Nitric Oxide and Infection

Nitric Oxide and Infection

Edited by

Ferric C. Fang

University of Colorado Health Sciences Center
Denver, Colorado

Kluwer Academic / Plenum Publishers
New York, Boston, Dordrecht, London, Moscow

Library of Congress Cataloging-in-Publication Data

Nitric oxide and infection / edited by Ferric C. Fang.
p. cm.
Includes bibliographical references and index.
ISBN 0-306-46147-1
1. Nitric oxide--Physiological effect. 2. Nitric oxide--Pathophysiology. 3. Infection. 4. Nitric oxide--Metabolism. 5. Immune response--Regulation. I. Fang, Ferric C.
[DNLM: 1. Nitric Oxide--immunology. 2. Nitric Oxide--metabolism. 3. Infection--metabolism. 4. Nitric-Oxide Synthase. QU 54 N7307 1999]
QP535.N1N542 1999
616.07'9--dc21
DNLM/DLC
for Library of Congress 99-30078
CIP

ISBN 0-306-46147-1

233 Spring Street, New York, N.Y. 10013

10 9 8 7 6 5 4 3 2 1

A C.I.P. record for this book is available from the Library of Congress.

Printed in the United States of America

Contributors

TAKAAKI AKAIKE • Department of Microbiology, Kumamoto University School of Medicine, Kumamoto 860, Japan

NICHOLAS M. ANSTEY • Menzies School of Health Research and Royal Darwin Hospital, Casuarina, Darwin NT0811, Northern Territory, Australia

NEIL R. BASTIAN • Department of Internal Medicine, Division of Infectious Diseases, University of Utah School of Medicine, Salt Lake City, Utah 84132

NIGEL BENJAMIN • Department of Clinical Pharmacology, St. Bartholomew's and the Royal London School of Medicine and Dentistry, London EC1M 6BQ, United Kingdom

CHRISTIAN BOGDAN • Institute of Clinical Microbiology, Immunology, and Hygiene, Friedrich-Alexander-University of Erlangen-Nuremberg, D-91054 Erlangen, Germany

KENNETH S. BOOCKVAR • Department of Medicine, Cornell University Medical Center, New York, New York 10021

JOHN CHAN • Departments of Medicine and Microbiology and Immunology, Albert Einstein College of Medicine, Bronx, New York 10467

JOHN A. COOK • Tumor Biology Section, Radiation Biology Branch, National Cancer Institute, Bethesda, Maryland 20892

WILLIAM DeGRAFF • Tumor Biology Section, Radiation Biology Branch, National Cancer Institute, Bethesda, Maryland 20892

MARY ANN DeGROOTE • Departments of Medicine, Pathology, and Microbiology, University of Colorado Health Sciences Center, Denver, Colorado 80262

ANDREAS DIEFENBACH • Institute of Clinical Microbiology, Immunology, and Hygiene, Friedrich-Alexander-University of Erlangen-Nuremberg, D-91054 Erlangen, Germany

CHARLES A. DINARELLO • Division of Infectious Diseases, University of Colorado Health Sciences Center, Denver, Colorado 80262

ROELF DYKHUIZEN • Department of Clinical Pharmacology, St. Bartholomew's and the Royal London School of Medicine and Dentistry, London EC1M 6BQ, United Kingdom, and Department of Medicine and Therapeutics, University of Aberdeen Medical School, Aberdeen AB9 2ZD, United Kingdom

FERRIC C. FANG • Departments of Medicine, Pathology, and Microbiology, University of Colorado Health Sciences Center, Denver, Colorado 80262

JOANNE FLYNN • Departments of Molecular Genetics and Biochemistry and Medicine, University of Pittsburgh School of Medicine, Pittsburgh, Pennsylvania 15261

RAYMOND FOUST III • The Institute for Environmental Medicine and Department of Biochemistry and Biophysics, University of Pennsylvania School of Medicine, Philadelphia, Pennsylvania 19104

BENJAMIN GASTON • University of Virginia Health Sciences Center, Charlottesville, Virginia 22908

MADHURA GOLE • The Institute for Environmental Medicine and Department of Biochemistry and Biophysics, University of Pennsylvania School of Medicine, Philadelphia, Pennsylvania 19104

ANDREW J. GOW • The Institute for Environmental Medicine and Department of Biochemistry and Biophysics, University of Pennsylvania School of Medicine, Philadelphia, Pennsylvania 19104

DONALD L. GRANGER • Department of Medicine, University of Utah Medical Center, Salt Lake City, Utah 84132

BENJAMIN HEMMENS • Institute for Pharmacology and Toxicology, Karl-Franzens University Graz, A-8010 Graz, Austria

JOHN B. HIBBS, JR. • Department of Internal Medicine, Division of Infectious Diseases, University of Utah School of Medicine, Salt Lake City, Utah 84132, and Veterans Affairs Medical Center, Salt Lake City, Utah 84148

HARRY ISCHIROPOULOS • The Institute for Environmental Medicine and Department of Biochemistry and Biophysics, University of Pennsylvania School of Medicine, Philadelphia, Pennsylvania 19104; *present address*: Stokes Research Institute, Children's Hospital of Philadelphia, Philadelphia, Pennsylvania 19104

MURALI KRISHNA • Tumor Biology Section, Radiation Biology Branch, National Cancer Institute, Bethesda, Maryland 20892

ROGER L. KURLANDER • Department of Clinical Pathology, National Institutes of Health, Bethesda, Maryland 20892

FOO Y. LIEW • Department of Immunology and Centre for Rheumatic Diseases, University of Glasgow, Glasgow G11 6NT, United Kingdom

STUART A. LIPTON • CNS Research Institute, Harvard Medical School, and Brigham and Women's Hospital, Boston, Massachusetts 02115

CHARLES J. LOWENSTEIN • Division of Cardiology, Department of Medicine, Johns Hopkins University School of Medicine, Baltimore, Maryland 21205

HIROSHI MAEDA • Department of Microbiology, Kumamoto University School of Medicine, Kumamoto 860, Japan

STUART MALCOLM • The Institute for Environmental Medicine and Department of Biochemistry and Biophysics, University of Pennsylvania School of Medicine, Philadelphia, Pennsylvania 19104

MITRA MAYBODI • Department of Ophthalmology, Washington University School of Medicine, St. Louis, Missouri 63110

BERND MAYER • Institute for Pharmacology and Toxicology, Karl-Franzens University Graz, A-8010 Graz, Austria

IAIN B. McINNES • Department of Immunology and Centre for Rheumatic Diseases, University of Glasgow, Glasgow G11 6NT, United Kingdom

JAMES B. MITCHELL • Tumor Biology Section, Radiation Biology Branch, National Cancer Institute, Bethesda, Maryland 20892

HEIKO MÜHL • Division of Infectious Diseases, University of Colorado Health Sciences Center, Denver, Colorado 80262, and Institute for General Pharmacology and Toxicology, Clinic of Johann Wolfgang Goethe University, D-60590 Frankfurt, Germany

ISABELLE P. OSWALD • INRA, Laboratory of Pharmacology-Toxicology, 31931 Toulouse Cedex 9, France

ROBERTO PACELLI • Tumor Biology Section, Radiation Biology Branch, National Cancer Institute, Bethesda, Maryland 20892

REBECCA M. POSTON • Embrex Corporation, Research Triangle Park, North Carolina 27709

DARYL D. REES • Centre for Clinical Pharmacology, University College London, London WC1E 6JJ, United Kingdom

MARTIN RÖLLINGHOFF • Institute of Clinical Microbiology, Immunology, and Hygiene, Friedrich-Alexander-University of Erlangen-Nuremberg, D-91054 Erlangen, Germany

MARTA SAURA • Division of Cardiology, Department of Medicine, Johns Hopkins University School of Medicine, Baltimore, Maryland 21205

W. MICHAEL SCHELD • Department of Medicine, University of Virginia, Charlottesville, Virginia 22908

JONATHAN S. STAMLER • Howard Hughes Medical Institute and Duke University Medical Center, Durham, North Carolina 27710

CHRISTOPH THIEMERMANN • The William Harvey Research Institute, St. Bartholomew's and the Royal London School of Medicine and Dentistry, London EC1M 6BQ, United Kingdom

GREGORY TOWNSEND • Department of Medicine, University of Virginia, Charlottesville, Virginia 22908

ANDRÉS VAZQUEZ-TORRES • Departments of Medicine, Pathology, and Microbiology, University of Colorado Health Sciences Center, Denver, Colorado 80262

YORAM VODOVOTZ • Cardiology Research Foundation and Medlantic Research Institute, Washington, D.C. 20010

J. BRICE WEINBERG • Division of Hematology and Oncology, Veterans Affairs and Duke University Medical Centers, Durham, North Carolina 27705

DAVID A. WINK • Tumor Biology Section, Radiation Biology Branch, National Cancer Institute, Bethesda, Maryland 20892

GILLIAN WRAY • The William Harvey Research Institute, St. Bartholomew's and the Royal London School of Medicine and Dentistry, London EC1M 6BQ, United Kingdom

CARLOS ZARAGOZA • Division of Cardiology, Department of Medicine, Johns Hopkins University School of Medicine, Baltimore, Maryland 21205

Preface

The past decade has witnessed a remarkable transformation in our understanding of the pathogenesis of infectious diseases. The deceptively simple molecule nitric oxide (NO) has been discovered to play important roles in numerous biological processes, casting many aspects of infection in a new light. Some researchers initially attempted to determine whether NO is "friend" or "foe" to the infected host. However, this overly simplistic view failed to account for the diverse mechanisms by which NO or its congeners can both contribute to host defense and mediate pathological processes, oftentimes simultaneously. This volume will attempt to synthesize a rapidly expanding scientific literature describing the biosynthesis of NO, mechanisms of its biological actions, and its complex roles in specific infectious settings, concluding with speculations on the future application of this newly acquired knowledge for the treatment of infectious diseases. The intended audience of this book includes microbiologists, molecular geneticists, biochemists, immunologists, cell biologists, physiologists, pharmacologists and pharmaceutical researchers, physicians, and other scientists with an interest in mechanisms of infectious disease pathogenesis. Of course, interested laypersons and curious family members are also welcome.

When I was first approached by Michael Hennelly of Plenum Press about editing a comprehensive book on NO and infection, I had the only sensible response to such an ambitious task—"no way!" However, on further reflection I acknowledged that such a book might represent a useful and novel contribution to scientists in a variety of disciplines, but only if the leading investigators in the field could be persuaded to contribute. To my surprise and pleasure, nearly all said yes. At this point, I realized I was stuck. My more selfish rationale for undertaking this project was the realization that I could learn a lot by editing this book, and that prediction has happily been confirmed. I hope that you, the reader, will be similarly rewarded.

Acknowledgments

I feel genuinely privileged for the opportunity to be studying host–pathogen interactions at such an interesting and important time in science. In addition to my patient publisher and the many talented contributors to this volume, I would like to specifically acknowledge some of my most influential mentors and colleagues: Steve Libby, Don Guiney, Don Helinski, Nancy Buchmeier, Don Granger, Charlie Davis, Carl Nathan, Larry Keefer, John Hibbs, Jonathan Stamler, Joshua Fierer, Elizabeth Ziegler, Skip Foster, Stanley Maloy, Stanley Falkow, Jim Imlay, Mike Spector, Ross Durland, Joe McCord, Mary Dinauer, Irwin Fridovich, Ron Gill, Mike Vasil, Charles Dinarello, Traci Testerman, and Andrés Vazquez-Torres. Appreciation is also due Mr. Rios, my eighth-grade biology teacher, and Mr. Thille, who showed me how to take apart bugs in the third grade—little did he know that I would still be doing it now. I want to thank Chip Schooley, Bob Schrier, Ron Lepoff, Laz Gerschenson, Randy Holmes, Nancy Madinger, Jan Monahan, the members of my laboratory, and the University of Colorado Health Sciences Center for providing me with a superb environment in which to live and work. Of course, I am grateful to the National Institutes of Health and the United States Department of Agriculture for their continuing (I hope) support. And last but certainly not least, I must thank my wonderful family who provide a continuing reminder of what is truly important in life. This book is dedicated to them.

Contents

Part A. Introduction

Part B. Historical Aspects of Nitric Oxide

Part C. Biological Roles of Nitric Oxide

3. *Biochemistry of Nitric Oxide*

Benjamin Gaston and Jonathan S. Stamler

4. *Enzymology of Nitric Oxide Biosynthesis*

Benjamin Hemmens and Bernd Mayer

16. ***Nitric Oxide in Schistosomiasis***
Isabelle P. Oswald

17. ***Nitric Oxide in Leishmaniasis: From Antimicrobial Activity to Immunoregulation***

Christian Bogdan, Martin Röllinghoff, and Andreas Diefenbach

18. ***Nitric Oxide in Viral Myocarditis***
Charles J. Lowenstein, Marta Saura, and Carlos Zaragoza

19. ***Nitric Oxide in Influenza***
Takaaki Akaike and Hiroshi Maeda

Part A

Introduction

CHAPTER 1

An Overview of Nitric Oxide in Infection

FERRIC C. FANG

> The universality of the deep chemistry of living things is indeed a fantastic and beautiful thing.
>
> Richard P. Feynman (Feynman, 1998)

1. Introduction

With a molecular mass of just 30 daltons, nitric oxide (NO) is certainly one of the smallest biological mediators in existence. For many years, this tiny molecule was of principal medical concern as a noxious constituent of automobile exhaust and cigarette smoke, while its more interesting biological properties were overlooked. From 1966 to 1985, there were fewer than 300 publications concerning NO in the entire MEDLINE-referenced medical literature. The subsequent explosion of scientific interest in NO is truly remarkable, reflected by nearly 20,000 NO-related papers over the past 10 years with more than 4000 of these reports published during the last year alone.

2. Historical Aspects of Nitric Oxide

In Chapter 2 of this book, John Hibbs and Neil Bastian provide their personal perspective on the events that transformed a minor component of polluted air into *Science* magazine's 1992 "molecule of the year" (Koshland, 1992). Several lines of apparently unrelated investigation serendipitously converged to yield unexpected insights. One of these initial investigations focused on the metabolism of nitrate in rodents and humans, stimulated by concerns about conversion of dietary nitrates

FERRIC C. FANG • Departments of Medicine, Pathology, and Microbiology, University of Colorado Health Sciences Center, Denver, Colorado 80262.

Nitric Oxide and Infection, edited by Fang. Kluwer Academic/Plenum Publishers, New York, 1999.

into carcinogenic nitrosamines by intestinal bacteria. Surprisingly, mammals were found to excrete significantly more nitrate than they ingest; it is now appreciated that this nitrate is an oxidative product of NO derived from NO synthase (NOS). Independently, pharmacologists and physiologists were discovering that NO is responsible for the ability of endothelial cells to control vascular tone. Shortly after Drs. Hibbs and Bastian submitted their chapter, the 1998 Nobel Prize in Medicine was awarded to Robert Furchgott, Ferid Murad, and Louis Ignarro for their discoveries concerning NO as a signaling molecule in the cardiovascular system. In parallel studies, biochemists reported that interferon-γ and lipopolysaccharide can induce enzymatic formation of NO from L-arginine by murine macrophages, and neurobiologists were learning that NO can function as a neurotransmitter. With the recognition that cell-derived NO could inhibit or kill tumor cells and microbes, the stage was set for the systematic investigation of NO's astonishingly diverse roles in infection.

3. Biological Roles of Nitric Oxide

Much of the biological versatility of NO relates to its ability to exist in various redox forms and congeners, each possessing distinctive reactivity. The biochemistry of NO, including critical reactions of nitrogen oxides with oxygen species, transition metals, carbon, nitrogen, and sulfur groups, is reviewed by Benjamin Gaston and Jonathan Stamler in Chapter 3. The availability of reactants and the specific redox environment can dictate whether NO plays a predominantly cytotoxic or physiologic signaling role.

Most biologically relevant NO is enzymatically generated from L-arginine, molecular oxygen, and NADPH, producing $NO^{\bullet}$ (NO radical), L-citrulline, and water. FMN, FAD, heme, tetrahydrobiopterin (H_4B), and calmodulin are required as cofactors. As detailed in Chapter 4 by Benjamin Hemmens and Bernd Mayer, the enzymology of NO synthesis is now fairly well appreciated. nNOS (NOS1) and eNOS (NOS3) were first discovered in neurons and endothelial cells, respectively, and generate relatively low fluxes of NO; hence, they are sometimes referred to as "constitutive" or cNOS isoforms. iNOS (NOS2) was first identified in cytokine-stimulated macrophages, and is alternatively known as the "inducible" isoform. In addition to neurons, endothelial cells, and macrophages, it is now recognized that NO can be synthesized by an enormous range of cell types including neutrophils, hepatocytes, mesangial cells, fibroblasts, chondrocytes, islet cells, myocytes, keratinocytes, and various epithelial cells.

Cytokine regulation of NO production by iNOS, reviewed by Heiko Mühl and Charles Dinarello in Chapter 5, is exceedingly important for understanding the production and overproduction of NO in infectious diseases. Several proinflammatory cytokines (e.g., IL-1, IL-2, TNFα, IFNγ, IFNα) have been implicated in iNOS induction, while counterregulatory factors (e.g., IL-4, IL-10, TGFβ, α-MSH) appear to exert a negative influence. Of great practical importance, a number of anti-inflammatory and immunosuppressive drugs such as nonsteroidal anti-inflammatory drugs, glucocorticoids, and cyclosporins have significant effects on iNOS expression or activity, suggesting that conventional notions of their mechanisms of action need to be reconsidered in a broader context.

Knowledge of cytokine regulation of NO production remains incomplete at this time, in part because much of our present understanding derives from work with rodent macrophages, and it now appears that rodents and humans control iNOS expression via quite different regulatory pathways. The production of NO by human mononuclear phagocytes was in fact held in serious question for many years, resulting from the failure of human monocyte-derived macrophages to demonstrate iNOS expression or NO generation under experimental conditions in which rodent cells elaborate copious quantities of NO. Brice Weinberg painstakingly reviews this controversy in Chapter 6. It is now abundantly clear from the more than 100 recent publications documenting iNOS mRNA, protein activity, and NO-related biological actions in human mononuclear cells, that reports of NO's absence from human macrophages (Schneemann *et al.*, 1993) were premature. Although the signals required to induce iNOS expression from normal human macrophages *in vitro* remain incompletely understood, mononuclear cells from patients with inflammatory conditions such as tuberculosis, malaria, AIDS, rheumatoid arthritis, and hepatitis consistently show evidence of enhanced NO production. It remains for future investigations to establish the biological importance and regulation of human macrophage-derived NO.

Of the potentially detrimental actions of NO during infection, among the most important is vascular collapse. In Chapter 7, Daryl Rees discusses the cardiovascular actions of NO, and specifically the role of NO in both maintaining basal vascular tone and altering vascular tone in response to local or systemic pathological states. This action is a consequence of interactions of NO and heme-containing guanylyl cyclase that increase cGMP levels to induce vasorelaxation. *In vitro* studies also implicate NO in the reversible myocardial depression associated with sepsis (Finkel *et al.*, 1992), although the clinical importance of this phenomenon is not yet established. NOS inhibition in patients with sepsis actually reduces cardiac output (Petros *et al.*, 1994) as a result of increased afterload, suggesting that NO-related vascular effects are of greater hemodynamic significance than direct effects on the myocardium.

Most other deleterious effects of NO during infection fall under the general rubric of cytotoxicity. Some of the same actions that make NO a potent antimicrobial mediator can also result in collateral damage to host tissues. In Chapter 8, Andrew Gow, Harry Ischiropoulos, and colleagues describe the ability of NO or its congeners to cause cytotoxicity via modification of proteins, injury to mitochondria, oxidation of membranes, and direct or indirect DNA damage. These effects may account for some of the tissue injury and organ dysfunction that occurs during infectious and other inflammatory states. Products arising from the interaction of NO with reactive oxygen species, such as peroxynitrite ($ONOO^-$), have been particularly implicated in NO-related cytotoxicity. NO and its congeners can also promote or inhibit apoptosis (programmed cell death) in experimental systems (Nicotera *et al.*, 1997). NO effects on apoptosis are not extensively discussed in this volume because their relevance to infection is presently unclear, but such actions could be important in NO-related tissue injury (see Chapter 21) or immunoregulatory phenomena. In any event, NO appears to play a major role as a mediator of immunopathological sequelae during certain infections including influenza (Chapter 19) and pertussis (Heiss *et al.*, 1994). Such observations suggest that NO inhibition may be beneficial to infected patients in specific situations.

However, NO clearly plays beneficial roles during infection as well. Paradoxically, NO can ameliorate oxygen-related cytotoxicity under specific conditions, possibly by scavenging oxidant species, terminating lipid peroxidation reactions, or inducing the expression of antioxidant systems. The chemical basis of these actions is reviewed in Chapter 9 by David Wink, James Mitchell, and co-workers. In one model of neuronal cytotoxicity, NO-dependent S-nitrosylation protected cells from death, while formation of NO-derived oxidant species results in neuronal destruction (Lipton *et al.*, 1993). Whether NO plays a cytotoxic or cytoprotective role is therefore highly dependent on its local redox environment and the balance between its various reactivities.

Immunomodulatory actions of NO, as reviewed by Iain McInnes and F. Y. Liew in Chapter 10, may be of great importance to the infected host. NO can regulate the formation of inflammatory mediators and maintain the integrity of the microcirculation by controlling platelet aggregation and leukocyte adhesion. Of course, NO and its derivatives are also important as antimicrobial effectors, possessing activity against an astonishingly broad range of parasitic, fungal, bacterial, and viral pathogens. As discussed by Nigel Benjamin and Roelf Dykhuizen in Chapter 11, this action is likely to play a significant role in mucosal and cutaneous innate immunity, in which a significant portion of the NO appears to arise from chemical reduction rather than from enzymatic sources. Reduction of nitrate to nitrite by commensal bacteria, with subsequent acidification of the nitrite in the stomach or on the skin, may provide an important first line of epithelial host defense.

The antimicrobial actions of NO are mechanistically complex, as Mary Ann De Groote and I describe in Chapter 12. In some cases, a direct antimicrobial effect of NO or its congeners has been demonstrated *in vitro*, while in other instances such an activity has been inferred by an increase in organism burden following inhibition of NO synthesis in tissue culture or animal models. Molecular targets of reactive nitrogen intermediates responsible for their antimicrobial activity include DNA, membranes, and reactive thiols, metals, amines, or aromatic residues of proteins. Microbes in turn may employ a variety of strategies to resist host-derived NO, including the production of scavengers, detoxifying enzymes, resistant targets, repair systems, or inhibitors.

4. Nitric Oxide in Specific Infections

Just as important scientific insights are often revealed in the details, so the analysis of specific infections has underscored some of the most significant biological aspects of NO. For example, overproduction of NO appears to be intimately involved in the dramatic hypotension and vasoplegia characteristic of septic shock. In Chapter 13, Gillian Wray and Christoph Thiemermann discuss experimental and clinical evidence to suggest that selective iNOS inhibition can help to resolve hemodynamic instability in sepsis.

In Chapter 14, John Chan and JoAnne Flynn discuss a very different role of NO in tuberculosis. The ability of NO to exert antimycobacterial activity *in vitro* and the requirement of NO synthesis for an effective host response in experimental murine tuberculosis have been well demonstrated by several investigators. Upregulation of NO production is also observed in human tuberculosis, but its functional significance in this setting is not yet established. Intriguingly, a possible role of NO in the maintenance of *Mycobacterium tuberculosis* latency has been suggested. With an estimated one-third of the world's population currently infected with the tubercle bacillus, this is an issue of no mean importance.

Malaria is another leading cause of infectious morbidity and mortality worldwide, and Nicholas Anstey and colleagues present intriguing data in Chapter 15 to suggest that NO plays a salutary role in human malaria. NOS expression in peripheral blood mononuclear cells correlated inversely with clinical severity in Tanzanian children with falciparum malaria (Anstey *et al.*, 1997), and the authors suggest that immunomodulatory and vascular actions of NO are more likely than direct antiparasitic activity of NO to be responsible for this effect.

In Chapter 16, Isabelle Oswald describes evidence that NO contributes to host defense in schistosomiasis. Endothelial cells can mediate NO-dependent antimicrobial activity against larval schistosomes, which reside within the vasculature, illustrating that host defense is not a function restricted to cells of the immune system. It is of particular interest that chronic schistostomiasis is associated with an enhanced risk of bladder and liver cancer, and DNA damage from chronic overproduction of NO-related species might be an important contributory factor. The mutational frequency of p53 genes analyzed from such tumors is characteristic of NO-related genotoxicity, lending credence to this hypothesis (Warren *et al.*, 1995). Increased NOS expression in chronic viral hepatitis might similarly contribute to the development of hepatocellular carcinoma (Kane *et al.*, 1997).

Evidence supporting an important role of NO as an immunomodulator and antimicrobial effector in leishmaniasis is reviewed by Christian Bogdan and co-workers in Chapter 17. Compelling data supporting a role of NO in the maintenance of microbial latency are presented, with the demonstration of persistent iNOS expression in latent murine *Leishmania major* infection and prompt reactivation of disease following administration of an iNOS inhibitor.

In Chapter 18, Charles Lowenstein and colleagues provide evidence to suggest that NO is also essential for host antiviral defenses in coxsackievirus myocarditis. NOS inhibition in this model exacerbated viral replication and tissue destruction.

However, a useful counterpoint is provided by Takaaki Akaike and Hiroshi Maeda, who examine a different and very important aspect of NO in Chapter 19. In a murine model of influenza pneumonitis, NO production appears to result in respiratory pathology and mortality without contributing to host defense. The prevention of lethality by NOS inhibition (Akaike *et al.*, 1996) strongly suggests that such strategies should be considered and further investigated in certain viral infections characterized by severe tissue injury. Unquestionably, therapeutic strategies targeting NO will have to consider the specific and sometimes dramatically different pathophysiological roles of reactive nitrogen intermediates in different infectious settings.

Studies of NO in bacterial meningitis, reviewed by Gregory Townsend and Michael Scheld in Chapter 20, illustrate the difficulty in sorting out the pathophysiological roles of NO. Experimental evidence suggests that NO is an important mediator of neuronal damage but is also required for the maintenance of cerebral perfusion. The use of highly isoform-specific NOS inhibitors may ultimately help to separate these actions.

The possible importance of NO in neurological infection is not limited to bacteria. In Chapter 21, Stuart Lipton discusses possible mechanisms by which NO may contribute to the central neurological sequelae of human immunodeficiency virus (HIV) infection. Ongoing clinical trials hope to establish whether this will provide a useful therapeutic target in patients with AIDS dementia.

As the final specific example, Kenneth Boockvar, Donald Granger, and colleagues provide a helpful discussion of the benefits of NO production during experimental listeriosis in Chapter 22. It is perhaps not coincidental that *Listeria monocytogenes* is an intracellular pathogen, as are so many of the pathogens for which NO appears to play an antimicrobial role (see Chapter 12, Table 2). This may relate to the abundance of NO-scavenging substances in the extracellular environment. Notably, NOS inhibition has a dramatic effect on the course of infection in primary listeriosis, but no impact in immune animals, emphasizing the role of NO in innate nonspecific immunity.

5. Future Directions

From this brief overview, it is evident that NO can play both beneficial and detrimental roles in the infected host. While vascular collapse and tissue injury may contribute to the morbidity and mortality of infection, the immunoregulatory, microcirculatory, cytoprotective, and antimicrobial effects of NO can be essential for host survival. As Andrés Vazquez-Torres and I conclude in Chapter 23, the therapeutic challenge before us is to develop means of selectively inhibiting excessive NO production and delivering NO to sites of infection or inadequate perfusion. Encouraging preliminary observations and our rapidly growing understanding of NO's complex roles during infection suggest that such approaches will ultimately succeed.

References

Akaike, T., Noguchi, Y., Ijiri, S., Setoguchi, K., Suga, M., Zheng, Y. M., Dietzschold, B., and Maeda, H., 1996, Pathogenesis of influenza virus-induced pneumonia: Involvement of both nitric oxide and oxygen radicals, *Proc. Natl. Acad. Sci. USA* **93:**2448–2453.

Anstey, N. M., Weinberg, J. B., Hassanali, M. Y., Mwaikambo, E. D., Manyenga, D., Misukonis, M. A., Arnelle, D. R., Hollis, D., McDonald, M. I., and Granger, D. L., 1996, Nitric oxide in Tanzanian children with malaria: Inverse relationship between malaria severity and nitric oxide production/nitric oxide synthase type 2 expression, *J. Exp. Med.* **184:**557–567.

Feynman, R. P., 1998, *The Meaning of It All: Thoughts of a Citizen Scientist,* Addison-Wesley, Reading, Mass.

Finkel, M. S., Oddis, C. V., Jacob, T. D., Watkins, S. C., Hattler, B. G., and Simmons, R. L., 1992, Negative inotropic effects of cytokines on the heart mediated by nitric oxide, *Science* **257:**387–389.

Heiss, L. N., Lancaster, J. R., Jr., Corbett, J. A., and Goldman, W. E., 1994, Epithelial autotoxicity of nitric oxide: Role in the respiratory cytopathology of pertussis, *Proc. Natl. Acad. Sci. USA* **91:**267–270.

Kane, J. M., Shears, L. L., Hierholzer, C., Ambs, S., Billiar, T. R., and Posner, M. C., 1997, Chronic hepatitis C virus infection in humans—Induction of hepatic nitric oxide synthase and proposed mechanisms for carcinogenesis, *J. Surg. Res.* **69:**321–324.

Koshland, D. E., Jr., 1992, The molecule of the year, *Science* **258:**1861.

Lipton, S. A., Choi, Y.-B., Pan, Z.-H., Lei, S. Z., Chen, H.-S. V., Sucher, N. J., Loscalzo, J., Singel, D. J., and Stamler, J. S., 1993, A redox-based mechanism for the neuroprotective and neurodestructive effects of nitric oxide and related nitroso-compounds, *Nature* **364:**626–632.

Nicotera, P., Brune, B., and Bagetta, G., 1997, Nitric oxide: Inducer or suppressor of apoptosis? *Trends Pharmacol. Sci.* **18:**189–190.

Petros, A., Lamb, G., Leone, A., Moncada, S., Bennett, D., and Vallance, P., 1994, Effects of a nitric oxide synthase inhibitor in humans with septic shock, *Cardiovasc. Res.* **28:**34–39.

Schneemann, M., Schoedon, G., Hofer, S., Blau, N., Guerrero, L., and Schaffner, A., 1993, Nitric oxide synthase is not a constituent of the antimicrobial armature of human mononuclear phagocytes, *J. Infect. Dis.* **167:**1358–1363.

Warren, W., Biggs, P. J., el-Baz, M., Ghoneim, M. A., Stratton, M. R., and Venitt, S., 1995, Mutations in the p53 gene in schistosomal bladder cancer: A study of 92 tumours from Egyptian patients and a comparison between mutational spectra from schistosomal and non-schistosomal urothelial tumours, *Carcinogenesis* **16:**1181–1189.

Part B

Historical Aspects of Nitric Oxide

CHAPTER 2

The Discovery of the Biological Synthesis of Nitric Oxide

JOHN B. HIBBS, JR. and NEIL R. BASTIAN

1. Introduction

Essential contributions to the discovery of the biological synthesis of nitric oxide (NO) occurred in three disparate lines of research: 1) studies of the role of macrophages as cytotoxic effector cells in innate resistance and cell-mediated immunity; 2) investigation of nitrosamine-induced carcinogenesis; and 3) studies of endothelium-dependent vascular relaxation. The clearest picture of the precursor, products, inhibitors, enzymology, and biological significance of this novel biochemistry emerged from the combined results of studies by investigators from all three areas when they independently converged on unexpected observations from very different points of departure. The discovery of the biological synthesis of NO illustrates how science benefits when an exchange of information occurs between several different disciplines of investigation that each have solved part of a broad and very complex biological problem. This seems to us to be a good example of the inherent interconnected character of biology as well as of scientific effort.

To review the discovery of the biological synthesis of NO we will return to the three disparate lines of research mentioned above and examine how, in 1987/1988, the aggregate experimental results of investigators working in different scientific fields, launched the rapidly moving and multidisciplinary field of NO biology.

JOHN B. HIBBS, JR. • Department of Internal Medicine, Division of Infectious Diseases, University of Utah School of Medicine, Salt Lake City, Utah 84132, and Veterans Affairs Medical Center, Salt Lake City, Utah 84148. *NEIL R. BASTIAN* • Department of Internal Medicine, Division of Infectious Diseases, University of Utah School of Medicine, Salt Lake City, Utah 84132.

Nitric Oxide and Infection, edited by Fang. Kluwer Academic/Plenum Publishers, New York, 1999.

Our group working in the area of innate resistance and cell-mediated immunity (CMI) used a bioassay to study an interaction between macrophage effector cells and neoplastic target cells. Macrophages activated by cytokines and/or microbial products such as lipopolysaccharide (LPS) caused, by an unknown mechanism, a reproducible pattern of redox enzyme inhibition as well as target cell cytostasis.

A second bioassay was used by several groups of investigators to study an interaction between endothelial and vascular smooth muscle cells. Agonists such as acetylcholine or bradykinin stimulated endothelial cells to release an unknown vasodilatory substance that activated soluble guanylyl cyclase in smooth muscle cells and caused muscle relaxation. The unknown vasodilatory substance was termed endothelium-dependent relaxation factor (EDRF) by Robert Furchgott.

The third line of research was carried out by investigators at MIT who were studying the toxicological consequences of nitrosamine ingestion. Metabolic balance studies firmly established the endogenous synthesis of inorganic nitrogen oxides by both rodents and humans. Further research by the same group extended nitrogen oxide synthesis to murine cells *in vitro*. However, the source and the biological significance of the nitrogen oxides measured by the MIT investigators *in vivo* (Green *et al.*, 1981a,b) and *in vitro* (Stuehr and Marletta, 1985) were not known at the time.

We review here the research completed by the end of 1988, which represents the foundation for what has become the multidisciplinary field of $NO^{\bullet}$ biochemistry, physiology, pathophysiology, and enzymology. We will also briefly discuss experimental work carried out in 1989 and 1990 that directly led to purification of the first NOS isoform, the neural NO synthase (nNOS or type 1 NOS). One of us was personally involved in a portion of the work to be described. We hope this has not interfered with our effort to be objective and to relate the events as they unfolded.

2. Converging Lines of Investigation through 1987

2.1. Studies of Activated Macrophages in Innate and Cell-Mediated Immunity

Early in the 1970s, Hibbs and colleagues discovered that murine macrophages activated by a cell-mediated immune (CMI) response acquired the ability to express nonspecific cytotoxicity for neoplastic target cells (Hibbs *et al.*, 1971, 1972a–d; Hibbs, 1973). Peritoneal macrophages from mice with chronic intracellular infection [either *Toxoplasma gondii* or *Mycobacterium bovis* strain bacillus Calmette–Guérin (BCG)] or macrophages from mice inoculated with killed

Mycobacterium butyricum (present in Freund's adjuvant) were found to be cytotoxic for syngeneic, allogeneic, and xenogeneic neoplastic cells *in vitro*. The results also showed that macrophages must undergo a functional change so as to express nonspecific cytotoxicity for target cells *in vitro*. Normal resident peritoneal macrophages, or normal macrophages elicited with sterile nonimmunogenic inflammatory stimulants such as 10% peptone or thioglycollate broth, are not cytotoxic for neoplastic cells (Hibbs *et al.*, 1971, 1972a–d; Hibbs, 1973). It was clear that signals generated during the development of cellular immunity to intracellular pathogens in mice caused functional modification of host macrophages. This modification resulted in acquisition by peritoneal macrophages of the ability to express nonspecific cytotoxicity for tumor cells when cultured *in vitro*. Independently, Alexander and Evans (1971) showed that murine macrophages treated with LPS *in vitro* are nonspecifically cytotoxic for syngeneic and allogeneic lymphoma cells by a nonphagocytic mechanism. Together, these results demonstrated that rodent macrophages are functionally modified by signals generated during a CMI response *in vivo* or by LPS *in vitro* to express cytotoxicity for neoplastic cells. The cytotoxic effect is nonphagocytic, requires close macrophage–target cell contact, and is immunologically nonspecific at the effector level. Susceptibility to cytotoxicity is independent of target cell antigens. Therefore, the cytotoxic activated macrophage–neoplastic cell coculture system provided a useful bioassay for examining the biochemical mechanisms of this newly recognized activated macrophage effector cell–tumor target cell interaction.

In 1977, Hibbs and colleagues identified the biochemical signals that induced expression of activated macrophage-mediated cytotoxicity (Hibbs *et al.*, 1977; Chapman and Hibbs, 1977; Weinberg *et al.*, 1978). In this work, a sequence of macrophage differentiation was elucidated, demonstrating that expression of nonspecific cytotoxicity for neoplastic cells by activated macrophages requires a priming signal, interferon-γ (IFNγ), and a second signal such as bacterial LPS (Chapman and Hibbs, 1977; Hibbs *et al.*, 1977; Weinberg *et al.*, 1978). These observations were confirmed by work in other laboratories (Russell *et al.*, 1977; Ruco and Meltzer, 1978). Later, these same signals were shown to induce expression of the high-output immune/inflammatory NO synthase (iNOS or NOS2).

The first metabolic perturbation of neoplastic target cells induced by cytotoxic activated macrophages was independently discovered by Keller (1973) and by Krahenbuhl and Remington (1974). They observed that cytotoxic activated macrophages cause the rapid onset of target cell cytostasis and inhibition of DNA synthesis without inducing lysis of the target cells. A series of investigations by Hibbs and colleagues were later carried out to identify other metabolic lesions induced in target cells by cytotoxic activated macrophages (Granger *et al.*, 1980; Hibbs *et al.*, 1984; Drapier and Hibbs, 1986). These lesions, as well as inhibition of DNA synthesis, were later shown to be caused by cytokine-induced NO synthesis.

Experiments showed that neoplastic target cells cocultivated with cytokine-activated macrophages develop inhibition of mitochondrial respiration, but the glycolytic pathway remains functional and the target cells remain viable (Granger *et al.*, 1980). Activated macrophage-induced inhibition of mitochondrial respiration was shown by Granger and Lehninger (1982) to result from inhibition of the two proximal oxidoreductases of the mitochondrial electron transport system [NADH:ubiquinone oxidoreductase (Complex I) and succinate:ubiquinone oxidoreductase (Complex II)]. Remarkably, other enzymes of the mitochondrial electron transport chain were not affected (Granger and Lehninger, 1982). These findings and earlier work published by Weinberg and Hibbs (1977) led Hibbs and co-workers to test the hypothesis that there is a link between the activated macrophage cytotoxic mechanism and intracellular iron metabolism. They reported in 1984 that cytotoxic activated macrophages cause a major perturbation of iron homeostasis in neoplastic target cells (Hibbs *et al.*, 1984). Viable target cells lose a significant portion of their intracellular iron ($>70\%$). Concurrently, a characteristic pattern of activated macrophage-induced metabolic inhibition is observed, including inhibition of both DNA replication and mitochondrial respiration. Drapier and Hibbs (1986) then discovered that enzymes with iron-dependent catalytic activity, particularly those with [4Fe–4S] centers, are targets of the cytokine-activated macrophage cytotoxic effector mechanism. Neoplastic cells cocultivated with cytokine-activated macrophages develop rapid inhibition of mitochondrial aconitase, which contains a [4Fe–4S] prosthetic group essential for catalytic activity. Removal of a labile iron atom from the [4Fe–4S] center by a cytotoxic activated macrophage-mediated mechanism was shown to be causally related to aconitase inhibition (Drapier and Hibbs, 1986). Inhibition of enzymes with [4Fe–4S] centers provided a mechanistic explanation for the earlier discovery made by Granger and Lehninger (1982), that cytokine-activated macrophages induce inhibition of mitochondrial respiration via specific inhibition of Complex I and Complex II, each of which contains [4Fe–4S] centers. These results indicated that iron and enzymes with [4Fe–4S] clusters are targets of the activated macrophage cytotoxic effector mechanism.

During the course of these metabolic studies, Hibbs and colleagues observed that cytotoxic activated macrophages require L-arginine in the culture medium to cause cytostasis and metabolic lesions in neoplastic target cells (Hibbs *et al.*, 1987a). D-Arginine cannot substitute for L-arginine, and the L-arginine analogue N^{ω}-monomethyl-L-arginine (L-NMMA) was found to potently inhibit induction of the observed metabolic changes (Hibbs *et al.*, 1987a). Hibbs and co-workers then discovered that L-arginine is directly converted to L-citrulline without loss of the guanidino carbon atom. They also observed that L-NMMA prevents the synthesis of L-citrulline from L-arginine (Hibbs *et al.*, 1987b) and prevents the development of cytostasis as well as the activated macrophage-mediated inhibition of mitochondrial respiration in neoplastic target cells (Hibbs *et al.*, 1987a,b). See Fig. 1 for a schematic depiction of experimental flow in the discipline of innate resistance and CMI.

In 1985, Dennis Stuehr and Michael Marletta reported that thioglycollate-elicited LPS-stimulated peritoneal macrophages from C_3H/HeN mice synthesize nitrite (NO_2^-, 60%) and nitrate (NO_3^-, 40%) (Stuehr and Marletta, 1985). This observation provided a clue to the fate of the imino nitrogen atom lost from the guanidino group of L-arginine when cytotoxic activated macrophages synthesize L-citrulline from L-arginine. It seemed very likely that the nitrite and nitrate identified by Stuehr and Marletta (1985) were products of the same pathway producing the L-citrulline from L-arginine identified by Hibbs *et al.* (1987a), and this indeed proved to be the case. Hibbs and colleagues reported early in 1987 that cytotoxic activated macrophages synthesize nitrite and L-citrulline in a coordinated manner from L-arginine, but not D-arginine, and that L-NMMA inhibits the synthesis of both products as well as the expression of cytotoxicity by activated macrophages (Hibbs *et al.*, 1987b). These experiments provided the first demonstration of an enzymatic reaction directly coupling synthesis of L-citrulline from L-arginine to oxidation of a terminal guanidino nitrogen atom of L-arginine, an activity later shown to be related to iNOS. In addition, L-NMMA was identified as a specific inhibitor of this enzymatic reaction. Also shown was a consistent correlation between the activity of this pathway and the expression of cytotoxicity by activated macrophages (Hibbs *et al.*, 1987a,b) (see Fig. 1). Later in 1987, Iyengar, Stuehr, and Marletta confirmed that a terminal guanidino nitrogen atom of L-arginine is the precursor of nitrite and nitrate synthesized by activated macrophages, and that L-citrulline is a product of the enzymatic reaction (Iyengar *et al.*, 1987). Therefore, in 1987, the enzymatic synthesis of nitrogen oxides from L-arginine was established, a potent and nontoxic inhibitor of this synthesis (L-NMMA) was identified, and the biological significance of cytokine-induced high output nitrogen oxide synthesis from L-arginine was elucidated [$NO^\bullet$ or other reactive nitrogen oxides were the effector molecules causing inhibition of mitochondrial respiration, aconitase activity, and DNA synthesis (cytostasis) in neoplastic target cells of activated macrophages]. The iron nitrosylating properties of $NO^\bullet$ or other nitrogen oxides provided a mechanistic explanation for the reproducible pattern of cytotoxic activated macrophage-mediated metabolic inhibition that had been elucidated earlier by Hibbs and colleagues.

2.2. Toxicological and Metabolic Studies Leading to Investigation of LPS- and Cytokine-Stimulated Macrophages

Stuehr and Marletta, working at MIT, discovered that LPS-treated RAW 264.7 cells synthesize nitrogen oxides (Stuehr and Marletta, 1985). This observation was an extension of earlier nitrogen metabolic balance studies carried out by Steven Tannenbaum and colleagues, which showed that mammals excrete more nitrogen

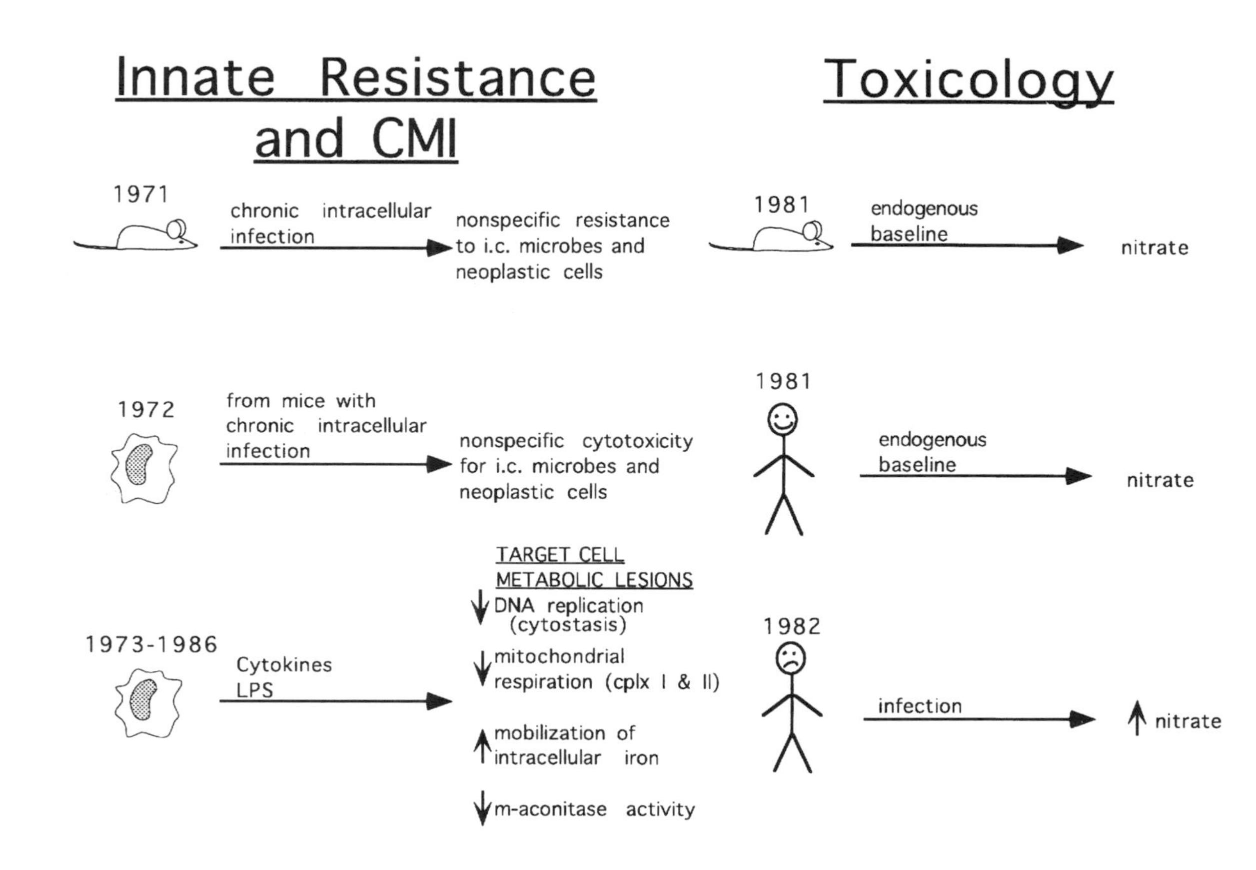

Innate Resistance and CMI
Toxicology
1971
chronic intracellular infection
nonspecific resistance to i.c. microbes and neoplastic cells
1981
endogenous baseline
nitrate
1972
from mice with chronic intracellular infection
nonspecific cytotoxicity for i.c. microbes and neoplastic cells
1981
endogenous baseline
nitrate
1973-1986
Cytokines LPS
TARGET CELL METABOLIC LESIONS
DNA replication (cytostasis)
mitochondrial respiration (cplx I & II)
mobilization of intracellular iron
m-aconitase activity
1982
infection
nitrate

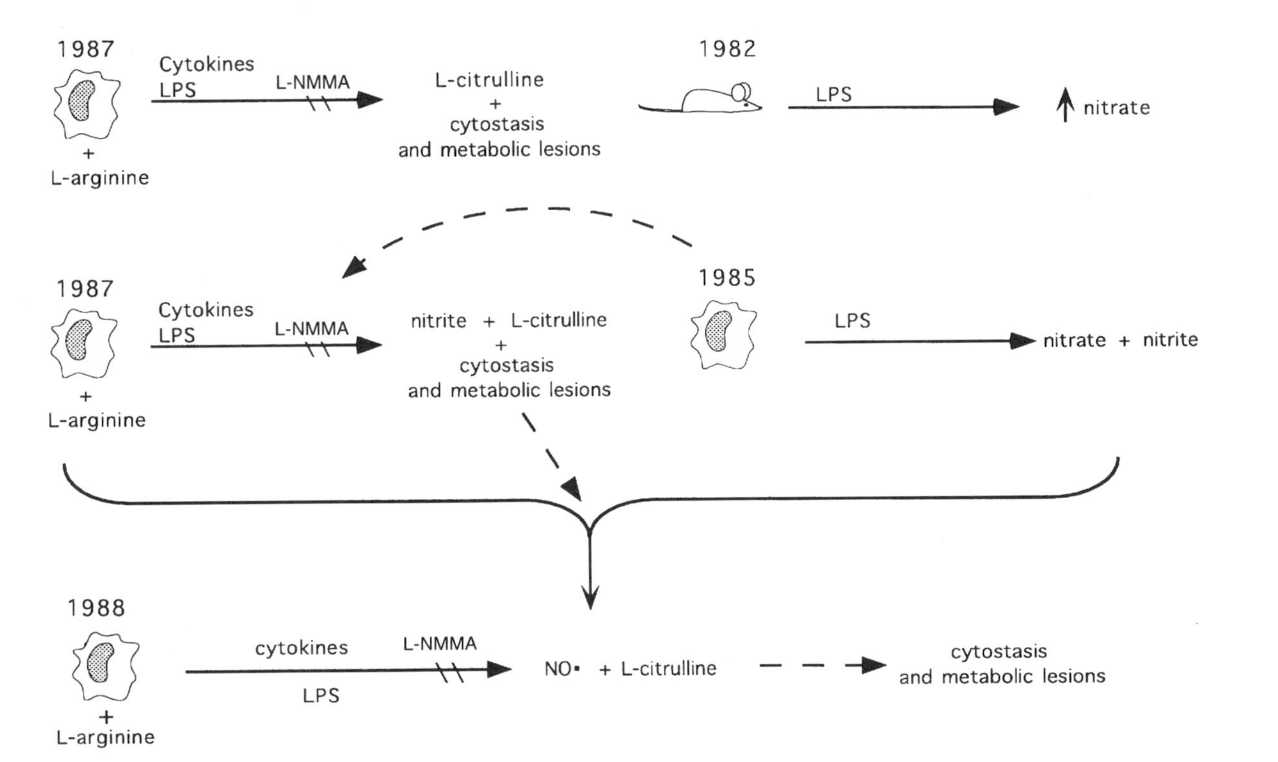

FIGURE 1. Experimental flow in the fields of innate resistance/cell-mediated immunity and toxicology. Work in the two fields developed independently until 1986 when the observation that LPS induced murine macrophages to synthesize nitrogen oxides published by workers in the toxicology field (Stuehr and Marletta, 1985) provided the information needed by workers in the innate resistance/cell-mediated immunity field to elucidate the precursor and products of the enzymatic reactions synthesizing nitrogen oxides (Hibbs *et al.*, 1987,a,b). The latter investigators were aware that expression of activated macrophage cytotoxicity was dependent on the direct conversion of L-arginine to L-citrulline and that the cytotoxic reaction and citrulline synthesis was inhibited by N^{ω}-monomethyl-L-arginine (L-NMMA). The observation of Stuehr and Marletta (1985) provided a clue to the fate of the terminal guanidino nitrogen atom that was lost when activated macrophages directly synthesized L-citrulline from L-arginine. The two scientific fields reached common ground again in 1988 when Hibbs *et al.* (1988) and Marletta *et al.* (1988) independently used methods that measured NO distinct from other nitrogen oxides, for the first time.

oxides than they ingest. This finding was in fact first noted by Mitchell *et al.* (1916), but this early research was neglected until 1981 when Tannenbaum and co-workers showed that both mice and humans maintained on a known low-nitrate diet (<180 μmole/day per human subject) continue to excrete relatively high concentrations of urinary nitrate (~800 μmole/day per human subject) (Green *et al.*, 1981a,b). Serendipitously, Tannenbaum's group observed an unexpected increase in nitrate biosynthesis during the course of a metabolic balance study in healthy adult humans when an individual on a monitored low-nitrate diet developed fever and diarrhea (Wagner and Tannenbaum, 1982). Urinary nitrate levels increased sixfold in this individual during the course of the illness. As a follow-up to this initial finding, they demonstrated that *Escherichia coli* LPS greatly enhances urinary excretion of nitrate in rats (Wagner and Tannenbaum, 1982; Wagner *et al.*, 1983a). Using an oral dose of [^{15}N]ammonium acetate, they showed that the enhanced urinary excretion of nitrate in LPS-treated rats is the result of increased nitrate synthesis. Tannenbaum and co-workers also carried out metabolic balance studies in humans using ^{15}N-labeled nitrate (Wagner *et al.*, 1983b), finding that the half-life of orally administered nitrate is approximately 5 h, and its volume of distribution is about 30% of body weight. Daily endogenous biosynthesis of nitrate was estimated to be ~1 mmole/day. Taken together, these important findings clearly established that mammals endogenously synthesize nitrate. In addition, they provided a link between induction of inflammation and increased endogenous synthesis of nitrate by humans and rodents.

Marletta and Stuehr moved the investigation of inorganic nitrogen oxide synthesis by mammals to an experimental tissue culture system. They reported in 1985 that LPS treatment induces murine peritoneal macrophages to synthesize nitrite and nitrate, which are released into the culture medium (Stuehr and Marletta, 1985). Furthermore, infection of mice with the BCG strain of *Mycobacterium bovis* induces production of high levels of nitrogen oxides in the tissues as determined by measurement of urinary nitrate excretion (Stuehr and Marletta, 1985). This significant discovery showed that LPS, which induces nitrogen oxide production *in vivo* (Wagner and Tannenbaum, 1982; Wagner *et al.*, 1983a), also induces murine macrophages to synthesize nitrogen oxides *in vitro* (Stuehr and Marletta, 1985). It was also demonstrated in this study that infection of mice with a facultative intracellular bacterium (BCG) results in high tissue production of nitrogen oxides (Stuehr and Marletta, 1985). In a subsequent study published in 1987, Stuehr and Marletta showed that two signals, IFNγ and bacterial LPS, are required for induction of high-output nitrogen oxide synthesis by macrophages (Stuehr and Marletta, 1987a,b). Hibbs and colleagues had previously shown that these identical signals are required to induce macrophage cytotoxicity for neoplastic target cells and intracellular pathogens (Chapman and Hibbs, 1977; Hibbs *et al.*, 1977; Weinberg *et al.*, 1978). Marletta, Stuehr, Tannenbaum, and colleagues also demonstrated that murine macrophages synthesizing nitrogen oxides are capable of nitrosamine formation (Miwa *et al.*, 1987). They confirmed that L-arginine is the precursor molecule for nitrogen oxide synthesis in macrophages activated by treatment with IFNγ and LPS (Iyengar *et al.*, 1987), reproducing the findings of Hibbs and colleagues published earlier that year (Hibbs *et al.*, 1987a,b). Thus, in 1987, the work of Hibbs and colleagues and that of Stuehr and Marletta had joined to become the study of a common biochemistry. (See Fig. 1 for a schematic depiction of experimental flow in the discipline of toxicology.)

2.3. Vascular Pharmacology and Physiology Studies

The discovery of the role of NO• in the cardiovascular system began with investigations of cyclic GMP and guanylyl cyclase. Subsequent work would demonstrate that the heme prosthetic group of soluble guanylyl cyclase is the primary target for NO•, derived from both pharmacological and biological sources. Both Ferid Murad and Louis Ignarro began working in this field in the early 1970s when cyclic GMP was known to exist as an intracellular biochemical messenger, similar to the better known cyclic AMP, but with no known functions. In 1977, Murad and colleagues demonstrated that the clinically important vasodilators sodium nitroprusside and nitroglycerine, as well as reagent NO•, activate guanylyl cyclase (Arnold *et al.*, 1977; Katsuki *et al.*, 1977; Murad *et al.*, 1978). They also suggested that NO• was the probable activator of guanylyl cyclase pharmacologically generated from nitrovasodilators, as well as from nitrite, *N*-nitroso compounds, azide, and other related molecules. This important observation suggested that the likely mechanism of action of nitrovasodilators was release of NO• and activation of guanylyl cyclase, and established that NO• is a pharmacologically active molecule.

In 1979, Ignarro and co-workers provided the first demonstration that reagent NO• is a vasorelaxant, and that its mechanism of action is attributable to the second-messenger actions of cyclic GMP (Gruetter *et al.*, 1979). In this and in a subsequent study, strong evidence was presented that sodium nitroprusside, organic nitrates, and other nitrovasodilators elevate cyclic GMP and relax vascular smooth muscle by releasing NO• (Gruetter *et al.*, 1979; Ignarro *et al.*, 1981). While studying the mechanism of action of nitrovasodilators, Ignarro's group observed the intermediate formation of chemically labile *S*-nitrosothiols which decompose to liberate NO• (Ignarro *et al.*, 1981). They were also the first to demonstrate that sodium nitroprusside and reagent NO• inhibit platelet aggregation, and that this effect is related to activation of platelet guanylyl cyclase (Mellion *et al.*, 1981).

The mechanism of guanylyl cyclase activation by NO• was also defined during the late 1970s and early 1980s. Building on the observations of Murad and colleagues that NO• is the likely final common activator of guanylyl cyclase by nitrovasodilators, Craven and DeRubertis discovered that heme is required for the activation of guanylyl cyclase by NO• (DeRubertis *et al.*, 1978; Craven and

DeRubertis, 1978). Their work showed that guanylyl cyclase becomes catalytically active at the same time an enzyme-associated heme–nitrosyl complex is formed (Craven *et al.*, 1979). Ignarro's group went on to show that NO• interacts with enzyme-bound heme iron to force a structural reconfiguration of the heme, resulting in a protoporphyrin IX-like binding interaction with the guanylyl cyclase apoprotein (Ignarro *et al.*, 1982a,b; Ohlstein *et al.*, 1982; Wolin *et al.*, 1982).

Prior to 1980, it was assumed by most researchers that vascular smooth muscle tissues are relaxed by neurotransmitters acting directly on smooth muscle cells. However, in 1980, Furchgott and Zawadzki showed that acetylcholine-induced relaxation of blood vessels depends on the presence of endothelial cells. When the endothelial layer was stripped from the vessels, smooth muscle relaxation no longer occurred. Using an arrangement of two strips of rabbit aorta, one with and the other without an intact endothelial layer, Furchgott and Zawadzki (1980) showed that binding of acetylcholine to receptors on endothelial cells causes the release of a substance that readily diffuses into the adjacent vascular smooth muscle tissue, causing it to relax. Because this factor had not been chemically identified, it became known as *endothelium dependent relaxing factor* (Cherry *et al.*, 1982). Shortly thereafter, Murad and colleagues showed that endothelium-dependent relaxation of vascular smooth muscle is induced through the formation of cyclic GMP and mediated by cyclic GMP-dependent protein phosphorylation (Rapoport and Murad, 1983; Rapoport *et al.*, 1983a,b). Subsequent work by Furchgott's laboratory confirmed the close relationship between EDRF, nitrovasodilators, and guanylyl cyclase activation (Martin *et al.*, 1985, 1986). In 1986, Salvador Moncada and co-workers showed that the endothelium relaxing properties of EDRF and prostacyclin can be separated (Gryglewski *et al.*, 1986a), and that superoxide anion inactivates EDRF (Gryglewski *et al.*, 1986b).

The identification of EDRF by Furchgott and Zawadzki (1980) was an essential step in the eventual demonstration of endogenous nitrogen oxide production by the cardiovascular system. Experimental results from the study of guanylyl cyclase/cyclic GMP chemistry and nitrovasodilator pharmacology provided the background information needed to create a suspicion that EDRF could be a nitrogen oxide. All evidence available in 1986 suggested that the chemical and biological properties of reagent NO•, pharmacologically generated nitrogen oxides, and EDRF are remarkably similar.

In 1987, the laboratories of Moncada (Palmer *et al.*, 1987) and Ignarro (Ignarro *et al.*, 1987a,b) independently published the results of experiments that suggested that EDRF is either NO• or a closely related nitrogen oxide. However, the methods used did not unequivocally establish that NO• was the nitrogen oxide produced by vascular endothelial cells (see Nathan, 1992, for an explanation). Also in 1987, Moncada and colleagues showed that agonist-induced nitrogen oxide synthesis by endothelial cells inhibits human platelet aggregation in a cyclic GMP-dependent manner (Radomski *et al.*, 1987a–d). This extended the earlier results of Ignarro and co-workers, who demonstrated the same cyclic GMP-dependent inhibition of platelet aggregation in response to reagent NO• and nitrovasodilators (Mellion *et al.*, 1981). Therefore, in 1987, investigators in the field of vascular pharmacology and physiology provided experimental evidence that characterized EDRF as either NO• or a closely related nitrogen oxide. (See Fig. 2 for a schematic representation of experimental flow in the discipline of vascular physiology.)

3. NO• Synthesis from L-Arginine Biochemically Unifies Studies of Endothelium-Dependent Relaxation and the Activated Macrophage Cytotoxic Reaction in 1988

The discovery that nitrogen oxide synthesis from L-arginine by activated macrophages was the effector mechanism causing cytostasis and metabolic lesions in neoplastic target cells was published in early 1987 (Hibbs *et al.*, 1987a,b). This

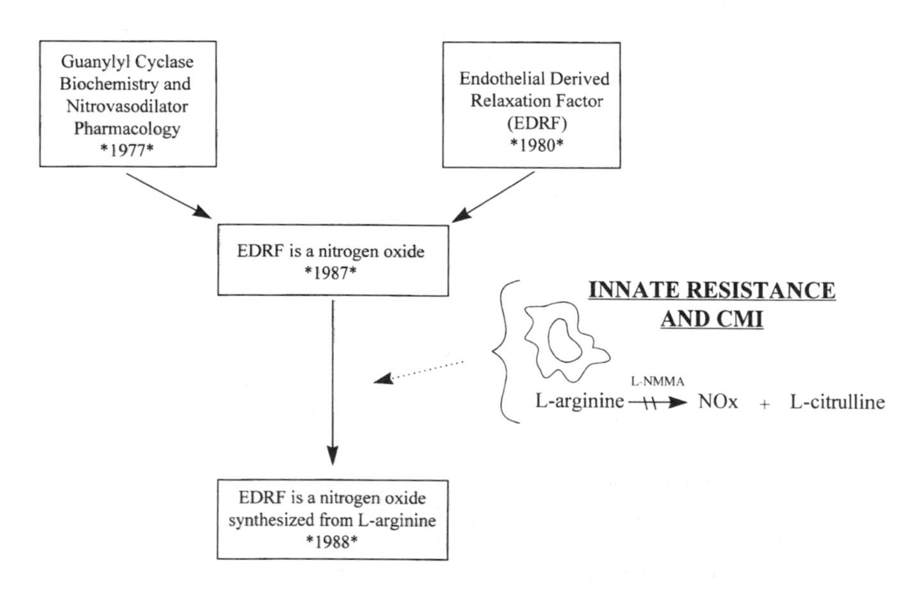

FIGURE 2. Experimental flow in the field of vascular pharmacology and physiology leading to the conclusion in 1987 that EDRF is a nitrogen oxide (Ignarro *et al.*, 1987a,b; Palmer *et al.*, 1987) and in 1988 that EDRF is a nitrogen oxide synthesized from L-arginine (Palmer *et al.*, 1988). Also shown is work from the field of innate resistance/cell-mediated immunity (Hibbs *et al.*, 1987a,b) which provided information that enabled elucidation of similar enzymatic activity synthesizing nitrogen oxides in the vascular system (Palmer *et al.*, 1988).

work occurred independently and without knowledge of the investigations that culminated in the discovery that an NO-like nitrogen oxide is synthesized in the cardiovascular system published later in 1987 (Palmer *et al.*, 1987; Ignarro *et al.*, 1987a,b). Likewise it is almost certain, the discovery that a nitrogen oxide was the effector molecule causing endothelium dependent vascular relaxation occurred independently and without knowledge of work carried out in the toxicology field or in the study of the role of macrophages in CMI/innate resistance. However, in 1987 the independent work carried out in studies of the activated macrophage effector mechanism causing cytostasis and metabolic lesions had a major influence on the previously independent studies of endothelium dependent vascular relaxation. The finding that nitrogen oxides were synthesized from a terminal guanidino nitrogen atom of L-arginine by activated macrophages (Hibbs *et al.*, 1987b) was used by investigators in the field of vascular physiology (Fig. 2). This resulted in the demonstration in 1988 that the enzymatic activity generating nitrogen oxides in endothelial cells was strikingly similar to the origin of nitrogen oxides in activated macrophages (Palmer *et al.*, 1988a). This exchange of information had important consequences for the basic understanding of the biological synthesis of nitrogen oxides in the cardiovascular system. In a second report, Moncada and colleagues observed that L-NMMA inhibits both the generation of $NO^{\bullet}$ by endothelial cells in culture and the endothelium-dependent relaxation of rabbit aortic rings (Palmer *et al.*, 1988b). Both of these effects were reversed by L-arginine. Likewise, Stuehr, Nathan, and co-workers showed that agonist-induced relaxation of guinea pig pulmonary artery is strongly inhibited by L-NMMA, and that this inhibition can be immediately and completely reversed by L-arginine but not D-arginine (Sakuma *et al.*, 1988). These results suggested a close identity of the effector mechanism being studied in two very different bioassays: relaxation of vascular smooth muscle cells by endothelial cell-derived nitrogen oxide synthesis and induction of neoplastic cell cytostasis and other metabolic lesions by activated macrophage-derived nitrogen oxide synthesis. In both bioassays, the nitrogen oxide released by the effector cell was synthesized from a terminal guanidino nitrogen atom of L-arginine and the synthesis was inhibited by L-NMMA. In addition, in both bioassays the nitrogen oxide released by the effector cell stimulated (endothelial cells) or inhibited (activated macrophages) target cell enzymes with iron-dependent catalytic activity.

Two important differences were immediately apparent between nitrogen oxide synthesis from L-arginine by activated macrophages and nitrogen oxide synthesis by endothelial cells. Nitrogen oxide synthesis by macrophages is cytokine inducible and high output, while nitrogen oxide synthesis by endothelial cells is constitutive (agonist inducible) and low output (Hibbs *et al.*, 1990). On the other hand, similarities also existed. Both processes target iron or iron-containing enzymes. High-output nitrogen oxide synthesis from L-arginine by cytokine-activated macrophages induces a reproducible pattern of inhibition of certain enzymes with non-heme-iron-dependent catalytic activity in neoplastic target cells. Likewise, low-output nitrogen oxide synthesis from L-arginine by agonist-stimulated endothelial cells targets the heme prosthetic group of soluble guanylyl cyclase, increasing cyclic GMP levels and triggering vascular smooth muscle relaxation.

4. Other Advances in NO$^\bullet$ Biology Published in 1988

Hibbs *et al.* (1988) as well as Marletta *et al.* (1988) simultaneously reported the first evidence that mammalian cells synthesize NO$^\bullet$, as distinct from other nitrogen oxides, from a terminal guanidino nitrogen atom of L-arginine (Fig. 1). Hibbs and co-workers used the volatility of NO$^\bullet$ to separate it from other nitrogen oxides prior to oxidation and measurement (Hibbs *et al.*, 1988). Marletta *et al.* (1988), like Moncada's group (Palmer *et al.*, 1988a,b), used a chemiluminescence assay. However, in contrast to Moncada's group, Marletta and colleagues used conditions in which other nitrogen oxides in the sample (particularly nitrate and nitrite) were not reduced to NO$^\bullet$ prior to the measurement of NO$^\bullet$. The synthesis of NO$^\bullet$ per se from L-arginine by mammalian cells was thus established by these reports of Hibbs' and Marletta's groups in 1988. In addition, Marletta's group studied NOS activity in a 100,000 *g* supernatant prepared from cytokine-treated RAW 264.7 cells, and made the important observation that NADPH is a cosubstrate for NO$^\bullet$ synthesis from L-arginine (Marletta *et al.*, 1988). Therefore, in 1988, two groups of investigators demonstrated the synthesis of NO$^\bullet$ by an enzymatic activity later shown to be mediated by iNOS.

Also in 1988, Garthwaite *et al.* (1988) reported the synthesis of nitrogen oxide/EDRF activity by rat cerebellar cells. They showed that L-glutamate, by acting on *N*-methyl-D-aspartate (NMDA) receptors of cerebellar cells, induces calcium-dependent release of a nitrogen oxide/EDRF-like factor that increases cyclic GMP levels in the cerebellar cells. This suggested a role for NO$^\bullet$ as either a neurotransmitter or a modulator of neurotransmission, and demonstrated a parallel between nitrogen oxide synthesis by cells of the vascular system (endothelial cells) and by cells of the central nervous system (cerebellar cells). Nitrogen oxide synthesis by both cell types was constitutive (agonist-induced), low output, and functioned as an intercellular messenger by activating soluble guanylyl cyclase. Garthwaite and colleagues also observed that agonist-induced activation of nitrogen oxide synthesis is calcium dependent, linking induction of low-output NO$^\bullet$/EDRF synthesis by neural cells to activation of the previously well-studied NMDA type of glutamate receptor. Thus, also in 1988, Garthwaite and colleagues were the first to identify the activity later shown to be mediated by a third NOS isoform, nNOS (NOS1).

Several other reports published in 1988 further explained the properties of endothelium-derived NO•. Ignarro and co-workers provided additional pharmacological evidence suggesting that EDRF is NO• (Ignarro *et al.*, 1988a,b). They showed that acetylcholine-induced NO•/EDRF is not distinguishable from reagent NO• when either is exposed to superoxide anion, superoxide dismutase, oxyhemoglobin, or when compared using a cascade superfusion bioassay. These results confirmed earlier studies that suggested EDRF and NO• possess very similar pharmacological properties as relaxants of vascular smooth muscle. Moncada and colleagues provided evidence of a reciprocal relationship between NO• and prostacyclin synthesis (Doni *et al.*, 1988), and showed that acetylcholine induces vasodilation in the isolated rabbit heart via the release of NO• or a related nitrogen oxide (Amezcua *et al.*, 1988). Moncada's group also demonstrated that reagent NO•, like reagent prostacyclin, prevents platelet activation during isolation, washing, and storage (Radomski *et al.*, 1988).

Additional observations published in 1988 strengthened the link between cytokine-induced synthesis of NO• from L-arginine and earlier studies of the immunology and biochemistry of cytokine-induced macrophage cytotoxicity. Two groups of investigators, Drapier, Wietzerbin, and Hibbs (1988) and Ding, Nathan, and Stuehr (1988), independently showed that tumor necrosis factor (TNFα) is a potent second signal in the induction of high-output NO• synthesis from L-arginine in IFNγ-treated macrophages. In addition, Drapier *et al.* (1988) demonstrated that microbial products such as muramyl dipeptide and LPS act by inducing TNFα synthesis. TNFα is the actual physiological cosignal induced by microbial products and the final intermediary that induces high-output NO• synthesis by IFNγ-primed murine macrophages. This observation extended earlier studies by Hibbs and co-workers describing the IFNγ and LPS requirement for induction of activated macrophage cytotoxicity (Chapman and Hibbs, 1977; Hibbs *et al.*, 1977; Weinberg *et al.*, 1978) and explained how innate resistance is activated when microbes invade tissues (Drapier *et al.*, 1988). In addition, the findings of this study were important in understanding the pathogenesis of high-output NO• production from L-arginine in patients with septic shock (see Chapters 7 and 13).

Drapier and Hibbs further showed that cytokine-activated macrophages develop inhibition of mitochondrial aconitase as well as Complex I and Complex II of the mitochondrial electron transport chain (Drapier and Hibbs, 1988). This pattern of NO•-mediated inhibition of redox enzymes was identical to that demonstrated earlier in neoplastic target cells of activated macrophages (Granger and Lehninger, 1982; Drapier and Hibbs, 1986).

Hibbs and colleagues were the first to show that cytokines induce nonmacrophage cell lines (murine EMT-6 mammary adenocarcinoma cells and murine embryo fibroblasts) to synthesize high-output NO• from L-arginine (Amber *et al.*, 1988a,b). Interleukin 1 was found to be a potent second signal for high-output NO•

synthesis by nonmacrophage cells, but not by macrophages. The pattern of metabolic inhibition caused by endogenous NO• synthesis in cytokine-treated EMT-6 cells was identical to that observed earlier in the neoplastic target cells of activated macrophages (Granger *et al.*, 1980; Granger and Lehninger, 1982; Hibbs *et al.*, 1984; Drapier and Hibbs, 1986) and in the activated macrophage effector cells themselves (Drapier and Hibbs, 1988). These experiments showed that cytokine-induced high-output NO• synthesis from L-arginine is not restricted to cells specialized for host defense such as macrophages, but is potentially an activity of most somatic cells involved in cell-mediated immune reactions, expanding the scope of innate resistance and cellular immunity by showing that somatic cells not specialized for host defense have the potential to control the replication of intracellular pathogens via cytokine-induced high-output NO• synthesis. The studies also revealed a novel mechanism for cytokine-induced and NO•-mediated autotoxicity, that is, high-output NO• synthesis by nonmacrophage somatic cells. Another report published in 1988 by Granger, Hibbs, and colleagues demonstrated that the L-arginine-dependent effector mechanism has potent cytostatic effects for the facultative intracellular fungal pathogen *Cryptococcus neoformans* (Granger *et al.*, 1988), providing the initial evidence of NO-related antimicrobial activity (see Chapter 12).

Therefore, in 1987–1988, the broad outline of the scope of NO• biology had become visible. A cytokine-inducible enzymatic activity producing high levels of NO• and L-citrulline had been defined in several mammalian cell phenotypes. Inducible high-output NO• synthesis caused inhibition of redox enzymes with nonheme iron-dependent catalytic activity and was inhibited by L-NMMA. High-output NO• synthesis appeared to be a component of innate resistance and cell-mediated immunity, but also appeared to have the potential for producing autotoxicity or autoimmunity if not properly regulated.

It was also demonstrated in 1988 that an agonist-activated constitutive low-output enzymatic activity synthesizing NO• from L-arginine was present in endothelial cells. The small amount of NO• produced by the agonist-activated endothelial enzyme diffuses to vascular smooth muscle cells where it combines with the heme prosthetic group of soluble guanylyl cyclase, stimulates cyclic GMP formation, and causes smooth muscle relaxation. In endothelial cells, L-NMMA inhibits the synthesis of NO• from L-arginine and also inhibits both the elevation of cyclic GMP and the relaxation of vascular smooth muscle. In addition, an agonist-activated and calcium-dependent constitutive low-output NO•/EDRF-synthesizing enzymatic activity was observed in cells from the cerebellum. This diffusible NO•-like factor was observed to elevate cyclic GMP in neural cells, indicating an important role in NMDA receptor-mediated neurotransmission and suggesting that agonist-activated low-output NO• synthesis from L-arginine might represent a general mechanism for stimulation of soluble guanylyl cyclase.

In 1988, work in all of the areas of biomedical science reviewed above (macrophage cytotoxicity, nitrogen metabolism and nitrosamine toxicology, vascular physiology/pharmacology, and neurophysiology) was unified in the study of a single biochemical reaction. Studies published in 1987–1988 provided the essential information that ultimately led to the purification of three NOS isoforms, the elucidation of NOS enzymology, and the extension of the physiological and pathophysiological significance of $NO^{\bullet}$ synthesis from L-arginine to many new fields of study.

5. Isolation of the First NOS Isoform

During 1989 and 1990, publications in the field of NO biology began to increase exponentially, and the first international meeting in this new area was organized by Salvador Moncada and held in London in 1989. For this period of intense activity and rapidly accumulating knowledge, we will restrict our review to studies that directly led to purification of the first NOS isoform, reported by Bredt and Snyder early in 1989.

In 1989, Garthwaite and colleagues, following up their 1988 study, demonstrated that L-NMMA inhibits the elevation of cyclic GMP induced by NMDA in rat cerebellar slices (Garthwaite *et al.*, 1989). L-Arginine but not D-arginine augments the response to NMDA and reverses the inhibition by L-NMMA. The results indicated that stimulation of NMDA receptors activates an enzyme that catalyzes the formation of $NO^{\bullet}$ from L-arginine. Moncada and co-workers then examined soluble enzymatic activity in a preparation of crude synaptosomal cytosol prepared from rat forebrain (Knowles *et al.*, 1989). Their results with soluble enzyme fractions from rat forebrain were very similar to those of Marletta and colleagues (Marletta *et al.*, 1988), who used the $100,000 \times g$ supernatant from cytokine-treated RAW 264.7 cell lysates. Both groups demonstrated that the enzymatic synthesis of nitrogen oxides and L-citrulline from L-arginine was dependent on NADPH. The function of the enzyme from rat brain was, as noted by Garthwaite and co-workers (Garthwaite *et al.*, 1988), also dependent on calcium. Moncada and co-workers went on to demonstrate that L-NMMA inhibits nitrogen oxide and L-citrulline formation as well as guanylyl cyclase stimulation by the soluble enzyme from rat forebrain (Knowles *et al.*, 1989). Bredt and Snyder then demonstrated that glutamate and NMDA stimulate the synthesis of L-citrulline from L-arginine in parallel with enhancement of cyclic GMP levels in slices of rat cerebellar tissue (Bredt and Snyder, 1989). L-NMMA blocks NMDA-dependent stimulation of both L-citrulline synthesis and cyclic GMP elevation. These results confirmed that enzymatic synthesis of nitrogen oxides mediates the stimulation of cyclic GMP formation in brain tissue by glutamate. Early in 1990, Bredt and

Snyder reported purification to homogeneity of the neural NOS (nNOS, NOS1) from rat cerebellum (Bredt and Snyder, 1990), using a 2′,5′-ADP Sepharose affinity column eluted with NADPH. The purified enzyme migrated as a single 150-kDa band on SDS/PAGE gels. In addition, Bredt and Snyder made the important discovery that nNOS is a calcium/calmodulin-requiring enzyme. Mayer *et al.* (1990) also purified a calcium/calmodulin-dependent NOS from porcine cerebellum, confirming the results of Bredt and Snyder. The neural NOS, like the endothelial NOS purified later, is a low-output enzyme. Total nitrogen oxide synthesis does not approach the minimum threshold for measurement with the Greiss reaction (>1 μM nitrite). Therefore, in the studies reviewed above, catalytic activity of nNOS in crude synaptosomal cytosol preparations or catalytic activity of purified nNOS was detected by measuring the conversion of [^{3}H]arginine to [^{3}H]citrulline. Purification of the nNOS isoform in 1990 can be viewed as the conclusion of the initial phase in the development of the broad multidisciplinary field of NO$^\bullet$ biology.

6. Conclusions

We have reviewed the experimental work reported through 1988 that provided the foundation for the new field of NO$^\bullet$ biology. In addition, we have described experiments carried out between 1988 and 1990 that led directly to the purification of the first NOS isoform. A wide range of biomedical scientists have extended and refined the core information published in the period through 1988, and the study of NO$^\bullet$ biology has entered many new fields and resulted in many new discoveries. It is clear that this work has had, and will continue to have, a major impact on our understanding of physiology and pathophysiology.

References

Alexander, P., and Evans, R., 1971, Endotoxin and double stranded RNA render macrophages cytotoxic, *Nature New Biol.* **232:**76–78.

Amber, I. J., Hibbs, J. B., Jr., Taintor, R. R., and Vavrin, Z., 1988a, The L-arginine dependent effector mechanism is induced in murine adenocarcinoma cells by culture supernatant from cytotoxic activated macrophages, *J. Leukoc. Biol.* **43:**187–192.

Amber, I. J., Hibbs, J. B., Jr., Taintor, R. R., and Vavrin, Z., 1988b, Cytokines induce an L-arginine dependent effector system in non-macrophage cells, *J. Leukoc. Biol.* **44:**58–65.

Amezcua, J. L., Dusting, G. J., Palmer, R. M., and Moncada, S., 1988, Acetylcholine induces vasodilatation in the rabbit isolated heart through the release of nitric oxide, the endogenous nitrovasodilator, *Br. J. Pharmacol.* **95:**830–834.

Arnold, W. P., Mittal, C. K., Katsuki, S., and Murad, F., 1977, Nitric oxide activates guanylate cyclase and increases guanosine 3′:5′-cyclic monophosphate levels in various tissue preparations, *Proc. Natl. Acad. Sci. USA* **74:**3203–3207.

Bredt, D. S., and Snyder, S. H., 1989, Nitric oxide mediates glutamate-linked enhancement of cGMP levels in the cerebellum, *Proc. Natl. Acad. Sci. USA* **86:**9030–9033.

Bredt, D. S., and Snyder, S. H., 1990, Isolation of nitric oxide synthetase, a calmodulin-requiring enzyme, *Proc. Natl. Acad. Sci. USA* **87:**682–685.

Chapman, H. A., Jr., and Hibbs, J. B., Jr., 1977, Modulation of macrophage tumoricidal capability by components of normal serum: A central role for lipid, *Science* **197:**282–285.

Cherry, P. D., Furchgott, R. F., Zawadzki, J. V., and Jothianandan, D., 1982, Role of endothelial cells in relaxation of isolated arteries by bradykinin, *Proc. Natl. Acad. Sci. USA* **79:**2106–2110.

Craven, P. A., and DeRubertis, F. R., 1978, Restoration of the responsiveness of purified guanylate cyclase to nitrosoguanidine, nitric oxide, and related activators by heme and hemeproteins, *J. Biol. Chem.* **253:**8433–8443.

Craven, P. A., DeRubertis, F. R., and Pratt, D. W.,1979, Electron spin resonance study of the role of $NO^{\bullet}$ catalase in the activation of guanylate cyclase by NaN_3 and NH_2OH: Modulation of enzyme responses by heme proteins and their nitrosyl derivatives, *J. Biol. Chem.* **254:**8213–8221.

DeRubertis, F. R., Craven, P. A., and Pratt, D. W., 1978, Electron spin resonance study of the role of nitrosyl-heme in the activation of guanylate cyclase by nitrosoguanidine and related agonists, *Biochem. Biophys. Res. Commun.* **83:**158–167.

Ding, A. H., Nathan, C. F., and Stuehr, D. J., 1988, Release of reactive nitrogen intermediates and reactive oxygen intermediates from mouse peritoneal macrophages, *J. Immunol.* **141:**2407–2414.

Doni, M. G., Whittle, B. J., Palmer, R. M., and Moncada, S., 1988, Actions of nitric oxide on the release of prostacyclin from bovine endothelial cells in culture, *Eur. J. Pharmacol.* **151:**19–25.

Drapier, J.-C., and Hibbs, J. B., Jr., 1986, Murine cytotoxic activated macrophages inhibit aconitase in tumor cells. Inhibition involves the iron-sulfur prosthetic group and is reversible, *J. Clin. Invest.* **78:**790–797.

Drapier, J.-C., and Hibbs, J. B., Jr., 1988, Differentiation of murine macrophages to express nonspecific cytotoxicity for tumor cells results in L-arginine-dependent inhibition of mitochondrial iron-sulfur enzymes in the macrophage effector cells, *J. Immunol.* **140:**2829–2838.

Drapier, J.-C., Wietzerbin, J., and Hibbs, J. B., Jr., 1988, Synergism of gamma-interferon and tumor necrosis factor to induce the L-arginine dependent effector mechanism of cytotoxicity in murine macrophages, *Eur. J. Immunol.* **18:**1587–1592.

Furchgott, R. F., and Zawadzki, J. V., 1980, The obligatory role of endothelial cells in the relaxation of arterial smooth muscle by acetylcholine, *Nature* **288:**373–376.

Garthwaite, J., Charles, S. L., and Chess-Williams, R., 1988, Endothelium-derived relaxing factor release on activation of NMDA receptors suggests role as intracellular messenger in brain, *Nature* **336:**385–388.

Garthwaite, J., Garthwaite, G., Palmer, R. M., and Moncada, S., 1989, NMDA receptor activation induces nitric oxide synthesis from L-arginine in rat brain slices, *Eur. J. Pharmacol.* **172:**413–416.

Granger, D. L., and Lehninger, A. I., 1982, Sites of inhibition of mitochondrial electron transport in macrophage-injured neoplastic cells, *J. Cell Biol.* **95:**527–535.

Granger, D. L., Taintor, R. R., Cook, J. L., and Hibbs, J. B., Jr., 1980, Injury of neoplastic cells by murine macrophages leads to inhibition of mitochondrial respiration, *J. Clin. Invest.* **65:**357–370.

Granger, D. L., Hibbs, J. B., Jr., Perfect, J. R., and Durack, D. T., 1988, Specific amino acid (L-arginine) requirement for the microbiostatic activity of murine macrophages, *J. Clin. Invest.* **81:**1129–1136.

Green, L. C., Tannenbaum, S. R., and Goldman, P., 1981a, Nitrate synthesis in the germfree and conventional rat, *Science* **212:**56–58.

Green, L. C., De Luzuriaga, K. R., Wagner, D. A., Rand, W., Istfan, N., Young, V. R., and Tannenbaum, S. R., 1981b, Nitrate biosynthesis in man, *Proc. Natl. Acad. Sci. USA* **78:**7764–7768.

Gruetter, C. A., Barry, B. K., McNamara, D. B., Gruetter, D. Y., Kadowitz, P. J., and Ignarro, L. J., 1979, Relaxation of bovine coronary artery and activation of coronary arterial guanylate cyclase by nitric oxide, nitroprusside and a carcinogenic nitrosamine, *J. Cyclic Nucleotide Res.* **5:**211–224.

Gryglewski, R. J., Moncada, S., and Palmer, R. M., 1986a, Bioassay of prostacyclin and endothelium-derived relaxing factor (EDRF) from porcine aortic endothelial cells, *Br. J. Pharmacol.* **87:**685–694.

Gryglewski, R. J., Palmer, R. M., and Moncada, S., 1986b, Superoxide anion is involved in the breakdown of endothelium-derived vascular relaxing factor, *Nature* **320:**454–456.

Hibbs, J. B., Jr., 1973, Macrophage nonimmunologic recognition: target cell factors related to contact inhibition, *Science* **180:**868–870.

Hibbs, J. B., Jr., Lambert, L. H., Jr., and Remington, J. S., 1971, Resistance to murine tumors conferred by chronic infection with intracellular protozoa, *Toxoplasma gondii* and *Besnotia jellisoni*, *J. Infect. Dis.* **124:**587–592.

Hibbs, J. B., Jr., Lambert, L. H., Jr., and Remington, J. S., 1972a, Possible role of macrophage mediated nonspecific cytotoxicity in tumour resistance, *Nature New Biol.* **235:**48–50.

Hibbs, J. B., Jr., Lambert, L. H., Jr., and Remington, J. S., 1972b, Control of carcinogenesis: A possible role for the activated macrophage, *Science* **177:**998–1000.

Hibbs, J. B., Jr., Lambert, L. H., Jr., and Remington, J. S., 1972c, *In vitro* nonimmunologic destruction of cells with abnormal growth characteristics by adjuvant activated macrophages, *Proc. Soc. Exp. Biol. Med.* **139:**1049–1052.

Hibbs, J. B., Jr., Lambert, L. H., Jr., and Remington, J. S., 1972d, Adjuvant induced resistance to tumor development in mice, *Proc. Soc. Exp. Biol. Med.* **139:**1053–1056.

Hibbs, J. B., Jr., Taintor, R. R., Chapman, H. A., Jr., and Weinberg, J. B., 1977, Macrophage tumor killing: Influence of the local environment, *Science* **197:**282–285.

Hibbs, J. B., Jr., Taintor, R. R., and Vavrin, Z., 1984, Iron depletion: Possible cause of tumor cell toxicity induced by activated macrophages, *Biochem. Biophys. Res. Commun.* **123:**716–723.

Hibbs, J. B., Jr., Vavrin, Z., and Taintor, R. R.,1987a, L-Arginine is required for expression of the activated macrophage effector mechanism causing selective metabolic inhibition in target cells, *J. Immunol.* **138:**550–565.

Hibbs, J. B., Jr., Taintor, R. R., and Vavrin, Z., 1987b, Macrophage cytotoxicity: Role for L-arginine deiminase and imino nitrogen oxidation to nitrite, *Science* **235:**473–476.

Hibbs, J. B., Jr., Taintor, R. R., Vavrin, Z., and Rachlin, E. M., 1988, Nitric oxide: A cytotoxic activated macrophage effector molecule, *Biochem. Biophys. Res. Commun.* **157:**87–94.

Hibbs, J. B., Jr., Taintor, R. R., Vavrin, Z., Granger, D. L., Drapier, J.-C., Amber, I. J., and Lancaster, J. R., Jr., 1990, Synthesis of nitric oxide from a terminal guanidino nitrogen atom of L-arginine: A molecular mechanism regulating cellular proliferation that targets intracellular iron, in: *Nitric Oxide from L-Arginine, a Bioregulatory System* (S. Moncada and E. A. Higgs, eds.), Elsevier, Amsterdam, pp. 189–223.

Ignarro, L. J., Lippton, H., Edwards, J. C., Baricos, W. H., Hyman, A. L., Kadowitz, P. J., and Gruetter, C. A., 1981, Mechanism of vascular smooth muscle relaxation by organic nitrates, nitrites, nitroprusside and nitric oxide: Evidence for the involvement of *S*-nitrosothiols as active intermediates, *J. Pharmacol. Exp. Ther.* **218:**739–749.

Ignarro, L. J., Wood, K. S., and Wolin, M. S., 1982a, Activation of purified soluble guanylate cyclase by protoporphyrin IX, *Proc. Natl. Acad. Sci. USA* **79:**2870–2873.

Ignarro, L. J., Degnan, J. N., Baricos, W. H., Kadowitz, P. J., and Wolin, M. S., 1982b, Activation of purified guanylate cyclase by nitric oxide requires heme: Comparison of heme-deficient, heme-reconstituted and heme-containing forms of soluble enzyme from bovine lung, *Biochim. Biophys. Acta* **718:**49–59.

Ignarro, L. J., Byrns, R. E., Buga, G. M., and Wood, K. S., 1987a, Endothelium-derived relaxing factor from pulmonary artery and vein possesses pharmacological and chemical properties that are identical to those for nitric oxide radical, *Circ. Res.* **61:**866–879.

Ignarro, L. J., Buga, G. M., Wood, K. S., Byrns, R. E., and Chaudhuri, G., 1987b, Endothelium-derived relaxing factor produced and released from artery and vein is nitric oxide, *Proc. Natl. Acad. Sci.*

USA **84:**9265–9269.
Ignarro, L. J., Buga, G. M., Byrns, R. E., Wood, K. S., and Chaudhuri, G., 1988a, Endothelium-derived relaxing factor and nitric oxide possess identical pharmacologic properties as relaxants of bovine arterial and venous smooth muscle, *J. Pharmacol. Exp. Ther.* **246:**218–226.
Ignarro, L. J., Byrns, R. E., Buga, G. M., Wood, K. S., and Chaudhuri, G., 1988b, Pharmacological evidence that endothelium-derived relaxing factor is nitric oxide: Use of pyrogallol and superoxide dismutase to study endothelium-dependent and nitric oxide-elicited vascular smooth muscle relaxation, *J. Pharmacol. Exp. Ther.* **244:**181–190.
Iyengar, R., Stuehr, D. J., and Marletta, M. A., 1987, Macrophage synthesis of nitrite, nitrate, and *N*-nitrosamines: Precursors and role of the respiratory burst, *Proc. Natl. Acad. Sci. USA* **84:**6369–6373.
Katsuki, S., Arnold, W., Mittal, C., and Murad, F., 1977, Stimulation of guanylate cyclase by sodium nitroprusside, nitroglycerin and nitric oxide in various tissue preparations and comparison to the effects of sodium azide and hydroxylamine, *J. Cyclic Nucleotide Res.* **3:**23–35.
Keller, R., 1973, Cytostatic elimination of syngeneic rat tumor cells *in vitro* by nonspecifically activated macrophages, *J. Exp. Med.* **138:**625–644.
Knowles, R. G., Palacios, M., Palmer, R. M. J., and Moncada, S., 1989, Formation of nitric oxide from L-arginine in the central nervous system: A transduction mechanism for stimulation of the soluble guanylate cyclase, *Proc. Natl. Acad. Sci. USA* **86:**5159–5162.
Krahenbuhl, J. L., and Remington, J. S., 1974, The role of activated macrophages in specific and nonspecific cytostasis of tumor cells, *J. Immunol.* **113:**507–516.
Marletta, M. A., Yoon, P. S., Iyengar, R., Leaf, C. D., and Wishnok, J. S., 1988, Macrophage oxidation of L-arginine to nitrite and nitrate: Nitric oxide is an intermediate, *Biochemistry* **27:**8706–8711.
Martin, W., Villani, G. M., Jothianandan, D., and Furchgott, R. F., 1985, Selective blockade of endothelium-dependent and glyceryl trinitrate-induced relaxation by hemoglobin and by methylene blue in the rabbit aorta, *J. Pharmacol. Exp. Ther.* **232:**708–716.
Martin, W., Furchgott, R. F., Villani, G. M., and Jothianandan, D., 1986, Phosphodiesterase inhibitors induce endothelium-dependent relaxation of rat and rabbit aorta by potentiating the effects of spontaneously released endothelium-derived relaxing factor, *J. Pharmacol. Exp. Ther.* **237:**539–547.
Mayer, B., John, M., and Bohme, E., 1990, Purification of a Ca^{2+}/calmodulin-dependent nitric oxide synthase from porcine cerebellum, *FEBS Lett.* **277:**215–219.
Mellion, B. T., Ignarro, L. J., Ohlstein, E. H., Pontecorvo, E. G., Hyman, A. L., and Kadowitz, P. J., 1981, Evidence for the inhibitory role of guanosine 3′,5′-monophosphate in ADP-induced human platelet aggregation, *Blood* **57:**946–955.
Mitchell, H. H., Shonle, H. A., and Grindley, H. S., 1916, The origin of the nitrates in the urine, *J. Biol. Chem.* **24:**461–490.
Miwa, M., Stuehr, D. J., Marletta, M. A., Wishnok, J. S., and Tannenbaum, S. R., 1987, Nitrosation of amines by stimulated macrophages, *Carcinogenesis* **8:**955–958.
Murad, F., Mittal, C. K., Arnold, W. P., Katsuki, S., and Kimura, H., 1978, Guanylate cyclase: Activation by azide, nitro compounds, nitric oxide, and hydroxyl radical and inhibition by hemoglobin and myoglobin, *Adv. Cyclic Nucleotide Res.* **9:**145–158.
Nathan, C., 1992, Nitric oxide as a secretory product of mammalian cells, *FASEB J.* **6:**3051–3064.
Ohlstein, E. H., Wood, K. S., and Ignarro, L. J., 1982, Purification and properties of heme-deficient hepatic soluble guanylate cyclase: Effects of heme and other factors on enzyme activation by NO, NO-heme and protoporphyrin IX, *Arch. Biochem. Biophys.* **218:**187–198.
Palmer, R. M., Ferrige, A. G., and Moncada, S., 1987, Nitric oxide release accounts for the biological activity of endothelium-derived relaxing factor, *Nature* **327:**524–526.
Palmer, R. M. J., Ashton, D. S., and Moncada, S., 1988a, Vascular endothelial cells synthesize nitric oxide from L-arginine, *Nature* **333:**664–666.

Palmer, R. M. J., Rees, D. D., and Moncada, S., 1988b, L-Arginine is the physiological precursor for the formation of nitric oxide in endothelium-dependent relaxation, *Biochem. Biophys. Res. Commun.* **153:**1251–1256.

Radomski, M. W., Palmer, R. M., and Moncada, S., 1987a, Comparative pharmacology of endothelium-derived relaxing factor, nitric oxide and prostacyclin in platelets, *Br. J. Pharmacol.* **92:**181–187.

Radomski, M. W., Palmer, R. M., and Moncada, S., 1987b, Endogenous nitric oxide inhibits human platelet adhesion to vascular endothelium, *Lancet* **2:**1057–1058.

Radomski, M. W., Palmer, R. M., and Moncada, S., 1987c, The anti-aggregating properties of vascular endothelium: Interactions between prostacyclin and nitric oxide, *Br. J. Pharmacol.* **92:**639–646.

Radomski, M. W., Palmer, R. M., and Moncada, S., 1987d, The role of nitric oxide and cGMP in platelet adhesion to vascular endothelium, *Biochem. Biophys. Res. Commun.* **148:**1482–1489.

Radomski, M. W., Palmer, R. M., Read, N. G., and Moncada, S., 1988, Isolation and washing of human platelets with nitric oxide, *Thromb. Res.* **50:**537–546.

Rapoport, R. M., and Murad, F., 1983, Agonist-induced endothelium-dependent relaxation in rat thoracic aorta may be mediated through cGMP, *Circ. Res.* **52:**352–357.

Rapoport, R. M., Draznin, M. B., and Murad, F., 1983a, Endothelium-dependent vasodilator- and nitrovasodilator-induced relaxation may be mediated through cyclic GMP formation and cyclic GMP-dependent protein phosphorylation, *Trans. Assoc. Am. Physicians* **96:**19–30.

Rapoport, R. M., Draznin, M. B., and Murad, F., 1983b, Endothelium dependent relaxation in rat aorta may be mediated through cyclic GMP-dependent protein phosphorylation, *Nature* **306:**274–276.

Ruco, L. P., and Meltzer, M. S., 1978, Macrophage activation for tumor cytotoxicity: Development of macrophage cytotoxic activity requires completion of a sequence of short-lived intermediary reactions, *J. Immunol.* **121:**2035–2042.

Russell, S. W., Doe, W. F., and McIntosh, A. T., 1977, Functional characterization of a stable, noncytolytic stage of macrophage activation in tumors, *J. Exp. Med.* **146:**1511–1520.

Sakuma, I., Stuehr, D. J., Gross, S. S., Nathan, C., and Levi, R., 1988, Identification of arginine as a precursor of endothelium-derived relaxing factor, *Proc. Natl. Acad. Sci. USA* **85:**8664–8667.

Stuehr, D. J., and Marletta, M. A., 1985, Mammalian nitrate biosynthesis: Mouse macrophages produce nitrite and nitrate in response to *Escherichia coli* lipopolysaccharide, *Proc. Natl. Acad. Sci. USA* **82:**7738–7742.

Stuehr, D. J., and Marletta, M. A., 1987a, Induction of nitrite/nitrate synthesis in murine macrophages by BCG infection, lymphokines or interferon-γ, *J. Immunol.* **139:**518–525.

Stuehr, D. J., and Marletta, M. A., 1987b, Synthesis of nitrite and nitrate in murine macrophage cell lines, *Cancer Res.* **47:**5590–5594.

Wagner, D. A., and Tannenbaum, S. R., 1982, Enhancement of nitrate biosynthesis by *Escherichia coli* lipopolysaccharide, in: *Nitrosamines and Human Cancer*, Banbury Report 12 (P.N. Magee, ed.), Cold Spring Harbor Press, Cold Spring Harbor, N.Y., pp. 437–441.

Wagner, D. A., Young, V. R., and Tannenbaum, S. R., 1983a, Mammalian nitrate biosynthesis: Incorporation of $^{15}NH_3$ into nitrate is enhanced by endotoxin treatment, *Proc. Natl. Acad. Sci. USA* **80:**4518–4521.

Wagner, D. A., Schultz, D. S., Deen, W. N., Young, V. R., and Tannenbaum, S. R., 1983b, Metabolic fate of an oral dose of ^{15}N-labelled nitrate in humans: Effect of diet supplementation with ascorbic acid, *Cancer Res.* **43:**1921–1925.

Weinberg, J. B., and Hibbs, J. B., Jr., 1977, Endocytosis of red blood cells or hemoglobin by activated macrophages inhibits their tumoricidal effect, *Nature* **269:**245–247.

Weinberg, J. B., Chapman, H. A., Jr., and Hibbs, J. B., Jr., 1978, Characterization of the effects of endotoxin on macrophage tumor cell killing, *J. Immunol.* **121:**72–80.

Wolin, M. S., Wood, K. S., and Ignarro, L. J., 1982, Guanylate cyclase from bovine lung: A kinetic analysis of the regulation of the purified soluble enzyme by protoporphyrin IX, heme and nitrosyl-heme, *J. Biol. Chem.* **257:**13312–13320.

Part C

Biological Roles of Nitric Oxide

CHAPTER 3

Biochemistry of Nitric Oxide

BENJAMIN GASTON and JONATHAN S. STAMLER

Host–pathogen interactions involve every redox form of atomic nitrogen (−3 through +5); electron transfer chemistry is commonplace. Eukaryotic nitric oxide synthase (NOS) is a "battery" driving this current of immune information and bioactivities.

1. Nitric Oxide Synthase and Nitrogen Oxides

1.1. Isoforms and Structure

All isoforms of NOS catalyze oxidation of the guanidino nitrogen of L-arginine to nitric oxide (NO). They are categorized as "constitutive," calcium- and calmodulin-dependent NOSs—there is a neuronal constitutive NOS (nNOS, NOS1) and an endothelial constitutive NOS (eNOS, NOS3)—or "inducible," and calcium-independent (iNOS, NOS2). iNOS is regulated primarily at the level of transcription, and is capable of higher output of NO; such output can have untoward effects. The three isoforms are over 80% conserved between mammalian species, but are distinct from one another with approximately 60% amino acid sequence homology (Nathan and Xie, 1994; Förstermann and Kleinert, 1995; Albakri and Stuehr, 1996; Crane *et al.*, 1997). iNOS is approximately 130 kDa, eNOS is approximately 150 kDa, and nNOS is approximately 165 kDa, with its larger size resulting from an N-terminal PDZ domain.

In fact, each isoform of NOS can be constitutive and each is inducible: Inducibility is a function of the stimulus rather than the gene product. Moreover,

BENJAMIN GASTON • University of Virginia Health Sciences Center, Charlottesville, Virginia 22908. ***JONATHAN S. STAMLER*** • Howard Hughes Medical Institute and Duke University Medical Center, Durham, North Carolina 27710.

Nitric Oxide and Infection, edited by Fang. Kluwer Academic/Plenum Publishers, New York, 1999.

"low-level" NO production is typical of human iNOS (Asano *et al.*, 1994; Mannick *et al.*, 1994); thus, cytotoxicity may be a consequence of unregulated NO synthesis rather than the amount of NO produced. Each NOS isoform is active as a dimer, and each has been identified in soluble and particulate cell fractions (Nathan and Xie, 1994). All monomers have a C-terminal reductase domain (~residues 600–1100) that binds NADPH and flavoproteins (FMN, FAD), and an N-terminal oxygenase domain (~residues 1–500) that contains heme, to which substrate O_2 binds, as well as sites for binding tetrahydrobiopterin (cofactor) and L-arginine (cosubstrate) (Albakri and Stuehr, 1996; Sennequier and Stuehr, 1996; Crane *et al.*, 1997). NOS domains are separated by an intervening calmodulin-binding region that controls electronic communication. Dimerization is also requisite for electron transport between the reductase flavoproteins and the oxygenase heme (Siddhanta *et al.*, 1996). The enzymology of NO biosynthesis will be discussed in greater detail in Chapter 4.

1.2. Cofactors and Substrates

One mole of O_2 and 1.5 NADPH equivalents are believed to be required per mole of NO produced by NOS (Nathan and Xie, 1994; Sennequier and Stuehr, 1996), although the stoichiometry is controversial (Schmidt *et al.*, 1997). Initially, electrons are transferred from NADPH to flavins on the reductase domains where calmodulin binding—tightly configured in iNOS, but largely dependent on calcium fluxes in nNOS and eNOS (Gachhui *et al.*, 1996)—allows electron transfer to the oxygenase domain. Electrons are also transferred to O_2, generating superoxide ($O_2^{-\bullet}$) and hydrogen peroxide (H_2O_2) (Heinzel *et al.*, 1992); indeed, some reactive oxygen generation is typical of all NOS isoforms under all conditions. In the absence of L-arginine, however, such "uncoupling" may be more excessive. In this regard, iNOS upregulation in response to endotoxin is accompanied by upregulation of argininosuccinate synthase and argininosuccinate lyase, which recycle L-citrulline to L-arginine, to ensure substrate availability, thus normalizing $O_2^{-\bullet}$ production (Nagasake *et al.*, 1996). Of note, under some circumstances, cellular arginases may compete with NOS for substrate (Ignarro *et al.*, 1997). Similarly, defects in either H_4-biopterin or glutathione biosynthesis inhibit NO synthesis. H_4-biopterin-forming cyclohydrolases may also be coinduced with NOS (Harbrecht *et al.*, 1997; Hussain *et al.*, 1997).

1.3. Nitric Oxide Synthase Regulation

Transcriptional, translational, and posttranslational mechanisms of NOS regulation have recently been reviewed (Nathan and Xie, 1994; Förstermann and Kleinert, 1995; Sase and Michel, 1997). Typically, transcriptional regulation involves cytokines such as interferon-γ (IFNγ) and interleukin-1β (IL-1β) effecting iNOS expression, in part through nuclear factor κB (NF-κB) binding. But transcriptional regulation of nNOS and eNOS is also important, underscoring the imprecision of the "constitutive" and "inducible" nomenclature. For example, transcription of eNOS is upregulated by shear stress and downregulated by transforming growth factor-β (TGFβ) (Harrison *et al.*, 1996). On the other hand, iNOS may be constitutively expressed (Kobzik *et al.*, 1993; Asano *et al.*, 1994; Mannick *et al.*, 1994) and regulated at the level of translation, as evidenced by TGFβ destabilization of iNOS mRNA. Vascular endothelial growth factor (VEGF) and tumor necrosis factor-α (TNFα) have similar destabilizing effects on eNOS mRNA, while IFNγ increases both the transcription and the stability of iNOS mRNA (Nathan and Xie, 1994). Cytokine regulation of iNOS expression is discussed in greater detail in Chapter 5.

Posttranslational modifications of NO synthases regulate their subcellular localization and function. Calcium ions bind reversibly to nNOS and eNOS via electrostatic interactions with the regulatory protein calmodulin, in what is principally an ionic bond. Several covalent modifications of the enzyme(s) have also been described. In particular, myristoylation and palmitoylation target eNOS to caveolae (Garcia-Cardeña *et al.*, 1997; Sase and Michel, 1997), whereas nNOS has adopted the use of N-terminal PDZ domains to attach itself to membranes (Chao *et al.*, 1996); protein–protein interactions dictate the subcellular localization of this isoform rather than protein acylation (Bredt, 1996). iNOS may also be found in the particulate fraction, but the determinants of its subcellular localization are not known. Phosphorylation of all three NOS isoforms has been reported, but the type or nature (serine, threonine, tyrosine) of the modification and the functional significance is controversial. NOS dimerization is subject to regulation as well (Nathan and Xie, 1994; Duval *et al.*, 1995; Albakri and Stuehr, 1996; Harbrecht *et al.*, 1997), but little is known of the importance of this mechanism in cells. There is some indication that certain splice variants are expressed that cannot dimerize; they may have a function, but cannot produce NO. NO itself can inhibit heme incorporation into the oxygenase domain (Albakri and Stuehr, 1996), coordinate to the heme directly (perhaps providing a level of tonic inhibition), and feed back to inhibit the enzyme through reaction with a thiol group on the protein ("feedback inhibition") (Patel *et al.*, 1996). Additional regulation through protein–protein interactions is proving to be a common mechanism of regulation. For example, PIN and CAPON (Jaffrey and Snyder, 1996; Jaffrey *et al.*, 1998) inhibit nNOS, and caveolin inhibits eNOS (Garcia-Cardeña *et al.*, 1997).

1.4. Nitrogen Oxide Redox Forms

NOS products have been variously characterized as NO ($NO^{\bullet}$, +2 oxidation), nitroxyl (NO^-, +1), or thiol-bound nitrosonium (NO^+, +3) (Fukuto *et al.*, 1992; Stamler *et al.*, 1992; Stamler, 1994; Arnelle and Stamler, 1995; Schmidt *et al.*,

1997). This redox spectrum is familiar to microbiologists and ecologists who appreciate that fixed nitrogen in the form of nitrate (NO_3^-, +5) may be reduced sequentially by denitrifying organisms to nitrite (NO_2^-, +3), NO (+2), and nitrous oxide (N_2O, +1, the product of NO^- protonation, dimerization, and dehydration), as well as to ammonia and biological amines (Arai *et al.*, 1996). Further, eukaryotic "redox catalysis" is not limited to NOS activity. Mammalian proteins can promote redox reactions among NO-related species (Stamler *et al.*, 1992; Stamler, 1994). Other reactive oxides of nitrogen are now appreciated to be of critical biological relevance, including peroxynitrite ($ONOO^-$), peroxynitrous acid (ONOOH), and dinitrogen trioxide (N_2O_3) (Beckman *et al.*, 1992; Wink *et al.*, 1994). Under biological conditions (particularly in the presence of CO_2), $ONOO^-$ may donate nitronium (NO_2^+, +5), a potent nitrating species (Beckman *et al.*, 1992; Ischiropoulos *et al.*, 1992).

2. Reactions of Nitrogen Oxides

2.1. Inorganic Reactions with Oxygen Species

NO is present in tissues in concentrations of 1–3000 nM under normal physiological conditions (Pinsky *et al.*, 1997). Its diffusion is constrained and its bioactivity modified by its reactivity toward several cellular components (Lancaster, 1994; Rubbo *et al.*, 1994). The reaction of $NO^\bullet$ with O_2, however, is third order ($k \approx 6 \times 10^6\ M^{-2}\ sec^{-1}$) (Rubbo *et al.*, 1994; Wink *et al.*, 1994) [Reactions (1) and (2)], which is too slow to be of major importance in the cytosol under normal physiological conditions, but may predominate where NO concentrations exceed 10 μM (namely, in membranes, where $NO^\bullet$–O_2 concentrate, and under inflammatory conditions). Ultimately, $NO^\bullet$–O_2 reactions produce NO_2^- by way of N_2O_3:

$$\begin{aligned} O_2 + NO^\bullet &\Rightarrow NO_2{}^\bullet \\ NO^\bullet + NO_2{}^\bullet &\Rightarrow N_2O_3 \end{aligned} \tag{1}$$

$$N_2O_3 + H_2O \Rightarrow NO^+ \cdots NO_2^- + H_2O \Leftrightarrow 2HNO_2 \tag{2}$$

NO reactions with hydroxyl radical ($HO^\bullet$) ($k \approx 10^{10}\ M^{-1}\ sec^{-1}$) and $O_2^{-\bullet}$ [Reaction (3); $k \approx 10^9\ M^{-1}\ sec^{-1}$] are considerably more rapid than with O_2 (Ischiropoulos *et al.*, 1992; Rubbo *et al.*, 1994; Miles *et al.*, 1996):

$$NO^\bullet + O_2^{-\bullet} \Rightarrow ONOO^- \tag{3a}$$

$$ONOO^- + H^+ + \Rightarrow HO^\bullet + NO_2{}^\bullet \tag{3b}$$

Though reaction of $NO^{\bullet}$ with $HO^{\bullet}$ may serve as an efficient detoxification route, $HO^{\bullet}$ is scavenged equally well by many other molecules that are more prevalent in cells. Accordingly, reactions of $NO^{\bullet}$ with O_2 and $O_2^{-\bullet}$ are thought to be of greater relevance *in vivo*. Reaction of $NO^{\bullet}$ with $O_2^{-\bullet}$ proceeds three times more rapidly than the enzymatic $O_2^{-\bullet}$ dismutation by superoxide dismutase. Nitric oxide can also terminate free radical lipid oxidation cascades by reaction with peroxy or alkoxy radicals (Rubbo *et al.*, 1994).

2.2. Reactions with Transition Metals

Rapid reactions with iron-containing proteins are among the most common interactions of nitrogen oxides in mammalian organisms. For example, heme proteins are thought to participate in the formation of NO_3^- (the end NO oxidation product) *in vivo*, classically involving a reaction of $NO^{\bullet}$ or NO_2^- within oxyferrous hemoglobin [Reactions (4) and (5)] (Ignarro *et al.*, 1993). It should be noted, however, that the reactions are incompletely understood and their importance *in vivo* is not known:

$$\text{Hgb–Fe}^{2+}\text{–O}_2 + \text{NO}^{\bullet} \Rightarrow \text{Hgb–Fe}^{3+} + \text{NO}_3^- \tag{4}$$

$$4\text{HbO}_2\text{Fe}^{2+} + 4\text{NO}_2^- + 4\text{H}^+ \Rightarrow 4\text{HbFe}_3^+ + \text{O}_2 + 2\text{H}_2\text{O} + 4\text{NO}_3^- \tag{5}$$

For example, NO bioactivity is often portrayed uniquely as requiring binding to unliganded or deoxy heme in guanylyl cyclase to effect functional change. However, many other metal centers, such as those of iron–sulfur clusters which do not bind O_2, are also of biological importance (Green *et al.*, 1991; Vanin *et al.*, 1997). Reactions of NO_x with protein iron–sulfur clusters can result in sulfur oxidation or reduction (Vanin *et al.*, 1997), or in the formation of iron–nitrosyl intermediates that appear able to catalyze either the formation or the decomposition of *S*-nitrosothiols (SNOs). In other words, metals can support NO group exchange with thiols to form SNOs. An example of this type of reaction is as follows:

$$\text{Fe}^+(\text{NO}^+)_2(\text{RS}^-)_2 + \text{H}^+ \rightarrow \text{Fe}^{2+} + \text{RSNO} + \text{NO}^{\bullet} + \text{RSH} \tag{6}$$

Nitrogen oxide–copper interactions are also of importance, particularly with regard to SNO bioactivities (Gordge *et al.*, 1995, 1996). For example, NO undergoes a series of redox reactions with copper proteins including charge transfers (forming NO^+ equivalents) in type I copper species (Gorren *et al.*, 1987; Musci *et al.*, 1991; Kobayashi and Shoun, 1995). Similar charge transfers are evident in iron-containing proteins, including myoglobin, hemoglobin, and cytochrome P450 (Wade and Castro, 1990).

2.3. Nitration and Nitros(yl)ation Reactions of C, N, and S Groups

2.3.1. Carbon

NO^+ or NO_2^+ attack makes the ring of an NADH molecule susceptible to subsequent nucleophilic attack by thiolate anion, exemplified by glyceraldehyde phosphate dehydrogenase (GAPDH) (Mohr *et al.*, 1996). Specifically, the transfer of protein-bound pyridine cofactor to the active-site thiol of GAPDH, leading to irreversible inactivation, is promoted by cytokine-stimulated NO production. Nitrosation also occurs readily at the 3-carbon position of tyrosine. The event appears to be common during inflammation *in vivo* (Haddad *et al.*, 1994) and may contribute to tissue injury when excessive. Several species, including $ONOO^-$, nitrogen dioxide ($NO_2{}^\bullet$) and NO_2^-–HOCl, promote phenolic carbon nitration. The reaction of $ONOO^-$ is catalyzed by Cu^{2+}–SOD (Beckman *et al.*, 1992; Ischiropoulos *et al.*, 1992), and that of NO_2^- is catalyzed by myeloperoxidase (Eiserich *et al.*, 1998).

2.3.2. Nitrogen

Nitrogen oxides react with primary amines, leading to deamination, and with secondary or tertiary amines to form potentially mutagenic nitrosoamines (Miwa *et al.*, 1987). The potential of this type of reaction for mutagenicity can be demonstrated in *Salmonella* species exposed to 50 parts per million ($\sim 2\,\mu M$) NO gas under aerobic conditions, where it results in cytidine-to-thymidine mutation (Wink *et al.*, 1991).

2.3.3. Sulfur

S-Nitrosylation reactions are favored over reactions with nitrogen, carbon, and oxygen (Stamler, 1994; Simon *et al.*, 1996). *S*-Nitrosothiols are present in human plasma, airway lining fluid, biliary fluid, macrophages, neutrophils, platelets, epithelial cells, and a variety of other sites relevant to immune function (Clancy and Abramson, 1992; Gaston *et al.*, 1993, 1998; Clancy *et al.*, 1994; Stamler, 1994; Freedman *et al.*, 1995; Jia *et al.*, 1996; Minamiyama *et al.*, 1996; Gaston and Fang, 1997; Hothersall *et al.*, 1997). Most recently, constitutive levels of *S*-nitrosoglutathione (GSNO) have been measured at $\sim 7\,\mu M$ in cerebellar extracts (Kluge *et al.*, 1997). The reaction by which SNO forms *in vitro* is dependent on [NO]. At nanomolar concentrations, the reaction with thiol exhibits first-order dependence on [NO] and [O_2] [Reaction (1)] (Wink *et al.*, 1994; DeMaster *et al.*, 1995; Gow *et al.*, 1997). The reaction is second order in [NO] at higher concentrations. SNO synthesis may also be catalyzed by metals *in vivo*. For example, iron- or copper-mediated NO^+ formation and transfer to thiol has been demonstrated (Kharitonov, 1995; Vanin *et al.*, 1997), and heme proteins support similar chemistry (Stamler *et al.*, 1992). Hemoglobin catalyzes SNO formation in an O_2-mediated allosteric transition that is coupled to NO oxidation and transfer from heme to thiol (Gow and Stamler, 1998). Further, SNOs are formed from $ONOO^-$ (Moro *et al.*, 1994; Kharitonov, 1995; Vanin *et al.*, 1997). Once formed, proteins and low-mass SNO may undergo transnitrosation reactions that facilitate their bioactivity, transmembrane transport, and metabolism (Meyer *et al.*, 1994; Scharfstein *et al.*, 1994; DeGroote *et al.*, 1995; Gordge *et al.*, 1995; Gaston and Fang, 1997). These NO group transfer reactions may be catalyzed by enzymes such as glutathione-*S*-transferase (Peng *et al.*, 1995; Ji *et al.*, 1996; Stamler *et al.*, 1997). Sites of *S*-nitrosylation in proteins may be identified by an acid–base motif (Stamler *et al.*, 1997b).

SNO bioactivities are more often inhibited than facilitated by homolytic breakdown to NO (Lipton *et al.*, 1993; Kowaluk and Fung, 1990; Gaston *et al.*, 1994; Gordge *et al.*, 1995, 1996; Stamler *et al.*, 1997a). In this sense, NO production from SNO may represent a metabolic route of degradation (Hausladen *et al.*, 1996; Singh, 1996). *In vivo*, NO is liberated by thiol and ascorbate (Hausladen and Stamler, 1998).

In vitro, five enzymes have been shown to catabolize GSNO. Two produce NO, including thioredoxin reductase (Nikitovic and Holmgren, 1996) and glutathione peroxidase (Hou *et al.*, 1996). A third, γ-glutamyl transpeptidase, facilitates NO release by cleaving GSNO to less stable *S*-nitrosocysteinylglycine (Askew *et al.*, 1995; DeGroote *et al.*, 1995; Hogg *et al.*, 1997; Gordge *et al.*, 1998). The fourth enzyme is glutathione-dependent formaldehyde dehydrogenase, whose major product is hydroxylamine (Jensen *et al.*, 1998), and the last is xanthine oxidase, which ultimately generates $ONOO^-$ (Trujillo *et al.*, 1998). There are also at least three more unidentified activities (Gaston and Fang, 1997; Hausladen and Stamler, 1998; Gordge *et al.*, 1998). However, the physiological relevance of any of these enzymatic pathways has not been clearly established.

Exposure *in vitro* to iron (Vanin *et al.*, 1997) or copper also leads to rapid breakdown of certain SNOs (Askew *et al.*, 1995; Gordge *et al.*, 1996; Hogg *et al.*, 1997; Gaston *et al.*, 1998). Evidence exists for additional metabolic pathways that control the fate of SNO and $ONOO^-$ (Sies *et al.*, 1997) in activated neutrophils (Clancy and Abramson, 1992; Park, 1996), platelets (Gordge *et al.*, 1996), airway epithelial cells (Gaston and Fang, 1997), *Salmonella* (DeGroote *et al.*, 1997), and

Escherichia coli (Hausladen *et al.*, 1996). Interestingly, Cu^+ may serve as a true catalyst in the presence of reducing equivalents (McAninly *et al.*, 1993):

$$\begin{aligned} &Cu^+ + RSNO \rightarrow [RSNO{\cdot}Cu]^+ \rightarrow RS^- + NO^{\cdot} + Cu^{2+} \\ &Cu^{2+} + RS^- \rightarrow Cu^+ + RS^{\cdot} \end{aligned} \tag{7}$$

The physiologic importance of this reaction, however, remains to be determined as free copper concentrations are negligible in biological systems. Perhaps it is relevant that the concentration of free copper is higher in acidic lysosomes, where Cu^{2+} concentrations are regulated by a NO_3^-–permeable chloride/bicarbonate exchanger (Alda and Garay, 1990), or that regulation of organelle NO_2^- flux [to form intracellular HNO_2 ($pK_a = 3.4$) and $NO^{\cdot}$] may be of relevance to innate immunity: Nathan (1995) suggests the possibility that pH- and copper-determined equilibria between SNO and other bioactive nitrogen oxides may be exploited by cells to achieve intracellular killing of parasitic organisms.

2.4. Prokaryotic Reduction Pathways

Nitrogen oxide reductive pathways in pathologically relevant bacteria and fungi are characterized as *dissimilatory*, in which nitrogen oxides are utilized as electron acceptors for anaerobic respiration in place of O_2, or *assimilatory*, in which the product ammonia is oxidatively incorporated into amino acids for protein synthesis (Jeter *et al.*, 1984; Haas *et al.*, 1990; Kobayashi and Shoun, 1995; Arai *et al.*, 1996). Thus, organisms colonizing the mammalian host—such as *Pseudomonas* or *Aspergillus* species in the cystic fibrosis airway, or *E. coli* in the normal colon—have the potential to establish an entire nitrogen redox cycle. Of note, bacterial nitric oxide reductase and nitrous oxide reductase show some resemblance to mitochondrial copper-containing cytochrome oxidases, and assimilatory nitrate reductases and fumarate nitrite reductase contain iron–sulfur complexes (Haas *et al.*, 1990) subject to feedback regulation by NO (Arai *et al.*, 1996). Further, denitrifying systems have recently been identified in eukaryotic cells (Kobayaski and Shoun, 1995).

3. Nitrogen Oxide Chemistry and Host–Pathogen Interactions

3.1. Cytotoxic Reactions

Nitrogen oxides can cause cell necrosis or apoptosis through free radical-mediated mechanisms and covalent modifications of proteins that impart an oxidative and/or nitrosative stress. The cytotoxicity of NO is critically dependent on the redox milieu (Rubbo *et al.*, 1994; Sexton *et al.*, 1994; Stamler, 1994). For example, NO$^\bullet$ and $ONOO^-$ may be more toxic to mammalian tissues under oxidizing conditions that deplete thiol reserves (Stamler, 1994). Further, $ONOO^-$ may undergo protonation ($pK_a \approx 6.8$) followed by homolytic cleavage [Reaction (3b)] to form $NO_2^\bullet$ and HO$^\bullet$ (or some reactive equivalent), to initiate of free radical cascades and oxidations, or reaction with $HCO_3^- - CO_2$ that promotes heterolytic nitration reactions. Potential targets subject to such covalent modification include carbohydrates (Beckman *et al.*, 1992), lipids (Rubbo *et al.*, 1994), and thiols (Wink *et al.*, 1994) (see Table I).

Many bacteria and other pathogenic microorganisms appear to be more sensitive to N_2O_3, $ONOO^-$, or SNO than to NO$^\bullet$ itself (see Chapter 12). For example, transnitrosation reactions mediated by SNOs that gain access to intracellular compartments can decrease the growth of *Salmonella typhimurium* (DeGroote *et al.*, 1995, 1996). Likewise, the toxicity of NO$^\bullet$ for *Salmonella* is greatly enhanced by $O_2^{-\bullet}$ (DeGroote *et al.*, 1997).

Both prokaryotic and eukaryotic defense systems may be compromised by excessive protein thiol nitrosylation (Luperchio *et al.*, 1996; Petit *et al.*, 1996; Hothersall *et al.*, 1997), or in other words, by nitrosative stress. Specifically, critical metabolic enzymes such as GAPDH (Stamler, 1994; Mohr *et al.*, 1996; Padgett and Whorton, 1997), ribonucleotide reductase (Roy *et al.*, 1995), glutathione-*S*-transferase (GST) (Clark and Debnam, 1988), glutathione peroxidase (GS-Px) (Asahi *et al.*, 1995), glutathione reductase (Becker *et al.*, 1995), and γ-glutamyl-cysteine synthase (Kuo *et al.*, 1996) may be inhibited by *S*-nitrosylation and *S*-oxidation. Nitrosants (nitrosating agents) are enzymatically metabolized (Hausladen and Stamler, 1998); GS-Px and GST may participate in nitrosant breakdown (Freedman *et al.*, 1995; Hogg *et al.*, 1997), and their inhibition may exacerbate injury (Clark and Debnam, 1988; Becker *et al.*, 1995; Hogg *et al.*, 1997) (Table II). Once thiol levels fall, additional nucleophilic targets become susceptible

TABLE I
NO-Related Species: Reaction and Effect

Reaction	$ONOO^-$-mediated nitration reactions	NO-mediated reactions with iron–sulfur clusters	*S*-Nitrosylation reactions	Oxidation of vicinal hiols (SNO, $ONOO^-$, NO_x)
Example	Inhibition of Mn-superoxide dismutase, aconitase	Inhibition of ribonucleotide reductase, and cytochrome complexes I and II	Activation of p47^{phox}, p21^{ras}, and OxyR; inhibition of neutrophil hexose monophosphate shunt	Inhibition of protein kinase C

TABLE II
S-Nitrosothiols and Glutathione Metabolic Enzymes

Enzyme	Inhibition by (S)NO: mechanism	$IC_{50} K_i$	SNO breakdown	K_M	SNO synthesis	K_M
γ-Glutamylcysteine synthetase	Decreased (rat hepatocyte) enzyme activity through thiol nitrosylation	Not available	—	—	—	—
γ-Glutamyltranspeptidase	—	—	Type II bovine kidney (to *S*-nitrosocysteinyl glycine), competitive with GSH	28 μM	—	—
Glutathione-*S*-transferase	Competitive inhibition[a]	~0.015 mM	—		Rat liver microsome (from GSH and amyl nitrite)[a]	0.45 mM (for amyl nitrite)
Glutathione reductase	a. Competitive[a] b. Active Cys63 nitrosylation[a] S → N rearrangement (irreversible inhibition)	0.5 or (K_i) 1 mM	—	—	—	—
Glutathione peroxidase	(Bovine) nitrosylation of vicinal thiols	2 μM	Platelets[a]	5.4 mM	—	—

[a] Specific for *S*-nitrosoglutathione.

to nitrosative attack. In particular, nitrosation of DNA bases results in deamination reactions that are likely to initiate programmed cell death.

When the geometry and proximity of dithiols are favorable, SNOs or $ONOO^-$ may promote oxidation to disulfide (Stamler, 1994; Arnelle and Stamler, 1995; Campbell *et al.*, 1996). This reaction is exemplified in the inhibitory effects of SNOs on protein kinase C (Gopalakrishna *et al.*, 1993) and *Yersinia enterocolitica* phosphotyrosine protein phosphatase (Caselli *et al.*, 1994), as well as in NO-mediated inhibition of NOS itself (Patel *et al.*, 1996). Protein function may be adversely affected if the redox state in the cell in general, and in the protein in particular, is not rapidly normalized. Oxidative and nitrosative stresses may synergize through additional molecular mechanisms (Hausladen and Stamler, 1998). In this regard, NO potentiates H_2O_2-mediated bacterial killing, and SNO synergizes with H_2O_2 in mediating DNA damage. Some of the complex mechanisms involved are catalyzed by transition metals on the one hand (Park and Kim, 1994; Pacelli *et al.*, 1995), whereas nitrosylated iron complexes may be directly cytotoxic (through radical and nitrosating reactions) on the other hand. Additionally, NO may interfere with cellular iron metabolism by destabilizing the iron–sulfur center of cytosolic aconitase (iron response element binding protein) (Stamler, 1994) or inactivating mitochondrial respiration through interaction with iron–sulfur enzyme complexes (Petit *et al.*, 1996; Karupiah and Harris, 1995). Prokaryotic heme iron-containing metabolic enzymes such as Anr and fumarate reductase may also be affected by NO (Haas *et al.*, 1990; Arai *et al.*, 1996). For example, nitrate reductase couples Fe^{2+}-to-Fe^{3+} oxidation with NO production, but the enzyme expels NO before the iron is rereduced, preventing autoinhibition (via formation of Fe^{2+}–NO) (Fülöp *et al.*, 1995). In the presence of higher concentrations of exogenous NO, this conformational balance may be upset.

3.2. Signaling Reactions

Nitrogen oxide signaling employs the same chemistry that produces toxicity, namely, nitrosylation and oxidation reactions. However, in signaling, selectivity and specificity are maintained. For example, NO-mediated oxidation of transcriptional activating proteins can be regulatory in the case of the iron cluster in SoxR (Demple, 1997) or destructive at the site of zinc finger clusters (Kröncke *et al.*, 1994). Alternatively, nitrosylation of the thiol in OxyR (Hausladen *et al.*, 1996) or the NADPH oxidase (Park, 1996) is regulatory. Of note, differential reactivity of $NO^\bullet$ and NO^+ (donating species) may account for some differences in cellular control mechanisms at the level of transcription. For example, $NO^\bullet$ stabilizes IκB (thereby inhibiting NF-κB) in endothelial cells (Peng *et al.*, 1995), while NO^+ donors or *S*-nitrosylation reactions can stimulate $p21^{ras}$, thereby activating NF-κB in T cells and monocytes (Lander *et al.*, 1993, 1995). Another redox-sensitive transcriptional activator, Zta, appears to be involved in NO-mediated inhibition of Epstein–Barr viral replication (Mannick *et al.*, 1994).

3.3. Prokaryotic Responses to Oxidative and Nitrosative Stress

Bacteria and other pathogens defend themselves against nitrosative and oxidative stress using many of the same reactions used for immune signaling and cytotoxic events in host cells (see Chapter 12). For example, *E. coli* utilizes a thiol nitrosylation reaction to activate genes under control of OxyR, thereby increasing expression of proteins that enhance cellular glutathione production and GSNO metabolism. OxyR is activated by SNO and H_2O_2, but not by $NO^\bullet$ or NO_2^-. Increased GSH production, in turn, protects the organism against many of the cytotoxic reactions described for SNO and for H_2O_2 (nitrosative and oxidative stress) (Pacelli *et al.*, 1995; Hausladen *et al.*, 1996). Similarly, partial SNO resistance in *Salmonella typhimurium* is afforded by the *metL* gene product, which participates in the homocysteine synthetic pathway and replenishes intracellular homocysteine, which may detoxify exogenous SNO (DeGroote *et al.*, 1996). Resistance to the cytostatic effects of GSNO is also found in *Salmonella* lacking dipeptide transport, which prevents *S*-nitroso-γ-glutamyl-cysteine from entering bacteria after being cleaved from GSNO at the cell surface (DeGroote *et al.*, 1995).

3.4. Eukaryotic Nitrosative Stress Responses

As in prokaryotic cells, eukaryotic nitrosative and oxidative stress responses involve metalloproteins and thiol groups. For example, metallothionein protects against NO-mediated DNA damage and cell death in NIH 3T3 cells (Schwarz *et al.*, 1995) and increased cellular GSH production mitigates NO-mediated cytotoxicity in fibroblasts, macrophages, and tumor cells (Petit *et al.*, 1995; Walker *et al.*, 1995; Hothersall *et al.*, 1997). Furthermore, upregulation of iNOS transcription in response to cytokine stimulation in hepatocytes is mediated by mitochondrial reactive oxygen intermediates (Duval *et al.*, 1995). In this regard, it is important to remember that $O_2^{-\bullet}$ and $NO^\bullet$ may sometimes "quench" one another's cytotoxic effects, depending on the pH, thiol concentration, and redox status of the environment (Rubbo *et al.*, 1994; Miles *et al.*, 1996), or may serve a physiological signaling function.

4. Summary

Production, transport, breakdown, and activity of $NO^\bullet$, NO_2^-, SNO, $ONOO^-$, iron–nitrosyl species, and other less well characterized nitrogen oxides are regulated both by redox reactions in various cellular compartments, and by transcriptional, translational, and post translational modifications of a wide spectrum of proteins. These diverse reactions, in turn, are used (1) as signals by host cells and in defense against parasitic organisms, (2) to upregulate and downregulate expression of inflammatory proteins, (3) to produce cytotoxic effects, and (4) in adaptation of cellular defenses. Understanding of these complex reactions will provide an important foundation for the development of new antimicrobial therapies.

References

Albakri, Q. A., and Stuehr, D. J., 1996, Intracellular assembly of inducible NO synthase is limited by nitric oxide-mediated changes in heme insertion and availability, *J. Biol. Chem.* **271**:5414–5421.

Alda, J. O., and Garay, R., 1990, Chloride (or bicarbonate)-dependent copper uptake through the anion exchanger in human red blood cells, *Am. J. Physiol.* **259**:C570–C576.

Arai, H., Kawasaki, S., Igarashi, Y., and Kodama, T., 1996, Arrangement and transcriptional regulation of the denitrification genes in *Pseudomonas aeruginosa*, in: *Molecular Biology of Pseudomonas* (T. Nakazoua, ed.), ASM Press, Washington, D.C., pp. 298–308.

Arnelle, D. R., and Stamler, J. S., 1995, NO^+, $NO^\bullet$, and NO^- donation by *S*-nitrosothiols: Implications for regulation of physiological functions by *S*-nitrosylation and acceleration of disulfide formation, *Arch. Biochem. Biophys.* **318**:279–285.

Asahi, M., Fujii, J., Suzuki, K., Seo, H. G., Kuzuya, T., Hori, M., Tada, M., Fujii, S., and Taniguchi, N., 1995, Inactivation of glutathione peroxidase by nitric oxide, *J. Biol. Chem.* **270**:21035–21039.

Asano, K., Chee, C. B. E., Gaston, B., Lilly, C. M., Gerard, C., Drazen, J. M., and Stamler, J. S., 1994, Constitutive and inducible nitric oxide synthase gene expression, regulation, and activity in human lung epithelial cells, *Proc. Natl. Acad. Sci. USA* **91**:10089–10093.

Askew, S. C., Butler, A. R., Flitney, F. W., Kemp, G. D., and Megson, I. L., 1995, Chemical mechanisms underlying the vasodilator and platelet anti-aggregating properties of *S*-nitroso-*N*-acetyl-DL-penicillamine and *S*-nitrosoglutathione, *Bioorg. Med. Chem.* **3**:1–9.

Becker, K., Gui, M., and Schirmer, R. H., 1995, Inhibition of human glutathione reductase by *S*-nitrosoglutathione, *Eur. J. Biochem.* **234**:472–478.

Beckman, J. S., Ischiropoulos, H., Zhu, L., van der Woerd, M., Smith, C., Chen, J., Harrison, J., Martin, J. C., and Tsia, M., 1992, Kinetics of superoxide dismutase- and iron catalyzed nitration of phenolics by peroxynitrite, *Arch. Biochem. Biophys.* **298**:438–445.

Bredt, D. S., 1996, Targeting nitric oxide to its targets, *Proc. Soc. Exp. Biol. Med.* **211**:41–48.

Campbell, D. L., Stamler, J. S., and Strauss, H., 1996, Redox modulation of L-type calcium channels in ferret ventricular myocytes—Dual mechanism regulation by nitric oxide and *S*-nitrosothiols, *J. Gen. Physiol.* **108**:277–293.

Caselli, A., Camici, G., Manao, G., Moneti, G., Pazzagli, L., Cappugi, G., and Ramponi, G., 1994, Nitric oxide causes inactivation of the low molecular weight phosphotyrosine protein phosphatase, *J. Biol. Chem.* **269**:24878–24882.

Chao, D. S., Gorospe, J. R., Brenman, J. E., Rafael, J. A., Peters, M. F., Froehner, S. C., and Hoffman, E. G., 1996, Selective loss of sarcolemmal nitric oxide synthase in Becker muscular dystrophy, *J. Exp. Med.* **184**:609–618.

Clancy, R. M., and Abramson, S. B., 1992, Novel synthesis of *S*-nitrosoglutathione and degradation by human neutrophils, *Anal. Biochem.* **204**:365–371.

Clancy, R. M., Levartovsky, D., Leszczynska-Piziak, J., Yegudin, J., and Abramson, S. B., 1994, Nitric oxide reacts with intracellular glutathione and activates the hexose monophosphate shunt in human neutrophils: Evidence for *S*-nitrosoglutathione as a bioactive intermediary, *Proc. Natl. Acad. Sci. USA* **91:**3680–3684.

Clark, A. G., and Debnam, P., 1988, Inhibition of glutathione S-transferases from rat liver by *S*-nitroso-L-glutathione, *Biochem. Pharmacol.* **37:**3199–3201.

Crane, B. R., Arvai, A. S., Gacchui, R., Wu, C., Ghosh, D. K., Getzoff, E. D., Stuehr, D. J., and Tainer, J. A., 1997, The structure of nitric oxide synthase oxygenase domain and inhibitor complexes, *Science* **278:**425–431.

DeGroote, M. A., Granger, D., Xu, Y., Campbell, G., Prince, R., and Fang, F. C., 1995, Genetic and redox determinants of nitric oxide cytotoxicity in a *Salmonella typhimurium* model, *Proc. Natl. Acad. Sci. USA* **92:**6399–6403.

DeGroote, M. A., Testerman, T., Xu, Y., Stauffer, G., and Fang, F. C., 1996, Homocysteine antagonism of nitric oxide-related cytostasis in *Salmonella typhimurium*, *Science* **272:**414–417.

DeGroote, M. A., Ochsner, U. A., Shiloh, M. U., Nathan, C., McCord, J. M., Dinauer, M. C., Libby, S. J., Vazquez-Torres, A., Xu, Y., and Fang, F. C., 1997, Periplasmic superoxide dismutase protects *Salmonella* from products of phagocyte NADPH-oxidase and nitric oxide synthase, *Proc. Natl. Acad. Sci. USA* **94:**13997–14001.

DeMaster, E. G., Quast, B. J., Redfern, B., and Nagasawa, H. T., 1995, Reaction of nitric oxide with the free sulfhydryl group of human serum albumin yields a sulfinic acid and nitrous oxide, *Biochemistry* **34:**11494–11499.

Demple, B., 1997, Study of redox-regulated transcription factors in prokaryotes, *Methods* **11:**267–278.

Duval, D. L., Sieg, D. J., and Billings, R. E., 1995, Regulation of hepatic nitric oxide synthase by reactive oxygen intermediates and glutathione, *Arch. Biochem. Biophys.* **316:**699–706.

Eiserich, J. P., Hristova, M., Cross, C. E., Jones, A. D., Freeman, B. A., Halliwell, B., and van der Vliet, A., 1998, Formation of nitric oxide-derived inflammatory oxidants by myeloperoxide in neutrophils, *Nature* **391:**393–397.

Förstermann,U., and Kleinert, H., 1995, Nitric oxide synthase: Expression and expressional control of the three isoforms, *Naunyn-Schmiedebergs Arch. Pharmacol.* **352:**351–364.

Freedman, J. E., Frei, B., Welch, G. N., and Loscalzo, J., 1995, Glutathione peroxidase potentiates the inhibition of platelet function by *S*-nitrosothiols, *J. Clin. Invest.* **96:**394–400.

Fukuto, J. M., Chiang, K., Hszieh, R., Wong, P., and Chaudhuri, G., 1992, The pharmacological activity of nitroxyl: A potent vasodilator with activity similar to nitric oxide and/or endothelium-derived relaxing factor, *J. Pharmacol. Exp. Ther.* **263:**546–551.

Fülöp, V., Moir, J. W. B., Ferguson, S. J., and Hajdu, J., 1995, The anatomy of a bifunctional enzyme: Structural basis for reduction of oxygen to water and synthesis of nitric oxide by cytochrome cd_1, *Cell* **81:**369–377.

Gachhui, R., Presta, A., Bentley, D. F., Abu-Soud, H. M., McArthur, R., Brudvig, G., Ghosh, D. K., and Stuehr, D. J., 1996, Characterization of the reductase domain of rat neuronal nitric oxide synthase generated in the methylotrophic yeast *Pichia pastoris*, *J. Biol. Chem.* **271:**20594–20602.

Garcia-Cardeña, G., Martasek, P., Liu, J., Fan, R., Masters, B. S. S., Lisanti, M. P., and Sessa, W.C., 1997, Inactivation of nitric oxide synthases by direct interactions with caveolins, *Jpn. J. Pharmacol.* **75:**5P.

Gaston, B., and Fang, K., 1997, *S*-Nitrosoglutathione breakdown by airway epithelial cells: A mechanism for *S*-nitrosothiol deficiency in the asthmatic airway, *Jpn. J. Pharmacol.* **75:**111P.

Gaston, B., Reilly, J., Drazen, J. M., Fackler, J., Ramdev, P., Arnelle, D., Mullins, M. E., Sugarbaker, D.., Chee, C., Singel, D. J., Loscalzo, J., and Stamler, J. S., 1993, Endogenous nitrogen oxides and bronchodilator nitrosothiols in human airways, *Proc. Natl. Acad. Sci. USA* **90:**10957–10961.

Gaston, B., Drazen, J. M., Jansen, A., Sugarbaker, D.., Loscalzo, J., Richards, W., and Stamler, J. S., 1994, Relaxation of human bronchial smooth muscle by *S*-nitrosothiols *in vitro*, *J. Pharmacol. Exp. Ther.* **268:**978–985.

Gaston, B., Sears, S., Woods, J., Hunt, J., Ponaman, M., McMahon, T., and Stamler, J. S., 1998, Bronchodilator *S*-nitrosothiol deficiency in asthmatic respiratory failure, *Lancet* **351:**1317–1319.

Gopalakrishna, R., Chen, Z. H., and Gundimeda, U., 1993, Nitric oxide and nitric oxide-generating agents induce a reversible inactivation of protein kinase C activity and phorbol ester binding, *J. Biol. Chem.* **268:**27180–27185.

Gordge, M. P., Meyer, D. J., Hothersall, J., Neild, G. H., Payne, N. N., and Dutra-Noronha, A., 1995, Copper chelation-induced reduction of the biological activity of *S*-nitrosothiols, *Br. J. Pharmacol.* **114:**1083–1089.

Gordge, M. P., Hothersall, J. S., Neild, G. H., and Dutra, A. A. N., 1996, Role of a copper (I)-dependent enzyme in the anti-platelet action of *S*-nitrosoglutathione, *Br. J. Pharmacol.* **119:**533–538.

Gordge, M. P., Addis, P., Noronha-Dutra, A., and Hothersall, J., 1998, Cell-mediated biotransformation of *S*-nitrosoglutathione, *Biochem. Pharmacol.* **55:**657–665.

Gorren, A. C. F., de Boer, E., and Wever, R., 1987, The reaction of nitric oxide with copper proteins and the photodissociation of copper–NO complexes, *Biochim. Biophys. Acta* **916:**38–47.

Gow, A. J., and Stamler, J. S., 1998, Reactions between nitric oxide and haemoglobin under physiological conditions, *Nature* **391:**169–174.

Gow, A. J., Buerk, D. G., and Ischiropoulos, H., 1997, A novel mechanism for the formulation of *S*-nitrosothiol *in vivo*, *J. Biol. Chem.* **272:**2841–2845.

Green, S. J., Nacy, C. A., and Meltzer, M. S., 1991, Cytokine-induced synthesis of nitrogen oxides in macrophages: A protective host response to *Leishmania* and other intracellular pathogens, *J. Leukoc. Biol.* **50:**93–103.

Haas, D., Gamper, M., and Zimmerman, A., 1990, Arginine network of *Pseudomonas aeruginosa:* Specific and global controls, in: *Pseudomonas: Biotransformations, Pathogenesis and Evolving Biotechnology* (S. Silver, ed.), American Society for Microbiology, Washington, D.C., pp. 303–316.

Haddad, Y., Pataki, G., Hu, P., Galiani, C., Beckman, J. S., and Matalon, S., 1994, Quantitation of nitrotyrosine levels in lung sections of patients and animals with acute lung injury, *J. Clin. Invest.* **94:**2407–2413.

Harbrecht, B. G., Di Silvio, M., Chough, V., Kim, Y. M., Simmons, R., and Billiar, T. M., 1997, Glutathione regulates nitric oxide synthase in cultured hepatocytes, *Ann. Surg.* **225:**76–87.

Harrison, D. G., Sayegh, H., Ohara, Y., Inoue, N., and Venema, R. C., 1996, Regulation of expression of the endothelial cell nitric oxide synthase, *Clin. Exp. Pharmacol. Physiol.* **23:**251–255.

Hausladen, A., and Stamler, J. S., 1999, Nitrosative stress, *Methods Enzymol.* **300:**389–395.

Hausladen, A., Privalle, C. T., Keng, T., DeAngelo, J., and Stamler, J. S., 1996, Nitrosative stress: Activation of the transcription factor OxyR, *Cell* **86:**719–729.

Heinzel, B., Mathias, J., Klatt, P., Bohme, E., and Mayer, B., 1992, Ca^{2+}/calmodulin-dependent formation of hydrogen peroxide by brain nitric oxide synthase, *Biochem. J.* **281:**627–630.

Heiss, L., Lancaster, J. R., Corbett, J. A., and Goldman, W. E., 1994, Epithelial autotoxicity of nitric oxide: Role in the respiratory cytopathology of pertussis, *Proc. Natl. Acad. Sci. USA* **91:**267–270.

Hogg, N., Singh, R. J., Konorev, E., Joseph, J., and Kalyanaraman, B., 1997, *S*-Nitroglutathione as a substrate for γ-glutamyl transpeptidase, *Biochem. J.* **323:**477–481.

Hothersall, J. S., Cunha, F. Q., Neild, G. H., and Noronha-Dutra, A. A., 1997, Induction of nitric oxide synthesis in J774 cells lowers intracellular glutathione: Effect of modulated glutathione redox status on nitric oxide synthase induction, *Biochem. J.* **322:**477–481.

Hou, Y., Guo, Z., Li, J., and Wang, P. G., 1996, Seleno compounds and glutathione peroxidase catalyzed decomposition of *S*-nitrosothiols, *Biochem. Biophys. Res. Commun.* **228:**88–93.

Hussain, N. A., Giaid, A., Dawiri, Q. E., Sakkal, D., Hattori, R., and Guo, Y., 1997, Expression of nitric oxide synthases and GTP cyclohydrolase I in the ventilatory and limb muscles during endotoxemia, *Am. J. Respir. Cell Mol. Biol.* **17:**173–180.

Ignarro, L. J., Fukuto, J. M., Griscavage, J. M., Rogers, N. E., and Byrns, R. E., 1993, Oxidation of nitric oxide in aqueous solution to nitrite but not nitrate: Comparison with enzymatically formed nitric oxide from L-arginine, *Proc. Natl. Acad. Sci.USA* **90:**8103–8107.

Ignarro, L., Buga, G., Wei, L., and Fukuto, J., 1997, Co-regulation of inducible NO synthase and arginase in endothelial and tumor cells, *Jpn. J. Pharmacol.* **75:**4P.

Ischiropoulos, H., Zhu, L., Chen, J., Tsai, M., Martin, J. C., Smith, C. D., and Beckman, J. S., 1992, Peroxynitrite-mediated tyrosine nitration catalyzed by superoxide dismutase, *Arch. Biochem. Biophys.* **298:**431–437.

Jaffrey, S. R., and Snyder, S. H., 1996, PIN: An associated protein inhibitor of neuronal nitric oxide synthase, *Science* **274:**774–777.

Jaffrey, S. R., Snowman,A. M., Eliasson, M. J. L., and Snyder, S. H., 1998, CAPON, a protein associated with neuronal nitric oxide synthase which regulates its interactions with PSD95, *Neuron* **20:**115–124.

Jensen, D., Belka, G., and Du Bois, G., 1998, *S*-Nitrosoglutathione is a substrate for rat alcohol dehydrogenase class III isozyme, *Biochem. J.* **331:**659–668.

Jeter, R. M., Sias, S. R., and Ingraham, J. L., 1984, Chromosomal location and function of genes affecting *Pseudomonas aeruginosa* nitrate assimilation, *J. Bacteriol.* **157:**673–677.

Ji, Y., Akerboom, T. P. M., and Sies, H., 1996, Microsomal formation of *S*-nitroglutathione from organic nitrites: Possible role of membrane-bound glutathione transferase, *Biochem. J.* **313:**377–380.

Jia, L., Bonaventura, C., Bonaventura, J., and Stamler, J. S., 1996, *S*-Nitrosohaemoglobin: A dynamic activity of blood involved in vascular control, *Nature* **380:**221–226.

Karupiah, G., and Harris, N., 1995, Inhibition of viral replication by nitric oxide and its reversal by ferrous sulfate and tricarboxylic acid cycle metabolites, *J. Exp. Med.* **181:**2171–2179.

Kharitonov, V., 1995, Kinetics of nitrosation of thiols by nitric oxide in the presence of oxygen, *J. Biol. Chem.* **270:**28158–28164.

Kluge, I., Gutteck-Amsler, U., Zollinger, M., and Do, K. Q., 1997, *S*-Nitrosoglutathione in rat cerebellum: Identification and quantification by liquid chromatography–mass spectrometry, *J. Neurochem.* **69:**2599–2607.

Kobayashi, M., and Shoun, H., 1995, The copper-containing dissimilatory nitrite reductase involved in the denitrifying system of the fungus *Fusarium oxysporum*, *J. Biol. Chem.* **270:**4146–4151.

Kobzik, L., Bredt, D. S., Lowenstein, C. J., Drazen, J., Gaston, B., Sugarbaker, D., and Stamler, J. S., 1993, Nitric oxide synthase in human and rat lung: Immunocytochemical and histochemical localization, *Am. J. Respir. Cell Mol. Biol.* **9:**371–377.

Kowaluk, E. A., and Fung, H. L., 1990, Spontaneous liberation of nitric oxide cannot account for *in vitro* vascular relaxation by *S*-nitrosothiols, *J. Pharmacol. Exp. Ther.* **255:**1256–1264.

Kröncke, K. D., Fechsel, K., Schmidt, T., Zenke, F. T., Dasting, I., Wesener, J. R., Betterman, H., Breunig, K. D., and Kolb-Bachofen, V., 1994, Nitric oxide destroys zinc-sulfur clusters inducing zinc release from metallothionein and inhibition of the zinc finger-type yeast transcription activator Lac9, *Biochem. Biophys. Res. Commun.* **200:**1105–1110.

Kuo, P. C., Abe, K. Y., and Schroeder, R. A., 1996, Interleukin-1-induced nitric oxide production modulates glutathione synthesis in cultured rat hepatocytes, *Am. J. Physiol.* **271:**C851–C862.

Lancaster, J. R., Jr., 1994, Simulation of the diffusion and reaction of endogenously produced nitric oxide, *Proc. Natl. Acad. Sci. USA* **91:**8137–8141.

Lander, H. M., Sehajpal, P., Levine, D. M., and Novogrodsky, A., 1993, Activation of human peripheral blood mononuclear cells by nitric oxide-generating compounds, *J. Immunol.* **150:**1509–1516.

Lander, H. M., Ogiste, J. S., Pearce, S. F. A., Levi, R., and Novogrodsky, A., 1995, Nitric oxide-stimulated guanine nucleotide exchange on $p21^{ras}$, *J. Biol. Chem.* **270:**7017–7020.

Lipton, S. A., Chol, Y. B., Pan, Z. H., Lel, S. Z., Chen, H. S. V., Sucher, N. J., Loscalzo, J., Singel, D. J., and Stamler, J. S., 1993, A redox-based mechanism for the neuroprotective and neurodestructive effects of nitric oxide and related nitroso-compounds, *Nature* **364:**626–632.

Luperchio, S., Tamir, S., and Tannenbaum, S. R., 1996, NO-induced oxidative stress and glutathione metabolism in rodent and human cells, *Free Radical Biol. Med.* **21:**513–519.

Mannick, J. B., Asano, K., Izumi, K., Kleff, E., and Stamler, J. S., 1994, Nitric oxide produced by human B lymphocytes inhibits apoptosis and Epstein–Barr virus reactivation, *Cell* **79:**1137–1146.

McAninly, J., Williams, D. L. H., Askew, S. C., Butler, A. R., and Russell, C., 1993, Metal ion catalysis in nitrosothiol (RSNO) decomposition, *J. Chem. Soc. Chem. Commun.* **1993:**1758–1759.

Meyer, D. J., Kramer, H., Ozer, N., Coles, B., and Ketterer, B., 1994, Kinetics and equilibria of *S*-nitrosothiol-thiol exchange between glutathione, cysteine, penicillamines and serum albumin, *FEBS Lett.* **345:**177–180.

Miles, A. M., Bohle, D. S., Glassbrenner, P. A., Hansert, B., Wink, D. A., and Grishma, M. B., 1996, Modulation of superoxide-dependent oxidation and hydroxylation reactions by nitric oxide, *J. Biol. Chem.* **271:**40–47.

Minamiyama, Y., Takemura, S., Koyama, K., Yu, H., Miyamoto, M., and Inoue, M., 1996, Dynamic aspects of glutathione and nitric oxide metabolism in endotoxemic rats, *Am. J. Physiol.* **271:**G575–G581.

Miwa, M., Stuehr, D. J., Marletta, M. A., Wishnok, J. S., and Tannenbaum, S. R., 1987, Nitrosation of amines by stimulated macrophages, *Carcinogenesis* **8:**955–958.

Mohr, S., Stamler, J. S., and Brune, B., 1994, Mechanism of covalent modification of glyceraldehyde-3-phosphate dehydrogenase at its active site thiol by nitric oxide, peroxynitrite and related nitrosating agents, *J. Biol. Chem.* **271:**223–227.

Mohr, S., Stamler, J. S., and Brune, B., 1996, Posttranslational modification of glyceraldehyde-3-phosphate dehydrogenase by *S*-nitrosylation and subsequent NADH attachment, *J. Biol. Chem.* **8:**4209–4214.

Moro, M. A., Darley-Usmar, U. M., Goodwin, D. A., Read, N. G., Zamora-Pino, R., Feelisch, M., Radomski, M. W., and Moncada, S., 1994, Paradoxical fate and biological action of peroxynitrite on human platelets, *Proc. Natl. Acad. Sci. USA* **91:**6702–6706.

Musci, G., Di Marco, S., di Patti, M. C. B., and Calabrese, L., 1991, Interaction of nitric oxide with ceruloplasmin lacking an EPR-detectable type 2 copper, *Biochemistry* **30:**9866–9872.

Nagasake, A., Gotoh, T., Takeya, M., Yu, Y., Takiguchi, M., Matsuzaki, H., Takatsuki, K., and Mori, M., 1996, Coinduction of nitric oxide synthase, argininosuccinate synthetase, and argininosuccinate lyase in lipopolysaccharide-treated rats, *J. Biol. Chem.* **271:**2658–2662.

Nathan, C., 1995, Natural resistance and nitric oxide, *Cell* **82:**873–876.

Nathan, C., and Xie, Q., 1994, Regulation of biosynthesis of nitric oxide, *J. Biol. Chem.* **269:**13725–13728.

Nikitovic, D., and Holmgren, A., 1996, *S*-Nitrosoglutathione is cleaved by the thioredoxin system with liberation of glutathione and redox regulating nitric oxide, *J. Biol. Chem.* **271:**19180–19185.

Pacelli, R., Wink, D. A., Cook, J. A., Krishna, M. C., DeGraff, W., Friedman, N., Tsokos, M., Samuni, A., and Mitchell, J. B., 1995, Nitric oxide potentiates hydrogen peroxide-induced killing of *Escherichia coli*, *J. Exp. Med.* **182:**1469–1479.

Padgett, C. M., and Whorton, A. R., 1997, Glutathione redox cycle regulates nitric oxide-mediated glyceraldehyde-3-phosphate dehydrogenase inhibition, *Am. J. Physiol.* **272:**C99–C108.

Park, J. W., 1996, Attenuation of $p47^{phox}$ membrane translocation as the inhibitory mechanism of *S*-nitrosothiol on the respiratory burst oxidase in human neutrophils, *Biochem. Biophys. Res. Commun.* **220:**31–35.

Park, J. W., and Kim, H. K., 1994, Strand scission in DNA induced by *S*-nitrosothiol with hydrogen peroxide, *Biochem. Biophys. Res. Commun.* **200:**966–972.

Patel, J. M., Zhang, J., and Block, E. R., 1996, Nitric oxide-induced inhibition of lung endothelial cell nitric oxide synthase via interaction with allosteric thiols: Role of thioredoxin in regulation of catalytic activity, *Am. J. Respir. Cell Biol.* **15:**410–419.

Peng, H. B., Libby, P., and Liao, J. K., 1995, Induction and stabilization of IκBα by nitric oxide mediates inhibition of NF-κB, *J. Biol. Chem.* **270:**14214–14219.

Petit, J. F., Nicaise, M., Lepoivre, M., Guissani, A., and Lemaire, G., 1996, Protection by glutathione against the antiproliferative effects of nitric oxide, *Biochem. Pharmacol.* **52:**205–212.

Pinsky, D. J., Patton, S., Mesaros, S., Brovkovych, V., Kubaszewski, E., Grunfeld, S., and Malinski, T., 1997, Mechanical transduction of nitric oxide synthesis in the beating heart, *Circ. Res.* **81:**372–379.

Roy, B., Lepoivre, M., Henry, Y., and Fontecave, M., 1995, Inhibition of ribonucleotide reductase by nitric oxide derived from thionitrites: Reversible modifications of both subunits, *Biochemistry* **34:**5411–5418.

Rubbo, H., Radi, R., Trujillo, M., Telleri, R., Kalyanaraman, B., Barnes, S., Kirk, M., and Freeman, B. A., 1994, Nitric oxide regulation of superoxide and peroxynitrite-dependent lipid peroxidation, *J. Biol. Chem.* **269:**26066–26075.

Sase, K., and Michel, T., 1997, Expression and regulation of endothelial nitric oxide synthase, *Trends Cardiovasc. Med.* **7:**28–37.

Scharfstein, J. S., Keaney, J. F., Jr., Silvka, A., Welch, G. N., Vita, J. A., Stamler, J. S., and Loscalzo, J., 1994, *In vivo* transfer of nitric oxide between a plasma protein-bound reservoir and low molecular weight thiols, *J. Clin. Invest.* **94:**1432–1439.

Schmidt, H. H., Hofmann, H., Schiendler, U., Shufenko, Z. S., Cunningham, D. D., and Feelisch, M., 1997, No NO$^\bullet$ from NO synthase, *Proc. Natl. Acad. Sci. USA* **93:**14492–14497.

Schwarz, M. A., Laxo, J. S., Yalowich, J. C., Allen, W. P., Whitmore, M., Bergonia, H. A., Tzeng, E., Billiar, T. R., Robbins, P. D., Lancaster, J. R., Jr., and Pitt, B. R., 1995, Metallothionein protects against the cytotoxic and DNA-damaging effects of nitric oxide, *Proc. Natl. Acad. Sci. USA* **92:**4452–4456.

Sennequier, N., and Stuehr, D. J., 1996, Analysis of substrate-induced electronic, catalytic, and structural changes in inducible NO synthase, *Biochemistry* **35:**5883–5892.

Sexton, D. J., Muruganandam, A., McKenney, D. J., and Mutus, B., 1994, Visible light photochemical release of nitric oxide from *S*-nitrosoglutathione: Potential photochemotherapeutic applications, *Photochem. Photobiol.* **59:**463–467.

Siddhanta, J., Wu, C., Abu-Soud, H. M., Zhang, J., Ghosh, D. K., and Stuehr, D. J., 1996, Heme iron reduction and catalysis by a nitric oxide synthase heterodimer containing one reductase and two oxygenase domains, *J. Biol. Chem.* **271:**7309–7312.

Sies, H., Sharov, V. S., Klotz, L. O., and Briviba, K., 1997, Glutathione peroxidase protects against peroxynitrite-mediated oxidations, *J. Biol. Chem.* **272:**27812–27817.

Simon, D. I., Mullins, M. E., Jia, L., Gaston, B., Singel, D. J., and Stamler, J. S., 1996, Polynitrosylated proteins: Characterization, bioactivity, and functional consequences, *Proc. Natl. Acad. Sci. USA* **93:**4736–4741.

Singh, R. J., Hogg, N., Joseph, J., and Kalyanaraman, B., 1996, Mechanism of nitric oxide release from *S*-nitrosothiols, *J. Biol. Chem.* **271:**18596–18603.

Stamler, J. S., 1994, Redox signaling: Nitrosylation and related target interactions of nitric oxide, *Cell* **78:**931–936.

Stamler, J. S., Singel, D. J., and Loscalzo, J., 1992, Biochemistry of nitric oxide and its redox-activated forms, *Science* **258:**1898–1902.

Stamler, J. S., Jia, L., En, J. P., McMahan, T. J., Demchenko, I. T., Bonaventura, J., Gernert, K., and Piantadosi, C. A., 1997a, Blood flow regulation by *S*-nitrosohemoglobin in the physiological oxygen gradient, *Science* **276:**2034–2037.

Stamler, J. S., Toone, E. J., Lipton, S. A., and Sucher, N. J., 1997b, (S)NO signals: Translocation, regulation, and a consensus motif, *Neuron* **18:**691–696.

Trujillo, M., Alvarez, M., Peluffo, G., Freeman, B., and Radi, R., 1998, Xanthine oxidase-mediated decomposition of *S*-nitrosothiols, *J. Biol. Chem.* **273:**7828–7834.

Vanin, A. F., Malenkova, I. V., and Serezhenkov, V. A., 1997, Iron catalyzes both decomposition and synthesis of *S*-nitrosothiols: Optical and electron paramagnetic resonance studies, *Nitric Oxide* **1**:191–203.

Wade, R. S., and Castro, C. E., 1990, Redox reactivity of iron(III) porphyrins and heme proteins with nitric oxide. Nitrosyl transfer to carbon, oxygen, nitrogen, and sulfur, *Chem. Res. Toxicol.* **3**:189–191.

Walker, M. W., Kinter, M. T., Roberts, R. J., and Spitz, D. R., 1995, Nitric oxide-induced cytotoxicity: Involvement of cellular resistance to oxidative stress and the role of glutathione in protection, *Pediatr. Res.* **37**:41–49.

Wink, D. A., Kadprzak, K. S., Maragos, C. M., Elespuru, R. K., Misra, M., Dunams, T. M., Cebula, T. A., Koch, W. H., Andrews, A. W., Allen, J. S., and Keefer, L. K., 1991, DNA deaminating ability and genotoxicity of nitric oxide and its progenitors, *Science* **254**:1001–1003.

Wink, D. A., Nims, R. W., Darbyshire, J. F., Christodoulou, D., Hanbauer, I., Cox, G. W., Laval, F., Laval, J., Cook, J. A., Krishna, M. C., DeGraff, W. G., and Mitchell, J. B., 1994, Reaction kinetics for nitrosation of cysteine and glutathione in aerobic nitric oxide solutions at neutral pH. Insights into the fate and physiological effects of intermediates generated in the NO/O_2 reaction, *Chem. Res. Toxicol.* **7**:519–525.

CHAPTER 4

Enzymology of Nitric Oxide Biosynthesis

BENJAMIN HEMMENS and BERND MAYER

1. Introduction to the Nitric Oxide Synthases

1.1. The NO Synthase Reaction

In mammals under normal physiological conditions, nitric oxide originates from the reaction shown in Fig. 1 (Palmer *et al.*, 1988). The nitrogen atom comes from the guanidino group of L-arginine (Palmer *et al.*, 1988; Sakuma *et al.*, 1988) and the oxygen atom from molecular oxygen (Kwon *et al.*, 1990; Leone *et al.*, 1991). N^G-hydroxy-L-arginine (NOHLA) has been identified as an intermediate (Stuehr *et al.*, 1991b), so that the reaction can be divided into two stages as shown.

1.2. NO Synthase Isoenzymes

The reaction is catalyzed by the enzyme nitric oxide synthase (NOS) (Griffith and Stuehr, 1995; Hemmens and Mayer, 1997; Stuehr, 1997). All known NOSs can be classified into three isoenzyme types. Table I gives an overview of the nomenclature used for these isoenzymes and their defining characteristics. This chapter will use the terms nNOS (neuronal NOS, NOS1), iNOS (inducible NOS, NOS2), and eNOS (endothelial NOS, NOS3). As far as is known, each NOS isoenzyme has the same catalytic mechanism, shares the same layout of catalytic domains within its amino acid sequences, and is homodimeric. The principal differences between them relate first to their tissue distribution and second to their modes of regulation. These two aspects are the main focus of the following discussion.

BENJAMIN HEMMENS and BERND MAYER • Institute for Pharmacology and Toxicology, Karl-Franzens University Graz, A-8010 Graz, Austria

Nitric Oxide and Infection, edited by Fang. Kluwer Academic/Plenum Publishers, New York, 1999.

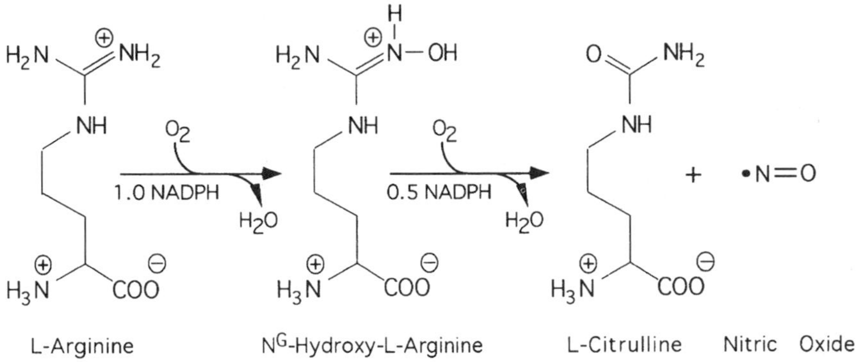

FIGURE 1. The nitric oxide synthase reaction.

1.3. Tissue Distribution and Physiological Roles

nNOS is expressed in neurons of both the central and peripheral nervous systems. In the CNS, NO is proposed to be involved in long-term potentiation (Snyder and Bredt, 1991; Garthwaite and Boulton, 1995; Son *et al.*, 1996; Wilson *et al.*, 1997). In peripheral neurons that innervate smooth muscle, NO is the neurotransmitter responsible for nonadrenergic noncholinergic relaxation of muscle (Rand and Li, 1995). eNOS is expressed in vascular endothelium. NO$^{\bullet}$ is the endothelium-derived relaxing factor that regulates blood pressure by relaxing adjacent smooth muscle cells (Marletta, 1989; Ignarro, 1990; Moncada *et al.*, 1991; Feelisch *et al.*, 1994). iNOS expression is induced in macrophages when they are activated by cytokines during an inflammatory response. NO is one of a number of toxic compounds produced by phagocytic cells to kill pathogens (Stuehr *et al.*, 1989; Tayeh and Marletta, 1989). This tissue specificity accounts for the major known roles of NO$^{\bullet}$; however, it is becoming clear that this classification is somewhat oversimplified and should not be interpreted too rigidly. It should be remembered that eNOS is present in blood vessels in the brain and can therefore influence events in the adjacent neurons; it has even been discovered within some neurons (Dinerman *et al.*, 1994; Son *et al.*, 1996; Wilson *et al.*, 1997) and cardiomyocytes (Balligand *et al.*, 1995a). iNOS is not only present in activated macrophages, but its synthesis can be induced in smooth (Busse and Mülsch, 1990) and cardiac muscle (Balligand *et al.*, 1993, 1995b), liver (Billiar *et al.*, 1990), and in glial cells (Murphy *et al.*, 1993). NO$^{\bullet}$ produced by macrophages can cause hypotension, most dramatically during septic shock (Petros *et al.*, 1994) (see Chapters 7 and 13).

TABLE I
NOS Isoenzymes

NOS isoform	Alternative terms	Human chromosome	Molecular mass	Distinctive properties	Subcellular localization	Tissue expression
Neuronal	Type I nNOS ncNOS bNOS	12	160 kDa	Ca^{2+} dependent, constitutively expressed	Binds to specific membrane proteins via an N-terminal PDZ domain	Neuronal cells, skeletal muscle
Endothelial	Type III eNOS ecNOS	7	134 kDA	Ca^{2+} dependent, constitutively expressed	Targets to the Golgi and to caveoli via N-terminal myristoylation and palmitoylation	Endothelial cells, epithelial cells, cardiomyocytes
Inducible	Type II iNOS macNOS	17	130 kDa	Ca^{2+} independent, induced by inflammatory stimuli (cytokines, LPS)	Soluble?	Macrophages, hepatocytes, astrocytes, smooth muscle cells (and many more)

1.4. Cofactors and Domain Layout

All NOSs contain bound FMN, FAD, heme, and tetrahydrobiopterin (Mayer *et al.*, 1990, 1991; Bredt *et al.*, 1991; Hevel *et al.*, 1991; Mülsch and Busse, 1991; Stuehr *et al.*, 1991a; Klatt *et al.*, 1992c; McMillan *et al.*, 1992; Stuehr and Ikeda-Saito, 1992). In early enzyme preparations the amounts of these molecules bound to the enzyme were somewhat variable, but the values obtained have converged gradually toward one equivalent each of FMN, FAD, and heme, and one-half equivalent of H_4B per polypeptide.

Figure 2 shows the common domain layout of the NOS isoenzymes. Each isoenzyme has an N-terminal domain of unique sequence and length. The homology between the isoenzymes begins sharply at a point equivalent to residue 300 in rat nNOS. Next comes the oxygenase domain with somewhat more than 400 residues. This section has no sequence similarity to other known proteins, and the recently obtained X-ray crystal structure (Crane *et al.*, 1997) has revealed a hitherto undescribed fold based largely on β-sheet structures. The oxygenase domain contains the binding sites for heme, L-arginine, and H_4B, as well as the intersubunit contacts of the native NOS homodimer. In the center of the sequence is a binding site for calmodulin (CaM) that is easily recognized as such by comparison with the CaM-binding sequences of many other proteins (Bredt *et al.*, 1991; Vorherr *et al.*, 1993; Zhang and Vogel, 1994; Anagli *et al.*, 1995; Venema *et al.*, 1996). The region to the C-terminal side of the CaM-binding site contains sequences homologous to nucleotide-binding domains in other proteins. This region is closely related to cytochrome P450 reductase and is called the *reductase domain*. The similarities with other proteins were recognized immediately from the first NOS sequence obtained (Bredt *et al.*, 1991) and allowed the FMN-, FAD-, and NADPH-binding domains to be located. The oxygenase and reductase domains have been shown to fold independently of each other, both by limited proteolysis of full-length NOS (Sheta *et al.*, 1994; McMillan and Masters, 1995; Ghosh *et al.*, 1996; Lowe *et al.*, 1996) and by expression as separate proteins (Ghosh *et al.*, 1995; Gachhui *et al.*, 1996).

1.5. Partial Reactions

The complete NOS reaction requires all of the cofactors previously mentioned. A variety of observations allow diferent stages of the reaction to be assigned to the various cofactors, and thus to different sites on the protein. Addition of L-arginine to the enzyme promotes a low-to-high spin shift of the heme, suggesting that the L-arginine binds close enough to the heme to displace distal ligands. Therefore, the actual synthesis of NOHLA and NO is believed to take place at the heme (Pufahl and Marletta, 1993).

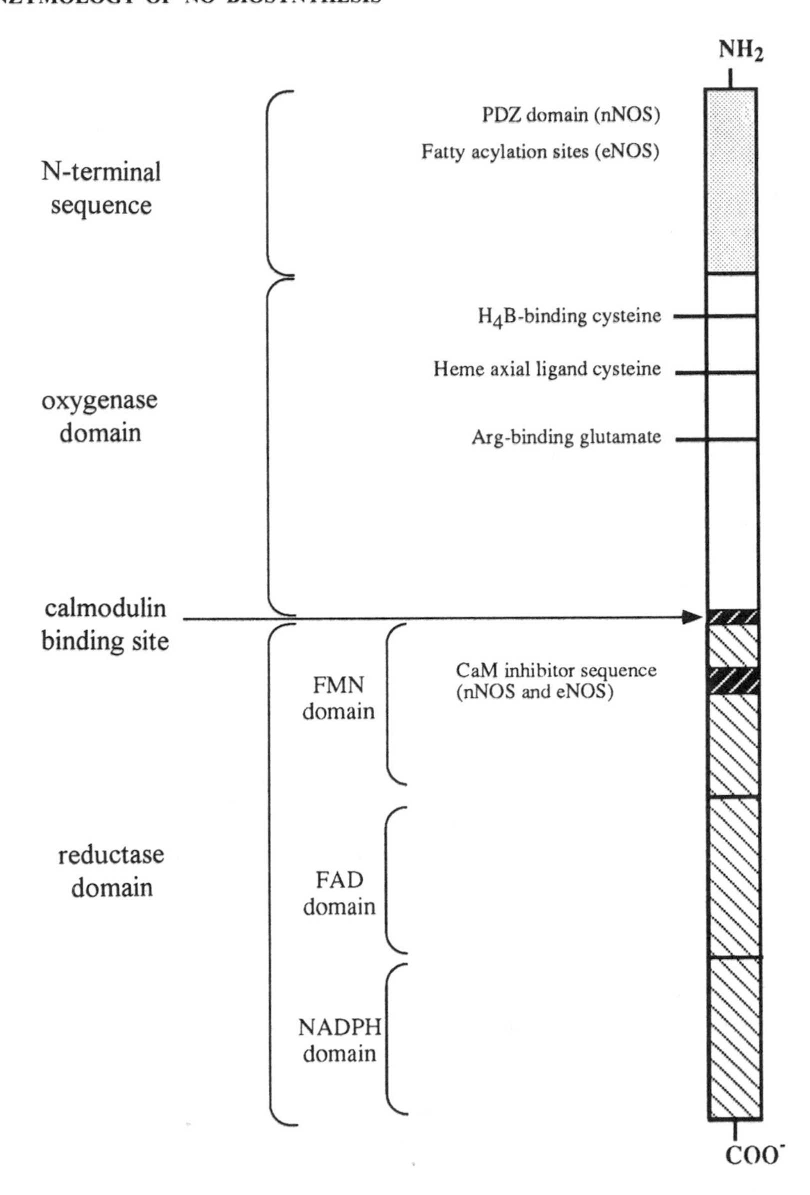

FIGURE 2. Domain layout of nitric oxide synthase. The PDZ domain in nNOS was identified by Hendriks (1995), the fatty acylation sites in eNOS by Busconi and Michel (1993), Sessa *et al.* (1993), Liu *et al.* (1995), and Garcia-Cardeña *et al.* (1996b), a cysteine residue important for H_4B binding by Cho *et al.* (1995), the heme proximal ligand cysteine by Chen *et al.* (1994), Richards and Marletta (1994), McMillan and Masters (1995), and Sari *et al.* (1996), and a glutamate residue important for L-arginine binding by Chen *et al.* (1997) and Gachhui *et al.* (1997). The function of the CaM inhibitory sequence was proposed by Gross *et al.* (1997).

The function of the reductase domain with its flavin cofactors is to transfer reducing equivalents from NADPH to the heme (Abu-Soud and Stuehr, 1993; Abu-Soud *et al.*, 1994). The correct attachment of the reductase domain to the oxygenase as found in the native protein appears to be essential for efficient coupling of their activities (Ghosh *et al.*, 1995; Chen *et al.*, 1996), although the isolated oxygenase domain can catalyze NO synthesis from NOHLA if supplied with hydrogen peroxide (Pufahl *et al.*, 1995). The reductase domain can also catalyze the reduction of other electron acceptors besides heme, such as oxidized cytochrome *c* (Klatt *et al.*, 1992a; Ghosh and Stuehr, 1995).

2. Regulation by Calcium/Calmodulin

2.1. Isoenzyme-Specific Activation Mechanisms

The activity of NOSs is completely dependent on the binding of CaM to its recognition sequence. The effect of CaM on the NOS reaction is mediated by the reductase domain: this is confirmed by the CaM dependence of reductase activity toward exogenous electron acceptors such as oxidized cytochrome *c* (Klatt *et al.*, 1992a; Gachhui *et al.*, 1996). To our knowledge, NOS is the only enzyme in which CaM controls this type of electron-transfer process.

Regulation by CaM is dramatically different between the constitutive isoenzymes nNOS and eNOS on the one hand, and inducible iNOS on the other. In both nNOS and eNOS, CaM binding and NOS activation are dependent on the Ca^{2+} concentration. The nNOS and eNOS proteins fold correctly without CaM, and bound CaM dissociates readily if the Ca^{2+} concentration is lowered. In contrast, purified iNOS is already fully active in the absence of added CaM or Ca^{2+}, and always has CaM irreversibly bound (Cho *et al.*, 1992). Reduction of Ca^{2+} concentrations to subphysiological levels neither induces dissociation of CaM nor significantly depresses iNOS enzyme activity. When iNOS was cloned and overexpressed, it would not fold correctly or incorporate heme without coexpression of CaM (Wu *et al.*, 1996). Thus, iNOS retains a need for bound CaM but is physiologically independent of Ca^{2+} concentration.

The mechanisms underlying these differences between iNOS and the other isoenzymes are now better understood as a result of two recent studies. One of these focused on the most obvious difference between the amino acid sequences of the enzymes: nNOS and eNOS have an insert of about 40 amino acids in the FMN domain that is absent from iNOS or from homologous FMN-binding domains in other proteins. It was discovered that peptides with the same sequence inhibit nNOS and eNOS, and cause dissociation of CaM from nNOS (Gross *et al.*, 1997). In several other CaM-dependent enzymes, a portion of the enzyme with sequence similarity to CaM binds to the CaM recognition sequence and must be displaced by CaM (Jarrett and Madhavan, 1991; Brickey *et al.*, 1994). It was proposed that the insert in the FMN domain of nNOS and eNOS could have an analogous function, and that its absence from iNOS might explain the tighter, Ca^{2+}-independent binding of CaM to this isoenzyme.

This insight has been nicely complemented by a sophisticated study in which the CaM-binding sites were swapped between proteins to construct chimeric NOS enzymes (Venema *et al.*, 1996). The iNOS chimera containing the CaM-binding site from eNOS bound CaM reversibly and was dependent on Ca^{2+}. Conversely, the eNOS chimera containing the CaM-binding sequence from iNOS was found to bind CaM irreversibly and was independent of Ca^{2+}. Therefore, the Ca^{2+}-independent CaM binding of iNOS is a function of the CaM-binding sequence itself and does not follow simply from the absence of the inhibitory insert. Intriguingly, though, Ca^{2+}-independent binding of CaM to this chimeric enzyme was not sufficient to cause enzyme activation. The enzyme *activity* showed a Ca^{2+} dependence similar to that of native eNOS. This key observation allows us to recognize that CaM binding and enzyme activation are separate events that happen to occur in the same Ca^{2+} range in nNOS and eNOS. The results can be combined to provide a model in which the binding of CaM to NOS is essentially determined by the CaM-binding site, but a conformational change of CaM on Ca^{2+} binding is necessary to displace the inhibitory insert and activate the enzyme. This requires that the inhibitory insert interact sterically with CaM, but without binding directly to the CaM-binding sequence: This was in fact proposed by Gross *et al.* (1997), because the inhibitory insert does not have sequence similarity to CaM. The absence of the inhibitory insert in iNOS results in Ca^{2+}-independent activation rather than Ca^{2+}-independent CaM binding.

Another CaM-related mode of regulation of eNOS, that has attracted increasing attention in the last couple of years, is its inhibition by the protein caveolin-1, a negative regulator of signal transduction. Caveolin is responsible for formation of caveolae, invaginations of the cell membrane where a variety of signalling proteins are gathered in an inactive form (Anderson, 1998). Caveolin inhibits eNOS with an IC_{50} of 1–3 μM, by interfering with electron transfer within the reductase domain (Ju *et al.*, 1997; Ghosh *et al.*, 1998). The inhibitions can be reversed by Ca^{2+}/CaM. In cells, a significant fraction of eNOS is bound to caveolae, and Ca^{2+} influx into the cells results in dissociation of the eNOS from the caveolin and its activation (Garcia-Cardeña *et al.*, 1997; Feron *et al.*, 1998). The sequestration of eNOS at the membrane by caveolin may be important in coupling eNOS activation to influx of Ca^{2+} from outside the cell rather than to increases in intracellular Ca^{2+} due to release from intracellular stores.

2.2. Physiological Importance of Calcium for NOS Regulation

Because iNOS is already fully active at the lowest Ca^{2+} concentrations known to occur physiologically, fluctuations in Ca^{2+} are almost certainly not important for its regulation *in vivo*. Instead, cytokine regulation of gene transcription provides the principal mechanism by which iNOS expression is controlled (see Chapter 5). In contrast, both nNOS and eNOS have the potential to be regulated by changes in intracellular Ca^{2+} concentration.

The importance of Ca^{2+} for regulation of eNOS in vascular endothelium has been apparent since pharmacological studies first implicated NO in endothelium-derived relaxation (Palmer *et al.*, 1988; Furchgott and Vanhoutte, 1989; Ignarro, 1989). Agents that cause $NO^{\bullet}$-mediated relaxation, such as acetylcholine and bradykinin, are calcium agonists; NO synthesis in endothelial cells is also stimulated by the calcium ionophore A23187. Activation of eNOS is not always associated with an increase in total intracellular Ca^{2+}, but recent observations with membrane-bound Ca^{2+} sensors suggest that local increases in Ca^{2+} near the plasma membrane may be sufficient for the response (Graier *et al.*, 1998). In addition, there is some evidence for Ca^{2+}-independent regulation involving mechanisms such as tyrosine phosphorylation of eNOS (Garcia-Cardeña *et al.*, 1996a).

In the brain, Ca^{2+} activation of nNOS forms a link between glutamate binding to postsynaptic NMDA-type receptors (that have a large Ca^{2+} conductance when glutamate is bound) and $NO^{\bullet}$ production (Garthwaite *et al.*, 1988; Garthwaite and Boulton, 1995). In peripheral nerves, the nitrergic transmission pathway is also triggered by an increase in intracellular Ca^{2+}, in this case associated with the opening of Ca^{2+} channels in response to the action potential (Rand and Li, 1995). For other neurotransmitters, presynaptic Ca^{2+} influx stimulates release from preexisting transmitter stores, but there is no store of the nitrergic transmitter (NO or an $NO^{\bullet}$ congener, perhaps an *S*-nitrosothiol). The most likely source of the transmitter is nNOS, activated by the Ca^{2+} influx.

3. Regulation by Tetrahydrobiopterin

3.1. An Allosteric Activator ... and More

An involvement of H_4B in NOS was discovered quite early in the process of purifying the enzyme from natural sources (Tayeh and Marletta, 1989; Mayer *et al.*, 1990; Werner-Felmayer *et al.*, 1990). Its role was initially somewhat controversial, first because the NOS reaction does not lead to stoichiometric oxidation of the pteridine in contrast to other H_4B-dependent enzyme reactions (Giovanelli *et al.*,

1991), and second because most NOS preparations contained no more than half a molar equivalent of H_4B per NOS polypeptide, although a certain amount of H_4B appeared to be irreversibly associated with the enzyme. Addition of exogenous H_4B to the purified enzyme resulted in a two to threefold activation (Klatt *et al.*, 1992b), indicating that H_4B at least had to be regarded as a significant allosteric activator. Subsequent results have confirmed this role, but also tend to indicate that the pteridine indeed has a direct chemical role in catalysis.

The puzzling presence of less than stoichiometric amounts of H_4B in NOS preparations has been resolved by the characterization of H_4B-free versions of the enzyme (Gorren *et al.*, 1996). These reveal two distinct phases of H_4B binding to NOS, interpreted as reflecting binding of one H_4B molecule to each subunit of the enzyme dimer, with strong negative cooperativity between the two sites. The first molecule binds with a dissociation constant in the low nanomolar range, and the second with a dissociation constant of a few micromolar. Therefore, when the enzyme is purified without adding extra H_4B to the buffer, the high-affinity site remains occupied leading to the usual H_4B content of half a mole per mole of polypeptide.

In line with the role of H_4B as an allosteric activator, two structural effects of its binding to NOS have been observed. The first of these concerns the environment of the heme. The heme group in NOS has a cysteine thiolate proximal ligand and consequently possesses spectral properties very similar to cytochrome P450s (Sono *et al.*, 1995). A well-known phenomenon in these enzymes is the transition of the heme iron from a low-spin to a high-spin state on substrate binding (e.g., see Fisher *et al.*, 1985). This has been shown for P450 enzymes to be correlated with the presence (in the low-spin state) or absence (in the high-spin state) of a sixth, distal heme ligand that is displaced on binding of substrate (Poulos *et al.*, 1986). A similar shift toward high spin occurs on addition of the L-arginine substrate to NOS (McMillan and Masters, 1993; Matsuoka *et al.*, 1994; Gerber and Ortiz DeMontellano, 1995; Roman *et al.*, 1995; Gorren *et al.*, 1996). However, not only L-arginine but also H_4B binding to NOS induces the transition from low spin to high spin (Gorren *et al.*, 1996). The recent crystal structures of dimeric iNOS oxygenase domain showed that while the L-arg, as expected, binds on the distal side of the heme, H_4B binds on the proximal side (Crane *et al.*, 1998). Thus the low-to-high spin drift stimulated by H_4B is not caused by direct steric displacement of the distal water, as might have been supposed, but by more subtle and indirect effects on the coordination properties of the heme iron and on the structure of the distal pocket. The high affinity of the first H_4B site in the enzyme dimer ensures that the heme in purified NOS is already predominantly high spin in the absence of substrate, unlike other P450s (Gorren *et al.*, 1996).

The other structural correlate of H_4B binding that has been characterized to date is a marked stabilization of the NOS dimer from dissociation by SDS (Klatt *et al.*, 1995; Rodriguez-Crespo *et al.*, 1996; Mayer *et al.*, 1997). The H_4B-free enzyme dissociates completely in 2% SDS at room temperature, whereas a substantial fraction of the dimeric enzyme (nearly 100% in the case of the pig brain enzyme, somewhat less for other forms) remains associated in the presence of saturating H_4B. This behavior was characterized by electrophoresis at low temperatures (LT-PAGE).

Besides these structural effects of H_4B, the outlines of its functional importance in the NOS reaction have also been significantly clarified. The central finding has been that NOS is completely unable to catalyze the oxidation of L-arginine to L-citrulline and $NO^{\bullet}$ in the absence of H_4B (Gorren *et al.*, 1996). The H_4B-free enzyme is at most able to catalyze NADPH-dependent production of reduced oxygen species (superoxide or peroxide) (Heinzel *et al.*, 1992; Pou *et al.*, 1992). Reductive oxygen activation is presumably a normal part of the NOS reaction mechanism, which then becomes stalled in the absence of H_4B, allowing the release of these reduced oxygen species into solution. If the H_4B-free enzyme is titrated with H_4B and assayed for both L-citrulline and peroxide formation, a switch is observed from peroxide formation to L-citrulline formation with increasing H_4B (Gorren *et al.*, 1996). Thus, H_4B plays an essential role in the coupling of NADPH oxidation to NO production. This apparent ability to influence the course of the reaction seems to hint at a direct chemical involvement of H_4B in catalysis. Recently the 4-amino analogue of H_4B was found to bind with similar affinity to that of H_4B, causing both the spin-state shift and stabilization of the dimer, but inhibiting the NOS reaction (Werner *et al.*, 1996; Mayer *et al.*, 1997). This may also be an indication that the allosteric effects of H_4B do not fully explain its role in NOS. The alternative is that H_4B may undergo a brief redox change, stabilizing some transition state during the reaction cycle. This may be very difficult to detect for obvious technical reasons, so at the present time the puzzle of the "essential" role of H_4B in NOS remains unsolved.

3.2. Physiological Relevance: Puzzles with Peroxynitrite

The negative cooperativity of H_4B binding to NOS, combined with the capacity of the H_4B-free enzyme to synthesize superoxide, raises something of a dilemma about the overall product of the NOS reaction. A study comparing the binding of DTT to the heme of nNOS (measured photometrically) with the kinetics of nNOS inhibition by DTT showed that an H_4B-free site can catalyze superoxide production if one of the H_4B sites in the dimer is occupied (Gorren *et al.*, 1997). To avoid superoxide formation by nNOS, it would appear necessary to saturate both H_4B sites. However, this would require the presence of a large amount of free H_4B, which autoxidizes to again produce superoxide (Mayer *et al.*, 1995). Superoxide reacts very rapidly with $NO^{\bullet}$ to form peroxynitrite ($ONOO^-$). Thus, with purified nNOS, formation of NO has only been detected when superoxide dismutase is added. Peroxynitrite, in contrast to NO, is highly cytotoxic and does not mediate the signaling functions of $NO^\bullet$ (see Chapter 8).

We have emphasized that these results were obtained with nNOS: In fact, iNOS and eNOS have very low "uncoupled" superoxide-producing activity when compared with nNOS (List *et al.*, 1997). Thus, nNOS is the isoenzyme most in danger of effectively producing peroxynitrite. This coincides with studies of mice carrying disrupted genes for either nNOS (Huang *et al.*, 1994) or eNOS (Huang *et al.*, 1996), which demonstrate that nNOS is responsible for brain damage during ischemia–reperfusion episodes, whereas eNOS has a protective effect under these conditions. It is intriguing to consider that nNOS might experience conditions during ischemia–reperfusion that switch its output from NO toward peroxynitrite.

Factors that could tip the activity of nNOS in this direction might include a deficiency of SOD, a depletion of reduced thiols such as GSH (which may be able to compete with superoxide for reaction with $NO^\bullet$), extremely low or high concentrations of H_4B, or excessive activation of the enzyme resulting from sustained high calcium concentrations. What is the evidence that these conditions occur in the ischemic brain? Confirmatory data are thus far available only for the last of these conditions: In the ischemic brain, extracellular glutamate concentrations are strongly elevated. Similar concentrations of glutamate are rapidly toxic to cultured neurons (for reviews see Choi and Rothman, 1990; Meldrum and Garthwaite, 1990; Bruno *et al.*, 1993; Choi, 1993; Bolanos *et al.*, 1997). Excess glutamate would be expected to lead to overactivation of NMDA receptors and, therefore, nNOS. The hypothesis that NOS is involved in glutamate neurotoxicity has its adherents (Dawson *et al.*, 1991), although there certainly are also NO-independent effects, and an effect of NO has not always been detected as this hypothesis would have predicted (Garthwaite and Garthwaite, 1994). Some of the apparent confusion may be related to an inability to distinguish between NOS isoenzymes; as mentioned earlier, an improvement of blood flow mediated by eNOS may tend to offset excitotoxic effects of nNOS. More detailed studies of the nNOS and eNOS knockout mice may help to disentangle these different effects.

With regard to H_4B, although it is fairly easy to determine the average concentration in brain homogenates, it is still very difficult to estimate the concentration in individual NOS-containing neurons. The best indication that H_4B can be limiting for NOS *in vivo* is the restoration of endothelium-dependent relaxation in blood vessels affected by a variety of cardiovascular disease states by increasing H_4B availability (Kinoshita *et al.*, 1997). However, the possible role of changes in H_4B concentration in the regulation of brain NOS is still an open question.

4. Regulation by Heme

4.1. Dual Role in Catalysis and Enzyme Assembly

The heme group of NOS is absolutely essential for NOS activity. It is not the purpose of the present review to discuss the catalytic role of the heme in detail (the basic issues are well presented by Griffith and Stuehr, 1995). We merely observe that many of the ideas on this aspect of NOS derive from analogies of much better studied P450 enzymes (e.g., see Mansuy *et al.*, 1995), while the experimental evidence on NOS itself is still rather fragmentary. For example, conversion of NOHLA to L-citrulline is catalyzed by other P450s (Clement *et al.*, 1993; Renaud *et al.*, 1993). The first step of the NOS reaction, from L-arginine to NOHLA, has also been suggested to follow a route similar to P450-catalyzed hydroxylations.

In addition to its catalytic function, heme plays a key role in the assembly of the native NOS structure. Dimerization of iNOS (Baek *et al.*, 1993), nNOS (Klatt *et al.*, 1996), and most recently eNOS (Venema *et al.*, 1997) has been shown to require heme. Heme-free, monomeric nNOS binds neither L-arginine nor H_4B. Apart from these aspects, heme-free enzymes appear to be structurally intact: Cytochrome *c* reductase activity, flavin content, and circular dichroism spectra are all similar to the native enzyme (Klatt *et al.*, 1996). nNOS dimerizes upon addition of heme alone (Klatt *et al.*, 1996), while iNOS requires the simultaneous addition of heme plus H4B (Baek *et al.*,1993), and both enzymes can be substantially reactivated (Baek *et al.*, 1993; Hemmens and Mayer, unpublished observations). Therefore, heme incorporation is essential to create the correct conformation of the dimeric contact points, as well as the substrate and H_4B binding sites.

4.2. Regulation by Heme Availability?

Albakri and Stuehr (1996) have presented intriguing results suggesting that iNOS activity in activated macrophages is limited by heme availability. Cytokine activation of the macrophage cell line RAW 264.7 was observed to induce synthesis of new iNOS protein, but less than half of the protein was assembled into active dimeric enzyme. In contrast, addition of L-NAME (which is metabolized to the NOS inhibitor *N*-methyl-L-arginine) resulted in full dimerization of the newly synthesized NOS, accompanied by a rise in intracellular free heme. It was concluded that NO produced by iNOS inhibits heme biosynthesis to levels that are limiting for NOS assembly. This could represent an important safety mechanism to prevent $NO^{\bullet}$ synthesis by macrophages from running out of control. A possible route for this feedback effect is suggested by the results of Hentze and colleagues, showing that $NO^{\bullet}$ can activate the iron regulatory protein, leading to a downregulation of protoporphyrin and heme biosynthesis (Weiss *et al.*, 1994; Pantopoulos and Hentze, 1995; Pantopoulos *et al.*, 1996). An effect of iron on NOS transcription has also been found, and it is proposed that these signaling pathways serve to match the supply of heme and NOS protein to each other (Weiss *et al.*, 1994, 1997; Hentze and Kuhn, 1996; Paraskeva and Hentze, 1996).

Whether these mechanisms are also important in neuronal or endothelial cells is still unclear. Since the level of NOS expression is lower in these cells than in macrophages, constitutive NO synthesis may not significantly perturb iron and porphyrin metabolism. A large proportion of heme-free NOS protein has not been observed in these cell types.

5. Regulation by Intracellular Targeting

5.1. N-Termini Are Isoenzyme Specific

Each NOS isoenzyme has an N-terminal extension of characteristic length and sequence. Experiments with truncated versions of each of the three isoenzymes suggest that these sequences are not directly involved in catalysis. Instead, for both nNOS and eNOS there is evidence that the N-terminal extensions are used to target the enzyme to particular sites within the cell.

5.2. nNOS Contains a PDZ Domain

The N-terminal sequence of nNOS contains a region homologous to a family of domains called *GLGF repeats* or *PDZ domains* (Hendriks, 1995). All proteins containing this motif are localized to cell–cell junctions. Studies with immunoprecipitation and expression of truncated forms of nNOS provide evidence that nNOS associates via this domain to the dystrophin complex in skeletal muscle, thus localizing nNOS to the sarcolemmal membrane (Brenman *et al.*, 1995). The protein partner of nNOS in this case has been proposed to be α-1 syntrophin. Another protein partner, PSD-95, was identified by the yeast two-hybrid methodology (Brenman *et al.*, 1996). PSD-95 is known to bind to the NMDA receptor. A peptide identical to the C-terminus of the 2B subtype of NMDA receptors competitively displaces PSD-95 from nNOS, suggesting that nNOS may be able to associate with this receptor type. This localization could have a function in the coupling of nNOS activation to NMDA receptor activation by glutamate.

5.3. eNOS Is Palmitoylated and Myristoylated

The N-terminal sequence of eNOS is also proposed to function in localizing the enzyme to membranes, but by means of fatty acylation rather than through protein–protein interactions. Gly-2 becomes myristoylated (Busconi and Michel, 1993; Sessa *et al.*, 1993), and Cys-15 and Cys-26 are palmitoylated (Liu *et al.*,

1995; Garcia-Cardeña *et al.*, 1996b). The eNOS enzyme with all three fatty acids attached is membrane bound, whereas enzyme lacking any of the fatty acyl moieties remains in the cytosol. The effects of partial modification are controversial. Fatty acylation can be influenced by receptor agonists and may be used to control the enzyme's location within the cell.

References

Abu-Soud, H. M., and Stuehr, D. J., 1993, Nitric oxide synthases reveal a novel role for calmodulin in controlling electron transfer, *Proc. Natl. Acad. Sci. USA* **90:**10769–10772.

Abu-Soud, H. M., Yoho, L. L., and Stuehr, D. J., 1994, Calmodulin controls neuronal nitric-oxide synthase by a dual mechanism—Activation of intra- and inter-domain electron transfer, *J. Biol. Chem.* **269:**32047–32050.

Albakri, Q. A., and Stuehr, D. J., 1996, Intracellular assembly of inducible NO synthase is limited by nitric oxide-mediated changes in heme insertion and availability, *J. Biol. Chem.* **271:**5414–5421.

Anagli, J., Hofmann, F., Quadroni, M., Vorherr, T., and Carafoli, E., 1995, The calmodulin-binding domain of the inducible (macrophage) nitric oxide synthase, *Eur. J. Biochem.* **233:**701–708.

Anderson, R. G. W., 1998, The caveolae membrane system, *Annu. Rev. Biochem.* **67:**199–225.

Baek, K. J., Thiel, B. A., Lucas, S., and Stuehr, D. J., 1993, Macrophage nitric oxide synthase subunits—Purification, characterization, and role of prosthetic groups and substrate in regulating their association into a dimeric enzyme, *J. Biol. Chem.* **268:**21120–21129.

Balligand, J. L., Kelly, R. A., Marsden, P. A., Smith, T. W., and Michel, T., 1993, Control of cardiac muscle cell function by an endogenous nitric oxide signaling system, *Proc. Natl. Acad. Sci. USA* **90:**347–351.

Balligand, J. L., Kobzik, L., Han, X. Q., Kaye, D. M., Belhassen, L., Ohara, D. S., Kelly, R. A., Smith, T. W., and Michel, T., 1995a, Nitric oxide-dependent parasympathctic signaling is due to activation of constitutive endothelial (type III) nitric oxide synthase in cardiac myocytes, *J. Biol. Chem.* **270:**14582–14586.

Balligand, J. L., Ungureanu-Longrois, D., Simmons, W. W., Kobzik, L., Lowenstein, C. J., Lamas, S., Kelly, R. A., Smith, T. W., and Michel, T., 1995b, Induction of NO synthase in rat cardiac microvascular endothelial cells by IL-1 beta and IFN-gamma, *Am. J. Physiol.* **268:**H1293–H1303.

Billiar, T. R., Curran, R. D., Stuehr, D. J., Stadler, J., Simmons, R. L., and Murray, S. A., 1990, Inducible cytosolic enzyme activity for the production of nitrogen oxides from L-arginine in hepatocytes, *Biochem. Biophys. Res. Commun.* **168:**1034–1040.

Bolanos, J. P., Almeida, A., Stewart, V., Peuchen, S., Land, J. M., Clark, J. B., and Heales, S. J., 1997, Nitric oxide-mediated mitochondrial damage in the brain: Mechanisms and implications for neurodegenerative diseases, *J. Neurochem.* **68:**2227–2240.

Bredt, D. S., Hwang, P. M., Glatt, C. E., Lowenstein, C., Reed, R. R., and Snyder, S. H., 1991, Cloned and expressed nitric oxide synthase structurally resembles cytochrome P-450 reductase, *Nature* **351:**714–718.

Brenman, J. E., Chao, D. S., Xia, H. H., Aldape, K., and Bredt, D. S., 1995, Nitric oxide synthase complexed with dystrophin and absent from skeletal muscle sarcolemma in Duchenne muscular dystrophy, *Cell* **82:**743–752.

Brenman, J. E., Chao, D. S., Gee, S. H., Mcgee, A. W., Craven, S. E., Santillano, D. R., Wu, Z. Q., Huang, F., Xia, H. H., Peters, M. F., Froehner, S. C., and Bredt, D. S., 1996, Interaction of nitric oxide synthase with the postsynaptic density protein PSD-95 and alpha 1-syntrophin mediated by PDZ domains, *Cell* **84:**757–767.

Brickey, D. A., Bann, J. G., Fong, Y. L., Perrino, L., Brennan, R. G., and Soderling, T. R., 1994, Mutational analysis of the autoinhibitory domain of calmodulin kinase II, *J. Biol. Chem.* **269:**29047–29054.

Bruno, V., Scapagnini, U., and Canonico, P. L., 1993, Excitatory amino acids and neurotoxicity, *Funct. Neurol.* **8:**279–292.

Busconi, L., and Michel, T., 1993, Endothelial nitric oxide synthase—N-terminal myristoylation determines subcellular localization, *J. Biol. Chem.* **268:**8410–8413.

Busse, R., and Mlsch, A., 1990, Induction of nitric oxide synthase by cytokines in vascular smooth muscle cells, *FEBS Lett.* **275:**87–90.

Chen, P. F., Tsai, A. L., and Wu, K. K., 1994, Cysteine-184 of endothelial nitric oxide synthase is involved in heme coordination and catalytic activity, *J. Biol. Chem.* **269:**25062–25066.

Chen, P. F., Tsai, A. L., Berka, V., and Wu, K. K., 1996, Endothelial nitric-oxide synthase—Evidence for bidomain structure and successful reconstitution of catalytic activity from two separate domains generated by a baculovirus expression system, *J. Biol. Chem.* **271:**14631–14635.

Chen, P. F., Tsai, A. L., Berka, V., and Wu, K. K., 1997, Mutation of Glu-361 in human endothelial nitric-oxide synthase selectively abolishes L-arginine binding without perturbing the behavior of heme and other redox centers, *J. Biol. Chem.* **272:**6114–6118.

Cho, H. J., Xie, Q. W., Calaycay, J., Mumford, R. A., Swiderek, K. M., Lee, T. D., and Nathan, C., 1992, Calmodulin is a subunit of nitric oxide synthase from macrophages, *J. Exp. Med.* **176:**599–604.

Cho, H. J., Martin, E., Xie, Q. W., Sassa, S., and Nathan, C., 1995, Inducible nitric oxide synthase: Identification of amino acid residues essential for dimerization and binding of tetrahydrobiopterin, *Proc. Natl. Acad. Sci. USA* **92:**11514–11518.

Choi, D. W., 1993, Nitric oxide—Foe or friend to the injured brain, *Proc. Natl. Acad. Sci. USA* **90:**9741–9743.

Choi, D. W., and Rothman, S. M., 1990, The role of glutamate neurotoxicity in hypoxic–ischemic neuronal death, *Annu. Rev. Neurosci.* **13:**171–182.

Clement, B., Schultze-Mosgau, M. H., and Wohlers, H., 1993, Cytochrome P450 dependent *N*-hydroxylation of a guanidine (debrisoquine), microsomal catalyzed reduction and further oxidation of the *N*-hydroxy-guanidine metabolite to the urea derivative—Similarity with the oxidation of arginine to citrulline and nitric oxide, *Biochem. Pharmacol.* **46:**2249–2267.

Crane, B. R., Arvai, A. S., Gachhui, R., Wu, C., Ghosh, D. K., Getzoff, E. D., Stuehr, D. J., and Tainer, J. A., 1997, The structure of nitric oxide synthase oxygenase domain and inhibitor complexes, *Science* **278:**425–431.

Crane, B. R., Arvai, A. S., Ghosh, D. K., Wu, C., Getzoff, E. D., Stuehr, D. J., and Tainer, J. A., 1988, Structure of nitric oxide synthease oxygenase dimer and pterin and substrate, *Science* **279:**2121–2126.

Dawson, V. L., Dawson, T. M., London, E. D., Bredt, D. S., and Snyder, S. H., 1991, Nitric oxide mediates glutamate neurotoxicity in primary cortical cultures, *Proc. Natl. Acad. Sci.* USA **88:**6368–6371.

Dinerman, J. L., Dawson, T. M., Schell, M. J., Snowman, A., and Snyder, S. H., 1994, Endothelial nitric oxide synthase localized to hippocampal pyramidal cells: Implications for synaptic plasticity, *Proc. Natl. Acad. Sci. USA* **91:**4214–4218.

Feelisch, W., Poel, M. T., Zamora, R., Deussen, A., and Moncada, S., 1994, Understanding the controversy over the identity of EDRF, *Nature* **368:**62–65.

Feron, O., Saldana, F., Michel, J. B., and Michel, T., 1988, The endothelial nitric-oxide synthease-caveolin regulatory cycle, *J. Biol. Chem.* **273:**3124–3128.

Fisher, M. T., Scarlata, S. F., and Sligar, S. G., 1985, High-pressure investigations of cytochrome P-450 spin and substrate binding equilibria, *Arch. Biochem. Biophys.* **240:**456–463.

Furchgott, R. F., and Vanhoutte, P. M., 1989, Endothelium-derived relaxing and contracting factors, *FASEB J.* **3:**2007–2018.

Gachhui, R., Presta, A., Bentley, D. F., Abu-Soud, H. M., McArthur, R., Brudvig, G., Ghosh, D. K., and Stuehr, D. J., 1996, Characterization of the reductase domain of rat neuronal nitric oxide synthase generated in the methylotrophic yeast *Pichia pastoris*—Calmodulin response is complete within the reductase domain itself, *J. Biol. Chem.* **271**:20594–20602.

Gachhui, R., Ghosh, D. K., Wu, C. Q., Parkinson, J., Crane, B. R., and Stuehr, D. J., 1997, Mutagenesis of acidic residues in the oxygenase domain of inducible nitric-oxide synthase identifies a glutamate involved in arginine binding, *Biochemistry* **36**:5097–5103.

Garcia-Cardeña, G., Fan, R., Stern, D. F., Liu, J. W., and Sessa, W. C., 1996a, Endothelial nitric oxide synthase is regulated by tyrosine phosphorylation and interacts with caveolin-1, *J. Biol. Chem.* **271**:27237–27240.

Garcia-Cardeña, G., Oh, P., Liu, J., Schnitzer, J. S., and Sessa, W. C., 1996b, Targeting of nitric oxide synthase to endothelial cell caveolae via palmitoylation: Implications for nitric oxide signaling, *Proc. Natl. Acad. Sci. USA* **93**:6448–6453.

Garcia-Cardeña, G., Martasek, P., Masters, B. S. S., Skidd, P. M., Couet, J., Li, S., Lisanti, M., and Sessa, W. C., 1997, Dissecting the interaction between nitric oxide synthase (NOS) and caveolin, *J. Biol. Chem.* **272**:25437–25440.

Garthwaite, G., and Garthwaite, J., 1994, Nitric oxide does not mediate acute glutamate neurotoxicity, nor is it neuroprotective, in rat brain slices, *Neuropharmacology* **33**:1431–1438.

Garthwaite, J., and Boulton, C. L., 1995, Nitric oxide signaling in the central nervous system, *Annu. Rev. Physiol.* **57**:683–706.

Garthwaite, J., Charles, S. L., and Chess-Williams, R., 1988, Endothelium-derived relaxing factor release on activation of NMDA receptors suggests role as intercellular messenger in the brain, *Nature* **336**:385–388.

Gerber, N. C., and Ortiz DeMontellano, P. R., 1995, Neuronal nitric oxide synthase—Expression in *Escherichia coli*, irreversible inhibition by phenyldiazene, and active site topology, *J. Biol. Chem.* **270**:17791–17796.

Ghosh, D. K., and Stuehr, D. J., 1995, Macrophage NO synthase: Characterization of isolated oxygenase and reductase domains reveals a head-to-head subunit interaction, *Biochemistry* **34**:801–807.

Ghosh, D. K., Abu-Soud, H. M., and Stuehr, D. J., 1995, Reconstitution of the second step in NO synthesis using the isolated oxygenase and reductase domains of macrophage NO synthase, *Biochemistry* **34**:11316–11320.

Ghosh, D. K., Abu-Soud, H. M., and Stuehr, D. J., 1996, Domains of macrophage NO synthase have divergent roles in forming and stabilizing the active dimeric enzyme, *Biochemistry* **35**:1444–1449.

Ghosh, S., Gachhui, R., Crooks, C., Wu, C., Lisanti, M. P., and Stuehr, D. J., 1998, Interaction between caveolin-1 and the reductase domain of endothelial nitric-oxide synthase. Consequences for catalysis, *J. Biol. Chem.* **273**:22267–22271.

Giovanelli, J., Campos, K. L., and Kaufman, S., 1991, Tetrahydrobiopterin, a cofactor for rat cerebellar nitric oxide synthase, does not function as a reactant in the oxygenation of arginine, *Proc. Natl. Acad. Sci. USA* **88**:7091–7095.

Gorren, A. C. F., List, B. M., Schrammel, A., Pitters, E., Hemmens, B., Werner, E. R., Schmidt, K., and Mayer, B., 1996, Tetrahydrobiopterin-free neuronal nitric oxide synthase: Evidence for two identical highly anticooperative pteridine binding sites, *Biochemistry* **35**:16735–16745.

Gorren, A. C. F., Schrammel, A., Schmidt, K., and Mayer, B., 1997, Thiols and neuronal nitric oxide synthase: Complex formation, competitive inhibition, and enzyme stabilization, *Biochemistry* **36**:4360–4366.

Graier, W. F., Paltauf-Doburzynska, J., Hill, B. J. F., Fleischhacker, E., Hoebel, B. G., Kostner, G. M., and Sturek, M., 1998, Submaximal stimulation of porcine endothelial cells causes focal Ca^{2+} elevation beneath the cell membrane, *J. Physiol. (London)* **506**:109–125.

Griffith, O. W., and Stuehr, D. J., 1995, Nitric oxides synthases: Properties and catalytic mechanism, *Annu. Rev. Physiol.* **57:**707–736.

Gross, S. S., Liu, Q., Jones, C. L., Weissman, B. A., Martasek, P., Roman, L. J., Masters, B. S. S., Smith, S. M. E., Irizzary, K., Patel, B., Morales, A. J., Harris, D. E., and Salerno, J. C., 1997, Identification of a unique regulatory domain in calcium-dependent NO synthases, *Jpn. J. Pharmacol.* **75** (Suppl. I):6P (abstract 13).

Heinzel, B., John, M., Klatt, P., Böhme, E., and Mayer, B., 1992, Ca^{2+}/calmodulin-dependent formation of hydrogen peroxide by brain nitric oxide synthase, *Biochem. J.* **281:**627–630.

Hemmens, B., and Mayer, B., 1997, Enzymology of nitric oxide synthases, in: *Methods in Molecular Biology*, Volume 100 (M. A. Titheradge, ed.), Humana Press, Totowa, N.J., pp. 1–32.

Hendriks, W., 1995, Nitric oxide synthase contains a discs-large homologous region (DHR) sequence motif, *Biochem. J.* **305:**687–688.

Hentze, M. W., and Kuhn, L. C., 1996, Molecular control of vertebrate iron metabolism: mRNA-based regulatory circuits operated by iron, nitric oxide, and oxidative stress, *Proc. Natl. Acad. Sci. USA* **93:**8175–8182.

Hevel, J. M., White, K. A., and Marletta, M. A., 1991, Purification of the inducible murine macrophage nitric oxide synthase—identification as a flavoprotein, *J. Biol. Chem.* **266:**22789–22791.

Huang, Z. H., Huang, P. L., Panahian, N., Dalkara, T., Fishman, M. C., and Moskowitz, M. A., 1994, Effects of cerebral ischemia in mice deficient in neuronal nitric oxide synthase, *Science* **265:**1883–1885.

Huang, Z. H., Huang, P. L., Ma, J. Y., Meng, W., Ayata, C., Fishman, M. C., and Moskowitz, M. A., 1996, Enlarged infarcts in endothelial nitric oxide synthase knockout mice are attenuated by nitro-L-arginine, *J. Cereb. Blood Flow Metab.* **16:**981–987.

Ignarro, L. J., 1989, Endothelium-derived nitric oxide: Actions and properties, *FASEB J.* **3:**31–36.

Ignarro, L. J., 1990, Biosynthesis and metabolism of endothelium-derived nitric oxide, *Annu. Rev. Pharmacol. Toxicol.* **30:**535–560.

Jarrett, H. W., and Madhavan, R., 1991, Calmodulin-binding proteins also have a calmodulin-like binding site within their structure. The flip-flop model, *J. Biol. Chem.* **266:**362–371.

Ju, H., Zou, R., Venema, V. J., and Venema, R. C., 1997, Direct interaction of endothelial nitric-oxide synthase and caveolin-1 inhibits synthase activity, *J. Biol. Chem.* **272:**18522–18525.

Kinoshita, H., Tsutsui, M., Milstien, S., and Katusic, Z. S., 1997, Tetrahydrobiopterin, nitric oxide and regulation of cerebral arterial tone, *Prog. Neurobiol.* **52:**295–302.

Klatt, P., Heinzel, B., John, M., Kastner, M., Böhme, E., and Mayer, B., 1992a, Ca^{2+}/calmodulin-dependent cytochrome c reductase activity of brain nitric oxide synthase, *J. Biol. Chem.* **267:**11374–11378.

Klatt, P., Heinzel, B., Mayer, B., Ambach, E., Werner-Felmayer, G., Wachter, H., and Werner, E. R., 1992b, Stimulation of human nitric oxide synthase by tetrahydrobiopterin and selective binding of the cofactor, *FEBS Lett.* **305:**160–162.

Klatt, P., Schmidt, K., and Mayer, B., 1992c, Brain nitric oxide synthase is a haemoprotein, *Biochem. J.* **288:**15–17.

Klatt, P., Schmidt, K., Lehner, D., Glatter, O., Bächinger, H. P., and Mayer, B., 1995, Structural analysis of porcine brain nitric oxide synthase reveals novel role of tetrahydrobiopterin and L-arginine in the formation of an SDS-resistant dimer, *EMBO J.* **14:**3687–3695.

Klatt, P., Pfeiffer, S., List, B. M., Lehner, D., Glatter, O., Werner, E. R., Schmidt, K., and Mayer, B., 1996, Characterization of heme-deficient neuronal nitric oxide synthase reveals role for heme in subunit dimerization and binding of amino acid substrate and tetrahydrobiopterin, *J. Biol. Chem.* **271:**7336–7342.

Kwon, N. S., Nathan, C. F., Gilker, C., Griffith, O. W., Matthews, D. E., and Stuehr, D. J., 1990, L-Citrulline production from L-arginine by macrophage nitric oxide synthase. The ureido oxygen derives from dioxygen, *J. Biol. Chem.* **265:**13442–13445.

Leone, A. M., Palmer, R. M. J., Knowles, R. G., Francis, P. L., Ashton, D. S., and Moncada, S., 1991, Constitutive and inducible nitric oxide synthases incorporate molecular oxygen into both nitric oxide and citrulline, *J. Biol. Chem.* **266:**23790–23795.

List, B. M., Klösch, B., Völker, C., Gorren, A. C. F., Sessa, W. C., Werner, E. R., Kukovetz, W. R., Schmidt, K., and Mayer, B., 1997, Characterization of bovine endothelial nitric oxide synthase as a homodimer with down-regulated uncoupled NADPH oxidase activity: Tetrahydrobiopterin binding kinetics and role of haem in dimerization, *Biochem. J.* **323:**159–165.

Liu, J. W., Garcia-Cardeña, G., and Sessa, W. C., 1995, Biosynthesis and palmitoylation of endothelial nitric oxide synthase: Mutagenesis of palmitoylation sites, cysteines-15 and/or -26, argues against depalmitoylation-induced translocation of the enzyme, *Biochemistry* **34:**12333–12340.

Lowe, P. N., Smith, D., Stammers, D. K., Riveros-Moreno, V., Moncada, S., Charles, I., and Boyhan, A., 1996, Identification of the domains of neuronal nitric oxide synthase by limited proteolysis, *Biochem. J.* **314:**55–62.

Mansuy, D., Boucher, J. L., and Clement, B., 1995, On the mechanism of nitric oxide formation upon oxidative cleavage of C=N(OH) bonds by NO-synthases and cytochromes P450, *Biochimie* **77:**661–667.

Marletta, M. A., 1989, Nitric oxide: Biosynthesis and biological significance, *Trends Biochem. Sci.* **14:**488–492.

Matsuoka, A., Stuehr, D. J., Olson, J. S., Clark, P., and Ikeda-Saito, M., 1994, L-Arginine and calmodulin regulation of the heme iron reactivity in neuronal nitric oxide synthase, *J. Biol. Chem.* **269:**20335–20339.

Mayer, B., John, M., and Böhme, E., 1990, Purification of a Ca^{2+}/calmodulin-dependent nitric oxide synthase from porcine cerebellum. Cofactor-role of tetrahydrobiopterin, *FEBS Lett.* **277:**215–219.

Mayer, B., John, M., Heinzel, B., Werner, E. R., Wachter, H., Schultz, G., and Böhme, E., 1991, Brain nitric oxide synthase is a biopterin- and flavin-containing multi-functional oxido-reductase, *FEBS Lett.* **288:**187–191.

Mayer, B., Klatt, P., Werner, E. R., and Schmidt, K., 1995, Kinetics and mechanism of tetrahydrobiopterin-induced oxidation of nitric oxide, *J. Biol. Chem.* **270:**655–659.

Mayer, B., Wu, C., Gorren, A. C. F., Pfeiffer, S., Schmidt, K., Clark, P., Stuehr, D. J., and Werner, E. R., 1997, Tetrahydrobiopterin binding to inducible nitric oxide synthase expressed in Escherichia coli. Heme spin shift and dimer stabilization by the potent pterin antagonist 4-amino-tetrahydrobiopterin, *Biochemistry* **36:**8422–8427.

McMillan, K., and Masters, B. S. S., 1993, Optical difference spectrophotometry as a probe of rat brain nitric oxide synthase heme–substrate interaction, *Biochemistry* **32:**9875–9880.

McMillan, K., and Masters, B. S. S., 1995, Prokaryotic expression of the heme- and flavin-binding domains of rat neuronal nitric oxide synthase as distinct polypeptides: Identification of the heme-binding proximal thiolate ligand as cysteine-415, *Biochemistry* **34:**3686–3693.

McMillan, K., Bredt, D. S., Hirsch, D. J., Snyder, S. H., Clark, J. E., and Masters, B. S. S., 1992, Cloned, expressed rat cerebellar nitric oxide synthase contains stoichiometric amounts of heme, which binds carbon monoxide, *Proc. Natl. Acad. Sci. USA* **89:**11141–11145.

Meldrum, B., and Garthwaite, J., 1990, Excitatory amino acid neurotoxicity and neurodegenerative disease, *Trends Pharmacol. Sci.* **11:**379–387.

Moncada, S., Palmer, R. M. J., and Higgs, E. A., 1991, Nitric oxide—Physiology, pathophysiology, and pharmacology, *Pharmacol. Rev.* **43:**109–142.

Mülsch, A., and Busse, R., 1991, Nitric oxide synthase in native and cultured endothelial cells—Calcium/calmodulin and tetrahydrobiopterin are cofactors, *J. Cardiovasc. Pharmacol.* **17:**S52–S56.

Murphy, S., Simmons, M. L., Agullo, L., Garcia, A., Feinstein, D. L., Galea, E., Reis, D. J., Minc-Golomb, D., and Schwartz, J. P., 1993, Synthesis of nitric oxide in CNS glial cells, *Trends Neurosci.* **16:**323–328.

Palmer, R. M. J., Ashton, D. S., and Moncada, S., 1988, Vascular endothelial cells synthesize nitric oxide from L-arginine, *Nature* **333:**664–666.

Pantopoulos, K., and Hentze, M. W., 1995, Nitric oxide signaling to iron-regulatory protein: Direct control of ferritin mRNA translation and transferrin receptor mRNA stability in transfected fibroblasts, *Proc. Natl. Acad. Sci. USA* **92:**1267–1271.

Pantopoulos, K., Weiss, G., and Hentze, M. W., 1996, Nitric oxide and oxidative stress (H_2O_2) control mammalian iron metabolism by different pathways, *Mol. Cell Biol.* **16:**3781–3788.

Paraskeva, E., and Hentze, M. W., 1996, Iron–sulphur clusters as genetic regulatory switches: The bifunctional iron regulatory protein-1, *FEBS Lett.* **389:**40–43.

Petros, A., Lamb, G., Leone, A., Moncada, S., Bennett, D., and Vallance, P., 1994, Effects of a nitric oxide synthase inhibitor in humans with septic shock, *Cardiovasc. Res.* **28:**34–39.

Pou, S., Pou, W. S., Bredt, D. S., Snyder, S. H., and Rosen, G. M., 1992, Generation of superoxide by purified brain nitric oxide synthase, *J. Biol. Chem.* **267:**24173–24176.

Poulos, T. L., Finzel, B. C., and Howard, A. J., 1986, Crystal structure of substrate-free *Pseudomonas putida* cytochrome P-450, *Biochemistry* **25:**5314–5322.

Pufahl, R. A., and Marletta, M. A., 1993, Oxidation of N^G-hydroxy-L-arginine by nitric oxide synthase—Evidence for the involvement of the heme in catalysis, *Biochem. Biophys. Res. Commun.* **193:**963–970.

Pufahl, R. A., Wishnok, J. S., and Marletta, M. A., 1995, Hydrogen peroxide-supported oxidation of *N*-G-hydroxy-L-arginine by nitric oxide synthase, *Biochemistry* **34:**1930–1941.

Rand, M. J., and Li, C. G., 1995, Nitric oxide as a neurotransmitter in peripheral nerves: Nature of transmitter and mechanism of transmission, *Annu. Rev. Physiol.* **57:**659–682.

Renaud, J. P., Boucher, J. L., Vadon, S., Delaforge, M., and Mansuy, D., 1993, Particular ability of liver P450s3A to catalyze the oxidation of N(omega)-hydroxyarginine to citrulline and nitrogen oxides and occurrence in NO synthases of a sequence very similar to the heme-binding sequence in P450s, *Biochem. Biophys. Res. Commun.* **192:**53–60.

Richards, M. K., and Marletta, M. A., 1994, Characterization of neuronal nitric oxide synthase and a C415H mutant, purified from a baculovirus overexpression system, *Biochemistry* **33:**14723–14732.

Rodriguez-Crespo, I., Gerber, N. C., and Ortiz DeMontellano, P. R., 1996, Endothelial nitric-oxide synthase—Expression in *Escherichia coli*, spectroscopic characterization, and role of tetrahydrobiopterin in dimer formation, *J. Biol. Chem.* **271:**11462–11467.

Roman, L. J., Sheta, E. A., Martasek, P., Gross, S. S., Liu, Q., and Masters, B. S. S., 1995, High-level expression of functional rat neuronal nitric oxide synthase in *Escherichia coli*, *Proc. Natl. Acad. Sci. USA* **92:**8428–8432.

Sakuma, I., Stuehr, D. J., Gross, S. S., Nathan, C., and Levi, R., 1988, Identification of arginine as a precursor of endothelium-derived relaxing factor, *Proc. Natl. Acad. Sci. USA* **85:**8664–8667.

Sari, M. A., Booker, S., Jaouen, M., Vadon, S., Boucher, J. L., Pompon, D., and Mansuy, D., 1996, Expression in yeast and purification of functional macrophage nitric oxide synthase. Evidence for cysteine-194 as iron proximal ligand, *Biochemistry* **35:**7204–7213.

Sessa, W. C., Barber, C. M., and Lynch, K. R., 1993, Mutation of N-myristoylation site converts endothelial cell nitric oxide synthase from a membrane to a cytosolic protein, *Circ. Res.* **72:**921–924.

Sheta, E. A., McMillan, K., and Masters, B. S. S., 1994, Evidence for a bidomain structure of constitutive cerebellar nitric oxide synthase, *J. Biol. Chem.* **269:**15147–15153.

Snyder, S. H., and Bredt, D. S., 1991, Nitric oxide as a neuronal messenger, *Trends Pharmacol. Sci.* **12:**125–128.

Son, H., Hawkins, R. D., Martin, K., Kiebler, M., Huang, P. L., Fishman, M. C., and Kandel, E. R., 1996, Long-term potentiation is reduced in mice that are doubly mutant in endothelial and neuronal nitric oxide synthase, *Cell* **87:**1015–1023.

Sono, M., Stuehr, D. J., Ikedasaito, M., and Dawson, J. H., 1995, Identification of nitric oxide synthase as a thiolate-ligated heme protein using magnetic circular dichroism spectroscopy—Comparison with cytochrome P-450-CAM and chloroperoxidase, *J. Biol. Chem.* **270:**19943–19948.

Stuehr, D. J., 1997, Structure–function aspects in the nitric oxide synthases, *Annu. Rev. Pharmacol. Toxicol.* **37:**339–359.

Stuehr, D. J., and Ikeda-Saito, M., 1992, Spectral characterization of brain and macrophage nitric oxide synthases—cytochrome-P-450-like hemeproteins that contain a flavin semiquinone radical, *J. Biol. Chem.* **267:**20547–20550.

Stuehr, D. J., Gross, S. S., Sakuma, I., Levi, R., and Nathan, C. F., 1989, Activated murine macrophages secrete a metabolite of arginine with the bioactivity of endothelium-derived relaxing factor and the chemical reactivity of nitric oxide, *J. Exp. Med.* **169:**1011–1020.

Stuehr, D. J., Cho, H. J., Kwon, N. S., Weise, M. F., and Nathan, C. F., 1991a, Purification and characterization of the cytokine-induced macrophage nitric oxide synthase: An FAD- and FMN-containing flavoprotein, *Proc. Natl. Acad. Sci. USA* **88:**7773–7777.

Stuehr, D. J., Kwon, N. S., Nathan, C. F., Griffith, O. W., Feldman, P. L., and Wiseman, J., 1991b, N omega-hydroxy-L-arginine is an intermediate in the biosynthesis of nitric oxide from L-arginine, *J. Biol. Chem.* **266:**6259–6263.

Tayeh, M. A., and Marletta, M. A., 1989, Macrophage oxidation of L-arginine to nitric oxide, nitrite, and nitrate. Tetrahydrobiopterin is required as a cofactor, *J. Biol. Chem.* **264:**19654–19658.

Venema, R. C., Sayegh, H. S., Kent, J. D., and Harrison, D. G., 1996, Identification, characterization, and comparison of the calmodulin-binding domains of the endothelial and inducible nitric oxide synthases, *J. Biol. Chem.* **271:**6435–6440.

Venema, R. C., Ju, H., Zou, R., Ryan, J. W., and Venema, V. J., 1997, Subunit interactions of endothelial nitric-oxide synthase—Comparisons to the neuronal and inducible nitric-oxide synthase isoforms, *J. Biol. Chem.* **272:**1276–1282.

Vorherr, T., Knopfel, L., Hofmann, F., Mollner, S., Pfeuffer, T., and Carafoli, E., 1993, The calmodulin binding domain of nitric oxide synthase and adenylyl cyclase, *Biochemistry* **32:**6081–6088.

Weiss, G., Werner-Felmayer, G., Werner, E. R., Grünewald, K., Wachter, H., and Hentze, M. W., 1994, Iron regulates nitric oxide synthase activity by controlling nuclear transcription, *J. Exp. Med.* **180:**969–976.

Weiss, G., Bogdan, C., and Hentze, M. W., 1997, Pathways for the regulation of macrophage iron metabolism by the anti-inflammatory cytokines IL-4 and IL-13, *J. Immunol.* **158:**420–425.

Werner, E. R., Pitters, E., Schmidt, K., Wachter, H., Werner-Felmayer, G., and Mayer, B., 1996, Identification of the 4-amino analogue of tetrahydrobiopterin as dihydropteridine reductase inhibitor and potent pteridine antagonist of rat neuronal nitric oxide synthase, *Biochem. J.* **320:**193–196.

Werner-Felmayer, G., Werner, E. R., Fuchs, D., Hausen, A., Reibnegger, G., and Wachter, H., 1990, Tetrahydrobiopterin-dependent formation of nitrite and nitrate in murine fibroblasts, *J. Exp. Med.* **172:**1599–1607.

Wilson, R. I., Yanovsky, J., Godecke, A., Stevens, D. R., Schrader, J., and Haas, H. L., 1997, Endothelial nitric oxide synthase and LTP, *Nature* **386:**338.

Wu, C. Q., Zhang, J. G., Abu-Soud, H., Ghosh, D. K., and Stuehr, D. J., 1996, High-level expression of mouse inducible nitric oxide synthase in *Escherichia coli* requires coexpression with calmodulin, *Biochem. Biophys. Res. Commun.* **222:**439–444.

Zhang, M. J., and Vogel, H. J., 1994, Characterization of the calmodulin-binding domain of rat cerebellar nitric oxide synthase, *J. Biol. Chem.* **269:**981–985.

CHAPTER 5

Cytokine Regulation of Nitric Oxide Production

HEIKO MÜHL and CHARLES A. DINARELLO

1. Introduction

In numerous animal models associated with increased production of proinflammatory cytokines, nitric oxide (NO) has been shown to affect severity of disease. NO can influence pathophysiology either as a cytotoxic component of nonspecific immune defense mechanisms (Fang, 1997) (see Chapters 8 and 12) or as a mediator that stimulates signal transduction pathways (Lander, 1997) (see Chapter 10). It is widely assumed that the macrophage type of inducible nitric oxide synthase (iNOS, NOS2) mediates most of the pathophysiological functions of NO. Competitive inhibition of NOS activity using L-arginine analogues such as N^G-monomethyl-L-arginine has been shown to reduce the severity of various inflammatory and autoimmune diseases (Table I).

Recently, it has been recognized that production of NO is a marker in human pathophysiology as well (Table II). With data from rodent systems supporting the idea of NO as a proinflammatory mediator, cytokine regulation and activation of iNOS might represent a novel target for anti-inflammatory intervention.

HEIKO MÜHL • Division of Infectious Diseases, University of Colorado Health Sciences Center, Denver, Colorado 80262, and Institute for General Pharmacology and Toxicology, Clinic of Johann Wolfgang Goethe University, D-60590 Frankfurt, Germany. ***CHARLES A. DINARELLO*** • Division of Infectious Diseases, University of Colorado Health Sciences Center, Denver, Colorado 80262.

Nitric Oxide and Infection, edited by Fang. Kluwer Academic / Plenum Publishers, New York, 1999.

TABLE I
Animal Models Supporting a Role for NO as a Mediator of Inflammatory Diseases

Zymosan-induced peritonitis and multiorgan failure (Cuzzocrea *et al.*, 1997)
Endotoxin-induced ocular inflammation (Goureau *et al.*, 1995)
Adjuvant arthritis (Connor *et al.*, 1995)
Carrageenan-induced pleural inflammation (Tracey *et al.*, 1995)
Experimental colitis (Neilly *et al.*, 1995)
Sephadex-induced lung edema (Andersson *et al.*, 1995)
Islet inflammation and diabetes (Reimers *et al.*, 1994)
Allergic airway inflammation (Miura *et al.*, 1996)
Immune-complex glomerulonephritis (Weinberg *et al.*, 1994)

2. Induction of iNOS Expression by Cytokines

Expression of iNOS is mediated by a variety of proinflammatory cytokines in a cell-specific manner. Regulation by endogenous cytokines and mediators mainly occurs at the level of gene expression. In most cases, changes in iNOS mRNA levels result in corresponding changes in iNOS protein expression.

2.1. Induction of iNOS by IL-1

In a variety of cell types, IL-1 is a prominent and sufficient stimulus for NO production via iNOS. These include human hepatocytes (Geller *et al.*, 1995), human and rat vascular smooth muscle cells (Junquero *et al.*, 1992; Kilbourn *et al.*, 1992), human chondrocytes (Palmer *et al.*, 1993), human astrocytes (Liu *et al.*, 1996), rat renal mesangial cells (Pfeilschifter and Schwarzenbach, 1990), rat microvascular endothelial cells (Bonmann *et al.*, 1997), rodent pancreatic β-cells (Corbett *et al.*, 1992), rat cardiac myocytes (LaPointe and Sitkins, 1996), rat fibroblasts (Jorens *et al.*, 1992), and rat myenteric neurons (Valentine *et al.*, 1996). Analysis of the iNOS 5′-flanking region has revealed nuclear factor κB

TABLE II
Human Diseases Associated with Increased Production of NO

Asthma (Kharitonov *et al.*, 1994)
Cystic fibrosis (Francoeur and Denis, 1995)
Ulcerative colitis and Crohn's disease (Kimura *et al.*, 1997)
Rheumatoid arthritis (Kaur and Halliwell, 1994)
Hepatic cirrhosis (Laffi *et al.*, 1995)
Systemic inflammatory response syndrome (Evans *et al.*, 1993)
Tuberculosis (Nicholson *et al.*, 1996)
HIV infection (Bukrinsky *et al.*, 1995)

(NF-κB) binding regions in the human, murine, and rat promoter sequences (Xie *et al.*, 1993; Beck and Sterzel, 1996; Eberhardt *et al.*, 1996; Nunokawa *et al.*, 1996). IL-1β induces NF-κB translocation and binding to the iNOS promoter (Kwon *et al.*, 1995). Moreover, inhibitors of NF-κB activation potently inhibit IL-1β-induced iNOS expression (Eberhardt *et al.*, 1994). Therefore, it is assumed that NF-κB activation is an essential step in the signaling cascade that results in IL-1-induced iNOS induction. In most cell types, IL-1 does not induce cell death. Therefore, in certain cells IL-1 is able to induce expression of iNOS over long periods of time. In rat mesangial cells, high levels of iNOS protein are detectable even after 48 h of exposure to IL-1β (H.M., unpublished results). Upregulation of iNOS expression by NO as observed in rat mesangial cells (Mühl and Pfeilschifter, 1995) and vascular smooth muscle cells (Mühl and Pfeilschifter, 1995; Boese *et al.*, 1996) may facilitate long-term expression of iNOS in these cells. This example of iNOS positive feedback regulation is in contrast to the negative feedback noted in other cell systems, such as RAW 264.7 macrophages stimulated by LPS/IFNγ (Weisz *et al.*, 1996). In addition to its ability to induce iNOS as a single stimulus, IL-1 acts as costimulus synergizing with TNFα or IFNγ for iNOS expression in a wide array of cell types, such as human intestinal epithelial cells (Linn *et al.*, 1997) and human mesangial cells (Nicolson *et al.*, 1993).

2.2. Induction of iNOS by TNFα

Similar to IL-1, TNFα is a stimulus for iNOS expression in a broad spectrum of cells. Accordingly, induction of iNOS is another example of the overlapping activities of IL-1 and TNFα. TNFα can act as a sole stimulus for induction of NO production via iNOS in many different cell types such as rat renal mesangial cells (Pfeilschifter and Schwarzenbach, 1990), rat hepatocytes (Geller *et al.*, 1993), and human neuroblastoma cells (Obregon *et al.*, 1997). Moreover, TNFα is a potent costimulus for iNOS expression, especially in combination with IL-1 or IFNγ, in a variety of cells including human mesangial cells (Nicolson *et al.*, 1993), human lung epithelial cells (Robbins *et al.*, 1994), and human keratinocytes (Sirsjo *et al.*, 1996).

2.3. Induction of iNOS by IL-2

IL-2 immunotherapy for treatment of cancer is accompanied by significant induction of NO synthesis (Hibbs *et al.*, 1992; Ochoa *et al.*, 1992). Studies with IL-2-treated mice have revealed iNOS expression in various tissues including endothelium and macrophages (Orucevic *et al.*, 1997). Interestingly, coadministration of IL-2 and the NOS inhibitor N^{G}-nitro-L-arginine-methylester to cancer patients significantly reduces development of one of the main side effects of IL-2 treatment in mice and humans, hypotension (Ochoa *et al.*, 1992), by inhibiting NO-dependent vasodilation (see Chapter 7).

2.4. Activation of NO Release from Human Monocytes via an IL-4/CD23 Pathway

It has been reported that IL-4 is able to induce release of NO from human monocytes (Dugas *et al.*, 1995). This observation has been linked to IL-4-induced expression of the low-affinity receptor for IgE, CD23. Activation of CD23 by IgE or anti-CD23 antibodies results in NO production by monocytes (Paul-Eugene *et al.*, 1995). Recent data have suggested that activation of NO release by the IL-4/CD23 pathway in human monocytic cells is mediated by constitutive endothelial NOS (Aubry *et al.*, 1997).

2.5. Induction of iNOS by IFNγ

In a broad spectrum of cell types including human intestinal epithelial cells (Linn *et al.*, 1997), human astrocytes (Liu *et al.*, 1996), human mesangial cells (Nicolson *et al.*, 1993), and rat cardiac myocytes (Hattori *et al.*, 1997), IFNγ is an important costimulus for iNOS expression. In some cell types like rat lung fibroblasts (Jorens *et al.*, 1992) or murine macrophages (Vodovotz *et al.*, 1994), IFNγ is sufficient to stimulate expression and activity of iNOS. Induction of iNOS has been associated with some antiviral (Karupiah *et al.*, 1993) and antitumor effects (Lavnikova *et al.*, 1993) of IFNγ.

The intracellular mechanisms leading to iNOS induction are best characterized in murine macrophages. Activation of iNOS by IFNγ is mediated by γ-activated sites and IFNγ-responsive elements in the iNOS promotor (Xie *et al.*, 1993; Martin *et al.*, 1994). The importance of interferon regulatory factor-1 in IFNγ induction of iNOS is underscored by a defect in iNOS expression observed in macrophages from mice homozygous for a targeted disruption of the interferon regulatory factor-1 gene (Kamijo *et al.*, 1994). Once induced by IFNγ, iNOS can remain active over prolonged periods of up to 1 week *in vitro* (Vodovotz *et al.*, 1994). LPS, another potent inducer of iNOS in most cells, mediates synergistic expression of iNOS in murine macrophages when added in combination with IFNγ (Xie *et al.*, 1993). However, LPS seems to simultaneously elicit a negative signal that limits IFNγ-induced NO release (Vodovotz *et al.*, 1994). Moreover, preincubation of cells with small quantities of LPS can inhibit subsequent IFNγ stimulation of NO production (Bogdan *et al.*, 1993).

2.6. iNOS Induction by IFNα/β

Recent evidence suggests that IFNα induces iNOS expression in human mononuclear cells. This pathway may be active *in vivo* as well as *in vitro*, as elevated levels of iNOS expression are detectable in patients with hepatitis C receiving IFNα (Sharara *et al.*, 1997). IFNα/β has also been found to be essential for NOS2 expression in murine leishmaniasis (Diefenbach *et al.*, 1998), although the reported effects of IFNα/β on iNOS expression in rodent macrophages (Zhang *et al.*, 1994; Kreil and Eibl, 1995; Zhou *et al.*, 1995; Lopez-Collazo *et al.*, 1998) or primary astrocytes (Stewart *et al.*, 1997) have been variable.

2.7. iNOS Induction by IL-12 and IL-18

IL-12 and the newly described cytokine IL-18 (formerly, *IFNγ-inducing factor*) are known for their ability to induce IFNγ production in T and NK cells (Dinarello *et al.*, 1998; Kohno and Kurimoto, 1998). Via this pathway, these cytokines are capable of inducing NO formation as demonstrated in murine peritoneal exudate cells (Zhang *et al.*, 1997).

3. Modulation of iNOS Expression

In view of the proinflammatory and cytotoxic properties of NO (see Chapter 8), control of iNOS expression is likely to play a critical role in limiting detrimental effects of NO in infection. The net level of iNOS protein production is dependent on the balanced expression of cytokines, growth factors, and vasoactive peptides that enhance or inhibit iNOS expression.

3.1. Modulation of iNOS Expression by Growth Factors and Vasoactive or Neuro-immunomodulatory Peptides

Inhibition of iNOS expression by TGFβ has been observed in murine peritoneal macrophages (Vodovotz *et al.*, 1993), rat renal mesangial cells (Pfeilschifter and Vosbeck, 1991), the murine insulin-producing β-cell line RINm5F (Mabley *et al.*, 1997), human vascular smooth muscle cells (Junquero *et al.*, 1992), and other cell types. It has become apparent that TGFβ interferes at multiple stages of iNOS expression. TGFβ reduces mRNA stability, decreases translational efficiency, and increases protein degradation of iNOS (Vodovotz *et al.*, 1993). Recently, the relevance of TGFβ in the control of iNOS activity has been confirmed *in vivo*. TGFβ$1^{-/-}$ mice have elevated iNOS expression and NO_x levels in serum (Vodovotz *et al.*, 1996), while mice overexpressing a cDNA coding for TGFβ1 in the liver show a significant reduction of serum nitrite/nitrate levels compared with wild-type mice after endotoxin-induced septic shock (Vodovotz *et al.*, 1998).

Insulinlike growth factor and platelet-derived growth factor are each potent inhibitors of cytokine-induced iNOS expression in rat vascular smooth muscle cells and mesangial cells (Schini *et al.*, 1992, 1994; Kunz *et al.*, 1997). Inhibition of NO release by platelet-derived growth factor can be reversed by calphostin C, suggesting that this inhibitory pathway involves protein kinase C (Kunz *et al.*, 1997). This is supported by the observation that activation of protein kinase C inhibits iNOS expression in rat vascular smooth muscle cells (Geng *et al.*, 1994; Nakayama *et al.*, 1994; Paul *et al.*, 1997), renal mesangial cells (Mühl and Pfeilschifter, 1994), and cardiac myocytes (LaPointe and Sitkins, 1996). Inhibition of cytokine-induced iNOS expression by the vasoactive peptides angiotensin II (Nakayama *et al.*, 1994), arginine vasopressin (Yamamoto *et al.*, 1997), and endothelin (Beck *et al.*, 1995) in rat vascular smooth muscle cells or mesangial cells is believed to be mediated at least partially by activation of protein kinase C. However, in macrophages (Jun *et al.*, 1994) or hepatocytes (Hortelano *et al.*, 1993), activation of protein kinase C can actually upregulate iNOS expression.

Differential cell type-specific effects on iNOS expression have been documented for basic fibroblast growth factor (bFGF). In rat vascular smooth muscle cells (Scott-Burden *et al.*, 1992) and mesangial cells (Kunz *et al.*, 1997), bFGF enhances NO release. Because NO can increase secretion of bFGF in these cells (Fukuo *et al.*, 1995), upregulation of iNOS by NO which has been observed in either cell type (Mühl and Pfeilschifter, 1995; Boese *et al.*, 1996) might be mediated by NO-induced bFGF in a positive feedback interaction. In contrast, a suppressive effect of bFGF on iNOS expression was reported for bovine retinal pigmented epithelial cells (Goureau *et al.*, 1993) and human microglial cells (Colasanti *et al.*, 1995).

The neuro-immunomodulatory peptide melanocyte-stimulating hormone (α-MSH) is a well-described anti-inflammatory mediator. The anti-inflammatory action of α-MSH is accompanied by inhibition of synthesis of proinflammatory cytokines like IL-1, TNFα, IL-6, or IL-8, along with increased production of anti-inflammatory IL-10 (Lipton and Catania, 1997). The observation that the peptide inhibits expression of iNOS in LPS/ IFNγ-stimulated RAW 264.7 macrophages is consistent with a role of α-MSH in reducing inflammation (Star *et al.*, 1995).

3.2. Inhibition of iNOS Expression by Cytokines

IL-10 is a well-characterized suppressor of macrophage function. However, IL-10 only modestly inhibits IFNγ-induced NO production in murine peritoneal macrophages, and has no effect on IFNγ/TNFα stimulation of NO formation in these cells (Bogdan *et al.*, 1991). Pretreatment with IL-10 effectively inhibits IFNγ induction of iNOS in the murine macrophage cell line J774 (Cunha *et al.*, 1992). In contrast, IL-10 upregulates iNOS expression in bone marrow-derived macrophages stimulated by IFNγ and TNFα (Corradin *et al.*, 1993). These apparently contradictory results suggest that IL-10 inhibition of IFNγ-induced iNOS might be reversed by TNFα.

Although IL-4 has been reported to stimulate eNOS in human mononuclear cells (see above), the cytokines IL-4 and IL-13 have been reported to inhibit cytokine-induced iNOS expression (Berkman *et al.*, 1996; Bogdan *et al.*, 1997). In addition, anti-inflammatory cytokines such as IL-4 or IL-10 may block NO synthesis by inducing arginase and altering the cellular L-arginine pool (Corraliza *et al.*, 1995).

3.3. Regulation of iNOS by the Cyclic AMP Signaling System

Cyclic AMP-elevating agents can induce iNOS as a sole stimulus in rat vascular smooth muscle cells (Koide *et al.*, 1993) and renal mesangial cells (Mühl *et al.*, 1994; Nusing *et al.*, 1996). In these and other cell types such as rat cardiac myocytes, cyclic AMP synergizes with cytokines for iNOS induction (Mühl *et al.*, 1994; Ikeda *et al.*, 1996a). Cyclic AMP has been observed to increase both gene transcription and iNOS mRNA stability (Kunz *et al.*, 1994). In rat mesangial cells, induction of gene transcription by cyclic AMP involves binding of CAAT/enhancer-binding protein (C/EBP) as well as of cyclic AMP-responsive element-binding protein (CREB) transcription factors to a corresponding C/EBP-response element in the rat iNOS promoter (Eberhardt *et al.*, 1998). Via the cyclic AMP signaling pathway, prostaglandins and β_2-adrenergic agonists (Mühl *et al.*, 1994) or mediators like adrenomedullin (Ikeda *et al.*, 1996b) can positively regulate iNOS. However, cyclic AMP inhibits iNOS expression in rat astrocytes (Pahan *et al.*, 1997), and conflicting results have been reported for macrophages. Cyclic AMP was found to stimulate NO production in rat peritoneal macrophages (Sowa, and Przewlocki, 1994) but was suppressive in the murine cell line J774 (Bulut *et al.*, 1993). Moreover, cyclic AMP reduces iNOS expression in rat liver Kupffer cells (Mustafa and Olson, 1998).

3.4. Suppression of iNOS Expression by Anti-Inflammatory and Immunosuppressive Drugs

Both salicylic acid and acetylsalicylic acid (aspirin) are anti-inflammatory compounds. Inhibition of prostaglandin synthesis is considered to be the prime mechanism for the anti-inflammatory action of these drugs. However, only aspirin is a potent inhibitor of cyclooxygenase (COX) activity. Therefore, it has been suggested that these agents may act via additional mechanisms. It is of interest that both agents are able to inhibit activation of the transcription factor NF-κB (Kopp and Gosh, 1994), and sodium salicylate can suppress expression of chemokines (for a review see Wu *et al.*, 1998). Aspirin or salicylic acid inhibits the expression of iNOS in murine macrophages stimulated by LPS/IFNγ (Amin *et al.*, 1995; Kepka-Lenhart *et al.*, 1996), rat hepatocytes stimulated by IL-1β (Sakitani *et al.*, 1997), and RINm5F cells stimulated by IL-1β (Kwon *et al.*, 1997). In RAW 264.7 macrophages, the inhibition of iNOS does not appear to be mediated by reduction of prostaglandin E_2 synthesis, as addition of exogenous prostaglandin E_2 cannot overcome inhibition of iNOS expression (Sakitani *et al.*, 1997). Interestingly, salicylates appear to affect iNOS expression in each of the three cell types on a posttranscriptional level; at least in hepatocytes and RINm5F cells, inhibition is not mediated by effects on NF-κB (Kwon *et al.*, 1997; Sakitani *et al.*, 1997).

Widely used nonsteroidal anti-inflammatory drugs (NSAIDs) like ibuprofen are of importance in the treatment of inflammatory diseases and pain. In addition, it has been reported recently that this group of drugs may slow progression of neurological conditions such as Alzheimer's disease (Rich *et al.*, 1995). In several experimental systems such as LPS/IFNγ-stimulated rat alveolar macrophages (Aeberhard *et al.*, 1995) or primary cerebellar glial cells (Stratman *et al.*, 1997), ibuprofen or other NSAIDs inhibit expression of iNOS. Moreover transcriptional activation of the human iNOS promoter is inhibited by ibuprofen (Kolyada *et al.*, 1996). Therefore, inhibition of iNOS might be an important contributory mechanism of action by ibuprofen or other NSAIDs.

Glucocorticoids are among the most potent anti-inflammatory drugs available for treatment of a variety of autoimmune and inflammatory diseases such as rheumatoid arthritis. Inhibition of release of proinflammatory cytokines and chemokines, as well as reduction of adhesion molecule expression, are well-characterized mechanisms of action of these compounds. Inhibition of iNOS expression by dexamethasone has been observed in a variety of cells, including vascular endothelial cells (Radomski *et al.*, 1990), rat mesangial cells (Pfeilschifter and Schwarzenbach, 1990), rat peritoneal neutrophils (McCall *et al.*, 1991), rat vascular smooth muscle cells (Marumo *et al.*, 1993), murine J774 macrophages (Baydoun *et al.*, 1993), and rat hepatocytes (De Vera *et al.*, 1997). The importance of glucocorticoids as inhibitors of cytokine-induced NO formation is suggested by potent suppression of iNOS induction *in vivo* (Knowles *et al.*, 1990). The inhibitory mechanisms of dexamethasone on iNOS induction and NO formation are cell type specific and stimulus dependent. For example, the effects of dexamethasone administered $\frac{1}{2}$h prior to stimulation differ markedly from those of dexamethasone administered 2 h poststimulation (Perrella *et al.*, 1994). In rat hepatocytes stimulated with a combination of IL-1β, TNFα, and IFNγ, dexamethasone inhibited iNOS mRNA accumulation by inhibiting NF-κB activation (De Vera *et al.*, 1997). Reduction at the level of mRNA was also noted in vascular endothelial cells stimulated with LPS/IFNγ, as well (Radomski *et al.*, 1990). Inhibition of IL-1β induction of iNOS by dexamethasone in cells such as rat aortic smooth muscle cells (Perrella *et al.*, 1994) and rat mesangial cells (Kunz *et al.*, 1996) appears to be more complex. In either cell type, dexamethasone inhibits gene transcription of iNOS. However, iNOS mRNA stability is simultaneously increased by dexamethasone. In mesangial cells, a potent suppression of NO formation is achieved by reduction of mRNA translation and increased degradation of iNOS protein (Kunz *et al.*, 1996). Additional mechanisms of reduction of NO formation by dexamethasone might involve inhibition of tetrahydrobiopterin synthesis (Simmons *et al.*, 1996; Pluess *et al.*, 1997) and interference with cytokine-induced cellular L-arginine uptake (Simmons *et al.*, 1996).

Cyclosporins are an important class of immunosuppressants used to prevent rejection in transplantation medicine. In addition to inhibiting T-cell signaling via calcineurin, these drugs can inhibit cytokine-induced iNOS expression in rat mesangial cells (Mühl *et al.*, 1993) and vascular smooth muscle cells (Marumo *et al.*, 1995). Nuclear run-on experiments and electrophoretic mobility shift assays have revealed that cyclosporin A reduces transcription of iNOS in mesangial cells at least in part by reduction of NF-κB binding (Kunz *et al.*, 1995). Interestingly, FK506, another inhibitor of calcineurin, has no effect on NO release from these cells (Mühl *et al.*, 1993; Marumo *et al.*, 1995), implying that the underlying signaling pathway responsible for iNOS inhibition is different from the well-characterized immunosuppressive action of the drug. Subcutaneous injection of cyclosporin A into rats can inhibit iNOS induction by LPS *in vivo* (Tack *et al.*, 1997). This inhibitory activity might contribute to side effects associated with cyclosporin A therapy.

4. Concluding Remarks

The essential concepts of this chapter are summarized in Fig. 1. iNOS is induced by proinflammatory cytokines and interferons in a broad spectrum of cell types. Whereas interferons seem to be of exceptional importance in monocytes/macrophages, predominantly IL-1 and TNFα trigger iNOS expression in cells including chondrocytes, hepatocytes, and astrocytes as well as renal mesangial cells and even neuronal cells.

Obviously, a variety of strategies that have the potential to control iNOS activity have evolved in biological systems, in order to prevent detrimental effects from high-output production of potentially cytotoxic molecular species. In this regard, TGFβ appears to be particularly important as an iNOS suppressor. However, additional modulators like IL-10 or cyclic AMP may also be important in certain cell types.

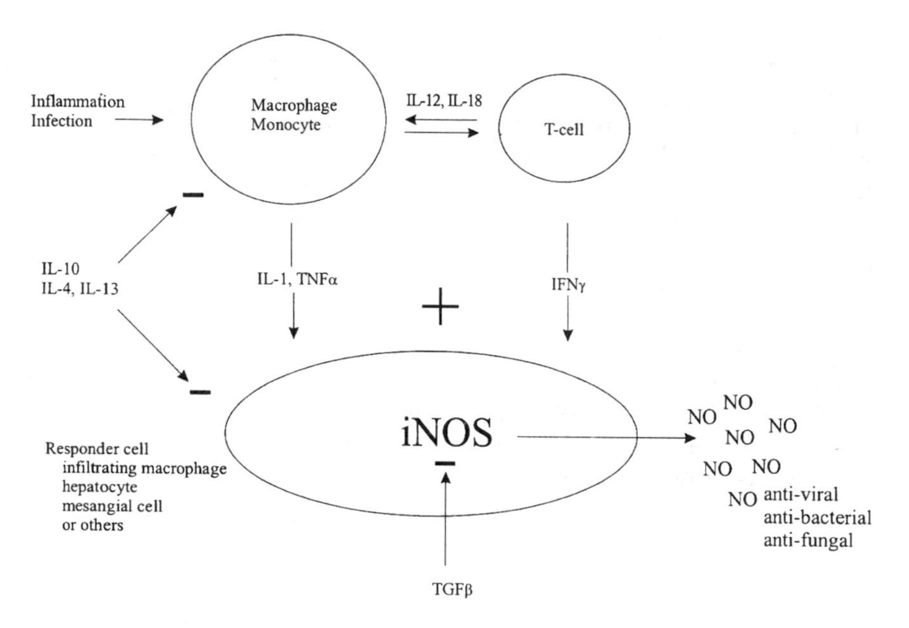

FIGURE 1. Cytokine regulation of NO production.

References

Aeberhard, E. E., Henderson, S. A., Arabolos, N. S., Griscavage, J. M., Catro, F. E., Barrett, C. T., and Ignarro, L. J., 1995, Nonsteroidal anti-inflammatory drugs inhibit expression of the inducible nitric oxide synthase gene, *Biochem. Biophys. Res. Commun.* **208:**1053–1059.

Amin, A. R., Vyas, P., Attur, M., Leszczynska-Piziak, J., Patel, I. R., Weissmann, G., and Abramson, S. B., 1995, The mode of action of aspirin-like drugs: Effect on inducible nitric oxide synthase, *Proc. Natl. Acad. Sci. USA* **92:**7926–7930.

Andersson, S. E., Kallstrom, L., Malm, M., Miller-Larsson, A., and Axelsson, B., 1995, Inhibition of nitric oxide synthase reduces Sephadex-induced oedema formation in the rat lung: Dependence on intact adrenal function, *Inflamm. Res.* **44:**418–422.

Aubry, J. P., Dugas, N., Lecoanet-Henchoz, S., Ouaaz, F., Zhao, H., Delfraissy, J. F., Graber, P., Kolb, J. P., Dugas, B., and Bonnefoy, J. Y., 1997, The 25-kDa soluble CD23 activates type III constitutive nitric oxide-synthase activity via CD11b and CD11c expressed by human monocytes, *J. Immunol.* **159:**614–622.

Baydoun, A. R., Bogle, R. G., Pearson, J. D., and Mann, G. E., 1993, Selective inhibition by dexamethasone of induction of NO synthase, but not of induction of L-arginine transport, in activated murine J774 cells, *Br. J. Pharmacol.* **110:**1401–1406.

Beck, K. F., and Sterzel, R. B., 1996, Cloning and sequencing of the proximal promoter of the rat iNOS gene: Activation of NFkappaB is not sufficient for transcription of the iNOS gene in rat mesangial cells, *FEBS Lett.* **394:**263–267.

Beck, K. F., Mohaupt, M. G., and Sterzel, R. B., 1995, Endothelin-1 inhibits cytokine-stimulated transcription of inducible nitric oxide synthase in glomerular mesangial cells, *Kidney Int.* **48:**1893–1899.

Berkman, N., Robichaud, A., Robbins, R. A., Roesems, G., Haddad, E. B., Barnes, P. J., and Chung, K. F., 1996, Inhibition of inducible nitric oxide synthase expression by interleukin-4 and interleukin-13 in human lung epithelial cells, *Immunology* **89:**363–367.

Boese, M., Busse, R., Mulsch, A., and Schini-Kerth, V., 1996, Effect of cyclic GMP-dependent vasodilators on the expression of inducible nitric oxide synthase in vascular smooth muscle cells: Role of cyclic AMP, *Br. J. Pharmacol.* **119:**707–715.

Bogdan, C., Vodovotz, Y., and Nathan, C., 1991, Macrophage deactivation by interleukin 10, *J. Exp. Med.* **174:**1549–1555.

Bogdan, C., Vodovotz, Y., Paik, J., Xie, Q. W., and Nathan, C., 1993, Traces of bacterial lipopolysaccharide suppress IFN-gamma-induced nitric oxide synthase gene expression in primary mouse macrophages, *J. Immunol.* **151:**301–309.

Bogdan, C., Thuring, H., Dlaska, M., Rollinghoff, M., and Weiss, G., 1997, Mechanism of suppression of macrophage nitric oxide release by IL-13: Influence of the macrophage population, *J. Immunol.* **159:**4506–4513.

Bonmann, E., Suschek, C., Spranger, M., and Kolb-Bachofen, V., 1997, The dominant role of exogenous or endogenous interleukin-1 beta on expression and activity of inducible nitric oxide synthase in rat microvascular brain endothelial cells, *Neurosci. Lett.* **230:**109–112.

Bukrinsky, M. I., Nottet, H. S., Schmidtmayerova, H., Dubrovsky, L., Flanagan, C. R., Mullins, M. E., Lipton, S. A., and Gendelman, H. E., 1995, Regulation of nitric oxide synthase activity in human immunodeficiency virus type 1 (HIV-1)-infected monocytes: Implications for HIV-associated neurological disease, *J. Exp. Med.* **181:**735–745.

Bulut, V., Severn, A., and Liew, F. Y., 1993, Nitric oxide production by murine macrophages is inhibited by prolonged elevation of cyclic AMP, *Biochem. Biophys. Res. Commun.* **195:**1134–1138.

Cetkovic-Cvrlje, M., Sandler, S., and Eizirik, D. L., 1993, Nicotinamide and dexamethasone inhibit interleukin-1-induced nitric oxide production by RINm5F cells without decreasing messenger ribonucleic acid expression for nitric oxide synthase, *Endocrinology* **133:**1739–1743.

Colasanti, M., Di Pucchio, T., Persichini, T., Sogos, V., Presta, M., and Lauro, G. M., 1995, Inhibition of inducible nitric oxide synthase mRNA expression by basic fibroblast growth factor in human microglial cells, *Neurosci. Lett.* **195:**45–48.

Connor, J. R., Manning, P. T., Settle, S. L., Moore, W. M., Jerome, G. M., Webber, R. K., Tjoeng, F. S., and Currie, M. G., 1995, Suppression of adjuvant-induced arthritis by selective inhibition of inducible nitric oxide synthase, *Eur. J. Pharmacol.* **273:**15–24.

Corbett, J. A., Wang, J. L., Sweetland, M. A., Lancaster, J. R., Jr., and McDaniel, M. L., 1992, Interleukin 1 beta induces the formation of nitric oxide by beta-cells purified from rodent islets of Langerhans. Evidence for the beta-cell as a source and site of action of nitric oxide, *J. Clin. Invest.* **90:**2384–2391.

Corradin, S. B., Fasel, N., Buchmuller-Rouiller, Y., Ransijn, A., Smith, J., and Mauel, J., 1993, Induction of macrophage nitric oxide production by interferon-gamma and tumor necrosis factor-alpha is enhanced by interleukin-10, *Eur. J. Immunol.* **23:**2045–2048.

Corraliza, I. M., Soler, G., Eichmann, K., and Modolell, M., 1995, Arginase induction by suppressors of nitric oxide synthesis (IL-4,IL-10 and PGE_2) in murine bone-marrow-derived macrophages, *Biochem. Biophys. Res. Commun.* **206:**667–673.

Cunha, F. Q., Moncada, S., and Liew, F. Y., 1992, Interleukin-10 (IL-10) inhibits the induction of nitric oxide synthase by interferon-gamma in murine macrophages, *Biochem. Biophys. Res. Commun.* **182:**1155–1159.

Cuzzocrea, S., Zingarelli, B., Sautebin, L., Rizzo, A., Crisafulli, C., Campo, G. M., Costantino, G., Calapai, G., Nava, F., DiRosa, M., and Caputi, A. P., 1997, Multiple organ failure following zymosan-induced peritonitis is mediated by nitric oxide, *Shock* **8:**268–275.

De Vera, M. E., Taylor, B. S., Wang, Q., Shapiro, R. A., Billiar, T. R., and Geller, D. A., 1997, Dexamethasone suppresses iNOS gene expression by upregulating I-kappa B alpha and inhibiting NF-kappa B, *Am. J. Physiol.* **273:**G1290–G1296.

Diefenbach, A., Schindler, H., Donhauser, N., Lorenz, E., Laskay, T., MacMicking, J., Rollinghoff, M., Gresser, I., and Bogdan, C., 1998, Type 1 interferon (IFN-alpha/beta) and type 2 nitric oxide synthase regulate the innate immune response to a protozoan parasite, *Immunity* **8:**77–87.

Dinarello, C. A., Novick, D., Puren, A. J., Fantuzzi, G., Shapiro, L., Mühl, H., Yoon, D.-Y., Reznikov, L. L., Kim, S.-H., and Rubinstein, M., 1998, Overview of interleukin-18: More than an interferon-γ inducing factor, *J. Leukoc. Biol.* **63:**658–664.

Dugas, B., Debre, P., and Moncada, S., 1995, Nitric oxide, a vital poison inside the immune and inflammatory network, *Res. Immunol.* **146:**664–670.

Eberhardt, W., Kunz, D., and Pfeilschifter, J., 1994, Pyrrolidine dithiocarbamate differentially affects interleukin 1 beta- and cAMP-induced nitric oxide synthase expression in rat renal mesangial cells, *Biochem. Biophys. Res. Commun.* **200:**163–170.

Eberhardt, W., Kunz, D., Hummel, R., and Pfeilschifter, J., 1996, Molecular cloning of the rat inducible nitric oxide synthase gene promoter, *Biochem. Biophys. Res. Commun.* **223:**752–756.

Eberhardt, W., Plüss, C., Hummel, R., and Pfeilschifter, J., 1998, Molecular mechanisms of inducible nitric oxide synthase gene expression by IL-1β and cAMP in rat mesangial cells, *J. Immunol.* **160:**4961–4969.

Evans, T., Carpenter, A., Kinderman, H., and Cohen, J., 1993, Evidence of increased nitric oxide production in patients with the sepsis syndrome, *Circ. Shock* **41:**77–81.

Fang, F. C., 1997, Mechanisms of nitric oxide-related antimicrobial activity, *J. Clin. Invest.* **99:**2818–2825.

Francoeur, C., and Denis, M., 1995, Nitric oxide and interleukin-8 as inflammatory components of cystic fibrosis, *Inflammation* **19:**587–598.

Fukuo, K., Inoue, T., Morimoto, S., Nakahashi, T., Yasuda, O., Kitano, S., Sasada, R., and Ogihara, T., 1995, Nitric oxide mediates cytotoxicity and basic fibroblast growth factor release in cultured vascular smooth muscle cells. A possible mechanism of neovascularization in atherosclerotic plaques, *J. Clin. Invest.* **95:**669–676.

Geller, D. A., Nussler, A. K., Di Silvio, M., Lowenstein, C. J., Shapiro, R. A., Wang, S. C., Simmons, R. L., and Billiar, T. R., 1993, Cytokines, endotoxin, and glucocorticoids regulate the expression of inducible nitric oxide synthase in hepatocytes, *Proc. Natl. Acad. Sci. USA* **90:**522–526.

Geller, D. A., De Vera, M. E., Russell, D. A., Shapiro, R. A., Nussler, A. K., Simmons, R. L., and Billiar, T. R., 1995, A central role for IL-1 beta in the in vitro and in vivo regulation of hepatic inducible nitric oxide synthase. IL-1 beta induces hepatic nitric oxide synthesis, *J. Immunol.* **155:**4890–4898.

Geng, Y. J., Wu, Q., and Hansson, G. K., 1994, Protein kinase C activation inhibits cytokine-induced nitric oxide synthesis in vascular smooth muscle cells, *Biochim. Biophys. Acta* **1223:**125–132.

Goureau, O., Lepoivre, M., Becquet, F., and Courtois, Y., 1993, Differential regulation of inducible nitric oxide synthase by fibroblast growth factors and transforming growth factor beta in bovine retinal pigmented epithelial cells: Inverse correlation with cellular proliferation, *Proc. Natl. Acad. Sci. USA* **90:**4276–4280.

Goureau, O., Bellot, J., Thillaye, B., Courtois, Y., and de Kozak, Y., 1995, Increased nitric oxide production in endotoxin-induced uveitis. Reduction of uveitis by an inhibitor of nitric oxide synthase, *J. Immunol.* **154:**6518–6523.

Hattori, Y., Nakanishi, N., and Kasai, K., 1997, Role of nuclear factor kappa B in cytokine-induced nitric oxide and tetrahydrobiopterin synthesis in rat neonatal cardiac myocytes, *J. Mol. Cell. Cardiol.* **29:**1585–1592.

Hibbs, J. B., Jr., Westenfelder, C., Taintor, R., Vavrin, Z., Kablitz, C., Baranowski, R. L., Ward, J. H., Menlove, R. L., McMurry, M. P., Kushner, J. P., and Samlowski, W. E., 1992, Evidence for cytokine-inducible nitric oxide synthesis from L-arginine in patients receiving interleukin-2 therapy, *J. Clin. Invest.* **89:**867–877.

Hortelano, S., Genaro, A. M., and Bosca, L., 1993, Phorbol esters induce nitric oxide synthase and increase arginine influx in cultured peritoneal macrophages, *FEBS Lett.* **320:**135–139.

Ikeda, U., Yamamoto, K., Ichida, M., Ohkawa, F., Murata, M., Iimura, O., Kusano, E., Asano, Y., and Shimada, K., 1996a, Cyclic AMP augments cytokine-stimulated nitric oxide synthesis in rat cardiac myocytes, *J. Mol. Cell. Cardiol.* **28:**789–795.

Ikeda, U., Kanbe, T., and Shimada, K., 1996b, Adrenomedullin increases inducible nitric oxide synthase in rat vascular smooth muscle cells stimulated with interleukin-1, *Hypertension* **27:**1240–1244.

Jorens, P. G., Van Overveld, F. J., Vermeire, P. A., Bult, H., and Herman, A. G., 1992, Synergism between interleukin-1 beta and interferon-gamma, an inducer of nitric oxide synthase, in rat lung fibroblasts, *Eur. J. Pharmacol.* **224:**7–12.

Jun, C. D., Choi, B. M., Hoon-Ryu, Um, J. Y., Kwak, H. J., Lee, B. S., Paik, S. G., Kim, H. M., and Chung, H. A. T., 1994, Synergistic cooperation between phorbol ester and IFN-gamma for induction of nitric oxide synthesis in murine peritoneal macrophages, *J. Immunol.* **153:**3684–3690.

Junquero, D. C., Scott-Burden, T., Schini, V. B., and Vanhoutte, P. M., 1992, Inhibition of cytokine-induced nitric oxide production by transforming growth factor-beta 1 in human smooth muscle cells, *J. Physiol. (London)* **454:**451–465.

Kamijo, R., Harada, H., Matsuyama, T., Bosland, M., Gerecitano, J., Shapiro, D., Le, J., Koh, S. I., Kimura, T.,Green, S. J., Mak, T. W., Taniguchi, T., and Vilcek, J., 1994, Requirement for transcription factor IRF-1 in NO synthase induction in macrophages, *Science* **263:**1612–1615.

Karupiah, G., Xie, Q. W., Buller, R. M., Nathan, C., Duarte, C., and MacMicking, J. D., 1993, Inhibition of viral replication by interferon-gamma-induced nitric oxide synthase, *Science* **261:**1445–1448.

Kaur, H., and Halliwell, B., 1994, Evidence for nitric oxide-mediated oxidative damage in chronic inflammation. Nitrotyrosine in serum and synovial fluid from rheumatoid patients, *FEBS Lett.* **350:**9–12.

Kepka-Lenhart, D., Chen, L. C., and Morris, S. M., Jr., 1996, Novel actions of aspirin and sodium salicylate: Discordant effects on nitric oxide synthesis and induction of nitric oxide synthase mRNA in a murine macrophage cell line, *J. Leukoc. Biol.* **59:**840–846.

Kharitonov, S. A., Yates, D., Robbins, R. A., Logan-Sinclair, R., Shinebourne, E. A., and Barnes, P.J., 1994, Increased nitric oxide in exhaled air of asthmatic patients, *Lancet* **343:**133–135.

Kilbourn, R. G., Gross, S. S., Lodato, R. F., Adams, J., Levi, R., Miller, L. L., Lachman, L. B., and Griffith, O. W., 1992, Inhibition of interleukin-1-alpha-induced nitric oxide synthase in vascular smooth muscle and full reversal of interleukin-1-alpha-induced hypotension by N omega-amino-L-arginine, *J. Natl. Cancer Inst.* **84:**1008–1016

Kimura, H., Miura, S., Shigematsu, T., Ohkubo, N., Tsuzuki, Y., Kurose, I., Higuchi, H., Akiba, Y., Hokari, R., Hirokawa, M., Serizawa, H., and Ishii, H., 1997, Increased nitric oxide production and inducible nitric oxide synthase activity in colonic mucosa of patients with active ulcerative colitis and Crohn's disease, *Dig. Dis. Sci.* **42:**1047–1054.

Knowles, R.G., Salter, M., Brooks, S. L., and Moncada, S., 1990, Anti-inflammatory glucocorticoids inhibit the induction by endotoxin of nitric oxide synthase in the lung, liver and aorta of the rat, *Biochem. Biophys. Res. Commun.* **172:**1042–1048.

Kohno, K., and Kurimoto, M., 1998, Interleukin 18, a cytokine which resembles IL-1 structurally and IL-12 functionally but exerts its effect independently of both, *Clin. Immunol. Immunopathol.* **86:**11–15.

Koide, M., Kawahara, Y., Nakayama, I., Tsuda, T., and Yokoyama, M., 1993, Cyclic AMP-elevating agents induce an inducible type of nitric oxide synthase in cultured vascular smooth muscle cells. Synergism with the induction elicited by inflammatory cytokines, *J. Biol. Chem.* **268:**24959–24966.

Koloyada, A. Y., Savikovsky, N., and Madias, N. E., 1996, Transcriptional regulation of the human iNOS gene in vascular-smooth muscle cells and macrophages: Evidence for tissue specificity, *Biochem. Biophys. Res. Commun.* **220:**600–605.

Kopp, E., and Ghosh, S., 1994, Inhibition of NF-κB by sodium salicylate and aspirin, *Science* **265:**956–959.

Kreil, T. R., and Eibl, M. M., 1995, Viral infection of macrophages profoundly alters requirements for induction of nitric oxide synthesis, *Virology* **212:**174–178.

Kunz, D., Mühl, H., Walker, G., and Pfeilschifter, J., 1994, Two distinct signaling pathways trigger the expression of inducible nitric oxide synthase in rat renal mesangial cells, *Proc. Natl. Acad. Sci. USA* **91:**5387–5391.

Kunz, D., Walker, G., Eberhardt, W., Nitsch, D., and Pfeilschifter, J., 1995, Interleukin 1 beta-induced expression of nitric oxide synthase in rat renal mesangial cells is suppressed by cyclosporin A, *Biochem. Biophys. Res. Commun.* **216:**438–446

Kunz, D., Walker, G., Eberhardt, W., and Pfeilschifter, J., 1996, Molecular mechanisms of dexamethasone inhibition of nitric oxide synthase expression in interleukin 1 beta-stimulated mesangial cells: Evidence for the involvement of transcriptional and posttranscriptional regulation, *Proc. Natl. Acad. Sci. USA* **93:**255–259.

Kunz, D., Walker, G., Eberhardt, W., Messmer, U. K., Huwiler, A., and Pfeilschifter, J., 1997, Platelet-derived growth factor and fibroblast growth factor differentially regulate interleukin 1beta- and cAMP-induced nitric oxide synthase expression in rat renal mesangial cells, *J. Clin. Invest.* **100:**2800–2809.

Kwon, G., Corbett, J. A., Rodi, C. P., Sullivan, P., and McDaniel, M. L., 1995, Interleukin-1 beta-induced nitric oxide synthase expression by rat pancreatic beta-cells: Evidence for the involvement of nuclear factor kappa B in the signaling mechanism, *Endocrinology* **136:**4790–4795.

Kwon, G., Hill, J. R., Corbett, J. A., and McDaniel, M. L., 1997, Effects of aspirin on nitric oxide formation and de novo protein synthesis by RINm5F cells and rat islets, *Mol. Pharmacol.* **52:**398–405.

Laffi, G., Foschi, M., Masini, E., Simoni, A., Mugnai, L., La Villa, G., Barletta, G., Mannaioni, P.F., and Gentilini, P., 1995, Increased production of nitric oxide by neutrophils and monocytes from cirrhotic patients with ascites and hyperdynamic circulation, *Hepatology* **22:**1666–1673.

Lander, H. M., 1997, An essential role for free radicals and derived species in signal transduction, *FASEB J.* **11:**118–124.

LaPointe, M. C., and Sitkins, J. R., 1996, Mechanisms of interleukin-1beta regulation of nitric oxide synthase in cardiac myocytes, *Hypertension* **27:**709–714.

Lavnikova, N., Drapier, J. C., and Laskin, D. L., 1993, A single exogenous stimulus activates resident rat macrophages for nitric oxide production and tumor cytotoxicity, *J. Leukoc. Biol.* **54:**322–328.

Linn, S. C., Morelli, P. J., Edry, I., Cottongim, S. E., Szabo, C., and Salzman, A. L., 1997, Transcriptional regulation of human inducible nitric oxide synthase gene in an intestinal epithelial cell line, *Am. J. Physiol.* **272:**G1499–G1508.

Lipton, J. M., and Catania, A., 1997, Anti-inflammatory actions of the neuroimmunomodulator α-MSH, *Immunol. Today* **18:**140–145.

Liu, J., Zhao, M. L., Brosnan, C. F., and Lee, S. C., 1996, Expression of type II nitric oxide synthase in primary human astrocytes and microglia: Role of IL-1beta and IL-1 receptor antagonist, *J. Immunol.* **157:**3569–3576.

Lopez-Collazo, E., Hortelano, S., and Bosca, L., 1998, Triggering of peritoneal macrophages with IFN-alpha/beta attenuates the expression of the inducible nitric oxide synthase through a decrease in NF-kappaB activation, *J. Immunol.* **160:**2889–2895.

Mabley, J. G., Cunningham, J. M., John, N., Di Matteo, M. A., and Green, I. C., 1997, Transforming growth factor beta 1 prevents cytokine-mediated inhibitory effects and induction of nitric oxide synthase in the RINm5F insulin-containing beta-cell line, *J. Endocrinol.* **155:**567–575.

Martin, E., Nathan, C., and Xie, Q. W., 1994, Role of interferon regulatory factor 1 in induction of nitric oxide synthase, *J. Exp. Med.* **180:**977–984.

Marumo, T., Nakaki, T., Nagata, K., Miyata, M., Adachi, H., Esumi, H., Suzuki, H., Saruta, T., and Kato, R., 1993, Dexamethasone inhibits nitric oxide synthase mRNA induction by interleukin-1 alpha and tumor necrosis factor-alpha in vascular smooth muscle cells, *Jpn. J. Pharmacol.* **63:**361–367.

Marumo, T., Nakai, T., Hishikawa, K., Suzuki, H., Kato, R., and Saruta, T., 1995, Cyclosporin A inhibits nitric oxide synthase induction in vascular smooth muscle cells, *Hypertension* **25:**764–768.

McCall, T. B., Palmer, R. M., and Moncada, S., 1991, Induction of nitric oxide synthase in rat peritoneal neutrophils and its inhibition by dexamethasone, *Eur. J. Immunol.* **21:**2523–2527.

Miura, M., Ichinose, M., Kageyama, N., Tomaki, M., Takahashi, T., Ishikawa, J., Ohuchi, Y., Oyake, T., Endoh, N., and Shirato, K., 1996, Endogenous nitric oxide modifies antigen-induced microvascular leakage in sensitized guinea pig airways, *J. Allergy Clin. Immunol.* **98:**144–151.

Mühl, H., and Pfeilschifter, J., 1994, Possible role of protein kinase C-epsilon isoenzyme in inhibition of interleukin 1 beta induction of nitric oxide synthase in rat renal mesangial cells, *Biochem. J.* **303:**607–612.

Mühl, H., and Pfeilschifter, J., 1995, Amplification of nitric oxide synthase expression by nitric oxide in interleukin 1 beta-stimulated rat mesangial cells, *J. Clin. Invest.* **95:**1941–1946.

Mühl, H., Kunz, D., Rob, P., and Pfeilschifter, J., 1993, Cyclosporin derivatives inhibit interleukin 1 beta induction of nitric oxide synthase in renal mesangial cells, *Eur. J. Pharmacol.* **249:**95–100.

Mühl, H., Kunz, D., and Pfeilschifter, J., 1994, Expression of nitric oxide synthase in rat glomerular mesangial cells mediated by cyclic AMP, *Br. J. Pharmacol.* **112:**1–8.

Mustafa, S. B., and Olson, M. S., 1998, Expression of nitric-oxide synthase in rat Kupffer cells is regulated by cAMP, *J. Biol. Chem.* **273:**5073–5080.

Nakayama, I., Kawahara, Y., Tsuda, T., Okuda, M., and Yokoyama, M., 1994, Angiotensin II inhibits cytokine-stimulated inducible nitric oxide synthase expression in vascular smooth muscle cells, *J. Biol. Chem.* **269:**11628–11633.

Neilly, P.J., Kirk, S. J., Gardiner, K. R., Anderson, N. H., and Rowlands, B. J., 1995, Manipulation of the L-arginine-nitric oxide pathway in experimental colitis, *Br. J. Surg.* **82:**1188–1191.

Nicholson, S., Bonecini-Almeida, M. da G., Lapa e Silva, J. R., Nathan, C., Xie, Q.-W., Mumford, R. W., Weidner, J. R., Calaycay, J., Geng, J., Boechat, N., Linhares, C., Rom, W., and Ho, J. L., 1996, Inducible nitric oxide synthase in pulmonary alveolar macrophages from patients with tuberculosis, *J. Exp. Med.* **183:**2293–2302.

Nicolson, A. G., Haites, N. E., McKay, N. G., Wilson, H. M., MacLeod, A. M., and Benjamin, N., 1993, Induction of nitric oxide synthase in human mesangial cells, *Biochem. Biophys. Res. Commun.* **193:**1269–1274.

Nunokawa, Y., Oikawa, S., and Tanaka, S., 1996, Human inducible nitric oxide synthase gene is transcriptionally regulated by nuclear factor-kappaB dependent mechanism, *Biochem. Biophys. Res. Commun.* **223:**347–352.

Nusing, R.M., Klein, T., Pfeilschifter, J., and Ullrich, V., 1996, Effect of cyclic AMP and prostaglandin E2 on the induction of nitric oxide- and prostanoid-forming pathways in cultured rat mesangial cells, *Biochem. J.* **313:**617–623.

Obregon, E., Punzon, M. C., Gonzalez-Nicolas, J., Fernandez-Cruz, E., Fresno, M., and Munoz-Fernandez, M. A., 1997, Induction of adhesion/differentiation of human neuroblastoma cells by tumour necrosis factor-alpha requires the expression of an inducible nitric oxide synthase, *Eur. J. Neurosci.* **9:**1184–1193.

Ochoa, J. B., Curti, B., Peitzman, A. B., Simmons, R. L., Billiar, T. R., Hoffman, R., Rault, R., Longo, D. L., Urba, W. J., and Ochoa, A. C., 1992, Increased circulating nitrogen oxides after human tumor immunotherapy: Correlation with toxic hemodynamic changes, *J. Natl. Cancer Inst.* **84:**864–867.

Orucevic, A., Hearn, S., and Lala, P. K., 1997, The role of active inducible nitric oxide synthase expression in the pathogenesis of capillary leak syndrome resulting from interleukin-2 therapy in mice, *Lab. Invest.* **76:**53–65.

Pahan, K., Namboodiri, A. M., Sikh, F. G., Smith, B. T., and Singh, I., 1997, Increasing cAMP attenuates induction of inducible nitric-oxide synthase in rat primary astrocytes, *J. Biol. Chem.* **272:**7786–7791.

Palmer, R. M., Hickery, M. S., Charles, I. G., Moncada, S., and Bayliss, M. T., 1993, Induction of nitric oxide synthase in human chondrocytes, *Biochem. Biophys. Res. Commun.* **193:**398–405.

Paul, A., Doherty, K., and Plevin, R., 1997, Differential regulation by protein kinase C isoforms of nitric oxide synthase induction in RAW 264.7 macrophages and rat aortic smooth muscle cells, *Br. J. Pharmacol.* **120:**940–946.

Paul-Eugene, N., Kolb, J. P., Sarfati, M., Arock, M., Ouaaz, F., Debre, P., Mossalayi, D. M., and Dugas, B., 1995, Ligation of CD23 activates soluble guanylate cyclase in human monocytes via an L-arginine-dependent mechanism, *J. Leukoc. Biol.* **57:**160–167.

Perrella, M. A., Yoshizumi, M., Fen, Z., Tsai, J.-C., Hsieh, C.-M., Kourembanas, S., and Lee, M.-E., 1994, Transforming growth factor-β1, but not dexamethasone, down-regulates nitric-oxide synthase mRNA after its induction by interleukin-1β in rat smooth muscle cells, *J. Biol. Chem.* **269:**14595–14600.

Pfeilschifter, J., and Schwarzenbach, H., 1990, Interleukin 1 and tumor necrosis factor stimulate cGMP formation in rat renal mesangial cells, *FEBS Lett.* **273:**185–187.

Pfeilschifter, J., and Vosbeck, K., 1991, Transforming growth factor beta 2 inhibits interleukin 1 beta- and tumour necrosis factor alpha-induction of nitric oxide synthase in rat renal mesangial cells, *Biochem. Biophys. Res. Commun.* **175:**372–379.

Pluess, C., Werner, E. R., Wachter, H., and Pfeilschifter, J., 1997, Differential effect of dexamethasone on interleukin 1beta- and cyclic AMP-triggered expression of GTP cyclohydrolase I in rat renal mesangial cells, *Br. J. Pharmacol.* **122:**534–538.

Radomski, M. W., Palmer, R. M., and Moncada, S., 1990, Glucocorticoids inhibit the expression of an inducible, but not the constitutive, nitric oxide synthase in vascular endothelial cells, *Proc. Natl. Acad. Sci. USA* **87:**10043–10047.

Reimers, J. I., Bjerre, U., Mandrup-Poulsen, T., and Nerup, J., 1994, Interleukin 1 beta induces diabetes and fever in normal rats by nitric oxide via induction of different nitric oxide synthases, *Cytokine* **6:**512–520.

Rich, J. B., Rasmusson, D. X., Folstein, M. F., Carson, K. A., Kawas, C., and Brandt, J., 1995, Nonsteroidal anti-inflammatory drugs in Alzheimer's disease, *Neurology* **45:**51–55.

Robbins, R. A., Barnes, P. J., Springall, D. R., Warren, J. B., Kwon, O. J., Buttery, L. D., Wilson, A. J., Geller, D. A., and Polak, J. M., 1994, Expression of inducible nitric oxide in human lung epithelial cells, *Biochem. Biophys. Res. Commun.* **203:**209–218.

Sakitani, K., Kitade, H., Inoue, K., Kamiyama, Y., Nishizawa, M., Okumura, T., and Ito, S., 1997, The anti-inflammatory drug sodium salicylate inhibits nitric oxide formation induced by interleukin-1beta at a translational step, but not at a transcriptional step, in hepatocytes, *Hepatology* **25:**416–420.

Schini, V. B., Durante, W., Elizondo, E., Scott-Burden, T., Junquero, D. C., Schafer, A. I., and Vanhoutte, P. M., 1992, The induction of nitric oxide synthase activity is inhibited by TGF-beta 1, PDGFAB and PDGFBB in vascular smooth muscle cells, *Eur. J. Pharmacol.* **216:**379–383.

Schini, V. B., Catovsky, S., Schray-Utz, B., Busse, R., and Vanhoutte, P. M., 1994, Insulin-like growth factor I inhibits induction of nitric oxide synthase in vascular smooth muscle cells, *Circ. Res.* **74:**24–32.

Scott-Burden, T., Schini, V. B., Elizondo, E., Junquero, D. C., and Vanhoutte, P. M., 1992, Platelet-derived growth factor suppresses and fibroblast growth factor enhances cytokine-induced production of nitric oxide by cultured smooth muscle cells. Effects on cell proliferation, *Circ. Res.* **71:**1088–1100.

Sharara, A. I., Perkins, D. J., Misukonis, M. A., Chan, S. U., Dominitz, J. A., and Weinberg, J. B., 1997, Interferon (IFN)-alpha activation of human blood mononuclear cells *in vitro* and *in vivo* for nitric oxide synthase (NOS) type 2 mRNA and protein expression: Possible relationship of induced NOS2 to the anti-hepatitis C effects of IFN-alpha *in vivo*, *J. Exp. Med.* **186:**1495–1502.

Simmons, W. W., Ungureanu-Longrois, D., Smith, G. K., Smith, T. W., and Kelly, R. A., 1996, Glucocorticoids regulate inducible nitric oxide synthase by inhibiting tetrahydrobiopterin synthesis and L-arginine transport, *J. Biol. Chem.* **271:**23928–23937.

Sirsjo, A., Karlsson, M., Gidlof, A., Rollman, O., and Torma, H., 1996, Increased expression of inducible nitric oxide synthase in psoriatic skin and cytokine-stimulated cultured keratinocytes, *Br. J. Dermatol.* **134:**643–648.

Sowa, G., and Przewlocki, R., 1994, cAMP analogues and cholera toxin stimulate the accumulation of nitrite in rat peritoneal macrophage cultures, *Eur. J. Pharmacol.* **266:**125–129.

Star, R. A., Rajora, N., Huang, J., Stock, R. C., Catania, A., and Lipton, J. M., 1995, Evidence of autocrine modulation of macrophage nitric oxide synthase by α-melanocyte-stimulating hormone, *Proc. Natl. Acad. Sci. USA* **92:**8016–8020.

Stewart, V. C., Giovannoni, G., Land, J. M., McDonald, W. I., Clark, J. B., and Heales, S. J., 1997, Pretreatment of astrocytes with interferon-alpha/beta impairs interferon-gamma induction of nitric oxide synthase, *J. Neurochem.* **68:**2547–2551.

Stratman, N. C., Carter, D. B., and Sethy, V. H., 1997, Ibuprofen: Effect on inducible nitric oxide synthase, *Brain Res. Mol. Brain Res.* **50:**107–112.

Tack, I., Marin-Castano, E., Bascands, J. L., Pecher, C., Ader, J. L., and Giorlami, J. P., 1997, Cyclosporine A-induced increase in glomerular cyclic GMP in rats and the involvement of the endothelin B receptor, *Br. J. Pharmacol.* **121:**433–440.

Tracey, W. R., Nakane, M., Kuk, J., Budzik, G., Klinghofer, V., Harris, R., and Carter, G., 1995, The nitric oxide synthase inhibitor, L-N^G-monomethylarginine, reduces carrageenan-induced pleurisy in the rat, *J. Pharmacol. Exp. Ther.* **273:**1295–1299.

Valentine, J. F., Tannahill, C. L., Stevenot, S. A., Sallustio, J. E., Nick, H. S., and Eaker, E. Y., 1996, Colitis and interleukin 1beta up-regulate inducible nitric oxide synthase and superoxide dismutase in rat myenteric neurons, *Gastroenterology* **111:**56–64.

Vodovotz, Y., Bogdan, C., Paik, J., Xie, Q. W., and Nathan, C., 1993, Mechanisms of suppression of macrophage nitric oxide release by transforming growth factor beta, *J. Exp. Med.* **178:**605–613.

Vodovotz, Y., Kwon, N. S., Pospischil, M., Manning, J., Paik, J., and Nathan, C., 1994, Inactivation of nitric oxide synthase after prolonged incubation of mouse macrophages with IFN-gamma and bacterial lipopolysaccharide, *J. Immunol.* **152:**4110–4118.

Vodovotz, Y., Geiser, A. G., Chesler, L., Letterio, J. J., Campbell, A., Lucia, M. S., Sporn, M. B., and Roberts, A. B., 1996, Spontaneously increased production of nitric oxide and aberrant expression of the inducible nitric oxide synthase *in vivo* in the transforming growth factor beta 1 null mouse, *J. Exp. Med.* **183:**2337–2342.

Vodovotz, Y., Kopp, J. B., Takeguchi, H., Shrivastav, S., Coffin, D., Lucia, M. S., Mitchell, J. B., Webber, R., Letterio, J., Wink, D., and Roberts, A. B., 1998, Increased mortality, blunted production of nitric oxide, and increased production of TNF-alpha in endotoxemic TGF-beta1 transgenic mice, *J. Leukoc. Biol.* **63:**31–39.

Weinberg, J. B., Granger, D. L., Pisetsky, D. S., Seldin, M. F., Misukonis, M. A., Mason, S. N., Pippen, A. M., Ruiz, P., Wood, E. R., and Gilkeson, G. S., 1994, The role of nitric oxide in the pathogenesis of spontaneous murine autoimmune disease: Increased nitric oxide production and nitric oxide synthase expression in MRL-lpr/lpr mice, and reduction of spontaneous glomerulonephritis and arthritis by orally administered N^G-monomethyl-L-arginine, *J. Exp. Med.* **179:**651–660.

Weisz, A., Cicatiello, L., and Esumi, H., 1996, Regulation of the mouse inducible-type nitric oxide synthase gene promoter by interferon-gamma, bacterial lipopolysaccharide and N^G-monomethyl-L-arginine, *Biochem. J.* **316**:209–215.

Xie, Q. W., Whisnant, R., and Nathan, C., 1993, Promoter of the mouse gene encoding calcium-independent nitric oxide synthase confers inducibility by interferon gamma and bacterial lipopolysaccharide, *J. Exp. Med.* **177**:1779–1784.

Yamamoto, K., Ikeda, U., Okada, K., Saito, T., and Shimada, K., 1997, Arginine vasopressin inhibits nitric oxide synthesis in cytokine-stimulated vascular smooth muscle cells, *Hypertens. Res.* **20**:209–216.

Zhang, T., Kawakami, K., Qureshi, M. H., Okamura, H., Kurimoto, M., and Saito, A., 1997, Interleukin-12 (IL-12) and IL-18 synergistically induce the fungicidal activity of murine peritoneal exudate cells against *Cryptococcus neoformans* through production of gamma interferon by natural killer cells, *Infect. Immun.* **65**:3594–3599.

Zhang, X., Alley, E. W., Russell, S. W., and Morrison, D. C., 1994, Necessity and sufficiency of beta interferon for nitric oxide production in mouse peritoneal macrophages, *Infect. Immun.* **62**:33–40.

Zhou, A., Chen, Z., Rummage, J. A., Jiang, H., Kolosov, M., Stewart, C. A., and Leu, R. W., 1995, Exogenous interferon-gamma induces endogenous synthesis of interferon-alpha and -beta by murine macrophages for induction of nitric oxide synthase, *J. Interferon Cytokine Res.* **15**:897–904.

CHAPTER 6

Human Mononuclear Phagocyte Nitric Oxide Production and Inducible Nitric Oxide Synthase Expression

J. BRICE WEINBERG

1. Introduction

Nitric oxide (NO) plays important roles in physiology and pathology, as detailed elsewhere in this volume. This small molecule regulates smooth muscle tone, functions as a neurotransmitter, regulates cellular proliferation, and protects the host against neoplasia and infection (Moncada and Higgs, 1993). NO may also mediate deleterious effects; for example, it appears to be important in inflammation, carcinogenesis, aging, and neurotoxicity (Moncada and Higgs, 1993; Bredt and Snyder, 1994; Clancy and Abramson, 1995). NO is produced from L-arginine by the actions of NO synthases (NOS), a family of enzymes encoded by separate genes (Nathan and Xie, 1994; Michel and Feron, 1997) (see Chapter 4). While neuronal NOS (nNOS, NOS1) and endothelial NOS (eNOS, NOS3) are produced constitutively and controlled in large part by changes in intracellular calcium concentrations, inducible NOS (NOS2, NOS2) is expressed mainly by mononuclear phagocytes, hepatocytes, chondrocytes, and smooth muscle cells (Nathan and Xie, 1994; MacMicking *et al.*, 1997). Activity of NOS2 is controlled primarily by regulation of mRNA transcription and translation. Under suitable conditions, NOS2 produces very high levels of NO.

Much of the work that has examined NOS2 regulation and NO production by mononuclear phagocytes (monocytes and macrophages) has been performed

---

J. BRICE WEINBERG • Division of Hematology and Oncology, Veterans Affairs and Duke University Medical Centers, Durham, North Carolina 27705.

Nitric Oxide and Infection, edited by Fang. Kluwer Academic / Plenum Publishers, New York, 1999.

using mouse or rat peritoneal macrophages, or mouse macrophage cell lines. In the late 1980s and early 1990s, investigators attempting to extend observations from rodent cells to human mononuclear phagocytes had difficulty demonstrating high-level NOS2 expression and NO production by these human cells (Schneemann *et al.*, 1993; Denis, 1994; Albina, 1995), even though researchers were able to demonstrate that humans produce NO and the NO metabolites nitrite/nitrate during infection or treatment with IL-2 (Green *et al.*, 1981; Ochoa *et al.*, 1991, 1992; Hibbs *et al.*, 1992), and showed that this NO was derived from L-arginine (Hibbs *et al.*, 1992). With improvements in techniques and reagents (e.g., specific antibodies), recent studies have more convincingly documented NOS2 expression and NO production by human mononuclear phagocytes with certain treatments *in vitro* and *in vivo*, particularly during certain disease states. In several instances, there are no satisfactory explanations for varying results from apparently identical experiments performed in different research laboratories, with some showing induction of NO production and NOS2 expression while others have failed to find evidence of NOS2 expression despite the use of comparable techniques. The purpose of this chapter is to review the literature regarding NOS2 expression and NO production by human mononuclear phagocytes.

Bioassays have been used to determine NO production by human mononuclear phagocytes. These have included assays demonstrating inhibition of platelet aggregation, induction of smooth muscle relaxation, or inhibition of cellular proliferation. Some of these bioassays have been coupled with use of NO scavengers (e.g., hemoglobin) or NOS enzyme inhibitors such as L-arginine analogues [e.g., N^G-monomethyl-L-arginine (L-NMMA)] to demonstrate that NO was responsible for the observed effects. Certain problems may arise in the chemical measurement of NO and its metabolites (Feelisch and Stamler, 1996). In oxygen-containing environments, NO is converted within seconds to nitrite and nitrate in approximate equimolar concentrations (Stamler *et al.*, 1992). Chemiluminescence measurements of NO in solution may underestimate amounts of NO formed because of the short life span of the molecule. Nitrite and nitrate are generally stable and unreactive at neutral pH. Nitrite can be relatively easily measured spectrophotometrically using the Griess reaction (Green *et al.*, 1982; Granger *et al.*, 1995). Nitrate is generally measured after conversion of nitrate to nitrite with nitrate reductase and subsequent detection with the Griess reaction. In the presence of hemoglobin (or other heme-containing compounds), nitrite is converted to nitrate; thus, studies in which only nitrite is measured may underestimate the amount of NO formed. NO reacting with low- and high-molecular-weight thiols may escape detection by conventional measurements of nitrite/nitrate (Stamler *et al.*, 1992). Also, NO reacting with superoxide to form peroxynitrite can result in nitrotyrosine formation with free tyrosine or tyrosine-containing proteins

(Beckman *et al.*, 1994a), which too would be missed by conventional assays of nitrite/nitrate. Some tissue culture media (e.g., most preparations of RPMI 1640) are supplemented with nitrate, and sera used in tissue culture may contain variable amounts of nitrite or nitrate. Thus, investigators must measure the nitrite/nitrate content of their cell-free culture medium with serum to avoid making erroneous determinations.

Despite these possible problems, *in vitro* studies in which investigators have carefully measured NO (by chemiluminescent or amperometric techniques), nitrite, or nitrite plus nitrate (NOx) generally appear to have accurately reflected cellular NO production. In some of these studies, cells or tissues obtained from research subjects have been used to detect NOS enzyme activity or NOS2 antigen content by immunofluorescent/immunohistochemical techniques and immunoblot analyses. Because of potential problems with antibody specificity in some studies, immunoblot assays (which give not only semiquantitative positive or negative results, but also antigen molecular mass) have provided more convincing information.

In this review, I discuss studies in which human blood and tissue cells have been explanted and examined either without *in vitro* manipulation, or following *in vitro* culture and treatment. I do not include studies in which investigators measured only serum, plasma, gas, or tissue fluid NO or NO metabolites. Some studies of blood cells have used purified monocytes, while others have used mononuclear cells (MNC—monocytes, lymphocytes, and variable numbers of platelets) isolated by centrifugation over ficoll–Hypaque. Some of the studies have used immune assays to identify cells (e.g., anti-CD14 or anti-CD68 antibodies to identify mononuclear phagocytes). Most investigators have assumed that monocytes in the MNC were the likely sources of NO, although reports have also described NO production and NOS2 expression by human Epstein–Barr virus-transformed B-lymphocyte cell line cells (Mannick *et al.*, 1994) and transformed human T-cell line cells (Mannick *et al.*, 1997). In addition, normal B and T lymphocytes have been noted to express NOS3 mRNA as detected by RT-PCR (Reiling *et al.*, 1996), so it is possible that lymphocytes in the MNC fraction might also produce NO. Likewise, there are reports that human platelets contain NOS2 and can produce NO (Radomski *et al.*, 1990; Malinski *et al.*, 1993; Chen and Metha, 1996). However, significant levels of NO production by lymphocytes and platelets (relative to mononuclear phagocytes) are unlikely (Weinberg *et al.*, 1995).

I also discuss reports analyzing tissues taken from humans and analyzed by immunocytology or immunohistology, *in situ* hybridization, or RT-PCR for NOS2 mRNA expression. I evaluate reports of NOS2 expression and NO expression by cells from patients with various pathological disorders. Finally, discussions of NO production and NOS2 expression by human myeloid leukemia cell lines are included.

2. Cytokine Activation of Human Mononuclear Phagocyte NO Production

2.1. Spontaneous NO Production and NOS2 Expression

Table I summarizes reports analyzing NO production and NOS2 expression in human mononuclear phagocytes. The papers are listed alphabetically by author for sequential years. Most investigators have noted that MNC and mononuclear phagocytes do not spontaneously produce NO or express NOS. However, some have presented evidence of NO production or NOS2 expression in MNC or mononuclear phagocytes from apparently normal individuals, without the need for special treatment *in vitro*. It is important to note that cell manipulation by preparation with centrifugation and other procedures and with culture in plastic vessels with media and sera may in itself "activate" cells. Salvemini *et al.* (1989) showed that human PMN (polymorphonuclear neutrophils) and MNC release a factor that blocks thrombin-induced platelet aggregation. The inhibition was abrogated by oxyhemoglobin and L-arginine analogues, and was enhanced by superoxide dismutase (SOD). Although they did not measure NO, their data suggested that the inhibiting factor was NO. Hunt and Goldin (1992) noted that normal monocytes spontaneously generated nitrite with *in vitro* culture. This nitrite production was stimulated by LPS, and inhibited by L-NMMA. Middleton *et al.* (1993a,b) found that unstimulated human MNC caused relaxation of rat colonic smooth muscle, and this effect was inhibited by L-NMMA, hemoglobin, or the guanylyl synthase inhibitor methylene blue; SOD enhanced the smooth relaxation, and granulocytes exerted a comparable effect. Martin and Edwards (1993) found that, with time in culture, human monocytes exhibit an increase in nitrite production and an increase in tumoricidal activity. The nitrite production and tumor cell killing were inhibited by L-NMMA.

In studies of the effects of ethanol on bone marrow cell growth, Wickramasinghe and Hasan (1993) found that human bone marrow macrophages produced low levels of nitrite; this production was inhibitable by L-NMMA. While ethanol inhibited thymidine and leucine incorporation into bone marrow cells, it did not influence nitrite production, and L-NMMA did not influence the ethanol effect. On the other hand, Petit *et al.* (1993) found that spontaneous monocyte-mediated antitumor activity was not inhibited by L-arginine analogues, and that tumoristatic monocytes did not produce nitrite.

Using RT-PCR, Chu *et al.* (1995) studied NOS2 mRNA expression in alveolar macrophages from a normal subject. NOS2 mRNA was found in untreated, freshly isolated alveolar macrophages that were allowed to adhere to plastic. There was evidence of structural diversity in the 5′ untranslated region of the mRNA, with the use of multiple transcription initiation sites. Eissa *et al.* (1996) showed by RT-PCR techniques that normal human alveolar macrophages contained NOS2 mRNA, and

TABLE I
NO Production and NOS2 Expression Analysis in Human Mononuclear Phagocytes[a,b]

Author	Cell type	*In vitro* Rx	Assay	Detected	Comment
1989					
Salvemini *et al.* (1989)	MNC	None	Inhibit platelet aggregation	Yes	
Schmidt *et al.* (1989)	HL-60	VD_3, fMLP	Chemiluminescence	No	HL-60 cells differentiated to monocyte-like cells did not produce NO, while neutrophil-like cells did
1990					
Cameron *et al.* (1990)	Alv Mac, Perit Mac	IFNγ	Nitrite, nitrate, L-Arg to L-Cit	No	No evidence of NO involvement in anticryptococcal effects
1991					
Denis (1991)	Mo	*M. avium*, TNF, GM-CSF	Nitrite	Yes	
Sherman *et al.* (1991)	Alv Mac	LPS, IFNγ	Nitrite, L-Cit	Yes	No increase in nitrite, but increased L-Cit with IFNγ treatment
1992					
Harwix *et al.* (1992)	Mo	LPS, IFNγ	Nitrite	No	No evidence of NO generation in tumoricidal Mo
Hunt and Goldin (1992)	Mo	None	Nitrite	Yes	More in Mo from patients with alcoholic hepatitis
Muñoz-Fernández *et al.* (1992)	Mo	IFNγ, TNF	Nitrite	Yes	Anti-*Trypanosoma cruzi* activity inhibited by L-NMMA

(*Continued*)

TABLE I (*Continued*)

Author	Cell type	*In vitro* Rx	Assay	Detected	Comment
Murray and Teitelbaum (1992)	Mo	IFNγ	Nitrite	No	No inhibition of antimicrobial effect by L-NMMA; also no NO production by Mo from patients receiving IFNγ *in vivo*
Padgett and Pruett (1992)	Mo	LPS, IFNγ, opsonized zymosan, SEB, PMA	Nitrite	No	
Summersgill *et al.* (1992)	HL-60	IFNγ	Nitrite	No	No inhibition of HL-60-mediated stasis of *Legionella pneumophilia* by L-NMMA
1993					
Belenky *et al.* (1993)	Mo	fMLP	Chemotaxis inhibition by L-NMMA	Yes	L-Arg analogues inhibited chemotaxis to fMLP
Ben-Efraim *et al.* (1993)	Perit Mac	LPS, PMA, IND	Nitrite	No	
Bermudez (1993)	Mo	TNF, GM-CSF, IFNγ, *Listeria monocytogenes*, *M. avium*	Nitrite	No	No inhibition of antimicrobial effect by L-NMMA or arginase
Condino-Neto *et al.* (1993)	MNC	None	Inhibit platelet aggregation	Yes	Inhibition blocked by an L-Arg analogue
Keller *et al.* (1993)	BM-derived Mac	IL-3, M-CSF, GM-CSF	Nitrite	No	Tumoricidal effect not inhibited by L-NMMA
Kobzik *et al.* (1993)	Alv Mac	None	Histology with diaphorase and anti-NOS2 Ab	Yes	More expression in those from patients with inflammation

Martin and Edwards (1993)	Mo	Culture	Nitrite	Yes	Tumoricidal effect inhibited by L-NMMA
Naotunne *et al.* (1993)	MNC	Malaria extracts	Inhibition of antimalarial action by L-NMMA	Yes	Inhibition of MNC antimalarial effects by L-NMMA
Petit *et al.* (1993)	Mo	None	Nitrite	No	No inhibition of antitumor activity by L-Arg analogues
Sakai and Milstien (1993)	MNC	LPS, IFNγ	Nitrite	No	No effect of adding BH_4 or sepiapterin
Schneemann *et al.* (1993)	Mo	LPS, IFNγ, GM-CSF, TNF, IL-2, PPD, *Listeria*, *Moraxella*	Nitrite, L-Arg consumption, L-Cit production, NOS activity	No	No benefit of adding BH_4
Wickramasinghe and Hasan (1993)	BM Mac	None	Nitrite	Yes	Inhibited by L-NMMA
Middleton *et al.* (1993a,b)	MNC	None	Smooth muscle relaxation	Yes	Relaxation inhibited by L-NMMA, Hb, or methylene blue
1994					
Barnewall and Rikihisa (1994)	Mo, THP-1	IFNγ, *Ehrlichia chaffeensis*, PMA	Nitrite	No	Mo-mediated anti-*Ehrlichia* effects not dependent on NO production
Beckman *et al.* (1994)	Arterial atheromata Mac	None	Immunohistology for nitrotyrosine	Yes	Nitrotyrosine as evidence of peroxynitrite (and NO) formation *in situ*
Bo *et al.* (1994)	Brain Mac	None	RT-PCR	Yes	NOS2 mRNA noted in brain tissue from patients with multiple sclerosis
DeMaria *et al.* (1994)	Mo	Anti-CD69 Ab, LPS, IFNγ	Nitrite and nitrate, tumor cytotoxicity	Yes	Inhibited by L-NMMA; LPS and IFNγ without effect

(Continued)

TABLE I (*Continued*)

Author	Cell type	*In vitro* Rx	Assay	Detected	Comment
Dumarey *et al.* (1994)	MNC	Live *M. avium*, LPS, IFNγ, TNF	Nitrite and nitrate	Yes	Inhibited by L-NMMA; only a certain strain effective
Essery *et al.* (1994)	MNC	LPS, SEB	Nitrite	Yes	
Gyan *et al.* (1994)	MNC	IFNγ	Nitrite	Yes	Nitrite production and antimalarial activity inhibited by L-NMMA
Haddad *et al.* (1994)	Alv Mac	None	Immunocytology, immunohistology	Yes	Nitrotyrosine as evidence of peroxynitrite (and NO) formation *in situ* in patients with acute lung injury
Kolb *et al.* (1994)	Mo	IL-4, IFNγ	Nitrite	Yes	Sequential IL-4 then IFNγ treatment increased nitrite; inhibited by L-NMMA
Leibovich *et al.* (1994)	Mo	LPS	Nitrite and nitrate	Yes	Inhibited by L-Arg analogues
Martin and Edwards (1994)	Mo	IFNγ	Nitrite	Yes	Increase with time in culture; no augmentation by IFNγ
Mautino *et al.* (1994)	Mo	IL-4	Nitrite	Yes	Heterogeneous production; allergic subjects higher; inhibited by L-NMMA
Paul-Eugene *et al.* (1994)	MNC	IL-4	Nitrite	Yes	NMMA inhibited IL-4-induced IgE production
Pietraforte *et al.* (1994)	Mo	gp120	Nitrite and spin trapping (DMPO)	Yes	Inhibited by L-Arg analogues
Reiling *et al.* (1994)	Mo, U937, THP-1, Mono-Mac6	LPS, IFNγ	RT-PCR	Yes	NOS3 mRNA in "resting" Mo and leukemia cells, and NOS2 mRNA in treated Mo
Thomsen *et al.* (1994)	Gynecological cancer MNC	None	L-Arg to L-Cit, immunohistology, immunoblot	No	NOS2 in tumor cells, but not in leukocytes
Tracey *et al.* (1994)	Alv Mac	None	Immunohistology	Yes	NOS2 antigen in Mac from patients with bronchiectasis and pneumonia

Tufano *et al.* (1994)	Mo	*Yersinia enterocolitica* porins	Nitrite	Yes	Some LPS contamination in porin
Zembala *et al.* (1994)	Mo	Selected tumor cells, IL-2, LPS, TNF, IFNγ	Nitrite	Yes	Inhibited by L-Arg analogues; no effect of IL-2, LPS, TNF, IFNγ
1995					
Bagasra *et al.* (1995)	Brain Mac	None	Northern blot; RT-*in situ*-PCR, immunohistology	Yes	NOS2 mRNA and nitrotyrosine in Mac of patients with multiple sclerosis and absent in "control" brains
Bose and Farnia (1995)	MNC	IFNγ, TNF, IL-1	Nitrite	Yes	
Bukrinsky *et al.* (1995)	Mo	M-CSF, HIV-1 infection, IFNγ, LPS; coculture with astroglial cells	Nitrite, RT-PCR, EPR (Fe-DETC)	Yes	All cells cultured with M-CSF, and then infected/treated with other additives; inhibited by L-NMMA
(Chu *et al.* (1995)	Alv Mac	IFNγ, IL-1, TNF, IL-6	RT-PCR	Yes	Structural diversity in the 5′ untranslated region of mRNA
Criado-Jimenez *et al.* (1995)	MNC	None	Nitrite	Yes	Higher activity in cells from cirrhosis patients; L-Arg analogue inhibited
Eue *et al.* (1995)	U937	Alkylphosphocholine, LPS, PMA	Nitrite	Yes	Rapid release of nitrite; LPS and PMA potentiated effects
Kim *et al.* (1995)	Mo	LPS	Nitrite, nitrate	Yes	Higher NO production in PBMC from trauma patients; some reduction of NO production by IL-13
Kooy *et al.* (1995)	Alv Mac	None	Immunohistology for nitrotyrosine	Yes	Nitrotyrosine as evidence of peroxynitrite (and NO) formation *in situ* in patients with acute lung injury
Kumar *et al.* (1995)	Mo	LPS, PPD, PMA, latex spheres	Nitrite, L-Cit	Yes	Higher in cells from patients with tuberculosis pretreatment

(*Continued*)

TABLE I (*Continued*)

Author	Cell type	*In vitro* Rx	Assay	Detected	Comment
Laffi *et al.* (1995)	Mo	None	L-Arg to L-Cit, inhibition of platelet aggregation	Yes	Higher activity in cells from patients with cirrhosis; inhibited by L-Arg analogues
Lecoanet-Henchoz *et al.* (1995)	Mo	sCD23, anti-CD11b, anti-CD11c	Nitrite, nitrate	Yes	Inhibited by L-Arg analogue
Masini *et al.* (1995)	Mo	None	L-Arg to L-Cit, inhibition of platelet aggregation	Yes	Higher in Mo from patients with cirrhosis; inhibited by L-NMMA
Paul-Eugene *et al.* (1995a)	Mo	IL-4, IgE immune complexes	Nitrite	Yes	Inhibited by L-NMMA
Paul-Eugene *et al.* (1995b)	Mo	IL-4, sCD23, IFNγ	Nitrite, L-Cit	Yes	Inhibited by L-Arg analogue or anti-CD23
Paul-Eugene *et al.* (1995c)	PBMC	IL-4	Inhibition of IL-4-induced increase in IL-4 and sCD23 production	Yes	Phenomenon blocked by L-NMMA
Perez-Mediavilla *et al.* (1995)	MNC	ECM peptides	Nitrite, immuno-cytology	Yes	Inhibited by L-NMMA
Perez-Perez *et al.* (1995)	THP-1	*H. pylori*, LPS	Nitrite	No	
Sakurai *et al.* (1995)	Synovial Mac	None	Nitrite, immunohistology, immunoblot, RT-PCR	Yes	RA and inflammatory osteoarthritis patients; inhibited by L-Arg analogue

Siedlar *et al.* (1995)	Mo	DeTa tumor cells	Nitrite	Yes	DeTa tumor-induced nitrite formation blocked by antibodies against MHC class I or II, CD44, LFA-3 (CD58), VLA-β1 (CD29)
Thomsen *et al.* (1995)	Breast cancer Mac	None	Anti-NOS2 Ab immunocytology, nitrite and nitrate, L-Arg to L-Cit	Yes	NOS2 associated with CD68$^+$ Mac by immunohistology
Vouldoukis *et al.* (1995)	Mo	anti-CD23 Ab, IgE IC, IL-4, IFNγ	Nitrite, immunoblot, RT-PCR, L-Arg to L-Cit	Yes	Inhibited by L-NMMA
Weinberg *et al.* (1995)	Mo, Perit Mac	LPS, IFNγ, TNF, IL-1, IL-2, IL-4, GM-CSF, IL-7, IL-6, VD3, PMA, Con A, PHA, A23187, bacteria, mycobacteria, HIV-1	Nitrite, nitrate, immunocytology, immunoblot, RT-PCR, Northern blot, RNase protection, L-Arg to L-Cit	Yes	Generally low levels (Mac > Mo) compared with mouse Mac; no effect of adding BH_4; mRNA detected only by RT-PCR
Wildhirt *et al.* (1995)	Myocardial Mac	None	Immunohistology	Yes	NOS2 in infarcted myocardium colocalized with Mac
Zinetti *et al.* (1995)	MNC, THP-1	LPS	Nitrite; L-NMMA- or Hb- or Mb-inhibitable LPS-induced TNF secretion and inhibition of THP-1 proliferation	Yes	No nitrite production noted, but indirect evidence of NO production
1996					
Anstey *et al.* (1996)	MNC	None	Immunoblot, serum and urine nitrite/nitrate	Yes	MNC NOS2 expression in normal Tanzanian children; increased NOS2 in asymptomatic and mild malaria; markedly decreased NOS2 in cerebral malaria

(*Continued*)

TABLE I (*Continued*)

Author	Cell type	*In vitro* Rx	Assay	Detected	Comment
Buttery *et al.* (1996)	Atherosclerotic plaque Mac	None	Immunohistology, immunoblot, *in situ* hybridization, nitrotyrosine	Yes	Nitrotyrosine as evidence of peroxynitrite (and NO) formation *in situ*
Chen *et al.* (1996)	THP-1	PMA, silica, LPS	Nitrite, Northern blot	Yes	PMA as "priming" agent; inhibited by L-NMMA or allopurinol
Condino-Neto *et al.* (1996)	MNC	IFNγ *in vivo*, LPS *in vitro*	Nitrite, nitrate, inhibition of platelet aggregation	No	
Dias-Da-Motta *et al.* (1996)	PBMC	PMA, zymosan	Inhibition of platelet aggregation	Yes	Inhibited by L-Arg analogue; production noted only in presence of SOD; more activity in cells from patients with sickle-cell disease
Dugas *et al.* (1996)	Mo	Anti-CD23 antibody	NMMA-induced inhibition of anti-CD23 antibody-induced IL-10 production	Yes	
Eissa *et al.* (1996)	Alv Mac	None	RT-PCR	Yes	Alternative splicing of mRNA
Kashem *et al.* (1996)	Kidney Mac, Mo	IFNγ, TNF	Immunohistology, RT-PCR	Yes	Associated with $CD68^+$ Mac in kidneys with IgA nephropathy and proliferative glomerulonephritis; NOS2 mRNA expression in normal Mo treated *in vitro* with IFNγ + TNF
Kawase *et al.* (1996)	HL-60	NaF, VD_3, PGE_2	Nitrite, immunoblot	Yes	Prevention of VD_3 + NaF induction of NOS2 by IND

Liu *et al.* (1996)	Fetal microglial cells	IL-1, TNF, IFNγ, LPS, TGFβ, IL-10, HIV-1 gp160	Nitrite, immunocytology, Northern blot, immunoblot	No	
López-Moratalla *et al.* (1996)	Mo	Peptides from thyroid autoantigens	Immunocytology	Yes	NOS2 in freshly isolated Mo from Graves'; increased activity induced by *in vitro* treatment with peptides from thyroid autoantigens
Magazine *et al.* (1996)	Mo	Morphine	NO-specific amperometric probe	Yes	Inhibited by L-Arg analogues or naloxone
Mannick *et al.* (1996)	Gastric MNC	None	Immunohistology for NOS2 and nitrotyrosine	Yes	NOS2 expression by gastric MNC in patients with *Helicobacter pylori* gastritis; number of positive cells decreased after antibiotic, β-carotene, or ascorbate treatment
McInnes *et al.* (1996)	Synovial Mac	SEB	Nitrite, RT-PCR, immunohistology	Yes	NOS2 in synovial Mac and synovial fibroblasts
McLachlan *et al.* (1996)	Mo	Dehydroepiandrosterone, LPS	Nitrite	Yes	
Nicholson *et al.* (1996)	Alv Mac	None	Immunohistology, immunoblot, RT-PCR	Yes	Increased in cells from patients with tuberculosis
Singer *et al.* (1996)	Colonic Mac	None	Immunohistology for NOS2 and nitrotyrosine, RT-PCR	Yes	Found only in inflammatory mucosal areas in patients with ulcerative colitis, Crohn's disease, and diverticulitis
St. Clair *et al.* (1996)	Mo, MNC	LPS, IFNγ	Nitrite, nitrate, immunoblot, L-Arg to L-Cit	Yes	Increased L-Arg to L-Cit activity and NOS2 Ag in freshly isolated cells from patients with RA; increased responsiveness of MNC of RA patients to IFNγ *in vitro;* inhibited by L-NMMA
Stefano *et al.* (1996)	Mo	Anandamide (tetrahydrocannabinol derivative)	NO-specific amperometric probe	Yes	Inhibited by L-Arg analogues and a cannabinoid antagonist

(*Continued*)

TABLE I (*Continued*)

Author	Cell type	*In vitro* Rx	Assay	Detected	Comment
Wang *et al.* (1996)	Mo; pleural and Perit Mac	LPS	Nitrite	Yes	More NO production if adherent to plastic; more NO production from tissue Mac from patients with cancer
Weyand *et al.* (1996)	Arterial Mac	None	Immunohistology	Yes	NOS2 in intimal Mac in giant cell arteritis
1997					
Amin *et al.* (1997)	Mo, HL-60 cells, U937 cells	TNF, IL-1, LPS	L-Arg to L-Cit, immunoblot, RT-PCR, Northern blot	Yes	NOS2 mRNA noted, but negative by immunoblot and L-Arg to L-Cit assay
Aubry *et al.* (1997)	Mo	sCD23, anti-CD11b, anti-CD11c	L-Arg to L-Cit, immunoblot, RT-PCR	Yes	Inhibited by L-NMMA; stimulated expression of NOS3 (NOS2 not studied)
Bagasra *et al.* (1997)	Brain Mac	None	RT-*in situ*-PCR, immunohistology	No	Absence of NOS2 mRNA and nitrotyrosine in brains of patients with AIDS
DeGroot *et al.* (1997)	Brain Mac	None	Immunohistology, nitrite	Yes	In brains of patients with multiple sclerosis, Mac positive for both NOS2 and "cNOS"; isolated Mac produced NO
Eis *et al.* (1997)	Fetal membrane	None	Immunohistology	Yes	NOS2 in $CD14^+$ fetal membrane Mac
Goto *et al.* (1997)	U937	Transfection with HTLV-I *tax*; IFNγ	RT-PCR, nitrite, nitrate	Yes	Additive effects of the transfection and IFNγ treatment
Grabowski *et al.* (1997)	Synovial Mac	None	Immunohistology	Yes	NOS2 associated chiefly with $CD68^+$ Mac by immunohistology; more NOS2 in RA vs. osteoarthritis; no NOS2 noted in tissues from normal subjects (hip fractures)

Hooper *et al.* (1997)	Brain Mac	None	Immunohistology, RT-*in situ*-PCR	Yes	NOS2 in brain Mac from multiple sclerosis patients
Ikeda *et al.* (1997)	Bowel Mac	None	Immunohistology	Yes	Increased NOS2 expression in ulcerative colitis
King *et al.* (1997)	Mo, U937, THP-1	Endothelin-1	NO-specific amperometric probe	Yes	Effect blocked by ET_B antagonist
Lafond-Walker *et al.* (1997)	Coronary artery Mac	None	Immunohistology, *in situ* hybridization	Yes	Only in accelerated graft arteriosclerosis in transplanted hearts; associated with $CD68^+$ Mac
Lammas *et al.* (1997)	Mo	ATP, BCG	Inhibition of Mo or BCG killing by L-Arg analogues	No	No inhibition of Mo or BCG killing by L-Arg analogues
Lopez-Guerrero *et al.* (1997)	U937	PMA, HSV-1	Nitrite	Yes	Inhibited by L-NMMA
Lopez-Guerrero *et al.* (1997)	U937	None	Nitrite	Yes	More NO production by parvovirus infection-resistant cells; inhibited by L-NMMA
McDermott *et al.* (1997)	Alv Mac	None	Immunohistology for NOS2 and nitrotyrosine	Yes	Lung transplant patients with obliterative bronchiolitis
Moilanen *et al.* (1997)	Foreign body Mac (joint prostheses)	None	L-Arg to L-Cit, Immunohistology, RT-PCR	Yes	NOS2 associated with $CD68^+$ Mac
Myatt *et al.* (1997)	Placental Mac (Hofbauer cells)	None	Immunohistology, RT-PCR	Yes	NOS2 associated with $CD14^+$ Mac
Nozaki *et al.* (1997)	Alv Mac	BCG	Immunocytology and immunoblot for NOS2 and nitro-tyrosine, RT-PCR	Yes	Produced by Alv Mac from patients with pulmonary fibrosis after *in vitro* challenge with BCG; L-NMMA inhibited Mac-mediated BCG killing

(*Continued*)

TABLE I (*Continued*)

Author	Cell type	*In vitro* Rx	Assay	Detected	Comment
Polack *et al.* (1997)	Mo	LPS	NMMA inhibition of LPS-induced increased tissue factor	Yes	—
Roman *et al.* (1997)	U937	sCD23, anti-CD11b, anti-CD11c	L-Arg to L-Cit, immunoblot, RT-PCR	Yes	Inhibited by L-NMMA; NOS3 expression stimulated
Saha *et al.* (1997)	PBMC	Monophosphoryl lipid A	Nitrite, immunocytology, flow cytometry	Yes	Monophosphoryl lipid A stimulated but LPS did not; inhibited by L-Arg analogue
Schneemann *et al.* (1997)	MNC	IL-4, IFNγ	Nitrite, L-Arg and L-Cit levels	No	
Seitzer *et al.* (1997)	Mo-derived multinucleated giant cells	*Nippostrongylus brasiliensis*	RT-PCR	Yes	NOS2 mRNA noted in 15–21% of single giant cells
Sharara *et al.* (1997)	Mo, MNC	IFNα *in vitro* and *in vivo*	Nitrite, nitrate, immunoblot, L-Arg to L-Cit, RT-PCR	Yes	Induction of NOS2 activity, mRNA, and Ag content by *in vitro* treatment of normal Mo, or *in vivo* treatment of hepatitis C patients; correlation of antiviral activity of IFNα with degree of NOS2 induction
Snell *et al.* (1997)	Mo	Polyribonucleotides, IFNγ, IFNα, IL-4	Nitrite	Yes	No effect of LPS, IFNγ, IFNα, and IL-4 alone, but potentiation of polyribonucleotide effect; inhibited by L-NMMA

ter Steege *et al.* (1997)	Bowel Mac	None	Immunohistology	Yes	Increased NOS2 and nitrotyrosine in CD14/CD68^{+} Mac celiac disease
Vitek *et al.* (1997)	Mo	Polyinosinic-poly-cytidylic acid and apolipoprotein E	Nitrite	Yes	Amyloid beta peptide inhibited the apolipoprotein E-induced increase
Vouldoukis *et al.* (1997)	Mo	Anti-CD23 Ab	Nitrite	Yes	IL-10 and IL-4 inhibited killing of *Leishmania*
Watkins *et al.* (1997)	Mac around loosened bone hip prostheses	None	*In situ* hybridization, immunohistology	Yes	No NOS2 in synovial cells
Wilcox *et al.* (1997)	Vessel Mac	None	*In situ* hybridization, immunohistology	Yes	Mac in atherosclerotic lesions expressed both NOS2 and nNOS
Zarlingo *et al.* (1997)	Placental Mac	None	Immunohistology, immunoblot	Yes	
1998					
Ambs *et al.* (1998)	Tumor Mac	None	L-Arg to L-Cit, Immunohistology for NOS2 and nitrotyrosine, immunoblot, RT-PCR	Yes	NOS2 in tumors (more in adenomas than in carcinomas), but low in normal tissue; present in MNC, tumor cells, and endothelial cells; nitrotyrosine in MNC
Luoma *et al.* (1998)	Arterial Mac	None	Immunohistology	Yes	NOS2 and nitrotyrosine associated with Mac

(Continued)

TABLE I (*Continued*)

Author	Cell type	*In vitro* Rx	Assay	Detected	Comment
Majano *et al.* (1998)	Liver MNC	None	Immunohistology, *in situ* hybridization	Yes	In chronic active hepatitis B or C patients, NOS2 protein and mRNA noted in hepatocytes, with only mRNA noted in MNC in hepatitis
Perkins *et al.* (1998)	Mo, MNC	None	Immunoblot, RT-PCR, L-Arg to L-Cit	Yes	Increased L-Arg to L-Cit activity, NOS2 Ag, NOS2 mRNA in Mo and MNC from patients with RA; anti-TNF antibody treatment *in vivo* reduced the increased NOS activity and NOS2 Ag expression

[a] Citations are listed alphabetically within each year.
[b] Abbreviations: Alv, alveolar; BH_4, tetrahydrobiopterin; BM, bone marrow; Con A, concanavalin A; EPR, electron paramagnetic resonance; fMLP, *N*-formyl methionylleucylphenalanine; IND, indomethacin; L-Arg to L-Cit, assay measuring conversion of L-arginine to L-citrulline; L-NMMA, N^G-monomethyl-L-arginine; Mac, macrophage; MNC, mononuclear cell; Mo, monocyte; Perit, peritoneal; PHA, phytohemagglutin; PMA, phorbol myristate acetate; PPD, purified protein derivative; RA, rheumatoid arthritis; SEB, staphylococcal enterotoxin B; VD_3, 1,25-dihydroxyvitamin D_3.

showed evidence of extensive mRNA alternative splicing. These authors postulated that alternative splicing might function to regulate levels of expression of functional enzyme.

2.2. Cytokines, Growth Factors, and Lipopolysaccharide

Based on experiences using rodent macrophages (see Chapter 5), investigators have tried to determine if various cytokines and microbial extracts would activate human mononuclear phagocytes for NO production and NOS2 expression. Most have focused on IFNγ, TNFα, and LPS. Even though Granger *et al.* (1988) had shown earlier that mouse macrophages require L-arginine to inhibit growth of *Cryptococcus neoformans*, these investigators later reported that human alveolar and peritoneal macrophages from normal individuals mediated fungistasis by an L-arginine-independent metabolism (Cameron *et al.*, 1990). In the latter study, they could find no evidence that the human cells generated NO or L-citrulline from L-arginine. Treatment of the cells with IFNγ *in vitro* enhanced fungistasis, but had no effect on L-arginine metabolism. They also noted that fungistasis was not inhibited by L-NMMA.

In contrast to the report of Cameron *et al.* (1990), Denis (1991) found that monocyte-derived macrophages (monocytes cultured for 7 days) from normal individuals treated *in vitro* with TNFα and GM-CSF had enhanced ability to restrict growth of virulent *Mycobacterium avium* and to kill avirulent *M. avium*. The killing was inhibited by L-NMMA. While treatment of the cells with TNFα and GM-CSF did not enhance nitrite formation, the treated cells produced nitrite when inoculated with *M. avium*. Sherman *et al.* (1991) noted that LPS and IFNγ enhanced nitrite and L-citrulline production by mouse macrophages, but human alveolar macrophages treated with LPS and IFNγ produced only L-citrulline. However, coculture with *Pneumocystis carinii* caused a slight increase in both nitrite and L-citrulline formation by the human macrophages. Muñoz-Fernández *et al.* (1992) demonstrated that human monocytes treated with IFNγ, TNFα, or both cytokines had increased ability to destroy *Trypanosoma cruzi* and increased production of nitrite. L-NMMA inhibited nitrite production, and nitrite production correlated with the trypanocidal activity.

However, some investigators subsequently reported their inability to detect NO production by human monocytes. Harwix *et al.* (1992) showed that cultured monocytes treated with IFNγ and LPS are tumoricidal, but did not produce nitrite. Murray and Teitelbaum (1992) showed that human monocytes treated *in vitro* with IFNγ displayed antimicrobial effects toward *Toxoplasma gondii, Chlamydia psittaci*, and *Leishmania donovani*, but these antimicrobial effects were not modified by L-NMMA or by depletion of L-arginine from the medium with arginase. Normal monocytes did not produce nitrite; likewise, treatment of AIDS patients with IFNγ *in vivo* or treatment of normal monocytes *in vitro* with

IFNγ ± LPS *in vitro* did not generate cells capable of producing nitrite (Murray and Teitelbaum, 1992). Similarly, Padgett and Pruett (1992) found that human monocytes cultured for 2 weeks failed to produce nitrite *in vitro*. Ben-Efraim *et al.* (1993) found no evidence that human peritoneal macrophages with or without PMA or indomethacin treatment could produce NO, although the cells displayed anti-tumor cell activity *in vitro*. Bermudez (1993) found that the antimicrobial actions of human monocytes toward *M. avium* or *Listeria monocytogenes* were not inhibited by L-NMMA or arginase, and that TNFα, GM-CSF, and IFNγ did not induce nitrite production. Keller *et al.* (1993) noted that human bone marrow-derived macrophages (cultured with IL-3, M-CSF, and GM-CSF) were tumoricidal, but the tumoricidal effect was not inhibited by L-NMMA and was not associated with nitrite generation.

Sakai and Milstien (1993) observed that human MNC did not produce nitrite after treatment with IFNγ and LPS, and elevation of MNC biopterin levels by adding biopterin or sepiapterin did not render the cells capable of producing nitrite. Schneemann *et al.* (1993) found that human mononuclear phagocytes did not produce nitrite, consume L-arginine, produce L-citrulline, or display NOS activity after treatment with LPS, IFNγ, GM-CSF, TNFα, IL-2, PPD, *Listeria*, or *Moraxella*. As Sakai and Milstien (1993) had noted, adding biopterin did not enable the cells to produce nitrite, although the autoxidation of tetrahydrobiopterin to produce superoxide (see Chapter 4) is a potentially confounding factor in these studies. Barnewall and Rikihisa (1994) showed that culture of human monocytes or THP-1 cells with IFNγ and *Ehrlichia chaffeensis* did not stimulate nitrite production, although growth of the *Ehrlichia* was inhibited by treatment of the cells with IFNγ or PMA.

Following this wave of negative reports, additional investigators were able to provide some confirmation of the original observations of inducible NO production by human MNC *in vitro*. Essery *et al.* (1994) showed that treatment of human MNC with staphylococcal enterotoxin B augmented nitrite formation. Gyan *et al.* (1994) discovered that human MNC cultured with IFNγ produced increased amounts of nitrite and inhibited growth of *Plasmodium falciparum*. Furthermore, the anti-parasitic effect could be partially blocked by L-NMMA. Leibovich *et al.* (1994) found that treatment of monocytes with LPS stimulated their production of angiogenic activity, and this increase in activity paralleled an increase in production of nitrite and nitrate. The production of the angiogenic activity and nitrite/nitrate was reduced when the cells were cultured in medium low in L-arginine or in the presence of L-NMMA. Martin and Edwards (1994) noted an increase in monocyte production of nitrite with increasing time in culture (up to 9 days). This effect was not altered by IFNγ, although IFNγ augmented the monocyte-mediated tumor cytotoxicity. Zembala *et al.* (1994) observed that normal human monocytes cultured with LPS, IFNγ, IL-2, and TNFα did not produce nitrite, but coculture of the monocytes with a human colorectal cell line (DeTa) caused the monocytes to produce nitrite; this production was inhibited by L-arginine analogues. Stimulation of nitrite production by DeTa cells was also inhibited by antibodies directed against CD44, CD29 (VLA-β_1), CD58 (LFA-3), and MHC class I or II antigen (Siedlar *et al.*, 1995). Bose and Farnia (1995) found that human monocytes treated with IFNγ, TNF, and IL-1 (and subsequently "triggered" with LPS) produced nitrite. As reported by Martin and Edwards (1994), the cells studied by Bose and Farnia (1995) produced higher levels of nitrite with increasing time in culture. Reiling *et al.* (1994) used RT-PCR analysis of untreated human monocytes, THP-1 cells, U937 cells, and Mono-Mac6 cells to document the presence of mRNA for a "constitutive" isoform of NOS (cNOS). On stimulation with LPS and IFNγ, levels of NOS2 mRNA became detectable and cNOS levels diminished.

Bukrinsky *et al.* (1995) studied human monocytes cultured for 7 days with M-CSF, then infected with HIV-1. Uninfected cells produced little or no NO, but those infected with HIV-1 produced significant amounts. Treatment of these cells with LPS or TNFα, or coculture of the cells with astroglial cells, further enhanced NO production, while IL-4 or L-NMMA decreased the production. NO production was assessed by measurement of nitrite and electron paramagnetic resonance (EPR) detection of NO "trapped" by iron-diethyldithiocarbamate (Fe-DETC). Also, NOS2 mRNA was demonstrated in infected and stimulated monocytes by RT-PCR (Bukrinsky *et al.*, 1995).

Weinberg *et al.* (1995) showed that LPS and/or IFNγ induced normal human monocytes and peritoneal macrophages to express low levels of NOS2 mRNA (as detected by RT-PCR). Immunofluorescence and immunoblot analyses demonstrated that IFNγ also induced detectable levels of NOS2 antigen. Production of nitrite/nitrate by human peritoneal macrophages was induced by IFNγ and LPS, as well as increased levels of NOS enzymatic activity in both monocytes and macrophages (Weinberg *et al.*, 1995). However, a large number of other agents in various combinations with LPS and IFNγ were tested for the ability to induce high-level nitrite/nitrate production, and none was effective; these included GM-CSF, IL-1, IL-2, IL-4, IL-7, IL-6, 1,25-vitamin D3, PMA, the calcium ionophore A23187, and the lectins concanavalin A and phytohemagglutinin. Likewise, live or heat-killed *M. avium*, *M. tuberculosis*, *Listeria monocytogenes*, *Candida albicans*, *Staphylococcus epidermidis*, HIV-1, culture in 3–50% human, dog, or fetal bovine serum (heated or unheated), or supplementation with excess L-arginine, NADPH, sepiapterin, or biopterin each failed to induce the cells to produce nitrite/nitrate. In experiments mixing lysates of murine and human cells or using neutralizing antibodies against TGFβ1, these investigators could show no evidence of an endogenous inhibitor of NOS expression or function in the human monocytes and macrophages (Weinberg *et al.*, 1995).

Zinetti *et al.* (1995) found that L-NMMA, hemoglobin, or myoglobin would inhibit LPS-induced TNFα secretion by MNC and THP-1 cells, even though they could detect no evidence of nitrite formation. They interpreted their findings as evidence of regulation of TNFα production by endogenous NO. Condino-Neto *et al.* (1996) studied patients with chronic granulomatous disease who had received IFNγ for 6 months. Although the authors had previously reported evidence of L-arginine-dependent NO production by neutrophils and MNC from such patients (Condino-Neto *et al.*, 1993), IFNγ treatment *in vivo* did not appear to enhance this production. Kashem *et al.* (1996) demonstrated by immunohistology that treatment of normal monocytes *in vitro* with a combination of TNF and IFN-γ caused expression of NOS2 mRNA.

St. Clair *et al.* (1996) showed that freshly isolated monocytes and MNC from patients with active rheumatoid arthritis (RA) had increased NOS enzyme activity and NOS2 antigen detection on immunoblot, when compared with cells from normal subjects. When MNC from patients with RA were cultured with IFNγ, they exhibited increased production of nitrite/nitrate, while those of normal subjects were not altered by this cytokine. L-NMMA inhibited expression of the NOS activity, and levels of enzyme activity and antigen were positively correlated with the severity of arthritis. Wang *et al.* (1996) noted that monocytes, pleural macrophages, and peritoneal macrophages produced nitrite following LPS treatment *in vitro*. Tissue macrophages from patients with malignancy were found to produce higher levels of nitrite than those from normal individuals. Amin *et al.* (1997) reaffirmed that normal monocytes did not express NOS2 antigen or have NOS enzyme activity. However, using RT-PCR, they demonstrated that monocytes expressed NOS2 mRNA. NOS2 mRNA could also be detected by Northern analysis of monocytes and U937 cells.

Polack *et al.* (1997) observed that L-NMMA inhibited LPS-induced monocyte tissue factor expression. Saha *et al.* (1997) noted that monophosphoryl lipid A, an LPS derivative with reduced toxicity, stimulated NO production by PBMC under *in vitro* conditions in which LPS itself is nonstimulatory; NOS2 protein, enzyme activity, and nitrite production were detected in this study. Snell *et al.* (1997) showed that the synthetic polyribonucleotides poly I:C (polyinosinic-polycytidylic acid) or poly I stimulated production of nitrite by human monocytes, an effect enhanced by pretreatment of the cells with LPS, IFNγ, IFNα, or IL-4. Nitrite production under these conditions was inhibited by L-NMMA.

While others focused on IFNγ, Sharara *et al.* (1997) demonstrated that IFNα functioned as an effective activator of human monocyte NO production and NOS2 expression when used either *in vitro* or *in vivo*. In rodent macrophages, exogenous IFNα cannot activate macrophages for NO production (Ding *et al.*, 1988), but macrophage-synthesized IFNα can augment NO production in an autocrine fashion (Zhou *et al.*, 1995). Sharara and coworkers noted that IFNα induced normal monocytes to produce nitrite/nitrate *in vitro*, as well as to express NOS2 antigen and mRNA, and display increased NOS enzyme activity (Fig. 1). NOS activity was inhibited by L-NMMA. In patients with hepatitis C, administration of IFNα *in vivo* increased NOS enzyme activity and caused the appearance of NOS2 antigen and mRNA in MNC, whereas cells from hepatitis C patients not receiving IFNα did not

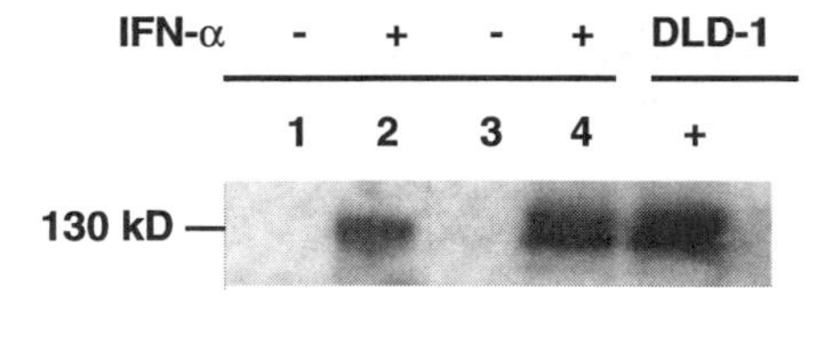

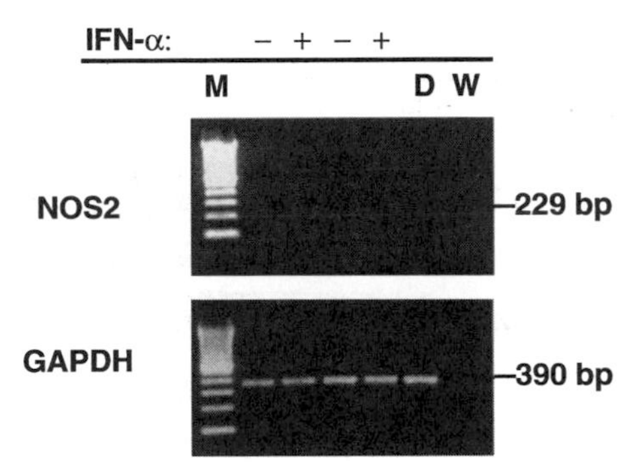

FIGURE 1. (Top) Immunoblot analyses of extracts of mononuclear cells treated *in vitro* with IFNα Cells were cultured with ("+") or without 500 units/ml of IFNα ("−") for 3 days. Extracts were obtained, and equivalent amounts of cellular protein were analyzed. The antibody "anti-MacNOS" was used. Cells from two normal individuals were analyzed. Analysis of the first is shown in lanes 1 and 2, and that of the second in lanes 3 and 4. DLD-1 "+" signifies extracts from the human colon cancer cell line DLD-1 treated with IFNγ, IL-1, and TNF *in vitro*. Results demonstrate that IFNα treatment induces NOS2 antigen expression.

(Bottom) Reverse transcriptase-polymerase chain reaction analysis of mRNA of mononuclear cells treated *in vitro* with IFNα. Cells from two normal individuals were cultured with ("+") or without 500 units/ml of IFNα ("−") for 12 h. RNA was extracted and analyzed for NOS2 and GAPDH mRNA. "M" signifies molecular weight markers, "D" cells of the human colon cancer cell line DLD-1 treated with IFNγ, IL-1, and TNF, and "W" distilled water.

Reproduced from *The Journal of Experimental Medicine*, 1997, Vol. 186, pp. 1495–1502 by copyright permission of The Rockefeller University Press.

express NOS2 antigen or mRNA (Fig. 2). In patients receiving IFNα, the degree of induction of NOS2 correlated significantly with the degree of improvement of their hepatitis. The investigators in this study speculated that enhanced NO production induced by IFNα might account for the development of "autoimmune" illnesses with inflammation similar to rheumatoid arthritis and systemic lypus enrythematosus noted in some patients treated with IFNα (Vial and Descotes, 1995). Notably, the work of Sharara and colleagues appears to be the only study to date that has demonstrated activation of human monocytes/MNC for NO production and NOS2 expression both *in vitro* and *in vivo* using a defined, purified agent (IFNα).

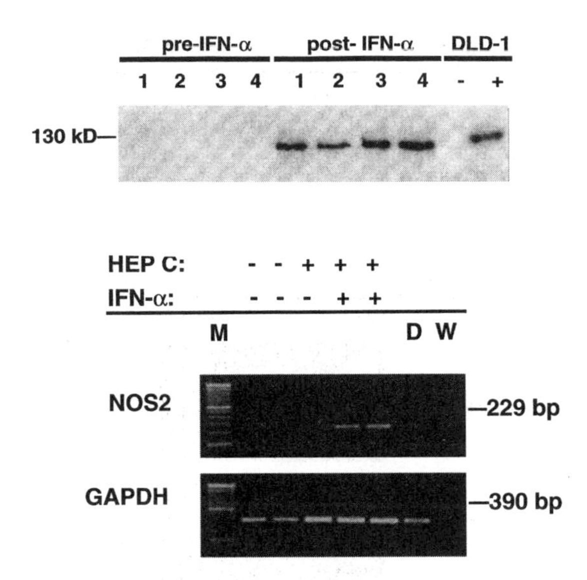

FIGURE 2. (Top) Immunoblot analyses of extracts of blood mononuclear cells from hepatitis C patients before and after IFNα treatment. Equivalent amounts of cellular protein were analyzed in each lane. The antibody "1E8-B8" was used. Samples from patients 1 to 4 were collected before IFNα treatment ("pre-IFN-α") or after receiving IFNα *in vivo* ("post-IFN-α"). Controls for human NOS2 were from the human colon cancer cell line DLD-1 without ("−") or with ("+") treatment with IFNγ, IL-1, and TNF *in vitro*.

(Bottom) Reverse transcriptase-polymerase chain reaction analysis of mononuclear cells from normal subjects, patients with hepatitis C, and patients with hepatitis C treated *in vivo* with IFNα. Cells were isolated, frozen, extracted, and analyzed. Cells from two normal subjects ("HEP C −" and "IFN-α −"), one patient with hepatitis C ("HEP C +" and "IFN-α −"), and two patients with hepatitis C on treatment with IFNα ("HEP C+" and "IFN-α+") were analyzed. "M" signifies molecular weight markers, "D" cells of the human colon cancer cell line DLD-1 treated with IFNγ, IL-1, and TNF, and "W" distilled water.

3. IgE and CD23 Activation of Human Mononuclear Phagocyte NO Production

A series of papers have indicated that IgE and CD23 (the low-affinity receptor for IgE) play a role in activating monocytes for NOS2 expression and NO production. Kolb *et al.* (1994) found that monocytes treated with IL-4 followed by IFNγ produce nitrite, and this production can be inhibited by L-NMMA. Mautino *et al.* (1994) noted that IL-4 augmented nitrite production by monocytes from some human donors, with the degree of enhancement varying among different donors. Subjects could be divided into "low producers" and "high producers"—IL-4 augmented nitrite production by cells from "low producers", while it decreased production by cells from "high producers". Allergic subjects appeared to have increased production (Mautino *et al.*, 1994). Paul-Eugene *et al.* (1994) observed that IL-4 stimulated IgE and nitrite production by MNC, with both nitrite and IgE production inhibitable by L-NMMA. Lecoanet-Henchoz *et al.* (1995) noted that soluble CD23 or anti-CD11b with anti-CD11c enhanced nitrite and nitrate formation by human monocytes via an L-NMMA-inhibitable mechanism. These investigators suggested that CD11b and CD11c served as receptors for CD23.

In related work, Paul-Eugene *et al.* (1995a–c) showed that ligation of monocyte CD23 with IgE immune complexes stimulated nitrite formation that could be further enhanced by IL-4, that L-NMMA blocked IL-4-induced increases in production of IL-4 and soluble CD23, and that IL-4 along with soluble CD23 and IFNγ induced nitrite and L-citrulline production by PBMC. Vouldoukis *et al.* (1995) showed that engagement of CD23 with IgE immune complexes or anti-CD23 antibody caused an increase in nitrite formation and *Leishmania major* killing (see also Chapter 17). IgE immune complexes, anti-CD23 antibody, and IFNγ enhanced nitrite formation and the ability of cell lysates to convert L-arginine to L-citrulline. While IL-4 treatment had little or no effect on NOS2 mRNA expression in the monocytes, sequential treatment with IL-4 and anti-CD23 induced expression of NOS2. In general, the abilities to express/produce NOS/NO correlated with abilities to kill parasites (Vouldoukis *et al.*, 1995), suggesting that this mechanism might be relevant to host defense against parasitic infections. Although the addition of TNFα did not enhance nitrite formation or *Leishmania* killing, anti-TNFα antibody inhibited NO production and parasite killing, providing evidence that endogenously produced TNFα was important in this process.

Aubry *et al.* (1997) noted that soluble CD23, anti-CD11b, and anti-CD11c treatment activated monocytes to express "constitutive" NOS3 as determined by enzyme activity, immunoblot, and RT-PCR (Aubry *et al.*, 1997). NOS activity was inhibited by EGTA and L-arginine analogues, but NOS2 was not examined in this study. Soluble CD23 enhanced monocyte cGMP content and TNFα production, and these effects were inhibited by L-NMMA. In a follow-up to their earlier study (Vouldoukis *et al.*, 1995), Vouldoukis *et al.* (1997) showed that recombinant IL-10 negatively regulated nitrite formation and *L. major* killing by monocytes activated by anti-CD23 antibody or IFNγ treatment. Contradicting many prior studies, Schneemann *et al.* (1997) did not find evidence that nitrite generation in IL-4/IFNγ-treated monocytes involved the NOS pathway. These investigators were unable to demonstrate L-arginine consumption, L-citrulline production, or synthesis of tetrahydrobiopterin in their cultures. They suggested that the nitrite measured was derived from nitrate in the culture medium or serum, possibly generated by nitrate reductase activity of IL-4/IFNγ, although this would not account for the ability of L-NMMA to inhibit nitrite formation in the earlier studies.

4. Activation of Human Mononuclear Phagocyte NO Production by Microbes

4.1. Mycobacteria

Denis and co-workers were the first to report that microbes could activate human monocytes for NO production (Denis, 1991), but this observation has been somewhat controversial. As mentioned earlier, these investigators noted that monocyte-derived macrophages treated *in vitro* with TNFα and GM-CSF had an enhanced ability to restrict the growth of virulent *Mycobacterium avium*, but nitrite production required the presence of the mycobacterium (Denis, 1991). In contrast, Bermudez (1993) found that the antimicrobial actions of human monocytes toward *M. avium* or *Listeria monocytogenes* was *not* inhibited by L-NMMA or arginase, and TNFα, GM-CSF, and IFNγ, with or without the presence of bacteria, failed to cause nitrite production. Dumarey *et al.* (1994) appeared to confirm Denis's observation that infection of human monocytes with live, virulent *M. avium* could induce nitrite production; LPS, IFNγ, and TNFα alone had no effect on nitrite formation in these studies, and virulent strains of *M. tuberculosis* or avirulent *M. avium* also had no effect. However, a subsequent report could not demonstrate induction of NO production by human monocytes or macrophages with inoculation of live or heat-killed *M. avium* complex (Weinberg *et al.*, 1995), although these investigators did not coincubate cytokines with the mononuclear phagocytes and mycobacteria.

4.2. Parasites

Sherman *et al.* (1991) noted that while LPS and IFNγ enhanced nitrite and L-citrulline production by mouse macrophages, these stimuli enhanced only L-citrulline formation by human alveolar macrophages unless the cells were cocultured with *Pneumocystis carinii,* which resulted in a slight increase in both nitrite and L-citrulline. Naotunne *et al.* (1993) reported that supernatants of human MNC cultured with extracts of freeze-thawed malarial parasites could inactivate *Plasmodium* gametocytes incubated with blood MNC. The gametocytes were subsequently unable to infect mosquitoes, and this effect could be inhibited by L-NMMA. Seitzer *et al.* (1997) studied multinucleated giant cells formed *in vitro* after culture of human MNC with larvae of *Nippostrongylus brasiliensis*. They demonstrated by RT-PCR analysis of single giant cells that 10 of 55 cells examined (18%) contained NOS2 mRNA, as well as mRNA for IL-1, TNFα, and IL-6. All examined cells exhibited NADPH diaphorase activity as well, with cells adherent to nematodes staining most strongly.

4.3. Viruses

Some researchers have noted that HIV-1 or components of HIV-1 activated human monocyte NO production. Pietraforte *et al.* (1994) found that HIV-1 envelope gp120 increased nitrite production by cultured monocytes. This production was inhibited by L-arginine analogues, and production of NO could also be demonstrated by EPR spin-trapping with 5,5-dimethyl-1-pyrroline-*N*-oxide (DMPO). Bukrinsky *et al.* (1995) studied human monocytes after culture for 7 days with M-CSF and subsequent infection with HIV-1. Uninfected cells produced little or no NO, but those infected with HIV-1 produced significant amounts. Treatment of infected cells with LPS or TNFα, or coculture with astroglial cells further enhanced NO formation, while IL-4 or L-NMMA decreased the NO production. NO production was assessed by both measurement of nitrite and EPR detection of NO trapped by Fe-DETC. Also, NOS2 mRNA was detected in infected and stimulated monocytes by RT-PCR (Bukrinsky *et al.*, 1995). However, other investigators have failed to detect significant increases in monocyte NO production after infection with HIV-1 *in vitro* (Weinberg *et al.*, 1995).

4.4. Bacteria

Other than the equivocal findings with mycobacteria discussed earlier, bacteria have not generally been shown to stimulate NO production by human MNC *in vitro*. As noted, Bermudez (1993) showed that *L. monocytogenes* did not enhance monocyte nitrite formation, and Schneemann *et al.* (1993) did not detect nitrite production, L-arginine consumption, L-citrulline production, or NOS activity after treatment of human mononuclear phagocytes with heat-killed *Listeria* or *Moraxella*. Weinberg *et al.* (1995) evaluated *L. monocytogenes*, *C. albicans*, *S. epidermidis*, *M. avium* complex, and *M. tuberculosis* for the ability to activate monocytes for NO production, and reported uniformly negative results. On the other hand, Tufano *et al.* (1994) reported that porins extracted from *Yersinia enterocolitica* (along with a small amount of LPS contamination) stimulated nitrite formation by cultured monocytes.

5. Activation of Human Mononuclear Phagocyte NO Production by Miscellaneous Agents

A variety of other agents or materials have been tested for the ability to stimulate normal human mononuclear phagocyte NO production. Although Belenky *et al.* (1993) did not measure NO production, they did demonstrate that L-arginine analogues attenuated chemotactic peptide fMLP-induced chemotaxis and fMLP-induced increases in cAMP in human monocytes, suggesting indirectly that NO production by these cells modulated their function. DeMaria *et al.* (1994) found that treatment of monocytes with an antibody directed against the activation antigen CD69 triggered production of nitrite and nitrate. The antibody enhanced monocyte-mediated cytotoxicity for tumor cells, an effect (along with nitrite production) inhibited by L-NMMA. Perez-Mediavilla *et al.* (1995) demonstrated that oligopeptides from certain extracellular matrix proteins enhanced nitrite production and NOS2 protein expression by human monocytes, and the nitrite production was inhibitable by L-NMMA. McLachlan *et al.* (1996) have noted that dehydroepiandrosterone (DHEA) stimulated nitrite formation by normal human monocytes, with synergistic effects provided by the addition of LPS.

Magazine *et al.* (1996) used an NO-specific amperometric probe to demonstrate that morphine enhanced NO formation. The NO production rate was rapid, with enhanced NO elaboration being detected within minutes of adding morphine to the cultures, and increased NO production was blocked by naloxone or by L-NMMA. Stefano *et al.* (1996) found that normal human monocytes produced NO when treated with the tetrahydrocannabinol derivative anandamide. As observed with morphine, NO production occurred rapidly, and L-arginine analogues inhibited production. The cannabinoid antagonist SR 141716A (but not the morphine receptor antagonist naloxone) also blocked the NO-inducing effect.

King *et al.* (1997) reported that endothelin-1 induced rapid release of NO (measured by a NO-specific amperometric probe) from MNC, and decreased adherence of MNC to endothelial cells. Vitek *et al.* (1997) demonstrated that cultured human monocytes pretreated with poly I:C produced greater quantities of nitrite in the presence of apolipoprotein E, an effect inhibited by the presence of amyloid beta peptide. Not all reports using miscellaneous agents have been positive; Lammas *et al.* (1997) have found that human monocyte-mediated killing of BCG organisms induced by treatment with ATP was *not* inhibited by L-arginine analogues.

6. Human Disorders Associated with Mononuclear Phagocyte NOS2 Expression and NO Production

The studies discussed above have principally dealt with normal cells treated *in vitro* with various agents in an attempt to stimulate NO production and NOS expression. However, there is generally more convincing evidence of NO production when human mononuclear phagocytes are "activated" *in vivo*. Some reports have documented NOS2 expression in monocytes and macrophages using immunohistological or immunocytological staining with specific antibodies, or have employed RT-PCR or *in situ* hybridization to demonstrate NOS2 mRNA within tissue mononuclear phagocytes. Others have examined cells taken from blood or various tissues. Table II summarizes reports in which investigators have

TABLE II
Human Disorders Associated with Mononuclear Phagocyte NOS2 Expression[a]

Disorder	Study	Cell	Assay	Comment
Liver disease				
Alcoholic hepatitis	Hunt and Goldin (1992)	Mo	Nitrite production *in vitro*	None by those from normal individuals; no treatment
Alcoholic cirrhosis	Criado-Jimenez *et al.* (1995)	MNC	Nitrite	Blocked by L-Arg analogue
Alcoholic cirrhosis with ascites	Laffi *et al.* 1995)	Mo	L-Arg to L-Cit, inhibition of platelet aggregation	Inhibited by L-Arg analogues
Cirrhosis	Masini *et al.* (1995)	Mo	L-Arg to L-Cit, inhibition of platelet aggregation	Higher in Mo from patients with cirrhosis; inhibited by L-NMMA
Hepatitis C patients	Sharara *et al.* (1997)	Mo, MNC	Nitrite, nitrate, immunoblot, L-Arg to L-Cit, RT-PCR	Induction of NOS2 activity and Ag content by *in vivo* treatment of hepatitis C patients; correlation of antiviral activity of IFNα with degree of NOS2 induction
Chronic viral hepatitis	Majano *et al.* (1998)	MNC	Immunohistology, *in situ* hybridization	In chronic active hepatitis B or C patients, NOS2 protein and mRNA noted in hepatocytes, with only mRNA noted in MNC in hepatitis
Lung disease				
Lung inflammation	Kobzik *et al.* (1993)	Alv Mac	NOS2 immunohistology	—
Acute lung injury (ARDS)	Haddad *et al.* (1994)	Alv Mac	Nitrotyrosine immunohistology	Presumed NO formation leading to nitrotyrosine
Lung inflammation (bronchiectasis and bronchopneumonia)	Tracey *et al.* (1994)	Alv Mac	NOS2 immunohistology	No staining in cells from normal lung
Acute lung injury (ARDS, sepsis, pneumonia)	Kooy *et al.* (1995)	Alv Mac	Nitrotyrosine immunohistology	Presumed NO formation leading to nitrotyrosine

(*Continued*)

TABLE II (*Continued*)

Disorder	Study	Cell	Assay	Comment
Pulmonary tuberculosis	Kumar *et al.* (1995)	Mo	Nitrite	Higher in cells from patients with tuberculosis pretreatment; enhanced *in vitro* by LPS, PPD, PMA, latex spheres
Pulmonary tuberculosis	Nicholson *et al.* (1996)	Alv Mac	Immunohistology, immunoblot, RT-PCR	Increased NOS2 in cells from patients with tuberculosis
Obliterative bronchiolitis in lung transplant patients	McDermott *et al.* (1997)	Alv Mac	Immunohistology for NOS2 and nitrotyrosine	Colocalization of NOS2 and peroxynitrite with Mac
Pulmonary fibrosis	Nozaki *et al.* (1997)	Alv Mac	Immunocytology and immunoblot for NOS2 and nitrotyrosine, RT-PCR	Produced by Alv Mac from patients with pulmonary fibrosis after *in vitro* challenge with BCG; L-NMMA inhibited Mac-mediated BCG killing
Cardiovascular disease				
Atherosclerosis	Beckman *et al.* (1994b)	Vessel Mac	Nitrotyrosine immunohistology	Presumed NO formation leading to nitrotyrosine
Myocardial infarct	Wildhirt *et al.* (1995)	Myocardial Mac	Immunohistology	NOS2 in infarcted myocardium colocalized with Mac
Atherosclerosis	Buttery *et al.* (1996)	Atherosclerotic plaque Mac	Immunohistology, immunoblot, *in situ* hybridization, nitrotyrosine	Nitrotyrosine associated with Mac
Giant cell arteritis	Weyand *et al.* (1996)	Arterial Mac	Immunohistology	NOS2 in intimal Mac in giant cell arteritis
Accelerated arteriosclerosis in transplanted hearts	Lafond-Walker *et al.* (1997)	Coronary artery Mac	Immunohistology, *in situ* hybridization	NOS2 associated with Mac
Atherosclerosis	Wilcox *et al.* (1997)	Vessel Mac	*In situ* hybridization, immunohistology	No NOS2 in normal vessels; in atherosclerotic vessels, Mac expressed NOS2 and NOS1
Atherosclerosis	Luoma *et al.* (1998)	Arterial Mac	Immunohistology	NOS2 and nitrotyrosine associated with Mac

Allergic disease				
Hay fever, asthma	Mautino *et al.* (1994)	Mo	Nitrite	IL-4 treatment
Arthritis				
RA and inflammatory osteoarthritis	Sakurai *et al.* (1995)	Synovial Mac	Nitrite, immunohistology, immunoblot, RT-PCR	Inhibited by L-Arg analogue
RA and osteoarthritis	McInnes *et al.* (1996)	Synovial Mac	Nitrite, RT-PCR, immunohistology	NOS2 in synovial Mac and synovial fibroblasts
RA	St. Clair *et al.* (1996)	Mo, MNC	Nitrite, nitrate, immunoblot, L-Arg to L-Cit	Increased L-Arg to L-Cit activity and NOS2 Ag in freshly isolated cells from patients with RA; increased responsiveness of MNC of RA patients to IFNγ *in vitro;* inhibited by L-NMMA
RA	Grabowski *et al.* (1997)	Synovial Mac	Immunohistology	NOS2 associated chiefly with CD68^{+} Mac by immunohistology; more NOS2 in RA vs. osteoarthritis; no NOS2 noted in tissues from normal subjects (hip fractures)
RA	Perkins *et al.* (1998)	Mo, MNC	Immunoblot, RT-PCR, L-Arg to L-Cit	Increased L-Arg to L-Cit activity, NOS2 Ag, NOS2 mRNA in MNC from patients with RA; anti-TNF antibody treatment *in vivo* reduced the increased NOS activity and NOS2 Ag expression
Inflammatory response to foreign body (joint prostheses)	Moilanen *et al.* (1997)	Foreign body Mac	Immunohistology, RT-PCR	NOS2 associated with Mac
Loosened bone prosthesis	Watkins *et al.* (1997)	Mac around loosened bone hip prostheses	*In situ* hybridization, immunohistology	No NOS2 in synovial cells
Cancer				
Breast cancer	Thomsen *et al.* (1995)	Breast cancer Mac	Anti-NOS2 Ab immunocytology, nitrite and nitrate, L-Arg to L-Cit	NOS2 associated with CD68^{+} Mac by immunohistology

(Continued)

TABLE II (*Continued*)

Disorder	Study	Cell	Assay	Comment
Various metastatic cancers to pleural or peritoneal spaces	Wang *et al.* (1996)	Pleural or Perit Mac	Nitrite	More NO production by Mac from patients with cancer
Colon cancer	Ambs *et al.* (1998)	Colon cancer MNC	L-Arg to L-Cit, Immunohistology for NOS2 and nitrotyrosine, immunoblot, RT-PCR	NOS2 in tumors (more in adenomas than in carcinomas), but low in normal tissue; present in MNC, tumor cells, and endothelial cells; nitrotyrosine in MNC
Parasitic disease				
African children: normal and malarious	Anstey *et al.* (1996)	MNC	Immunoblot	MNC NOS2 expression in normal Tanzanian children; increased NOS2 in asymptomatic and mild malaria; markedly decreased NOS2 in cerebral malaria
Renal disease				
IgA nephropathy and proliferative glomerulonephritis	Kashem *et al.* (1996)	Kidney Mac, Mo	Immunohistology, RT-PCR	NOS2 associated with Mac
Bowel disease				
Crohn's disease, ulcerative colitis, diverticulitis	Singer *et al.* (1996)	Bowel Mac	Immunohistology, RT-PCR	Found associated with Mac only at sites of inflammation
Ulcerative colitis	Ikeda *et al.* (1997)	Bowel Mac	Immunohistology	Increased NOS2 in areas of inflammation
Celiac disease	ter Steege *et al.* (1997)	Bowel Mac	Immunohistology	Increased NOS2 and nitrotyrosine in Mac in disease sites

Helicobacter pylori gastritis	Mannick *et al.* (1996)	Gastric Mac	Immunohistology for NOS2 and nitrotyrosine	NOS2 expression by gastric MNC in patients with *Helicobacter pylori* gastritis; number of positive cells decreased after antibiotic, β-carotene, or ascorbate treatment
Multiple sclerosis				
	Bo *et al.* (1994)	Brain Mac	RT-PCR	NOS2 NADPH diaphorase activity detectable in brains of MS patients
	Bagasra *et al.* (1995)	Brain Mac	Immunohistology, RT-*in situ*-PCR	NOS2 mRNA and nitrotyrosine in Mac of patients with multiple sclerosis and absent in "control" brains
	DeGroot *et al.* (1997)	Brain Mac	Immunohistology; nitrite	In brains of MS patients, Mac positive for both NOS2 and "cNOS"; isolated Mac produced NO
	Hooper *et al.* (1997)	Brain Mac	Immunohistology, RT-in situ-PCR	NOS2 in brain Mac from MS patients
Miscellaneous conditions				
Trauma	Kim *et al.* (1995)	Mo	Nitrite, nitrate	Higher NO production in PBMC from trauma patients; some reduction of NO production by IL-13
Sickle-cell disease	Dias-Da-Motta *et al.* (1996)	PBMC	Inhibition of platelet aggregation	Inhibited by L-Arg analogue; production noted only in presence of SOD; more activity in cells from patients with sickle-cell disease
Graves' disease (thyrotoxicosis)	López-Moratalla *et al.* (1996)	Mo	Immunocytology	NOS2 in freshly isolated Mo from Graves'; increased activity induced by *in vitro* treatment with peptides from thyroid autoantigens
Crohn's disease, ulcerative colitis, diverticulitis	Singer *et al.* (1996)	Bowel Mac	Immunohistology, RT-PCR	Found associated with Mac only at sites of inflammation

(*Continued*)

TABLE II *(Continued)*

Disorder	Study	Cell	Assay	Comment
Pregnancy	Eis *et al.* (1997)	Fetal membrane Mac	Immunohistology	NOS2 associated with membrane Mac; greater NOS2 expression in preterm labor vs. term labor
Pregnancy	Myatt *et al.* (1997)	Placental Mac (Hofbauer cells)	Immunohistology, RT-PCR	NOS2 associated with Mac
Pregnancy	Zarlingo *et al.* (1997)	Placental Mac	Immunohistology, immunoblot	—

[a]For abbreviations, see Table I.

examined mononuclear phagocytes from patients with various disease states for NO production and NOS2 expression.

6.1. Hepatic Disease

When Hunt and Goldin (1992) observed that normal monocytes spontaneously generated nitrite during *in vitro* culture, they also noted that monocytes from alcoholic patients with or without liver disease exhibited higher spontaneous nitrite production than those from normal subjects. Similarly, Criado-Jiminez *et al.* (1995) found that MNC from alcoholic subjects produced nitrite, and those from subjects with cirrhosis produced even higher levels. In both studies, the nitrite production was inhibited by L-arginine analogues. Laffi *et al.* (1995) have helped to confirm and extend these findings by documenting that monocytes from patients with alcoholic cirrhosis and ascites spontaneously converted L-arginine to L-citrulline and inhibited thrombin-induced platelet aggregation. Masini *et al.* (1995) also found that monocytes from individuals with cirrhosis produced more NO than those from normal subjects.

Sharara *et al.* (1997) studied MNC from individuals with hepatitis C who were or were not undergoing treatment with IFNα. In the absence of IFNα therapy, cells from hepatitis C patients did not display NOS2 antigen or mRNA, and their NOS enzyme activities were comparable to those of normal individuals. However, administration of IFNα increased NOS enzyme activity and caused the appearance of NOS2 antigen and mRNA in MNC. In patients receiving IFNα, the induction of NOS2 correlated significantly with the degree of improvement in their hepatitis. In another study, Majano *et al.* (1998) used immunohistological methods and *in situ* hybridization to analyze liver biopsy specimens from patients with chronic active viral hepatitis, alcoholic hepatitis, and cholestasis. Large quantities of NOS2 protein and mRNA were noted within hepatocytes of patients with hepatitis B or hepatitis C, along with small amounts of NOS2 mRNA in mononuclear cells. NOS2 protein was not detected with mononuclear cells, but the patients were not receiving IFNα treatment. Liver tissue from patients without liver disease or with nonviral liver disease contained very little detectable NOS2.

6.2. Pulmonary Disease

In a histological and cytological study of surgically resected human lung tissue, Kobzik *et al.* (1993) found that human alveolar macrophages contained NADPH diaphorase activity and NOS2 antigen, most markedly in areas of inflammation. Tracey *et al.* (1994) used immunohistological techniques to demonstrate NOS2 expression in alveolar macrophages from patients with bronchiectasis or pneumonia. In contrast, no cells from normal lung were found to contain NOS2 antigen. Haddad *et al.* (1994), employed immunohistological and immunocytochemical stains to show that human alveolar macrophages from patients with acute lung injury (adult respiratory distress syndrome) contain nitrotyrosine. Although tissue from patients without lung injury contained small amounts of nitrotyrosine, there was twice as much in samples from patients with acute lung injury. The presence of nitrotyrosine was presumptive evidence that NO had been formed, possibly reacting with superoxide to form peroxynitrite, which in turn caused tyrosine nitration (Beckman *et al.*, 1994a).

In studies of autopsy tissues from patients with acute lung injury, Kooy *et al.* (1995) identified nitrotyrosine in alveolar macrophages, alveolar epithelium, lung interstitium, and proteinaceous alveolar exudate. Samples from patients with sepsis-induced diffuse alveolar damage had extensive staining of the endothelium and subendothelial tissues. McDermott *et al.* (1997) demonstrated the presence of alveolar macrophages containing NOS2 and nitrotyrosine in lung tissue from transplant patients with obliterative bronchiolitis; NOS2 and nitrotyrosine were also seen in neutrophils, airway epithelium, and vascular endothelium of these specimens, but little or no reactivity was observed in control lungs. Nozaki *et al.* (1997) documented that alveolar macrophages from patients with idiopathic pulmonary fibrosis contain NOS2 mRNA, NOS2 antigen, and nitrotyrosine following inoculation with BCG *in vitro*. Alveolar macrophages from these patients were able to kill BCG organisms, and this killing was inhibited by L-NMMA. Alveolar macrophages from patients with lung cancer or nonmalignant pulmonary nodules did not express NOS2 or nitrotyrosine, and did not kill BCG.

In studies of monocytes from patients with pulmonary tuberculosis, Kumar *et al.* (1995) found that monocytes stimulated *in vitro* with LPS, PPD, PMA, or latex spheres produced nitrite and L-citrulline. Nitrite and L-citrulline production were higher in subjects before their tuberculosis had been treated. Nicholson *et al.* (1996) found that alveolar macrophages from patients with tuberculosis expressed NOS2 antigen (by immunocytological staining and immunoblot with an antibody highly specific for human NOS2) and NOS2 mRNA (by RT-PCR). Sixty-five percent of alveolar macrophages from all 11 patients with untreated tuberculosis expressed NOS2, but only 10% of alveolar macrophages from 5 normal subjects detectably expressed the antigen. Alveolar macrophages from patients with other inflammatory disorders such as pneumonia, cancer, and sarcoidosis also contained NOS2.

6.3. Cardiovascular Disease

In immunohistological and immunoblot studies of human arterial atherosclerotic plaques, Beckman *et al.* (1994b) found extensive nitration of protein tyrosines associated with macrophages. Buttery *et al.* (1996) also showed that human atherosclerotic lesions contained NOS2 and nitrotyrosine in association with macrophages, foam cells, and smooth muscle cells. These investigators used *in situ* hybridization to demonstrate evidence of NOS2 mRNA, and immunoblot analysis to confirm that the NOS2 protein isoform was present. The presence of nitrotyrosine suggested that peroxynitrite was formed in the lesions, and might play a role in the pathogenesis of atherosclerosis. By applying *in situ* hybridization and immunohistological methods, Wilcox *et al.* (1997) demonstrated that NOS1 and NOS2 were not expressed in normal human vessels, but endothelial cells expressed NOS1. In contrast, atherosclerotic vessels exhibited decreased NOS3 expression in the vicinity of the lesions, and increased expression of NOS and NOS2 within a variety of cells (including macrophages). Luoma *et al.* (1998) have demonstrated that macrophages within atherosclerotic lesions expressed NOS2 and contained nitrotyrosine; high levels of SOD expression were also noted within the same lesions.

Lafond-Walker *et al.* (1997) found NOS2 antigen and mRNA in macrophages within transplanted hearts undergoing accelerated graft arteriosclerosis. NOS2 was noted in the neointima from seven of ten transplanted vessels with accelerated graft arteriosclerosis, but was absent from five arteries with atherosclerosis and from two normal coronary arteries. This study noted no relationship between NOS2 expression and levels of the immunosuppressive drug cyclosporine A.

Wildhirt *et al.* (1995) found NOS2 associated with macrophages in areas of myocardial infarction in patients 7 and 25 days after infarct. Weyand *et al.* (1996) detected NOS2 in intimal macrophages within arteries from patients with giant cell arteritis. Interestingly, adventitial $CD68^+$ macrophages expressing TGFβ were positive for IL-6 and IL-1, but negative for NOS2, while $CD68^+$ macrophages expressing NOS2 were negative for TGFβ and positive for 72-kDa collagenase. Nonmacrophage cells (probably smooth muscle cells) in the inflammatory lesions also expressed NOS2.

6.4. Allergic Disease

As noted earlier, several researchers have demonstrated the importance of IgE and CD23 in the control of monocyte NO production. Mautino *et al.* (1994) noted that IL-4 enhanced nitrite production by monocytes from different donors to varying degrees. Cells from atopic subjects (i.e., those with hay fever or asthma) appeared to have increased *in vitro* production of NO.

6.5. Rheumatological Disease

Numerous studies have demonstrated proinflammatory effects of NO in animal models of arthritis (reviewed by Clancy and Abramson, 1995). There are now also compelling data to document enhanced expression of NOS2 and production of NO by mononuclear phagocytes from human subjects with inflammatory arthritis. In a study of human synovial cells from patients with inflammatory arthritides, Sakurai *et al.* (1995) showed that synovial macrophages produced nitrite, contained NOS2 antigen (by immunohistology and immunoblot), and expressed NOS2 mRNA. NOS2 expression and nitrite production were noted in cells from both RA and inflammatory osteoarthritis, and could be inhibited by L-arginine analogues. McInnes *et al.* (1996) studied synovial membrane cells from patients with osteoarthritis and RA. Although they detected increased nitrite production in cultures of these tissues, and confirmed NOS2 expression by synovial macrophages, the most abundant NOS2-expressing cells were synovial fibroblasts. Other researchers noted in immunohistological studies of resected joint specimens that RA patients have large numbers of NOS2-containing $CD68^+$ macrophages in synovial lining areas (Grabowski *et al.*, 1997); chondrocytes, fibroblasts, and smooth muscle cells in these specimens also contained small amounts of NOS2. Tissues from osteoarthritis patients were found to have lower levels of NOS2 expression, and tissue from patients without arthritis (e.g., hip fracture) had no NOS2 expression (Grabowski *et al.*, 1997).

St. Clair *et al.* (1996) found that freshly isolated monocytes and MNC from patients with active RA had increased NOS enzyme activity and NOS2 antigen expression by immunoblot (Fig. 3), when compared with those of normal subjects. When MNC were cultured with IFNγ, only the cells from RA patients responded with increased production of nitrite/nitrate, and L-NMMA inhibited the NOS activity. Levels of NOS2 enzyme activity and NOS2 antigen were positively correlated with the severity of arthritis. Investigators from this group also studied RA patients receiving treatment with the chimeric, monoclonal anti-TNFα antibody cA2 (Perkins *et al.*, 1998), which has been recently found to induce a dramatic clinical improvement in the majority of RA patients treated. Perkins and associates confirmed increased blood MNC expression of NOS enzyme and NOS2 antigen in patients with RA, and further demonstrated that these MNC had increased expression of NOS2 mRNA (RT-PCR). They found that cells from patients receiving anti-TNFα antibody treatment 4 weeks earlier had a decrease in MNC NOS2 antigen and NOS activity overexpression. Antibody-induced reductions in NOS activity and NOS2 antigen expression correlated significantly with reduction in the number of painful joints. The authors proposed that antibody-induced decreases in NOS overexpression may account (wholly or in part) for treatment-related clinical improvement.

A study by Moilanen *et al.* (1997) found that foreign body macrophages in the granulomatous, pseudosynovial membrane adjacent to loosened joint prostheses contained NOS2 in 10 of 13 cases examined. CD23 was not detectable. Calcium-independent NOS enzyme activity was detected in 12 of 13 specimens, and RT-PCR analysis revealed NOS2 mRNA in 3 of 3 tissue samples studied (normal blood monocytes were negative). In a comparable study, Watkins *et al.* (1997) demonstrated by immunohistology and *in situ* histochemistry that human macrophages at the interfascial membrane and pseudocapsule surrounding failed prosthetic hip

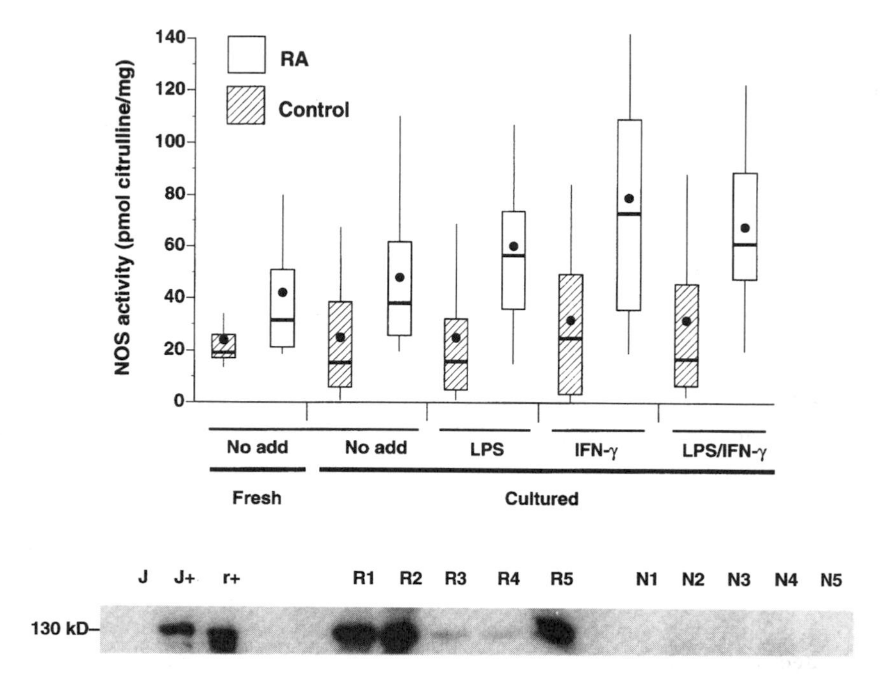

FIGURE 3. (Top) NOS activity in freshly isolated and cultured mononuclear cells from control subjects and RA patients. Blood mononuclear cells were prepared and extracts from freshly isolated cells ("Fresh") or from cells cultured 5 days with no additions, with 1 μg/ml LPS alone, with 500 U/ml IFNγ alone, or with 500 U/ml IFNγ and 1 μg/ml LPS were assayed for NOS activity (ability to convert L-arginine to L-citrulline). Assays were done as six replicates for each subject. Results are shown as medians (horizontal bar), means (circle), the interquartile range (box), and the 10th to 90th percentile range (vertical lines). There were 20 control subjects and 25 RA patients. Using the Wilcoxon Rank Sum test, RA patients' NOS activities differed significantly from control subjects' NOS activities in the categories "Fresh" ($p<0.003$), "No add" ($p<0.005$), "LPS" ($p<0.002$), "IFN-γ" ($p<0.002$), and "LPS/IFN-γ" ($p<0.002$). In analyses of cultured cells, the within-group comparison was significant for RA patients ($p<0.001$), but not for control subjects. Pairwise comparisons for cells from RA patients revealed significant differences for treatments that included IFNγ ["No add" versus "IFN-γ" ($p<0.003$) and "No add" versus "LPS/IFN-γ" ($p<0.003$)].

(Bottom) Immunoblot analysis of mononuclear cells from normal controls and RA patients for NOS2 expression. Blood mononuclear cells were isolated and extracts were analyzed for NOS2 antigen content using an NOS2-specific mouse monoclonal anti-NOS2 antibody. The NOS2 antigen has a molecular mass of approximately 130 kDa. Extracts from the murine macrophage cell line J774 and RAW 264 were used as negative and positive controls. Forty micrograms of protein from the extracts was used in each lane. Parallel gels and blots done using isotype-specific control immunoglobulin showed no reactivity. J = J774 control; J+ = J774 cells cultured with LPS + IFNγ; r+ = RAW 264 cells cultured with LPS + IFNγ; R1–5 = samples from five RA patients; N1–5 = samples from five normal (control) subjects.

Reproduced from *The Journal of Experimental Medicine*, 1996, Vol. 184, pp. 1173–1178 by copyright permission of The Rockefeller University Press.

joints contained NOS2; these investigators noted phagocytosed polyethylene debris within several NOS2-positive macrophages.

6.6. Neoplastic Disease

Thomsen *et al.* noted that macrophages within human breast cancer specimens produced nitrite and nitrate and contained NOS2 (Thomsen *et al.*, 1995). There was a general positive relationship between the grade of malignancy and the amount of NOS content. Ambs and co-workers examined resected specimens from individuals with colon adenomas and adenocarcinomas for NOS activity and NOS2 expression (Ambs *et al.*, 1998). They found calcium-dependent NOS enzyme activity in normal colon tissue, but levels were lower in adenomas and carcinoma tissues (possibly indicative of a general decrease in endothelial cells and autonomic neurons in colon tumors). However, levels of calcium-dependent NOS (NOS2) were much higher in adenomas and carcinomas, with very low levels in normal tissues adjacent to the tumors. Calcium-independent NOS activity in tumors decreased with increasing stage of the tumor, with the lowest activities being noted in metastatic tumors. Immunoblot analysis and RT-PCR detected NOS2 mRNA in the tumor tissue. Normal colon epithelium, colon cancer cells, and MNC expressed NOS2 antigen, while only MNC (and rare PMN) expressed nitrotyrosine (Ambs *et al.*, 1998).

In a study of gynecological tumors (ovarian, endometrial, and mixed mesodermal), Thomsen and associates noted high levels of NOS enzyme activity in tumor cells, but not in normal gynecological tissues (Thomsen *et al.*, 1994). Immunohistology showed that NOS2 antigen was in tumor cells, but not in normal tissue. Other workers showed that central nervous tumors (astrocytoma, meningioma, Schwannoma, ependymoma, medulloblastoma, and mixed glioma) had NOS activity and expressed NOS1 and NOS2. In general, the highest levels of expression were in the tumors with the highest histologic tumor grades (Cobbs *et al.*, 1995). Hematopoietic cells in the tumor tissue were not reported to express NOS or NADPH diaphorase activity.

6.7. Parasitic Disease

Anstey *et al.* (1996) studied children from Tanzania (see Chapter 15) to determine the influence of malaria on MNC NOS2 expression and NO production, and to determine whether these parameters related to disease severity. They found that urine and plasma levels of nitrite/nitrate (corrected for renal impairment) correlated inversely with disease severity, with highest levels in subclinical infection and lowest levels in fatal cerebral malaria. Likewise, blood MNC NOS2 antigen was detectable by immunoblot in control children and those with subclinical infection, but was undetectable in all but one subject with cerebral malaria. Quantitated MNC NOS2 antigen levels paralleled nitrite/nitrate levels, and were inversely related to disease severity. Levels of IL-10, a cytokine known to suppress NO synthesis, increased with disease severity. The authors hypothesized that high IL-10 levels in severe disease might decrease NOS expression and NO production. Based on their data, they also suggested that NO had a protective (rather than a pathological) role in African children with malaria. It is important to note that MNC from healthy control African children of this study exhibited basal expression of NOS2 and relatively high levels of plasma and urine nitrite/nitrate (Anstey *et al.*, 1996). Studies in American adults have only rarely found NOS2 antigen expression in MNC or monocytes from normal individuals (Weinberg *et al.*, 1995; St. Clair *et al.*, 1996; Sharara *et al.*, 1997). Anstey's group postulated that this "constitutive" expression of NOS2 may be related to subclinical malaria or other infections, or to genetic differences in the control of NOS2 expression.

6.8. Renal Disease

Kashem *et al.* (1996) demonstrated by immunohistology that renal macrophages from patients with IgA or proliferative nephropathy contained NOS2 antigen. They also found that kidney biopsy samples from these patients contained NOS2 mRNA, as determined by RT-PCR, whereas normal kidney samples expressed neither NOS2 antigen nor mRNA.

6.9. Gastrointestinal Disease

Using immunohistology and RT-PCR, Singer *et al.* (1996) demonstrated NOS2 antigen and mRNA in samples of colonic epithelium from patients with inflammatory bowel disease. NOS2 was observed in association with macrophages from areas of inflammation in patients with ulcerative colitis, Crohn's disease, and diverticulitis. Areas of inflammation also contained nitrotyrosine, suggesting that peroxynitrite may have been present. NOS2 was noted in neutrophils located in the colon lumen and within crypt abscesses, and these cells may have contributed to the formation of nitrotyrosine via a myeloperoxidase-dependent mechanism (Eiserich *et al.*, 1998). Ikeda *et al.* (1997) also demonstrated increased NOS2 expression in colonic lesions of patients with ulcerative colitis. Increased expression was seen only in individuals with active colitis, and was not found in patients with nonspecific colitis, ischemic colitis, or infectious colitis. These investigators also noted serum nitrite/nitrate levels to be approximately twofold higher in patients with ulcerative colitis than in control subjects.

ter Steege *et al.* (1997) used immunohistology to demonstrate increased NOS2 and nitrotyrosine within small bowel macrophages in celiac disease. NOS2 was found in 10 of 11 cases of celiac disease, compared with only 1 of 7 controls, while nitrotyrosine in association with NOS2-positive cells was noted in 5 of 6 cases, compared with 0 of 6 controls.

6.10. Multiple Sclerosis

Using RT-PCR, Bo *et al.* (1994) identified macrophages expressing NOS2 mRNA and associated NADPH diaphorase activity within demyelinated regions of brain tissue from patients with multiple sclerosis (MS). Bagasra *et al.* (1995) also employed RT-PCR to demonstrate NOS2 mRNA along with nitrotyrosine in the brains of all seven patients with MS, while the three control brains lacked these substances. DeGroot *et al.* (1997) showed both NOS2 and a constitutive NOS isoform within brain tissue from patients with MS, and were able to detect nitrite production by brain macrophages *in vitro*. Hooper *et al.* (1997) likewise confirmed the presence of NOS2 mRNA and protein in brain macrophages from patients with MS.

6.11. Miscellaneous Disease

Condino-Neto *et al.* (1993) noted that MNC from patients with chronic granulomatous disease (cells incapable of producing superoxide) inhibited thrombin-induced platelet aggregation; this inhibition was blocked by an L-arginine analogue. However, these investigators did not specifically examine NOS expression. López-Moratalla *et al.* (1996) found that freshly isolated monocytes from patients with Graves' disease spontaneously expressed NOS2 antigen. Monocyte NOS2 antigen content could be increased further by treatment of the cells *in vitro* with peptides from thyroid autoantigens (thyrotropin receptor, thyroid peroxidase, and thyroglobulin). Kim *et al.* (1995) noted that monocytes obtained from patients with trauma produced more NO after LPS treatment *in vitro* than those from normal individuals; IL-13 was able to decrease this production. Dias-Da-Motta *et al.* (1996) reported that NO produced by PBMC from patients with sickle-cell anemia inhibited platelet aggregation if the assays were performed in the presence of SOD, but the results were not significantly different than cells from normal individuals.

A study by Myatt *et al.* (1997) revealed that NOS2 was present and localized within placental $CD14^+$ macrophages (Hofbauer cells), but there was no difference observed in specimens from normotensive, preeclamptic, or intrauterine growth-restricted pregnancies. NOS2 mRNA was also detected, and in some cases NOS2 protein was also seen within syncytiotrophoblast and vascular endothelium. Zarlingo *et al.* (1997) detected NOS2 staining in placental villous stromal macrophages of humans and other species, as well as within placental syncytio-trophoblasts and vascular endothelial cells. Eis *et al.* (1997) demonstrated NOS2 within decidual macrophages and other cell types. These researchers found greater

Table III
NO Production and NOS2 Expression Analysis in Human Leukemia Cell Line Cells[a]

Study	Cell type	*In vitro* Rx	Assay	Detected	Comment
Schmidt *et al.* (1989)	HL-60	VD_3, fMLP	Chemiluminescence	No	HL-60 cells differentiated to monocytelike cells did not produce NO
Summersgill *et al.* (1992)	HL-60	IFNγ	Nitrite	No	No inhibition of HL-60 mediated stasis of *Legionella pneumophila* by L-NMMA
Barnewall and Rikihisa (1994)	Mo, THP-1	IFNγ, *Ehrlichia chaffeensis*	Nitrite	No	Mo-mediated anti-*Ehrlichia* effects not dependent on NO production
Reiling *et al.* (1994)	Mo, U937, THP-1, Mono-Mac 6	LPS, IFNγ	RT-PCR	Yes	eNOS mRNA in "resting" Mo and leukemia cells, and iNOS mRNA in treated Mo
Eue *et al.* (1995)	U937	Alkylphosphocholine, LPS, PMA	Nitrite	Yes	Rapid release of nitrite; potentiation of effects by LPS and PMA
Perez-Perez *et al.* (1995)	THP-1	*H. pylori*, LPS	Nitrite	No	—
Zinetti *et al.* (1995)	MNC, THP-1	LPS	NMMA-inhibitable LPS-induced TNF secretion and inhibition of THP-1 proliferation	Yes	—
Chen *et al.* (1996)	THP-1	PMA, silica, LPS	Nitrite, Northern blot	Yes	PMA as "priming" agent; inhibited by L-NMMA or allopurinol
Kawase *et al.* (1996)	HL-60	NaF, VD_3, PGE_2	Nitrite, immunoblot	Yes	Prevention of VD_3 + NaF induction of iNOS by IND
Rajora *et al.* (1996)	THP-1	IFNγ, LPS, TNF, IL4, anti-CD23 Ab	Nitrite	Yes	Slight inhibition by α-MSH

(*Continued*)

Table III (*Continued*)

Study	Cell type	*In vitro* Rx	Assay	Detected	Comment
Amin *et al.* (1997)	Mo, HL-60 cells, U937 cells	TNF, IL-1, LPS	L-Arg to L-Cit, immunoblot, RT-PCR, Northern blot	Yes	iNOS mRNA noted, but negative by immunoblot and L-Arg to L-Cit assay
Goto *et al.* (1997)	U937	Transfection with HTLV-1 *tax;* IFNγ	RT-PCR, nitrite, nitrate	Yes	Additive effects of the transfection and IFNγ treatment
King *et al.* (1997)	Mo, U937, THP-1	Endothelin-1	NO-specific amperometric probe	Yes	—
Lopez-Guerrero and Alonso (1997)	U937	PMA, HSV-1	Nitrite	Yes	Inhibited by L-NMMA
Lopez-Guerrero *et al.* (1997)	U937	None	Nitrite	Yes	More NO production by parvovirus infection-resistant cells; inhibited by L-NMMA
Roman *et al.* (1997)	U937	sCD23, anti-CD11b, anti-CD11c	L-Arg to L-Cit, immunoblot, RT-PCR	Yes	Inhibited by L-NMMA; eNOS expression stimulated

[a] For abbreviations: see Table I.

intensity of NOS2 staining in membranes of patients with preterm labor compared with those not in labor. There was no NOS2 staining in amnion epithelium or chorion trophoblast.

7. Use of Human Leukemia Cell Lines to Study Human Mononuclear Phagocyte NO Production

Human myeloid leukemia cell lines have been useful in the study of differentiation of hematopoietic precursors to more mature monocytic and neutrophilic cells (Koeffler, 1986; Collins, 1987; Auwerx, 1991; Steube *et al.*, 1997). While some investigators have differentiated these cells by various means into more "mature" "macrophages," these cells clearly differ in numerous ways from normal human mononuclear phagocytes. Nevertheless, factors such as convenience, uniformity of the cell populations, ease in obtaining large numbers of cells, and others make myeloid leukemia cell lines an attractive model system for study. Table III summarizes reports in which investigators have used the acute myeloid cell line HL-60, and the monoblastic cell lines U937, THP-1, and Mono-Mac6 for studies of NO production and NOS expression. These publications document both positive and negative reports of NO production by leukemia cell lines.

Schmidt *et al.* (1989) demonstrated that HL-60 cells differentiated to monocytic cells and treated with fMLP did not produce detectable NO. However, they also noted that HL-60 cells differentiated to neutrophil-like cells with cAMP and treated with fMLP *did* produce measurable amounts of NO. This production was inhibited by L-canavanine. Summersgill *et al.* (1992) noted that HL-60 myeloid leukemia cells treated with IFNγ were bacteriostatic for *Legionella pneumophila*, but produced no detectable nitrite.

Barnewall and Rikihisa (1994) showed that culture of human monocytes or THP-1 cells with IFNγ and *Ehrlichia chaffeensis* did not stimulate nitrite production, although growth of the *Ehrlichia* was inhibited by treatment of cells with IFNγ or PMA. Reiling *et al.* (1994) demonstrated by RT-PCR that THP-1 cells, U937 cells, and Mono-Mac6 cells displayed NOS2 mRNA on stimulation with LPS and IFNγ. Eue *et al.* (1995) showed that U937 cells rapidly produced nitrite after treatments with alkylphosphocholine, and this production could be augmented by LPS or PMA. Perez-Perez *et al.* (1995) observed that THP-1 cells stimulated with *Helicobacter pylori in vitro* did not produce nitrite. Although Zinetti *et al.* (1995) found that L-NMMA, hemoglobin, or myoglobin inhibited LPS-induced TNFα secretion by THP-1 cells, they could not detect nitrite formation. They postulated that endogenously produced NO modulated TNF production.

Chen *et al.* (1996) reported that THP-1 cells produced nitrite and expressed NOS2 mRNA after "priming" with PMA, followed by treatment with LPS and silica. Nitrite production was inhibited by either L-NMMA or allopurinol. Kawase *et al.* (1996) noted that HL-60 cells treated with sodium fluoride and 1,25-dihydroxyvitamin D_3 expressed NOS2 protein and produced nitrite, as well as PGE_2, IL-1, IL-6, and TNFα. Addition of indomethacin to the cultures blocked production of nitrite, and adding PGE_2 countered the inhibitory effects of indomethacin. Rajora *et al.* (1996) discovered that treatment with IFNγ and LPS did not enhance nitrite formation by THP-1 cells. However, pretreatment of these cells with IL-4 followed by anti-CD23 antibody, IFNγ, and TNFα caused an increase in nitrite formation, which could be slightly inhibited by α-melanocyte-stimulating hormone.

Amin *et al.* (1997) showed that HL-60 cells did not express NOS2 antigen or have NOS enzyme activity. However, using RT-PCR they were able to demonstrate that HL-60 cells (along with monocytes, neutrophils, and Jurkat cells) expressed NOS2 mRNA. NOS2 mRNA could also be detected by Northern analysis of monocytes and U937 cells. King *et al.* (1997) noted that endothelin-1 induced rapid release of NO from THP-1 cells. Endothelin-1 also caused rounding of THP-1 cells adherent to fibronectin-coated plates, and an L-arginine analogue prevented this effect. NO release could be blocked by BQ-788, an antagonist of ET_B receptors. Lopez-Guerrero and Alonso (1997) reported that U937 cells were induced to produce nitrite after treatment with PMA. Cells treated with PMA were susceptible to infection with herpes simplex virus type 1 (HSV-1), and HSV-1 infection further increased nitrite production. Although L-NMMA inhibited nitrite production, this did not alter the HSV-1 infection. Another publication from the same group noted that certain clones of U937 cells selected for resistance to parvovirus infection produced nitrite and superoxide constitutively, with both the nitrite production and resistance to parvovirus inhibited by L-NMMA (Lopez-Guerrero *et al.*, 1997).

Roman *et al.* (1997) noted that soluble CD23, anti-CD11b antibody, or anti-CD11c antibody induced U937 cells to express NOS enzyme activity, NOS3 antigen, and NOS3 mRNA (RT-PCR), but NOS2 was not evaluated. NOS enzyme activity was inhibited by either EGTA or L-NMMA. Finally, Goto *et al.* (1997) described increased expression of NOS2 mRNA and nitrite/nitrate production in U937 cells transfected with the HTLV-1 *tax* gene. Treatment of control cells with IFNγ increased NOS2 mRNA expression and nitrite/nitrate production, and this enhancement was greater in *tax*-transfected cells. The authors speculated that increased NOS2 expression in transfected cells was mediated by effects of *tax* on NF-κB.

8. Summary and Conclusions

Despite lingering controversy, it is evident from a detailed review of the literature that human mononuclear phagocytes can be stimulated by various means both *in vitro* and *in vivo* to express NOS2 and produce NO. It is difficult to quantitatively compare levels of NO production and NOS2 expression by human and murine mononuclear phagocytes. However, based on several studies in which both murine and human cells have been examined in parallel under identical conditions, human cells appear to produce less NO and express lower levels of NOS2 than do murine cells. This might help to explain the innate resistance of mice to certain human pathogens (e.g., *Mycobacterium tuberculosis*). Despite the apparent species differences, it is very likely that human mononuclear phagocyte-generated NO is important in certain pathological states (e.g., in resistance to infection and mediation of inflammation), and possibly under normal physiological conditions as well. Pharmacological modulation of mononuclear phagocyte NO production should prove to be a useful therapeutic option in some disease states.

References

Albina, J. E., 1995, On the expression of nitric oxide synthase by human macrophages—why no NO? *J. Leukoc. Biol.* **58:**643–649.

Ambs, S., Merriam, W. G., Bennett, W. P., Felleybosco, E., Ogunfusika, M. O., Oser, S. M., Klein, S., Shields, P. G., Billiar, T. R., and Harris, C. C., 1998, Frequent nitric oxide synthase-2 expression in human colon adenomas—implication for tumor angiogenesis and colon cancer progression, *Cancer Res.* **58:**334–341.

Amin, A. R., Attur, M., Vyas, P., Leszczynska-Piziak, J., Levartovsky, D., Rediske, J., Clancy, R. M., Vora, K. A., and Abramson, S. B., 1997, Expression of nitric oxide synthase in human peripheral blood mononuclear cells and neutrophils, *J. Inflamm.* **47:**190–205.

Anstey, N. M., Weinberg, J. B., Hassanali, M., Mwaikambo, E. D., Manyenga, D., Misukonis, M. A., Arnelle, D. R., Hollis, D., McDonald, M. I., and Granger, D. L., 1996, Nitric oxide in Tanzanian children with malaria. Inverse relationship between malaria severity and nitric oxide production/nitric oxide synthase type 2 expression, *J. Exp. Med.* **184:**557–567.

Aubry, J. P., Dugas, N., Lecoanet-Henchoz, S., Ouaaz, F., Zhao, H. X., Delfraissy, J. F., Graber, P., Kolb, J. P., Dugas, B., and Bonnefoy, J. Y., 1997, The 25-kDa soluble CD23 activates type III constitutive nitric oxide-synthase activity via CD11b and CD11c expressed by human monocytes, *J. Immunol.* **159:**614–622.

Auwerx, J., 1991, The human leukemia cell line, THP-1: A multifaceted model for the study of monocyte–macrophage differentiation, *Experientia (Basel)* **47:**22–31.

Bagasra, O., Bobroski, F., Sarker, A., Bagasra, A., Saikumari, P., and Pomerantz, R. J., 1997, Absence of the inducible form of nitric oxide synthase in the brains of patients with the acquired immunodeficiency syndrome, *J. Neurovirology* **3:**153–167.

Bagasra, O., Michaels, F. H., Zheng, Y. M., Bobroski, L. E., Spitsin, S. V., Fu, Z. F., Tawadros, R., and Koprowski, H., 1995, Activation of the inducible form of nitric oxide synthase in the brains of patients with multiple sclerosis, *Proc. Natl. Acad. Sci. USA* **92:**12041–12045.

Barnewall, R. E., and Rikihisa, Y., 1994, Abrogation of gamma interferon-induced inhibition of *Ehrlichia chaffeensis* infection in human monocytes with iron-transferrin, *Infect. Immun.* **62:**4804–4810.

Beckman, J. S., Chen, J., Ischiropoulos, H., and Crow, J.P., 1994a, Oxidative chemistry of peroxynitrite, *Methods Enzymol.* **233:**229–240.

Beckman, J. S., Ye, Y. Z., Anderson, P. G., Chen, J., Accavitti, M. A., Tarpey, M. M., and White, C. R., 1994b, Extensive nitration of protein tyrosines in human atherosclerosis detected by immunohistochemistry, *Biol. Chem. Hoppe-Seyler* **375:**81–88.

Belenky, S. N., Robbins, R. A., and Rubinstein, I., 1993, Nitric oxide synthase inhibitors attenuate human monocyte chemotaxis *in vitro*, *J. Leukoc. Biol.* **53:**498–503.

Ben-Efraim, S., Tak, C., Fieren, M. J. W. A., Romijn, J. C., Beckmahn, I., and Bonta, I. L., 1993, Activity of human peritoneal macrophages against a human tumor: Role of tumor necrosis factor-α, PGE_2 and nitrite, *in vitro* studies, *Immunol. Lett.* **37:**27–33.

Bermudez, L. E., 1993, Differential mechanisms of intracellular killing of *Mycobacterium avium* and *Listeria monocytogenes* by activated human and murine macrophages. The role of nitric oxide, *Clin. Exp. Immunol.* **91:**277–281.

Bo, L., Dawson, T. M., Wesselingh, S., Mork, S., Choi, S., Kong, P. A., Hanley, D., and Trapp, B. D., 1994, Induction of nitric oxide synthase in demyelinating regions of multiple sclerosis brains, *Ann. Neurol.* **36:**778–786.

Bose, M., and Farnia, P., 1995, Proinflammatory cytokines can significantly induce human mononuclear phagocytes to produce nitric oxide by a cell maturation-dependent process, *Immunol. Lett.* **48:**59–64.

Bredt, D. S., and Snyder, S. H., 1994, Nitric oxide: A physiologic messenger molecule, *Annu. Rev. Biochem.* **63:**175–195.

Bukrinsky, M. I., Nottet, H., Schmidtmayerova, H., Dubrovsky, L., Flanagan, C. R., Mullins, M. E., Lipton, S. A., and Gendelman, H. E., 1995, Regulation of nitric oxide synthase activity in human immunodeficiency virus type 1 (HIV-1)-infected monocytes—Implications for HIV-associated neurological disease, *J. Exp. Med.* **181:**735–745.

Buttery, L. D. K., Springall, D. R., Chester, A. H., Evans, T. J., Standfield, N., Parums, D. V., Yacoub, M. H., and Polak, J. M., 1996, Inducible nitric oxide synthase is present within human atherosclerotic lesions and promotes the formation and activity of peroxynitrite, *Lab. Invest.* **75:**77–85.

Cameron, M. L., Granger, D. L., Weinberg, J. B., Kozumbo, W. J., and Koren, H. S., 1990, Human alveolar and peritoneal macrophages mediate fungistasis independently of L-arginine oxidation to nitrite or nitrate, *Am. Rev. Respir. Dis.* **142:**1313–1319.

Chen, F., Kuhn, D. C., Gaydos, L. J., and Demers, L. M., 1996, Induction of nitric oxide and nitric oxide synthase mRNA by silica and lipopolysaccharide in PMA-primed THP-1 cells, *APMIS* **104:**176–182.

Chen, L. Y., and Metha, J. L., 1996, Further evidence of the presence of constitutive and inducible nitric oxide synthase isoforms in human platelets, *J. Cardiovasc. Pharmacol.* **27:**154–158.

Chu, S. C., Wu, H. P., Banks, T. C., Eissa, N. T., and Moss, J., 1995, Structural diversity in the 5′-untranslated region of cytokine-stimulated human inducible nitric oxide synthase mRNA, *J. Biol. Chem.* **270:**10625–10630.

Clancy, R. M., and Abramson, S. B., 1995, Nitric oxide—A novel mediator of inflammation, *Proc. Soc. Exp. Biol. Med.* **210:**93–101.

Cobbs, C. S., Brenman, J. E., Aldape, K. D., Bredt, D. S., and Israel, M. A., 1995, Expression of nitric oxide synthase in human central nervous system tumors, *Cancer Res.* **55:**727–730.

Collins, S. J., 1987, The HL-60 promyelocytic leukemia cell line: Proliferation, differentiation, and cellular oncogene expression, *Blood* **70:**1233–1244.

Condino-Neto, A., Muscara, M. N., Grumach, A. S., Carneiro-Sampaio, M. M., and DeNucci, G., 1993, Neutrophils and mononuclear cells from patients with chronic granulomatous disease release nitric oxide, *Br. J. Clin. Pharmacol.* **35:**485–490.

Condino-Neto, A., Muscara, M. N., Bellinatipires, R., Carneiro-Sampaio, M. M. S., Brandao, A. C., Grumach, A. S., and DeNucci, G., 1996, Effect of therapy with recombinant human interferon-gamma on the release of nitric oxide by neutrophils and mononuclear cells from patients with chronic granulomatous disease, *J. Interferon Cytokine Res.* **16:**357–364.

Criado-Jimenez, M., Rivas-Cabanero, L., Martin-Oterino, J. A., Lopez-Novoa, J. M., and Sanchez-Rodriguez, A., 1995, Nitric oxide production by mononuclear leukocytes in alcoholic cirrhosis, *J. Mol. Med.* **73:**31–33.

DeGroot, C. J. A., Ruuls, S. R., Theeuwes, J. W. M., Dijkstra, C. D., and Vandervalk, P., 1997, Immunocytochemical characterization of the expression of inducible and constitutive isoforms of nitric oxide synthase in demyelinating multiple sclerosis lesions, *J. Neuropathol. Exp. Neurol.* **56:**10–20.

DeMaria, R., Cifone, M. G., Trotta, R., Rippo, M. R., Festuccia, C., Santoni, A., and Testi, R., 1994, Triggering of human monocyte activation through CD69, a member of the natural killer cell gene complex family of signal transducing receptors, *J. Exp. Med.* **180:**1999–2004.

Denis, M., 1991, Tumor necrosis factor and granulocyte macrophage-colony stimulating factor stimulate human macrophages to restrict growth of virulent *Mycobacterium avium* and to kill avirulent *M. avium:* Killing effector mechanism depends on the generation of reactive nitrogen intermediates, *J. Leukoc. Biol.* **49:**380–387.

Denis, M., 1994, Human monocytes/macrophages: NO or no NO? *J. Leukoc. Biol.* **55:**682–684.

Dias-Da-Motta, P., Arruda, V. R., Muscara, M. N., Saad, S. T., DeNucci, G., Costa, F. F., and Condino-Neto, A., 1996, The release of nitric oxide and superoxide anion by neutrophils and mononuclear cells from patients with sickle cell anaemia, *Br. J. Haematol.* **93:**333–340.

Ding, A. H., Nathan, C. F., and Stuehr, D. J., 1988, Release of reactive nitrogen intermediates and reactive oxygen intermediates from mouse peritoneal macrophages. Comparison of activating cytokines and evidence for independent production, *J. Immunol.* **141:**2407–2412.

Dugas, N., Vouldoukis, I., Becherel, P., Arock, M., Debre, P., Tardieu, M., Mossalayi, D. M., Delfraissy, J. F., Kolb, J. P., and Dugas, B., 1996, Triggering of CD23b antigen by anti-CD23 monoclonal antibodies induces interleukin-10 production by human macrophages, *Eur. J. Inmmunol.* **26:**1394–1398.

Dumarey, C. H., Labrousse, V., Rastogi, N., Vargaftig, B. B., and Bachelet, M., 1994, Selective *Mycobacterium avium*-induced production of nitric oxide by human monocyte-derived macrophages, *J. Leukoc. Biol.* **56:**36–40.

Eis, A. L. W., Brockman, D. E., and Myatt, L., 1997, Immunolocalization of the inducible nitric oxide synthase isoform in human fetal membranes, *Am. J. Reprod. Immunol.* **38:**289–294.

Eiserich, J. P., Hristova, M., Cross, C. E., Jones, A. D., Freeman, B. A., Halliwell, B., and van der Vliet, A., 1998, Formation of nitric oxide-derived inflammatory oxidants by myeloperoxidase in neutrophils, *Nature* **391:**393–397.

Eissa, N. T., Strauss, A. J., Haggerty, C. M., Choo, E. K., Chu, S. C., and Moss, J., 1996, Alternative splicing of human inducible nitric-oxide synthase mRNA. Tissue-specific regulation and induction by cytokines, *J. Biol. Chem.* **271:**27184–27187.

Essery, S. D., Saadi, A. T., Twite, S. J., Weir, D. M., Blackwell, C. C., and Busuttil A., 1994, Lewis antigen expression on human monocytes and binding of pyrogenic toxins, *Agents Actions* **41:**108–110.

Eue, I., Zeisig, R., and Arndt, D., 1995, Alkylphosphocholine-induced production of nitric oxide and tumor necrosis factor alpha by U937 cells, *J. Cancer Res. Clin. Oncol.* **121:**350–356.

Feelisch, M., and Stamler, J. S., 1996, *Methods in Nitric Oxide Research*, Wiley, New York.

Goto, H., Nakamura, T., Shirabe, S., Ueki, Y., Nishiura, Y., Furuya, T., Tsujino, A., Nakane, S., Eguchi, K., and Nagataki, S., 1997, Up-regulation of NOS2 mRNA expression and increased production of NO in human monoblast cell line U937 transfected by HTLV-I *tax* gene, *Immunobiology* **197:**513–521.

Grabowski, P. S., Wright, P. K., Vanthof, R. J., Helfrich, M. H., Ohshima, H., and Ralston, S. H., 1997, Immunolocalization of inducible nitric oxide synthase in synovium and cartilage in rheumatoid arthritis and osteoarthritis, *Br. J. Rheumatol.* **36:**651–655.

Granger, D. L., Hibbs, J. B., Jr., Perfect, J. R., and Durack, D. T., 1988, Specific amino acid (L-arginine) requirement for the microbiostatic activity of murine macrophages, *J. Clin. Invest.* **81:**1129–1136.

Granger, D. L., Taintor, R. R., Boockvar, K. S., and Hibbs, J. B., Jr., 1995, Determination of nitrate and nitrite in biological samples using bacterial nitrate reductase coupled with the Griess reaction, *Methods: A Companion to Methods in Enzymology* **7:**78–83.

Green, L. C., de Luzuriaga, K. R., Wagner, D. A., Rand, W., Istfan, N., Young, V. R., and Tannenbaum, S. R., 1981, Nitrate biosynthesis in man, *Proc. Natl. Acad. Sci. USA* **78:**7764–7768.

Green, L. C., Wagner, D. A., Glogowski, J., Skipper, P. L., Wishnok, J. S., and Tannenbaum, S. R., 1982, Analysis of nitrate, nitrite, and [^{15}N]nitrate in biological fluids, *Anal. Biochem.* **126:**131–138.

Gyan, B., Troye-Blomberg, M., Perlmann, P., and Bjorkman, A., 1994, Human monocytes cultured with and without interferon-gamma inhibit *Plasmodium falciparum* parasite growth in vitro via secretion of reactive nitrogen intermediates, *Parasite Immunol.* **16:**371–375.

Haddad, I. Y., Pataki, G., Hu, P., Galliani, C., Beckman, J. S., and Matalon, S., 1994, Quantitation of nitrotyrosine levels in lung sections of patients and animals with acute lung injury, *J. Clin. Invest.* **94:**2407–2413.

Harwix, S., Andreesen, R., Ferber, E., and Schwamberger, G., 1992, Human macrophages secrete a tumoricidal activity distinct from tumour necrosis factor-α and reactive nitrogen intermediates, *Res. Immunol.* **143:**89–94.

Hibbs, J. B., Jr., Westenfelder, C., Taintor, R., Vavrin, Z., Kablitz, C., Baranowski, R. L., Ward, J. H., Menlove, R. L., McMurry, M. P., Kushner, J. P., and Samlowski, W. E., 1992, Evidence for cytokine-inducible nitric oxide synthesis from L-arginine in patients receiving interleukin-2 therapy, *J. Clin. Invest.* **89:**867–877.

Hooper, D. C., Bagasra, O., Marini, J. C., Zborek, A., Ohnishi, S. T., Kean, R., Champion, J. M., Sarker, A. B., Bobroski, L., Farber, J. L., Akaike, T., Maeda, H., and Koprowski, H., 1997, Prevention of experimental allergic encephalomyelitis by targeting nitric oxide and peroxynitrite: Implications for the treatment of multiple sclerosis, *Proc. Natl. Acad. Sci. USA* **94:**2528–2533.

Hunt, N. C., and Goldin, R. D., 1992, Nitric oxide production by monocytes in alcoholic liver disease, *J. Hepatol.* **14:**146–150.

Ikeda, I., Kasajima, T., Ishiyama, S., Shimojo, T., Takeo, Y., Nishikawa, T., Kameoka, S., Hiroe, M., and Mitsunaga, A., 1997, Distribution of inducible nitric oxide synthase in ulcerative colitis, *Am. J. Gastroenterol.* **92:**1339–1341.

Kashem, A., Endoh, M., Yano, N., Yamauchi, F., Nomoto, Y., and Sakai, H., 1996, Expression of inducible-NOS in human glomerulonephritis—The possible source is infiltrating monocytes/macrophages, *Kidney Int.* **50:**392–399.

Kawase, T., Oguro, A., Orikasa, M., and Burns, D. M., 1996, Characteristics of NaF-induced differentiation of HL-60 cells, *J. Bone Miner. Res.* **11:**1676–1687.

Keller, R., Keist, R., Joller, P., and Groscurth, P., 1993, Mononuclear phagocytes from human bone marrow progenitor cells; morphology, surface phenotype, and functional properties and activated cells, *Clin. Exp. Immunol.* **91:**176–182.

Kim, C., Schinkel, C., Fuchs, D., Stadler, J., Walz, A., Zedler, S., von Donnersmarck, G. H., and Faist, E., 1995, Interleukin-13 effectively down-regulates the monocyte inflammatory potential during traumatic stress, *Arch. Surg.* **130:**1330–1336.

King, J. M., Srivastava, K. D., Stefano, G. B., Bilfinger, T. V., Bahou, W. F., and Magazine, H. I., 1997, Human monocyte adhesion is modulated by endothelin B receptor-coupled nitric oxide release, *J. Immunol.* **158:**880–886.

Kobzik, L., Bredt, D. S., Lowenstein, C. J., Drazen, J., Gaston, B., Sugarbaker, D., and Stamler, J. S.,

1993, Nitric oxide synthase in human and rat lung: Immunocytochemical and histochemical localization, *Am. J. Respir. Cell. Mol. Biol.* **9:**371–377.

Koeffler, H. P., 1986, Human acute myeloid leukemia lines: Models of leukemogenesis, *Semin. Hematol.* **23:**223–236.

Kolb, J. P., Paul-Eugene, N., Damais, C., Yamaoka, K., Drapier, J.-C., and Dugas, B., 1994, Interleukin-4 stimulates cGMP production by IFN-γ-activated human monocytes. Involvement of the nitric oxide synthase pathway, *J. Biol. Chem.* **269:**9811–9816.

Kooy, N. W., Royall, J. A., Ye, Y. Z., Kelly, D. R., and Beckman, J. S., 1995, Evidence for in vivo peroxynitrite production in human acute lung injury, *Am. J. Respir. Crit. Care Med.* **151:**1250–1254.

Kumar, V., Jindal, S. K., and Ganguly, N. K., 1995, Release of reactive oxygen and nitrogen intermediates from monocytes of patients with pulmonary tuberculosis, *Scand. J. Clin. Lab. Invest.* **55:**163–169.

Laffi, G., Foschi, M., Masini, E., Simoni, A., Mugnai, L., Lavilla, G., Barletta, G., Mannaioni, P. F., and Gentilini, P., 1995, Increased production of nitric oxide by neutrophils and monocytes from cirrhotic patients with ascites and hyperdynamic circulation, *Hepatology* **22:**1666–1673.

Lafond-Walker, A., Chen, C. L., Augustine, S., Wu, T. C., Hruban, R. H., and Lowenstein, C. J., 1997, Inducible nitric oxide synthase expression in coronary arteries of transplanted human hearts with accelerated graft arteriosclerosis, *Am. J. Pathol.* **151:**919–925.

Lammas, D. A., Stober, C., Harvey, C. J., Kendrick, N., Panchalingam, S., and Kumararatne, D. S., 1997, ATP-induced killing of mycobacteria by human macrophages is mediated by purinergic P2Z(p2X7) receptors, *Immunity* **7:**433–444.

Lecoanet-Henchoz, S., Gauchat, J. F., Aubry, J. P., Graber, P., Life, P., Paul-Eugene, N., Ferrua, B., Corbi, A. L., Dugas, B., Platerzyberk, C., and Bonnefoy, J. Y., 1995, CD23 regulates monocyte activation through a novel interaction with the adhesion molecules CD11b-Cd18 and CD11c-CD18, *Immunity* **3:**119–125.

Leibovich, S. J., Polverini, P. J., Fong, T. W., Harlow, L. A., and Koch, A. E., 1994, Production of angiogenic activity by human monocytes requires an L-arginine/nitric oxide-synthase-dependent effector mechanism, *Proc. Natl. Acad. Sci. USA* **91:**4190–4194.

Liu, J., Zhao, M. L., Brosnan, C. F., and Lee, S. C., 1996, Expression of type II nitric oxide synthase in primary human astrocyte and microglie: role of IL-1 beta and IL-1 receptor antogonist, *J. Immunol.* **157:**3569–3576.

Lopez-Guerrero, J. A., and Alonso, M. A., 1997, Nitric oxide production induced by herpes simplex virus type 1 does not alter the course of the infection in human monocytic cells, *J. Gen. Virol.* **78:**1977–1980.

Lopez-Guerrero, J. A., Rayet, B., Tuynder, M., Rommelaere, J., and Dinsart, C., 1997, Constitutive activation of U937 promonocytic cell clones selected for their resistance to parvovirus H-1 infection, *Blood* **89:**1642–1653.

López-Moratalla, N., Calleja, A., Gonzalez, A., Perez-Mediavilla, L. A., Aymerich, M. S., Burrel, M. A., and Santiago, E., 1996, Inducible nitric oxide synthase in monocytes from patients with Graves' disease, *Biochem. Biophys. Res. Commun.* **226:**723–729.

Luoma, J. S., Stralin, P., Marklund, S. L., Hiltunen, T. P., Sarkioja, T., and Ylaherttuala, S., 1998, Expression of extracellular SOD and NOS2 in macrophages and smooth muscle cells in human and rabbit atherosclerotic lesions—Colocalization with epitopes characteristic of oxidized LDL and peroxynitrite-modified proteins, *Arterio. Thromb. Vasc. Biol.* **18:**157–167.

MacMicking, J., Xie, Q. W., and Nathan, C., 1997, Nitric oxide and macrophage function, *Annu. Rev. Immunol.* **15:**323–350.

Magazine, H. I., Liu, Y., Bilfinger, T. V., Fricchione, G. L., and Stefano, G. B., 1996, Morphine-induced conformational changes in human monocytes, granulocytes, and endothelial cells and in invertebrate immunocytes and microglia are mediated by nitric oxide, *J. Immunol.* **156:**4845–4850.

Majano, P. L., García-Monzón, C., López-Cabrera, M., Lara-Pezzi, E., Fernández-Ruiz, E., García-Iglesias, C., Borque, M. J., and Moreno-Otero, R., 1998, Inducible nitric oxide synthase expression in chronic viral hepatitis. Evidence for a virus-induced gene upregulation, *J. Clin. Invest.* **101:**1343–1352.

Malinski, T., Radomski, M. W., Taha, Z., and Moncada, S., 1993, Direct electrochemical measurement of nitric oxide released from human platelets, *Biochem. Biophys. Res. Commun.* **194:**960–965.

Mannick, E. E., Bravo, L. E., Zarama, G., Realpe, J. L., Zhang, X. J., Ruiz, B., Fortham, E. T., Mera, R., Miller, M. J., and Correa, P., 1996, Inducible nitric oxide synthase, nitrotyrosine, and apoptosis, in Helicobacter pylori gastritis: effects of antibiotics and antioxidants, *Cancer Res.* **56:**3238–3243.

Mannick, J. B., Asano, K., Izumi, K., Kieff, E., and Stamler J. S., 1994, Nitric oxide produced by human B lymphocytes inhibits apoptosis and Epstein–Barr virus reactivation, *Cell* **79:**1137–1146.

Mannick, J. B., Miao, X. Q., and Stamler, J. S., 1997, Nitric oxide inhibits Fas-induced apoptosis, *J. Biol. Chem.* **272:**24125–24128.

Martin, J. H., and Edwards, S. W., 1993, Changes in mechanisms of monocyte/macrophage-mediated cytotoxicity during culture. Reactive oxygen intermediates are involved in monocyte-mediated cytotoxicity, whereas reactive nitrogen intermediates are employed by macrophages in tumor cell killing, *J. Immunol.* **150:**3478–3486.

Martin, J. H., and Edwards, S. W., 1994, Interferon-gamma enhances monocyte cytotoxicity via enhanced reactive oxygen intermediate production. Absence of an effect on macrophage cytotoxicity is due to failure to enhance reactive nitrogen intermediate production, *Immunology* **81:**592–597.

Masini, E., Mugnai, L., Foschi, M., Laffi, G., Gentilini, P., and Mannaioni, R. F., 1995, Changes in the production of nitric oxide and superoxide by inflammatory cells in liver cirrhosis, *Int. Arch. Allergy Immunol.* **107:**197–198.

Mautino, G., Paul-Eugene, N., Chanez, P., Vignola, A. M., Kolb, J. P., Bousquet, J., and Dugas, B., 1994, Heterogeneous spontaneous and interleukin-4-induced nitric oxide production by human monocytes, *J. Leukoc. Biol.* **56:**15–20.

McDermott, C. D., Gavita, S. M., Shennib, H., and Giaid, A., 1997, Immunohistochemical localization of nitric oxide synthase and the oxidant peroxynitrite in lung transplant recipients with obliterative bronchiolitis, *Transplantation* **64:**270–274.

McInnes, I. B., Leung, B. P., Field, M., Wei, X. Q., Huang, F.-P., Sturrock, R. D., Kinninmonth, A., Weidner, J., Mumford, R., and Liew, F. Y., 1996, Production of nitric oxide in the synovial membrane of rheumatoid and osteoarthritis patients, *J. Exp. Med.* **184:**1519–1524.

McLachlan, J. A., Serkin, C. D., and Bakouche, O., 1996, Dehydroepiandrosterone modulation of lipopolysaccharide-stimulated monocyte cytotoxicity, *J. Immunol.* **156:**328–335.

Michel, T., and Feron, O., 1997, Nitric oxide synthases—Which, where, how, and why, *J. Clin. Invest.* **100:**2146–2152.

Middleton, S. J., Cuthbert, A. W., Shorthouse, M., and Hunter, J. O., 1993a, Nitric oxide affects mammalian distal colonic smooth muscle by tonic neural inhibition, *Br. J. Pharmacol.* **108:**974–979.

Middleton, S. J., Shorthouse, M., and Hunter, J. O., 1993b, Relaxation of distal colonic circular smooth muscle by nitric oxide derived from human leucocytes, *Gut* **34:**814–817.

Moilanen, E., Moilanen, T., Knowles, R., Charles, I., Kadoya, Y., al-Saffar, N., Revell, P. A., and Moncada, S., 1997, Nitric oxide synthase is expressed in human macrophages during foreign body inflammation, *Am. J. Pathol.* **150:**881–887.

Moncada, S., and Higgs, A., 1993, The L-arginine–nitric oxide pathway, *N. Engl. J. Med.* **329:**2002–2012.

Muñoz-Fernández, M. A., Fernández, M. A., and Fresno, M., 1992, Activation of human macrophages for the killing of intracellular *Trypanosoma cruzi* by TNF-alpha and IFN-gamma through a nitric oxide-dependent mechanism, *Immunol. Lett.* **33:**35–40.

Murray, H. W., and Teitelbaum, R. F., 1992, L-Arginine-dependent reactive nitrogen intermediates and the antimicrobial effect of activated human mononuclear phagocytes, *J. Infect. Dis.* **165:**513–517.

Myatt, L., Eis, A. L. W., Brockman, D. E., Kossenjans, W., Greer, I., and Lyall, F., 1997, Inducible (type II) nitric oxide synthase in human placental villous tissue of normotensive, pre-eclamptic and intrauterine growth-restricted pregnancies, *Placenta* **18:**261–268.

Naotunne, T. S., Karunaweera, N. D., Mendis, K. N., and Carter, R., 1993, Cytokine-mediated inactivation of malarial gametocytes is dependent on the presence of white blood cells and involves reactive nitrogen intermediates, *Immunology* **78:**555–562.

Nathan, C. and Xie, Q.-W., 1994, Regulation of biosynthesis of nitric oxide, *J. Biol. Chem.* **269:**13725–13728.

Nicholson, S., Bonecini-Almeida, M. D. G., Silva, L. E., Jr., Nathan, C., Xie, Q.-W., Mumford, R., Weidner, J. R., Calaycay, J., Geng, J., Boechat, N., Linhares, C., Rom, W., and Ho, J. L., 1996, Inducible nitric oxide synthase in pulmonary alveolar macrophages from patients with tuberculosis, *J. Exp. Med.* **183:**2293–2302.

Nozaki, Y., Hasegawa, Y., Ichiyama, S., Nakashima, I., and Shimokata, K., 1997, Mechanism of nitric oxide-dependent killing of *Mycobacterium bovis* BCG in human alveolar macrophages, *Infect. Immun.* **65:**3644–3647.

Ochoa, J. B., Udekwu, A. O., Billiar, T. R., Curran, R. D., Cerra, F. B., Simmons, R. L., and Peitzman, A. B., 1991, Nitrogen oxide levels in patients after trauma and during sepsis, *Ann. Surg.* **214:**621–626.

Ochoa, J. B., Curti, B., Peitzman, A. B., Simmons, R. L., Billiar, T. R., Hoffman, R., Rault, R., Longo, D. L., Urba, W. J., and Ochoa, A. C., 1992, Increased circulating nitrogen oxides after human tumor immunotherapy: Correlation with toxic hemodynamic changes, *J. Natl. Cancer Inst.* **84:**864–867.

Padgett, E. L., and Pruett, S. B., 1992, Evaluation of nitrite production by human monocyte-derived macrophages, *Biochem. Biophys. Res. Commun.* **286:**775–781.

Paul-Eugene, N., Kolb, J. P., Damais, C., Yamaoka, K., and Dugas, B., 1994, Regulatory role of nitric oxide in the IL-4-induced IgE production by normal human peripheral blood mononuclear cells, *Lymphokine Cytokine Res.* **13:**287–293.

Paul-Eugene, N., Kolb, J. P., Sarfati, M., Arock, M., Ouaaz, F., Debre, P., Mossalayi, D. M., and Dugas, B., 1995a, Ligation of CD23 activates soluble guanylate cyclase in human monocytes via an L-arginine-dependent mechanism, *J. Leukoc. Biol.* **57:**160–167.

Paul-Eugene, N., Mossalayi, D., Sarfati, M., Yamaoka, K., Aubry, J. P., Bonnefoy, J. Y., Dugas, B., and Kolb, J. P., 1995b, Evidence for a role of Fc epsilon RII/CD23 in the IL-4-induced nitric oxide production by normal human mononuclear phagocytes, *Cell. Immunol.* **163:**314–318.

Paul-Eugene, N., Pene, J., Bousquet, J., and Dugas, B., 1995c, Role of cyclic nucleotides and nitric oxide in blood mononuclear cell IgE production stimulated by IL-4, *Cytokine* **7:**64–69.

Perez-Mediavilla, L. A., Lopez-Zabalza, M. J., Calonge, M., Montuenga, L., López-Moratalla, N., and Santiago, E., 1995, Inducible nitric oxide synthase in human lymphomononuclear cells activated by synthetic peptides derived from extracellular matrix proteins, *FEBS Lett.* **357:**121–124.

Perez-Perez, G. I., Shepherd, V. L., Morrow, J. D., and Blaser, M. J., 1995, Activation of human THP-1 cells and rat bone marrow-derived macrophages by *Helicobacter pylori* lipopolysaccharide, *Infect. Immun.* **63:**1183–1187.

Perkins, D. J., St. Clair, W. E., Misukonis, M. A., and Weinberg, J. B., 1998, Reduction of NOS2 overexpression in rheumatoid arthritis patients treated with anti-TNF-alpha monoclonal antibody (cA2), *Arth. Rheum.* **41:**2205–2210.

Petit, J. F., Phan-Bich, L., Lemaire, G., Martinache, C., and Lopez, M., 1993, During their differentiation into macrophages, human monocytes acquire cytostatic activity independent of NO and TNF alpha, *Res. Immunol.* **144:**277–280.

Pietraforte, D., Tritarelli, E., Testa, U., and Minetti, M., 1994, Gp120 HIV envelope glycoprotein increases the production of nitric oxide in human monocyte-derived macrophages, *J. Leukoc. Biol.* **55:**175–182.

Polack, B., Pernod, G., Barro, C., and Doussiere, J., 1997, Role of oxygen radicals in tissue factor induction by endotoxin in blood monocytes, *Haemostasis* **27:**193–200.

Radomski, M. W., Palmer, R. M., and Moncada, S., 1990, Characterization of the L-arginine:nitric oxide pathway in human platelets, *Br. J. Pharmacol.* **101:**325–328.

Rajora, N., Ceriani, G., Catania, A., Star, R. A., Murphy, M. T., and Lipton, J. M., 1996, Alpha-MSH production, receptors, and influence on neopterin in a human monocyte/macrophage cell line, *J. Leukoc. Biol.* **59:**248–253.

Reiling, N., Ulmer, A. J., Duchrow, M., Ernst, M., Flad, H.-D., and Hauschildt, S., 1994, Nitric oxide synthase: mRNA expression of different isoforms in human monocytes/macrophages, *Eur. J. Immunol.* **24:**1941–1944.

Reiling, N., Kroncke, R., Ulmer, A. J., Gerdes, J., Flad, H.-D., and Hauschildt, S., 1996, Nitric oxide synthase—Expression of the endothelial, Ca^{2+}/calmodulin-dependent in human B and T lymphocytes, *Eur. J. Immunol.* **26:**511–516.

Roman, V., Dugas, N., Abadie, A., Amirand, C., Zhao, H., Dugas, B., and Kolb, J. P., 1997, Characterization of a constitutive type III nitric oxide synthase in human U937 monocytic cells—Stimulation by soluble CD23, *Immunology* **91:**643–648.

Saha, D. C., Astiz, M. E., Lin, R. Y., Rackow, E. C., and Eales, L. J., 1997, Monophosphoryl lipid A stimulated up-regulation of nitric oxide synthase and nitric oxide release by human monocytes *in vitro*, *Immunopharmacology* **37:**175–184.

Sakai, N., and Milstien, S., 1993, Availability of tetrahydrobiopterin is not a factor in the inability to detect nitric oxide production by human macrophages, *Biochem. Biophys. Res. Commun.* **193:**378–383.

Sakurai, H., Kohsaka, H., Liu, M. F., Higashiyama, H., Hirata, Y., Kanno, K., Saito, I., and Miyasaka, N., 1995, Nitric oxide production and inducible nitric oxide synthase expression in inflammatory arthritides, *J. Clin. Invest.* **96:**2357–2363.

Salvemini, D., de Nucci, G., Gryglewski, R. J., and Vane, J. R., 1989, Human neutrophils and mononuclear cells inhibit platelet aggregation by releasing a nitric oxide-like factor, *Proc. Natl. Acad. Sci. USA* **86:**6328–6332.

Schmidt, H. H. H. W., Seifert, R., and Böhme, E., 1989, Formation and release of nitric oxide from human neutrophils and HL-60 cells induced by a chemotactic peptide, platelet activating factor and leukotriene B_4, *FEBS Lett.* **244:**357–360.

Schneemann, M., Schoedon, G., Hofer, S., Blau, N., Guerrero, L., and Schaffner, A., 1993, Nitric oxide synthase is not a constituent of the antimicrobial armature of human mononuclear phagocytes, *J. Infect. Dis.* **167:**1358–1363.

Schneemann, M., Schoedon, G., Linscheid, P., Walter, R., Blau, N., and Schaffner, A., 1997, Nitrite generation in interleukin-4-treated human macrophage cultures does not involve the nitric oxide synthase pathway, *J. Infect. Dis.* **175:**130–135.

Seitzer, U., Scheeltoellner, D., Toellner, K. M., Reiling, N., Haas, H., Galle, J., Flad, H.-D., and Gerdes, J., 1997, Properties of multinucleated giant cells in a new *in vitro* model for human granuloma formation, *J. Pathol.* **182:**99–105.

Sharara, A. I., Perkins, D. J., Misukonis, M. A., Chan, S. U., Dominitz, J. A., and Weinberg, J. B., 1997, Interferon-alpha activation of human mononuclear cells *in vitro* and *in vivo* for nitric oxide synthase type 2 mRNA and protein expression. Possible relationship of induced NOS2 to the anti-hepatitis C effects of IFN-α *in vivo*, *J. Exp. Med.* **186:**1495–1502.

Sherman, M. P., Loro, M. L., Wong, V. Z., and Tashkin, D. P., 1991, Cytokine- and *Pneumocystis carinii*-induced L-arginine oxidation by murine and human pulmonary alveolar macrophages, *J. Protozool.* **38:**234S–236S.

Siedlar, M., Marcinkiewicz, J., and Zembala, M., 1995, MHC class I and class II determinants and some adhesion molecules are engaged in the regulation of nitric oxide production *in vitro* by human monocytes stimulated with colon carcinoma cells, *Clin. Immunol. Immunopathol.* **77**:380–384.

Singer, I. I., Kawka, D. W., Scott, S., Weidner, J. R., Mumford, R. A., Riehl, T. E., and Stenson, W. F., 1996, Expression of inducible nitric oxide synthase and nitrotyrosine in colonic epithelium in inflammatory bowel disease, *Gastroenterology* **111**:871–885.

Snell, J. C., Chernyshev, O., Gilbert, D. L., and Colton, C. A., 1997, Polyribonucleotides induce nitric oxide production by human monocyte-derived macrophages, *J. Leukoc. Biol.* **62**:369–373.

Stamler, J. S., Singel, D. J., and Loscalzo, J., 1992, Biochemistry of nitric oxide and its redox-activated forms, *Science* **258**:1898–1902.

St. Clair, E. W., Wilkinson, W. E., Lang, T., Sanders, L., Misukonis, M. A., Gilkeson, G. S., Pisetsky, D. S., Granger, D. L., and Weinberg, J. B., 1996, Increased expression of blood mononuclear cell nitric oxide synthase type 2 in rheumatoid arthritis patients, *J. Exp. Med.* **184**:1173–1178.

Stefano, G. B., Liu, Y., and Goligorsky, M. S., 1996, Cannabinoid receptors are coupled to nitric oxide release in invertebrate immunocytes, microglia, and human monocytes, *J. Biol. Chem.* **271**:19238–19242.

Steube, K. G., Teepe, D., Meyer, C., Zaborski, M., and Drexler, H. G., 1997, A model system in haematology and immunology—The human monocytic cell line Mono-Mac-1, *Leuk. Res.* **21**:327–335.

Summersgill, J. T., Powell, L. A., Buster, B. L., Miller, R. D., and Ramirez, J. A., 1992, Killing of *Legionella pneumophila* by nitric oxide in gamma-interferon-activated macrophages, *J. Leukoc. Biol.* **52**:625–629.

ter Steege, J., Buurman, W., Arends, J. W., and Forget, P., 1997, Presence of inducible nitric oxide synthase, nitrotyrosine, CD68, and CD14 in the small intestine in celiac disease, *Lab. Invest.* **77**:29–36.

Thomsen, L. L., Lawton, F. G., Knowles, R. G., Beesley, J. E., Riveros-Moreno, V., and Moncado, S., 1994, Nitric oxide synthase in human gynecological cancer, *Cancer Res.* **54**:1352–1354.

Thomsen, L. L., Miles, D. W., Happerfield, L., Bobrow, L. G., Knowles, R. G., and Moncada, S., 1995, Nitric oxide synthase activity in human breast cancer, *Br. J. Cancer* **72**:41–44.

Tracey, W. R., Xue, C., Klinghofer, V., Barlow, J., Pollock, J. S., Forstermann, U., and Johns, R. A., 1994, Immunochemical detection of inducible NO synthase in human lung, *Am. J. Physiol.* **266**:L722–L727.

Tufano, M. A., Rossano, F., Catalanotti, P., Liguori, G., Marinelli, A., Baroni, A., and Marinelli, P., 1994, Properties of *Yersinia enterocolitica* porins: Interference with biological functions of phagocytes, nitric oxide production and selective cytokine release, *Res. Microbiol.* **145**:297–307.

Vial, T., and Descotes, J., 1995, Immune-mediated side-effects of cytokines in humans, *Toxicology* **105**:31–57.

Vitek, M. P., Snell, J., Dawson, H., and Colton, C. A., 1997, Modulation of nitric oxide production in human macrophages by apolipoprotein-E and amyloid-beta peptide, *Biochem. Biophys. Res. Commun.* **240**:391–394.

Vouldoukis, I., Riverosmoreno, V., Dugas, B., Ouaaz, F., Becherel, P., Debre, P., Moncada, S., and Mossalayi, M. D., 1995, The killing of *Leishmania major* by human macrophages is mediated by nitric oxide induced after ligation of the Fc-epsilon-RII/CD23 surface antigen, *Proc. Natl. Acad. Sci. USA* **92**:7804–7808.

Vouldoukis, I., Becherel, P. A., Riverosmoreno, V., Arock, M., Dasilva, O., Debre, P., Maizer, D., and Mossalayi, M. D., 1997, Interleukin-10 and interleukin-4 inhibit intracellular killing of *Leishmania infantum* and *Leishmania major* by human macrophages by decreasing nitric oxide generation, *Eur. J. Immunol.* **27**:860–865.

Wang, C. L., Su, M. H., Chao, T. Y., Shaio, M. F., and Yang, K. D., 1996, When do human macrophages release nitric oxide? Variable effects of certain *in vitro* cultural and *in vivo* resident conditions, *Proc. Natl. Sci. Counc. Repub. China B* **20**:65–70.

Watkins, S. C., Macaulay, W., Turner, D., Kang, R., Rubash, H. E., and Evans, C. H., 1997, Identification of inducible nitric oxide synthase in human macrophages surrounding loosened hip prostheses, *Am. J. Pathol.* **150:**1199–1206.

Weinberg, J. B., Misukonis, M. A., Shami, P. J., Mason, S. N., Sauls, D. L., Dittman, W. A., Wood, E. R., Smith, G. K., McDonald, B., Bachus, K. E., Haney, A. F., and Granger, D. L., 1995, Human mononuclear phagocyte inducible nitric oxide synthase (NOS2). Analysis of NOS2 mRNA, NOS2 protein, biopterin, and nitric oxide production by blood monocytes and peritoneal macrophages, *Blood* **86:**1184–1195.

Weyand, C. M., Wagner, A. D., Bjornsson, J., and Goronzy J. J., 1996, Correlation of the topographical arrangement and the functional pattern of tissue-infiltrating macrophages in giant cell arteritis, *J. Clin. Invest.* **98:**1642–1649.

Wickramasinghe, S. N., and Hasan, R., 1993, Possible role of macrophages in the pathogenesis of ethanol-induced bone marrow damage, *Br. J. Haematol.* **83:**574–579.

Wilcox, J. N., Subramanian, R. R., Sundell, C. L., Tracey, W. R., Pollock, J. S., Harrison, D. G., and Marsden, P. A., 1997, Expression of multiple isoforms of nitric oxide synthase in normal and atherosclerotic vessels, *Arterio. Thromb. Vasc. Biol.* **17:**2479–2488.

Wildhirt, S. M., Dudek, R. R., Suzuki, H., and Bing, R. J., 1995, Involvement of inducible nitric oxide synthase in the inflammatory process of myocardial infarction, *Int. J. Cardiol.* **50:**253–261.

Zarlingo, T. J., Eis, A. L. W., Brockman, D. E., Kossenjans, W., and Myatt, L., 1997, Comparative localization of endothelial and inducible nitric oxide synthase isoforms in haemochorial and epitheliochorial placentae, *Placenta* **18:**511–520.

Zembala, M., Siedlar, M., Marcinkiewicz, J., and Pryjma, J., 1994, Human monocytes are stimulated for nitric oxide release in vitro by some tumor cells but not by cytokines and lipopolysaccharide, *Eur. J. Immunol.* **24:**435–439.

Zhou, A. Q., Chen, Z. F., Rummage, J. A., Jiang, H., Kolosov, M., Kolosova, I., Stewart, C. A., and Leu, R. W., 1995, Exogenous interferon-gamma induces endogenous synthesis of interferon-alpha and -beta by murine macrophages for induction of nitric oxide synthase, *J. Interferon Cytokine Res.* **15:**897–904.

Zinetti, M., Fantuzzi, G., Delgado, R., Di Santo, E., Ghezzi, P., and Fratelli, M., 1995, Endogenous nitric oxide production by human monocytic cells regulates LPS-induced TNF production, *Eur. Cytokine Network* **6:**45–48.

CHAPTER 7

Cardiovascular Actions of Nitric Oxide

DARYL D. REES

1. Introduction

As described in Chapter 2, the search for the identity of the labile vasodilator endothelium-derived relaxing factor (EDRF) (Furchgott and Zawadzki, 1980) helped lead to the discovery in the vasculature of an enzyme, nitric oxide (NO) synthase, that generates NO from one of the terminal guanidino nitrogen atoms of the semiessential amino acid L-arginine (Palmer *et al.*, 1988a,b). Because NO is soluble in both lipid and water and diffuses freely within and between cells, it can transmit signals between cells or from one part of a cell to another. Furthermore, NO has a physiological half-life of only a few seconds and reacts with oxygen free radicals (Moncada *et al.*, 1991). In the blood, hemoglobin inactivates NO by binding it to form nitrosohemoglobin and by catalyzing the oxidation of NO to nitrite and nitrate, resulting in the formation of methemoglobin (Wennmalm *et al.*, 1992). NO can also form complexes with other heme-containing proteins such as soluble guanylyl cyclase, accounting for many of its physiological actions (Ignarro, 1990), or can react with plasma constituents including thiols, albumin, and a variety of other proteins (Stamler *et al.*, 1992) (see also Chapter 3). Thus, the fate and actions of NO in the cardiovascular system depend on its local environment.

As detailed in Chapter 4, the NO synthases have three isoforms: neuronal NO synthase (nNOS, NOS1), immunologically induced NO synthase (iNOS, NOS2), and endothelial NO synthase (eNOS, NOS3). The isoforms of NO synthase are large (125–155 kDa) dimeric enzymes, containing a reductase domain that shuttles electrons from NADPH, FAD, and FMN to the oxygenase domain where oxidation of L-arginine occurs (Bredt *et al.*, 1991; White and Marletta, 1992; Stuehr, 1997),

---

DARYL D. REES • Centre for Clinical Pharmacology, University College London, London, WC1E 6JJ, United Kingdom.

Nitric Oxide and Infection, edited by Fang. Kluwer Academic / Plenum Publishers, New York, 1999.

forming NO• and L-citrulline as coproducts. The eNOS and nNOS isoforms are regulated by calmodulin and require an elevation of intracellular calcium for activation. In contrast, the iNOS isoform binds calmodulin tightly so that its activity appears functionally independent of the intracellular calcium concentration. The activities of eNOS and nNOS may also be affected by phosphorylation or various posttranslational modifications that regulate localization within the cell (Michel and Feron, 1997; Stuehr, 1997).

All NOS isoforms may be inhibited by naturally occurring or synthetic analogues of arginine that compete for the arginine-binding site. Endogenous analogues include N^G-monomethyl-L-arginine (L-NMMA) (Hibbs *et al.*, 1987) and N^G, N^G-dimethylarginine (asymmetric dimethylarginine, ADMA) (MacAllister *et al.*, 1994), and synthetic analogues include N^G-nitro-L-arginine methyl ester (L-NAME) (Rees *et al.*, 1990a) and *N*-[3-(aminomethyl)benzyl] acetamidine (1400W) (Garvey *et al.*, 1997). The most widely studied of these, L-NMMA, is equipotent for all three isoforms and also competes with arginine for entry into the cell via the y^+ cationic amino acid transporter system (Bogle *et al.*, 1995). Although there is close homology between NOS isoforms, selective inhibitors have been developed with selectivity for iNOS over eNOS conferred by substituting the guanidino function of arginine with an amidine group (Garvey *et al.*, 1997). L-NMMA and other inhibitors of NOS have been used extensively as probes to characterize the physiological and pathophysiological roles of NO in the cardiovascular system. It is now evident that NO generated by the endothelium, nerves, and smooth muscle plays a fundamental role in the physiology and pathophysiology of vessel tone and blood pressure.

2. Physiological Functions of NO in the Vasculature

2.1. NO as a Vasodilator

In 1980, Furchgott and Zawadzki demonstrated that the relaxation of rabbit aorta in response to acetylcholine (ACh) is entirely dependent on the presence of an intact endothelial cell layer. NO was later shown to account for this endothelium-dependent relaxation, and the release of NO from the endothelium has since been demonstrated in arteries, arterioles, veins, and venules from a wide range of species including humans, both *in vitro* and *in vivo* (Ignarro *et al.*, 1987; Palmer *et al.*, 1988b; Furchgott and Vanhoutte, 1989; Rees *et al.*, 1989a,b; Vallance *et al.*, 1989a,b). Furthermore, the vasorelaxant properties of many hormones and autocoids including bradykinin, substance P, and serotonin have been shown to be endothelium dependent (Furchgott, 1983; Moncada *et al.*, 1991), as is "flow-dependent dilatation" (Griffith *et al.*, 1987). Emerging evidence suggests that the ACh-induced relaxation of resistance vessels of many tissues is only partially mediated by NO (Nagao *et al.*, 1992; Parsons *et al.*, 1994). Indeed, ACh does not induce endothelium-dependent relaxation of the aorta in mice with a targeted deletion in the eNOS gene (Huang *et al.*, 1995), but does induce vasodilatation of the mesenteric bed and a fall in blood pressure (Rees *et al.*, unpublished data). NO dependency of ACh-induced vasorelaxation appears to decrease with reduction in vessel size, a factor that may explain conflicting data in the literature. Whether other "endothelium-dependent" mediators exhibit these characteristics remains to be established.

Pharmacological inhibition of NOS with substrate analogues such as L-NMMA not only impairs the response to "endothelium-dependent" dilators, particularly in conduit vessels, but also causes an endothelium-dependent vasoconstriction of isolated arteries, arterioles, and to a lesser extent, veins in many species including humans (Rees *et al.*, 1989a,b; Vallance *et al.*, 1989a,b; Calver *et al.*, 1993). Similarly, administration of L-NMMA *in vivo* causes widespread vasoconstriction and elevation in blood pressure. Examination of the aorta *ex vivo* from animals treated with L-NMMA shows a reduced release of NO from this tissue (Rees *et al.*, 1989b). In humans, infusion of L-NMMA into the brachial artery of healthy volunteers reduces resting blood flow by about 40–50% (Vallance *et al.*, 1989a), and its administration intravenously increases systemic vascular resistance and blood pressure (Haynes *et al.*, 1993). This indicates that continuous generation of NO maintains resistance vessels in a dilated state, and that the cardiovascular system as a whole is in a state of active vasodilatation in both animals and humans (Rees *et al.*, 1989b; Moncada *et al.*, 1991; Calver *et al.*, 1993; Navarro *et al.*, 1994).

Mice with a targeted deletion in the eNOS gene are hypertensive (Fig. 1), indicating that absence of eNOS activity throughout development is not fully compensated for by changes in sympathetic or other activity that regulates blood pressure and vascular tone, and possibly suggesting that NO controls the baroreceptor set point (Huang *et al.*, 1995). Treatment with L-NMMA, even at high doses, has only a small effect on the blood pressure of eNOS mutant mice but elevates the blood pressure of wild-type animals to levels approximating those of the eNOS knockout animals (Rees *et al.*, unpublished data). This confirms that the generalized vasoconstriction and elevation in blood pressure produced by L-NMMA is related largely to inhibition of eNOS.

Although the release of NO by ACh appears to decrease from conduit to resistance vessels, this is not the case for basal NO release. Basal NO-mediated dilatation is somewhat less in veins, possibly reflecting differences in the local chemical or physical environment between arteries and veins (Calver *et al.*, 1993). Basal NO-mediated dilatation appears to be greater in women than in men and is further enhanced during normal pregnancy (Williams *et al.*, 1997), possibly because of an estrogen-dependent increase in expression of eNOS and nNOS.

The endothelial layer of the vascular wall may be viewed as a signal transducer, detecting physical and chemical stimuli and altering NO synthesis

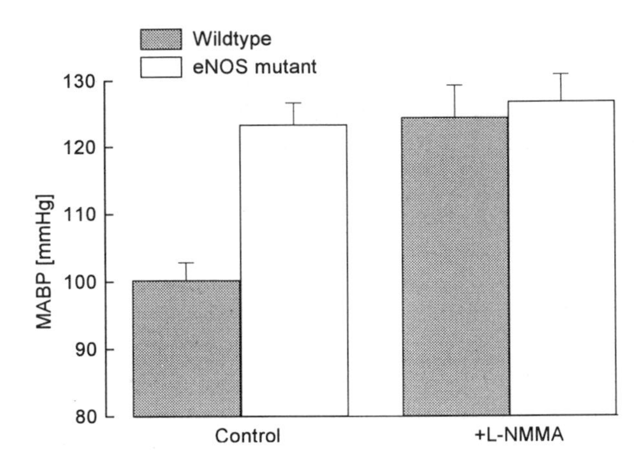

FIGURE 1. Mean arterial blood pressure is higher in mice lacking eNOS. Results are shown for eNOS mutant mice and their wild-type counterparts. Administration of L-NMMA (100 mg/kg per h) over 12 h has no effect on eNOS mutant mice but elevates blood pressure of wild-type mice to that of the eNOS mutant. Each point is the mean ± S.E.M. of five to seven animals.

accordingly. The release of NO proceeds at a basal, background rate via activation of eNOS by the normal resting calcium concentration within the endothelial cell. This is increased further by shear stress resulting from blood flowing across the endothelium. The shear deformation of the endothelial cell directly opens ion channels leading to the extrusion of potassium, allowing entry of calcium into the cell (Lansman *et al.*, 1987; Olesen *et al.*, 1988). In addition, vasoactive mediators such as ACh, bradykinin, and serotonin act on specific cell surface receptors. Either type of stimulus elevates intracellular calcium, thus activating eNOS to generate increased amounts of NO (Fig. 2). The active vasodilatation is counterbalanced by active vasoconstriction mediated by the sympathetic nervous system and circulating vasoconstrictor agents (Vallance, 1996). In addition, certain vasoconstrictor agents such as norepinephrine can stimulate the release of NO to modify the overall level of constriction produced (Cocks and Angus,1983).

Although ACh has been used extensively to investigate endothelium-dependent relaxation, its physiological relevance is unclear, as neuronally derived ACh is believed to be destroyed very rapidly *in vivo* by pseudocholinesterases. However, it has been suggested that ACh can be generated within the endothelial cell and hence may have a role in endothelial NO generation (Parnavelas *et al.*, 1985). Other mediators, such as serotonin released from aggregating platelets or bradykinin and substance P generated locally, may have a physiological role by stimulating NO release from the endothelium and altering local blood flow.

NO formed by the endothelial cell diffuses to the underlying smooth muscle where it binds to the heme moiety of the soluble guanylyl cyclase, causing a

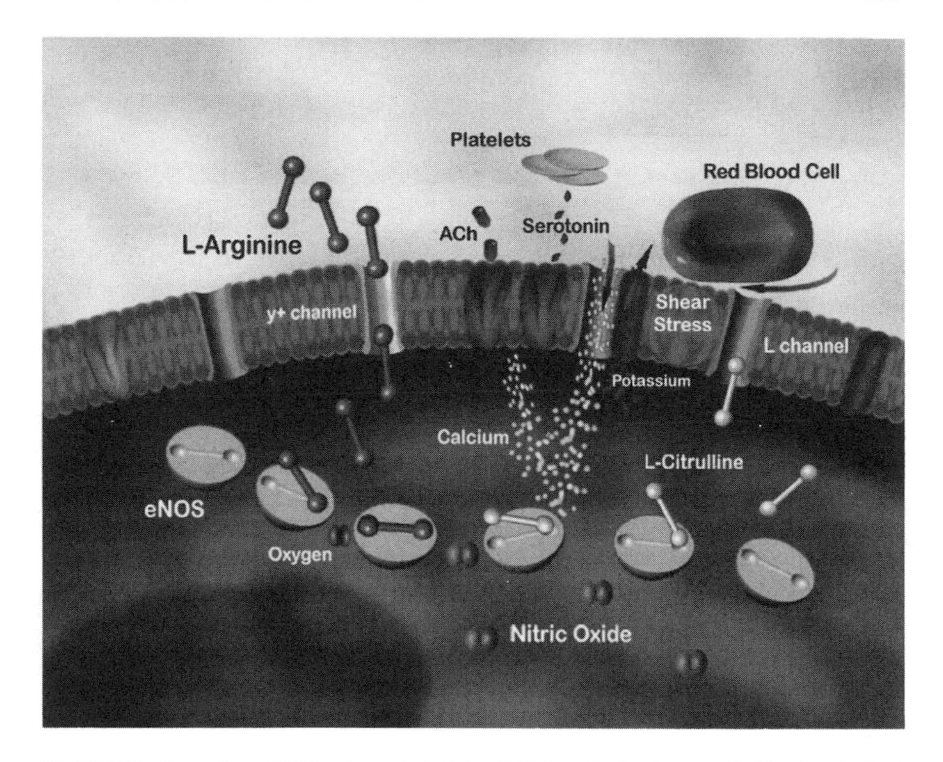

FIGURE 2. Generation of NO by the endothelial cell. NO production is controlled by the physical action of pulsatile flow and shear stress of circulating blood and by the chemical action of vasoactive mediators such as serotonin derived from platelets and acetylcholine (ACh). The resultant elevation of intracellular calcium activates the endothelial NO synthase (eNOS). L-Arginine enters the cell via the y^+ basic amino acid transporter system and combines with the substrate binding site of eNOS. Molecular oxygen is also a substrate for this reaction, which proceeds via the formation of N^G-L-hydroxyarginine and results in the formation of NO with L-citrulline as the coproduct (see text for further details).

conformational change in the enzyme. This increases the activity of the enzyme ~400-fold, and the resultant enhanced synthesis of cyclic guanosine monophosphate (cGMP) in the smooth muscle cell leads to activation of cGMP-dependent protein kinases, calcium sequestration, and vasodilatation (Ignarro, 1990).

Certain blood vessels are innervated by nitrergic nerves, which produce NO via the action of nNOS. NO generated in this manner may also contribute to the control of vessel tone, with NO acting as a direct vasodilator or a neuromodulator altering the release of other transmitters (Gustafsson *et al.*, 1990; Toda and Okamura, 1990). In the cerebral circulation, nitrergic nerves may be important in vasoneuronal coupling, i.e., the process by which blood flow is increased to active areas of the brain (Toda, 1995). In the corpus cavernosum, NO released from nitrergic nerves mediates relaxation of the smooth muscle and increases blood flow, leading to penile erection (Rajfer *et al.*, 1992).

2.2. Other Physiological Actions of NO in the Cardiovascular System

In addition to its effects on smooth muscle within the blood vessel, NO also affects blood cells. NO prevents the adhesion of platelets and white cells to endothelium, inhibits the aggregation of platelets, and induces disaggregation of aggregated platelets (Radomski, 1996). Platelets themselves produce NO, which may act as a negative feedback mechanism to inhibit platelet aggregation and adhesion (Radomski *et al.*, 1990a). The antiaggregatory/antiadhesive effects of NO are mediated by activation of soluble guanylyl cyclase and an elevation in cGMP concentrations, which lead to a decrease in intracellular calcium. NO also alters the expression of adhesion molecules (IIb/IIIa and P selectin) on the surface of the endothelium and circulating cells (Radomski, 1996). Basal release of NO can decrease adhesion of polymorphonuclear leukocytes (PMNs) to the vascular endothelium, at least in part by inhibiting endothelial expression of adhesion molecules including intercellular adhesion molecule-1 (ICAM-1) to the neutrophil ligand CD11a/CD18 (Kanwar and Kubes, 1995). At high concentrations, NO inhibits vascular smooth muscle cell growth and may induce apoptosis (programmed cell death). Taken together, these actions of NO are thought to contribute to vascular homeostasis—inhibition of atherogenesis and prevention of vessel occlusion.

Endothelial NOS may also play a role in maintaining microvascular permeability and integrity. Studies have shown that inhibition of eNOS increases microvascular fluid and protein flux. Microbial products and inflammatory mediators such as platelet-activating factor (PAF) acutely increase intestinal vascular permeability that can be augmented by pretreatment with L-NMMA or reversed by concurrent administration of NO donors (Whittle, 1995). The processes by which low levels of NO maintain microvascular integrity could relate to its ability to inhibit leukocyte adhesion to the vascular endothelium by reacting with superoxide ($O_2^{-}\bullet$). In the absence of NO, it has been suggested that $O_2^{-}\bullet$ may activate mast cell degranulation that promotes leukocyte adhesion to the endothelium (Kanwar and Kubes, 1995).

NO may have a role in controlling the function of the heart. When generated by the coronary vasculature, NO increases coronary blood flow supplying the myocardium. Although NOS inhibitors have little effect on basal ventricular contractility, they markedly increase the inotropic effect of the β-agonist isoproterenol (Balligand *et al.*, 1993a). This suggests that β-adrenergic stimulation activates NOS to generate NO, which subsequently attenuates ventricular contractility. Moreover, L-NMMA can inhibit the negative inotropic effects of inflammatory cytokines on isolated papillary muscles *in vitro* (Finkel *et al.*, 1992).

3. Pathophysiology of NO

Given the fundamental role of NO in controlling the cardiovascular system, its implication in a number of cardiovascular disorders has been almost inevitable. A reduction in basal NO synthesis or actions may lead to an increase in vessel tone, elevated blood pressure, and thrombus formation, whereas an overproduction of NO may lead to excessive vasodilatation, hypotension, vascular leakage, and possibly cellular metabolic disruption. Infection or inflammation can lead to both types of disruption of NO production.

3.1. Enhanced Generation of NO

Exposure to certain microbes, microbial products, or cytokines results in the induction of iNOS in phagocytic and other cells. Once expressed, iNOS is fully active even at the low basal concentrations of calcium present in cells. Although initially described as a mechanism of macrophage cytotoxicity important for host defense (Hibbs, *et al.*, 1987; Hibbs, 1992), it is now clear that the capacity to express iNOS in response to infection and specific inflammatory stimuli exists in virtually every cell type. The combination of cytokines and microbial products required to activate iNOS and the subsequent time course of expression vary according to the type of cell, the experimental conditions, and the host species (Fig. 3; Rees *et al.*, 1990b; Salter *et al.*, 1991; Rees, 1995). Gram-negative bacterial endotoxin, gram-positive wall fragments including lipoteichoic acid, and cytokines such as TNFα, IL-1, and IFNγ can induce the enzyme (Rees *et al.*, 1990b; Radomski *et al.*, 1990b) whereas IL-4, IL-8, IL-10, and TGFβ inhibit induction (McCall *et al.*, 1992; Schini *et al.*, 1992); more information regarding cytokine regulation of NO synthesis is provided in Chapter 5. Elevated concentrations of NO can negatively "feed back" to the enzyme and inhibit its activity, providing an additional regulatory mechanism (Assreuy *et al.*, 1993).

Isolated blood vessels treated with endotoxin or certain cytokines show a time-dependent expression of iNOS over several hours, which begins after a lag period of approximately 2 h (depending on the species). This is accompanied by an increase in tissue cGMP levels, a progressive vasorelaxation, and a hyporesponsiveness to vasoconstrictor agents over the same time interval (Fig. 4; Rees *et al.*, 1990b; Stoclet *et al.*, 1995); each of these changes can be prevented by treatment with inhibitors of NOS, inhibitors of protein synthesis (e.g., cycloheximide), or glucocorticoids, but reversal of the vascular disturbances can only be achieved with NOS inhibition (Rees *et al.*, 1990b). Thus, the increased production of NO functionally antagonizes the effect of vasoconstrictor agents and plays a major role in microbial/cytokine-induced vasoplegia. iNOS is expressed in both the endothelium and the smooth muscle layers, although the latter, with its greater mass, is the major source of increased NO (Rees *et al.*, 1990b, Stoclet *et al.*, 1995).

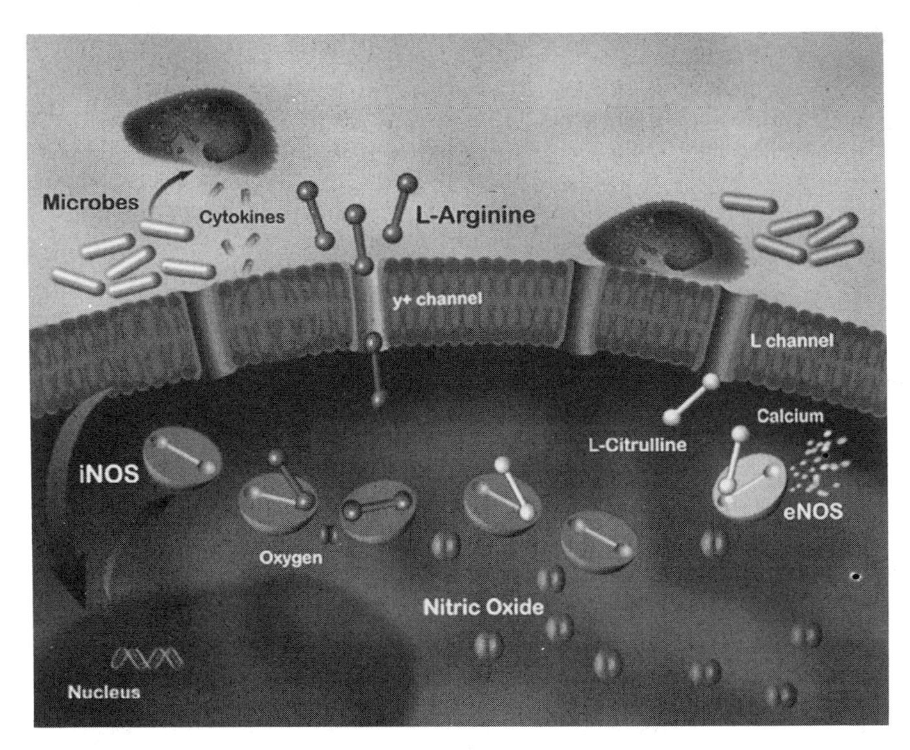

FIGURE 3. Microbial products and/or cytokines activate inducible NO synthase (iNOS) in a wide variety of cells. iNOS-producing cells include vascular endothelial cells (above) and smooth muscle cells. Expression of this enzyme requires *de novo* protein synthesis over several hours and large quantities of NO are produced. The iNOS isoform binds calmodulin tightly so that its activity appears functionally independent of intracellular calcium concentration (see text for further details).

In vivo studies in mice have demonstrated that administration of bacterial endotoxin stimulates the rapid release of TNFα and IL-6, which declines within 2–3 h (Sheehan *et al.,* 1989, Silva *et al.,* 1990; Rees *et al.*, 1998). A similar cytokine profile has been observed in patients with septic shock (Damas *et al.,* 1992) and in a report describing the experimental administration of endotoxin in man (Taveira da Silva *et al.,* 1993) (see also Chapter 13 for a detailed discussion of NO in sepsis). Endotoxin induces the expression of iNOS in tissues including the heart (Salter *et al.*, 1991; Mitchell *et al.,* 1993; Rees *et al.*, 1998), beginning approximately 2–4 h after administration and generally resolving by 24 h (Knowles *et al.*, 1990; Salter *et al.*, 1991; Mitchell *et al.*, 1993; Cunha *et al.*, 1994). iNOS expression is accompanied by an increase in plasma concentrations of nitrite/nitrate over a similar time period in close association with a progressive decline in blood pressure and a significantly reduced vasopressor response to noradrenaline, suggesting a

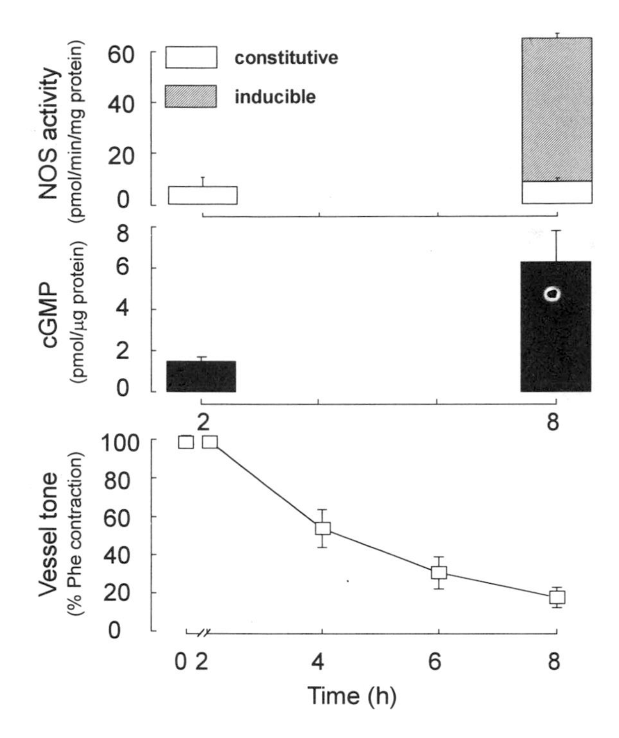

FIGURE 4. LPS activation of iNOS. Endotoxin (LPS from *Salmonella typhi*, 100 ng/ml) activates the expression of the inducible NO synthase (NOS) and elevates cGMP in isolated rat aortae and decreases the tone of the vessels precontracted with phenylphrine (Phe) over a similar time course ($n = 10$). Reproduced with permission from Rees *et al.* (1990b).

similar induction profile in both heart and blood vessels (Fig. 5; Rees *et al.*, 1998). Thus, in the mouse, as in other species (Thiemermann and Vane, 1990; Wright *et al.*, 1992) including humans (Ochoa *et al.*, 1991; Evans *et al.*, 1993), increased generation of NO by iNOS underlies the hypotension and hyporesponsiveness to vasoconstrictor agents in endotoxic shock and may contribute to associated cardiac dysfunction. This is further confirmed by observations in mice lacking iNOS (Fig. 6), which demonstrate a diminished blood pressure response to endotoxin when compared with wild-type animals (MacMicking *et al.*, 1995; Rees *et al.*, 1998).

Gram-positive bacterial cell wall fragments including lipoteichoic acid and peptidoglycan appear to induce a response similar to endotoxin (Zembowicz and Vane, 1992; Cunha *et al.*, 1993). Other bacteria such as *Mycobacterium bovis* (strain BCG) (Stuehr and Marletta, 1987) and heat-inactivated *Corynebacterium parvum* (Billiar *et al.*, 1992; Rees *et al.*, 1995) induce iNOS after a latent period of 1–2 days, and the induction can persist over a period of weeks. *C. parvum* induces a sequential and differential induction of NOS, first occurring in macrophages and

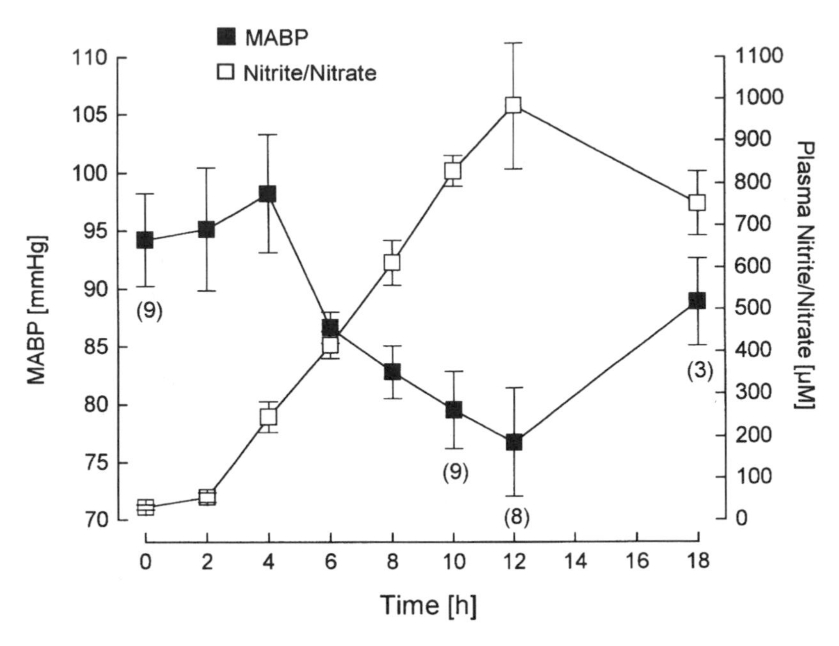

FIGURE 5. Effects of LPS on blood pressure and NO_x production. Endotoxin (*Escherichia coli* O26:B6, 12.5 mg/kg i.v.) induces a fall in mean arterial blood pressure (MABP) and elevates the concentrations of plasma nitrite/nitrate in the conscious mouse. Each point is the mean ± S.E.M. of three to nine animals, where (*n*) represents the number of survivors. Reproduced with permission from Rees *et al.* (1998).

hepatocytes, and subsequently in the spleen, heart, and aorta. This suggests that bacteria are initially sequestered by macrophages and Kupffer cells in the liver, with phagocyte activation and subsequent generation of cytokines stimulating the induction of NOS in other tissues. Induction in these tissues depends on their reponsiveness to different cytokines. In the vasculature, iNOS expression occurs 12–20 days after the administration of *C. parvum*, coinciding with the development of mild but measurable hypotension. In contrast, the time course of changes in plasma nitrite/nitrate predominantly appears to reflect the production of NO by hepatocytes and macrophages, in which NOS activity is greatest. The hypotension observed following administration of *C. parvum* is not as severe as that induced by endotoxin, yet the maximum levels of NOS induction in the vessel wall appear to be greater (Fig. 7) (Rees *et al.*, 1995, 1998), suggesting that the cardiovascular system may be able to adapt and compensate for the slow onset of iNOS expression in the vasculature following *C. parvum* administration.

Administration of live *Pseudomonas aeruginosa* in conscious sheep leads to systemic vasodilatation without an elevation of plasma nitrate/nitrite (Meyer *et al.*, 1992). The ability of *Pseudomonas* to convert nitrite to nitrous oxide (N_2O) might

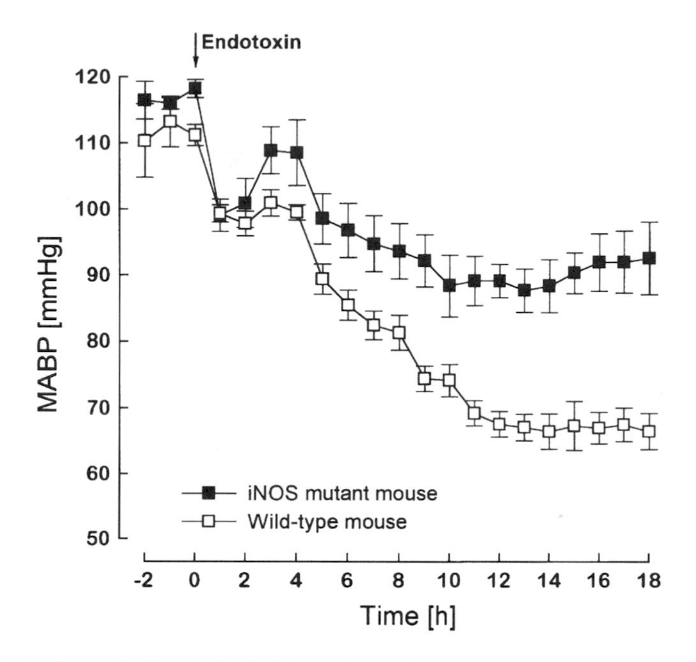

FIGURE 6. Effects of LPS on iNOS-deficient mice. Endotoxin (*Escherichia coli* O26:B6, 3 mg/kg i.v.)-induced fall in mean arterial blood pressure (MABP) in conscious wild-type mice is reduced in mice lacking inducible NO synthase (iNOS). Each point is the mean ± S.E.M. of four to five animals. Reproduced with permission from Rees *et al.* (1998).

conceivably account for this unanticipated phenomenon (Braun and Zumft, 1991). The ability of certain microbes to convert nitrite and nitrate to other products might also explain the lower plasma concentrations of nitrite/nitrate in human septic shock and in animal models challenged with live organisms, when compared with the responses to endotoxin or heat-inactivated organisms. These studies further suggest that plasma concentrations of nitrite/nitrate may not always accurately reflect vascular iNOS activity. On the other hand, administration of live *Streptococcus pyogenes* to mice does result in an elevation of plasma nitrite/nitrate concentrations accompanied by a gradual fall in blood pressure over a period of 48 h, although iNOS expression in the vasculature has yet to be analyzed in this model (Rees *et al.*, 1997). Furthermore, administration of live *Escherichia coli* for 2 h to conscious baboons stimulates the rapid appearance of TNFα and IL-6 in plasma, with levels declining over 24 h in parallel with plasma endotoxin concentrations. Plasma concentrations of nitrite/nitrate are also elevated in this model, and reach a maximum at 12 h in association with a fall in systemic vascular resistance (SVR). High concentrations of plasma nitrate at 48 h are associated with

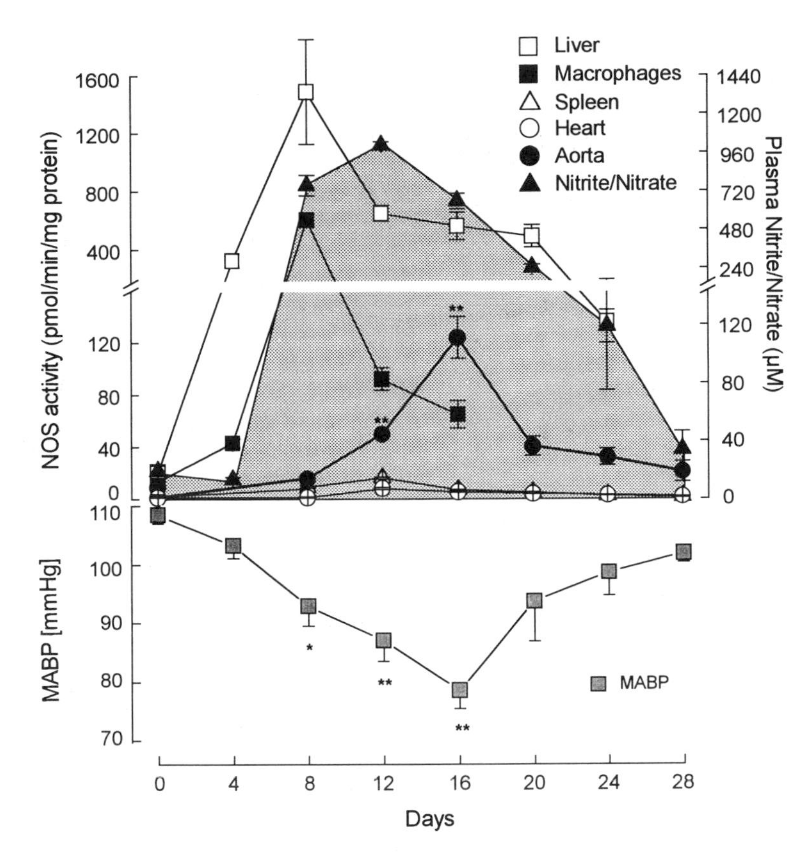

FIGURE 7. Time course of the sequential induction of the inducible NO synthase (iNOS). iNOS expression was measured in liver, macrophages, spleen, heart, and aorta ($n = 3-10$). Elevation of plasma concentrations of nitrite/nitrate ($n = 4$, shaded area) and associated changes in mean arterial blood pressure (MABP; $n = 4-10$) followed a single administration of *Corynebacterium parvum* (100 mg/kg i.p.). Day 0 represents control, untreated group. $^{*}p<0.05$; $^{**}p<0.01$. Reproduced with permission from Rees (1995).

a low SVR and increased mortality, while low plasma nitrate concentations at 48 h are associated with a higher SVR and survival (Redl *et al.*, 1997).

Thus, studies with microbial products or live organisms suggest that the profile of NOS induction *in vivo* may vary depending on the microbial strain, with a time course lasting from days to weeks; this time course may determine the severity of the cardiovascular collapse following infection. These observations provide substantial evidence that overproduction of NO accounts for the vasodilatation characteristic of septic shock (Vallance and Moncada, 1993; Rees *et al.*, 1998) and the hypotension induced by cytokine therapy (Hibbs *et al.*, 1992). Increased NO synthesis can also be seen in anaphylactic and hemorrhagic shock, and may play a role in some local inflammatory conditions of the heart, including acute myocarditis (see Chapter 18), postmyocardial infarction, dilated cardiomyopathy, and heart transplant rejection (de Belder *et al.*, 1995). Expression of iNOS is associated with myocardial depression, measurable experimentally as a decrease in baseline and isoproterenol-stimulated contractility of myocardial preparations (Balligand, *et al.*, 1993b).

Additional vascular and metabolic effects of NO deserve mention here. Microbial products and cytokines lead to the release of procoagulants such as tissue factor and induce the expression of endothelial adhesion molecules leading to increased thrombogenicity and disseminated intravascular coagulation (DIC). Increased generation of NO by the blood vessel may act as a counterbalance, attenuating the hypercoagulable state and protecting the microvasculature (Radomski, 1996). Although microbial products and inflammatory mediators acutely increase microvascular permeability, the expression of iNOS in a variety of tissues following administration of endotoxin *in vivo* can dramatically augment vascular leakage. The processes by which excessive NO produces an increase in microvascular permeability are unclear. Cytoskeletal derangements may be involved (Salzman *et al.*, 1995) and formation of peroxynitrite ($ONOO^-$) from $NO^\bullet$ and $O_2^{-\bullet}$ has been implicated in the cellular injury (Beckman *et al.*, 1990). Indeed, stimuli that lead to induction of iNOS often lead to increased generation of $O_2^{-\bullet}$ and thus formation of $ONOO^-$. Increased generation of $NO^\bullet$, $ONOO^-$, or other NO congeners can suppress cellular respiration (Brown *et al.*, 1993; Lizasoain *et al.*, 1996) and hence oxygen consumption by a direct action on the iron–sulfur centered proteins such as cytochrome *c* oxidase and complexes I–III of the respiratory chain, and inactivate enzymes involved in the citric acid cycle and DNA synthesis (Hibbs *et al.*, 1990). Inhibition of respiration leads to a transition from aerobic to anaerobic glucose metabolism with reduced ATP generation and excessive lactate formation. Reduced oxygen utilization, if persistent, can lead to cell damage or death. Thus, overproduction of NO represents a common mechanism by which cytokine responses to microbial products lead to the excessive vasodilatation, myocardial depression, increased microvascular permeability, reduced oxygen extraction, and multiple organ failure observed in septic shock.

While iNOS expression occurs readily in rodent cells or tissues exposed to microbial products and/or cytokines, the induction of functionally active iNOS has sometimes been difficult to demonstrate in human macrophages *in vitro* and appears to require a different combination of stimuli (Paul-Eugene, *et al.*, 1995) (this topic is discussed in much greater detail in Chapter 6). Nevertheless, there is good evidence for overproduction of NO in several human clinical conditions including septic shock and dilated cardiomyopathy (Ochoa *et al.*, 1991; Evans *et al.*, 1993; Petros *et al.*,1994).

3.2. Reduced Generation of NO

Virtually every cardiovascular risk factor—e.g., hypertension, diabetes, hyperlipidemia, tobacco use, hyperhomocyst(e)inemia—seems to be associated with a reduction in basal or stimulated NO-mediated dilatation. There is also an association between infection or inflammation and acute cardiovascular events (Nieminen *et al.*, 1993; Syrjanen, 1993). An antecedent febrile respiratory infection is a major risk factor for stroke in young and middle-aged adults (Syrjanen *et al.*, 1988), and transient endotoxemia often occurs postoperatively during a period of increased incidence of myocardial infarction and stroke (Baigrie *et al.*, 1993).

The link between infection and a reduction in basal or stimulated NO-mediated dilatation appears to result from "endothelial stunning" by bacterial toxins and inflammatory cytokines, an effect that can persist for several weeks (Bhagat *et al.*, 1996). The endothelial dysfunction appears to be independent of the profound cardiovascular changes observed in septic shock and remains long after the hemodynamic alterations appear to have stabilized. Possible mechanisms include an effect on eNOS mRNA stability, alterations in eNOS itself, alterations of other mediators such as prostaglandins (Yoshizumi *et al.*, 1993; Bhagat *et al.*, 1996), downregulation of eNOS by NO, or increased reaction of NO with $O_2^{-\bullet}$ (Mugge *et al.*, 1991). Endothelial injury and reduced NO production following infection or inflammation may predispose to vasospasm, thrombosis, vessel occlusion, and possibly hypertension, if prolonged.

4. Pharmacology of NO in the Vasculature

4.1. Agents that Increase NO

In situations characterized by the impaired activity or production of NO, it may be desirable to mimic or enhance the physiological generation of NO, or to administer NO itself. When inhaled at 5–80 ppm, NO gas has been shown to reverse persistent pulmonary hypertension of the newborn, pulmonary hypertension induced by hypoxia or after surgery, and chronic pulmonary hypertension (Pepke-Zaba *et al.*, 1991; Frostell *et al.*, 1993). The beneficial effects of NO last throughout the inhalation period, and in some cases persist after termination of treatment.

The adult respiratory distress syndrome (ARDS) represents the pulmonary manifestation of a global inflammatory process, in which widespread induction of iNOS appears to occur in the injured lung and may be an early contributor to its pathogenesis. It is characterized by alveolar edema and a loss of hypoxic pulmonary vasoconstriction, producing a marked increase in ventilation–perfusion mismatch (Singh and Evans, 1997). Inhalation of NO alleviates the pulmonary hypertension and hypoxemia because NO is distributed preferentially to ventilated portions of the lung, increasing local blood flow and improving the ventilation–perfusion ratio. Studies have shown that even very low concentrations of inhaled NO, comparable to those in the atmosphere (~ 0.1 ppm) or generated in the nasal cavity (Lundberg *et al.*, 1994), can be beneficial in ARDS (Gerlach *et al.*, 1993; Rossaint *et al.*, 1993). In the upper respiratory tract, endogenous release of NO appears to be important in maintaining ciliary function and may provide a natural sterilizing system for the mucosa, thus reducing susceptibility to lower respiratory tract infection (see also Chapter 11). This natural barrier is lost in intubated patients and may contribute to their increased susceptibility to infection. It is conceivable that very low concentrations of inhaled NO may provide a means of limiting the susceptibility to infection in these patients.

NO donors (nitrovasodilators) have been in therapeutic use as antianginal agents for over a century. These compounds behave as prodrugs that exert their pharmacological actions after their metabolism to NO by enzymatic and nonenzymatic processes (Feelisch, 1993). Cardiovascular conditions commonly treated with NO donors include stable and unstable angina, coronary vasospasm, myocardial infarction, and congestive heart failure. Scientific interest in the L-arginine:NO pathway has given rise to novel therapeutic indications for NO donors in which tissue selectivity may be required. For example, *S*-nitrosoglutathione has significant antiplatelet effects at doses that barely cause vasodilatation (Radomski *et al.*, 1992). Selective targeting to platelets without accompanying hypotension may be of use in thrombotic disorders, in which the NO donor may be used alone or in combination with other antithrombotic agents. In particular, a platelet-selective NO donor may be of use in restoring the hemostatic–thrombotic balance of sepsis-induced DIC.

4.2. Agents that Decrease NO

Inhibition of the synthesis of NO may be desirable in cardiovascular/inflammatory disorders such as septic shock. Nonselective NOS inhibitors such as L-NMMA can restore loss of vessel tone and prevent hyporesponsiveness to vasoconstrictor agents in endotoxin-treated vessels *in vitro* (Rees *et al.*, 1990b; Stoclet *et al.*, 1995). Furthermore, L-NMMA improves the hemodynamic disturbance (Fig. 8) seen in experimental models of endotoxin shock (Thiemermann, 1994; Stoclet *et al.*, 1995, Rees *et al.*, 1998), but its effects on tissue damage and mortality remain controversial. Adverse responses have been observed in studies in which a high bolus dose of the NOS inhibitor was used ($\geq$30 mg/kg), or when L-NMMA was administered at the same time as endotoxin. Indeed, under these circumstances L-NMMA has precipitated glomerular thrombosis (Shultz and Raij, 1992) and enhanced vascular leak (Whittle, 1995) in the rat. In contrast, administration of NOS inhibitors several hours after endotoxin (i.e., at the time of iNOS expression) can reverse the vascular leak (Whittle, 1995). Thus, the time of

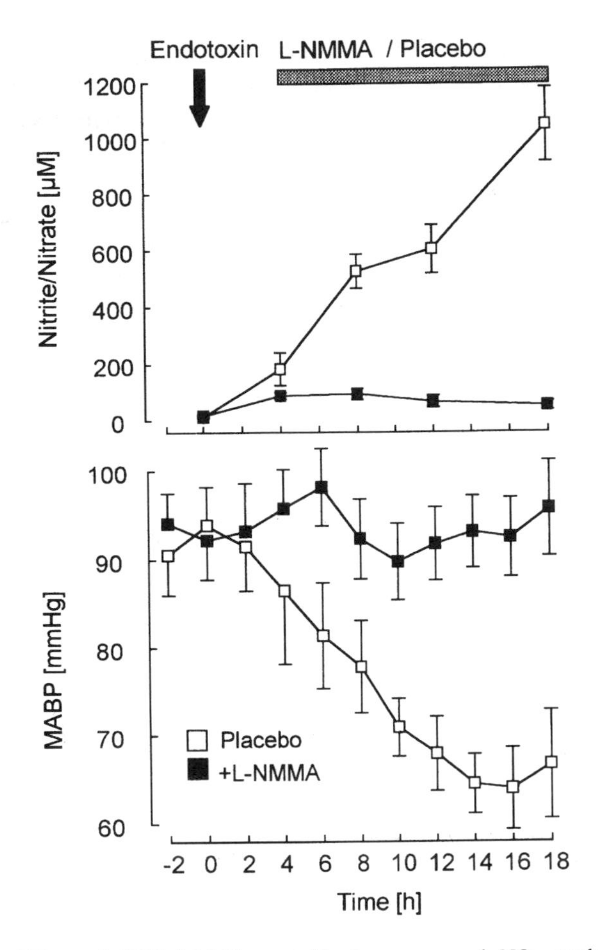

FIGURE 8. Effects of NOS inhibition on blood pressure and NO_x production. L-NMMA (10 mg/kg/per hr i.v.) administered as a continuous infusion inhibits the elevation of concentrations of plasma nitrite and nitrate and the fall in mean arterial blood pressure (MABP) in the conscious mouse following endotoxin (*Escherichia coli* O26:B6, 6 mg/kg i.v.) treatment. Each point is the mean ± S.E.M. of four to six animals. Reproduced with permission from Rees *et al.* (1998).

administration and dose of the NOS inhibitor appear to be crucial; early or complete inhibition may worsen outcome (Wright *et al.*, 1992; Cobb *et al.*, 1995), while reduction of NO overproduction appears to be protective (Wright *et al.*, 1992; Rees, *et al.*, 1998). This dual action suggests that some NO is required to maintain normal homeostasis and is protective early in the course of endotoxemia, whereas the subsequent vast overproduction is largely damaging. Adverse effects of L-NMMA in various animal models of septic shock may also relate to inadequate fluid resuscitation, a necessary intervention because of the vascular leak. Failure to do so can lead to a hypodynamic state, with reduced cardiac output and peripheral vasoconstriction seen in several canine studies (Cobb, *et al.*, 1995). Inhibition of NOS in this situation, resulting in further vasoconstriction, would be anticipated to have deleterious consequences. Indeed, after appropriate fluid resuscitation, a continuous infusion of a NOS inhibitor can reverse the systemic vasodilatation induced by live *Pseudomonas aeruginosa* in sheep (Meyer *et al.*, 1992) or by live *E. coli* in baboons (Redl *et al.*, 1997). In the latter model of septic shock, L-NMMA reversed the elevation of plasma nitrite/nitrate concentrations and improved survival from 29% to 75% (Redl *et al.*, 1997).

In the heart, microbial products and cytokines stimulate the expression of iNOS, which contributes to the myocardial depression observed in septic shock (Finkel *et al.*, 1992). Reduced isoproterenol-stimulated contractility of isolated myocytes can be restored by NOS inhibitors (Balligand, *et al.*, 1993b). However, NOS inhibitors administered to reverse low blood pressure and low systemic vascular resistance in experimental septic shock actually decrease cardiac output (Kilbourn *et al.*, 1995; Grover *et al.*, 1995). The reduction in heart rate and contractility, probably resulting from baroreceptor reflex action and increased afterload, reduces myocardial function. Thus, treatment of septic shock with a NOS inhibitor to restore blood pressure may require additional inotropic support so as to limit a potentially detrimental reduction in cardiac output.

L-NMMA has been used in clinical studies for the treatment of septic shock; additional discussion of NOS inhibition in sepsis can be found in Chapter 13. Phase I and Phase II studies for L-NMMA in 32 and 312 patients, respectively, demonstrate that this non-isoform-specific NOS inhibitor can effectively restore blood pressure in patients with septic shock, enabling conventional vasopressor therapy to be reduced and/or removed with inotropic support adjusted as clinically appropriate. The studies also reported no apparent adverse effects on several indices of organ function (Grover *et al.*, 1995). Although these studies were too small to detect changes in mortality, they suggest that L-NMMA has the potential to be utilized as a treatment for septic shock in a carefully controlled clinical setting. Ongoing clinical studies are using NOS inhibitors to combat the hypotension following cytokine therapy in cancer patients. Whether these compounds will have therapeutic benefit in this and other low blood pressure states such as anaphylactic and hemorrhagic shock remains to be determined.

Selective inhibitors of iNOS are in development and may have a more useful therapeutic role by blocking excessive pathological NO generation without affecting the physiological generation of NO from eNOS and nNOS. One such compound is 7W93, which has no effect on normal blood pressure in a conscious mouse model of endotoxin shock, but reverses an endotoxin-induced decline in blood pressure (Rees *et al.*, manuscript in preparation). Encouragingly, 7W93 improves survival when administered as a continuous infusion either prior to or after the endotoxin-induced fall in blood pressure. Interestingly, high doses of

7W93 administered during the shock phase increase blood pressure above normal levels, suggesting that even selective inhibitors of iNOS will need to be carefully titrated. This may be related to elevated levels of endogenous catecholamines following immunological activation (Benedict and Grahame-Smith, 1978; Jones and Romano, 1989) and suggests that a small increase in NO production can be beneficial in counterbalancing an increase in sympathetic activation. Other selective inhibitors of iNOS such as *S*-methyl-isothiourea have also demonstrated a survival benefit in a mouse model of shock, and *N*-[3-(aminomethyl)benzyl] acetamidine (1400W) has shown beneficial hemodynamic effects in a rat model of endotoxin shock (Wray *et al.*, 1998). Although clinical use of these compounds is limited by their unfavorable toxicological profile, it seems likely that selective inhibitors of iNOS will ultimately have a greater therapeutic index and more widespread clinical utility than nonselective inhibitors.

With NO implicated in microbial killing (see Chapter 12), there is the possibility that use of NOS inhibitors as a treatment to restore cardiovascular homeostasis in septic shock may lead to uncontrolled infection. Indeed, infection with the protozoal parasite *Leishmania major* in L-NMMA-treated mice or iNOS mutant mice shows greater parasite number than in untreated or wild-type mice (Wei *et al.*, 1995). In contrast, Lingnau *et al.* (1996) have shown a decrease in *Pseudomonas* in the blood following L-NMMA treatment in a sheep model of septic shock. Further studies are required to determine the antimicrobial role of NO in humans.

Glucocorticoids inhibit the induction of iNOS by suppressing transcription of iNOS mRNA, but have no effect on iNOS activity once the enzyme is expressed (Radomski *et al.*, 1990b; Rees *et al.*, 1990b; Geller *et al.*, 1993). Accordingly, clinical data have shown that steroids are only effective if administered before or at the onset of shock and are therefore more effective at preventing rather than treating the condition. The inhibitory effect on iNOS induction occurs within the therapeutic range and may contribute to the clinical effects of these drugs in chronic as well as acute inflammatory disorders. Glucocorticoids also prevent the reduction in basal or stimulated NO-mediated dilatation ("endothelial stunning") following bacterial endotoxin or cytokine administration (Bhagat *et al.*, 1996). Other clinically used drugs that inhibit NOS include antifungal imidazoles such as clotrimazole. These drugs do not block expression of the enzyme, but rather appear to inactivate the NOS protein (Bogle *et al.*, 1994).

Since the discovery of NO as a biologically active molecule in 1987, it has become clear that NO plays a fundamental role in the physiology and pathophysiology of the cardiovascular system. Some clinically useful drugs are now known to act on the L-arginine:NO pathway, and new uses for these agents are being explored. Novel drugs to activate or inhibit the pathway are becoming available and will doubtless have a major impact on the future treatment of cardiovascular disorders, particularly those related to infection.

References

Assreuy, J., Cunha, F. Q., Liew, F. Y., and Moncada, S., 1993, Feedback inhibition of nitric oxide synthase activity by nitric oxide, *Br. J. Pharmacol.* **108:**833–837.

Baigrie, R. J., Lamont, P. M., Whiting, S., and Morris, P. J., 1993, Portal endotoxin and cytokine responses during abdominal aortic surgery, *Am. J. Surg.* **166:**248–251.

Balligand, J. L., Kelly, R. A., Marsden, P. A., Smith, T. W., and Michel, T., 1993a, Control of cardiac muscle cell function by an endogenous nitric oxide signaling system, *Proc. Natl. Acad. Sci. USA* **90:**347–351.

Balligand, J. L., Ungureanu, D., Kelly, R. A., Kobzik, L., Pimental, D., Michel, T., and Smith, T. W., 1993b, Abnormal contractile function due to induction of nitric oxide synthesis in rat cardiac myocytes follows exposure to activated macrophage-conditioned medium, *J. Clin. Invest.* **91:**2314–2319.

Beckman, J. S., Beckman, T. W., Chen, J., Marshall, P. A., and Freeman, B. A., 1990, Apparent hydroxyl radical production by peroxynitrite: Implications for endothelial injury from nitric oxide and superoxide, *Proc. Natl. Acad. Sci. USA* **87:**1620–1624.

Benedict, C. R., and Grahame-Smith, D. G., 1978, Plasma noradrenaline and adrenaline concentrations and dopamine-β-hydroxylase activity in patients with shock due to septicaemia, trauma and haemorrhage, *Q. J. Med.* **185:**1–20.

Bhagat, K., Moss, R., Collier, J., and Vallance, P., 1996, Endothelial "stunning" following a brief exposure to endotoxin: A mechanism to link infection and infarction? *Cardiovasc. Res.* **32:**822–829.

Billiar, T. R., Curran, R. D., Harbrecht, B. G., Stadler, J., Williams, D. L., Ochoa, J. B., Di Silvio, M., Simmons, R. L., and Murray, S. A., 1992, Association between synthesis and release of cGMP and nitric oxide biosynthesis by hepatocytes, *Am. J. Physiol.* **262:**C1077–C1082.

Bogle, R. G., Whitley, G. S., Soo, S. C., Johnstone, A. P., and Vallance, P., 1994, Effect of antifungal imidazoles on mRNA levels and enzyme activity of inducible nitric oxide synthase, *Br. J. Pharmacol.* **111:**1257–1261.

Bogle, R. G., MacAllister, R. J., Whitley, G. S., and Vallance, P., 1995, Induction of N^G-monomethyl-L-arginine uptake: A mechanism for differential inhibition of NO synthases? *Am. J. Physiol.* **269:**C750–C756.

Braun, C., and Zumft, W. G., 1991, Marker exchange of the structural genes for nitric oxide reductase blocks the denitrification pathway of *Pseudomonas stutzeri* at nitric oxide, *J. Biol. Chem.* **266:**22785–22788.

Bredt, D. S., Hwang, P. M., Glatt, C. E., Lowenstein, C., Reed, R. R., and Snyder, S. H., 1991, Cloned and expressed nitric oxide synthase structurally resembles cytochrome P-450 reductase, *Nature* **351:**714–718.

Brown, G. C., Bolanos, J. P., Heales, S. J., and Clark, J. B., 1995, Nitric oxide produced by activated astrocytes rapidly and reversibly inhibits cellular respiration, *Neurosci. Lett.* **193:**201–204.

Calver, A., Collier, J., and Vallance, P., 1993, Nitric oxide and cardiovascular control, *Exp. Physiol.* **78:**303–326.

Cobb, J. P., Natanson, C., Quezado, Z. M., Hoffman, W. D., Koev, C. A., Banks, S., Correa, R., Levi, R., Elin, R. J., Hosseini, J. M., and Danner, R. L., 1995, Differential hemodynamic effects of L-NMMA in endotoxemic and normal dogs, *Am. J. Physiol.* **268:**H1634–H1642.

Cocks, T. M., and Angus, J. A., 1983, Endothelium-dependent relaxation of coronary arteries by noradrenaline and serotonin, *Nature* **305:**627–630.

Cunha, F. Q., Moss, D. W., Leal, L. M., Moncada, S., and Liew, F. Y., 1993, Induction of macrophage parasiticidal activity by *Staphylococcus aureus* and exotoxins through the nitric oxide synthesis pathway, *Immunology* **78:**563–567.

Cunha, F. Q., Assreuy, J., Moss, D. W., Rees, D. D., Leal, L. M., Moncada, S., Carrier, M., O'Donnell,

C. A., and Liew, F. Y., 1994, Differential induction of nitric oxide synthase in various organs of the mouse during endotoxaemia: Role of TNF-alpha and IL-1-beta, *Immunology* **81:**211–215.

Damas, P., Ledoux, D., Nys, M., Vrindts, Y., De Groote, D., Franchimont, P., and Lamy, M., 1992, Cytokine serum level during severe sepsis in human IL-6 as a marker of severity, *Ann. Surg.* **215:**356–362.

de Belder, A. J., Radomski, M. W., Martin, J. F., and Moncada, S., 1995, Nitric oxide and the pathogenesis of heart muscle disease, *Eur. J. Clin. Invest.* **25:**1–8.

Evans, T., Carpenter, A., Kinderman, H., and Cohen, J., 1993, Evidence of increased nitric oxide production in patients with the sepsis syndrome, *Circ. Shock* **41:**77–81.

Feelisch, M., 1993, Biotransformation to nitric oxide of organic nitrates in comparison to other nitrovasodilators, *Eur. Heart J.* **14:**123–132.

Finkel, M. S., Oddis, C. V., Jacob, T. D., Watkins, S. C., Hattler, B. G., and Simmons, R. L., 1992, Negative inotropic effects of cytokines on the heart mediated by nitric oxide, *Science* **257**:387–389.

Frostell, C. G., Blomqvist, H., Hedenstierna, G., Lundberg, J., and Zapol, W. M., 1993, Inhaled nitric oxide selectively reverses human hypoxic pulmonary vasoconstriction without causing systemic vasodilation, *Anesthesiology* **78:**427–435.

Furchgott, R. F., 1983, Role of endothelium in responses of vascular smooth muscle, *Circ. Res.* **53:**557–573.

Furchgott, R. F., and Vanhoutte, P. M., 1989, Endothelium-derived relaxing and contracting factors, *FASEB J.* **3:**2007–2018.

Furchgott, R. F., and Zawadzki, J. V., 1980, The obligatory role of endothelial cells in the relaxation of arterial smooth muscle by acetylcholine, *Nature* **288:**373–376.

Garvey, E. P., Oplinger, J. A., Furfine, E. S., Kiff, R. J., Laszlo, F., Whittle, B. J., and Knowles, R. G., 1997, 1400W is a slow, tight binding, and highly selective inhibitor of inducible nitric-oxide synthase *in vitro* and *in vivo*, *J. Biol. Chem.* **272:**4959–4963.

Geller, D. A., Nussler, A. K., Di Silvio, M., Lowenstein, C. J., Shapiro, R. A., Wang, S. C., Simmons, R. L., and Billiar, T. R., 1993, Cytokines, endotoxin, and glucocorticoids regulate the expression of inducible nitric oxide synthase in hepatocytes, *Proc. Natl. Acad. Sci. USA* **90:**522–526.

Gerlach, H., Pappert, D., Lewandowski, K., Rossaint, R., and Falke, K. J., 1993, Long-term inhalation with evaluated low doses of nitric oxide for selective improvement of oxygenation in patients with adult respiratory distress syndrome, *Intensive Care Med.* **19:**443–449.

Griffith, T. M., Edwards, D. H., Davies, R. L., Harrison, T. J., and Evans, K. T., 1987, EDRF coordinates the behaviour of vascular resistance vessels, *Nature* **329:**442–445.

Grover, R., Zaccardelli, D., Colice, G., Guntupalli, K., Watson, D., and Vincent, J.-L., 1995, The cardiovascular effects of 546C88 in human septic shock, *Intensive Care Med.* **21:**S21.

Gustafsson, L. E., Wiklund, C. U., Wiklund, N. P., Persson, M. G., and Moncada, S., 1990, Modulation of autonomic neuroeffector transmission by nitric oxide in guinea pig ileum, *Biochem. Biophys. Res. Commun.* **173:**106–110.

Haynes, W. G., Noon, J. P., Walker, B. R., and Webb, D. J., 1993, Inhibition of nitric oxide synthesis increases blood pressure in healthy humans, *J. Hypertens.* **11:**1375–1380.

Hibbs, J. B., Jr., 1992, Overview of cytotoxic mechanisms and defence of the intracellular environment against microbes, in: *The Biology of Nitric Oxide:Enzymology, Biochemistry and Immunology* (S. Moncada, M. A. Marletta, J. B. Hibbs, Jr., and E. A. Higgs, eds.), Portland Press, London, pp. 201–206.

Hibbs, J. B., Jr., Vavrin, Z., and Taintor, R. R., 1987, L-Arginine is required for expression of the activated macrophage effector mechanism causing selective metabolic inhibition in target cells, *J. Immunol.* **138:**550–565.

Hibbs, J. B., Jr., Taintor, R. R., Vavrin, Z., Granger, D. L., Drapier, J.-C., Amber, I. J., and Lancaster, J. R., Jr., 1990, Synthesis of nitric oxide from a terminal guanidino nitrogen atom of L-arginine: A molecular mechanism regulating cellular proliferation that targets intracellular iron, in: *Nitric Oxide from L-Arginine, a Bioregulatory System* (S. Moncada and E. A. Higgs, eds.), Elsevier, Amsterdam, pp. 189–223.

Hibbs, J. B., Jr., Westenfelder, C., Taintor, R., Vavrin, Z., Kablitz, C., Baranowski, R. L., Ward, J. H., Menlove, R. L., McMurry, M. P., Kushner, J. P., and Samlowski, W. E., 1992, Evidence for cytokine-inducible nitric oxide synthesis from L-arginine in patients receiving interleukin-2 therapy, *J. Clin. Invest.* **89:**867–877.

Huang, P. L., Huang, Z., Mashimo, H., Bloch, K. D., Moskowitz, M. A., Bevan, J. A., and Fishman, M. C., 1995, Hypertension in mice lacking the gene for endothelial nitric oxide synthase, *Nature* **377:**239–242.

Ignarro, L. J., 1990, Haem-dependent activation of guanylate cyclase and cyclic GMP formation by endogenous nitric oxide: A unique transduction mechanism for transcellular signaling, *Pharmacol. Toxicol.* **67:**1–7.

Ignarro, L. J., Buga, G. M., Wood, K. S., Byrns, R. E., and Chaudhuri, G., 1987, Endothelium-derived relaxing factor produced and released from artery and vein is nitric oxide, *Proc. Natl. Acad. Sci. USA* **84:**9265–9269.

Jones, S. B., and Romano, F. D., 1989, Dose- and time-dependent changes in plasma catecholamines in response to endotoxin in conscious rats, *Circ. Shock* **28:**59–68.

Kanwar, S., and Kubes, P., 1995, Nitric oxide is an antiadhesive molecule for leukocytes, *New Horiz.* **3:**93–104.

Kilbourn, R. G., Fonseca, G. A., Griffith, O. W., Ewer, M., Price, K., Striegel, A., Jones, E., and Logothetis, C. J., 1995, N^G-methyl-L-arginine, an inhibitor of nitric oxide synthase, reverses interleukin-2-induced hypotension, *Crit. Care Med.* **23:**1018–1024.

Knowles, R. G., Merrett, M., Salter, M., and Moncada, S., 1990, Differential induction of brain, lung and liver nitric oxide synthase by endotoxin in the rat, *Biochem. J.* **270:**833–836.

Lansman, J. B., Hallam, T. J., and Rink, T. J., 1987, Single stretch-activated ion channels in vascular endothelial cells as mechanotransducers? *Nature* **325:**811–813.

Lingnau, W., McGuire, R., Dehring, D. J., Traber, L. D., Linares, H. A., Nelson, S. H., Kilbourn, R. G., and Traber, D. L., 1996, Changes in regional hemodynamics after nitric oxide inhibition during ovine bacteremia, *Am. J. Physiol.* **270:**R207–R216.

Lizasoain, I., Moro, M. A., Knowles, R. G., Darley Usmar, V., and Moncada, S., 1996, Nitric oxide and peroxynitrite exert distinct effects on mitochondrial respiration which are differentially blocked by glutathione or glucose, *Biochem. J.* **314:**877–880.

Lundberg, J. O., Weitzberg, E., Nordvall, S. L., Kuylenstierna, R., Lundberg, J. M., and Alving, K., 1994, Primarily nasal origin of exhaled nitric oxide and absence in Kartagener's syndrome, *Eur. Respir. J.* **7:**1501–1504.

MacAllister, R. J., Fickling, S. A., Whitley, G. S., and Vallance, P., 1994, Metabolism of methylarginines by human vasculature; implications for the regulation of nitric oxide synthesis, *Br. J. Pharmacol.* **112:**43–48.

MacMicking, J. D., Nathan, C., Hom, G., Chartrain, N., Fletcher, D. S., Trumbauer, M., Stevens, K., Xie, Q.-W., Sokol, K., Hutchinson, N., Chen, H., and Mudgett, J. S., 1995, Altered responses to bacterial infection and endotoxic shock in mice lacking inducible nitric oxide synthase, *Cell* **81:**641–650.

McCall, T. B., Palmer, R. M., and Moncada, S., 1991, Induction of nitric oxide synthase in rat peritoneal neutrophils and its inhibition by dexamethasone, *Eur. J. Immunol.* **21:**2523–2527.

Meyer, J., Traber, L. D., Nelson, S., Lentz, C. W., Nakazawa, H., Herndon, D. N., Noda, H., and Traber, D. L., 1992, Reversal of hyperdynamic response to continuous endotoxin administration by inhibition of NO synthesis, *J. Appl. Physiol.* **73:**324–328.

Michel, T., and Feron, O., 1997, Nitric oxide synthases: Which, where, how, and why? *J. Clin. Invest.* **100:**2146–2152.

Mitchell, J. A., Kohlhaas, K. L., Sorrentino, R., Warner, T. D., Murad, F., and Vane, J. R., 1993, Induction by endotoxin of nitric oxide synthase in the rat mesentery: Lack of effect on action of vasoconstrictors, *Br. J. Pharmacol.* **109:**265–270.

Moncada, S., Palmer, R. M., and Higgs, E. A., 1991, Nitric oxide: Physiology, pathophysiology, and pharmacology, *Pharmacol. Rev.* **43:**109–142.

Mugge, A., Lopez, J. A., Piegors, D. J., Breese, K. R., and Heistad, D. D., 1991, Acetylcholine-induced vasodilatation in rabbit hindlimb *in vivo* is not inhibited by analogues of L-arginine, *Am. J. Physiol.* **260:**H242–H247.

Nagao, T., Illiano, S., and Vanhoutte, P. M., 1992, Heterogeneous distribution of endothelium-dependent relaxations resistant to N^G-nitro-L-arginine in rats, *Am. J. Physiol.* **263:**H1090–H1094.

Navarro, J., Sanchez, A., Saiz, J., Ruilope, L. M., Garcia Estan, J., Romero, J. C., Moncada, S., and Lahera, V., 1994, Hormonal, renal, and metabolic alterations during hypertension induced by chronic inhibition of NO in rats, *Am. J. Physiol.* **267:**R1516–R1521.

Nieminen, M. S., Mattila, K., and Valtonen, V., 1993, Infection and inflammation as risk factors for myocardial infarction, *Eur. Heart J.* **14:**12–16.

Ochoa, J. B., Udekwu, A. O., Billiar, T. R., Curran, R. D., Cerra, F. B., Simmons, R. L., and Peitzman, A. B., 1991, Nitrogen oxide levels in patients after trauma and during sepsis, *Ann. Surg.* **214:**621–626.

Olesen, S. P., Clapham, D. E., and Davies, P. F., 1988, Haemodynamic shear stress activates a K^+ current in vascular endothelial cells, *Nature* **331:**168–170.

Palmer, R. M., Ashton, D. S., and Moncada, S., 1988a, Vascular endothelial cells synthesize nitric oxide from L-arginine, *Nature* **333:**664–666.

Palmer, R. M., Rees, D. D., Ashton, D. S., and Moncada, S., 1988b, L-Arginine is the physiological precursor for the formation of nitric oxide in endothelium-dependent relaxation, *Biochem. Biophys. Res. Commun.* **153:**1251–1256.

Parnavelas, J. G., Kelly, W., and Burnstock, G., 1985, Ultrastructural localization of choline acetyltransferase in vascular endothelial cells in rat brain, *Nature* **316:**724–725.

Parsons, S. J., Hill, A., Waldron, G. J., Plane, F., and Garland, C. J., 1994, The relative importance of nitric oxide and nitric oxide-independent mechanisms in acetylcholine-evoked dilatation of the rat mesenteric bed, *Br. J. Pharmacol.* **113:**1275–1280.

Paul-Eugene, N., Kolb, J. P., Sarfati, M., Arock, M., Ouaaz, F., Debre, P., Mossalayi, D. M., and Dugas, B., 1995, Ligation of CD23 activates soluble guanylate cyclase in human monocytes via an L-arginine-dependent mechanism, *J. Leukoc. Biol.* **57:**160–167.

Pepke-Zaba, J., Higenbottam, T. W., Dinh Xuan, A. T., Stone, D., and Wallwork, J., 1991, Inhaled nitric oxide as a cause of selective pulmonary vasodilatation in pulmonary hypertension, *Lancet* **338:**1173–1174.

Petros, A., Lamb, G., Leone, A., Moncada, S., Bennett, D., and Vallance, P., 1994, Effects of a nitric oxide synthase inhibitor in humans with septic shock, *Cardiovasc. Res.* **28:**34–39.

Radomski, M. W., 1996, Nitric oxide: Biological mediator, modulator and effector, *Ann. Med.* **27:**321–329.

Radomski, M. W., Palmer, R. M., and Moncada, S., 1990a, An L-arginine/nitric oxide pathway present in human platelets regulates aggregation, *Proc. Natl. Acad. Sci. USA* **87:**5193–5197.

Radomski, M. W., Palmer, R. M., and Moncada, S., 1990b, Glucocorticoids inhibit the expression of an inducible, but not the constitutive, nitric oxide synthase in vascular endothelial cells, *Proc. Natl. Acad. Sci. USA* **87:**10043–10047.

Radomski, M. W., Rees, D. D., Dutra, A., and Moncada, S., 1992, *S*-nitroso-glutathione inhibits platelet activation *in vitro* and *in vivo*, *Br. J. Pharmacol.* **107:**745–749.

Rajfer, J., Aronson, W. J., Bush, P. A., Dorey, F. J., and Ignarro, L. J., 1992, Nitric oxide as a mediator of relaxation of the corpus cavernosum in response to nonadrenergic, noncholinergic neurotransmission, *N. Engl. J. Med.* **326:**90–94.

Redl, H., Schlag, G., Gasser, H., Davies, J., Rees, D. D., and Grover, R., 1997, Treatment with NO synthase inhibitor, 546C88 twelve hours after start of *E. coli* bacteraemia is beneficial in a baboon model of septic shock, *Shock* **8:**51.

Rees, D. D., 1995, Role of nitric oxide in the vascular dysfunction of septic shock, *Biochem. Soc. Trans.* **23:**1025–1029.

Rees, D. D., Palmer, R. M., Hodson, H. F., and Moncada, S., 1989a, A specific inhibitor of nitric oxide formation from L-arginine attenuates endothelium-dependent relaxation, *Br. J. Pharmacol.* **96:**418–424.

Rees, D. D., Palmer, R. M., and Moncada, S., 1989b, Role of endothelium-derived nitric oxide in the regulation of blood pressure, *Proc. Natl. Acad. Sci. USA* **86:**3375–3378.

Rees, D. D., Cellek, S., Palmer, R. M., and Moncada, S., 1990a, Dexamethasone prevents the induction by endotoxin of a nitric oxide synthase and the associated effects on vascular tone: An insight into endotoxin shock, *Biochem. Biophys. Res. Commun.* **173:**541–547.

Rees, D. D., Palmer, R. M., Schulz, R., Hodson, H. F., and Moncada, S., 1990b, Characterization of three inhibitors of endothelial nitric oxide synthase *in vitro* and *in vivo*, *Br. J. Pharmacol.* **101:**746–752.

Rees, D. D., Monkhouse, J. E., Morren, D., Davies, N., and Moncada, S., 1997, Nitric oxide induction and inhibition in the conscious mouse, *Shock* **8:**72.

Rees, D. D., Monkhouse, J. E., Cambridge, D., and Moncada, S., 1998, Nitric oxide and the haemodynamic profile of endotoxin shock in the conscious mouse, *Br. J. Pharmacol.* **124:**540–546.

Rossaint, R., Falke, K. J., Lopez, F., Slama, K., Pison, U., and Zapol, W. M., 1993, Inhaled nitric oxide for the adult respiratory distress syndrome, *N. Engl. J. Med.* **328:**399–405.

Salter, M., Knowles, R. G., and Moncada, S., 1991, Widespread tissue distribution, species distribution and changes in activity of Ca^{2+}-dependent and Ca^{2+}-independent nitric oxide synthases, *FEBS Lett.* **291:**145–149.

Salzman, A. L., Menconi, M. J., Unno, N., Ezzell, R. M., Casey, D. M., Gonzalez, P. K., and Fink, M. P., 1995, Nitric oxide dilates tight junctions and depletes ATP in cultured Caco-2BBe intestinal epithelial monolayers, *Am. J. Physiol.* **268:**G361–G373.

Schini, V. B., Durante, W., Elizondo, E., Scott Burden, T., Junquero, D. C., Schafer, A. I., and Vanhoutte, P. M., 1992, The induction of nitric oxide synthase activity is inhibited by TGF-beta 1, PDGFAB and PDGFBB in vascular smooth muscle cells, *Eur. J. Pharmacol.* **216:**379–383.

Sheehan, K. C. F., Ruddle, N. H., and Schreiber, R. D., 1989, Generation and characterization of hamster monoclonal antibodies that neutralize murine tumour necrosis factors, *J. Immunol.* **142:**3884–3893.

Shultz, P. J., and Raij, L., 1992, Endogenously synthesized nitric oxide prevents endotoxin-induced glomerular thrombosis, *J. Clin. Invest.* **90:**1718–1725.

Silva, A. T., Bayston, K. F., and Cohen, J., 1990, Prophylactic and therapeutic effects of a monoclonal antibody to tumour necrosis factor-α in experimental gram-negative shock, *J. Infect. Dis.* **162:**421–427.

Singh, S., and Evans, T. W., 1997, Nitric oxide, the biological mediator of the decade: Fact or fiction? *Eur. Respir. J.* **10:**699–707.

Stamler, J. S., Simon, D. I., Osborne, J. A., Mullins, M. E., Jaraki, O., Michel, T., Singel, D. J., and Loscalzo, J., 1992, *S*-nitrosylation of proteins with nitric oxide: Synthesis and characterization of biologically active compounds, *Proc. Natl. Acad. Sci. USA* **89:**444–448.

Stoclet, J. C., Schott, C. A., Schneider, F., Berton, C., and Paya, D., 1995, Induction by endotoxin of nitric oxide production in vascular smooth cells, in: *Shock, Sepsis and Organ Failure—Nitric Oxide* (G. Schlagg and H. Redl, eds.), Springer-Verlag, Berlin, pp. 102–117.

Stuehr, D. J., 1997, Structure–function aspects in the nitric oxide synthases, *Annu. Rev. Pharmacol. Toxicol.* **37:**339–359.

Stuehr, D. J., and Marletta, M. A., 1987, Induction of nitrite/nitrate synthesis in murine macrophages by BCG infection, lymphokines, or interferon-gamma, *J. Immunol.* **139:**518–525.

Syrjanen, J., 1993, Infection as a risk factor for cerebral infarction, *Eur. Heart J.* **14:**17–19.

Syrjanen, J., Valtonen, V. V., Iivanainen, M., Kaste, M., and Huttunen, J. K., 1988, Preceding infection as an important risk factor for ischaemic brain infarction in young and middle aged patients, *Br. Med. J.* **296:**1156–1160.

Taveira da Silva, A. M., Kaulbach, H. C., Chuidian, F. S., Lambert, D. R., Suffredini, A. F. and Danner, R. L., 1993, Shock and multiple organ dysfunction after self administration of *Salmonella* endotoxin, *N. Engl. J. Med.* **328:**1457–1460.

Thiemermann, C., 1994, The role of the L-arginine:nitric oxide pathway in circulatory shock, *Adv. Pharmacol.* **28:**45–79.

Thiemermann, C., and Vane, J., 1990, Inhibition of nitric oxide synthesis reduces the hypotension induced by bacterial lipopolysaccharides in the rat in vivo, *Eur. J. Pharmacol.* **182:**591–595.

Toda, N., 1995, Nitric oxide and the regulation of cerebral arterial tone, in: *Nitric Oxide in the Nervous System* (S. Vincent, ed.), Academic Press, Orlando, pp. 207–225.

Toda, N., and Okamura, T., 1990, Mechanism underlying the response to vasodilator nerve stimulation in isolated dog and monkey cerebral arteries, *Am. J. Physiol.* **259:**H1511–H1517.

Vallance, P., 1996, Control of the human cardiovascular system by nitric oxide, *J. Hum. Hypertens.* **10:**377–381.

Vallance, P., and Moncada, S., 1993, Role of endogenous nitric oxide in septic shock, *New Horiz.* **1:**77–86.

Vallance, P., Collier, J., and Moncada, S., 1989a, Nitric oxide synthesised from L-arginine mediates endothelium dependent dilatation in human veins in vivo, *Cardiovasc. Res.* **23:**1053–1057.

Vallance, P., Collier, J., and Moncada, S., 1989b, Effects of endothelium-derived nitric oxide on peripheral arteriolar tone in man, *Lancet* **2:**997–1000.

Wei, X. Q., Charles, I. G., Smith, A., Ure, J., Feng, G. J., Huang, F. P., Xu, D., Muller, W., Moncada, S., and Liew, F. Y., 1995, Altered immune responses in mice lacking inducible nitric oxide synthase, *Nature* **375:**408–411.

Wennmalm, A., Benthin, G., and Petersson, A. S., 1992, Dependence of the metabolism of nitric oxide (NO) in healthy human whole blood on the oxygenation of its red cell haemoglobin, *Br. J. Pharmacol.* **106:**507–508.

White, K. A., and Marletta, M. A., 1992, Nitric oxide synthase is a cytochrome P-450 type hemoprotein, *Biochemistry* **31:**6627–6631.

Whittle, B. J., 1995, Nitric oxide in physiology and pathology, *Histochem. J.* **27:**727–737.

Williams, D. J., Vallance, P. J. T., Neild, G. H., Spencer, J. A. D., and Imms, F. J., 1997, Nitric oxide-mediated vasodilation in human pregnancy, *Am. J. Physiol.* **272:**H748–H752.

Wray, G. M., Millar, C. G., Hinds, C. J., and Thiemermann, C., 1998, Selective inhibition of the activity of inducible nitric oxide synthase prevents the circulatory failure, but not the organ injury/dysfunction, caused by endotoxin, *Shock* **9:**329–335.

Wright, C. E., Rees, D. D., and Moncada, S., 1992, Protective and pathological roles of nitric oxide in endotoxin shock, *Cardiovasc. Res.* **26:**48–57.

Yoshizumi, M., Perrella, M. A., Burnett, J. C., Jr., and Lee, M. E., 1993, Tumor necrosis factor downregulates an endothelial nitric oxide synthase mRNA by shortening its half-life, *Circ. Res.* **73:**205–209.

Zembowicz, A., and Vane, J. R., 1992, Induction of nitric oxide synthase activity by toxic shock syndrome toxin 1 in a macrophage–monocyte cell line, *Proc. Natl. Acad. Sci. USA* **89:**2051–2055.

CHAPTER 8

Biochemical Regulation of Nitric Oxide Cytotoxicity

ANDREW J. GOW, RAYMOND FOUST III, STUART MALCOLM, MADHURA GOLE, and HARRY ISCHIROPOULOS

Nitric oxide (NO•) has been strongly implicated as a molecular mediator of tissue injury and organ system dysfunction associated with infection. Cytotoxic actions of NO• are also relevant to understanding NO•-related antimicrobial activity (Chapter 12). This chapter will describe the various reactive pathways by which NO• can react with biomolecules and the implications of these reactions for NO•-derived cytotoxicity.

1. Reactivity of Nitric Oxide

NO• is a simple diatomic molecule in which the highest occupied orbital contains an unpaired electron; hence, it is a free radical. Despite this designation, NO• is relatively stable and is freely diffusable across biological membranes. Although NO• is a structurally simple molecule, it may be involved in over 30 different biological functions involving all major organ systems. The highly complex reactivity of NO• with biological targets may explain its functional diversity. Biochemically, we know of at least four different reactive pathways for NO• that may occur in biological systems (Fig. 1). NO• possesses an ability to

ANDREW J. GOW, RAYMOND FOUST III, STUART MALCOLM, MADHURA GOLE, and HARRY ISCHIROPOULOS • The Institute for Environmental Medicine and Department of Biochemistry and Biophysics, University of Pennsylvania School of Medicine, Philadelphia, Pennsylvania 19104; *present address for HI:* Stokes Research Institute, Children's Hospital of Philadelphia, Philadelphia, Pennsylvania 19104

Nitric Oxide and Infection, edited by Fang. Kluwer Academic / Plenum Publishers, New York, 1999.

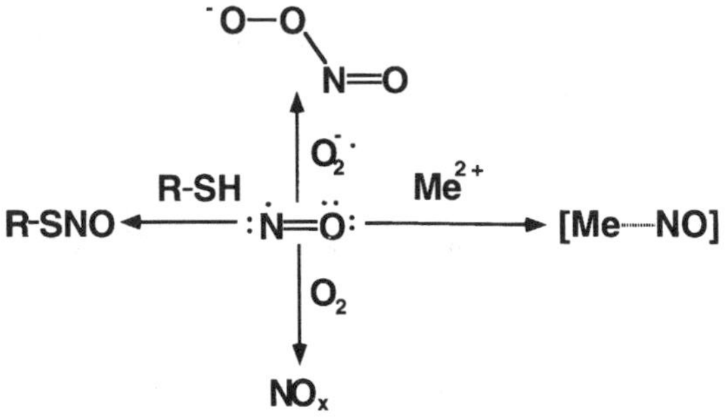

FIGURE 1. Potential reactive pathways for nitric oxide. R-SH, reduced thiol residues; Me, transition metals such as Fe, Cu, and Zn; NO_x, products of autoxidation of nitric oxide including N_2O_3 and N_2O_4.

donate electrons and it is this capacity that forms the basis of the classical physiological action of NO•, the activation of guanylyl cyclase (Murad, 1994). NO• binds to the iron of the heme moiety (Cassoly and Gibson, 1975), forming transition metal–nitrosyl complexes. This reaction occurs very rapidly, with a rate constant of the order of $10^7\,M^{-1}sec^{-1}$, and has been implicated in the regulation of a number of enzymes including guanylyl cyclase and aconitase (Drapier and Hibbs, 1996; Hausladen and Fridovich, 1996; Kennedy *et al.*, 1997).

Metal–nitrosyl complexes, however, are not stable end products of NO• metabolism. These complexes have been shown to decay releasing various redox congeners of NO• such as NO^+ (Vanin *et al.*, 1996), NO^- (Gow and Stamler, 1998), and NO_3^- (Eich *et al.*, 1996). The decay products of metal–nitrosyl complexes are dependent on the redox state of the metal and the presence of secondary reactants, such as oxygen. Both the varying redox state congeners of NO• and the alternative redox states of the transition metals that result from the breakdown of nitrosyl complexes may produce cytotoxic effects. For instance, $NO^{\bullet +}$ is a potent nitrosative species and hence may cause damaging nitrosation in a wide range of targets. Therefore, it is necessary for the concentration of NO• to be tightly controlled, and such control is most likely a result of the reaction of NO• with thiols.

Reduced thiols are the most abundant of all potential targets for NO• and thus provide a likely pathway for controlling NO• concentration. A number of different reaction mechanisms have been proposed for *S*-nitrosothiol synthesis, although many of these have depended on the autoxidation of NO• to form higher oxides of nitrogen (Keshive *et al.*, 1996). Such reactions are unlikely under physiological conditions as they are second order with respect to NO• and therefore very slow ($\sim 10^{-12}\,M^{-1}\,sec^{-1}$) at submicromolar concentrations of NO•. Recent experimental evidence was provided for the formation of *S*-nitrosothiol from the direct interaction of NO• with thiol in the presence of an electron acceptor (Gow *et al.*,

1997). This reaction mechanism, in combination with metal-catalyzed reactions such as those involving dinitrosyl–iron complexes (Mülsch *et al.*, 1992) and direct interaction with thiyl radical, provides physiological pathways for *S*-nitrosothiol formation. As the intracellular concentration of reduced thiol can be as high as 15 mM, this pathway constitutes a major target for $NO^{\bullet}$ reactivity. *S*-nitrosothiols appear to be bioactive compounds with a wide variety of effects (Stamler *et al.*, 1992). They have been implicated in signal transduction via regulation of ion channels such as the NMDA receptor (Kashii *et al.*, 1996) and the ryanodine receptor (Xu *et al.*, 1998), and via control of enzymes such as guanylyl cyclase and the hexose monophosphate shunt (Murad, 1994; Clancy *et al.*, 1997); specific interactions will be discussed in greater detail below.

$NO^{\bullet}$ does not directly mediate lipid peroxidation reactions. However, it has been implicated as both a prooxidant and an antioxidant in superoxide-, hydrogen peroxide-, and peroxynitrite-mediated lipid peroxidations (Rubbo *et al.*, 1994; Rubbo and Freeman, 1996). The prooxidant effects of $NO^{\bullet}$ are thought to occur via direct interaction between $NO^{\bullet}$ and reactive oxygen species to form more potent secondary oxidants. The antioxidant effects of $NO^{\bullet}$ are likely to be caused by interaction with alkoxyl and peroxyl radical intermediates in lipid peroxidation, thus terminating lipid radical chain propagation reactions (Rubbo *et al.*, 1995a).

Based on the known rate constants, the fastest reactions for nitric oxide are combinations with other free radicals. The reaction of $NO^{\bullet}$ with superoxide, which occurs with a second-order rate constant of approximately $6 \times 10^9\,M^{-1}sec^{-1}$, produces peroxynitrite, a stronger two-electron oxidant than either $NO^{\bullet}$ or superoxide alone* (Koppenol, 1996). It is capable of oxidizing thiol residues to sulfenic and sulfonic acids (Quijano *et al.*, 1997), oxidizing DNA to generate strand breaks (see below), oxidizing unsaturated fatty acyl chains to generate LO(O)NO (Rubbo *et al.*, 1994), and nitrating certain amino acid side chains such as tyrosine (Crow and Beckman, 1995). Peroxynitrite has also been shown to react with carbon dioxide (Denicola *et al.*, 1996; Gow *et al.*, 1996a). This interaction partially inhibits the oxidative capacity of peroxynitrite, but catalyzes the nitrative capacity. The mechanism of carbon dioxide-catalyzed nitration is unclear, but generation of 3-nitrotyrosine has been shown to inhibit protein function and alter signal transduction (Gow *et al.*, 1996b).

2. $NO^{\bullet}$-Mediated Protein Modification

As we have discussed, three of the primary routes of $NO^{\bullet}$ reactivity involve transition metals, superoxide, and reduced thiols. Each of these pathways relates to

* $E^{\circ}(ONOO^-/NO_2^{\bullet}) = 1.4$ V, $E^{\circ}(NO_2^{\bullet}/NO_2^-) = 0.99$ V.

protein modification and alteration of function. These variant routes of NO• interaction determine whether NO• will be either cytoprotective or cytotoxic depending on the conditions and cell type. For instance, the binding of NO• to the heme moiety of guanylyl cyclase activates the enzyme, resulting in the synthesis of cGMP and a physiological response (Murad, 1994). However, the interaction of nitrogen oxides with iron–sulfur centers can lead to their disruption and hence the inhibition of enzymes such as aconitase that are dependent on these centers (Cooper and Brown, 1995; Kennedy *et al.*, 1997).

The formation of *S*-nitrosothiols can lead to the modification of protein function, such as the closing of the NMDA receptor calcium channel (Lei *et al.*, 1992), or the inactivation of glyceraldehyde-3-phosphate dehydrogenase (Mohr *et al.*, 1994), which may exert a cytoprotective effect by altering intracellular conditions, e.g., reducing intracellular calcium and glycolytic rate. Reaction with thiols may also be protective by preventing inappropriate interaction with metal centers and superoxide anion.

Significant protein modification can occur via peroxynitrite or by the interaction of peroxynitrite with carbon dioxide. The interaction of peroxynitrite with carbon dioxide results in the formation of a potent nitrating agent. Thus, the most probable protein modification resulting from peroxynitrite/carbon dioxide formation is tyrosine nitration (Denicola *et al.*, 1996; Gow *et al.*, 1996a). Treatment of biological samples, such as plasma, with nitrating agents has shown that certain proteins are more susceptible to nitration, implying that there are selective targets for nitration (Gow *et al.*, 1996b). Previously, we have shown that nitrotyrosine formation inhibits tyrosine phosphorylation and may therefore interrupt signal transduction pathways (Gow *et al.*, 1996b). In addition, nitration may target proteins for degradation. Finally, nitration has been shown *in vitro* to inhibit the function of a wide variety of proteins. Recent evidence has shown that certain proteins are nitrated and functionally inhibited *in vivo*. These proteins include Mn-superoxide dismutase (MacMillan-Crow *et al.*, 1996), tyrosine hydroxylase, and cytoskeletal proteins such as actin and neurofilament-L (Boota *et al.*, 1996; Crow *et al.*, 1997).

3. NO• and Mitochondria

Mitochondria play a critical role in controlling the intracellular environment by generating energy, maintaining redox potential, and determining the intracellular calcium concentration. Therefore, alterations in mitochondrial homeostasis will strongly influence cellular function. The mitochondrion is a major source of oxygen radicals, reflecting reduction of oxygen by the electron transport chain. In addition, the mitochondrion appears to be a critical target for both NO•- and peroxynitrite-mediated cellular effects. We recently demonstrated that peroxynitrite-mediated delayed endothelial cell death occurs via inhibition of mitochondrial function (Gow *et al.*, 1998). Mitochondria represent a complex target site for NO• reactivity, possessing membrane, protein, and DNA. Clearly, damage to the mitochondrial membrane will result in loss of transmembrane potential, a fall in the reductive capacity of the cell, uncoupling of the electron transport chain, and leakage of calcium. Oxidation of mitochondrial DNA has also been shown to correlate with injury. In addition, there is a wide variety of protein targets for both NO• and peroxynitrite.

Both NO• and peroxynitrite have been shown to alter mitochondrial electron transport function, albeit by different mechanisms (Heales *et al.*, 1994; Hu *et al.*, 1994; Bolanos *et al.*, 1995, 1996; Richter *et al.*, 1995; Kennedy *et al.*, 1997). NO• has been demonstrated to inhibit cytochrome oxidase, and there is also a potential for interaction with iron–sulfur proteins (although this is not thought to occur physiologically). Peroxynitrite inhibits complexes I–III (Lemasters *et al.*, 1987; Bolanos *et al.*, 1995) and induces a cyclosporin A-sensitive calcium efflux (Hu *et al.*, 1994). Peroxynitrite has also been shown to induce calcium efflux from isolated liver mitochondria by oxidation of critical thiols in a manner that induces pyridine nucleotide-linked calcium release, a pathway inhibited by cyclosporin A (Packer and Murphy, 1995). Within neuronal cells, the intracellular level of glutathione, and hence reduced thiol, has been shown to be critical in determining the effectiveness of NO• inhibition of mitochondrial function (Bolanos *et al.*, 1996). The NO•- or peroxynitrite-induced uncoupling of the mitochondrial electron transport chain results in an increase in superoxide and hydrogen peroxide production (Bolanos *et al.*, 1995; Quijano *et al.*, 1997). The increase in superoxide can be further amplified by the inactivation of Mn-superoxide dismutase strategically located inside the mitochondria where superoxide production occurs (Thompson, 1995). Recently, Mn-superoxide dismutase was found to be nitrated and inactivated in rejected human transplanted kidney tissues (MacMillan-Crow *et al.*, 1996). Overall, it appears that mitochondria represent a critical target for nitric oxide and peroxynitrite. Collapse of the mitochondrial membrane potential and a decline in energy production and reduced equivalents represent early events in peroxynitrite-induced cell death.

4. Mechanisms of NO•-Induced DNA Damage

NO• itself is considered to have no direct effect on DNA. However, both peroxynitrite and the higher oxides of nitrogen have been shown to modify this macromolecule (Wink *et al.*, 1991; Tamir and Tannenbaum, 1996). Exposure of DNA to higher oxides of nitrogen results in the deamination of purines and pyrimidines. Peroxynitrite exposure of isolated DNA results in strand breaks and modification of purine bases (Zingarelli *et al.*, 1996). In addition, novel adducts such as 8-nitroguanine and 4,5-nitrosooxy-2′-deoxyguanosine have been detected on exposure of DNA to peroxynitrite (Yermilov *et al.*, 1995; Douki *et al.*, 1996). Although these modified bases have not been measured *in vivo*, they can be detected after exposure of human epidermal keratinocytes to peroxynitrite (Spencer *et al.*, 1996). Activation of poly-ADP ribosyl transferase in response to DNA strand breaks from intracellularly generated peroxynitrite has been reported, and the activation of this enzyme has been shown to lead to cell death (Szabo *et al.*, 1996, 1997; Zingarelli *et al.*, 1996).

NO•-related DNA damage can occur in ways other than direct attack by one of its reactive intermediates. NO• has been shown to inhibit DNA repair processes (Laval *et al.*, 1997). Employing NO• donors and T4 DNA, Graziewicz described an inhibition of DNA ligase by reactive oxygen species generated from NO• (Graziewicz *et al.*, 1996). This may account for the increased incidence of single-strand breaks in NO•-exposed cells. There are a number of potential nuclear targets for NO•-derived oxidants. Regulation of the cellular redox environment is critical in preventing the reaction of these species with DNA.

5. NO• and Cell Injury

NO• has been implicated in a wide variety of disease states including neurodegeneration and neuronal injury (Schulz *et al.*, 1995), ischemic injury, damage to the pulmonary surfactant system (Haddad *et al.*, 1994), and inflammation-related tissue damage (Rubbo *et al.*, 1995b). However, as stated previously, NO• has also been postulated to have a physiological role in virtually every organ system. We propose that the balance between physiological and pathological roles is maintained by the alternative reactivities outlined above, and summarized in Fig. 2. In this model, low physiological fluxes of NO• primarily interact with thiols and heme iron within the intracellular environment. Although these are not the fastest reactions for NO•, thiols and heme groups provide the primary targets as a result of their relatively high intracellular concentration. These interactions result in physiological modulation of cellular function. For example, binding to the heme iron of guanylyl cyclase results in enzyme activation and a cascade of reactions resulting from cGMP production.

Nitrosylation of thiol residues can also result in altered protein function, and these reactions form the basis of some of the non-cGMP-mediated effects of NO•. *S*-nitrosylation inhibits Ca^{2+} flow through the NMDA receptor (Lipton and Stamler, 1994), inhibits the glycolytic function of glyceraldehyde-3-phosphate dehydrogenase (Mohr *et al.*, 1994), activates the ryanodine receptor (Xu *et al.*, 1998), and activates p21ras (Lander *et al.*, 1996). Clearly some of these actions can be contradictory—inhibition of the NMDA receptor will reduce cytoplasmic Ca^{2+} concentrations, whereas activation of the ryanodine receptor increases cytoplasmic

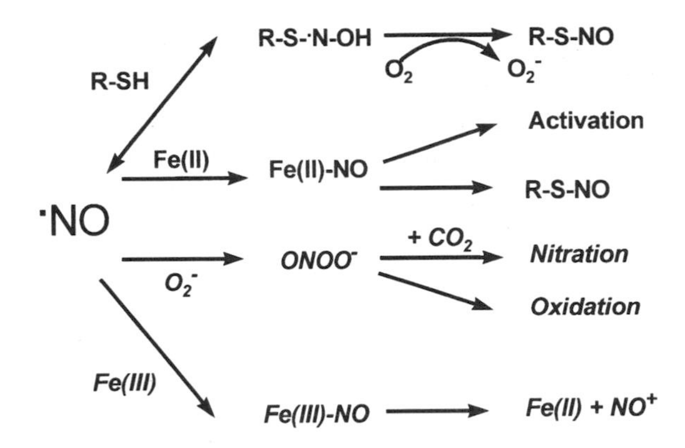

FIGURE 2. Primary and secondary targets for nitric oxide in the cellular environment. Primary reactions are shown in bold, secondary reactions in italics.

Ca^{2+}. Therefore, the physiological response elicited by NO• exposure depends on the specific NO•–target interaction. The complexity of this reactive network is complicated even further when the importance of both protein nitrosylation and the degree of nitrosylation are considered. For example, the ryanodine receptor is reversibly activated by low levels of nitrosylation. However, when greater than one quarter of the receptor thiols are oxidized (e.g., mediated by nitrosylation), activation becomes irreversible (Xu *et al.*, 1998). Therefore, a mechanism for physiological control of protein function can also result in pathological consequences.

Nitrosylation itself may not be the final protein modification. The glycolytic enzyme glyceraldehyde-3-phosphate dehydrogenase possesses a highly reactive thiol that is very readily nitrosylated. Nitrosylation leads to a subsequent protein modification, ADP-ribosylation (Dimmeler *et al.*, 1992). In other settings, nitrosylation may facilitate sulfhydryl oxidation (Stamler, 1995; Becker *et al.*, 1998). In these ways, nitrosylation can act as a signal for alternative reactions to occur. Thus, it can be seen that various biological functions can be modulated by two very simple chemical properties of NO•: formation of *S*-nitrosyl and Fe-nitrosyl groups.

In Fig. 1, two other reactive pathways for NO• are indicated: combination with superoxide to produce peroxynitrite, and autoxidation to higher oxides of nitrogen. These two reactive pathways can occur under normal physiological conditions, but their contribution to the overall consumption of NO• under such conditions is minimal. In the case of superoxide, the low concentration of superoxide in biological systems results in a low rate of peroxynitrite formation. Autoxidation of NO• is bimolecular with respect to NO• and is thus very unlikely at concentrations of NO• in the micromolar and lower range. However, under certain pathological conditions these reactions become more relevant.

Increases in the flux of NO• augment the relevancy of its secondary reactions. Once the rate of NO• exposure exceeds the capacity of the intracellular thiols and heme groups to consume it, resulting either from increased NO• production or from depletion of reduced thiols and hemes, the concentration of NO• available for secondary reactions will increase. When a cell is exposed to higher concentrations of NO•, the rate of production of peroxynitrite and higher oxides of nitrogen increases. The concentration of higher oxides of nitrogen increases directly as a function of NO• concentration, as this reaction is limited only by NO• concentration. However, the formation of peroxynitrite is dependent equally on both NO• and superoxide concentrations, either of which can be rate limiting. Therefore, increases in NO• concentration alone will result in an increased rate of peroxynitrite production, but not in increased peroxynitrite concentration. It is important to note that although peroxynitrite production is dependent on both NO• and superoxide concentrations, it is not necessary for these reactants to be present in equimolar concentrations.

An increase in the production of secondary metabolites of NO•, such as peroxynitrite and higher oxides of nitrogen, can result in cytotoxicity via a number of mechanisms. The most likely pathway of cellular injury via nitrogen oxides is DNA damage. These oxides can induce both chemical and mutational toxicity. Peroxynitrite has been implicated in a wide variety of potential toxic processes including nitrotyrosine formation (Crow and Beckman, 1995), poly-ADP ribosyl transferase activation (Szabo *et al.*, 1996), mitochondrial dysfunction (Gow *et al.*, 1998), depletion of ATP, damage to zinc fingers (Crow *et al.*, 1995), disruption of iron–sulfur centers (Kennedy *et al.*, 1997), depletion of intracellular glutathione (Walker *et al.*, 1995; Lizasoain *et al.*, 1996; Thom *et al.*, 1997), and DNA modification (Yermilov *et al.*, 1996).

Therefore, a number of disease states that involve the induction of NO• production may result in cytotoxicity mediated by secondary reactions of NO•. Disease states such as inflammation (Kaur and Halliwell, 1994), sepsis (Doughty *et al.*, 1996; Groeneveld *et al.*, 1997), arteriosclerosis (Buttery *et al.*, 1996), respiratory distress syndrome (Malcolm *et al.*, 1997), carbon monoxide poisoning (Ischiropoulos *et al.*, 1996), bronchopulmonary dysplasia (Banks *et al.*, 1997), and shock (Zingarelli *et al.*, 1997) each result in an increased production of NO• which can induce these secondary reactions. This is confirmed by the demonstration of 3-nitrotyrosine in many of these conditions, indicating that secondary reactions of NO• are occurring. Cytotoxicity via these secondary reactions can be increased by the presence of other reactants. For instance, the mutant Cu,Zn-superoxide dismutase found in familial amyotrophic lateral sclerosis catalyzes the nitration of tyrosine residues by peroxynitrite, and may thereby increase the cytotoxic potential of peroxynitrite production (Crow *et al.*, 1997).

Disease states in which the concentrations of primary reactants are reduced may result in an increased occurrence of secondary NO• reactions. This is of greatest relevance with regard to the intracellular concentration of reduced thiol. The high concentration of free thiol is critical to its action as a primary reactant for $NO^{\bullet}$; therefore, any reduction in reduced thiol concentration will lead to decreased formation of *S*-nitrosothiol and increased intracellular $NO^{\bullet}$. Because thiols are closely correlated to the reductive capacity of the cell, the latter is compromised following loss of reduced thiols. This situation occurs whenever a cell is exposed to oxidative stress, such as during ischemia–reperfusion injury. Thus, one mechanism of oxidative stress-related cytotoxicity may occur via increased toxicity of $NO^{\bullet}$. Intracellular reduced thiol concentration decreases with aging, and many diseases associated with aging (such as Parkinson's disease and amyotrophic lateral sclerosis) may result from a gradual increase in the toxicity of $NO^{\bullet}$ (Schulz and Beal, 1995; Yoritaka *et al.*, 1996).

In conclusion, although $NO^{\bullet}$ is a simple molecule, it is capable of a number of different reactivities. The local environment of $NO^{\bullet}$, which controls the balance between these various reactivities, is critical in determining whether $NO^{\bullet}$ is beneficial or injurious.

References

Banks, B. A., Ischiropoulos, H., McClelland, M., Ballard, P. L., and Ballard, R. A., 1997, Plasma 3-nitrotyrosine is elevated in premature infants who develop bronchopulmonary dysplasia, *Pediatrics* **101:**870–874.

Becker, K., Savvides, S. N., Keese, M., Schirmer, R. H., and Karplus, P. A., 1998, Enzyme inactivation through sulfhydryl oxidation by physiologic NO-carriers, *Nature Struct. Biol.* **5:**267–271.

Bolanos, J. P., Heales, S. J., Land, J. M., and Clark, J. B., 1995, Effect of peroxynitrite on the mitochondrial respiratory chain: Differential susceptibility of neurones and astrocytes in primary culture, *J. Neurochem.* **64:**1965–1972.

Bolanos, J. P., Heales, S. J., Peuchen, S., Barker, J. E., Land, J. M., and Clark, J. B., 1996, Nitric oxide-mediated mitochondrial damage: A potential neuroprotective role for glutathione, *Free Radical Biol. Med.* **21:**995–1001.

Boota, A., Zar, H., Kim, Y., Johnson, B., Pitt, B., and Davies, P., 1996, IL-1β stimulates superoxide and delayed peroxynitrite production by pulmonary vascular smooth muscle cells, *Am. J. Physiol.* **271:**L932–L938.

Buttery, L. D., Springall, D. R., Chester, A. H., Evans, T. J., Stanfield, E. N., Parums, A. V., Yacoub, M. H., and Polak, J. M., 1996, Inducible nitric oxide is present in human atherosclerotic lesions and promotes the formation and activity of peroxynitrite, *Lab. Invest.* **75:**77–85.

Cassoly, R., and Gibson, Q. H., 1975, Conformation, co-operativity and ligand binding in human hemoglobin, *J. Mol. Biol.* **91:**301–313.

Clancy, R. M., Abramson, S. B., Kohne, C., and Rediske, J., 1997, Nitric oxide attenuates cellular hexose monophosphate shunt response to oxidants in articular chondrocytes and acts to promote oxidant injury, *J. Cell. Physiol.* **172:**183–191.

Cooper, C. E., and Brown, G. C., 1995, The interactions between nitric oxide and brain nerve terminals as studied by electron paramagnetic resonance, *Biochem. Biophys. Res. Commun.* **212:**404–412.

Crow, J. P., and Beckman, J. S., 1995, Reactions between nitric oxide, superoxide, and peroxynitrite: Footprints of peroxynitrite *in vivo*, *Adv. Pharmacol.* **34:**17–43.

Crow, J. P., Beckman, J. S., and McCord, J. M., 1995, Sensitivity of the essential zinc-thiolate moiety of yeast alcohol dehydrogenase to hypochlorite and peroxynitrite, *Biochemistry* **34:**3544–3552.

Crow, J. P., Sampson, J. B., Zhuang, Y., Thompson, J. A., and Beckman, J. S., 1997, Decreased zinc affinity of amyotrophic lateral sclerosis-associated superoxide dismutase mutants leads to enhanced catalysis of tyrosine nitration by peroxynitrite, *J. Neurochem.* **69:**1936–1944.

Denicola, A., Freeman, B. A., Trujillo, M., and Radi, R., 1996, Peroxynitrite reaction with carbon dioxide/bicarbonate: Kinetics and influence on peroxynitrite-mediated oxidations, *Arch. Biochem. Biophys.* **333:**49–58.

Dimmeler, S., Lottspeich, F., and Brune, B., 1992, Nitric oxide causes ADP-ribosylation and inhibition of glyceraldehyde-3-phosphate dehydrogenase, *J. Biol. Chem.* **267:**16771–16774.

Doughty, L. A., Kaplan, S. S., and Carcillo, J. A., 1996, Inflammatory cytokine and nitric oxide responses in pediatric sepsis and organ failure, *Crit. Care Med.* **24:**1137–1143.

Douki, T., Cadet, J., and Ames, B., 1996, An adduct between peroxynitrite and 2′-deoxyguanosine: 4,5-dihydro-5-hydroxy-4-(nitrosooxy)-2′-deoxyguanosine, *Chem. Res. Toxicol.* **9:**3–7.

Drapier, J. C., and Hibbs, J. B., Jr., 1996, Aconitases: A class of metalloproteins highly sensitive to nitric oxide synthesis, *Methods Enzymol.* **269:**26–36.

Eich, R. F., Li, T., Lemon, D. D., Doherty, D. H., Curry, S. R., Aitken, J. F., Mathews, A. J. X., Johnson, K. A., Smith, R. D., Phillips, G. N., Jr., and Olson, J. S., 1996, Mechanism of NO-induced oxidation of myoglobin and hemoglobin, *Biochemistry* **35:**6976–6983.

Gow, A. J., and Stamler, J. S., 1998, Nitric oxide and hemoglobin reactions under physiological conditions, *Nature* **391:**169–173.

Gow, A., Duran, D., Thom, S. R., and Ischiropoulos, H., 1996a, Carbon dioxide enhancement of peroxynitrite-mediated protein tyrosine nitration, *Arch. Biochem. Biophys.* **333:**42–48.

Gow, A. J., Duran, D., Malcolm, S., and Ischiropoulos, H., 1996b, Effects of peroxynitrite-induced protein modifications on tyrosine phosphorylation and degradation, *FEBS Lett.* **385:**63–66.

Gow, A. J., Buerk, D. G., and Ischiropoulos, H., 1997, A novel reaction mechanism for the formation of *S*-nitrosothiol *in vivo*, *J. Biol. Chem.* **272:**2841–2845.

Gow, A. J., Thom, S. R., and Ischiropoulos, H., 1998, Nitric oxide and peroxynitrite-mediated pulmonary cell death, *Am. J. Physiol.* **274:**L112–L118.

Graziewicz, M., Wink, D. A., and Laval, F., 1996, Nitric oxide inhibits DNA ligase activity: Potential mechanisms for NO-mediated DNA damage, *Carcinogenesis* **17:**2501–2505.

Groeneveld, P. H., Kwappenberg, K. M., Langermans, J. A., Nibbering, P. H., and Curtis, L., 1997, Relation between pro- and anti-inflammatory cytokines and the production of nitric oxide (NO) in severe sepsis, *Cytokine* **9:**138–142.

Haddad, I. Y., Crow, J. P., Hu, P., Ye, Y., Beckman, J., and Matalon, S., 1994, Concurrent generation of nitric oxide and superoxide damages surfactant protein A, *Am. J. Physiol.* **267:**L242–L249.

Hausladen, A., and Fridovich, I., 1996, Measuring nitric oxide and superoxide: Rate constants for aconitase reactivity, *Methods Enzymol.* **269:**37–41.

Heales, S. J., Bolanos, J. P., Land, J. M., and Clark, J. B., 1994, Trolox protects mitochondrial complex IV from nitric oxide-mediated damage in astrocytes, *Brain Res.* **668:**243–245.

Hu, P., Ischiropoulos, H., Beckman, J. S., and Matalon, S., 1994, Peroxynitrite inhibition of oxygen consumption and sodium transport in alveolar type II cells, *Am. J. Physiol.* **266:**L628–L634.

Ischiropoulos, H., Beers, M. F., Ohnishi, S. T., Fisher, D., Garner, S. E., and Thom, S. R., 1996, Nitric oxide production and perivascular nitration in brain after carbon monoxide poisoning in the rat, *J. Clin. Invest.* **97:**2260–2267.

Kashii, S., Mandai, M., Kikuchi, M., Honda, Y., Tamura, Y., Kaneda, K., and Akaike, A., 1996, Dual actions of nitric oxide in *N*-methyl-D-aspartate receptor-mediated neurotoxicity in cultured retinal neurons, *Brain Res.* **711:**93–101.

Kaur, H., and Halliwell, B., 1994, Evidence for nitric oxide-mediated oxidative damage in chronic inflammation. Nitrotyrosine in serum and synovial fluid from rheumatoid patients, *FEBS Lett.* **350:**9–12.

Kennedy, M. C., Antholine, W. E., and Beinert, H., 1997, An EPR investigation of the products of the reaction of cytosolic and mitochondrial aconitases with nitric oxide, *J. Biol. Chem.* **272:**20340–20347.

Keshive, M., Singh, S., Wishnok, J. S., Tannenbaum, S. R., and Deen, W. M., 1996, Kinetics of *S*-nitrosation of thiols in nitric oxide solutions, *Chem. Res. Toxicol.* **9:**988–993.

Koppenol, W. H., 1996, Thermodynamics of reactions involving nitrogen–oxygen compounds, *Methods Enzymol.* **268:**7–12.

Lander, H. M., Jacovina, A. T., Davis, R. J., and Tauras, J. M., 1996, Differential activation of mitogen-activated protein kinases by nitric oxide-related species, *J. Biol. Chem.* **271:**19705–19709.

Laval, F., Wink, D. A., and Laval, J., 1997, A discussion of mechanisms of NO genotoxicity: Implication of inhibition of DNA repair proteins, *Rev. Physiol. Biochem. Pharmacol.* **131:**175–191.

Lei, S. Z., Pan, Z. H., Aggarwal, S. K., Chen, H. S., Hartman, J., Sucher, N. J., and Lipton, S. A., 1992, Effect of nitric oxide production on the redox modulatory site of the NMDA receptor–channel complex, *Neuron* **8:**1087–1099.

Lemasters, J. J., DiGuiseppi, J., Nieminen, A., and Herman, B., 1987, Blebbing, free calcium and mitochondrial membrane potential preceding cell death in hepatocytes, *Nature* **325:**78–81.

Lipton, S. A., and Stamler, J. S., 1994, Actions of redox-related congeners of nitric oxide at the NMDA receptor, *Neuropharmacology* **33:**1229–1233.

Lizasoain, I., Moro, M. A., Knowles, R. G., Darley-Usmar, V., and Moncada, S., 1996, Nitric oxide and peroxynitrite exert distinct effects on mitochondrial respiration which are differentially blocked by glutathione or glucose, *Biochem. J.* **314:**877–880.

MacMillan-Crow, L. A., Crow, J. P., Kerby, J. D. X., Beckman, J. S., and Thompson, J. A., 1996, Nitration and inactivation of manganese superoxide dismutase in chronic rejection of human renal allografts, *Proc. Natl. Acad. Sci. USA* **93:**11853–11858.

Malcolm, S., Foust, R., III, and Ischiropoulos, H., 1997, Biomarkers of oxidative stress in ARDS, in: *Acute Respiratory Distress Syndrome: Cellular and Molecular Mechanisms and Clinical Management* (S. Matalon, ed.), Plenum Press, New York, pp. 395–400.

Mohr, S., Stamler, J. S., and Brune, B., 1994, Mechanism of covalent modification of glyceraldehyde-3-phosphate dehydrogenase at its active site thiol by nitric oxide, peroxynitrite and related nitrosating agents, *FEBS Lett.* **348:**223–227.

Mülsch, A., Mordvintcev, P., Vanin, A. F., and Busse, R., 1992, The potent vasodilating and guanylyl cyclase activating dinitrosyl-iron(II) complex is stored in a protein-bound form in vascular tissue and is released by thiols, *FEBS Lett.* **294:**252–256.

Murad, F., 1994, Regulation of cytosolic guanylyl cyclase by nitric oxide: The NO–cyclic GMP signal transduction system, *Adv. Pharmacol.* **26:**19–33.

Packer, M. A., and Murphy, M. P., 1995, Peroxynitrite formed by simultaneous nitric oxide and superoxide generation causes a cyclosporin A-sensitive mitochondrial calcium efflux and depolarisation, *Eur. J. Biochem.* **234:**231–239.

Quijano, C., Alvarez, B., Gatti, R. M., Augusto, O., and Radi, R., 1997, Pathways of peroxynitrite oxidation of thiol groups, *Biochem. J.* **322:**167–173.

Richter, C., Gogvadze, V., Laffranchi, R., Schlapbach, R., Schweizer, M., Suter, M., Walter, P., and Yaffee, M., 1995, Oxidants in mitochondria: From physiology to diseases, *Biochim. Biophys. Acta* **1271:**67–74.

Rubbo, H., and Freeman, B. A., 1996, Nitric oxide regulation of lipid oxidation reactions: Formation and analysis of nitrogen-containing oxidized lipid derivatives, *Methods Enzymol.* **269:**385–394.

Rubbo, H., Radi, R., Trujillo, M., Telleri, R., Kalyanaraman, B., Barnes, S., Kirk, M., and Freeman, B. A., 1994, Nitric oxide regulation of superoxide and peroxynitrite-dependent lipid peroxidation.

Formation of novel nitrogen-containing oxidized lipid derivatives, *J. Biol. Chem.* **269:**26066–26075.

Rubbo, H., Parthasarathy, S., Barnes, S., Kirk, M., Kalyanaraman, B., and Freeman, B. A., 1995a, Nitric oxide inhibition of lipoxygenase-dependent liposome and low-density lipoprotein oxidation: Termination of radical chain propagation reactions and formation of nitrogen-containing oxidized lipid derivatives, *Arch. Biochem. Biophys.* **324:**15–25.

Rubbo, H., Tarpey, M., and Freeman, B. A., 1995b, Nitric oxide and reactive oxygen species in vascular injury, *Biochem. Soc. Symp.* **61:**33–45.

Schulz, J. B., and Beal, M. F., 1995, Neuroprotective effects of free radical scavengers and energy repletion in animal models of neurodegenerative disease, *Ann. N.Y. Acad. Sci.* **765:**100–118.

Schulz, J. B., Matthews, R. T., Muqit, M. M., Browne, S. E., and Beal, M. F., 1995, Inhibition of neuronal nitric oxide synthase by 7-nitroindazole protects against MPTP-induced neurotoxicity in mice, *J. Neurochem.* **64:**936–939.

Spencer, J. P., Wong, J., Jenner, A., Arouma, O. I., Cross, C. E., and Halliwell, B., 1996, Base modification and strand breakage in isolated calf thymus DNA and human skin epidermal keratinocytes exposed to peroxynitrite or 3-morpholinosydnonimine, *Chem. Res. Toxicol.* **9:**1152–1158.

Stamler, J. S., 1995, *S*-Nitrosothiols and the bioregulatory actions of nitrogen oxides through reactions with thiol groups, *Curr. Top. Microbiol.* **196:**19–36.

Stamler, J. S., Simon, D. I., Osborne, J. A., Mullins, M. E., Jaraki, O., Michel, T., Singel, D., and Loscalzo, J., 1992, *S*-Nitrosylation of proteins with nitric oxide: Synthesis and characterization of biologically active compounds, *Proc. Natl. Acad. Sci. USA* **89:**444–448.

Szabo, C., Zingarelli, B., and Salzman, A. L., 1996, Role of poly-ADP ribosyltransferase activation in the vascular contractile and energetic failure elicited by exogenous and endogenous nitric oxide and peroxynitrite, *Circ. Res.* **78:**1051–1063.

Szabo, C., Ferrer-Sueta, G., Zingarelli, B., Southan, G. J., Salzman, A. L., and Radi, R., 1997, Mercaptoethylguanidine and guanidine inhibitors of nitric-oxide synthase react with peroxynitrite and protect against peroxynitrite-induced oxidative damage, *J. Biol. Chem.* **272:**9030–9036.

Tamir, S., and Tannenbaum, S. R., 1996, The role of nitric oxide ($NO^•$) in the carcinogenic process, *Biochim. Biophys. Acta* **1288:**F31–F36.

Thom, S. R., Kang, M., Fisher, D., and Ischiropoulos, H., 1997, Release of glutathione from erythrocytes and other markers of oxidative stress in carbon monoxide poisoning, *J. Appl. Physiol.* **82:**1424–1432.

Thompson, G. B., 1995, Apoptosis in the pathogenesis and treatment of disease, *Science* **267:**1456–1462.

Vanin, A. F., Stukan, R. A., and Manukhina, E. B., 1996, Physical properties of dinitrosyl iron complexes with thiol-containing ligands in relation with their vasodilator activity, *Biochim. Biophys. Acta* **1295:**5–12.

Walker, M. W., Kinter, M. T., Roberts, R. J., and Spitz, D. R., 1995, Nitric oxide-induced cytotoxicity: Involvement of cellular resistance to oxidative stress and the role of glutathione in protection, *Pediatr. Res.* **37:**41–49.

Wink, D. A., Kasprzak, K. S., Maragos, C. M., Elespuru, R. K., Misra, M., Dunams, T. M., Cebula, T. A., Koch, W. H., Andrews, A. W., Allen, J. S., and Keefer, L. K., 1991, DNA deaminating ability and genotoxicity of nitric oxide and its progenitors, *Science* **254:**1001–1003.

Xu, L., Eu, J. P., Meissner, G., and Stamler, J. S., 1998, Activation of the cardiac calcium release channel (ryanodine receptor) by poly *S*-nitrosylation, *Science* **279:**234–237.

Yermilov, V., Rubio, J., and Ohshima, H., 1995, Formation of 8-nitroguanine in DNA treated with peroxynitrite in vitro and its rapid removal from DNA by depurination, *FEBS Lett.* **376:**207–210.

Yermilov, V., Yoshie, Y., Rubio, J., and Ohshima, H., 1996, Effects of carbon dioxide/bicarbonate on induction of DNA single-strand breaks and formation of 8-nitroguanine, 8-oxoguanine and base-propenal mediated by peroxynitrite, *FEBS Lett.* **399:**67–70.

Yoritaka, A., Hattori, N., Uchida, K., Tanaka, M., Stadtman, E. R., and Mizuno, Y., 1996, Immunohistochemical detection of 4-hydroxynonenal protein adducts in Parkinson disease, *Proc. Natl. Acad. Sci. USA* **93:**2696–2701.

Zingarelli, B., O'Connor, M., Wong, H., Salzman, A. L., and Szabo, C., 1996, Peroxynitrite-mediated DNA strand breakage activates poly-adenosine diphosphate ribosyl synthetase and causes cellular energy depletion in macrophages stimulated with bacterial lipopolysaccharide, *J. Immunol.* **156:**350–358.

Zingarelli, B., Day, B. J., Crapo, J. D., Salzman, A. L., and Szabo, C., 1997, The potential role of peroxynitrite in the vascular contractile and cellular energetic failure in endotoxic shock, *Br. J. Pharmacol.* **120:**259–267.

CHAPTER 9

Cytoprotective Effects of NO against Oxidative Injury

DAVID A. WINK, YORAM VODOVOTZ, WILLLIAM DeGRAFF, JOHN A. COOK, ROBERTO PACELLI, MURALI KRISHNA, and JAMES B. MITCHELL

1. Introduction

Research spanning nearly two decades has revealed that free radicals play major physiological and pathophysiological roles. Reactive oxygen species (ROS) derived from peroxide and superoxide oxidize key cellular molecules, with important effects in numerous disease states (Ames *et al.*, 1981, 1993; Halliwell and Gutteridge, 1989). Another free radical, nitric oxide ($NO^{\bullet}$), has been ascribed with both physiological roles essential for maintaining homeostasis (Ignarro, 1990; Moncada *et al.*, 1991) and the ability to damage specific cellular targets through the formation of reactive nitrogen oxide species (RNOS) (Gross and Wolin, 1995; Wink *et al.*, 1996b) (reviewed in Chapter 8). Further complexity is introduced by studies showing that NO can *protect* against oxidative stress mediated by ROS (Wink *et al.*, 1994; Gupta *et al.*, 1997), while other reports suggest that NO *augments* oxidative stress-mediated toxicity (Pacelli *et al.*, 1995; Hata *et al.*, 1996; Yamada *et al.*, 1996). These apparently conflicting results raise the issue of whether NO formation in the presence of ROS is beneficial or deleterious; this issue is important in determining the extent to which either oxygen or nitrogen free radical

---

DAVID A. WINK, WILLLIAM DeGRAFF, JOHN A. COOK, ROBERTO PACELLI, MURALI KRISHNA, and JAMES B. MITCHELL • Tumor Biology Section, Radiation Biology Branch, National Cancer Institute, Bethesda, Maryland 20892. ***YORAM VODOVOTZ*** • Cardiology Research Foundation and Medlantic Research Institute, Washington, D.C. 20010.

Nitric Oxide and Infection, edited by Fang. Kluwer Academic / Plenum Publishers, New York, 1999.

production should be modulated in different clinical situations. In this chapter, we will discuss the chemical and biological basis for these apparently contradictory observations.

2. NO and Peroxide Cytotoxicity

Hydrogen peroxide (H_2O_2) mediates oxidation of biological molecules, which can result in tissue damage. While NO does *not* react chemically with H_2O_2 (Wink *et al.*, 1993), it can protect cells against toxicity mediated by H_2O_2 (Wink *et al.*, 1993, 1994, 1995a,b, 1996a; Kim *et al.*, 1995; Gupta *et al.*, 1997). Lung fibroblasts exposed to increasing concentrations of H_2O_2 exhibited marked increases in cytotoxicity (Wink *et al.*, 1993). Surprisingly, the presence of NONOates, a class of compounds that release $NO^{\bullet}$ in a controlled manner over specific time periods (Keefer *et al.*, 1996), resulted in protection against the cytotoxicity of H_2O_2 (Table I) (Wink *et al.*, 1993). Pre- or posttreatment with these NO donors did not result in protection; in fact, nitrite, the by-product of the decomposition of $NO^{\bullet}$, increases the cytotoxicity of H_2O_2. Similar observations have been made in neuronal (Wink *et al.*, 1993), hepatoma (Wink *et al.*, 1994), and endothelial cells (Chang *et al.*, 1996; Gupta *et al.*, 1997). Other reports suggest that NO derived from endothelial cells is involved in the protection against damage to vascular smooth muscle mediated by H_2O_2 (Linas and Repine, 1997).

Protective effects of NO are not restricted to NONOates, but an extensive examination of these compounds reveals some important differences among *S*-nitrosothiols, sodium nitroprusside, and molsidomines (Farias-Eisner *et al.*, 1996; Wink *et al.*, 1996a,c). Studies have shown that, like the NONOates, compounds containing *S*-nitroso functional groups also protect against H_2O_2-mediated toxicity (Table II) (Wink *et al.*, 1996a). However, compounds such as 3-morpholinosydnonimine (SIN-1) and sodium nitroprusside (SNP) increase the toxicity of H_2O_2 (Table II) (Farias-Eisner *et al.*, 1996; Wink *et al.*, 1996a). Angeli's salt (AS;

TABLE I

Effect of NO Donors on Agents that Induce Oxidative Stress[a]

	H_2O_2	Xanthine oxidase	*t*-Butyl peroxide
Fibroblasts	+[a]	+	+
Hepatoma	nd[b]	+	nd
Neuronal	+	+	nd
Endothelial	+	nd	nd
E. coli	−	nd	nd

[a] +, protective; −, augments cytotoxicity.
[b] nd, no data available.

$Na_2N_2O_3$), a compound similar to the NONOates which donates nitroxyl (NO^-) instead of NO•, significantly potentiates the toxicity of H_2O_2 (Table II) (Wink *et al.*, 1996a). These results demonstrate that different redox species of NO can influence the toxicity of ROS differently from NO• itself, and raise the possibility of differential modulation of the toxicity of H_2O_2.

The difference exhibited by the various NO donors may be explained by the effect of the different NO donors on cellular antioxidant defenses, as well as by the amount and flux of NO produced during the experiment. One of the major cellular defenses against H_2O_2 is consumption by the enzymes glutathione (GSH) peroxidase and catalase (Halliwell and Gutteridge, 1989). When the kinetics for the disappearance of H_2O_2 were examined in the presence of the different NO donors, it was noted that several of the compounds inhibit the cellular consumption of H_2O_2 to varying degrees. In these studies, SNP, ($Et_2N[N(O)NO]Na$) (DEA/NO), AS, and *S*-nitroso-*N*-acetylpenicillamine (SNAP) all increased the amount of time required to decompose 0.75 mM H_2O_2 by as much as 30–200% (Wink *et al.*, 1996a). In the case of SIN-1 and *S*-nitrosoglutathione (GSNO), the consumption of H_2O_2 was retarded by as much as 400% (Wink *et al.*, 1996a). Thus, the enhancement of H_2O_2-mediated toxicity by AS and SIN-1 might be explained partially by the inhibition of H_2O_2 consumption. However, this cannot be the sole mechanism by which NO enhances or protects against H_2O_2, as GSNO, SNAP, and DEA/NO also decrease the rate of decomposition of H_2O_2, yet are cytoprotective.

Furthermore, an examination of different NO donors and products has revealed different effects on intracellular levels of GSH. Exposure of V79 cells to 1 mM nitrite, SNAP, SIN-1, GSNO, DEA/NO, or AS results in varying degrees of depletion of intracellular GSH (Table II). Intracellular GSH was evaluated as previously described (Tietze, 1969; Cook *et al.*, 1997). Exposure to SNAP, GSNO,

TABLE II
Effect of NO Donor Compounds

NO donor	Cytotoxicity[a]	NO detection[b]	Effect on H_2O_2 consumption (%)[c]	Effect on GSH (intracellular %)[d]
DEA/NO	+	+	–	–25%
PAPA/NO	+	+	nd	nd
SNAP	+	+	–	–25%
GSNO	+	+	nd	nd
SIN–1	–	–	–	–90%
SNP	–	–	ne	nd
AS	–	–	–	–85%

[a]+, protective; –, augments cytotoxicity.
[b]+, presence of detectable NO; –, lack of NO detection.
[c]–, decrease in peroxide consumption rate; nd, no data available; ne, no effect.
[d]–, decrease in intracellular GSH; nd, no data available.

or DEA/NO results in only a modest decrease (<30%), after which the levels of GSH recover rapidly. However, SIN-1 and AS decrease intracellular GSH levels by as much as 85%. Nitrite (1 mM) decreases the GSH levels in these cells by 50% after a 1-h exposure.

The other major explanation for the difference in protective effects among the various chemical NO donors may reflect the actual flux of NO produced by each compound, as assessed electrochemically with an electrode selective for NO. The temporal profiles of NO release by the different compounds demonstrate that different amounts of NO are released over time (Wink *et al.*, 1996a). Both the NONOates and the *S*-nitrosothiol complexes, which protect against H_2O_2 toxicity, release NO over the time course of exposure to H_2O_2. However, SIN-1, SNP, and AS do not produce measurable NO (<1 μM) under these experimental conditions, coincident with a lack of protection against H_2O_2 (Wink *et al.*, 1996a).

SNP appears to increase the toxicity of ROS by yet other mechanisms. The chemistry mediated by SNP can result in formation of chemical species other than NO, such as cyanide (CN^-) and iron. Desferrioxamine (DF) completely protects cells from H_2O_2, yet DF only partially protects against the toxicity mediated by SNP combined with H_2O_2 (Wink *et al.*, 1996a). This discrepancy may be accounted for by the enhanced CN^- released from SNP. Monocytes and polymorphonuclear leukocytes have been shown to facilitate the release of CN^- from SNP, a phenomenon believed to mediated by H_2O_2. The authors suggested that a transition metal complex with a labile ligand could then further oxidize substrates via Fenton-type catalysis (Campbell *et al.*, 1993). Additional evidence supporting this hypothesis was reported by Imlay *et al.* (1988), who showed that bacteria became more sensitive to H_2O_2 in the presence of CN^-. The fact that DF completely protects against the toxicity of CN^- suggests that metal–peroxide reactions are required to initiate cytotoxicity. Thus, the DF-insensitive enhancement of H_2O_2-mediated toxicity by SNP might be attributed to an iron complex that cannot be bound by DF; such a complex could catalyze the Fenton oxidation chemistry of cellular molecules.

Freeman and co-workers have investigated the effect of NO on xanthine oxidase (XO)-mediated lipid peroxidation and found that NO acts as an antioxidant (Rubbo *et al.*, 1995). We have also examined the effect of NO on organic hydroperoxide-mediated toxicity, thought to be mediated by oxidation of lipophilic membranes (Wink *et al.*, 1995a). Our studies further illustrate the importance of the amounts of NO present over time vis-à-vis the exposure to oxidants. DEA/NO, whose half-life is about 2 min, does not protect against either *t*-butyl hydroperoxide or cumene hydroperoxide. ($NH_3^+(C_3H_6)(N[N(O)NO]^-(C_3H_7))$) (PAPA/NO), a NONOate with a half-life of 15 min, markedly protects against both *t*-butyl hydroperoxide and cumene hydroperoxide (Table I). The different effects of these two NONOates on cytotoxicity are attributable to the timing of NO delivery. For a given iso-survival curve, exposure to organic peroxides requires up to 2 hr; in contrast, the half-life is only 10–15 min for H_2O_2 (Wink *et al.*, 1995a). Because alkyl hydroperoxides require more time to penetrate cells and exert their damage, more sustained fluxes of NO have greater protective efficacy.

Several potential mechanisms may be involved in the protection against organic hydroperoxide-mediated toxicity by NO. Intracellular metalloproteins such as those containing heme moieties react quickly with organic peroxides to form hypervalent complexes. These complexes can decompose and release intracellular iron, which in turn can catalyze damage to macromolecules such as DNA. Nitric oxide can show near diffusion-controlled rate constants with these hypervalent metalloproteins, which may restore these oxidized species to the ferric form (Fig. 1). The reduction of metallo-oxo-proteins prevents both their oxidative chemistry and their decomposition to release intracellular iron (Kanner *et al.*, 1991; Wink *et al.*, 1994; Gorbunov *et al.*, 1995), thus limiting intracellular damage mediated by oxidative stress.

Although NO can protect against the toxicity of H_2O_2 to mammalian cells, the opposite effect is observed when the target is *Escherichia coli.* H_2O_2, delivered either as a bolus or through the enzymatic activity of XO, exhibits only modest bactericidal activity (Pacelli *et al.*, 1995). However, simultaneous exposure to both H_2O_2 and NO, the latter delivered either as gas or by a NONOate complex, increases bactericidal activity by four orders of magnitude. Addition of either catalase or superoxide dismutase demonstrates that $NO^{\bullet}/H_2O_2$ are the chemical species responsible for this bactericidal activity. The combination of $NO^{\bullet}$ and H_2O_2 would appear to be ideally suited for killing *E. coli* while exerting a protective effect on the host. This mechanism may hold true for other species of bacteria as well, albeit with different kinetics. In a recent study, staphylococcal killing by $O_2^{-\bullet}$ was abrogated by NO at early time points, yet NO helped to sustain killing for longer time intervals. Maximal killing depended on the timing of exposure to $NO^{\bullet}$, H_2O_2, and $O_2^{-\bullet}$ (Kaplan *et al.*, 1996). These findings may explain why NO and ROS are produced by immune effector cells at different times following exposure to different pathogens.

The diametrically opposite responses of mammalian cells and some prokaryotes to the combination of NO/H_2O_2 may reflect their different cellular structures and complements of metalloproteins. Bacteria utilize iron–sulfur clusters to a greater extent than do mammalian cells, and these types of proteins are especially susceptible to degradation mediated by NO or RNOS (Drapier and Bouton, 1996; Hentze and Kuhn, 1996). In *E. coli,* decomposition of iron complexes which appears to occur in close proximity to the cytoplasm. This relative lack of compartmentalization may allow iron to bind and oxidize DNA. However, because of the organellar structure of mammalian cells, metal labilization may be limited to the cytoplasm and mitochondria. In this cellular arrangement, metals would be required to travel large distances to reach the nucleus and bind to DNA.

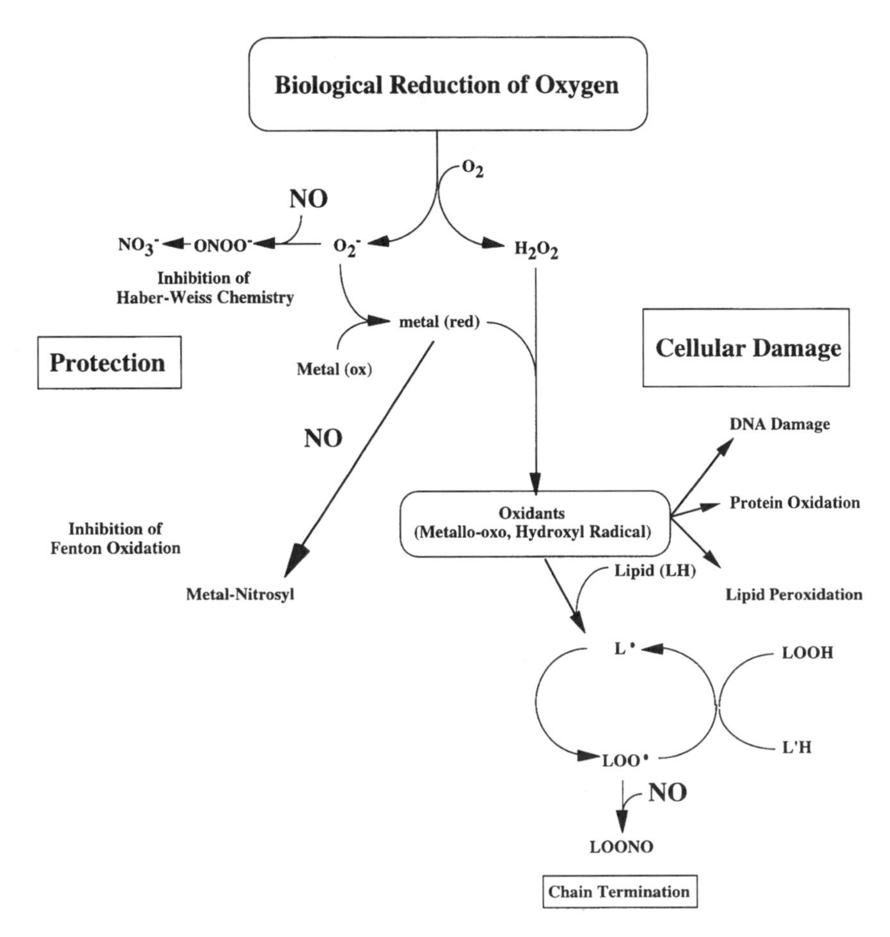

FIGURE 1. Cytoprotective mechanisms of NO against oxidative stress.

3. The Effect of $NO\cdot/O_2^-\cdot$ on Cytotoxicity

Treatment of cells with a bolus of peroxynitrite results in both bacterial (Zhu *et al.*, 1992) and mammalian cell death (see Pryor and Squadrito, 1996, for review). However, treatment with the combination of XO and NO donors, believed to generate peroxynitrite *in situ*, is not necessarily toxic to various cell types. In fact, as discussed above, the presence of NO can actually protect cells against toxicity mediated by ROS (Wink *et al.*, 1993, 1994, 1995a,b, 1996a; Kim *et al.*, 1995; Gupta *et al.*, 1997). Other studies have indicated that ovarian carcinoma cells exposed to 5 mM SIN-1 do not suffer appreciable toxicity (Farias-Eisner *et al.*, 1996). These results suggest that peroxynitrite formation *in situ* is not necessarily toxic to mammalian cells (Wink *et al.*, 1996a), and that there may be a distinct difference between treating cells with bolus millimolar concentrations of peroxynitrite, as opposed to the simultaneous production of $NO^{\bullet}$ and $O_2^{-\bullet}$.

This discrepancy may be explained by the peroxynitrite concentrations actually achieved. Zhu *et al.* (1992) reported that high concentrations of bolus peroxynitrite are required to penetrate cells. Although the simultaneous generation of $NO^{\bullet}$ and $O_2^{-\bullet}$ results in the formation of peroxynitrite, the short lifetime of this compound in solution does not permit the accumulation of sufficiently high concentrations of peroxynitrite to penetrate the cell. Therefore, the amount of peroxynitrite that can cross the cell membrane under biologically relevant conditions—and thus the contribution of peroxynitrite to toxicological mechanisms—may be low.

Another factor that may affect the toxicity mediated by peroxynitrite is its reaction with $NO^{\bullet}$ to form $NO_2^{\bullet}$ and N_2O_3. Peroxynitrite can be formed from the $NO^{\bullet}/O_2^{-\bullet}$ reaction, which in turn can react with excess $NO^{\bullet}$ to form $NO_2^{\bullet}$ and eventually N_2O_3 (Wink *et al.*, 1997):

$$ONOO^- + H^+ + NO^{\bullet} \rightarrow NO_2^{\bullet} + HNO_2$$

$$NO_2^{\bullet} + NO^{\bullet} \rightarrow N_2O_3$$

$$N_2O_3 + H_2O \rightarrow HNO_2$$

These reactions suggest that excess $NO^{\bullet}$ can moderate the oxidative chemistry of peroxynitrite, in addition to that of Fenton-type reactions (Fig. 2).

Additional cytoprotective mechanisms may also be involved. Pretreatment with SNAP can induce cross-resistance of isolated hepatocytes to subsequent treatment with H_2O_2 (Kim *et al.*, 1995). However, this protection is inhibitable by cycloheximide, suggesting that the nitrogen oxide may be inducing the synthesis of antioxidant protein(s).

4. Conclusions

In mammalian cells, it appears that NO can provide cytoprotective effects against chemical insults by agents that generate oxidative stress. This antioxidant effect may be important to minimize tissue injury by ROS-dependent processes required for the destruction of various pathogenic microorganisms. Although peroxynitrite may have potent oxidizing properties, the conversion to less toxic RNOS by NO may serve to control the chemistry of peroxynitrite *in vivo*.

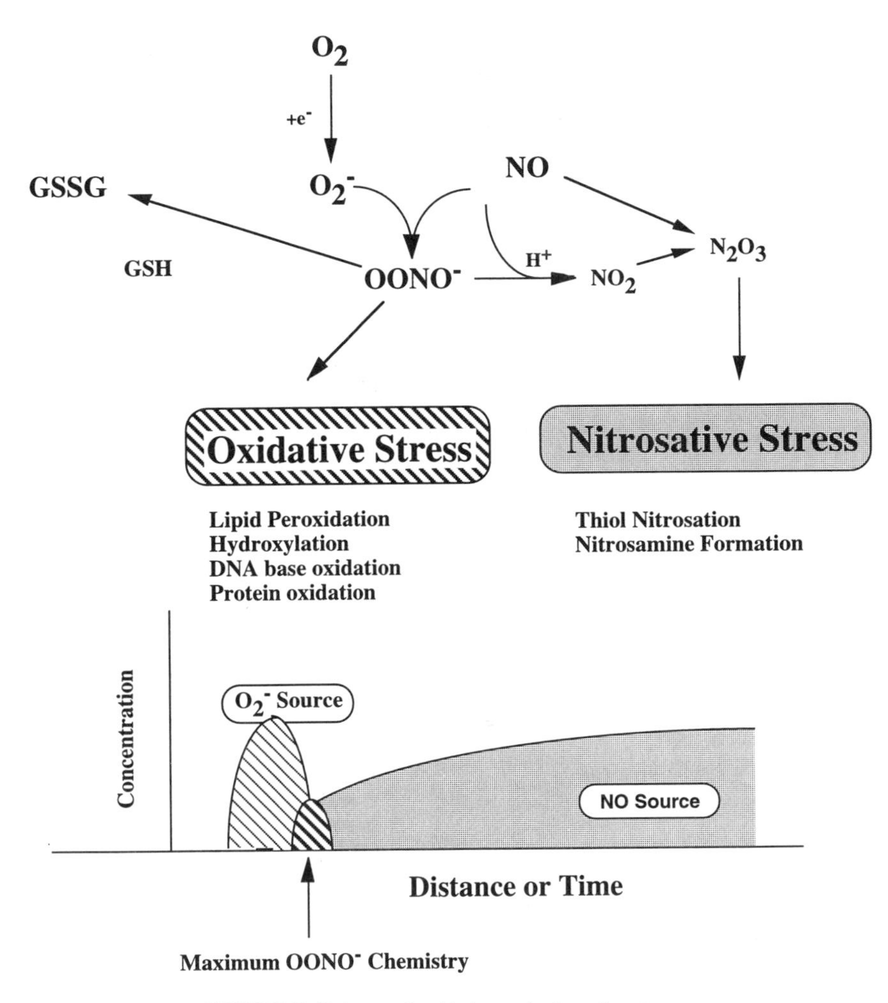

FIGURE 2. Balance of oxidative and nitrosative stress.

References

Ames, B. N., Cathcart, R., Schwiers, E., and Hochstein, P., 1981, Uric acid provides an antioxidant defense in humans against oxidant- and radical-causing aging and cancer: A hypothesis, *Proc. Natl. Acad. Sci. USA* **78:**6858–6862.

Ames, B. N., Shigenaga, M. K., and Hagen, T. M., 1993, Oxidants, antioxidants, and the degenerative diseases of aging, *Proc. Natl. Acad. Sci. USA* **90:**7915–7922.

Campbell, J. M., McCrae, F., Reglinski, J., Wilson, R., Smith, W. E., and Sturrock, R. D., 1993, The interaction of sodium nitroprusside with peripheral white blood cells in vitro: A rationale for cyanide release *in vivo*, *Biochim. Biophys. Acta* **1156:**327–333.

Chang, J., Rao, N. V., Markewitz, B. A., Hoidal, J. R., and Michael, J. R., 1996, Nitric oxide donor prevents hydrogen peroxide-mediated endothelial cell injury, *Am. J. Physiol.* **270:**L931–L940.

Cook, J. A., Krishna, M. C., Pacelli, R., DeGraff, W., Liebmann, J., Russo, A., Mitchell, J. B., and Wink, D. A., 1997, Nitric oxide enhancement of melphalan-induced cytotoxicity, *Br. J. Cancer* **76:**325–334.

Drapier, J.-C., and Bouton, C., 1996, Modulation by nitric oxide of metalloprotein regulatory activities, *BioEssays* **18:**1–8.

Farias-Eisner, R., Chaudhuri, G., Aeberhard, E., and Fukuto, J. M., 1996, The chemistry and tumoricidal activity of nitric-oxide hydrogen-peroxide and the implications to cell resistance susceptibility, *J. Biol. Chem.* **271:**6144–6151.

Gorbunov, N. V., Osipov, A. N., Day, B. W., Zayas-Rivera, B., Kagan, V. E., and Elsayed, N. M., 1995, Reduction of ferrylmyoglobin and ferrylhemoglobin by nitric oxide: A protective mechanism against ferryl hemoprotein-induced oxidations, *Biochemistry* **34:**6689–6699.

Gross, S. S., and Wolin, M. S., 1995, Nitric oxide: Pathophysiological mechanisms, *Annu. Rev. Physiol.* **57:**737–769.

Gupta, M. P., Evanoff, V., and Hart, C. M., 1997, Nitric oxide attenuates hydrogen peroxide-mediated injury to porcine pulmonary artery endothelial cells, *Am. J. Physiol.* **272:**L1133–L1141.

Halliwell, B., and Gutteridge, J. M. C., 1989, *Free Radicals in Biology and Medicine,* Clarendon Press, Oxford, pp. 416–509.

Hata, Y., Ota, S., Hiraishi, H., Terano, A., and Ivey, K. J., 1996, Nitric oxide enhances cytotoxicity of cultured rabbit gastric mucosal cells induced by hydrogen peroxide, *Biochim. Biophys. Acta* **1290:**257–260.

Hentze, M. W., and Kuhn, L. C., 1996, Molecular control of vertebrate iron metabolism: mRNA-based regulatory circuits operated by iron, nitric oxide, and oxidative stress, *Proc. Natl. Acad. Sci. USA* **93:**8175–8182.

Ignarro, L. J., 1990, Biosynthesis and metabolism of endothelium-derived nitric oxide, *Annu. Rev. Pharmacol. Toxicol.* **30:**535–560.

Imlay, J. A., Chin, S. M., and Linn, S., 1988, Toxic DNA damage by hydrogen peroxide through the Fenton reaction *in vivo* and *in vitro*, *Science* **240:**640–642.

Kanner, J., Harel, S., and Granit, R., 1991, Nitric oxide as an antioxidant, *Arch. Biochem. Biophys.* **289:**130–136.

Kaplan, S. S., Lancaster, J. R., Basford, R. E., and Simmons, R. L., 1996, Effect of nitric oxide on staphylococcal killing and interactive effect with superoxide, *Infect. Immun.* **64:**69–76.

Keefer, L. K., Nims, R. W., Davies, K. W., and Wink, D. A., 1996, NONOates (diazenolate-2-oxides) as nitric oxide dosage forms, *Methods Enzymol.* **268:**281–294.

Kim, Y. M., Bergonia, H., and Lancaster, J. R., Jr., 1995, Nitrogen oxide-induced autoprotection in isolated rat hepatocytes, *FEBS Lett.* **374:**228–232.

Linas, S. L., and Repine, J. E., 1997, Endothelial cells protect vascular smooth muscle cells from H_2O_2 attack, *Am. J. Physiol* **272:**F767–F773.

Moncada, S., Palmer, R. M. J., and Higgs, E. A., 1991, Nitric oxide: Physiology, pathophysiology, and pharmacology, *Pharmacol. Rev.* **43:**109–142.

Pacelli, R., Wink, D. A., Cook, J. A., Krishna, M. C., DeGraff, W., Friedman, N., Tsokos, M., Samuni, A., and Mitchell, J. B., 1995, Nitric oxide potentiates hydrogen peroxide-induced killing of *Escherichia coli*, *J. Exp. Med.* **182:**1469–1479.

Pryor, W. A., and Squadrito, G. L., 1996, The chemistry of peroxynitrite and peroxynitrous acid: Products from the reaction of nitric oxide with superoxide, *Am. J. Physiol.* **268:**L699–L721.

Rubbo, H., Parthasarathy, S., Barnes, S., Kirk, M., Kalyanaraman, B., and Freeman, B. A., 1995, Nitric oxide inhibition of lipoxygenase-dependent liposome and low-density lipoprotein oxidation: Termination of radical chain propagation reactions and formation of nitrogen-containing oxidized lipid derivatives, *Arch. Biochem. Biophys.* **324:**15–25.

Tietze, F., 1969, Enzymic method for quantitative determination of nanogram amounts of total and oxidized glutathione. Application to mammalian blood and other tissues, *Anal. Biochem.* **27:**502–522.

Wink, D. A., Hanbauer, I., Krishna, M. C., DeGraff, W., Gamson, J., and Mitchell, J. B., 1993, Nitric oxide protects against cellular damage and cytotoxicity from reactive oxygen species, *Proc. Natl. Acad. Sci. USA* **90:**9813–9817.

Wink, D. A., Hanbauer, I., Laval, F., Cook, J. A., Krishna, M. C., and Mitchell, J. B., 1994, Nitric oxide protects against the cytotoxic effects of reactive oxygen species, *Ann. N. Y. Acad. Sci.* **738:**265–278.

Wink, D. A., Cook, J. A., Krishna, M. C., Hanbauer, I., DeGraff, W., Gamson, J., and Mitchell, J. B., 1995a, Nitric oxide protects against alkyl peroxide-mediated cytotoxicity: Further insights into the role nitric oxide plays in oxidative stress, *Arch. Biochem. Biophys.* **319:**402–407.

Wink, D. A., Cook, J. A., Pacelli, R., Liebmann, J., Krishna, M. C., and Mitchell, J. B., 1995b, Nitric oxide (NO) protects against cellular damage by reactive oxygen species, *Toxicol. Lett.* **82–83:**221–226.

Wink, D. A., Cook, J., Pacelli, R., DeGraff, W., Gamson, J., Liebmann, J., Krishna, M., and Mitchell, J. B., 1996a, Effect of various nitric oxide-donor agents on peroxide mediated toxicity. A direct correlation between nitric oxide formation and protection, *Arch. Biochem. Biophys.* **331:**241–248.

Wink, D. A., Grisham, M., Mitchell, J. B., and Ford, P. C., 1996b, Direct and indirect effects of nitric oxide. Biologically relevant chemical reactions in biology of NO, *Methods Enzymol.* **268:**12–31.

Wink, D. A., Hanbauer, I., Grisham, M. B., Laval, F., Nims, R. W., Laval, J., Cook, J. A., Pacelli, R., Liebmann, J., Krishna, M. C., Ford, P. C., and Mitchell, J. B., 1996c, The chemical biology of NO. Insights into regulation, protective and toxic mechanisms of nitric oxide, *Curr. Top. Cell. Regul.* **34:**159–187.

Wink, D. A., Cook, J. A., Kim, S., Vodovotz, Y., Pacelli, R., Krishna, M. C., Russo, A., Mitchell, J. B., Jourd'heuil, D., Miles, A. M., and Grisham, M. B., 1997, Superoxide modulates the oxidation and nitrosation of thiols by nitric oxide derived reactive intermediates, *J. Biol. Chem.* **272:**11147–11151.

Yamada, M., Momose, K., Richelson, E., and Yamada, M., 1996, Sodium nitroprusside-induced apoptotic cellular death via production of hydrogen peroxide in murine neuroblastoma N1E-115, *J. Pharmacol. Toxicol. Methods* **35:**11–17.

Zhu, L., Gunn, C., and Beckman, J. S., 1992, Bactericidal activity of peroxynitrite, *Arch. Biochem. Biophys.* **298:**452–457.

CHAPTER 10

Immunomodulatory Actions of Nitric Oxide

IAIN B. McINNES and FOO Y. LIEW

Although NO was first described as the elusive endothelium-derived relaxing factor (EDRF) (Ignarro *et al.*, 1987; Palmer *et al.*, 1987), many studies have subsequently established that NO production and responsiveness occurs in many cells implicated in inflammation, innate host defense, and the evolution of antigen-specific immune responses. This chapter will review the role of NO in the regulation of these protective responses.

1. Regulation of NO Synthesis during Immune Responses

As discussed in Chapter 4, three isoforms of NO synthase (NOS) have been identified and their enzymology extensively studied (Bredt and Snyder, 1994; Nathan and Xie, 1994). Constitutively expressed endothelial NOS (eNOS, NOS3) and neuronal NOS (nNOS, NOS1) are capable of rapid-onset, short-lived generation of low concentrations of NO (together termed cNOS). Inducible NOS (iNOS, NOS2), in contrast, is present in cells only after specific upregulation, which requires novel protein synthesis, but thereafter generates high concentrations of NO over prolonged periods. NOS isoforms share 30–40% homology with cytochrome P450 reductase (CPR), with consensus sequences for redox-active cofactors including NADPH, FAD, and FMN. During activation, NOS forms dimers in the presence of tetrahydrobiopterin, heme, and L-arginine. In contrast to cNOS, which is calcium dependent, iNOS tightly binds calmodulin at a basic, hydrophobic site,

IAIN B. McINNES and FOO Y. LIEW • Department of Immunology and Centre for Rheumatic Diseases, University of Glasgow, Glasgow G11 6NT, United Kingdom.

Nitric Oxide and Infection, edited by Fang. Kluwer Academic / Plenum Publishers, New York, 1999.

and iNOS enzyme activity is independent of ambient calcium concentration (Cho *et al.*, 1992).

iNOS was first cloned from murine macrophages and subsequently from human hepatocytes and chondrocytes (reviewed by Charles *et al.*, 1993; Geller *et al.*, 1993; Nathan and Xie, 1994), but has not yet been cloned from human macrophages. Human iNOS, encoded on chromosome 17, shares ~50% homology with cNOS and 80% homology with murine iNOS. Widespread tissue distribution of iNOS has been reported, with expression observed in human keratinocytes, hepatocytes, osteoblasts/osteoclasts, chondrocytes, uterine smooth muscle cells, mesangial cells, dermal fibroblasts, neutrophils, and respiratory epithelial cells. Expression in human tumors has also been detected, including colorectal adenocarcinoma and glioblastoma. Considerable controversy has surrounded attempts to demonstrate the presence and activity of iNOS in human macrophages, a topic discussed in greater detail in Chapter 6. Whereas some authors have demonstrated NO production or L-arginine-/NO-dependent activity (e.g., Denis, 1991; Zembala *et al.*, 1994; Burkrinsky *et al.*, 1995; Dugas *et al.*, 1995), others have been unable to detect any evidence of iNOS activity (Schneemann *et al.*, 1993). The required stimuli for iNOS upregulation in human macrophages appear to differ from those in rodents, and *in vitro* NO production has been approximately an order of magnitude lower than that observed in rodent macrophages. Whether this represents a functionally significant difference in the role of NO in the generation of immune responses in rodents compared with humans remains unclear. Nevertheless, the widespread tissue distribution of iNOS confers on host tissue cells of either species the ability to contribute to regulation of immune responses through high-output NO generation.

The effect of NO as an immunomodulator is intimately linked to regulation of its own synthesis, and numerous feedback loops are likely to exist. Modulation of immune responses requires rapid elaboration of immunoactive mediators. Because NO may not be readily stored in a biologically active form, its concentration in tissues and thus its contribution to immunity is regulated through NOS activity. eNOS generates NO at picomolar concentrations in response to local vasoactive mediators such as bradykinin, thrombin, histamine, acetylcholine, and 5-hydroxytryptamine, to cytokines such as IL-1β and endothelin-3, or to physical factors, including shear stress or increased blood flow (reviewed by Lyons, 1995). Whether eNOS output may be further upregulated is unclear. Evidence for induction of a novel isoform resembling nNOS by IL-1β and LPS in osteoarthritis chondrocytes has been reported, suggesting that the delineation between low- and high-output NOS on the basis of calcium dependency alone may be oversimplified (Amin *et al.*, 1995).

The predominant source of NO in inflammatory lesions, however, is iNOS. Given the ubiquitous effects of NO in inflammatory lesions, it might be anticipated that many factors control iNOS expression and activity (Fig. 1). Thus, iNOS may be

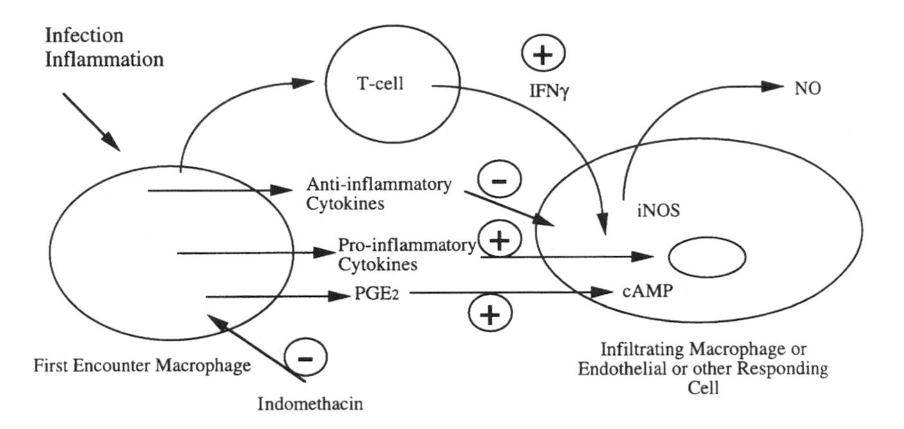

FIGURE 1. Regulation of NO production.

activated *in vitro* by cytokines, microbial products (particularly LPS and superantigen toxins), picolinic acid, cAMP-elevating agents, and physical factors such as UV light or trauma (reviewed by Nathan and Xie, 1994; Lyons, 1995). Cytokines appear to exert the major regulatory influence *in vivo* (reviewed in detail in Chapter 5). IFNγ is a potent inducer of NO production by rodent macrophages and endothelial cells, in synergy with LPS (Liew and Cox, 1991). IL-1β and TNFα also increase iNOS expression in many cells, either alone or in synergistic combination (Liew, 1994). Cytokine requirements vary with the species and tissue origin of cells. Thus, rodent smooth muscle cells respond either to IL-1β alone, or in synergy with IFNγ or TNFα, whereas human vascular smooth muscle cells require a combination of LPS, IFNγ, and TNFα for NO production. Similarly, rodent hepatocytes respond to LPS alone, whereas human hepatocytes require a combination of LPS, IFNγ, TNFα, and IL-2 (reviewed by Geller *et al.*, 1993; Liew, 1994; Nathan and Xie, 1994; Lyons, 1995).

Multiple inhibitory mediators have been described. TGFβ, IL-4, IL-8, IL-10 (indirectly through effects on TNFα production), IL-13, MIP-1α, epidermal growth factor (EGF), platelet-derived growth factor (PDGF), and fibroblast growth factor (FGF) each oppose iNOS activation (reviewed by Liew, 1994; Nathan and Xie, 1994; Lyons, 1995). Species and tissue specificity appear to be important. TGFβ inhibits rodent macrophage and endothelial iNOS expression, but enhances NO production in Swiss 3T3 fibroblasts (Gilbert and Herschman, 1993). Moreover, IL-10 has been shown to increase iNOS activity in avian osteoclasts (Sunyer *et al.*, 1996), as has IL-4 in human macrophages (Dugas *et al.*, 1995). The temporal sequence of ligand binding appears important, as preexposure of macrophages to LPS suppresses subsequent IFNγ-induced NO production. Such observations emphasize the difficulties attached to extrapolation between species and cell types. Furthermore, it has only recently been appreciated that iNOS may be constitutively present in human tissues, such as lung, retina, skeletal muscle, kidney, or CNS, in the absence of specific activating factors (reviewed by Nathan and Xie, 1994). This suggests that, in addition to a postulated responsive role in inflammation, iNOS may be involved in normal physiological regulation.

Engagement of class II MHC by allospecific monoclonal antibodies or by bacterial superantigens [toxic shock syndrome toxin (TSST)-1, staphylococcal enterotoxin B (SEB)] in the presence of syngeneic lymphocytes increases NO production by macrophages (Isobe and Nakashima, 1992; Tao and Stout, 1993). The requirement for T cells in this model can be replaced by exogenous IFNγ (our unpublished observations). Similarly, activation of macrophage NO synthesis follows cross-linking of CD69 by antibody. Cell–cell contact between macrophages and T lymphocytes of both Th1 and Th2 subsets leads to iNOS expression mediated in part through CD40/CD40 ligand and LFA-1/ICAM-1 (e.g., Stout *et al.*, 1996). Thus, homo- or heterotypic cell contact can induce NO synthesis in an inflammatory lesion. The relative contribution of such mechanisms in the context of high levels of cytokine production is currently unclear.

Glucocorticoids, which are potent immunosuppressive agents *in vivo*, inhibit iNOS-dependent activity (Di Rosa *et al.*, 1990). The rate of iNOS transcription is reduced, and although mRNA is briefly stabilized, translation is significantly retarded and degradation of iNOS protein is enhanced (Kunz *et al.*, 1996). NF-κB p50 or p65 nuclear translocation is unaffected, but NF-κB and not AP-1 binding in the iNOS promoter region is prevented (Kleinert *et al.*, 1996). L-N^{ω}-substituted arginines also inhibit NOS activity in biological systems. L-N^{ω}-monomethylarginine (L-NMMA) is commonly used when no isoform specificity is necessary. Although a completely isoform-specific inhibitor has yet to be identified, L-N^{ω}-nitroarginine (L-NNA) exhibits relative specificity for cNOS, as does L-N^{ω}-aminoarginine for iNOS. Aminoguanidine and *N*-iminoethyl-L-lysine are additional relatively iNOS-specific inhibitors used in animal models. However, mice in which the nNOS, iNOS, and eNOS genes have been specifically inactivated have now been generated, greatly facilitating the evaluation of the specific contribution of individual isoforms in different biological systems *in vivo* (Huang *et al.*, 1995; MacMicking *et al.*, 1995; Nelson *et al.*, 1995; Wei *et al.*, 1995).

2. Functional Consequences of NO Production during Immune Responses

The extensive tissue distribution and wide range of potential regulatory factors indicate roles for NO at multiple levels in host defense.

2.1. NO in Acute Inflammation

By virtue of its EDRF activity, NO can induce vasodilatation through relaxation of vascular smooth muscle, leading to erythema and increased local temperature (reviewed by Schmidt and Walter, 1994). Data from murine dextran- and carrageenin-induced models of inflammation indicate that NO also induces clinically detectable edema formation, through alteration of endothelial permeability. Thus, two features of the classical inflammatory response are regulated by NO. A further level of complexity lies in the interaction of iNOS with constitutive and inducible isoforms of cyclooxygenase and their products, the prostaglandins.

NO inhibits platelet aggregation as a function of its cardioprotective role, through production of cGMP. Subsequent studies of ischemia–reperfusion in mesenteric vessels and myocardium have indicated that NO also reduces neutrophil adhesion through CD11/CD18-, ICAM-1-, and P-selectin-dependent pathways, and by scavenging reactive oxygen intermediates that enhance adhesion (Kubes, *et al.*, 1993). These data indicate that NO can modify cellular recruitment, and in particular the crucial interaction between leukocyte and endothelium that is critical for the evolution of cell-mediated responses.

2.2. T-Cell Activation and Cytokine Synthesis

NO exerts biphasic effects on T-lymphocyte responsiveness. Initial observations described inhibition of *in vitro* antigen- or mitogen-driven T-cell proliferation, either by NO donors or by macrophage-derived NO in cocultures (Merryman *et al.*, 1993). Some antiproliferative effects of NO may be attributable to inhibitory effects on Janus kinases (Duhe *et al.*, 1998). Subsequently, it was established that low-dose NO significantly enhances peripheral blood lymphocyte activation, measured by PHA-induced proliferation, increased glucose uptake, increased NF-κB binding activity, and activation of protein tyrosine kinase $p56^{lck}$ (Lander *et al.*, 1993). Moreover, *in vitro* and *in vivo,* L-arginine enhances lymphocyte proliferation, and increases NK-cell and lymphokine-activated killer activity (Park *et al.*, 1991). Of most interest, however, has been the potential for modulating effects on functional maturation of T lymphocytes. T cells may be segregated on the basis of the predominant cytokines expressed within inflammatory lesions. Thus, IL-12, IFNγ, and TNFα synthesis is associated with local, cell-mediated Th1-type responses, whereas IL-4, IL-5, and IL-10 expression is more often associated with humoral immunity and hypersensitivity. Recent studies in murine T-cell clones have established that NO preferentially inhibits Th1 clonal proliferation to antigen, but has no effect on Th2 clones (Liew *et al.*, 1991; Taylor-Robinson *et al.*, 1994; Wei *et al.*, 1995). Moreover, proliferative responses by spleen cells to mitogen and to staphylococcal superantigens are diminished in iNOS-deficient mice (Wei *et al.*, 1995; McInnes *et al.*, 1998). Thus, the local concentration of NO and the maturational phenotype together influence the modulatory effect of NO on T cells. Amplification of the cellular immune response may be further modified by NO through induction of apoptosis (Albina *et al.*, 1993), and through altered T-cell recirculation and tissue ingress. In particular, P-selectin, which is downregulated by peroxynitrite *in vitro*, has recently been shown to recruit Th1 rather than Th2 cells to inflamed skin or joints in murine inflammatory models (Lefer *et al.*, 1997). One can speculate that through such pathways, local NO synthesis may further modify the nature of the T-cell compartment within an inflammatory lesion.

Whereas a clear effect of NO on Th1-cell cytokine production has been demonstrated in several systems, it remains unclear whether Th2 cells are directly influenced by NO synthesis. Whole spleen cell cultures responding to *Listeria monocytogenes* produce enhanced IFNγ in the presence of iNOS inhibitors, which can be reversed by the addition of NO donors (Xiong *et al.*, 1996). These data concur with similar observations made *in vivo*. iNOS-deficient mice generate exaggerated Th1 responses with increased IFNγ generation during infection with *Staphylococcus aureus* (McInnes *et al.*, 1998) and *Leishmania major* (Wei *et al.*, 1995). Moreover, spleen cells from iNOS-deficient mice synthesize high levels of IFNγ a priori, indicating that NO is normally required to regulate Th1-type responses. Initial reports suggested that IL-2 synthesis by T lymphocytes, like that of IFNγ, was inhibited by NO (Liew *et al.*, 1991). Such studies provided a molecular explanation for macrophage-mediated T-cell suppression. However, recent *in vitro* studies using a picryl chloride-driven T-cell/macrophage coculture system found no evidence of NO-mediated inhibition of IL-2 synthesis by macrophages. Rather, such a role was established for cyclooxygenase products, specifically PGE_2, as indomethacin enhances IL-2 synthesis (Marcinkiewicz *et al.*, 1996). In the same system, high concentrations of NO donors were able to reduce IL-2 synthesis, confirming previous observations. Thus, the local concentration of NO in tissues may determine the ultimate effect of NO *in vivo*. Moreover, NO may itself modify prostaglandin synthesis through direct effects on cyclooxygenase (e.g., Manfield *et al.*, 1996; Habib *et al.*, 1997), suggesting the presence of a complex regulatory network *in vivo*.

Similar uncertainty surrounds the role of NO in regulation of Th2-cell secretion. Although suppression of IL-4 production by Th2 clones has been observed (Nukaya *et al.*, 1995), it appears more likely that NO exerts either neutral (Marcinkiewicz *et al.*, 1996) or positive (Chang *et al.*, 1997) effects in the overall development of Th2-type immunity. Suppression of IFNγ by NO may lead to enhanced IL-4 production through Th1/Th2 cross-regulation. Direct enhancement of IL-4 production by Th2 clones and by EL4 T-lymphoma cells stimulated with mitogen has also been reported, in association with weak activation of the IL-4 promoter site and inhibition of IL-2 transcription (Chang *et al.*, 1997). In contrast, no enhancement of IL-4 was observed using the picryl chloride T-cell model *in vitro* (Marcinkiewicz *et al.*, 1996), and we have not observed an increase in IL-4 synthesis in wild-type compared with iNOS-deficient mice (McInnes *et al.*, 1998, and additional unpublished observations). Further investigations using antigen-driven Th2 responses will be required to clarify this issue *in vivo*. This may have particular clinical relevance, as therapeutic administration of NO (e.g., during pulmonary compromise) could conceivably lead to disadvantageous Th2 enhancement (Barnes and Liew, 1995).

2.3. Monokine Production

NO is also implicated in the regulation of monokine production. Increased TNFα production from PBMC exposed to NO donors has been detected, although the specific cellular origin of TNFα was not specified (Lander *et al.*, 1993). Production of cytokines, including IL-6 and TNFα, by purified blood monocytes/macrophages or macrophage cell lines has been variously reported to be suppressed or enhanced in the presence of exogenous NO donors (Deakin *et al.*, 1995; McInnes *et al.*, 1996). The mechanism underlying these observations is unknown, but may reflect modification of transcription factors such as NF-κB. We have recently demonstrated that NO induces TNFα synthesis by synovial tissues from rheumatoid arthritis patients, indicating that NO regulation of monokines is likely to have pathological relevance (McInnes *et al.*, 1996). Complex feedback loops allow the effector function of NO to overlap with its immunomodulatory role to finely tune immune responses.

3. NO and Immune Responses *in Vivo*

The availability of relatively isoform-selective NOS inhibitors and iNOS-deficient mice has facilitated investigation of the role of NO in immune regulation *in vivo*. Such studies rarely separate the effector and regulatory functions of NO. While recognizing this functional complexity for NO in any given model system, it is nevertheless informative to consider the outcome of altered NO production in a range of models.

3.1. Immune Regulation during Infection

Early studies detected increased nitrate generation during septicemia. NO has now been implicated in the response to a large number of organisms, including intracellular bacteria, fungi, protozoa, helminths, and viruses (considered in more detail in Chapter 12). In most cases, microbicidal activity is demonstrable *in vitro*, where it is inhibited by L^{ω}-arginine analogues, and is enhanced by addition of macrophage activating factors, such as IFNγ or LPS. Normally *in vivo*, T cells and macrophages cooperate to regulate NO synthesis through cytokine production. Thus, host responses to *Leishmania major* are dependent on the generation of an effective Th1 response, in which IL-12 and IFNγ production leads to NO-mediated resistance. Parasite killing activity is demonstrable in splenic macrophages *in vitro*, and treatment of infected mice with iNOS inhibitors increases lesion size and parasitic load (Liew *et al.*, 1990). Moreover, *L. major* infection in iNOS-deficient mice is of increased severity and mortality, despite the presence of an enhanced Th1 response, indicating that NO is critical in host defense against this organism (Wei *et al.*, 1995) (see Chapter 17). Similar obervations have been made during *Listeria monocytogenes* infection (MacMicking *et al.*, 1995) (see Chapter 22). However, NO-mediated microbial killing can also proceed in the absence of T cells. *L. monocytogenes*-infected SCID mice exhibit increased mortality and enhanced recovery of viable *Listeria* from spleens after treatment with aminoguanidine. IFNγ production by NK cells is sufficient to confer resistance (Beckerman *et al.*, 1993). Thus, at least in rodent models, the production of NO by activated macrophages constitutes a principal component of the antimicrobial armamentarium.

To further address the complex interaction of NO synthesis, antimicrobial activities, and immunoregulation, we recently explored the effect of iNOS deficiency on murine staphylococcal infection (McInnes *et al.*, 1998). The role of NO in gram-positive infection has been poorly defined. NO has been implicated in *S. aureus* killing by cytoplasts from human neutrophils (Malewista *et al.*, 1996), and NO donors are bactericidal for *S. aureus* in a cell-free system, although the time course of bacterial killing is delayed compared with that mediated by reactive oxygen intermediates (Kaplan *et al.*, 1996). These findings suggest an antimicrobial role for NO. However, *in vivo* injection of certain staphylococcal exotoxins, such as TSST-1 or SEB, leads to a T-lymphocyte-mediated shock syndrome in BALB/c mice, which can be exacerbated by exogenous NOS inhibitors, suggesting that NO-dependent T-cell regulation might be an important regulatory pathway (Florquin and Goldman, 1996). In contrast, staphylococcal cell-wall components, such as peptidoglycan and lipoteichoic acid, synergistically induce multiorgan failure in rats by an NO-mediated mechanism (De Kimpe *et al.*, 1995). Such studies have not adequately separated the relative contribution of constitutive and inducible NO isoforms. We observed that iNOS-deficient mice develop more severe septicemia and arthritis than wild-type littermate controls, and exhibit significantly increased mortality (McInnes *et al.*, 1998). The increased disease severity is correlated with increased IFNγ production *in vivo* and *in vitro*. Mortality in this model is related to T-cell overactivity (Zhao *et al.*, 1995), and it is likely that the immune dysregulation in iNOS-deficient animals contributes significantly to their severe immunopathology. NOS inhibition has also resulted in altered pathology and cytokine profiles in models of sterile inflammation (Hogaboam *et al.*, 1997). These observations clearly demonstrate that the balance of effector and regulatory function by NO can have important clinical significance.

3.2. Regulation of Autoimmune Responses by NO

The evidence reviewed above clearly implicates NO as an integral component of protective host immune responses. That NO is similarly involved in deleterious autoimmune responses, either as an aggressive or a protective component, has been explored in several animal model systems. Murine disease resembling insulin-dependent diabetes mellitus (IDDM) occurs following inoculation with streptozotocin. NO modifies cytokine production within the pancreatic islets and inhibition of NO production using NOS inhibitors leads to delayed onset of disease, with attenuation of the pancreatic inflammatory infiltrate (Lukic *et al.*, 1991; Kolb and Kolb-Bachofen, 1992). Similarly, in the genetically predisposed non-obese diabetic (NOD) mouse model, transfer of NOD mouse spleen cells induces diabetes in irradiated recipients. The onset of disease can be significantly delayed by aminoguanidine treatment (Corbett *et al.*, 1993). These findings implicate NO as an aggressor in IDDM pathogenesis. However, its role in experimental allergic encephalomyelitis (EAE) is more complex. NO production is upregulated in EAE, and iNOS levels have been reported to correlate with disease severity (Okuda *et al.*, 1995). Although NOS inhibitors were found to inhibit clinical signs and progression of EAE in mice and rats (Zielasek *et al.*, 1995; Zhao *et al.*, 1996), paradoxical aggravation of EAE following administration of L^{ω}-arginine analogues (Ruuls *et al.*, 1996) or in $iNOS^{-/-}$ mice (Fenyk-Melody *et al.*, 1998) has also been described. Suppression or aggravation of EAE by NOS inhibitors may depend on the mode of disease induction, or on the choice of inhibitor in T-cell-induced or myelin basic protein-induced EAE. Such discrepancies again emphasize the "double-edged" effector function of NO as an immunosuppressor or neurotoxin, depending on subtle variables of immunogen, inhibitor dosage, and regimen.

NO has been implicated in regulation of immune-complex-mediated disease. In pulmonary alveolitis induced by intratracheal injection of preformed immune complexes, NOS inhibitors reduce the severity of pulmonary hemorrhage and edema formation. Similar inhibition of dermal vasculitis has also been observed. A major component of this model is dependent on intact complement function, suggesting that NO synthesis may interact with complement to mediate pathology (Mulligan *et al.*, 1991). Graft-versus-host disease (GVHD) in mice resembles the early stages of inflammatory bowel disease (IBD) or gut hypersensitivity syndromes such as celiac disease. $(CBA \times BALB/c)F_1$ recipients of CBA spleen cells develop GVHD, which can be significantly retarded by L-NMMA treatment, with preservation of intestinal architecture and reduced density of intraepithelial lymphocyte infiltration (Garside *et al.*, 1992). However, it is unclear whether this effect operates primarily through immunoregulatory modification, or via hemodynamic effects in the mesenteric vasculature. NO production has been detected in human IBD, indicating a possible role in human disease pathogenesis (Broughton-Evans *et al.*, 1993). However, altered epithelial permeability as found in IBD results in increased exposure to bacteria and bacterial products within the lamina propria, with the potential for enhanced local NO production and consequent immunomodulation. Whether NO is ultimately protective or detrimental in this setting is therefore unclear.

Evidence for a role of NO in immune responses during inflammatory arthritis has been derived from several animal studies. Adjuvant arthritis in rats bears histopathological similarity to rheumatoid arthritis. iNOS is detectable in synovial membranes, and elevated levels of urinary and plasma nitrite are maximal after 14 days. Continuous administration of NOS inhibitors prevents or attenuates the clinical severity of arthritis, normalizes weight gain, reduces the acute-phase response, and retards erosive articular destruction (Ialenti *et al.*, 1993; Stefanovic-Racic *et al.*, 1994). Treatment during adjuvant priming alone is sufficient to confer a reduction in disease severity, and anti-mycobacterial antigen-specific T-cell responses are suppressed in treated rats. Similar data have been obtained in streptococcal cell wall (SCW)-induced arthritis in rats, in which L-NMMA inhibits the onset and progression of arthritis (McCartney-Francis *et al.*, 1993). Administration of NOS inhibitors to MRL-MP-*lpr*/*lpr* mice suppresses the development of renal pathology and attenuates clinical and histological evidence of arthritis (Weinberg *et al.*, 1994). NO synthesis is closely linked to IL-12 production in this model, suggesting that cytokine modulation by NO may constitute a positive feedback loop that culminates in end-organ damage (Huang *et al.*, 1996). In addition to these data implicating NO generation in articular pathology in rodents, we have recently demonstrated that NO is produced by macrophages and fibroblasts within the synovial membrane of patients with rheumatoid arthritis. Such NO synthesis may enhance TNFα production, a cytokine that is critical to disease pathogenesis (McInnes *et al.*, 1996). These observations demonstrate a mechanism whereby mesenchymal cells, such as synoviocytes, can contribute to immune regulation and immunopathology through the elaboration of factors such as NO. The widespread expression of iNOS strongly suggests that this will be a general phenomenon, although future confirmatory studies in other tissues are required.

4. Conclusions

The above observations clearly establish the production and importance of NO during immune responses in a variety of both antigen- and non-antigen-driven host responses *in vivo* and *in vitro* (Table I). Complex feedback loops have evolved whereby NO may mediate both effector and regulatory roles. This tight balance renders therapeutic intervention more difficult. Thus, although iNOS inhibition appears attractive as an immunomodulatory target, careful estimation of its net effects on tissue pathology will be required. This will be particularly important in

TABLE I
Immunomodulatory Effects of NO

Vasodilatation
Alteration of endothelial permeability
Inhibition of platelet aggregation
Inhibition of neutrophil adhesion
Scavenging of reactive oxygen intermediates
Bifunctional effects on lymphocyte proliferation
Modulation of NK and LAK cell activity
Activation of cyclooxygenase
Regulation of cytokine production (e.g., TNFα)

treatment of complex inflammatory diseases, such as the inflammatory arthropathies, in which multiple conflicting effects of NO in host tissues might be anticipated.

ACKNOWLEDGMENTS. The authors acknowledge the support of the Wellcome Trust, the Nuffield Foundation, and the Robertson Trust.

References

Albina, J. E., Cui, S., Mateo, R. B., and Reichner, J. S., 1993, Nitric oxide mediated apoptosis in murine peritoneal macrophages, *J. Immunol.* **150:**5080–5085.

Amin, A. R., Di Cesare, P. E., Vyas, P., Attur, M., Tzeng, E., Billiar, T. R., Stuchin, S. A., and Abramson, S. B., 1995, The expression and regulation of nitric oxide synthase in human osteoarthritis affected chondrocytes: Evidence for up-regulated neuronal NOS activity, *J. Exp. Med.* **182:**2097–2102.

Barnes, P. J., and Liew, F. Y., 1995, Nitric oxide and asthmatic inflammation, *Immunol. Today* **16:**128–130.

Beckerman, K. P., Rogers, J. A., Corbett, R. D., Schreiber, M. L., and Unanue, E. R., 1993, Release of nitric oxide during the T cell independent pathway of macrophage activation. Its role in resistance to *Listeria monocytogenes*, *J. Immunol.* **150:**888–895.

Bredt, D. S., and Snyder, S. H., 1994, Nitric oxide: A physiological messenger molecule, *Annu. Rev. Biochem.* **63:**175–195.

Broughton-Evans, N. K., Evans, S. M., Hawkey, C. J., Cole, A. T., Balsitis, M., Whittle, B. J. R., and Moncada, S., 1993, Nitric oxide synthase activity in ulcerative colitis and Crohn's disease, *Lancet* **342:**338–340.

Burkrinsky, M. I., Nottet, H. S. L. M., Schmidtmayerova, N., Dubrovsky, L., Mullins, M. E., Lipton, S. A., and Gendelman, H. E., 1995, Regulation of nitric oxide activity in HIV-infected monocytes: Implications for HIV associated neurological disease, *J. Exp. Med.* **181:**735–745.

Chang, R.-H., Lin Feng, M.-H., Liu, W.-H., and Lai, M.-Z., 1997, Nitric oxide increases interleukin-4 expression in T lymphocytes, *Immunology* **90:**364–369.

Charles, I. G., Palmer, R. J., Hickery, M. S., Bayliss, M. T., Chubb, A. P., Hall, V. S., Moss, D. W., and

Moncada, S., 1993, Cloning, characterization and expression of a cDNA encoding an inducible nitric oxide synthase from human chondrocytes, *Proc. Natl. Acad. Sci. USA* **90:**11419–11423.

Cho, H. J., Xie, Q. W., Calalcay, J., Mumford, R. A., Lee, T. D., and Nathan, C., 1992, Calmodulin is a subunit of nitric oxide synthase from macrophages, *J. Exp. Med.* **176:**599–604.

Corbett, J. A., Mikhael, A., Shimizu, J., Frederick, K., Misko, T. P., McDaniel, M. L., Kanagawa, O., and Unanue, E. R., 1993, Nitric oxide production in islets from nonobese diabetic mice. Aminoguanidine-sensitive and -resistant stages in the immunological diabetic process, *Proc. Natl. Acad. Sci. USA* **90:**8992–8995.

Deakin, A. M., Payne, A. N., Whittle, B. J. R., and Moncada, S., 1995, The modulation of IL-6 and TNFα release by nitric oxide following stimulation of J774 cells with LPS and IFNγ, *Cytokine* **7:**408–416.

De Kimpe, S. J., Kengatharan, M., Thiermann, C., and Vane, J. R., 1995, The cell wall components peptidoglycan and lipoteichoic acid from *Staphylococcus aureus* act in synergy to cause shock and multiple organ failure, *Proc. Natl. Acad. Sci. USA* **92:**10359–10363.

Denis, M., 1991, Tumour necrosis factor and granulocyte colony stimulating factor stimulate human macrophages to restrict growth of virulent *Mycobacterium avium* and to kill avirulent *M. avium:* Killing effector mechanism depends on generation of reactive nitrogen intermediates, *J. Leukoc. Biol.* **49:**380–387.

Di Rosa, M., Radomski, M., Carnuccio, R., and Moncada, S., 1990, Glucocorticoids inhibit the induction of nitric oxide synthesis in macrophages, *Biochem. Biophys. Res. Commun.* **172:**1246–1252.

Dugas, B., Djavad Mossalayi, M., Damais, C., and Kolb, J. P., 1995, Nitric oxide production by human monocytes: Evidence for a role of CD23, *Immunol. Today* **16:**574–580.

Duhe, R. J., Evans, G. A., Erwin, R. A., Kirken, R. A., Cox, G. W., and Farrar, W. L., 1998, Nitric oxide and thiol redox regulation of Janus kinase activity, *Proc. Natl. Acad. Sci. USA* **95:**126–131.

Fenyk-Melody, J. E., Garrison, A. E., Brunnert, S. R., Weidner, J. R., Shen, F., Shelton, B. A., and Mudgett, J. S., 1998, Experimental autoimmune encephalomyelitis is exacerbated in mice lacking the NOS2 gene, *J. Immunol.* **160:**2940–2946.

Florquin, S., and Goldman, M., 1996, Immunoregulatory mechanisms of T cell dependent shock induced by a bacterial superantigen in mice, *Infect. Immun.* **64:**3443–3445.

Garside, P., Hutton, A., Severn, A., Liew, F. Y., and Mowat, A. M., 1992, Nitric oxide mediates intestinal pathology in graft versus host disease, *Eur. J. Immunol.* **22:**2141–2145.

Geller, D. A., Nussler, A. K., Di Silvio, M., Lowenstein, C. J., Shapiro, R. A., Wang, S. C., Simmons, R. I., and Billiar, T. R., 1993, Cytokines, endotoxin, and glucocorticoids regulate the expression of inducible nitric oxide synthase in hepatocytes, *Proc. Natl. Acad. Sci. USA* **90:**522–526.

Gilbert, R. S., and Herschman, H. R., 1993, TGF beta differentially modulates the iNOS gene in distinct cell types. *Biochem. Biophys. Res. Commun.* **195:**380–384

Habib, A., Bernard, C., Lebret, M., Creminon, C., Esposito, B., Tedgui, A., and Maclouf, J., 1997, Regulation of the expression of cyclooxygenase-2 by nitric oxide in rat peritoneal macrophages, *J. Immunol.* **158:**3845–3851.

Hogaboam, C. M., Chensue, S. W., Steinhauser, M. L., Huffnagle, G. B., Lukacs, N. W., Strieter, R. M., and Kunkel, S. L., 1997, Alteration of the cytokine phenotype in an experimental lung granuloma model by inhibiting nitric oxide, *J. Immunol.* **159:**5585–5593.

Huang, F.-P., Feng, G.-J., Lindop, G., Stott, D., and Liew, F. Y., 1996, The role of IL-12 and nitric oxide in the development of spontaneous autoimmune disease in MRL/MP-lpr/lpr mice, *J. Exp. Med.* **183:**1447–1459.

Huang, P. L., Huang, Z., Mashimo, H., Bloch, K. D., Moskowitz, M. A., Bevan, J. A., and Fishman, M. C., 1995, Hypertension in mice lacking the gene for endothelial nitric oxide synthase, *Nature* **377:**239–242.

Ialenti, A., Moncada, S., and Di Rosa, M., 1993, Modulation of adjuvant arthritis by endogenous nitric oxide, *Br. J. Pharmacol.* **110:**701–706.

Ignarro, L. J., Buga, G. M., Wood, K. S., Byrns, R. E., and Chaudhuri, G., 1987, Endothelium derived relaxation factor produced and released from arteries and veins in nitric oxide, *Proc. Natl. Acad. Sci. USA* **84:**9265–9269

Isobe, K., and Nakashima, J., 1992, Feedback suppression of staphylococcal enterotoxin-stimulated T-lymphocyte proliferation by macrophages through inductive nitric oxide synthesis, *Infect. Immun.* **60:**4832–4837.

Kaplan, S. S., Lancaster, J. R., Basford, R. E., and Simmons, R. L., 1996, Effect of nitric oxide on staphylococcal killing and interactive effect with superoxide, *Infect. Immun.* **64:**69–76.

Kleinert, H., Euchenhofer, C., Ihrigbiedert, I., and Forstermann, U., 1996, Glucocorticoids inhibit the induction of iNOS by downregulating cytokine induced activity of transcription factor nuclear factor-κB, *Mol. Pharmacol.* **49:**15–21

Kolb, H., and Kolb-Bachofen, V., 1992, Nitric oxide: A pathogenetic factor in autoimmunity, *Immunol. Today* **13:**157–160.

Kubes, P., Kanwar, S., Niu, X., and Gaboury, J. P., 1993, Nitric oxide synthesis inhibition induces leukocyte adhesion via superoxide and mast cells, *FASEB J.* **7:**1293–1299.

Kunz, D., Walker, G., Eberhardt, W., and Pfeilschifter, J., 1996, Molecular mechanisms of dexamethasone inhibition of nitric oxide synthase expression in IL-1 stimulated mesangial cells: Evidence for the involvement of transcriptional and posttranscriptional regulation, *Proc. Natl. Acad. Sci. USA* **93:**255–259.

Lander, H. M., Sehajpal, P., Levine, D. M., and Novogrodsky, A., 1993, Activation of human peripheral blood cells by nitric oxide generating compounds, *J. Immunol.* **150:**1509–1516.

Lefer, D. J., Scalia, R., Campbell, B., Nossuli, T., Hayward, R., Salamon, M., Grayson, J., and Lefer, A. M., 1997, Peroxynitrite inhibits leukocyte endothelial cell interactions and protects against ischaemia perfusion injury in rats, *J. Clin. Invest.* **99:**684–691.

Liew, F. Y., 1994, Regulation of nitric oxide synthesis in infectious and autoimmune diseases, *Immunol. Lett.* **43:**95–98.

Liew, F. Y., and Cox, F. E. G., 1991, Non specific defence mechanism: The role of nitric oxide, *Immunol. Today* **12:**A17–A21.

Liew, F. Y., Millot, S., Parkinson, C., Palmer, R. M. J., and Moncada, S., 1990, Macrophage killing of *Leishmania* parasite *in vivo* is mediated by nitric oxide from L-arginine, *J. Immunol.* **144:**4794–4797.

Liew, F. Y., Li, Y., Severn, A., Millot, S., Schmidt, J., Salter, M., and Moncada, S., 1991, A possible novel pathway of regulation by murine T helper type-2 cells of a Th1 cell activity via the modulation of the induction of nitric oxide synthase in macrophages, *Eur. J. Immunol.* **21:**2489–2494.

Lukic, M. L., Stosic-Grujicic, S., Ostojic, N., Chan, W. L., and Liew, F. Y., 1991, Inhibition of nitric oxide generation affects the induction of diabetes by streptozocin in mice, *Biochem. Biophys. Res. Commun.* **178:**913–920.

Lyons, C. R., 1995, The role of nitric oxide in inflammation, *Adv. Immunol.* **60:**323–360.

MacMicking, J. D., Nathan, C., Hom, G., Chartrain, N., Fletcher, D. S., Trumbauer, M., Stevens, K., Xie, Q. W., Sokol, K., Hutchinson, N., Chen, H., and Mudgett, J. S., 1995, Altered responses to bacterial infection and endotoxic shock in mice lacking inducible nitric oxide synthase, *Cell* **81:**641–650.

Malewista, S. E., Montgomery, R. R., and Van Blaricom, G., 1996, Evidence for nitrogen intermediates in killing of staphylococci in human neutrophil cytoplasts, *J. Clin. Invest.* **90:**631–636.

Manfield, L., Jang, D., and Murrell, G. A. C., 1996, Nitric oxide enhances cyclooxygenase activity in articular cartilage, *Inflamm. Res.* **45:**254–258.

Marcinkiewicz, J., Grabowska, A., and Chain, B. M., 1996, Is there a role for nitric oxide in regulation of T cell secretion of IL-2? *J. Immunol.* **156:**4617–4621.

McCartney-Francis, N., Allen, J. B., Mizel, D. E., Albina, J. E., Xie, Q., Nathan, C. F., and Wahl, S.,

1993, Suppression of arthritis by an inhibitor of nitric oxide synthase, *J. Exp. Med.* **178:**749–754.
McInnes, I. B., Leung, B. P., Field, M., Huang, F.-P., Wei, X. Q., Sturrock, R. D., Kinninmonth, A., Mumford, R. A., and Liew, F. Y., 1996, Nitric oxide production in the synovial membranes of rheumatoid and osteoarthritis patients, *J. Exp. Med.* **184:**1519–1524.
McInnes, I. B., Leung, B. P., Wei, X. Q., Gemmell, C. G., and Liew, F. Y., 1998, Septic arthritis following *Staphylococcus aureus* infection in mice lacking inducible nitric oxide synthase, *J. Immunol.* **160:**308–315.
Merryman, P. F., Clancy, R. M., He, X. Y., and Abramson, S. B., 1993, Modulation of human T cell responses by nitric oxide and its derivative *S*-nitrosoglutathione, *Arthritis Rheum.* **36:**1414–1422.
Mulligan, M. S., Heirel, J. M., Marletta, M. A., and Ward, P. A., 1991, Tissue injury caused by deposition of immune complexes is L-arginine dependent, *Proc. Natl. Acad. Sci. USA* **88:**6338–6342.
Nathan, C., and Xie, Q.-W., 1994, Regulation of biosynthesis of nitric oxide, *J. Biol. Chem.* **269:**13725–13728.
Nelson, R. J., Demas, G. E., Huang, P. L., Fishman, M. C., Dawson, V. L., Dawson, T. M., and Snyder, S. H., 1995, Behavioural abnormalities in male mice lacking neuronal nitric oxide synthase, *Nature* **378:**383–386.
Nukaya, I., Takagi, K., Kawabe, T., and Suketa, Y., 1995, Suppression of cytokine production by Th2 cells by nitric oxide in comparison with Th1 cells, *Microbiol. Immunol.* **39:**709–714.
Okuda, Y., Nakatsuji, Y., Fujimura, H., Esumi, H., Ogura, T., and Yanagihara, T., 1995, Expression of the inducible isoform of nitric oxide synthase in the CNS of mice correlates with severity of actively induced EAE, *J. Neuroimmunol.* **62:**103–112.
Palmer, R. M. J., Ferrige, A. G., and Moncada, S., 1987, Nitric oxide release accounts for the biologic activity of endothelium derived relaxing factor, *Nature* **327:**524–526.
Park, K. G. M., Hayes, P. D., Garlick, P. J., Sewell, H., and Eremin, O., 1991, Stimulation of lymphocyte natural cytotoxicity by L-arginine, *Lancet* **337:**645–646.
Ruuls, S. R., Van Der Linden, S., Sontrop, K., Huitinga, I., and Dijkstra, C. D., 1996, Aggravation of experimental allergic encephalomyelitis (EAE) by administration of nitric oxide (NO) synthase inhibitors, *Clin. Exp. Immunol.* **103:**467–474.
Schmidt, H. H. H., and Walter, U., 1994, NO at work, *Cell* **78:**919–925.
Schneemann, M., Schoedon, G., Hoefer, S., Blau, N., Guerrero, L., and Schaffner, A., 1993, Nitric oxide synthase is not a constituent of the antimicrobial armature of human mononuclear phagocytes, *J. Infect. Dis.* **167:**1358–1363.
Stefanovic-Racic, M., Meyers, K., Meschter, C., Coffey, J. W., Hoffman, R. A., and Evans, C. H., 1994, N-Monomethyl arginine, an inhibitor of nitric oxide synthase, suppresses the development of adjuvant arthritis in rats, *Arthritis Rheum.* **37:**1062–1069.
Stout, R., Suttles, J., Xu, J., Grewal, I. S., and Flavell, R. A., 1996, Impaired T cell mediated macrophage activation in CD40 ligand-deficient mice, *J. Immunol.* **156:**8–11.
Sunyer, T., Rothe, L., Jiang, X., Osdoby, P., and Collin-Osdoby, P., 1996, Proinflammatory agents, IL-8 and IL-10, upregulate inducible nitric oxide synthase expression and nitric oxide production in avian osteoclast like cells, *J. Cell. Biochem.* **60:**469–483.
Tao, X., and Stout, R., 1993, T cell mediated cognate signalling of nitric oxide production by macrophages. Requirements for macrophage activation by plasma membranes isolated from T cells, *Eur. J. Immunol.* **23:**2916–2921.
Taylor-Robinson, A. W., Liew, F. Y., Severn, A., Xu, D., McSorley, S., Garside, P., Padron, J., and Phillips, R. S., 1994, Regulation of the immune response by nitric oxide differentially produced by T helper type 1 and T helper type 2 cells, *Eur. J. Immunol.* **24:**980–984.
Wei, X. Q., Charles, I., Smith, A., Ure, J., Feng, G. J., Huang, F. P., Xu, D., Muller, W., Moncada, S., and Liew, F. Y., 1995, Altered immune responses in mice lacking inducible nitric oxide synthase, *Nature* **375:**408–411.

Weinberg, J. B., Granger, D. L., Pisetsky, D. S., Seldin, M. J., Misukonis, M. A., Mason, S. N., Pippen, A. M., Ruiz, P., Wood, E. R., and Gilkeson, G. S., 1994, The role of nitric oxide in the pathogenesis of spontaneous murine autoimmune disease expression in MRL-*lpr/lpr* mice, and reduction of spontaneous glomerulonephritis and arthritis by orally administered N^G-monomethyl-L-arginine, *J. Exp. Med.* **179:**651–660.

Xiong, H., Kawamura, I., Nishibori, T., and Mitsuyama, M., 1996, Suppression of IFNγ production from *Listeria monocytogenes* specific T cells by endogenously produced nitric oxide, *Cell. Immunol.* **172:**118–125.

Zembala, M., Siedlar, M., Marcinkiewicz, J., and Pryjma, J., 1994, Human macrophages are stimulated for nitric oxide release in vitro by some tumour cell lines but not by cytokines and lipopolysaccharide, *Eur. J. Immunol.* **24:**435–439.

Zhao, W., Tilton, R. G., Corbett, J. A., McDaniel, M. L., Misko, T. P., and Williamson, J. R., 1996, Experimental allergic encephalomyelitis in the rat is inhibited by aminoguanidine, an inhibitor of nitric oxide synthase, *J. Neuroimmunol.* **64:**123–133.

Zhao, Y.-X., Abdelnour, A., Kalland, T., and Tarkowski, A., 1995, Overexpression of the T cell receptor Vβ3 in transgenic mice increases mortality during infection by enterotoxin A producing *Staphylococcus aureus*, *Infect. Immun.* **63:**4463–4469.

Zielasek, J., Jung, S., Gold, R., Liew, F. Y., Toyka, K. V., and Hartung, H. P., 1995, Administration of nitric oxide synthase inhibitors in experimental autoimmune neuritis and experimental autoimmune encephalomyelitis, *J. Neuroimmunol.* **59:**81–88.

CHAPTER 11

Nitric Oxide and Epithelial Host Defense

NIGEL BENJAMIN and ROELF DYKHUIZEN

1. Introduction

It is now clear that nitric oxide (NO) synthesis by mammalian cells contributes to host defense against a number of pathogenic microorganisms (detailed in Chapter 12). The mechanism of NO production for the purpose of microbial killing has almost universally been considered to be via a five-electron oxidation of L-arginine that is accomplished by the inducible form of the nitric oxide synthase (NOS) enzyme (Chapter 4). In this chapter we consider an alternative mechanism for the generation of NO, the enzymatic and chemical reduction of nitrate (NO_3^-), and provide evidence that this system may be important in the protection of humans against pathogenic organisms.

Nitrate is ubiquitous in nature, partly because it is a very thermodynamically stable molecule. However, nitrate is used by plants and certain bacteria as a source of nitrogen for incorporation into protein as amine groups (RNH_2). In this conversion, plants have developed a range of enzymes to accomplish the required eight-electron reduction, using energy derived from photosynthesis. Green, leafy plants such as lettuce often contain large amounts of nitrate, especially if they are grown under low light conditions (Cantliffe, 1972). Most other food products have a relatively low content of nitrate and nitrite (Table I).

There has been some concern about dietary nitrate as a potential precursor to carcinogenic molecules in the gastrointestinal tract (Spiegelhalder *et al.*, 1976; Tannenbaum *et al.*, 1976; Green, 1995). When swallowed, nitrate is rapidly

---

NIGEL BENJAMIN • Department of Clinical Pharmacology, St. Bartholomew's and the Royal London School of Medicine and Dentistry, London EC1M 6BQ, United Kingdom. ***ROELF DYKHUIZEN*** • Department of Clinical Pharmacology, St. Bartholomew's and the Royal London School of Medicine and Dentistry, London EC1M 6BQ United Kingdom, and Department of Medicine and Therapeutics, University of Aberdeen Medical School, Aberdeen AB9 2ZD, United Kingdom.

Nitric Oxide and Infection, edited by Fang. Kluwer Academic / Plenum Publishers, New York, 1999.

TABLE I

Contribution (%) of Various Foodstuffs to Dietary Intake of Nitrate and Nitrite.[a]

	Cured meats	Fresh meats	Vegetables	Fruits and juices	Baked goods and cereals	Milk products	Water
NO_3^-	1.6	0.8	87	6	1.6	0.2	2.6
NO_2^-	39	7.7	16	1.3	34	1.3	1.3

[a] Adapted from Committee on Nitrite and Alternative Curing Agents in Food (1981).

absorbed. At least 25% is concentrated in the salivary glands by an uncharacterized mechanism, so that the nitrate concentration of saliva exceeds that of plasma by at least tenfold. Nitrate is rapidly reduced to nitrite (NO_2^-) in the mouth by mechanisms that will be discussed below. Saliva containing large amounts of nitrite is acidified in the normal stomach to produce nitrous acid, which could potentially nitrosate amines to form *N*-nitrosamines (Tannenbaum *et al.*, 1974) that are known to be potent carcinogens in experimental systems (Crampton, 1980). From this theoretical understanding of nitrate metabolism, a number of studies have been performed to examine the relationship between nitrate intake and cancer (particularly gastric cancer) in humans. In general, it has been found that there is either no relationship or an inverse relationship, i.e., individuals with a high nitrate intake have a lower rate of cancer (Forman *et al.*, 1985; Al-Dabbagh *et al.*, 1986; Knight *et al.*, 1990). Similarly, in animal studies it has been generally impossible to demonstrate an increased risk of cancer (or any other adverse effect) when nitrate intake is increased (Vittozzi, 1992).

The interest in nitrate metabolism stimulated studies in humans that confirmed a discovery originally made in 1916 (Mitchell *et al.*, 1916), that mammals (including humans) synthesize inorganic nitrate (Tannenbaum *et al.*, 1978; Tannenbaum, 1979; Green *et al.*, 1981a,b; see also Chapter 2). Even on a nitrate-free diet, considerable concentrations of nitrate can be detected in plasma (approximately 30 μM) and urine (approximately 800 μmole/24 hr). It was also found that nitrate has a long half-life of 5–8 hr (Wagner *et al.*, 1983), which seems to reflect efficient (80%) nitrate reabsorption from the renal tubules by an active transport mechanism (Kahn *et al.*, 1975). It is now believed that endogenous nitrate synthesis derives from NOS enzymes acting on L-arginine (Hibbs *et al.*, 1992). The NO formed is rapidly oxidized to nitrate when it encounters superoxide or oxidized hemoglobin. It is still unclear whether all endogenous nitrate synthesis derives from this route, as the enrichment of urinary nitrate with ^{15}N is only about one-half of the steady state of [^{15}N]arginine enrichment following prolonged infusion of ^{15}N-labeled arginine (Macallan *et al.*, 1997). This may indicate that nitrate also derives from another source, or that the intracellular enrichment of labeled arginine is less than that in the plasma because of transamination reactions.

The peculiar metabolism of nitrate, characterized by renal salvage, salivary concentration, and conversion to nitrite in the mouth, has led us to consider that oxides of nitrogen in the mouth and stomach may be produced by a functional mechanism to provide host defense against ingested pathogens (Benjamin *et al.*, 1994).

2. Oral Nitrate Reduction

Although Tannenbaum and his colleagues considered that salivary bacteria may be reducing nitrate to nitrite, Sasaki and Matano (1979) actually demonstrated that this activity is localized almost entirely on the surface of the tongue in humans. They suggested that the nitrate reductase enzyme was most likely to be a mammalian nitrate reductase. Using a rat tongue preparation, we also found that the dorsal surface of the tongue possesses very high nitrate reductase activity, which is confined to the posterior two-thirds (Duncan *et al.*, 1995) (Fig. 1). However, microscopic analysis of the tongue surface revealed a dense population of gram-negative and gram-positive bacteria, 80% of which showed marked *in vitro* nitrate-reducing activity.

Our suspicion that the nitrate reduction was being accomplished by bacteria was strengthened by the observation that in rats bred in a germ-free environment and lacking colonization by bacteria, no nitrate-reducing activity could be demonstrated on the tongue. Furthermore, treatment of healthy

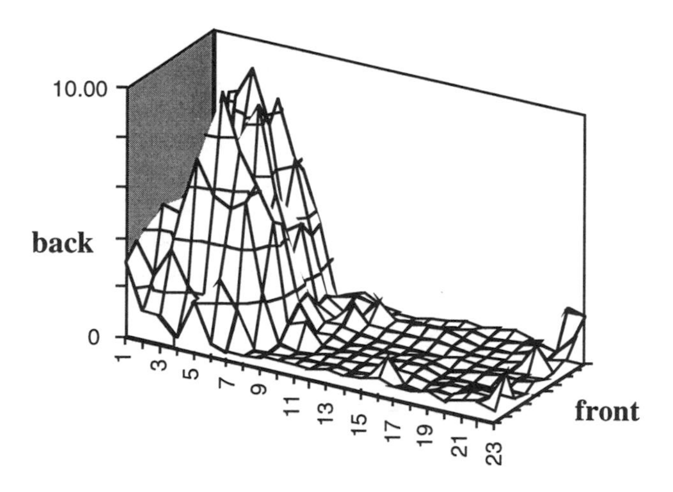

FIGURE 1. Nitrate reductase activity of dorsal surface of rat tongue. Data expressed as micromoles nitrite formed per hour per square millimeter; $n = 7$.

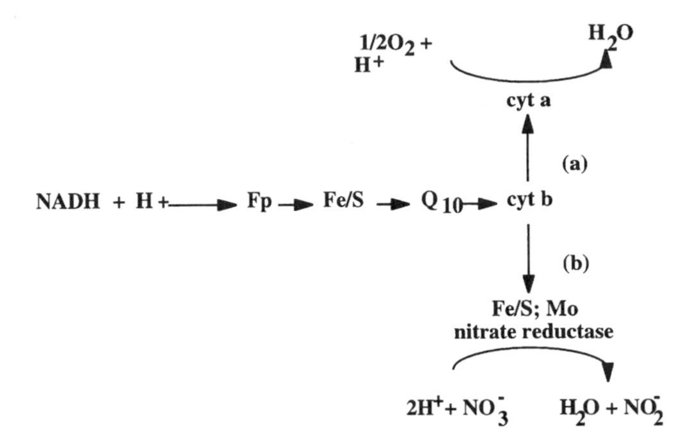

FIGURE 2. Facultative anaerobic bacteria can divert electrons from NADH to molecular oxygen (a) or to nitrate ion (b) when oxygen is not available. In the latter case, nitrite is produced as a by-product.

volunteers with the broad-spectrum antibiotic amoxicillin reduces salivary nitrite concentrations (Dougall *et al.*, 1995).

Although we have not been able to characterize the organisms in normal human tongues (this would require a deep biopsy, as the majority of the bacteria are at the bottom of the papillary clefts of the tongue surface), the most commonly found nitrite-producing organisms in the rat (Li *et al.*, 1997) are *Staphylococcus sciuri,* followed by *Staphylococcus intermedius, Pasteurella* spp., and *Streptococcus* spp. Both morphometric quantification of bacteria on tongue sections and enumeration of culturable bacteria reveal an increase in the density of bacteria toward the posterior tongue.

We now believe that these organisms represent true symbionts, and that the mammalian host actively encourages the growth of nitrite-forming organisms on the surface of the tongue. These bacteria are facultative anaerobes that use nitrate instead of oxygen as an electron acceptor for oxidation of carbon compounds to derive energy under microaerobic conditions (Fig. 2). Nitrite represents a waste product of this process from the standpoint of the bacteria, but can be utilized elsewhere by the mammalian host for its antimicrobial potential.

3. Acidification of Nitrite—Production of NO in the Mouth and Stomach

Nitrite formed on the tongue surface can be acidified in two ways. It can be swallowed into the acidic stomach, or it may encounter the periodontal acid environment provided by organisms such as *Lactobacillus* and *Streptococcus mutans* that are implicated in caries production. Acidification of nitrite produces nitrous acid (HNO_2), which has an acid dissociation constant of 3.2; in the normal fasting stomach (pH 1–2), complete conversion to nitrous acid will occur:

$$
\begin{aligned}
&NO_2^- + H^+ \leftrightarrow HNO_2 \\
&\quad 2HNO_2 \leftrightarrow N_2O_3 + H_2O \leftrightarrow NO + NO_2 \\
&\quad 3HNO_2 \leftrightarrow 2NO + NO_3^- + H^+ + H_2O
\end{aligned}
$$

Nitrous acid is unstable and will spontaneously decompose to nitric oxide (NO) and nitrogen dioxide (NO_2). Under reducing conditions, more NO will be formed than NO_2. Lundberg *et al.* (1994b) were the first to show a very high concentration of NO in gas expelled from the stomachs of healthy volunteers, which increases when nitrate intake is increased and diminishes when stomach acidification is impaired with the proton pump inhibitor omeprazole. We have conducted further studies on the amount of NO produced following ingestion of inorganic nitrate, measured directly during nasogastric intubation of healthy human volunteers. Following ingestion of 1 mmole of inorganic nitrate (the quantity of nitrate found in a large portion of lettuce), a pronounced increase in stomach headspace gas NO ensues, peaking at about 1 hr and persisting above control levels for at least 6 hr (McKnight *et al.*, 1997) (Fig. 3). The concentration of NO measured in the headspace gas of the stomach during these experiments would be lethal if inhaled continuously for as little as 20 min.

The concentration of NO in the stomach is in fact even much higher than would be expected from the concentration of nitrite in saliva and the measured pH in the gastric lumen. *In vitro* studies suggest that these concentrations of nitrite and acid would generate about one-tenth of the NO that is actually measured (McKnight, Smith, and Benjamin, unpublished data). It is therefore postulated that a reducing agent such as ascorbic acid (which is actively secreted into the stomach) (Sobala *et al.*, 1989, 1991; Schorah *et al.*, 1991) or reduced thiol (which is present in high concentrations in the gastric mucosa) contributes to the enhanced NO production.

We were surprised to find that NO is also generated in the oral cavity from salivary nitrite (Duncan *et al.*, 1995), because saliva is generally neutral or slightly alkaline. A possible mechanism for NO production in saliva is acidification at the gingival margins, as noted above. It will be important to determine if this is the case, because NO formed in this way may be able to inhibit the growth of organisms that generate acid. Such a mechanism could help to explain the importance of saliva in protection from dental caries. As in the stomach, acidification of saliva results in larger quantities of NO production than would be expected from the concentration of nitrite present. Saliva contains ascorbate

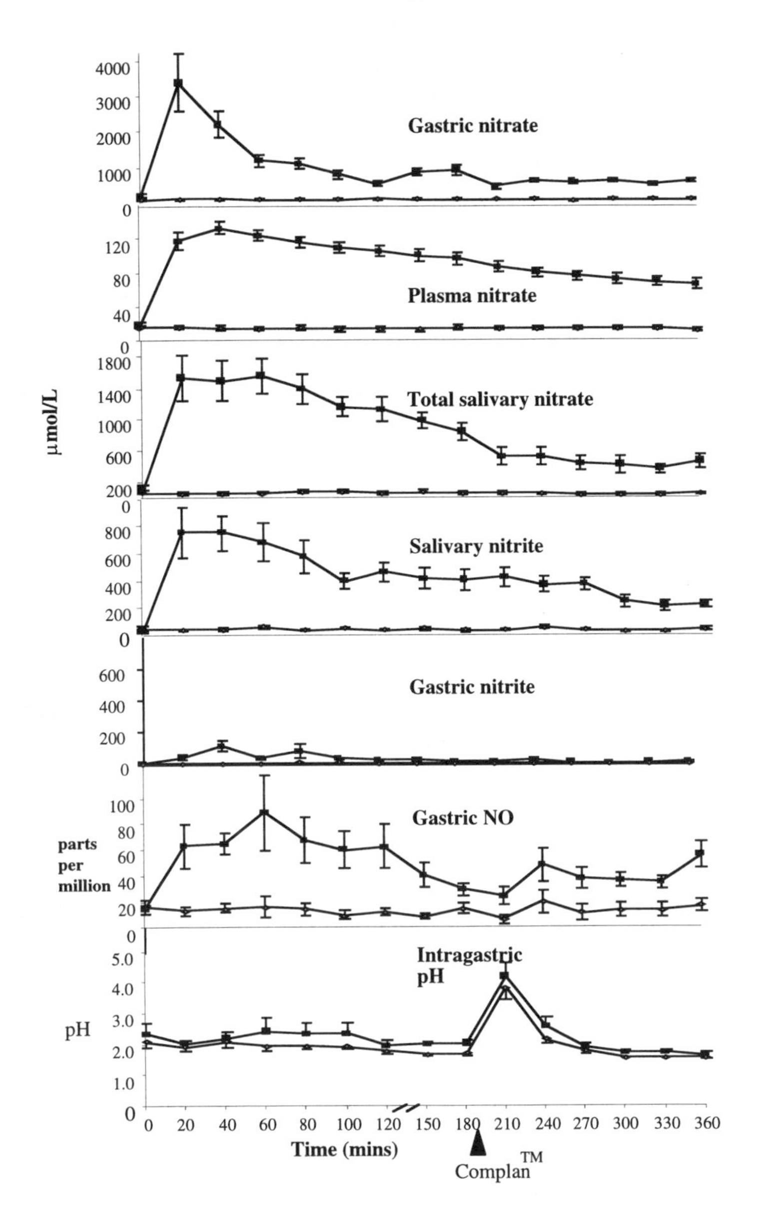

FIGURE 3. Fate of orally administered nitrate in healthy volunteers. Two millimoles of inorganic potassium nitrate or potassium chloride (control) was administered orally to ten healthy volunteers. Nitrate and nitrite were measured in blood, saliva, and gastric juice and NO was measured in gastric headspace gas. A liquid meal (Complan) was given at 180 min.

(Leggott *et al.*, 1986), and treatment with ascorbic acid oxidase partly reduces NO levels toward predicted values. It is likely that additional agents present in normal saliva also augment NO production.

A further explanation may be provided by L-arginine-derived NO, which also appears to contribute to the production of NO in the mouth and stomach. Jones-Carson *et al.* (1995) have shown that $\gamma\delta$ T cells can stimulate macrophage NO production and anticandidal activity *in vitro*, and that depletion of these lymphocytes *in vivo* enhances the susceptibility of mice to oral and orogastric mucosal *Candida* infection. This enhanced susceptibility is associated with reduced iNOS mRNA expression in mucosal tissues.

4. NO Synthesis in the Skin

By using a chemiluminescent NO detector, we have been able to demonstrate generation of NO by normal human skin (Weller *et al.*, 1996). Because NO has the ability to diffuse readily across membranes, we initially believed that we were measuring NO manufactured by eNOS that had escaped from vascular endothelium to the skin surface. However, we found that the release of NO from the hand was not affected by the NOS antagonist monomethyl-arginine infused into the brachial artery of healthy volunteers in amounts sufficient to maximally reduce forearm blood flow (Fig. 4). Furthermore, application of inorganic nitrite substantially elevated skin NO synthesis (Fig. 5).

This finding, along with the observations that cutaneous NO release is enhanced by acidity and reduced by antibiotic therapy, makes it likely that NO is being formed by nitrite reduction in the skin, as it is in the mouth and stomach. Normal human perspiration contains approximately 5 μM nitrite, and this concentration is precisely the amount predicted necessary to generate the observed amount of NO release. The source of nitrite is presently unclear, but is likely to reflect reduction of sweat nitrate by commensal skin bacteria, which are known to produce the nitrate reductase enzyme.

This observation has led us to the hypothesis that skin NO synthesis may also constitute a host defense mechanism to protect against pathogenic skin infections, especially fungi. The release of NO is enhanced following licking of the skin, reflecting the large quantities of nitrite in saliva, which may help to explain why animals and humans have an instinctive urge to lick their wounds (Benjamin *et al.*, 1997). We have also shown that the application of inorganic nitrite and an organic acid is effective in the treatment of patients with tinea pedis ("athlete's foot") (Weller *et al.*, 1998).

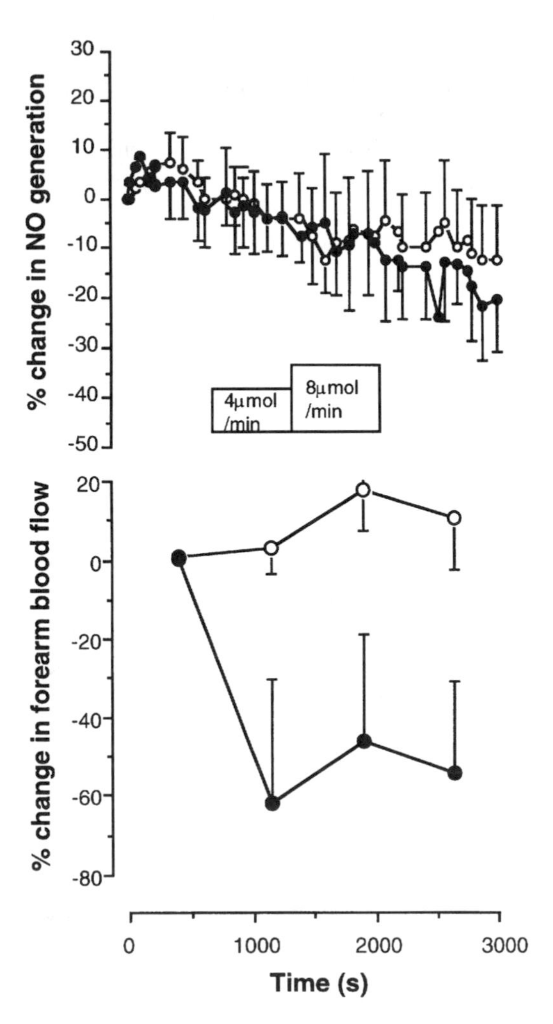

FIGURE 4. Effect of brachial artery infusion of L-NMMA in healthy volunteers. Despite a marked decrease in blood flow when monomethyl arginine was infused (filled symbols), there was no significant change in NO release from the skin surface.

5. Antimicrobial Activity of Acidified Nitrite

Inorganic nitrates have been used as food preservatives for centuries (Binkert and Kolari, 1975). It has subsequently become clear that nitrate itself is generally nonreactive with organic molecules, and has to be chemically or enzymatically reduced to nitrite in order to be effective as an antimicrobial agent whose potency is enhanced in an acidic environment. As well as its beneficial effect in limiting the

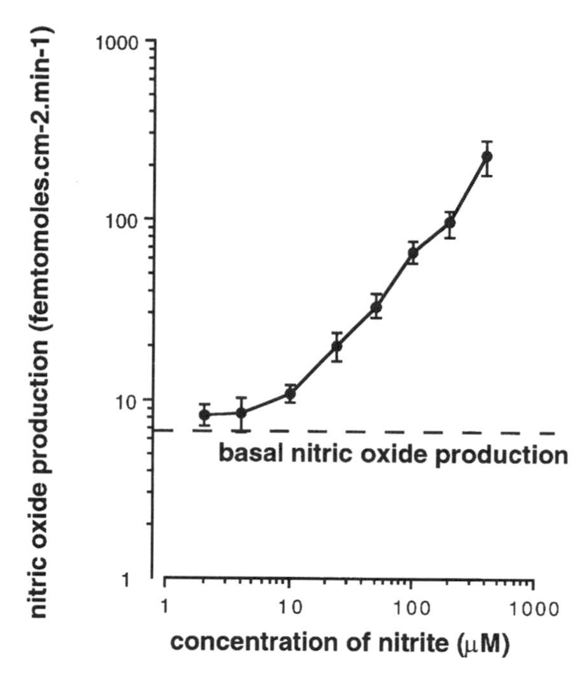

FIGURE 5. Application of inorganic nitrite increases skin nitric oxide production. The concentration of nitrite in healthy sweat is approximately 4 μM. Note that both axes are logarithmic scales.

growth of serious pathogens such as *Clostridium botulinum* (Reddy *et al.*, 1983), nitrite also provides the benefit of rendering muscle tissue a bright pink by the formation of nitrosomyoglobin.

The exact molecular species responsible for microbial killing in mammalian cells that synthesize NO is not completely clear (Fang, 1997). Indeed, different organisms appear to be susceptible to different reactive nitrogen species. Acidification of nitrite results in a complex mixture of nitrogen oxides, as well as nitrous acid. The additional stress of acidification may also help to make microorganisms more susceptible to nitrogen oxides. Nitrous acid, dinitrogen trioxide, and nitrogen dioxide are each effective nitrosating agents (NO^+ donors) (Williams, 1988). Nitrosation may occur at the microbial cell surface or intracellularly, and can involve intermediates such as *S*-nitrosothiols, which are also good NO^+ donors. Reduced thiols are in high concentration in gastric mucosa, and will inevitably be nitrosated in the presence of nitrite and acid. Thiocyanate is also present in saliva, and chloride ions are in high concentration in the stomach. Each of these anions will catalyze nitrosation reactions to form additional reactive intermediates that may add to the toxicity of acidified nitrite (Williams, 1988).

Many human pathogens that cause gastrointestinal disease are remarkably resistant to acid alone. Incubation of *Candida albicans* at pH 1 for 1 hr has no detectable effect on its subsequent growth. However, addition of nitrite to the acid incubation medium at concentrations found in saliva results in nearly complete sterilization of *C. albicans* cultures (Fig. 6).

Similarly, *E. coli* viability is markedly affected by the addition of nitrite to an incubation medium buffered to pH 3. As little as 10 μM nitrite will slow the growth of this organism, a significantly lower concentration than the 100 μM to 1 mM range typically found in saliva (depending on dietary nitrate intake). Common enteric pathogens such as *Salmonella typhimurium, Yersinia enterocolitica, Shigella sonnei*, and *E. coli* O157:H7 are also highly sensitive to the combination of acid and nitrite (Dykhuizen *et al.*, 1996a). Most of these organisms are not killed following exposure to pH 3 for 1 hr, but are susceptible to the addition of nitrite at a concentration normally found in saliva (Fig. 7).

Our investigations have demonstrated differing relative susceptibility of enteric bacterial species to acidified nitrite (listed from most to least susceptible): *Y. enterocolitica* > *S. enteritidis* > *S. typhimurium* = *S. sonnei* ($p < 0.05$). *E. coli* O157:H7 and *S. sonnei* are the most resistant of these organisms to acid; they survive exposure to pH 2.1 for 30 min, which kills the other bacteria tested. However, the growth of *E. coli* O157:H7 is inhibited by acid pH up to pH 4.2, while *Salmonella* and *Shigella* strains manage to maintain growth under these conditions

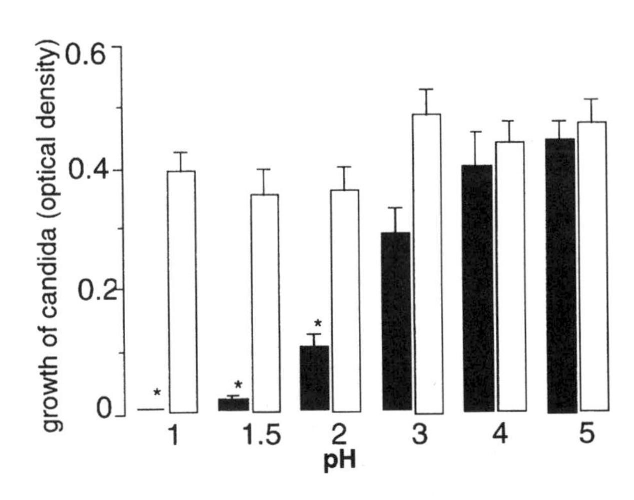

FIGURE 6. The effect of exposure to nitrite and differing hydrogen ion concentrations on the survival of *Candida albicans*. Open bars show growth of *C. albicans*, measured by optical density 9 hr following exposure to acid alone for 1 hr, while closed bars show growth following exposure to acid and 250 μM sodium nitrite. Asterisk denotes significant difference from control ($p < 0.05$, Mann–Whitney U test).

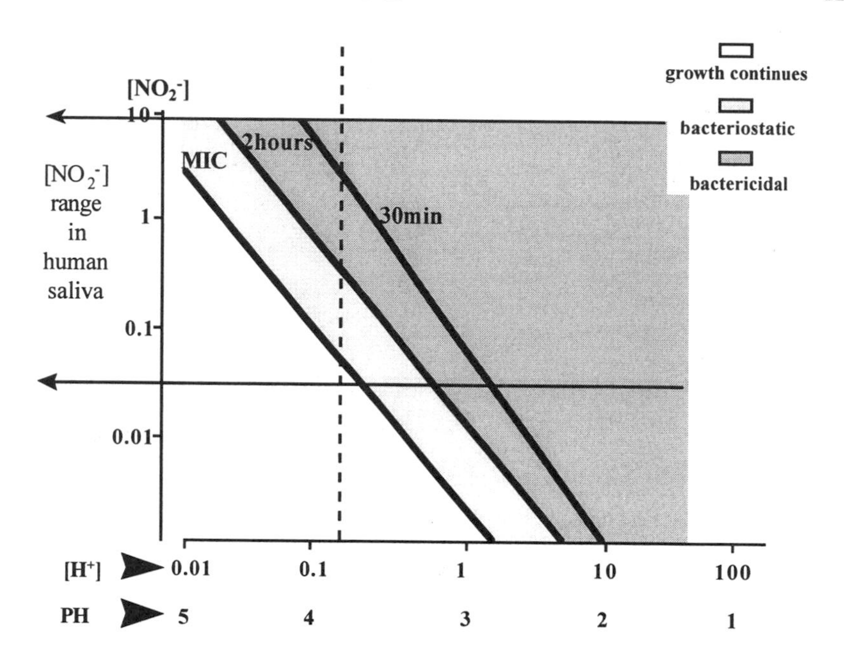

FIGURE 7. MBC_{30min}, MBC_{2hr}, and MIC µmole/ml for *Salmonella enteritidis* exposed to varying concentrations of acid and nitrite. At typical gastric pH values, increasing nitrite concentration from a low normal salivary concentration to a high normal value renders the acidified nitrite mixture bactericidal rather than bacteriostatic.

unless nitrite is present in the solution. It is conceivable that *E. coli* O157:H7 withstands relatively acidic conditions by limiting its growth, but this defensive strategy can be overcome by the concomitant presence of nitrite.

Perhaps not surprisingly, the important gastric pathogen *Helicobacter pylori* is resistant to the combination of nitrite and acid (Dykhuizen *et al.*, 1998), but the mechanism of this resistance is unknown. The generation of ammonia from urea via the urease enzyme may help *H. pylori* to locally neutralize acidity and limit nitrogen oxide formation. Alternatively, the organism may have developed specific biochemical mechanisms for protection, which could provide an attractive target for novel antimicrobial strategies.

6. NO Production in the Upper and Lower Airways

Exhaled air from healthy mammals (including humans) contains small concentrations of NO (Gustafsson *et al.*, 1991), which can be measured by chemiluminescent or mass spectrometric methods (Leone *et al.*, 1994). The normal concentration is approximately 20 ppb. Surprisingly, the nasopharynx produces prodigious amounts of NO (Lundberg *et al.*, 1995b), which seem to arise from the nasal sinuses (Lundberg *et al.*, 1995a) (a quick functional test of the chemiluminescence meter is to place the sampling tube inside one nostril, which typically causes the meter to register approximately 300 ppb). The source of NO from both the lungs and nasopharynx appears to be L-arginine via NOS, rather than from nitrate reduction. Treatment of children with antibiotics has no effect on this NO synthesis (Baraldi *et al.*, 1997). It seems likely that in addition to providing a continuous source of NO to aid ventilation–perfusion matching in the lungs, nasopharyngeal NO synthesis may have host defense antimicrobial functions. Reduced iNOS expression in the bronchial epithelium of patients with cystic fibrosis has been proposed as an explanation for the enhanced susceptibility of these patients to bacterial colonization (Meng *et al.*, 1998). However, as yet there has been little interest in the use of NO to prevent or treat infection in the upper and lower airways, as opposed to its effect in improving pulmonary gas exchange.

7. NO Production in the Lower Intestinal Tract

NO can be measured directly in the gas normally present in the human colon (Lundberg *et al.*, 1994a), and NO concentrations are increased in patients suffering from inflammatory bowel disease. In this setting, it is likely that the NO is being manufactured from L-arginine rather than from nitrate reduction.

In patients with infective gastroenteritis, it appears that NO may be formed in larger quantities than in other infective or inflammatory conditions, although it is not clear whether this synthesis originates from the gut itself. Early studies on nitrate balance in humans serendipitously found that diarrheal illness was associated with greatly increased endogenous nitrate synthesis in one of the volunteers (Green *et al.*, 1981a). Another study showed that plasma and urinary nitrate concentrations increase dramatically in infants with gastroenteritis. In this study, methemoglobin levels also increased, likely reflecting the reaction of NO with hemoglobin to produce methemoglobin and nitrate. Our studies also indicate that patients admitted to the hospital with infectious diarrhea have considerably elevated plasma nitrate concentrations, to a degree exceeding even that of patients with overwhelming septicemia (Dykhuizen *et al.*, 1995, 1996b; Neilly *et al.*, 1995) (Fig. 8). Elevated NO production does not seem to be pathogen specific, as similar increases in plasma nitrite and local NOS upregulation have also been described in patients with cholera (Janoff *et al.*, 1998).

These observations, along with recent studies showing little increase in NO synthesis in healthy volunteers administered typhoid vaccine (Macallan *et al.*, 1997) (which activates systemic immune responses) or in some patients with infections other than gastroenteritis, suggest that gut infections may be an

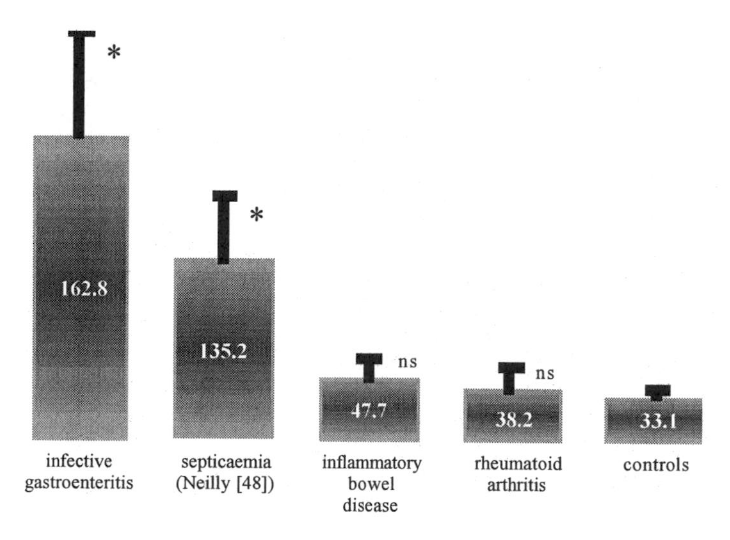

FIGURE 8. Elevation of plasma nitrate concentrations in infective gastroenteritis and other infectious and inflammatory conditions. Most subjects with gastroenteritis were less systemically ill than those with septicemia. Asterisks denote $P < 0.05$.

unusually potent stimulus for NO synthesis. We have considered potential alternative sources of nitrate synthesis in addition to L-arginine, but measurements of ^{15}N flux from labeled arginine to nitrate suggest that the NO is derived from NOS in this setting (Dykhuizen, Forte, and Benjamin, unpublished observations). The demonstration of iNOS upregulation in the rectal mucosa of patients with acute shigellosis (Islam *et al.*, 1997) and the ability of *Salmonella* to stimulate iNOS expression in IFNγ-treated intestinal epithelial cells (Salzman *et al.*, 1998) support this conclusion.

The most likely reason for marked NO synthesis in gastroenteritis is to mediate killing of gut pathogens. The large amounts of nitrate formed will greatly increase salivary nitrate and, hence, gastric NO synthesis by the enterosalivary circulation of nitrate detailed above. This could have the salutary effect of preventing fecal–oral recirculation of pathogens in infected individuals. However, it is also possible that this exuberant production of NO contributes to the pathophysiology and cytopathology associated with bacterial enteritis (Salzman, 1995) (see Chapter 8).

8. Conclusions

This chapter describes a novel and potentially important mechanism for host defense of epithelial sufaces: the production of reactive nitrogen oxides by the reduction of inorganic nitrate to nitrite and subsequent acidification. In addition, it is clear that enzyme-derived NO from NOS provides antimicrobial oxides of nitrogen at mucosal surfaces in the respiratory and gastrointestinal tracts. Whereas it is evident that acidified nitrite can inhibit or kill a variety of human gastrointestinal and cutaneous pathogens *in vitro*, definitive evidence for a physiological role of this mechanism is currently lacking (Green, 1995). Studies now under way will help to determine whether augmenting this system by increasing dietary nitrate intake can prevent gastroenteritis. Understanding the system of enterosalivary circulation of nitrate and subsequent production of nitrogen oxides may also lead to the development of new antimicrobial strategies by augmenting what appears to be a simple and effective epithelial host defense system.

References

Al-Dabbagh, S., Forman, D., Bryson, D., Stratton, I., and Doll, R., 1986, Mortality of nitrate fertiliser workers, *Br. J. Ind. Med.* **43:**507–515.

Baraldi, E., Azzolin, N. M., Biban, P., and Zacchello, F., 1997, Effect of antibiotic therapy on nasal nitric oxide concentration in children with acute sinusitis, *Am. J. Respir. Crit. Care Med.* **155:**1680–1683.

Benjamin, N., O'Driscoll, F., Dougall, H., Duncan, C., Smith, L., Golden, M., and McKenzie, H., 1994, Stomach NO synthesis, *Nature* **368:**502.

Benjamin, N., Pattullo, S., Weller, R., Smith, L., and Ormerod, A., 1997, Wound licking and nitric oxide, *Lancet* **349:**1776.

Binkert, E. F., and Kolari, O. E., 1975, The history and use of nitrate and nitrite in the curing of meat, *Food Cosmet. Toxicol.* **13:**655–661.

Cantliffe, D. J., 1972, Nitrate accumulation in vegetable crops as affected by photoperiod and light duration, *J. Am. Soc. Hortic. Sci.* **97:**414–418.

Committee on Nitrite and Alternative Curing Agents in Food, 1981, *The Health Effects of Nitrate, Nitrite, and N-Nitroso Compounds*, National Academy Press, Washington, D.C., pp. 5.41–5.52.

Crampton, R. F., 1980, Carcinogenic dose-related response to nitrosamines, *Oncology* **37:**251–254.

Dougall, H. T., Smith, L., Duncan, C., and Benjamin, N., 1995, The effect of amoxycillin on salivary nitrite concentrations: An important mechanism of adverse reactions? *Br. J. Clin. Pharmacol.* **39:**460–462.

Duncan, C., Dougall, H., Johnston, P., Green, S., Brogan, R., Leifert, C., Smith, L., Golden, M., and Benjamin, N., 1995, Chemical generation of nitric oxide in the mouth from the enterosalivary circulation of dietary nitrate, *Nature Med.* **1:**546–551.

Dykhuizen, R. S., Copland, M., Smith, C. C., Douglas, G., and Benjamin, N., 1995, Plasma nitrate concentration and urinary nitrate excretion in patients with gastroenteritis, *J. Infect.* **31:**73–75.

Dykhuizen, R. S., Frazer, R., Duncan, C., Smith, C. C., Golden, M., Benjamin, N., and Leifert, C., 1996a, Antimicrobial effect of acidified nitrite on gut pathogens: Importance of dietary nitrate in host defense, *Antimicrob. Agents Chemother.* **40:**1422–1425.

Dykhuizen, R. S., Masson, J., McKnight, G., Mowat, A. N., Smith, C. C., Smith, L. M., and Benjamin, N., 1996b, Plasma nitrate concentration in infective gastroenteritis and inflammatory bowel disease, *Gut* **39:**393–395.

Dykhuizen, R. S., Fraser, A., McKenzie, H., Golden, M., Leifert, C., and Benjamin, N., 1998, *Helicobactor pylori* is killed by nitrite under acidic conditions, *Gut* **42**(3):334–337.

Fang, F. C., 1997, Mechanisms of nitric oxide-related antimicrobial activity, *J. Clin. Invest.* **99:**2818–2825.

Forman, D., Al-Dabbagh, S., and Doll, R., 1985, Nitrate, nitrites and gastric cancer in Great Britain, *Nature* **313**:620–625.

Green, L. C., Ruiz de Luzuriaga, K., Wagner, D. A., Rand, W., Istfan, N., Young, V. R., and Tannenbaum, S. R., 1981a, Nitrate biosynthesis in man, *Proc. Natl. Acad. Sci. USA* **78**:7764–7768.

Green, L. C., Tannenbaum, S. R., and Goldman, P., 1981b, Nitrate synthesis in the germfree and conventional rat, *Science* **212**:56–58.

Green, S. J., 1995, Nitric oxide in mucosal immunity, *Nature Med.* **1**:515–517.

Gustafsson, L. E., Leone, A. M., Persson, M. G., Wiklund, N. P., and Moncada, S., 1991, Endogenous nitric oxide is present in the exhaled air of rabbits, guinea pigs and humans, *Biochem. Biophys. Res. Commun.* **181**:852–857.

Hibbs, J. B., Jr., Westenfelder, C., Taintor, R., Vavrin, Z., Kablitz, C., Baranowski, R. L., Ward, J. H., Menlove, R. L., McMurry, M. P., Kushner, J. P., and Samlowski, W. E., 1992, Evidence for cytokine-inducible nitric oxide synthesis from L-arginine in patients receiving interleukin-2 therapy, *J. Clin. Invest.* **89**:867–877.

Islam, D., Veress, B., Bardhan, P. K., Lindberg, A. A., and Christensson, B., 1997, *In situ* characterization of inflammatory responses in the rectal mucosae of patients with shigellosis, *Infect. Immun.* **65**:739–740.

Janoff, E. N., Hayakawa, H., Taylor, D. N., Fasching, C. E., Kenner, J. R., Jaimes, E., and Raij, L., 1997, Nitric oxide production during *Vibrio cholerae* infection, *Am. J. Physiol.* **273**(5 Pt 1):G1160–G1167.

Jones-Carson, J., Vazquez-Torres, A., van der Heyde, H. C., Warner, T., Wagner, R. D., and Balish, E., 1995, Gamma delta T cell-induced nitric oxide production enhances resistance to mucosal candidiasis, *Nature Med.* **1**(6):552–557.

Kahn, T., Bosch, J., Levitt, M. F., and Goldstein, M. H., 1975, Effect of sodium nitrate loading on electrolyte transport by the renal tubule, *Am. J. Physiol.* **229**:746–753.

Knight, T. M., Forman, D., Pirastu, R., Comba, P., Iannarilli, R., Cocco, P. L., Angotzi, G., Ninu, E., and Schierano, S., 1990, Nitrate and nitrite exposure in Italian populations with different gastric cancer rates, *Int. J. Epidemiol.* **19**:510–515.

Leggott, P. J., Robertson, P. B., Rothman, D. L., Murray, P. A., and Jacob, R. A., 1986, Response of lingual ascorbic acid test and salivary ascorbate levels to changes in ascorbic acid intake, *J. Dent. Res.* **65**:131–134.

Leone, A. M., Gustafsson, L. E., Francis, P. L., Persson, M. G., Wiklund, N. P., and Moncada, S., 1994, Nitric oxide is present in exhaled breath in humans: Direct GC-MS confirmation, *Biochem. Biophys. Res. Commun.* **201**:883–887.

Li, H., Duncan, C., Townend, J., Killham, K., Smith, L. M., Johnston, P., Dykhuizen, R., Kelly, D., Golden, M., Benjamin, N., and Leifert, C., 1997, Nitrate-reducing bacteria on rat tongues, *Appl. Environ. Microbiol.* **63**:924–930.

Lundberg, J. O., Hellstrom, P. M., Lundberg, J. M., and Alving, K., 1994a, Greatly increased luminal nitric oxide in ulcerative colitis, *Lancet* **344**:1673–1674.

Lundberg, J.O., Weitzberg, E., Lundberg, J. M., and Alving, K., 1994b, Intragastric nitric oxide production in humans: Measurements in expelled air, *Gut* **35**:1543–1546.

Lundberg, J. O., Farkas-Szallasi, T., Weitzberg, E., Rinder, J., Lidholm, J., Anggaard, A., Hokfelt, T., Lundberg, J. M., and Alving, K., 1995a, High nitric oxide production in human paranasal sinuses, *Nature Med.* **1**:370–373.

Lundberg, J. O., Lundberg, J. M., Settergren, G., Alving, K., and Weitzberg, E., 1995b, Nitric oxide, produced in the upper airways, may act in an 'aerocrine' fashion to enhance pulmonary oxygen uptake in humans, *Acta Physiol. Scand.* **155**:467–468.

Macallan, D. C., Smith, L. M., Ferber, J., Milne, E., Griffin, G. E., Benjamin, N., and McNurlan, M.A., 1997, Measurement of NO synthesis in humans by L-[$^{15}N_2$]arginine: Application to the response to vaccination, *Am. J. Physiol.* **272**:R1888–R1896.

McKnight, G. M., Smith, L. M., Drummond, R. S., Duncan, C. W., Golden, M., and Benjamin, N., 1997, Chemical synthesis of nitric oxide in the stomach from dietary nitrate in humans, *Gut* **40:**211–214.

Meng, Q. H., Springall, D. R., Bishop, A. E., Morgan, K., Evans, T. J., Habib, S., Bruenert, D. C., Gyi, K. M., Hodson, M. E., Yacoub, M. H., and Polak, J. M., 1998, Lack of inducible nitric oxide synthase in bronchial epithelium—A possible mechanism of susceptibility to infection in cystic fibrosis, *J. Pathol.* **184:**323–331.

Mitchell, H. H., Shonle, H. A., and Grindley, H. S., 1916, The origin of the nitrates in the urine, *J. Biol. Chem.* **24:**461–490.

Neilly, I. J., Copland, M., Haj, M., Adey, G., Benjamin, N., and Bennett, B., 1995, Plasma nitrate concentrations in neutropenic and non-neutropenic patients with suspected septicaemia, *Br. J. Haematol.* **89:**199–202.

Reddy, D., Lancaster, J. R., and Cornforth, D. P., 1983, Nitrite inhibition of *Clostridium botulinum:* Electron spin resonance detection of iron–nitric oxide complexes, *Science* **221:**769–770.

Salzman, A. L., 1995, Nitric oxide in the gut, *New Horiz.* **3:**352–364.

Salzman, A. L., Eavespyles, T., Linn, S. C., Denenberg, A. G., and Szabo, C., 1998, Bacterial induction of inducible nitric oxide synthase in cultured human intestinal epithelial cells, *Gastroenterology* **114:**93–102.

Sasaki, T., and Matano, K., 1979, Formation of nitrite from nitrate at the dorsum linguae, *J. Food Hyg. Soc. Jpn.* **20:**363–369.

Schorah, C. J., Sobala, G. M., Sanderson, M., Collis, N., and Primrose, J. M., 1991, Gastric juice ascorbic acid: Effects of disease and implications for gastric carcinogenesis, *Am. J. Clin. Nutr.* **53:**287S–293S.

Sobala, G. M., Schorah, C. J., Sanderson, M., Dixon, M. F., Tompkins, D. S., Godwin, P., and Axon, A. T. R., 1989, Ascorbic acid in the human stomach, *Gastroenterology* **97:**357–363.

Sobala, G. M., Pignatelli, B., Schorah, C. J., Bartsch, H., Sanderson, M., Dixon, M. F., King, R. F. G., and Axon, A. T. R., 1991, Levels of nitrite, nitrate, *N*-nitroso compounds, ascorbic acid and total bile acids in gastric juice of patients with and without precancerous conditions of the stomach, *Carcinogenesis* **12:**193–198.

Spiegelhalder, B., Eisenbrand, G., and Preussman, R., 1976, Influence of dietary nitrate on nitrite content of human saliva: Possible relevance to *in-vivo* formation of *N*-nitroso compounds, *Foods Cosmet. Toxicol.* **14:**545–548.

Tannenbaum, S. R., 1979, Nitrate and nitrite: Origin in humans, *Science* **205:**1332, 1334–1337.

Tannenbaum, S. R., Sinskey, A. J., Weisman, M., and Bishop, W., 1974, Nitrite in human saliva. Its possible relationship to nitrosamine formation, *J. Natl. Cancer Inst.* **53:**79–84.

Tannenbaum, S. R., Weisman, M., and Fett, D., 1976, The effect of nitrate intake on nitrite formation in human saliva, *Food Cosmet. Toxicol.* **14:**549–552.

Tannenbaum, S. R., Fett, D., Young, V. R., Land, P. D., and Bruce, W. R., 1978, Nitrite and nitrate are formed by endogenous synthesis in the human intestine, *Science* **200:**1487–1489.

Vittozzi, L., 1992, Toxicology of nitrates and nitrites, *Food Additives Contam.* **9:**579–585.

Wagner, D. A., Schultz, D. S., Deen, W. M., Young, V. R., and Tannenbaum, S. R., 1983, Metabolic fate of an oral dose of ^{15}N-labeled nitrate in humans: Effect of diet supplementation with ascorbic acid, *Cancer Res.* **43:**1921–1925.

Weller, R., Pattullo, S., Smith, L., Golden, M., Ormerod, A., and Benjamin, N., 1996, Nitric oxide is generated on the skin surface by reduction of sweat nitrate, *J. Invest. Dermatol.* **107:**327–331.

Weller, R., Ormerod, A. D., Hobson, R. P., and Benjamin, N. J., 1998, A randomized trial of acidified nitrite cream in the treatment of tinea pedis, *J. Am. Acad. Dermatol.* **38:**559–563.

Williams, D. H. L., 1988, *Nitrosation*, Cambridge University Press, London.

CHAPTER 12

Antimicrobial Properties of Nitric Oxide

MARY ANN DeGROOTE and FERRIC C. FANG

1. Introduction

The ability of nitric oxide (NO) to inhibit microbial pathogens has been appreciated for many years by the food processing industry, which routinely adds nitrite to meat during the curing process (Tarr, 1941; Shank *et al.*, 1962; Incze *et al.*, 1974; Pierson and Smoot, 1982; DeGiusti and DeVito, 1992). However, cell-derived NO-related antimicrobial activity generated from the enzymatic oxidation of L-arginine (Chapter 4) has been recognized only recently to be an important component of host defense. In addition to indirect effects resulting from modulation of immune responses (Chapter 10), it is now evident that NO or its congeners can exert direct inhibitory or lethal effects on microbial targets. This chapter will consider present knowledge concerning mechanisms of NO-related antimicrobial activity and relevant microbial defenses against NO-related cytotoxicity.

2. NO as an Endogenous Antimicrobial Mediator

Several independent lines of evidence have suggested a critical role of NO as an endogenous antimicrobial mediator. In experimental animals or humans, expression of the inducible NO synthase isoform (iNOS, NOS2) can be upregulated by cytokines associated with an effective host immune response to infection, such as IFNγ, TNFα, IL-1, and IL-2 (Hibbs *et al.*, 1992; MacMicking *et al.*, 1997a) (see also Chapter 5). Oxidation products of NO can be readily

---

MARY ANN DeGROOTE and FERRIC C. FANG • Departments of Medicine, Pathology, and Microbiology, University of Colorado Health Sciences Center, Denver, Colorado 80262.

Nitric Oxide and Infection, edited by Fang. Kluwer Academic / Plenum Publishers, New York, 1999.

detected from the plasma and urine of infected experimental animals and patients (Ochoa *et al.*, 1991; Hibbs *et al.*, 1992; Evans *et al.*, 1993; Anstey *et al.*, 1996; Dykhuizen *et al.*, 1996b; Wong *et al.*, 1996). Moreover, localized iNOS expression can be demonstrated directly at sites of infection (Gazzinelli *et al.*, 1993; Neilly *et al.*, 1995; Goldman *et al.*, 1996; Lowenstein *et al.*, 1996; Mannick *et al.*, 1996; Nicholson *et al.*, 1996; Rottenberg *et al.*, 1996; Stenger *et al.*, 1996; Adler *et al.*, 1997; Chambers *et al.*, 1997). Increased NOS expression has been associated with a good clinical outcome in infections such as malaria (Anstey *et al.*, 1996), and inhibition of NO production in many experimental models of infection results in enhanced microbial proliferation and increased mortality (Table I) (see also Chapters 14, 17, 18, 22). Similar effects on microbial proliferation can be seen when NOS inhibitors are added to infected phagocytic cells (Table I). Lastly, NO-generating compounds have been demonstrated to directly inhibit or kill microbes (Table I). An astonishingly broad range of pathogenic parasites, fungi, bacteria, and viruses have been found to be susceptible to inhibition or killing by NO (DeGroote and Fang, 1995), whether derived from activated cells or chemical donors (Table II).

NO has been particularly implicated in infections with obligate or facultative intracellular pathogens such as *Leishmania,* mycobacteria, chlamydia, rickettsia, and *Salmonella* (Feng and Walker, 1993; Mayer *et al.*, 1993; DeGroote *et al.*, 1996; Stenger *et al.*, 1996; MacMicking *et al.*, 1997b). This may reflect the ability of NO-scavenging substances such as hemoglobin to antagonize NO-related antimicrobial activity in extracellular compartments (Mabbott *et al.*, 1994; Kim *et al.*, 1996; Coulson *et al.*, 1998). However, recent observations suggest that NO may also play a role in infections with bacteria usually considered to be extracellular pathogens, such as *Pseudomonas aeruginosa* and *Klebsiella pneumoniae* (Gosselin *et al.*, 1995; Tsai *et al.*, 1997).

It is especially intriguing to consider whether NO might play a role in microbial latency (Granger *et al.*, 1993). Many intracellular microbial pathogens, including *Toxoplasma*, *Leishmania*, mycobacteria, and herpesviruses, are capable of establishing prolonged subclinical latent infection. Subsequent impairment of host defenses by immunosuppressive agents or intercurrent illness can allow reactivation of infection, but the molecular basis by which the host can indefinitely suppress microbial replication to maintain a state of latency has heretofore constituted one of the great mysteries of microbial pathogenesis. Recent observations that inhibition of NO production can result in rapid reactivation of *Leishmania major*, *Mycobacterium tuberculosis*, or Epstein–Barr virus infection in experimental models (Mannick *et al.*, 1994; Stenger *et al.*, 1996; MacMicking *et al.*, 1997b) strongly support the concept that NO plays a central role in many persistent or latent infections.

TABLE I
Nitric Oxide in Experimental Models of Infection

Exacerbation of infection by NOS inhibition

Evans *et al.* (1993), Green *et al.* (1993), Boockvar *et al.* (1994), Feng *et al.* (1994), Leitch and He (1994), Petray *et al.* (1994, 1995), Blasi *et al.* (1995), Brieland *et al.* (1995), Chan *et al.* (1995), Lovchik *et al.* (1995), MacMicking *et al.* (1995, 1997b), Vazquez-Torres *et al.* (1995), Wei *et al.* (1995), DeGroote *et al.* (1996, 1997), Fukatsu *et al.* (1996), Hayashi *et al.* (1996), Hiraoka *et al.* (1996), Lowenstein *et al.* (1996), Meli *et al.* (1996), Rajan *et al.* (1996), Stenger *et al.* (1996), Tucker *et al.* (1996), Nathan (1997), Scharton-Kersten *et al.* (1997), Tay and Welsh (1997), Tsai *et al.* (1997), Umezawa *et al.* (1997), Holscher *et al.* (1998), Maclean *et al.* (1998)

Enhancement of microbial proliferation by NOS inhibition in phagocytes

Granger *et al.* (1988, 1990), James and Glaven (1989), Adams *et al.* (1990, 1991), Green *et al.* (1990), Liew *et al.* (1990a,b), Denis, (1991a,b), Flesch and Kaufmann (1991), Mauel *et al.* (1991), Vincendeau and Daulouede (1991), Anthony *et al.* (1992), Chan *et al.* (1992), Fischer-Stenger and Marciano-Cabral (1992), Fortier *et al.* (1992), Gazzinelli *et al.* (1992), Lin and Chadee (1992), Munoz-Fernandez *et al.* (1992), Park and Rikihisa (1992), Summersgill *et al.* (1992), Vincendeau *et al.* (1992), Bermudez (1993), Cenci *et al.* (1993), Chao *et al.* (1993), Croen (1993), Cunha *et al.* (1993), Jiang *et al.* (1993), Kanazawa *et al.* (1993), Karupiah *et al.* (1993), Naotunne *et al.* (1993), Assreuy *et al.* (1994), Cillari *et al.* (1994), Gyan *et al.* (1994), Lane *et al.* (1994a,b), Lee *et al.* (1994), Melkova and Esteban (1994), Akarid *et al.* (1995), Blasi *et al.* (1995), Breummer and Stevens (1995), Didier (1995), Harris *et al.* (1995), Karupiah and Harris (1995), Norris *et al.* (1995), Nunoshiba *et al.* (1995), Peterson *et al.* (1995), Rementeria *et al.* (1995), Vouldoukis *et al.* (1995, 1997), Chen *et al.* (1996), Cowley *et al.* (1996), Pertile *et al.* (1996), Proudfoot *et al.* (1996), Vazquez-Torres *et al.* (1996), Akaki *et al.* (1997), Arias *et al.* (1997), Cogliati *et al.* (1997), DeGroote *et al.* (1997), Lin *et al.* (1997), Lopez-Guerrero *et al.* (1997), Miyagi *et al.* (1997), Mnaimneh *et al.* (1997), Nozaki *et al.* (1997), Sakiniene *et al.* (1997), Spithill *et al.* (1997), Thomas *et al.* (1997), Coulson *et al.* (1998), Gross *et al.* (1998), Hickman-Davis *et al.* (1998), Kudeken *et al.* (1998), Turco *et al.* (1998)

Killing or inhibition of microbes by NO-donor compounds

Tarr (1941), Shank *et al.* (1962), Incze *et al.* (1974), Moran *et al.* (1975), O'Leary and Solberg (1976), Yarbrough *et al.* (1980), Morris and Hansen (1981), Mancinelli and McKay (1983), Morris *et al.* (1984), Payne *et al.* (1990a,b), Alspaugh and Granger (1991), Rockett *et al.* (1991), DeGiusti and DeVito (1992), Zhu *et al.* (1992), Croen (1993), Denicola *et al.* (1993), Doi *et al.* (1993), Klebanoff (1993), Assreuy *et al.* (1994), Bohne *et al.* (1994), Mannick *et al.* (1994, 1995), O'Brien *et al.* (1994), Vespa *et al.* (1994), Virta *et al.* (1994), Akarid *et al.* (1995), Bi and Reiss (1995), Brunelli *et al.* (1995), DeGroote *et al.* (1995, 1996, 1997), Kawanishi, (1995), Kunert, (1995), Lopez-Jaramillo *et al.* (1995, 1998), Pacelli *et al.* (1995), Vazquez-Torres *et al.* (1995), Dykhuizen *et al.* (1996a), Guillemard *et al.* (1996), Hausladen *et al.* (1996), Johnson *et al.* (1996), Taylor *et al.* (1996), Ahmed *et al.* (1997), Akaki *et al.* (1997), Bourguignon *et al.* (1997), Cogliati *et al.* (1997), Fernandes and Assreuy (1997), Igietseme *et al.* (1997), Lane *et al.* (1997), Lemesre *et al.* (1997), Mauel and Ransijn (1997), Marcinkiewicz (1997), Rhoades and Orme (1997), Thomas *et al.* (1997), Xie *et al.* (1997), Zaragoza *et al.* (1997), Zhao *et al.* (1997), Chen *et al.* (1998), Crawford and Goldberg (1998), Kudeken *et al.* (1998), Nagata *et al.* (1998), Sanders *et al.* (1998)

TABLE II
Pathogenic Microbial Targets of Nitric Oxide

Viruses

- Coronavirus (Lane *et al.*, 1997)
- Coxsackievirus (Hiraoka *et al.*, 1996; Lowenstein *et al.*, 1996; Zaragoza *et al.*, 1997)
- Ectromelia virus (Karupiah *et al.*, 1993; Nathan, 1997)
- Encephalomyocarditis virus (Guillemard *et al.*, 1996)
- Epstein–Barr virus (Mannick *et al.*, 1994; Kawanishi, 1995)
- Friend leukemia virus (Akarid *et al.*, 1995)
- Herpes simplex virus (Croen, 1993; Karupiah *et al.*, 1993; Komatsu *et al.*, 1996; Adler *et al.*, 1997; Maclean *et al.*, 1998)
- Human immunodeficiency virus-1 (Mannick *et al.*, 1995)
- Japanese encephalitis virus (Lin *et al.*, 1997)
- Murine cytomegalovirus (Tay and Welsh, 1997)
- Parvovirus (Lopez-Guerrero *et al.*, 1997)
- Poliovirus (Komatsu *et al.*, 1996)
- Reovirus (Pertile *et al.*, 1996)
- Rhinovirus (Sanders *et al.*, 1998)
- Sindbis virus (Tucker *et al.*, 1996)
- Vaccinia virus (Karupiah *et al.*, 1993; Melkova and Esteban, 1994, 1995; Harris *et al.*, 1995; Karupiah and Harris, 1995; Rolph *et al.*, 1996)
- Vesicular stomatitis virus (Bi and Reiss, 1995; Komatsu *et al.*, 1996)

Bacteria

- *Bacillus cereus* (Morris and Hansen, 1981; Morris *et al.*, 1984)
- *Brucella abortus, B. suis* (Jiang *et al.*, 1993; Gross *et al.*, 1998)
- *Burkholderia pseudomallei* (Miyagi *et al.*, 1997)
- *Chlamydia trachomatis* (Mayer *et al.*, 1993; Chen *et al.*, 1996; Igietseme, 1996; Igietseme *et al.*, 1996, 1997)
- *Clostridium perfringens, C. sporogenes* (Shank *et al.*, 1962; Moran *et al.*, 1975; O'Leary and Solberg, 1976; Payne *et al.*, 1990a,b)
- *Ehrlichia risticii* (Park and Rikihisa, 1992)
- *Enterococcus faecium* (Incze *et al.*, 1974)
- *Escherichia coli* (Zhu *et al.*, 1992; Klebanoff, 1993; Virta *et al.*, 1994; Brunelli *et al.*, 1995; Nunoshiba *et al.*, 1995; Pacelli *et al.*, 1995; Dykhuizen *et al.*, 1996a; Hausladen *et al.*, 1996; Marcinkiewicz, 1997; Nagata *et al.*, 1998)
- *Francisella tularensis* (Anthony *et al.*, 1992; Fortier *et al.*, 1992; Green *et al.*, 1993; Cowley *et al.*, 1996)
- *Helicobacter pylori* (Nagata *et al.*, 1998)
- *Klebsiella pneumoniae* (Tsai *et al.*, 1997)
- *Legionella pneumophila* (Summersgill *et al.*, 1992; Brieland *et al.*, 1995; Rajagopalan-Levasseur *et al.*, 1996)
- *Listeria monocytogenes* (Beckerman *et al.*, 1993; Bermudez, 1993; Boockvar *et al.*, 1994; MacMicking *et al.*, 1995; Akaki *et al.*, 1997)
- *Micrococcus roseus, M. luteus* (Mancinelli and McKay, 1983)
- *Mycobacterium avium* complex (Denis, 1991b; Doi *et al.*, 1993; Akaki *et al.*, 1997; Zhao *et al.*, 1997)
- *Mycobacterium bovis* (Flesch and Kaufmann, 1991)

TABLE II (*Continued*)

Mycobacterium leprae (Green, 1995)
Mycobacterium tuberculosis (Denis, 1991a; Chan *et al.*, 1992, 1995, 1996; O'Brien *et al.*, 1994; Arias *et al.*, 1997; MacMicking *et al.*, 1997b; Rhoades and Orme, 1997)
Mycoplasma pulmonis (Hickman-Davis *et al.*, 1998)
Pseudomonas aeruginosa (Tarr, 1941; Gosselin *et al.*, 1995)
Rickettsia conorii, R. prowazekii (Feng and Walker, 1993; Feng *et al.*, 1994; Walker *et al.*, 1997; Turco *et al.*, 1998)
Salmonella typhimurium, S. enteritidis (Incze *et al.*, 1974; DeGroote *et al.*, 1995, 1996, 1997; Dykhuizen *et al.*, 1996a; Meli *et al.*, 1996; Umezawa *et al.*, 1997; Chen *et al.*, 1998; Crawford and Goldberg, 1998)
Shigella sonnei (Dykhuizen *et al.*, 1996a)
Staphylococcus aureus (Shank *et al.*, 1962; Mancinelli and McKay, 1983; Malawista *et al.*, 1992; Kaplan *et al.*, 1996; Sakiniene *et al.*, 1997)
Yersinia enterocolitica (DeGiusti and DeVito, 1992; Dykhuizen *et al.*, 1996a)

Fungi

Aspergillus fumigatus (Kunert, 1995)
Candida albicans (Cenci *et al.*, 1993; Blasi *et al.*, 1995; Lopez-Jaramillo *et al.*, 1995; Rementeria *et al.*, 1995; Vazquez-Torres *et al.*, 1995, 1996)
Cryptococcus neoformans (Granger *et al.*, 1988; Alspaugh and Granger, 1991; Lee *et al.*, 1994; Lovchik *et al.*, 1995; Xie *et al.*, 1997)
Epidermophyton floccosum (Lopez-Jaramillo *et al.*, 1995)
Histoplasma capsulatum (Lane *et al.*, 1994a,b; Brummer and Stevens, 1995)
Penicillium marneffei (Cogliati *et al.*, 1997; Kudeken *et al.*, 1998)
Trichophyton tonsurans, T. mentagrophytes (Lopez-Jaramillo *et al.*, 1995)

Parasites

Babesia bovis (Johnson *et al.*, 1996)
Brugia malayi (Rajan *et al.*, 1996; Taylor *et al.*, 1996; Thomas *et al.*, 1997)
Cryptosporidium parvum (Leitch and He, 1994)
Echinococcus multilocularis (Kanazawa *et al.*, 1993)
Encephalitozoon hellem, E. intestinalis, E. cuniculi (Didier, 1995; He *et al.*, 1996)
Entamoeba histolytica (Lin and Chadee, 1992)
Fasciola hepatica (Spithill *et al.*, 1997)
Giardia lamblia (Fernandes and Assreuy, 1997)
Leishmania major, L. enriettii, L. amazonensis, L. mexicana, L. chagasi, L. infantum (Green *et al.*, 1990; Liew *et al.*, 1990a,b, 1991; Mauel *et al.*, 1991; Cunha *et al.*, 1993; Evans *et al.*, 1993; Assreuy *et al.*, 1994; Cillari *et al.*, 1994; Vouldoukis *et al.*, 1995, 1997; Wei *et al.*, 1995; Augusto *et al.*, 1996; Proudfoot *et al.*, 1996; Stenger *et al.*, 1996; Bourguignon *et al.*, 1997; Lemesre *et al.*, 1997; Mauel and Ransijn, 1997; Lopez-Jaramillo *et al.*, 1998)
Naegleria fowleri (Fischer-Stenger and Marciano-Cabral, 1992)
Onchocerca lienalis (Taylor *et al.*, 1996)
Plasmodium falciparum, P. chabaudi, P. vinckii, P. vivax, P. berghei (Rockett *et al.*, 1991; Motard *et al.*, 1993; Naotunne *et al.*, 1993; Taylor-Robinson *et al.*, 1993; Gyan *et al.*, 1994; Mellouk *et al.*, 1994; Seguin *et al.*, 1994; Ahvazi *et al.*, 1995)
Schistosoma mansoni (James and Glaven, 1989; Oswald *et al.*, 1994; Wynn *et al.*, 1994; Ahmed *et al.*, 1997; Coulson *et al.*, 1998)

TABLE II (*Continued*)

Toxoplasma gondii (Adams *et al.*, 1990; Chao *et al.*, 1993; Gazzinelli *et al.*, 1993; Bohne *et al.*, 1994; Peterson *et al.*, 1995; Hayashi *et al.*, 1996; Khan *et al.*, 1997; Scharton-Kersten *et al.*, 1997)
Trypanosoma cruzi, T. brucei, T. musculi (Vincendeau and Daulouede, 1991; Gazzinelli *et al.*, 1992; Munoz-Fernandez *et al.*, 1992; Vincendeau *et al.*, 1992; Denicola *et al.*, 1993; Petray *et al.*, 1994; Vespa *et al.*, 1994; Norris *et al.*, 1995; Silva *et al.*, 1995; Bourguignon *et al.*, 1997; Mnaimneh *et al.*, 1997; Holscher *et al.*, 1998)

3. Interactions between Reactive Oxygen and Nitrogen Intermediates

Although iNOS and the NADPH oxidase responsible for the phagocyte respiratory burst are independently regulated, both enzyme systems can be stimulated during infection, enhancing the possibility of interaction between reactive nitrogen and oxygen species. Studies in mice (see Chapter 19) (Umezawa *et al.*, 1997) have suggested that iNOS may also synergistically interact with xanthine oxidase (XO), but XO appears not to be a significant source of superoxide in human phagocytes (Simmonds *et al.*, 1985). Simultaneous production of NO and superoxide can result in the formation of a complex variety of reactive molecular species (Fig. 1), each with distinctive reactivity, stability, diffusibility, and target specificity. In addition to NO radical (NO•), congeners such as peroxynitrite ($ONOO^-$), nitrogen dioxide (NO_2•), *S*-nitrosothiols (RSNO), dinitrogen trioxide (N_2O_3), and dinitrosyl iron complexes (DNIC) may contribute to antimicrobial activity. Interactions between reactive nitrogen and oxygen intermediates can provide synergistic antimicrobial activity through the production of cytotoxic species such as peroxynitrite (Brunelli *et al.*, 1995; DeGroote *et al.*, 1997) or hydroxyl radical (HO•) (Pacelli *et al.*, 1995). Alternatively, NO• can antagonize oxidant injury by terminating lipid peroxidation reactions (Freeman *et al.*, 1995) or stimulating the expression of antioxidant stress responses (Nunoshiba *et al.*, 1993); the interaction of superoxide (O_2^-•) and NO• may actually reduce the antimicrobial potency of NO• for certain microorganisms including *Cryptococcus neoformans* (Tohyama *et al.*, 1996), *L. major* (Assreuy *et al.*, 1994), and *Giardia lamblia* (Fernandes and Assreuy, 1997).

NO• appears to cross membranes and enter target cells readily. Although superoxide does not enter bacterial cells to a significant extent (Nunoshiba *et al.*, 1993), its product peroxynitrite can cross membranes (Marla *et al.*, 1997; Denicola *et al.*, 1998); reactivity with lipids and proteins may nevertheless limit the effective diffusion of peroxynitrite *in vivo*. *In situ* formation of peroxynitrite from O_2^-• and NO• may be particularly important with regard to membrane and periplasmic targets (DeGroote *et al.*, 1997). RSNO may be imported by specific microbial

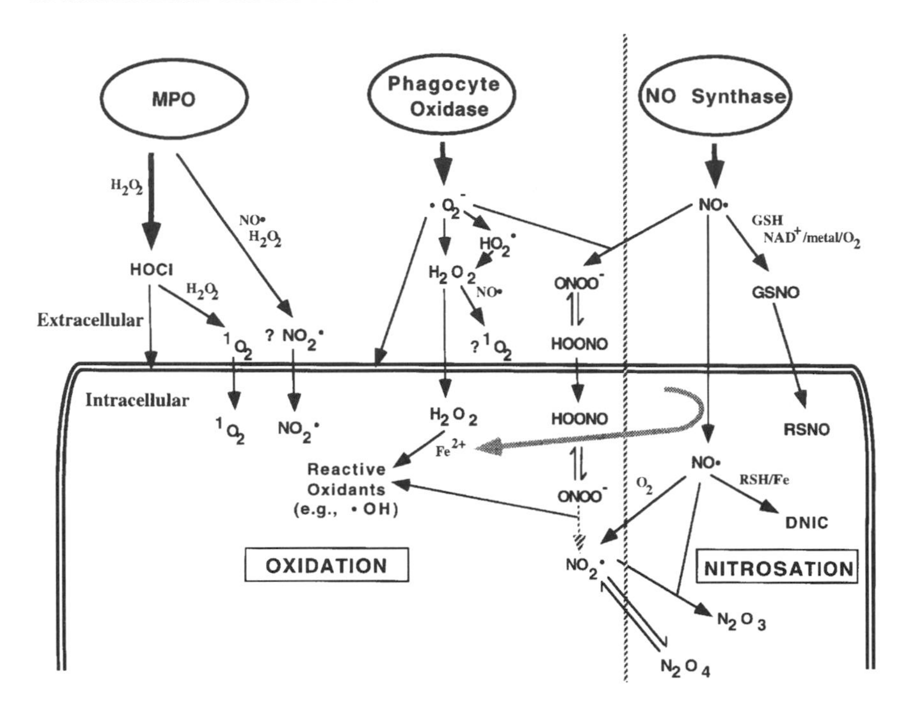

FIGURE 1. Potential interactions between phagocyte-derived reactive oxygen and nitrogen intermediates. Some possible reactions of products originating from nitric oxide synthase, phagocyte oxidase, and myeloperoxidase are shown in relation to a hypothetical microbe situated within a phagolysosome. "Extracellular" refers to the phagolysosomal compartment, and "intracellular" to the microbial cytosol. Chemical species are separated according to their predominant tendency toward oxidative or nitrosative reactivity. DNIC, dinitrosyl iron complexes; GSH, glutathione; GSNO, *S*-nitrosoglutathione; H_2O_2, hydrogen peroxide; HOCl, hypochlorous acid; HOONO, peroxynitrous acid; MPO, myeloperoxidase; NAD, nicotinamide adenine dinucleotide; $NO^•$, nitric oxide; $NO_2^•$, nitrogen dioxide; N_2O_3, dinitrogen trioxide; N_2O_4, dinitrogen tetroxide; O_2, molecular oxygen; $^1O_2$, singlet oxygen; $^•OH$, hydroxyl; $ONOO^-$, peroxynitrite; RSNO, *S*-nitrosothiol. Reproduced from Fang (1997), with permission of the Rockefeller University Press.

transport systems. Genetic analyses in *Salmonella typhimurium* indicate that *S*-nitrosoglutathione (GSNO) is processed by γ-glutamyl transpeptidase and taken up through the actions of an ATP-binding permease (Dpp) belonging to the ABC transporter family, which normally functions to import glutathione metabolites and other small peptides into the cell (Abouhamad *et al.*, 1991). Inactivation of the Dpp transporter completely abrogates the antimicrobial activity of GSNO for *Salmonella* (DeGroote *et al.*, 1995).

4. Microbial Targets of NO

Although many microbes have been shown indirectly or directly to be susceptible to NO (Table I), it is now evident that microbial species vary considerably in their susceptibility to various congeners of NO. For example, while $NO^{\bullet}$ has little antimicrobial activity for *E. coli* or *S. typhimurium*, it appears to be toxic for *Staphylococcus aureus, L. major*, or *G. lamblia* (Assreuy *et al.*, 1994; Brunelli *et al.*, 1995; DeGroote *et al.*, 1995; Kaplan *et al.*, 1996; Fernandes and Assreuy, 1997). In contrast, $ONOO^-$ is rapidly microbicidal for *E. coli* and *S. typhimurium* (Brunelli *et al.*, 1995; DeGroote *et al.*, 1995), but not for *S. aureus, L. major*, or *G. lamblia* (Assreuy *et al.*, 1994; Kaplan *et al.*, 1996; Fernandes and Assreuy, 1997). Some microbial species, such as *Mycobacterium avium*, appear to be relatively resistant to NO (Doi *et al.*, 1993; Zhao *et al.*, 1997), and susceptibility to NO can be quite variable even among strains of a single microbial species, as has been described for *M. tuberculosis* (O'Brien *et al.*, 1994; Rhoades and Orme, 1997). Additional interactions of NO and reactive oxygen species can augment antimicrobial activity via complex mechanisms; using a diethylamine adduct (DEA/NO) as an NO donor, Pacelli *et al.* (1995) demonstrated synergistic bactericidal interactions of hydrogen peroxide (H_2O_2) and $NO^{\bullet}$ exerted against *E. coli*. This synergy appeared to involve mobilization of redox-active transition metals, decreased cellular respiration, depletion of antioxidant thiols, and DNA damage, but did not involve $ONOO^-$ as an intermediate. RSNO and H_2O_2 similarly exert synergistic antibacterial activity (Marcinkiewicz, 1997). RSNO have their own distinctive spectrum of antimicrobial activity, exerting microbiostatic activity against *E. coli* or *S. typhimurium* (DeGroote *et al.*, 1995), and microbicidal activity against *S. aureus, L. major*, or *G. lamblia* (Assreuy *et al.*, 1994; Kaplan *et al.*, 1996; Fernandes and Assreuy, 1997). Although some of the latter activity may be attributable to homolytic release of $NO^{\bullet}$ by RSNO, Rockett *et al.* (1991) found the microbicidal activity of RSNO against *Plasmodium falciparum* to exceed that of $NO^{\bullet}$ by as much as 1000-fold. These observations, along with the studies in enteric bacteria (DeGroote *et al.*, 1995), suggest that RSNO possess a unique mechanism of action independent of their ability to release $NO^{\bullet}$.

5. Molecular Targets and Mechanisms of NO-Related Antimicrobial Activity

With multiple molecular species contributing to NO-related antimicrobial activity (Fig. 1), it is not surprising that multiple targets within microbial cells have been identified. Reactive nitrogen intermediates can modify DNA, proteins, and lipids (Fig. 2), acting both at the microbial surface and within the cell.

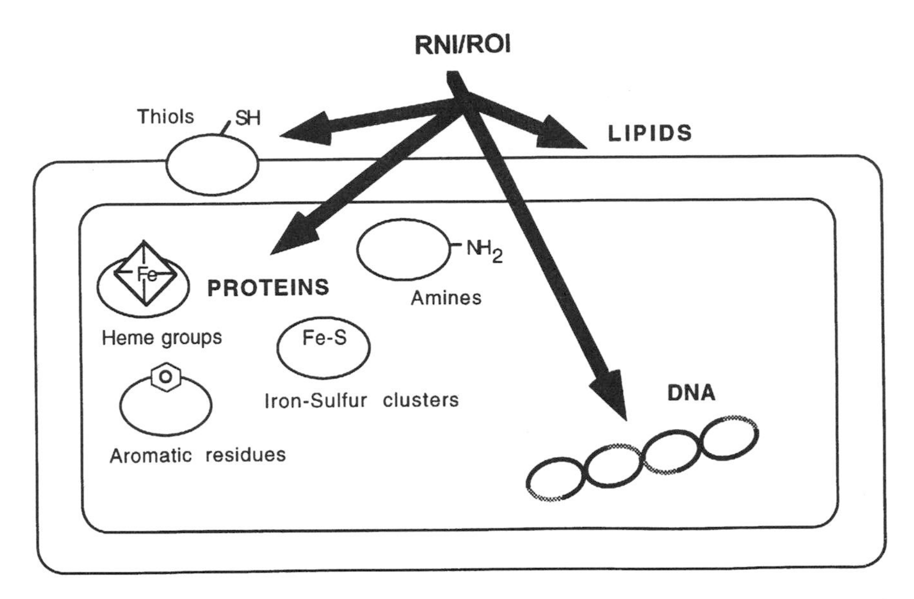

FIGURE 2. Microbial cellular targets of reactive oxygen and nitrogen intermediates. Reproduced from Fang (1997), with permission of the Rockefeller University Press.

5.1. DNA

In vitro studies have demonstrated the ability of NO to deaminate DNA (Wink *et al.*, 1991), probably acting via an *N*-nitrosating intermediate (e.g., N_2O_3) as deamination occurs only under aerobic conditions. $NO_2^{\bullet}$ or $ONOO^-$ is also capable of mediating DNA damage, generating a variety of alterations including abasic sites and strand breaks (Juedes and Wogan, 1996). It has been suggested that some DNA-modifying effects of NO congeners might result from interactions with DNA repair enzymes rather than from direct actions on DNA.

S. typhimurium mutants deficient in DNA repair systems have increased susceptibility to NO-donor compounds *in vitro* (DeGroote *et al.*, 1995), and sequence analysis of DNA alterations induced by NO donors implicates deamination as a mechanism (Maragos *et al.*, 1993). Although the biological significance of these observations remains to be established, restoration of virulence to DNA repair-deficient mutants by abrogation of reactive oxygen and nitrogen intermediate production in mice (Shiloh *et al.*, 1999) suggests that DNA may be an important target of NO in host–pathogen interactions.

5.2. Proteins

Reactive nitrogen species may react with multiple targets on proteins, including iron–sulfur clusters, heme groups, thiols, aromatic or phenolic residues, tyrosyl radicals, and amines (Fig. 2). Some NO congeners, such as $ONOO^-$ and $NO_2^\bullet$, can also cause nonspecific oxidative protein modifications (Ischiropoulos and Al-Mehdi, 1995). Initial studies of NO-related cytotoxicity in tumor cells revealed inactivation of enzymes containing iron–sulfur (Fe–S) clusters (e.g., aconitase, NADH dehydrogenase, succinate dehydrogenase) and efflux of iron–nitrosyl complexes, leading to the proposal that $NO^\bullet$ directly releases iron from Fe–S clusters and promotes iron depletion (Drapier *et al.*, 1991). Subsequent studies have suggested that NO congeners other than $NO^\bullet$ itself, such as $ONOO^-$, may be responsible for aconitase inactivation (Castro *et al.*, 1994; Hausladen and Fridovich, 1994), and the kinetics of NO-dependent effects on Fe–S clusters are more consistent with an indirect mechanism of action (Hentze and Kühn, 1996). Nevertheless, inactivation of metabolic enzymes involved in the tricarboxylic acid cycle and the electron transport chain required for cellular respiration may still constitute an important mechanism of NO-related antimicrobial activity. $ONOO^-$ has been shown to potently inactivate [4Fe–4S] dehydratases of *E. coli*, and releases iron in the process that could potentiate the toxicity of H_2O_2 (Keyer and Imlay, 1997). Inhibition of metabolic enzymes and transmembrane transporters may result in dissipation of electrochemical proton-motive force. Similarities in the cytotoxic effects of NO donors and metabolic inhibitors on *Schistosoma mansoni* larvae are consistent with the idea that common mechanistic pathways are involved (Ahmed *et al.*, 1997). NO-dependent antiparasitic activity for *Leishmania* spp. promastigotes and amastigotes can be reversed by iron and L-cysteine as well as components of the citric acid cycle, suggesting that the antimicrobial action of NO is dependent on its ability to inhibit Fe–S-containing metabolic enzymes (Lemesre *et al.*, 1997). High NO fluxes have also been shown to inhibit respiration reversibly in *E. coli* respiration (Pacelli *et al.*, 1995; Nagata *et al.*, 1998), and irreversibly in *H. pylori* (Nagata *et al.*, 1998).

NO is capable of coordinate interactions with transition metals such as Fe, Cu, Co, and Mn (Drapier, 1997). DNIC arising from the interaction of $NO^\bullet$ and iron are potent nitrosating species in their own right (Boese *et al.*, 1997). Electron paramagnetic resonance (EPR) studies have demonstrated the presence of DNIC in activated macrophages (Drapier, 1997) and in the plasma of patients with leishmaniasis (Augusto *et al.*, 1996), establishing their biological relevance to infectious diseases.

In eukaryotic cells, many important signaling properties of $NO^\bullet$ result from its interaction with heme iron in guanylyl cyclase (Murad, 1994). Other heme-protein targets of NO include hemoglobin, cytochrome P450, catalase, and NOS itself (Kim *et al.*, 1995). Although guanylyl cyclase is activated by NO, other NO–heme interactions can interfere with protein function; in the example of catalase, this might heighten susceptibility to oxidant stress, providing yet another potential mechanism for synergy between reactive nitrogen and oxygen species.

Ribonucleotide reductase, a highly conserved enzyme required for DNA synthesis, has been strongly implicated as a target of NO in tumor cells (Lepoivre *et al.*, 1990). Ribonucleotide reductase is a nonheme metalloenzyme, and quenching of a tyrosyl radical at the active site resulting in release of iron (Guittet *et al.*, 1998), as well as nitrosylation of thiol groups are believed to be responsible for enzyme inactivation. IFNγ- and NO-dependent inhibition of vaccinia virus replication can be partially abrogated by supplementation with deoxyribonucleosides, suggesting that ribonucleotide reductase inhibition contributes to the antiviral effect (Melkova and Esteban, 1995). However, supplementation with iron or tricarboxylic acid cycle metabolites can also partially restore viral replication, suggesting that NO is acting at multiple target sites including aconitase, NADH:ubiquinone oxidoreductase (complex I), and succinate ubiquinone oxidoreductase (complex II) of the mitochondrial electron transport chain (Karupiah and Harris, 1995). Observations in *Leishmania* also indicate a complex mechanism of NO toxicity, involving impaired respiration, decreased protein and DNA synthesis, impaired membrane transport, and an inhibition of glyceraldehyde-3-phosphate dehydrogenase and aconitase activity (Lemesre *et al.*, 1997; Mauel and Ransijn, 1997).

Elegant studies by Stamler (Stamler, 1994, 1995) and others have established that thiols are highly important protein targets of NO-mediated modification. Under physiologic conditions, *S*-nitrosylation (attachment or transfer of NO^+ to sulfhydryl groups) by nitrosating species such as RSNO, N_2O_3, or dinitrosyl–thiol–iron complexes is favored over *N*-nitrosylation reactions, although the latter may also occur (Boese *et al.*, 1997). Thiol nitrosylation can itself alter protein function, or facilitate further protein modification such as ADP-ribosylation, disulfide bond formation (Stamler, 1995), or sulfhydryl oxidation to sulfenic or sulfinic acid derivatives (Becker *et al.*, 1998). In *Bacillus cereus,* nitrosylation of surface thiols has been shown to be responsible for inhibition of spore outgrowth (Morris and Hansen, 1981; Morris *et al.*, 1984). In contrast, cytostasis resulting from nitrosylation of intracellular protein targets by RSNO has been suggested by observations in *S. typhimurium* (DeGroote *et al.*, 1995). *S*-Nitrosylation *in vitro* has been demonstrated to inactivate many proteins (Stamler, 1995), including glyceraldehyde-3-phosphate dehydrogenase and γ-glutamylcysteinyl synthetase. However, the identity of critical thiol targets responsible for NO-related antimicrobial activity remains uncertain.

$ONOO^-$ can mediate nitration of tyrosine residues (Chao *et al.*, 1994), disrupting signaling pathways involving tyrosine phosphorylation and altering protein function or stability. Myeloperoxidase also appears to be able to catalyze this protein modification in the presence of nitrite (NO_2^-) and H_2O_2 (Eiserich *et al.*, 1996). Specific antisera have been used to demonstrate the presence of nitrotyrosine *in situ* (Ye *et al.*, 1996). Nitrotyrosine has been found on the surface of *Staphylococcus aureus* following ingestion by human neutrophils (Evans *et al.*, 1996) and in human alveolar macrophages after ingestion of *Mycobacterium bovis*

BCG (Nozaki *et al.*, 1997). Although the NO-dependent tyrosine nitration was not essential for the killing of *S. aureus*, it was found to correlate with human phagocyte antimicrobial activity against BCG.

5.3. Lipids

NO-associated lipid alterations have been principally associated with $ONOO^-$, although $NO_2^{\bullet}$ can also induce lipid peroxidation (Halliwell *et al.*, 1992). $ONOO^-$-mediated peroxidation appears to be independent of iron (Rubbo *et al.*, 1994). Lipid peroxidation may contribute to NO-related antimicrobial activity, but is less likely to represent a major mechanism of damage to bacterial membranes, in which the presence of saturated and monounsaturated fatty acids limits the potential for radical chain propagation reactions.

6. Microbial Defenses against NO

6.1. Avoidance

Microbes have developed multiple mechanisms of protection against toxic effects of reactive nitrogen intermediates. Obviously, factors such as polysaccharide capsules, which interfere with phagocytosis (Cross and Kelly, 1990), can indirectly allow microbes to avoid exposure to phagocyte-derived NO. A more ingenious strategy has been described in *Francisella tularensis*, which gives rise to phase variants with altered LPS no longer capable of inducing NO production by murine macrophages (Cowley *et al.*, 1996). The failure to stimulate NO synthesis appears to facilitate intracellular replication of the bacteria. Alternatively, NO production may be inhibited directly; some experimental evidence suggests that hemozoin, a heme-containing pigment produced by *Plasmodium* spp., can suppress production of reactive nitrogen and oxygen species (Prada *et al.*, 1996). Interaction of *Cryptococcus neoformans* with macrophages also appears to inhibit NO synthesis (Kawakami *et al.*, 1997). Pathogenic *Yersinia* spp. secrete a tyrosine phosphatase (YopH) that subverts multiple functions of macrophages, including phagocytosis (Bliska *et al.*, 1991). Although specific effects of YopH on NO production have not yet been reported, observations in other experimental systems suggest that inhibition of tyrosine phosphorylation signaling pathways might negatively impact NO synthesis (Martiny *et al.*, 1996).

6.2. Stress Regulons

Microbial mechanisms of protection against NO appear to overlap to some extent with known antioxidant defense systems (Fig. 3). Studies in *Escherichia coli*

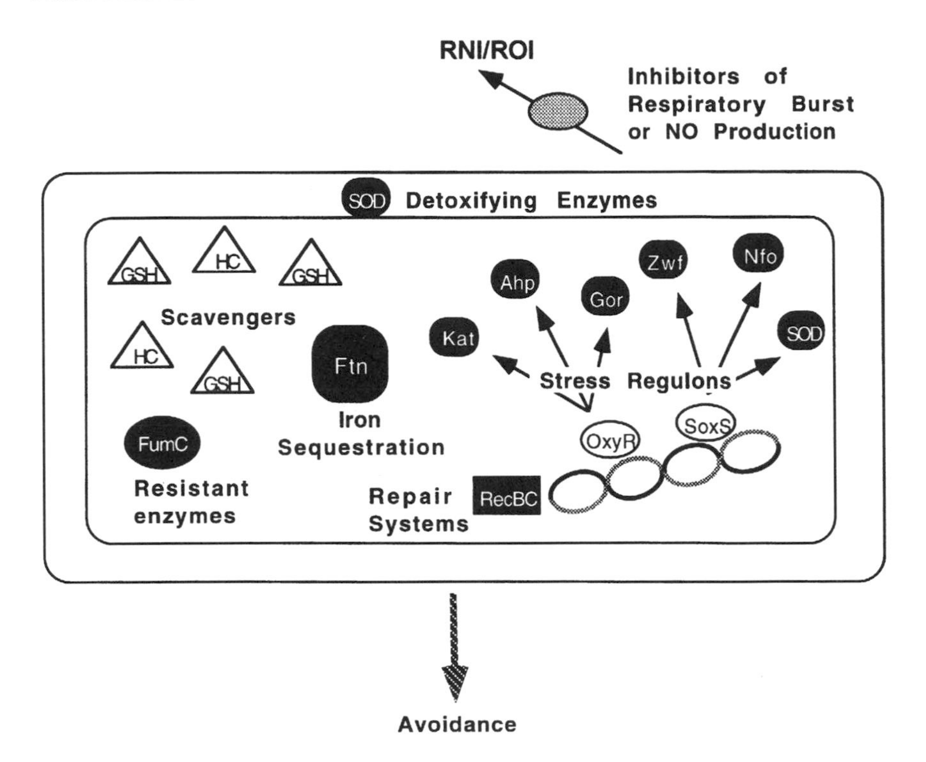

FIGURE 3. Microbial defenses against reactive oxygen and nitrogen intermediates. SOD, superoxide dismutase; GSH, glutathione; HC, homocysteine; FumC, resistant fumarase; Ftn, ferritin; Kat, catalase; Ahp, alkyl hydroperoxide reductase; Gor, glutathione reductase; Zwf, glucose-6-phosphate dehydrogenase; Nfo, endonuclease IV; OxyR, H_2O_2 response regulator; SoxS, superoxide response regulator; RecBC, exonuclease V. Reproduced from Fang (1997), with permission of the Rockefeller University Press.

have demonstrated that exposure to reactive nitrogen intermediates can induce the expression of some of the same genetic regulons previously shown to be elicited by oxidative stress. For example, the SoxRS two-component regulatory system initially was recognized to respond to redox-cycling agents which elevate intracellular $O_2^{-\bullet}$ (Tsaneva and Weiss, 1990), but subsequently was found to respond to NO as well (Nunoshiba *et al.*, 1993). SoxRS-regulated genes such as *zwf* (glucose-6-phosphate dehydrogenase), *sodA* (manganese superoxide dismutase), and *nfo* (endonuclease IV) may defend bacteria against NO by generating reducing equivalents (NADPH), limiting formation of $ONOO^-$, and excising damaged DNA bases. *E. coli* carrying a deletion of the *soxRS* genes was found to have increased susceptibility to NO-dependent antimicrobial activity of murine macrophages (Nunoshiba *et al.*, 1995), although analogous studies of an *S.*

typhimurium soxS mutant did not demonstrate this effect (Fang *et al.*, 1997). Another *E. coli* regulon, controlled by OxyR, was originally identified by its induction in response to H_2O_2 (Christman *et al.*, 1985), but more recently was found to play a role in resistance to RSNO (Hausladen *et al.*, 1996). *S*-Nitrosylation of the OxyR protein itself appears to result in its activation. Stimulation of OxyR-regulated gene expression by RSNO is most evident in glutathione-deficient *gshA* mutant bacteria, and confers increased resistance of *E. coli* to *S*-nitrosocysteine under aerobic conditions.

However, the role of antioxidant regulons in the NO resistance of pathogenic bacteria adapted to the intracellular environment remains to be established. *S. typhimurium* carrying mutations in *soxS* or *oxyR* are not attenuated for virulence (Miller *et al.*, 1989; Fang *et al.*, 1997; Taylor *et al.*, 1998), suggesting that additional systems may be involved in the NO resistance of this facultative intracellular pathogen. Preliminary studies in *M. tuberculosis* have revealed significant differences in the stress responses to reactive nitrogen and oxygen intermediates (Garbe *et al.*, 1996), but the regulatory mechanisms uniquely responsive to NO have not yet been characterized. A novel *M. tuberculosis* locus designated *noxR1* has recently been described to confer resistance to reactive nitrogen and oxygen intermediates when expressed in heterologous bacterial species (Ehrt *et al.*, 1997), but its mechanism of action is unknown.

6.3. Scavengers

Low-molecular-weight thiols play an important role scavenging reactive oxygen and nitrogen intermediates in microbial cells, as they do in mammalian cells. Intracellular thiol concentrations may help to account for differing susceptibility of bacterial species to NO. Staphylococci have a low thiol content (Newton *et al.*, 1996) and appear to be highly susceptible to $NO^{\bullet}$ *in vitro* (Kaplan *et al.*, 1996), whereas enteric bacteria have high concentrations of intracellular thiols and correspondingly high resistance to $NO^{\bullet}$ (Brunelli *et al.*, 1995; DeGroote *et al.*, 1995), unless glutathione (GSH) synthesis is prevented by mutation of the *gshB* gene. Glutathione-deficient *S. typhimurium* is hypersusceptible not only to $NO^{\bullet}$, but also to the $ONOO^-$ generator SIN-1 (3-morpholinosydnonimine hydrochloride) and to RSNO (Fang and DeGroote, unpublished data). Moreover, $NO^{\bullet}$ donors can reduce intracellular GSH levels in *E. coli*. Complex molecular interactions between RSNO and GSH have been analyzed *in vitro* (Singh *et al.*, 1996), and can produce a mixture of oxidized thiols, ammonia, and nitrite depending on the relative GSH concentration and presence of oxygen. It is possible that thiols in addition to GSH, such as mycothiol in *M. tuberculosis* (Newton *et al.*, 1996) and trypanothione in trypanosomes and *Leishmania* spp. (Fairlamb and Cerami, 1992), could perform similar scavenging functions. Other scavengers may also be important, but are less well characterized. For example, the phenazine pigment pyocyanin produced by *Pseudomonas aeruginosa* has been reported to inhibit NO *in vitro* by undergoing nitrosylation (Warren *et al.*, 1990), and a flavohemoglobin has been recently reported to protect *Salmonella* from reactive nitrogen species *in vitro* (Crawford and Goldberg, 1998), although the biological significance of these phenomena during infection remains to be determined.

Homocysteine is an interesting low-molecular-weight thiol that has been implicated in resistance to RSNO in *Salmonella* (DeGroote *et al.*, 1996). Mutant *S. typhimurium* strains selected for the phenotype of increased susceptibility to GSNO were found to have reduced synthesis of homocysteine resulting from a mutation in the *metL* gene (encoding aspartokinase-homoserine dehydrogenase). Addition of exogenous homocysteine abrogates GSNO hypersusceptibility of *metL* mutant strains under both aerobic and anaerobic conditions. *S. typhimurium* carrying a *metL* mutation has reduced survival in murine macrophages and reduced virulence in mice, which can be restored by inhibition of NOS; these observations suggest that interactions between RSNO and intracellular thiols such as homocysteine are relevant to the pathophysiology of infection. It is particularly intriguing to consider the possible role of antagonistic nitrogen oxide–homocysteine interactions in the pathogenesis of vascular disease and neural tube developmental defects associated with hyperhomocysteinemia in humans (McCully, 1969, 1996), which might provide a molecular connection between microbes and man.

6.4. Repair Systems

As mentioned earlier in this chapter, several lines of evidence indicate the importance of DNA as a target of reactive nitrogen intermediates and DNA repair systems as important defenses against NO-related cytotoxicity (Wink *et al.*, 1991; Nguyen *et al.*, 1992; Juedes and Wogan, 1996; Tamir *et al.*, 1996). The RecBCD exonuclease is a central participant in bacterial DNA repair processes involving homologous recombination, and helps to confer resistance to NO-donor compounds (DeGroote *et al.*, 1995). The attenuated virulence and intramacrophage survival of *recBC* mutant *S. typhimurium* may in part relate to NO-dependent DNA damage, although oxidative genotoxicity is clearly a major factor as well (Buchmeier *et al.*, 1993).

6.5. Detoxifying Enzymes

Finally, microbial enzymes can affect the redox transformation of NO by limiting the concentration of reactants. This has most clearly been demonstrated in the case of the periplasmic copper- and zinc-containing superoxide dismutase (Cu,Zn-SOD) of *S. typhimurium* (DeGroote *et al.*, 1997). Simultaneous production of $NO^{\bullet}$ and $O_2^{-\bullet}$ can produce toxic congeners such as $ONOO^{-}$, but diversion of

$O_2^{-\bullet}$ by Cu,Zn-SOD appears to limit such reactions and reduce NO-related cytotoxicity. Inactivation of the *sodC* gene encoding Cu,Zn-SOD greatly enhances the susceptibility of *Salmonella* to $NO^\bullet$ and $O_2^{-\bullet}$ *in vitro*, while reducing intramacrophage survival and virulence for mice. Abrogation of either $NO^\bullet$ or $O_2^{-\bullet}$ production can completely restore these phenotypes, demonstrating that NOS and the NADPH phagocyte oxidase can act synergistically to kill microbial pathogens, but that this action is antagonized by microbial SOD. Alkyl hydroperoxide reductase, encoded by the *ahpC* gene, has recently been implicated in the susceptibility of *Salmonella* and *M. tuberculosis* to reactive nitrogen intermediates *in vitro* (Chen *et al.*, 1998), but the physiologic significance of this observation has not been established; notably, *ahpC* is not required for *Salmonella* virulence in mice (Taylor *et al.*, 1998). Microbial systems limiting free iron availability (Crosa, 1997) might similarly control the rate of formation of DNIC, but scant experimental data to address this question are currently available.

7. Summary and Conclusions

Multiple molecular species account for NO-associated antimicrobial activity, which has been amply demonstrated *in vitro*, in tissue culture systems, and in experimental animal models. Interactions between reactive nitrogen and oxygen intermediates increase the potential range of reactive species formed, and in some cases can enhance antimicrobial potency (Pacelli *et al.*, 1995; DeGroote *et al.*, 1997; Umezawa *et al.*, 1997). NO congeners can exert cytotoxic actions at multiple cellular targets, including DNA, proteins, and lipids. Microbial pathogens correspondingly utlilize a variety of defenses against NO-related cytotoxicity, which include scavengers, expression of specific stress regulons, detoxifying enzymes, repair systems, and mechanisms of avoiding or subverting phagocytic cells.

To understand the physiologic significance of NO-related antimicrobial activity, it is essential to consider the relative importance of other NO-associated actions in the context of specific infectious states. Despite its remarkably broad spectrum of activity (Table I), NO is not an effective antimicrobial mediator for all pathogenic microbes, and in some infections dramatically contributes to morbidity (e.g., Chapters 8, 13, 19–21). Hence, inhibition of NO production can be detrimental in some infections (Evans *et al.*, 1993; Green *et al.*, 1993; Boockvar *et al.*, 1994; Feng *et al.*, 1994; Leitch and He, 1994; Petray *et al.*, 1994, 1995; Robertson *et al.*, 1994; Statman *et al.*, 1994; Blasi *et al.*, 1995; Chan *et al.*, 1995; Lovchik *et al.*, 1995; MacMicking *et al.*, 1995, 1997b; Vazquez-Torres *et al.*, 1995; Wei *et al.*, 1995; DeGroote *et al.*, 1996, 1997; Fukatsu *et al.*, 1996; Hayashi *et al.*, 1996; Hiraoka *et al.*, 1996; Lowenstein *et al.*, 1996; Meli *et al.*, 1996; Rajan *et al.*, 1996; Stenger *et al.*, 1996; Tucker *et al.*, 1996; Nathan, 1997; Scharton-Kersten *et al.*, 1997; Tay and Welsh, 1997; Tsai *et al.*, 1997; Umezawa *et al.*, 1997; Holscher *et al.*, 1998; Maclean *et al.*, 1998) and beneficial in others (Teale and Atkinson, 1992; Sternberg *et al.*, 1994; Akaike *et al.*, 1996; Kreil and Eibl, 1996; Adler *et al.*, 1997; Khan *et al.*, 1997; Nathan, 1997). The therapeutic challenge will be to selectively inhibit excessive NO production or to deliver NO equivalents to sites of infection in the appropriate clinical settings. Our ever-increasing understanding of NOs role as an antimicrobial mediator promises to provide important insights into host–pathogen interactions, and will hopefully lead to the development of novel therapeutic strategies in infection.

References

Abouhamad, W. N., Manson, M., Gibson, M. M., and Higgins, C. F., 1991, Peptide transport and chemotaxis in *Escherichia coli* and *Salmonella typhimurium:* Characterization of the dipeptide permease (Dpp) and the dipeptide-binding protein, *Mol. Microbiol.* **5:**1035–1047.

Adams, L. B., Hibbs, J. B., Jr., Taintor, R. R., and Krahenbuhl, J. L., 1990, Microbiostatic effect of murine-activated macrophages for *Toxoplasma gondii:* Role for synthesis of inorganic nitrogen oxides from L-arginine, *J. Immunol.* **144:**2725–2729.

Adams, L. B., Franzblau, S. G., Vavrin, Z., Hibbs, J. B., Jr., and Krahenbuhl, J. L., 1991, L-Arginine-dependent macrophage effector functions inhibit metabolic activity of *Mycobacterium leprae*, *J. Immunol.* **147:**1642–1646.

Adler, H., Beland, J. L., Del-Pan, N. C., Kobzik, L., Brewer, J. P., Martin, T. R., and Rimm, I. J., 1997, Suppression of herpes simplex virus type 1 (HSV-1)-induced pneumonia in mice by inhibition of inducible nitric oxide synthase (iNOS, NOS2), *J. Exp. Med.* **185:**1533–1540.

Ahmed, S. F., Oswald, I. P., Caspar, P., Hieny, S., Keefer, L., Sher, A., and James, S. L., 1997, Developmental differences determine larval susceptibility to nitric oxide-mediated killing in a murine model of vaccination against *Schistosoma mansoni*, *Infect. Immun.* **65:**219–226.

Ahvazi, B. C., Jacobs, P., and Stevenson, M. M., 1995, Role of macrophage-derived nitric oxide in suppression of lymphocyte proliferation during blood-stage malaria, *J. Leukoc. Biol.* **58:**23–31.

Akaike, T., Noguchi, Y., Ijiri, S., Setoguchi, K., Suga, M., Zheng, Y. M., Dietzschold, B., and Maeda, H., 1996, Pathogenesis of influenza virus-induced pneumonia: Involvement of both nitric oxide and oxygen radicals, *Proc. Natl. Acad. Sci. USA* **93:**2448–2453.

Akaki, T., Sato, K., Shimizu, T., Sano, C., Kajitani, H., Dekio, S., and Tomioko, H., 1997, Effector molecules in expression of the antimicrobial activity of macrophages against *Mycobacterium avium* complex—Roles of reactive nitrogen intermediates, reactive oxygen intermediates, and free fatty acids, *J. Leukoc. Biol.* **62:**795–804.

Akarid, K., Sinet, M., Desforges, B., and Gougerot-Pocidalo, M. A., 1995, Inhibitory effect of nitric oxide on the replication of a murine retrovirus *in vitro* and *in vivo*, *J. Virol.* **69:**7001–7005.

Alspaugh, J. A., and Granger, D. L., 1991, Inhibition of *Cryptococcus neoformans* replication by nitrogen oxides supports the role of these molecules as effectors of macrophage-mediated cytostasis, *Infect. Immun.* **59:**2291–2296.

Anstey, N. M., Weinberg, J. B., Hassanali, M. Y., Mwaikambo, E. D., Manyenga, D., Misukonis, M. A., Arnelle, D. R., Hollis, D., McDonald, M. I., and Granger, D. L., 1996, Nitric oxide in Tanzanian children with malaria: Inverse relationship between malaria severity and nitric oxide production/nitric oxide synthase type 2 expression, *J. Exp. Med.* **184:**557–567.

Anthony, L. S. D., Morrissey, P. J., and Nano, F. E., 1992, Growth inhibition of *Francisella tularensis* live vaccine strain by IFN-γ-activated macrophages is mediated by reactive nitrogen intermediates derived from L-arginine metabolism, *J. Immunol.* **148:**1829–1834.

Arias, M., Rojas, M., Zabaleta, J., Rodriguez, J. I., Paris, S. C., Barrera, L. F., and Garcia, L. F., 1997, Inhibition of virulent *Mycobacterium tuberculosis* by Bcg(R) and Bcg(S) macrophages correlates with nitric oxide production, *J. Infect. Dis.* **176:**1552–1558.

Assreuy, J., Cunha, F. Q., Epperlein, M., Noronha-Dutra, A., O'Donnell, C. A., Liew, F. Y., and Moncada, S., 1994, Production of nitric oxide and superoxide by activated macrophages and killing of *Leishmania major*, *Eur. J. Immunol.* **24:**672–676.

Augusto, O., Linares, E., and Giorgio, S., 1996, Possible roles of nitric oxide and peroxynitrite in murine leishmaniasis, *Braz. J. Med. Biol. Res.* **29:**853–862.

Becker, K., Savvides, S. N., Keese, M., Schirmer, R. H., and Karplus, P. A., 1998, Enzyme inactivation through sulfhydryl oxidation by physiologic NO-carriers, *Nature Struct. Biol.* **5:**267–271.

Beckerman, K. P., Rogers, H. W., Corbett, J. A., Schreiber, R. D., McDaniel, M. L., and Unanue, E. R., 1993, Release of nitric oxide during the T cell-independent pathway of macrophage activation. Its role in resistance to *Listeria monocytogenes*, *J. Immunol.* **150:**888–895.

Bermudez, L. E., 1993, Differential mechanisms of intracellular killing of *Mycobacterium avium* and *Listeria monocytogenes* by activated human and murine macrophages. The role of nitric oxide, *Clin. Exp. Immunol.* **91:**277–281.

Bi, Z., and Reiss, C. S., 1995, Inhibition of vesicular stomatitis virus infection by nitric oxide, *J. Virol.* **69:**2208–2213.

Blasi, E., Pitzurra, L., Puliti, M., Chimienti, A. R., Mazzolla, R., Barluzzi, R., and Bistoni, F., 1995, Differential susceptibility of yeast and hyphal forms of *Candida albicans* to macrophage-derived nitrogen-containing compounds, *Infect. Immun.* **63:**1806–1809.

Bliska, J. B., Guan, K. L., Dixon, J. E., and Falkow, S., 1991, Tyrosine phosphate hydrolysis of host proteins by an essential *Yersinia* virulence determinant, *Proc. Natl. Acad. Sci. USA* **88:**1187–1191.

Boese, M., Keese, M. A., Becker, K., Busse, R., and Mulsch, A., 1997, Inhibition of glutathione reductase by dinitrosyl-iron-dithiolate complex, *J. Biol. Chem.* **272:**21767–21773.

Bohne, W., Heesemann, J., and Gross, U., 1994, Reduced replication of *Toxoplasma gondii* is necessary for induction of bradyzoite-specific antigens: A possible role for nitric oxide in triggering stage conversion, *Infect. Immun.* **62:**1761–1767.

Boockvar, K. S., Granger, D. L., Poston, R. M., Maybodi, M., Washington, M. K., Hibbs, J. B., Jr., and Kurlander, R. L., 1994, Nitric oxide produced during murine listeriosis is protective, *Infect. Immun.* **62:**1089–1100.

Bourguignon, S. C., Alves, C. R., and Giovanni-De-Simone, S., 1997, Detrimental effect of nitric oxide on *Trypanosoma cruzi* and *Leishmania major* like cells, *Acta Trop.* **66:**109–118.

Brieland, J. K., Remick, D. G., Freeman, P. T., Hurley, M. C., Fantone, J. C., and Engleberg, N. C., 1995, *In vivo* regulation of replicative *Legionella pneumophila* lung infection by endogenous tumor necrosis factor alpha and nitric oxide, *Infect. Immun.* **63:**3253–3258.

Brummer, E., and Stevens, D. A., 1995, Antifungal mechanisms of activated murine bronchoalveolar or peritoneal macrophages for *Histoplasma capsulatum*, *Clin. Exp. Immunol.* **102:**65–70.

Brunelli, L., Crow, J. P., and Beckman, J. S., 1995, The comparative toxicity of nitric oxide and peroxynitrite to *Escherichia coli*, *Arch. Biochem. Biophys.* **316:**327–334.

Buchmeier, N., Lipps, C. J., So, M. H. Y., and Heffron, F., 1993, Recombination-deficient mutants of *Salmonella typhimurium* are avirulent and sensitive to the oxidative burst of macrophages, *Mol. Microbiol.* **7:**933–936.

Castro, L., Rodriguez, M., and Radi, R., 1994, Aconitase is readily inactivated by peroxynitrite, but not by its precursor, nitric oxide, *J. Biol. Chem.* **269:**29409–29415.

Cenci, E., Romani, L., Mencacci, A., Spaccapelo, R., Schiaffella, E., Puccetti, P., and Bistoni, F., 1993, Interleukin-4 and interleukin-10 inhibit nitric oxide-dependent macrophage killing of *Candida albicans*, *Eur. J. Immunol.* **23:**1034–1038.

Chambers, M. A., Marshall, B. G., Wangoo, A., Bune, A., Cook, H. T., Shaw, R. J., and Young, D. B.,

1997, Differential responses to challenge with live and dead *Mycobacterium bovis* bacillus Calmette-Guerin, *J. Immunol.* **158:**1742–1748.
Chan, J., Xing, Y., Magliozzo, R. S., and Bloom, B. R., 1992, Killing of virulent *Mycobacterium tuberculosis* by reactive nitrogen intermediates produced by activated macrophages, *J. Exp. Med.* **175:**1111–1122.
Chan, J., Tanaka, K., Carroll, D., Flynn, J., and Bloom, B. R., 1995, Effects of nitric oxide synthase inhibitors on murine infection with *Mycobacterium tuberculosis*, *Infect. Immun.* **63:**736–740.
Chan, J., Tian, Y., Tanaka, K. E., Tsang, M. S., Yu, K., Salgame, P., Carroll, D., Kress, Y., Teitelbaum, R., and Bloom, B. R., 1996, Effects of protein calorie malnutrition on tuberculosis in mice, *Proc. Natl. Acad. Sci. USA* **93:**14857–14861.
Chao, C. C., Anderson, W. R., Hu, S., Gekker, G., Martella, A., and Peterson, P. K., 1993, Activated microglia inhibit multiplication of *Toxoplasma gondii* via a nitric oxide mechanism, *Clin. Immunol. Immunopathol.* **67:**178–183.
Chao, C. C., Gekker, G., Hu, S., and Peterson, P. K., 1994, Human microglial cell defense against *Toxoplasma gondii*. The role of cytokines, *J. Immunol.* **152:**1246–1252.
Chen, B., Stout, R., and Campbell, W. F., 1996, Nitric oxide production: A mechanism of *Chlamydia trachomatis* inhibition in interferon-gamma-treated RAW264.7 cells, *FEMS Immunol. Med. Microbiol.* **14:**109–120.
Chen, L., Xie, Q. W., and Nathan, C., 1998, Alkyl hydroperoxidase reductase subunit C (AhpC) protects bacterial and human cells against reactive nitrogen intermediates, *Mol. Cell.* **1:** 795–805.
Christman, M. F., Morgan, R. W., Jacobson, F. S., and Ames, B. N., 1985, Positive control of a regulon for defenses against oxidative stress and some heat-shock proteins in *Salmonella typhimurium*, *Cell* **41:**753–762.
Cillari, E., Arcoleo, F., Dieli, M., D'Agostino, R., Gromo, G., Leoni, F., and Milano, S., 1994, The macrophage-activating tetrapeptide tuftsin induces nitric oxide synthesis and stimulates murine macrophages to kill *Leishmania* parasites *in vitro*, *Infect. Immun.* **62:**2649–2652.
Cogliati, M., Roverselli, A., Boelaert, J. R., Taramelli, D., Lombardi, L., and Viviani, M. A., 1997, Development of an *in vitro* macrophage system to assess *Penicillium marneffei* growth and susceptibility to nitric oxide, *Infect. Immun.* **65:**279–284.
Coulson, P. S., Smythies, L. E., Betts, C., Mabbott, N. A., Sternberg, J. M., Wei, X. G., Liew, F. Y., and Wilson, R. A., 1998, Nitric oxide produced in the lungs of mice immunized with the radiation-attenuated *Schistosoma* vaccine is not the major agent causing challenge parasite elimination, *Immunology* **93:**55–63.
Cowley, S. C., Myltseva, S. V., and Nano, F. E., 1996, Phase variation in *Francisella tularensis* affecting intracellular growth, lipopolysaccharide antigenicity and nitric oxide production, *Mol. Microbiol.* **20:**867–874.
Crawford, M. J., and Goldberg, D. E., 1998, Role for the *Salmonella* flavohemoglobin in protection from nitric oxide, *J. Biol. Chem.* **273:**12543–12547.
Croen, K. D., 1993, Evidence for antiviral effect of nitric oxide. Inhibition of herpes simplex virus type 1 replication, *J. Clin. Invest.* **91:**2446–2452.
Crosa, J. H., 1997, Signal transduction and transcriptional and posttranscriptional control of iron-regulated genes in bacteria, *Microbiol. Mol. Biol. Rev.* **61:**319–336.
Cross, A. S., and Kelly, N. M., 1990, Bacteria–phagocyte interactions: Emerging tactics in an ancient rivalry, *FEMS Microbiol. Immunol.* **2:**245–258.
Cunha, F. Q., Assreuy, J., Xu, D., Charles, I., Liew, F. Y., and Moncada, S., 1993, Repeated induction of nitric oxide synthase and leishmanicidal activity in murine macrophages, *Eur. J. Immunol.* **23:**1385–1388.
DeGiusti, M., and DeVito, E., 1992, Inactivation of *Yersinia enterocolitica* by nitrite and nitrate in food, *Food Additives Contamin.* **9:**405–408.

DeGroote, M. A., and Fang, F. C., 1995, NO inhibitions: Antimicrobial properties of nitric oxide, *Clin. Infect. Dis.* **21**(Suppl 2)**:**S162–S165.

DeGroote, M. A., Granger, D., Xu, Y., Campbell, G., Prince, R., and Fang, F. C., 1995, Genetic and redox determinants of nitric oxide cytotoxicity in a *Salmonella typhimurium* model, *Proc. Natl. Acad. Sci. USA* **92:**6399–6403.

DeGroote, M. A., Testerman, T., Xu, Y., Stauffer, G., and Fang, F. C., 1996, Homocysteine antagonism of nitric oxide-related cytostasis in *Salmonella typhimurium*, *Science* **272:**414–417.

DeGroote, M. A., Ochsner, U. A., Shiloh, M. U., Nathan, C., McCord, J. M., Dinauer, M. C., Libby, S. J., Vazquez-Torres, A., and Fang, F. C., 1997, Periplasmic superoxide dismutase protects *Salmonella* from products of phagocyte NADPH-oxidase and nitric oxide synthase, *Proc. Natl. Acad. Sci. USA* **94:**13997–14001.

Denicola, A., Rubbo, H., Rodriguez, D., and Radi, R., 1993, Peroxynitrite-mediated cytotoxicity to *Trypanosoma cruzi*, *Arch. Biochem. Biophys.* **304:**279–286.

Denicola, A., Souza, J. M., and Radi, R., 1998, Diffusion of peroxynitrite across erythrocyte membranes, *Proc. Natl. Acad. Sci. USA* **95:**3566–3571.

Denis, M., 1991a, Interferon-gamma-treated murine macrophages inhibit growth of tubercle bacilli via the generation of reactive nitrogen intermediates, *Cell. Immunol.* **132:**150–157.

Denis, M., 1991b, Tumor necrosis factor and granulocyte macrophage-colony stimulating factor stimulate human macrophages to restrict growth of virulent *Mycobacterium avium* and to kill avirulent *M. avium*, *J. Leukoc. Biol.* **49:**380–387.

Didier, E. S., 1995, Reactive nitrogen intermediates implicated in the inhibition of *Encephalitozoon cuniculi* (phylum Microspora) replication in murine peritoneal macrophages, *Parasite Immunol.* **17:**405–412.

Doi, T., Ando, M., Akaike, T., Suga, M., Sato, K., and Maeda, H., 1993, Resistance to nitric oxide in *Mycobacterium avium* complex and its implication in pathogenesis, *Infect. Immun.* **61:**1980–1989.

Drapier, J. C., 1997, Interplay between NO and [Fe-S] clusters: Relevance to biological systems, *Methods* **11:**319–329.

Drapier, J. C., Pellat, C., and Henry, Y., 1991, Generation of EPR-detectable nitrosyl-iron complexes in tumor target cells co-cultured with activated macrophages, *J. Biol. Chem.* **266:**10162–10167.

Dykhuizen, R. S., Frazer, R., Duncan, C., Smith, C. C., Golden, M., Benjamin, N., and Leifert, C., 1996a, Antimicrobial effect of acidified nitrite on gut pathogens: Importance of dietary nitrate in host defense, *Antimicrob. Agents Chemother.* **40:**1422–1425.

Dykhuizen, R. S., Masson, J., McKnight, G., Mowat, A. N., Smith, C. C., Smith, L. M., and Benjamin, N., 1996b, Plasma nitrate concentration in infective gastroenteritis and inflammatory bowel disease, *Gut* **39:**393–395.

Ehrt, S., Shiloh, M. U., Ruan, J., Choi, M., Gunzburg, S., Nathan, C., Xie, C. W., and Riley, L. W., 1997, A novel antioxidant gene from *Mycobacterium tuberculosis*, *J. Exp. Med.* **186:**1885–1896.

Eiserich, J. P., Cross, C. E., Jones, A. D., Halliwell, B., and Vandervliet, A., 1996, Formation of nitrating and chlorinating species by reaction of nitrite with hypochlorous acid—A novel mechanism for nitric oxide-mediated protein modification, *J. Biol. Chem.* **271:**19199–19208.

Evans, T. G., Thai, L., Granger, D. L., and Hibbs, J. B., Jr., 1993, Effect of *in vivo* inhibition of nitric oxide production in murine leishmaniasis, *J. Immunol.* **151:**907–915.

Evans, T. J., Buttery, L. D., Carpenter, A., Springall, D. R., Polak, J. M., and Cohen, J., 1996, Cytokine-treated human neutrophils contain inducible nitric oxide synthase that produces nitration of ingested bacteria, *Proc. Natl. Acad. Sci. USA* **93:**9553–9558.

Fairlamb, A. H., and Cerami, A., 1992, Metabolism and functions of trypanothione in the kinetoplastida, *Annu. Rev. Microbiol.* **46:**695–729.

Fang, F. C., 1997, Mechanisms of nitric oxide-related antimicrobial activity, *J. Clin. Invest.* **99:**2818–2825.

Fang, F. C., Vazquez-Torres, A., and Xu, Y., 1997, The transcriptional regulator SoxS is required for resistance of *Salmonella typhimurium* to paraquat but not for virulence in mice, *Infect. Immun.* **65:**5371–5375.

Feng, H. M., and Walker, D. H., 1993, Interferon-γ and tumor necrosis factor-α exert their antirickettsial effect via induction of synthesis of nitric oxide, *Am. J. Pathol.* **143:**1016–1023.

Feng, H. M., Popov, V. L., and Walker, D. H., 1994, Depletion of gamma interferon and tumor necrosis factor alpha in mice with *Rickettsia conorii*-infected endothelium: Impairment of rickettsicidal nitric oxide production resulting in fatal, overwhelming rickettsial disease, *Infect. Immun.* **62:**1952–1960.

Fernandes, P. D., and Assreuy, J., 1997, Role of nitric oxide and superoxide in *Giardia lamblia* killing, *Braz. J. Med. Biol. Res.* **30:**93–99.

Fischer-Stenger, K., and Marciano-Cabral, F., 1992, The arginine-dependent cytolytic mechanism plays a role in destruction of *Naegleria fowleri* amoebae by activated macrophages, *Infect. Immun.* **60:**5126–5131.

Flesch, I. E. A., and Kaufmann, S. H. E., 1991, Mechanisms involved in mycobacterial growth inhibition by gamma interferon-activated bone marrow macrophages: Role of reactive nitrogen intermediates, *Infect. Immun.* **59:**3231–3218.

Fortier, A. H., Polsinelli, T., Green, S. J., and Nacy, C. A., 1992, Activation of macrophages for destruction of *Francisella tularensis:* Identification of cytokines, effector cells, and effector molecules, *Infect. Immun.* **60:**817–825.

Freeman, B. A., White, C. R., Gutierrez, H., Paler-Martínez, A., Tarpey, M. M., and Rubbo, H., 1995, Oxygen radical–nitric oxide reactions in vascular diseases, *Adv. Pharmacol.* **34:**45–69.

Fukatsu, K., Saito, H., Fukushima, R., Lin, M. T., Inoue, T., Inaba, T., Furukawa, S., Han, I., and Muto, T., 1996, Effects of three inhibitors of nitric oxide synthase on host resistance to bacterial infection, *Inflamm. Res.* **45:**109–112.

Garbe, T. R., Hibler, N. S., and Deretic, V., 1996, Response of *Mycobacterium tuberculosis* to reactive oxygen and nitrogen intermediates, *Mol. Med.* **2:**134–142.

Gazzinelli, R. T., Oswald, I. P., Hieny, S., James, S. L., and Sher, A., 1992, The microbicidal activity of interferon-gamma-treated macrophages against *Trypanosoma cruzi* involves an L-arginine-dependent, nitrogen oxide-mediated mechanism inhibitable by interleukin-10 and transforming growth factor-beta, *Eur. J. Immunol.* **22:**2501–2506.

Gazzinelli, R. T., Eltoum, I., Wynn, T. A., and Sher, A., 1993, Acute cerebral toxoplasmosis is induced by in vivo neutralization of TNF-α and correlates with the down-regulated expression of inducible nitric oxide synthase and other markers of macrophage activation, *J. Immunol.* **151:**3672–3681.

Goldman, D., Cho, Y., Zhao, M., Casadevall, A., and Lee, S. C., 1996, Expression of inducible nitric oxide synthase in rat pulmonary *Cryptococcus neoformans* granulomas, *Am. J. Pathol.* **148:**1275–1282.

Gosselin, D., DeSanctis, J., Boule, M., Skamene, E., Matouk, C., and Radzioch, D., 1995, Role of tumor necrosis factor alpha in innate resistance to mouse pulmonary infection with *Pseudomonas aeruginosa*, *Infect. Immun.* **63:**3272–3278.

Granger, D. L., Hibbs, J. B., Jr., Perfect, J. R., and Durack, D. T., 1988, Specific amino acid (L-arginine) requirement for the microbiostatic activity of murine macrophages, *J. Clin. Invest.* **81:**1129–1136.

Granger, D. L., Hibbs, J. B., Jr., Perfect, J. R., and Durack, D. T., 1990, Metabolic fate of L-arginine in relation to the microbiostatic capability of murine macrophages, *J. Clin. Invest.* **85:**264–273.

Granger, D. L., Cameron, M. L., Lee-See, K., and Hibbs, J. B., Jr., 1993, Role of macrophage-derived nitrogen oxides in antimicrobial function, in: *Mononuclear Phagocytes in Cell Biology* (G. Lopez-Berestein and J. Klostergaard, eds.), CRC Press, Boca Raton, pp. 8–30.

Green, S. J., 1995, Nitric oxide in mucosal immunity, *Nature Med.* **1:**515–517.

Green, S. J., Meltzer, M. S., Hibbs, J. B., Jr., and Nacy, C. A., 1990, Activated macrophages destroy intracellular *Leishmania major* amastigotes by an L-arginine-dependent killing mechanism, *J. Immunol.* **144:**278–283.

Green, S. J., Nacy, C. A., Schreiber, R. D., Granger, D. L., Crawford, R. M., Meltzer, M. S., and Fortier, A. H., 1993, Neutralization of gamma interferon and tumor necrosis factor alpha blocks in vivo synthesis of nitrogen oxides from L-arginine and protection against *Francisella tularensis* infection in *Mycobacterium bovis* BCG-treated mice, *Infect. Immun.* **61:**689–698.

Gross, A., Spiesser, S., Terraza, A., Rouot, B., Caron, E., and Dornand, J., 1998, Expression and bactericidal activity of nitric oxide synthase in *Brucella suis*-infected murine macrophages, *Infect. Immun.* **66:**1309–1316.

Guillemard, E., Geniteau-Legendre, M., Kergot, R., Lemaire, G., Petit, J. F., Labarre, C., and Quero, A. M., 1996, Activity of nitric oxide-generating compounds against encephalomyocarditis virus, *Antimicrob. Agents Chemother.* **40:**1057–1059.

Guittet, O., Ducastel, B., Salem, J. S., Henry, Y., Rubin, H., Lemaire, G., and Lepoivre, M., 1998, Differential sensitivity of the tyrosyl radical of mouse ribonucleotide reductase to nitric oxide and peroxynitrite, *J. Biol. Chem.* **273:**22136–22144.

Gyan, B., Troye-Blomberg, M., Perlmann, P., and Bjorkman, A., 1994, Human monocytes cultured with and without interferon-gamma inhibit *Plasmodium falciparum* parasite growth *in vitro* via secretion of reactive nitrogen intermediates, *Parasite Immunol.* **16:**371–375.

Halliwell, B., Hu, M. L., Louie, S., Duvall, T. R., Tarkington, B. K., Motchnik, P., and Cross, C. E., 1992, Interaction of nitrogen dioxide with human plasma. Antioxidant depletion and oxidative damage, *FEBS Lett.* **313:**62–66.

Harris, N., Buller, R. M., and Karupiah, G., 1995, Gamma interferon-induced, nitric oxide-mediated inhibition of vaccinia virus replication, *J. Virol.* **69:**910–915.

Hausladen, A., and Fridovich, I., 1994, Superoxide and peroxynitrite inactivate aconitases, but nitric oxide does not, *J. Biol. Chem.* **269:**29405–29408.

Hausladen, A., Privalle, C. T., Keng, T., De Angelo, J., and Stamler, J. S., 1996, Nitrosative stress: Activation of the transcription factor OxyR, *Cell* **86:**719–729.

Hayashi, S., Chan, C. C., Gazzinelli, R., and Roberge, F. G., 1996, Contribution of nitric oxide to the host parasite equilibrium in toxoplasmosis, *J. Immunol.* **156:**1476–1481.

He, Q., Leitch, G. J., Visvesvara, G. S., and Wallace, S., 1996, Effects of nifedipine, metronidazole, and nitric oxide donors on spore germination and cell culture infection of the microsporidia *Encephalitozoon hellem* and *Encephalitozoon intestinalis*, *Antimicrob. Agents Chemother.* **40:**179–185.

Hentze, M. W., and Kühn, L. C., 1996, Molecular control of vertebrate iron metabolism: mRNA-based regulatory circuits operated by iron, nitric oxide, and oxidative stress, *Proc. Natl. Acad. Sci. USA* **93:**8175–8182.

Hibbs, J. B., Jr., Westenfelder, C., Taintor, R., Vavrin, Z., Kablitz, C., Baranowski, R. L., Ward, J. H., Menlove, R. L., McMurray, M. P., Kushner, J. P., and Samlowski, W., 1992, Evidence for cytokine-inducible nitric oxide synthesis from L-arginine in patients receiving interleukin-2 therapy, *J. Clin. Invest.* **89:**867–877.

Hickman-Davis, J. M., Lindsey, J. R., Zhu, S., and Matalon, S., 1998, Surfactant protein A mediates mycoplasmacidal activity of alveolar macrophages, *Am. J. Physiol.* **18:**L270–L277.

Hiraoka, Y., Kishimoto, C., Takada, H., Nakamura, M., Kurokawa, M., Ochiai, H., and Shiraki, K., 1996, Nitric oxide and murine coxsackievirus B3 myocarditis: Aggravation of myocarditis by inhibition of nitric oxide synthase, *J. Am. Coll. Cardiol.* **28:**1610–1615.

Holscher, C., Kohler, G., Muller, U., Mossmann, H., Schaub, G. A., and Brombacher, F., 1998, Defective nitric oxide effector functions lead to extreme susceptibility of *Trypanosoma cruzi*-infected mice deficient in gamma interferon receptor of inducible nitric oxide synthase, *Infect. Immun.* **66:**1208–1215.

Igietseme, J. U., 1996, The molecular mechanism of T-cell control of *Chlamydia* in mice: Role of nitric oxide, *Immunology* **87:**1–8.

Igietseme, J. U., Uriri, I. M., Hawkins, R., and Rank, R. G., 1996, Integrin-mediated epithelial–T cell interaction enhances nitric oxide production and increased intracellular inhibition of *Chlamydia*, *J. Leukoc. Biol.* **59:**656–662.

Igietseme, J. U., Uriri, I. M., Chow, M., Abe, E., and Rank, R. G., 1997, Inhibition of intracellular multiplication of human strains of *Chlamydia trachomatis* by nitric oxide, *Biochem. Biophys. Res. Commun.* **232:**595–601.

Incze, K., Farkas, J., Mihalys, V., and Zukal, E., 1974, Antibacterial effect of cysteine-nitrosothiol and possible precursors thereof, *Appl. Microbiol.* **27:**202–205.

Ischiropoulos, H., and Al-Mehdi, A. B., 1995, Peroxynitrite-mediated oxidative protein modifications, *FEBS Lett.* **364:**279–282.

James, S. L., and Glaven, J., 1989, Macrophage cytotoxicity against schistosomula of *Schistosoma mansoni* involves arginine-dependent production of reactive nitrogen intermediates, *J. Immunol.* **143:**4208–4212.

Jiang, X., Leonard, B., Benson, R., and Baldwin, C. L., 1993, Macrophage control of *Brucella abortus:* Role of reactive oxygen intermediates and nitric oxide, *Cell. Immunol.* **151:**192–196.

Johnson, W. C., Cluff, C. W., Goff, W. L., and Wyatt, C. R., 1996, Reactive oxygen and nitrogen intermediates and products from polyamine degradation are babesiacidal *in vitro*, *Ann. N.Y. Acad. Sci.* **791:**136–147.

Juedes, M. J., and Wogan, G. N., 1996, Peroxynitrite-induced mutation spectra of pSP189 following replication in bacteria and in human cells, *Mutat. Res.* **349:**51–61.

Kanazawa, T., Asahi, H., Hata, H., Mochida, K., Kagei, N., and Stadecker, M. J., 1993, Arginine-dependent generation of reactive nitrogen intermediates is instrumental in the *in vitro* killing of protoscoleces of *Echinococcus multilocularis* by activated macrophages, *Parasite Immunol.* **15:**619–623.

Kaplan, S. S., Lancaster, J. R., Jr., Basford, R. E., and Simmons, R. L., 1996, Effect of nitric oxide on staphylococcal killing and interactive effect with superoxide, *Infect. Immun.* **64:**69–76.

Karupiah, G., and Harris, N., 1995, Inhibition of viral replication by nitric oxide and its reversal by ferrous sulfate and tricarboxylic acid cycle metabolites, *J. Exp. Med.* **181:**2171–2179.

Karupiah, G., Xie, Q. W., Buller, R. M., Nathan, C., Duarte, C., and MacMicking, J. D., 1993, Inhibition of viral replication by interferon-gamma-induced nitric oxide synthase, *Science* **261:**1445–1448.

Kawakami, K., Zhang, T. T., Qureshi, M. H., and Saito, A., 1997, *Cryptococcus neoformans* inhibits nitric oxide production by murine peritoneal macrophages stimulated with interferon-gamma and lipopolysaccharide, *Cell. Immunol.* **180:**47–54.

Kawanishi, M., 1995, Nitric oxide inhibits Epstein–Barr virus DNA replication and activation of latent EBV, *Intervirology* **38:**206–213.

Keyer, K., and Imlay, J. A., 1997, Inactivation of dehydratase [4Fe–4S] clusters and disruption of iron homeostasis upon cell exposure to peroxynitrite, *J. Biol. Chem.* **272:**27652–27659.

Khan, I. A., Schwartzmann, J. D., Matsuura, T., and Kasper, L. H., 1997, A dichotomous role for nitric oxide during acute *Toxoplasma gondii* infection in mice, *Proc. Natl. Acad. Sci. USA* **94:**13955–13960.

Kim, Y.-M., Bergonia, H. A., Müller, C., Pitt, B. R., Watkins, W. D., and Lancaster, J. R. Jr., 1995, Nitric oxide and intracellular heme, *Adv. Pharmacol.* **34:**277–291.

Kim, Y. M., Hong, S. J., Billiar, T. R., and Simmons, R. L., 1996, Counterprotective effect of erythrocytes in experimental bacterial peritonitis is due to scavenging of nitric oxide and reactive oxygen intermediates, *Infect. Immun.* **64:**3074–3080.

Klebanoff, S. J., 1993, Reactive nitrogen intermediates and antimicrobial activity: Role of nitrite, *Free Radical Biol. Med.* **14:**351–360.

Komatsu, T., Bi, Z., and Reiss, C. S., 1996, Interferon-gamma induced type I nitric oxide synthase activity inhibits viral replication in neurons, *J. Neuroimmunol.* **68:**101–108.

Kreil, T. R., and Eibl, M. M., 1996, Nitric oxide and viral infection: NO antiviral activity against a flavivirus *in vitro*, and evidence for contribution to pathogenesis in experimental infection *in vivo*, *Virology* **219:**304–306.

Kudeken, N., Kawakami, K., and Saito, A., 1998, Different susceptibilities of yeasts and conidia of *Penicillium marneffei* to nitric oxide (NO)-mediated fungicidal activity of murine macrophages, *Clin. Exp. Immunol.* **112:**287–293.

Kunert, J., 1995, Effect of nitric oxide donors on survival of conidia, germination and growth of *Aspergillus fumigatus in vitro*, *Folia Microbiol.* **40:**238–244.

Lane, T. E., Otero, G. C., Wu-Hsieh, B. A., and Howard, D. H., 1994a, Expression of inducible nitric oxide synthase by stimulated macrophages correlates with their antihistoplasma activity, *Infect. Immun.* **62:**1478–1479.

Lane, T. E., Wu-Hsieh, B. A., and Howard, D. H., 1994b, Antihistoplasma effect of activated mouse splenic macrophages involves production of reactive nitrogen intermediates, *Infect. Immun.* **62:**1940–1945.

Lane, T. E., Paoletti, A. D., and Buchmeier, M. J., 1997, Disassociation between the *in vitro* and *in vivo* effects of nitric oxide on a neurotropic murine coronavirus, *J. Virol.* **71:**2202–2210.

Lee, S. C., Dickson, D. W., Brosnan, C. F., and Casadevall, A., 1994, Human astrocytes inhibit *Cryptococcus neoformans* growth by a nitric oxide-mediated mechanism, *J. Exp. Med.* **180:**365–369.

Leitch, G. J., and He, Q., 1994, Arginine-derived nitric oxide reduces fecal oocyst shedding in nude mice infected with *Cryptosporidium parvum*, *Infect. Immun.* **62:**5173–5176.

Lemesre, J. L., Sereno, D., Daulouede, S., Veyret, B., Brajon, N., and Vincendeau, P., 1997, *Leishmania* spp.: Nitric oxide-mediated metabolic inhibition of promastigote and axenically grown amastigote forms, *Exp. Parasitol.* **86:**58–68.

Lepoivre, M., Chenais, B., Yapo, A., Lemaire, G., Thelander, L., and Tenu, J.-P., 1990, Alterations of ribonucleotide reductase activity following induction of the nitrite-generating pathway in adenocarcinoma cells, *J. Biol. Chem.* **265:**14143–14149.

Liew, F. Y., Li, Y., and Millott, S., 1990a, Tumor necrosis factor-alpha synergizes with IFN-gamma in mediating killing of *Leishmania major* through the induction of nitric oxide, *J. Immunol.* **145:**4306–4310.

Liew, F. Y., Millott, S., Parkinson, C., Palmer, R. M., and Moncada, S., 1990b, Macrophage killing of *Leishmania* parasite *in vivo* is mediated by nitric oxide from L-arginine, *J. Immunol.* **144:**4794–4797.

Liew, F. Y., Li, Y., Moss, D., Parkinson, C., Rogers, M. V., and Moncada, S., 1991, Resistance to *Leishmania major* infection correlates with the induction of nitric oxide synthase in murine macrophages, *Eur. J. Immunol.* **21:**3009–3014.

Lin, J. Y., and Chadee, K., 1992, Macrophage cytotoxicity against *Entamoeba histolytica* trophozoites is mediated by nitric oxide from L-arginine, *J. Immunol.* **148:**3999–4005.

Lin, Y. L., Huang, Y. L., Ma, S. H., Yeh, C. T., Chiou, S. Y., Chen, L. K., and Liao, C. L., 1997, Inhibition of Japanese encephalitis virus infection by nitric oxide: Antiviral effect of nitric oxide on RNA virus replication, *J. Virol.* **71:**5227–5235.

Lopez-Guerrero, J. A., Rayet, B., Tuynder, M., Rommelaere, J., and Dinsart, C., 1997, Constitutive activation of U937 promonocytic cell clones selected for their resistance to parvovirus H-1 infection, *Blood* **89:**1642–1653.

Lopez-Jaramillo, P., Ruano, C., Rivera, J., Teran, E., and Moncada, S., 1995, Treatment of cutaneous mycosis with the NO donor *S*-nitroso-*N*-acetylpenicillamine (SNAP), *Endothelium* **3** (Suppl.)**:**s13 (#50).

Lopez-Jaramillo, P., Ruano, C., Rivera, J., Teran, E., Salazar-Irigoyen, R., Esplugues, J. V., and

Moncada, S., 1998, Treatment of cutaneous leishmaniasis with nitric-oxide donor, *Lancet* **351:**1176–1177.

Lovchik, J. A., Lyons, C. R., and Lipscomb, M. F., 1995, A role for gamma interferon-induced nitric oxide in pulmonary clearance of *Cryptococcus neoformans*, *Am. J. Respir. Cell Mol. Biol.* **13:**116–124.

Lowenstein, C. J., Hill, S. L., Lafond-Walker, A., Wu, J., Allen, G., Landavere, M., Rose, N. R., and Herskowitz, A., 1996, Nitric oxide inhibits viral replication in murine myocarditis, *J. Clin. Invest.* **97:**1837–1843.

Mabbott, N. A., Sutherland, I. A., and Sternberg, J. M., 1994, *Trypanosoma brucei* is protected from the cytostatic effects of nitric oxide under *in vivo* conditions, *Parasitol. Res.* **80:**687–690.

Maclean, A., Wei, X. Q., Huang, F. P., Alalem, U. A. H., Chan, W. L., and Liew, F. Y., 1998, Mice lacking inducible nitric-oxide synthase are more susceptible to herpes simplex virus infection despite enhanced Th1 cell responses, *J. Gen. Virol.* **79:**825–830.

MacMicking, J. D., Nathan, C., Hom, G., Chartrain, N., Fletcher, D. S., Trumbauer, M., Stevens, K., Xie, Q. W., Sokol, K., Hutchinson, N., Chen, H., and Mudgett, J. S., 1995, Altered responses to bacterial infection and endotoxic shock in mice lacking inducible nitric oxide synthase, *Cell* **81:**641–650.

MacMicking, J., Xie, Q. W., and Nathan, C., 1997a, Nitric oxide and macrophage function, *Annu. Rev. Immunol.* **15:**323–350.

MacMicking, J. D., North, R. J., LaCourse, R., Mudgett, J. S., Shah, S. K., and Nathan, C. F., 1997b, Identification of nitric oxide synthase as a protective locus against tuberculosis, *Proc. Natl. Acad. Sci. USA* **94:**5243–5248.

Malawista, S. E., Montgomery, R. R., and van Blaricom, G., 1992, Evidence for reactive nitrogen intermediates in killing of staphylococci by human neutrophil cytoplasts. A new microbicidal pathway for polymorphonuclear leukocytes, *J. Clin. Invest.* **90:**631–636.

Mancinelli, R. L., and McKay, C. P., 1983, Effects of nitric oxide and nitrogen dioxide on bacterial growth, *Appl. Environ. Microbiol.* **46:**198–202.

Mannick, E. E., Bravo, L. E., Zarama, G., Realpe, J. L., Zhang, X. J., Ruiz, B., Fontham, E. T., Mera, R., Miller, M. J., and Correa, P., 1996, Inducible nitric oxide synthase, nitrotyrosine, and apoptosis in *Helicobacter pylori* gastritis: Effect of antibiotics and antioxidants, *Cancer Res.* **56:**3238–3243.

Mannick, J. B., Asano, K., Izumi, K., Kieff, E., and Stamler, J. S., 1994, Nitric oxide produced by human B lymphocytes inhibits apoptosis and Epstein–Barr virus reactivation, *Cell* **79:**1137–1146.

Mannick, J. B., Stamler, J. S., Tang, E., and Finberg, R., 1995, Nitric oxide inhibits human immunodeficiency virus replication, *Endothelium* **3** (Suppl.):s13 (#52).

Maragos, C. M., Andrews, A. W., Keefer, L. K., and Elespuru, R. K., 1993, Mutagenicity of glyceryl trinitrate (nitroglycerin) in *Salmonella typhimurium*, *Mutat. Res.* **298:**187–195.

Marcinkiewicz, J., 1997, Nitric oxide and antimicrobial activity of reactive oxygen intermediates, *Immunopharmacology* **37:**35–41.

Marla, S. S., Lee, J., and Groves, J. T., 1997, Peroxynitrite rapidly permeates phospholipid membranes, *Proc. Natl. Acad. Sci. USA* **94:**14243–14248.

Martiny, A., Vannier-Santos, M. A., Borges, V. M., Meyer-Fernandes, J. R., Assreuy, J., Cunha e Silva, N. L., and de Souza, W., 1996, *Leishmania*-induced tyrosine phosphorylation in the host macrophage and its implication to infection, *Eur. J. Cell Biol.* **71:**206–215.

Mauel, J., and Ransijn, A., 1997, *Leishmania* spp.—Mechanisms of toxicity of nitrogen oxidation products, *Exp. Parasitol.* **87:**98–111.

Mauel, J., Ransijn, A., and Buchmuller-Rouiller, Y., 1991, Killing of *Leishmania* parasites in activated murine macrophages is based on an L-arginine-dependent process that produces nitrogen derivatives, *J. Leukoc. Biol.* **49:**73–82.

Mayer, J., Woods, M. L., Vavrin, Z., and Hibbs, J. B. Jr., 1993, Gamma interferon-induced nitric oxide production reduces *Chlamydia trachomatis* infectivity in McCoy cells, *Infect. Immun.* **61:**491–497.

McCully, K. S., 1969, Vascular pathology of homocysteinemia: Implications for the pathogenesis of atherosclerosis, *Am. J. Pathol.* **56:**111–128.

McCully, K. S., 1996, Homocysteine and vascular disease, *Nature Med.* **2:**386–389.

Meli, R., Raso, G. M., Bentivoglio, C., Nuzzo, I., Galdiero, M., and Di Carlo, R., 1996, Recombinant human prolactin induces protection against *Salmonella typhimurium* infection in the mouse: Role of nitric oxide, *Immunopharmacology* **34:**1–7.

Melkova, Z., and Esteban, M., 1994, Interferon-gamma severely inhibits DNA synthesis of vaccinia virus in a macrophage cell line, *Virology* **198:**731–735.

Melkova, Z., and Esteban, M., 1995, Inhibition of vaccinia virus DNA replication by inducible expression of nitric oxide synthase, *J. Immunol.* **155:**5711–5718.

Mellouk, S., Hoffman, S. L., Liu, Z. Z., de la Vega, P., Billiar, T. R., and Nussler, A. K., 1994, Nitric oxide-mediated antiplasmodial activity in human and murine hepatocytes induced by gamma interferon and the parasite itself: Enhancement by exogenous tetrahydrobiopterin, *Infect. Immun.* **62:**4043–4046.

Miller, S. I., Kukral, A. M., and Mekalanos, J. J., 1989, A two-component regulatory system (*phoP phoQ*) controls *Salmonella typhimurium* virulence, *Proc. Natl. Acad. Sci. USA* **86:**5054–5058.

Miyagi, K., Kawakami, K., and Saito, A., 1997, Role of reactive nitrogen and oxygen intermediates in gamma interferon-stimulated murine macrophage bactericidal activity against *Burkholderia pseudomallei*, *Infect. Immun.* **65:**4108–4113.

Mnaimneh, S., Geffard, M., Veyret, B., and Vincendeau, P., 1997, Albumin nitrosylated by activated macrophages possesses antiparasitic effects neutralized by anti-NO-acetylated-cysteine antibodies, *J. Immunol.* **158:**308–314.

Moran, D. M., Tannenbaum, S. R., and Archer, M. C., 1975, Inhibitor of *Clostridium perfringens* formed by heating sodium nitrite in a chemically defined medium, *Appl. Microbiol.* **30:**838–843.

Morris, S. L., and Hansen, J. N., 1981, Inhibition of *Bacillus cereus* spore outgrowth by covalent modification of a sulfhydryl group by nitrosothiol and iodoacetate, *J. Bacteriol.* **148:**465–471.

Morris, S. L., Walsh, R. C., and Hansen, J. N., 1984, Identification and characterization of some bacterial membrane sulfhydryl groups which are targets of bacteriostatic and antibiotic action, *J. Biol. Chem.* **259:**13590–13594.

Motard, A., Landau, I., Nussler, A., Grau, G., Baccam, D., Mazier, D., and Targett, G. A., 1993, The role of reactive nitrogen intermediates in modulation of gametocyte infectivity of rodent malaria parasites, *Parasite Immunol.* **15:**21–26.

Munoz-Fernandez, M. A., Fernandez, M., and Fresno, M., 1992, Activation of human macrophages for the killing of intracellular *Trypanosoma cruzi* by TNF-α and IFN-γ through a nitric oxide-dependent mechanism, *Immunol. Lett.* **33:**35–40.

Murad, F., 1994, Regulation of cytosolic guanylyl cyclase by nitric oxide: The NO–cyclic GMP signal transduction system, *Adv. Pharmacol.* **26:**19–33.

Nagata, K., Yu, H., Nishikawa, M., Kashiba, M., Nakamura, A., Sato, E. F., Tamura, T., and Inoue, M., 1998, *Helicobacter pylori* generates superoxide radicals and modulates nitric oxide metabolism, *J. Biol. Chem.* **273:**14071–14073.

Naotunne, T. S., Karunaweera, N. D., Mendis, K. N., and Carter, R., 1993, Cytokine-mediated inactivation of malarial gametocytes is dependent on the presence of white blood cells and involves reactive nitrogen intermediates, *Immunology* **78:**555–562.

Nathan, C., 1997, Inducible nitric oxide synthase: What difference does it make? *J. Clin. Invest.* **100:**2417–2423.

Neilly, I. J., Copland, M., Haj, M., Adey, G., Benjamin, N., and Bennett, B., 1995, Plasma nitrite concentrations in neutropenic and non-neutropenic patients with suspected septicaemia, *Br. J. Haematol.* **89:**199–202.

Newton, G. L., Arnold, K., Price, M. S., Sherrill, C., Delcardayre, S. B., Aharonowitz, Y., Cohen, G., Davies, J., Fahey, R. C., and Davis, C., 1996, Distribution of thiols in microorganisms: Mycothiol is a major thiol in most actinomycetes, *J. Bacteriol.* **178:**1990–1995.

Nguyen, T., Brunson, D., Crespi, C. L., Penman, B. W., Wishnok, J. S., and Tannenbaum, S. R., 1992, DNA damage and mutation in human cells exposed to nitric oxide in vitro, *Proc. Natl. Acad. Sci. USA* **89:**3030–3034.

Nicholson, S., Bonecini-Almeida, M. da G., Lapa e Silva, J. R., Nathan, C., Xie, Q.-W., Mumford, R., Weidner, J. R., Calaycay, J., Geng, J., Boechat, N., Linhares, C., Rom, W., and Ho, J. L., 1996, Inducible nitric oxide synthase in pulmonary alveolar macrophages from patients with tuberculosis, *J. Exp. Med.* **183:**2293–2302.

Norris, K. A., Schrimpf, J. E., Flynn, J. L., and Morris, S. M., Jr., 1995, Enhancement of macrophage microbicidal activity: Supplemental arginine and citrulline augment nitric oxide production in murine peritoneal macrophages and promote intracellular killing of *Trypanosoma cruzi*, *Infect. Immun.* **63:**2793–2796.

Nozaki, Y., Hasegawa, Y., Ichiyama, S., Nakashima, I., and Shimokata, K., 1997, Mechanism of nitric oxide-dependent killing of *Mycobacterium bovis* BCG in human alveolar macrophages, *Infect. Immun.* **65:**3644–3647.

Nunoshiba, T., deRojas-Walker, T., Wishnok, J. S., Tannenbaum, S. R., and Demple, B., 1993, Activation by nitric oxide of an oxidative-stress response that defends *Escherichia coli* against activated macrophages, *Proc. Natl. Acad. Sci. USA* **90:**9993–9997.

Nunoshiba, T., deRojas-Walker, T., Tannenbaum, S. R., and Demple, B., 1995, Roles of nitric oxide in inducible resistance of *Escherichia coli* to activated murine macrophages, *Infect. Immun.* **63:**794–798.

O'Brien, L., Carmichael, J., Lowrie, D. B., and Andrew, P. W., 1994, Strains of *Mycobacterium tuberculosis* differ in susceptibility to reactive nitrogen intermediates *in vitro*, *Infect. Immun.* **62:**5187–5190.

Ochoa, J. B., Udekwu, A. O., Billiar, T. R., Curran, R. D., Cerra, F. B., Simmons, R. L., and Peitzman, A. B., 1991, Nitrogen oxide levels in patients after trauma and during sepsis, *Ann. Surg.* **214:**621–626.

O'Leary, V., and Solberg, M., 1976, Effect of sodium nitrite inhibition on intracellular thiol groups and on the activity of certain glycolytic enzymes in *Clostridium perfringens*, *Appl. Environ. Microbiol.* **31:**208–212.

Oswald, I. P., Eltoum, I., Wynn, T. A., Schwartz, B., Caspar, P., Paulin, D., Sher, A., and James, S. L., 1994, Endothelial cells are activated by cytokine treatment to kill an intravascular parasite, *Schistosoma mansoni*, through the production of nitric oxide, *Proc. Natl. Acad. Sci. USA* **91:**999–1003.

Pacelli, R., Wink, D. A., Cook, J. A., Krishna, M. C., DeGraff, W., Friedman, N., Tsokos, M., Samuni, A., and Mitchell, J. B., 1995, Nitric oxide potentiates hydrogen peroxide-induced killing of *Escherichia coli*, *J. Exp. Med.* **182:**1469–1479.

Park, J., and Rikihisa, Y., 1992, L-arginine-dependent killing of intracellular *Ehrlichia risticii* by macrophages treated with gamma interferon, *Infect. Immun.* **60:**3504–3508.

Payne, M. J., Glidewell, C., and Cammack, R., 1990a, Interactions of iron-thiol-nitrosyl compounds with the phosphoroclastic system of *Clostridium sporogenes*, *J. Gen. Microbiol.* **136:**2077–2087.

Payne, M. J., Woods, L. F., Gibbs, P., and Cammack, R., 1990b, Electron paramagnetic resonance spectroscopic investigation of the inhibition of the phosphoroclastic system of *Clostridium sporogenes* by nitrite, *J. Gen. Microbiol.* **136:**2067–2076.

Pertile, T. L., Karaca, K., Sharma, J. M., and Walser, M. M., 1996, An antiviral effect of nitric oxide: Inhibition of reovirus replication, *Avian Dis.* **40:**342–348.

Peterson, P. K., Gekker, G., Hu, S., and Chao, C. C., 1995, Human astrocytes inhibit intracellular multiplication of *Toxoplasma gondii* by a nitric oxide-mediated mechanism, *J. Infect. Dis.* **171:**516–518.

Petray, P., Rottenberg, M. E., Grinstein, S., and Orn, A., 1994, Release of nitric oxide during the experimental infection with *Trypanosoma cruzi*, *Parasite Immunol.* **16:**193–199.

Petray, P., Castanos-Velez, E., Grinstein, S., Orn, A., and Rottenberg, M. E., 1995, Role of nitric oxide in resistance and histopathology during experimental infection with *Trypanosoma cruzi*, *Immunol. Lett.* **47:**121–126.

Pierson, M. D., and Smoot, L. A., 1982, Nitrite, nitrite alternatives, and the control of *Clostridium botulinum* in cured meats, *Crit. Rev. Food Sci. Nutr.* **17:**141–187.

Prada, J., Malinowski, J., Muller, S., Bienzle, U., and Kremsner, P. G., 1996, Effects of *Plasmodium vinckei* hemozoin on the production of oxygen radicals and nitrogen oxides in murine macrophages, *Am. J. Trop. Med. Hyg.* **54:**620–624.

Proudfoot, L., Nikolaev, A. V., Feng, G. J., Wei, W. Q., Ferguson, M. A., Brimacombe, J. S., and Liew, F. Y., 1996, Regulation of the expression of nitric oxide synthase and leishmanicidal activity by glycoconjugates of *Leishmania* lipophosphoglycan in murine macrophages, *Proc. Natl. Acad. Sci. USA* **93:**10984–10989.

Rajagopalan-Levasseur, P., Lecointe, D., Bertrand, G., Fay, M., and Gougerot-Pocidalo, M. A., 1996, Differential nitric oxide (NO) production by macrophages from mice and guinea pigs infected with virulent and avirulent *Legionella pneumophila* serogroup 1, *Clin. Exp. Immunol.* **104:**48–53.

Rajan, T. V., Porte, P., Yates, J. A., Keefer, L., and Shultz, L. D., 1996, Role of nitric oxide in host defense against an extracellular, metazoan parasite, *Brugia malayi*, *Infect. Immun.* **64:**3351–3353.

Rementeria, A., Garcia-Tobalina, R., and Sevilla, M. J., 1995, Nitric oxide-dependent killing of *Candida albicans* by murine peritoneal cells during an experimental infection, *FEMS Immunol. Med. Microbiol.* **11:**157–162.

Rhoades, E. R., and Orme, I. M., 1997, Susceptibility of a panel of virulent strains of *Mycobacterium tuberculosis* to reactive nitrogen intermediates, *Infect. Immun.* **65:**1189–1195.

Robertson, F. M., Offner, P. J., Ciceri, D. P., Becker, W. K., and Pruitt, B. A. J., 1994, Detrimental hemodynamic effects of nitric oxide synthase inhibition in septic shock, *Arch. Surg.* **129:**149–156.

Rockett, K. A., Awburn, M. M., Cowden, W. B., and Clark, I. A., 1991, Killing of *Plasmodium falciparum in vitro* by nitric oxide derivatives, *Infect. Immun.* **59:**3280–3283.

Rolph, M. S., Cowden, W. B., Medveczky, C. J., and Ramshaw, I. A., 1996, A recombinant vaccinia virus encoding inducible nitric oxide synthase is attenuated *in vivo*, *J. Virol.* **70:**7678–7685.

Rottenberg, M. E., Castanos-Velez, E., de Mesquita, R., Laguardia, O. G., Biberfeld, P., and Orn, A., 1996, Intracellular co-localization of *Trypanosoma cruzi* and inducible nitric oxide synthase (iNOS): Evidence for dual pathway of iNOS induction, *Eur. J. Immunol.* **26:**3203–3213.

Rubbo, H., Radi, R., Trujillo, M., Telleri, R., Kalyanaraman, B., Barnes, S., Kirk, M., and Freeman, B. A., 1994, Nitric oxide regulation of superoxide and peroxynitrite-dependent lipid peroxidation. Formation of novel nitrogen-containing oxidized lipid derivatives, *J. Biol. Chem.* **269:**26066–26075.

Sakiniene, E., Bremell, T., and Tarkowski, A., 1997, Inhibition of nitric oxide synthase (NOS) aggravates *Staphylococcus aureus* septicaemia and septic arthritis, *Clin. Exp. Immunol.* **110:**370–377.

Sanders, S. P., Siekierski, E. S., Porter, J. D., Richards, S. M., and Proud, D., 1998, Nitric oxide inhibits rhinovirus-induced cytokine production and viral replication in a human respiratory epithelial cell line, *J. Virol.* **72:**934–942.

Scharton-Kersten, T. M., Yap, G., Magram, J., and Sher, A., 1997, Inducible nitric oxide is essential for host control of persistent but not acute infection with the intracellular pathogen *Toxoplasma gondii*, *J. Exp. Med.* **185:**1261–1273.

Seguin, M. C., Klotz, F. W., Schneider, I., Weir, J. P., Goodbary, M., Slayter, M., Raney, J. J., Aniagolu, J. U., and Green, S. J., 1994, Induction of nitric oxide synthase protects against malaria in mice exposed to irradiated *Plasmodium berghei* infected mosquitoes: Involvement of interferon gamma and CD8+ T cells, *J. Exp. Med.* **180:**353–358.

Shank, J. L., Silliker, J. H., and Harper, R. H., 1962, The effect of nitric oxide on bacteria, *Appl. Microbiol.* **10:**185–189.

Shiloh, M. U., MacMicking, J., Nicholson, S., Brause, J., Potter, S., Fang, F. C., Marino, M., Old, L., Dinauer, M., and Nathan, C., 1999, Bactericidal activity of mice and macrophages deficient in both NADPH oxidase and inducible nitric oxide synthase, *Immunity* **10**:29–38.

Silva, J. S., Vespa, G. N., Cardoso, M. A., Aliberti, J. C., and Cunha, F. Q., 1995, Tumor necrosis factor alpha mediates resistance to *Trypanosoma cruzi* infection in mice by inducing nitric oxide production in infected gamma interferon-activated macrophages, *Infect. Immun.* **63:**4862–4867.

Simmonds, H. A., Goday, A., and Morris, G. S., 1985, Superoxide radicals, immunodeficiency and xanthine oxidase activity: Man is not a mouse, *Clin. Sci.* **68:**561–565.

Singh, S. P., Wishnok, J. S., Keshive, M., Deen, W. M., and Tannenbaum, S. R., 1996, The chemistry of the *S*-nitrosoglutathione/glutathione system, *Proc. Natl. Acad. Sci. USA* **93:**14428–14433.

Spithill, T. W., Piedrafita, D., and Smooker, P. M., 1997, Immunological approaches for the control of fasciolosis, *Int. J. Parasitol.* **27:**1221–1235.

Stamler, J. S., 1994, Redox signaling: Nitrosylation and related target interactions of nitric oxide, *Cell* **78:**931–936.

Stamler, J. S., 1995, *S*-Nitrosothiols and the bioregulatory actions of nitrogen oxides through reactions with thiol groups, *Curr. Top. Microbiol.* **196:**19–36.

Statman, R., Cheng, W., Cunningham, J. N., Henderson, J. L., Damiani, P., Siconolfi, A., Rogers, D., and Horovitz, J. H., 1994, Nitric oxide inhibition in the treatment of the sepsis syndrome is detrimental to tissue oxygenation, *J. Surg. Res.* **57:**93–98.

Stenger, S., Donhauser, N., Thuring, H., Rollinghoff, M., and Bogdan, C., 1996, Reactivation of latent leishmaniasis by inhibition of inducible nitric oxide synthase, *J. Exp. Med.* **183:**1501–1514.

Sternberg, J., Mabbott, N., Sutherland, I., and Liew, F. Y., 1994, Inhibition of nitric oxide synthesis leads to reduced parasitemia in murine *Trypanosoma brucei* infection, *Infect. Immun.* **62:**2135–2137.

Summersgill, J. T., Powell, L. A., Buster, B. L., Miller, R. D., and Ramirez, J. A., 1992, Killing of *Legionella pneumophila* by nitric oxide in IFN-γ-activated macrophages, *J. Leukoc. Biol.* **52:**625–629.

Tamir, S., Burney, S., and Tannenbaum, S. R., 1996, DNA damage by nitric oxide, *Chem. Res. Toxicol.* **9:**821–827.

Tarr, H. L. A., 1941, Bacteriostatic action of nitrates, *Nature* **147:**417–418.

Tay, C. H., and Welsh, R. M., 1997, Distinct organ-dependent mechanisms for the control of murine cytomegalovirus infection by natural killer cells, *J. Virol.* **71:**267–275.

Taylor, M. J., Cross, H. F., Mohammed, A. A., Trees, A. J., and Bianco, A. E., 1996, Susceptibility of *Brugia malayi* and *Onchocerca lienalis* microfilariae to nitric oxide and hydrogen peroxide in cell-free culture and from IFN gamma-activated macrophages, *Parasitology* **112:**315–322.

Taylor, P. D., Inchley, C. J., and Gallagher, M. P., 1998, The *Salmonella typhimurium* AhpC polypeptide is not essential for virulence in BALB/c mice but is recognized as an antigen during infection, *Infect. Immun.* **66:**3208–3217.

Taylor-Robinson, A. W., Phillips, R. S., Severn, A., Moncada, S., and Liew, F. Y., 1993, The role of Th1 and Th2 cells in a rodent malaria infection, *Science* **260:**1931–1934.

Teale, D. M., and Atkinson, A. M., 1992, Inhibition of nitric oxide synthesis improves survival in a murine peritonitis model of sepsis that is not cured by antibiotics alone, *J. Antimicrob. Chemother.* **30:**839–842.

Thomas, G. R., McCrossan, M., and Selkirk, M. E., 1997, Cytostatic and cytotoxic effects of activated macrophages and nitric oxide donors on *Brugia malayi*, *Infect. Immun.* **65:**2732–2739.

Tohyama, M., Kawakami, K., and Saito, A., 1996, Anticryptococcal effect of amphotericin B is mediated through macrophage production of nitric oxide, *Antimicrob. Agents Chemother.* **40:**1919–1923.

Tsai, W. C., Strieter, R. M., Zisman, D. A., Wilkowski, J. M., Bucknell, K. A., Chen, G. H., and

Standiford, T. J., 1997, Nitric oxide is required for effective innate immunity against *Klebsiella pneumoniae*, *Infect. Immun.* **65:**1870–1875.

Tsaneva, I. R., and Weiss, B., 1990, *soxR*, a locus governing a superoxide response regulon in *Escherichia coli* K-12, *J. Bacteriol.* **172:**4197–4205.

Tucker, P. C., Griffin, D. E., Choi, S., Bui, N., and Wesselingh, S., 1996, Inhibition of nitric oxide synthesis increases mortality in Sindbis virus encephalitis, *J. Virol.* **70:**3972–3977.

Turco, J., Liu, H., Gottlieb, S. F., and Winkler, H. H., 1998, Nitric oxide-mediated inhibition of the ability of *Rickettsia prowazekii* to infect mouse fibroblasts and mouse macrophagelike cells, *Infect. Immun.* **66:**558–566.

Umezawa, K., Akaike, T., Fujii, S., Suga, M., Setoguchi, K., Ozawa, A., and Maeda, H., 1997, Induction of nitric oxide synthesis and xanthine oxidase and their roles in the antimicrobial mechanism against *Salmonella typhimurium* infection in mice, *Infect. Immun.* **65:**2932–2940.

Vazquez-Torres, A., Jones-Carson, J., Warner, T., and Balish, E., 1995, Nitric oxide enhances resistance of SCID mice to mucosal candidiasis, *J. Infect. Dis.* **172:**192–198.

Vazquez-Torres, A., Jones-Carson, J., and Balish, E., 1996, Peroxynitrite contributes to the candidacidal activity of nitric oxide-producing macrophages, *Infect. Immun.* **64:**3127–3133.

Vespa, G. N., Cunha, F. Q., and Silva, J. S., 1994, Nitric oxide is involved in control of *Trypanosoma cruzi*-induced parasitemia and directly kills the parasite *in vitro*, *Infect. Immun.* **62:**5177–5182.

Vincendeau, P., and Daulouede, S., 1991, Macrophage cytostatic effect on *Trypanosoma musculi* involves an L-arginine-dependent mechanism, *J. Immunol.* **146:**4338–4343.

Vincendeau, P., Daulouede, S., Veyret, B., Darde, M. L., Bouteille, B., and Lemesre, J. L., 1992, Nitric oxide-mediated cytostatic activity on *Trypanosoma brucei gambiense* and *Trypanosoma brucei brucei*, *Exp. Parasitol.* **75:**353–360.

Virta, M., Karp, M., and Vuorinen, P., 1994, Nitric oxide donor-mediated killing of bioluminescent *Escherichia coli*, *Antimicrob. Agents Chemother.* **38:**2775–2779.

Vouldoukis, I., Riveros-Moreno, V., Dugas, B., Ouaaz, F., Becherel, P., Debre, P., Moncada, S., and Mossalayi, M. D., 1995, The killing of *Leishmania major* by human macrophages is mediated by nitric oxide induced after ligation of the Fc epsilon RII/CD23 surface antigen, *Proc. Natl. Acad. Sci. USA* **92:**7804–7808.

Vouldoukis, I., Becherel, P. A., Riveros-Moreno, V., Arock, M., da Silva, O., Debre, P., Mazier, D., and Mossalayi, M. D., 1997, Interleukin-10 and interleukin-4 inhibit intracellular killing of *Leishmania infantum* and *Leishmania major* by human macrophages by decreasing nitric oxide generation, *Eur. J. Immunol.* **27:**860–865.

Walker, D. H., Popov, V. L., Crocquet-Valdes, P. A., Welsh, C. J., and Feng, H. M., 1997, Cytokine-induced, nitric oxide-dependent, intracellular antirickettsial activity of mouse endothelial cells, *Lab. Invest.* **76:**129–138.

Warren, J. B., Loi, R., Rendell, N. B., and Taylor, G. W., 1990, Nitric oxide is inactivated by the bacterial pigment pyocyanin, *Biochem. J.* **266:**921–923.

Wei, X. Q., Charles, I. G., Smith, A., Ure, J., Feng, G. J., Huang, F. P., Xu, D., Muller, W., Moncada, S., and Liew, F. Y., 1995, Altered immune responses in mice lacking inducible nitric oxide synthase, *Nature* **375:**408–411.

Wink, D. A., Kasprzak, K. S., Maragos, C. M., Elespuru, R. K., Misra, M., Dunams, T. M., Cebula, T. A., Koch, W. H., Andrews, A. W., Allen, J. S., and Keefer, L. K., 1991, DNA deaminating ability and genotoxicity of nitric oxide and its progenitors, *Science* **254:**1001–1003.

Wong, H. R., Carcillo, J. A., Burckart, G., and Kaplan, S. S., 1996, Nitric oxide production in critically ill patients, *Arch. Dis. Child.* **74:**482–489.

Wynn, T. A., Oswald, I. P., Eltoum, I. A., Caspar, P., Lowenstein, C. J., Lewis, F. A., James, S. L., and Sher, A., 1994, Elevated expression of Th1 cytokines and nitric oxide synthase in the lungs of vaccinated mice after challenge infection with *Schistosoma mansoni*, *J. Immunol.* **153:**5200–5209.

Xie, Q. F., Kawakami, K., Kudeken, N., Zhang, T., Qureshi, M. H., and Saito, A., 1997, Different susceptibility of three clinically isolated strains of *Cryptococcus neoformans* to the fungicidal effects of reactive nitrogen and oxygen intermediates—Possible relationships with virulence, *Microbiol. Immunol.* **41:**725–731.

Yarbrough, J. M., Rake, J. B., and Eagon, R. G., 1980, Bacterial inhibitory effects of nitrite: Inhibition of active transport, but not of group translocation, and of intracellular enzymes, *Appl. Environ. Microbiol.* **39:**831–834.

Ye, Y. Z., Strong, M., Huang, Z. Q., and Beckman, J. S., 1996, Antibodies that recognize nitrotyrosine, *Methods Enzymol.* **269:**201–209.

Zaragoza, C., Ocampo, C. J., Saura, M., McMillan, A., and Lowenstein, C. J., 1997, Nitric oxide inhibition of coxsackievirus replication *in vitro*, *J. Clin. Invest.* **100:**1760–1767.

Zhao, B., Collins, M. T., and Czuprynski, C. J., 1997, Effects of gamma interferon and nitric oxide on the interaction of *Mycobacterium avium* subsp. *paratuberculosis* with bovine monocytes, *Infect. Immun.* **65:**1761–1766.

Zhu, L., Gunn, C., and Beckman, J. S., 1992, Bactericidal activity of peroxynitrite, *Arch. Biochem. Biophys.* **298:**452–457.

Part D

Nitric Oxide in Specific Infections

CHAPTER 13

Nitric Oxide in Sepsis

GILLIAN WRAY and CHRISTOPH THIEMERMANN

1. Septic Shock—An Introduction

The medical syndrome of shock can be defined as a *progressive failure of the circulation to provide blood and oxygen to vital organs of the body.* In clinical practice, the key symptom of shock is a severe fall in blood pressure that is often associated with the dysfunction or failure of several important organs including lung, kidney, liver, and brain. The most common cause of shock is the contamination of blood with bacteria (bacteremia), viruses, fungi, or parasites resulting in systemic infection and ultimately shock (septic shock). Other causes of shock include severe hemorrhage (hemorrhagic shock), trauma (traumatic shock), failure of the heart to maintain a sufficient cardiac output (cardiogenic shock), interruption of the innervation of blood vessels (neurogenic shock), and severe allergic reactions (anaphylactic shock). Septic shock, regardless of its etiology, is defined as a persistent systemic response to infection with hypotension despite adequate fluid replacement, resulting in impaired tissue perfusion and oxygen extraction (Parrillo, 1990). This definition of septic shock is independent of the presence or absence of a multiple organ dysfunction syndrome (MODS), which is defined as impaired organ function such that homeostasis cannot be maintained without intervention (Baue, 1993). Primary MODS is a direct result of a well-defined insult to a specific organ. Secondary MODS occurs as a consequence of an exaggerated host response, termed the *systemic inflammatory response syndrome* (SIRS). Current therapeutic approaches for septic shock include antimicrobial chemotherapy, volume replacement, inotropic and vasopressor support, oxygen therapy and mechanical

GILLIAN WRAY and CHRISTOPH THIEMERMANN • The William Harvey Research Institute, St. Bartholomew's and the Royal London School of Medicine and Dentistry, London EC1M 6BQ, United Kingdom.

Nitric Oxide and Infection, edited by Fang. Kluwer Academic / Plenum Publishers, New York, 1999.

ventilation as well as hemodialysis and hemofiltration. However, these interventions have failed to make a substantial impact on the high mortality associated with septic shock (Natanson *et al.*, 1994). Hence, septic shock remains the major cause of death in noncoronary intensive care units with an estimated mortality ranging between 50 and 80%. As shock is also by far the most common cause of prolonged admission to an intensive care unit, the clinical and socioeconomic importance of this illness is substantial. The great need to explore the pathophysiological events leading to circulatory failure, tissue ischemia, and MODS in septic shock is highlighted by the fact that numerous clinical trials evaluating the effects of potential novel therapeutic interventions in patients with septic shock have (at best) demonstrated a 5% reduction in 28-day mortality. Interestingly, in trials involving more than 500 patients, this benefit has consistently been demonstrated using a variety of drugs that antagonize various pathophysiological aspects of septic shock, e.g., antibodies against TNFα, IL-1 receptor antagonist, PAF receptor antagonists, to name but a few (Charles Nathanson, personal communication). This chapter reviews the role of endogenous nitric oxide (NO) production in the pathophysiology of septic shock in animals and humans, and discusses therapeutic approaches aimed at modulating the formation of NO in animal models of septic shock. Lastly, we review the results of a recently completed phase II clinical trial in which the effects and side effects of one NO synthase (NOS) inhibitor were investigated in 312 patients with septic shock.

2. Biosynthesis and Physiological Roles of NO

NO is generated from L-arginine by a family of enzymes collectively called *NO synthases* (see Chapter 4 for a more detailed discussion). The oxidation of one of the guanidino nitrogen atoms of the semiessential amino acid L-arginine by NOS generates NO and L-citrulline. The heme iron-dependent oxidation of L-arginine is coupled to the reductive activation of molecular oxygen, and requires reducing equivalents shuttled from the electron donor NADPH to heme through the flavins FAD and FMN. In addition to heme, flavins, and NADPH, NOS requires the presence of tetrahydrobiopterin, which appears to act as both an allosteric effector and a redox-active cofactor of the oxidation of L-arginine. Thus, NOS contains an oxygenase domain (containing the catalytic center) and a reductase domain. The synthesis of NO from L-arginine and molecular oxygen involves the generation of N^G-hydroxy-L-arginine and water and the subsequent oxidation of N^G-hydroxy-L-arginine in the presence of molecular oxygen to form NO, L-citrulline, and water. When generated, NO diffuses to adjacent cells where it activates soluble guanylate cyclase, resulting in the formation of cGMP, which in turn mediates many (but not all) of the effects of NO. NO is generated in many mammalian cells by at least three different isoforms of NOS. The NOS in endothelial cells (eNOS, NOS3) and neuronal cells (nNOS, NOS1) are expressed constitutively, and both enzymes require an increase in intracellular calcium for activation. Activation of macrophages and many other cells with proinflammatory cytokines or endotoxin results in the expression of a distinct isoform of NOS [inducible NOS (iNOS, NOS2)], the activity of which is functionally independent of changes in intracellular calcium (see Nathan, 1992; Dinerman *et al.*, 1993; Moncada and Higgs, 1993; Morris and Billiar, 1994; Thiemermann, 1994; Szabo and Thiemermann, 1995; and Chapter 5 for reviews).

Thus, it is not surprising that NO has many biological functions in the cardiovascular, nervous, and immune systems. For instance, activation of eNOS by shear stress results in a continuous release of picomolar amounts of NO that help to regulate blood pressure and organ blood flow by causing vasodilatation and opposing the effects of circulating catecholamines (see Chapter 7). NO also reduces the adhesion of platelets and polymorphonuclear leukocytes (PMNs) to the endothelium (Moncada and Higgs, 1993). The latter effect of NO is at least in part related to the prevention by NO of the expression of the adhesion molecules P-selectin and intercellular adhesion molecule (ICAM-1) on the surface of endothelial cells. In addition to preventing the adhesion of platelets to endothelial cells, NO can directly attenuate the activation of platelets. These effects of NO are associated with prevention of the expression of P-selectin on platelets, secretion of platelet granules, intracellular calcium flux, and binding of glycoprotein IIb/IIIa to fibrinogen (Loscalzo and Welch, 1995).

3. Role of NO in the Pathophysiology of Septic Shock

Since the discovery in 1990 that an enhanced formation of endogenous NO contributes to (1) the hypotension caused by endotoxin (Fig. 1) and TNFα (Kilbourn *et al.*, 1990a,b; Thiemermann and Vane, 1990), (2) the vascular hyporesponsiveness to vasoconstrictor agents (also termed *vasoplegia*) (Julou-Schaeffer *et al.*, 1990; Rees *et al.*, 1990), and (3) the protection of liver integrity in rodents with sepsis (Billiar *et al.*, 1990), there has been a steadily increasing interest in the role of NO in the pathophysiology of animals and humans with septic shock (Fig. 2). The overproduction of NO in animal models of circulatory shock is caused by a transient early activation of eNOS and a subsequent induction of iNOS activity, resulting in the formation of nanomolar amounts of NO in macrophages (host defense), vascular smooth muscle (hypotension, vascular hyporeactivity, maldistribution of blood flow), and parenchymal cells (Thiemermann, 1998). The ability of NOS inhibitors to attenuate the hypotension and vasoplegia caused by endotoxin in animals (see above), together with the resistance to endotoxin-induced hypotension in mice with inactivation of the iNOS gene (iNOS knockout mice) (MacMicking *et al.*, 1995, Wei *et al.*, 1995), supports the hypothesis that an

Potentially harmful

- Hypotension
- Hyporesponsiveness to pressor agents
- Myocardial dysfunction
- Formation of peroxynitrite → DNA strand breaks → PARS; Inhibition Na/K-ATPase; Tyrosine nitration; Direct damage to pulmonary surfactant
- Inhibition of mitochondrial respiration

FIGURE 1. Harmful effects of nitric oxide overproduction in sepsis or septic shock. These include possible mechanisms contributing to organ injury.

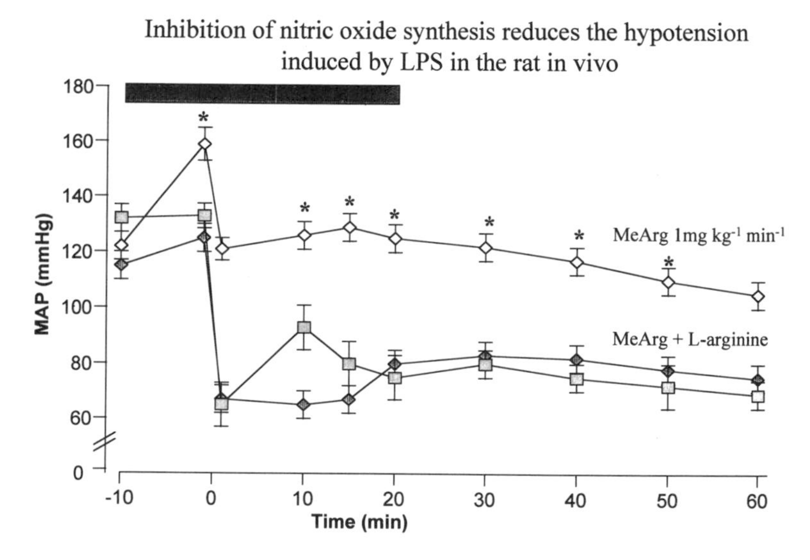

FIGURE 2. Effects of NOS inhibition on endotoxin-induced hypotension. The NOS inhibitor N^{G}-methyl-L-arginine (MeArg) attenuates the fall in mean arterial blood pressure (MAP) in response to *E. coli* lipopolysaccharide (LPS) in the anesthetized rat. Different groups of animals received either LPS only (squares, 15 mg/kg i.v. at time 0, $n = 9$), MeArg plus LPS (open diamonds, 1 mg/kg i.v., $n = 7$), or MeArg plus L-arginine and LPS (shaded diamonds, 6 mg/kg per min i.v., $n = 3$). The shaded bar indicates the infusion period of MeArg or MeArg plus L-Arg. Data are expressed as means $\pm$ S.E.M. of n observations. $*p < 0.05$, taken as statistically significant (one-way ANOVA).

TABLE I
Possible Effects of Administration of NOS Inhibitors in Septic Shock

Beneficial	Adverse
Increased blood pressure	Excessive vasoconstriction
Restores responsiveness to pressor agents	Pulmonary hypertension
Cardiac output return to baseline values	Fall in cardiac output
Decreased production of peroxynitrite	Increased platelet adhesiveness
Attenuation of inhibition of mitochondrial respiration	Increased neutrophil adhesion
Improved organ function	Worsened organ function
Improved survival	

overproduction of NO by iNOS contributes to the circulatory failure characteristic of septic shock. It is less clear whether increased formation of NO also contributes to the organ injury and dysfunction caused by endotoxin. The formation of NO by eNOS (and potentially also by iNOS) also exerts beneficial effects in shock including vasodilatation, immunoregulation (see Chapter 10), prevention of platelet and leukocyte adhesion, maintenance of microcirculatory blood flow (see Chapter 7), cytoprotection (see Chapter 9), and augmentation of host defense (see Chapter 12). Thus, it is not surprising that basic and clinical scientists have advocated the use of apparently contradictory therapeutic approaches including inhibition of NOS activity, enhancement of NO availability (NO donors, NO inhalation), or a combination of both approaches. The following paragraphs highlight some of the effects and side effects of inhibitors of NOS activity (Table I) in animal models of septic shock. For a more detailed review of (1) the many roles of NO in the pathophysiology of septic or other forms of shock, (2) the mechanisms leading to the induction of iNOS, and (3) a more detailed account of the chemistry and pharmacology (isoenzyme selectivity) of NOS inhibitors, the interested reader is referred to recent reviews of these topics (Morris and Billiar, 1994; Szabo and Thiemermann, 1995; Thiemermann *et al.*, 1995; Cobb and Danner, 1996; Southan and Szabo, 1996; Kilbourn *et al.*, 1997).

4. Inhibition of NOS Activity in Animal Models of Septic or Endotoxic Shock

Although there is good evidence that endotoxemia or sepsis in rodents results in the induction of iNOS in various tissues, leading to an increase in the plasma levels of nitrite/nitrate (from 20 up to 600 μM), there is limited information regarding the time course of iNOS induction, the degree of iNOS activity in tissues, or even the plasma levels of nitrite/nitrate in large animals (pig, dog, sheep, baboon) including humans with sepsis and septic shock. Clearly, sepsis or endotoxemia results in an increase in the plasma levels of nitrite/nitrate in these species. However, it appears that the rise in the plasma levels of nitrite/nitrate in humans with septic shock is less than in rodents. When evaluating the role of NO or elucidating the effects of NOS inhibitors in animal models of shock, one needs to consider that (1) many of the experimental models are nonresuscitated, hypodynamic models of shock, (2) the effects and side effects of nonselective inhibitors of NOS activity (see below) will vary greatly depending on the degree of iNOS induction in the species, and (3) any observed effects of the respective NOS inhibitor used will depend on the chosen dose regimen and timing of the intervention.

4.1. N^G-Methyl-L-arginine

The N-substituted L-arginine analogue, N^G-methyl-L-arginine (L-NMMA), was the first agent reported to inhibit NOS activity. L-NMMA is an endogenous substance present in the urine of both animals and humans (Carnigie *et al.*, 1977; Park *et al.*, 1988). Although L-NMMA inhibits all isoforms of NOS to a variable degree, it is a more potent inhibitor of iNOS than of eNOS activity in cultured cells (Gross *et al.*, 1990) and in the rat (Thiemermann *et al.*, 1995). Because L-NMMA is a competitive inhibitor of the binding of L-arginine to NOS, an excess of L-arginine can reverse the inhibition of NOS activity by L-NMMA. As L-NMMA is only a moderately selective inhibitor of iNOS activity, it is not entirely surprising that the effects of L-NMMA in models of shock vary from "very beneficial" (inhibition of iNOS activity) to "moderately beneficial with some adverse effects" (inhibition of eNOS activity masks the beneficial effects of iNOS inhibition) to "detrimental" (marked inhibition of eNOS activity). Clearly, the observed result is highly dependent on the dose of L-NMMA as well as on the model of shock (e.g., species, degree of iNOS induction). When given after the onset of hypotension, infusions of relatively low doses of L-NMMA (3 to 10 mg/kg per hr) have been convincingly demonstrated to exert beneficial hemodynamic effects in rodent (Thiemermann and Vane, 1990), ovine (Booke *et al.*, 1996), canine (Kilbourn *et al.*, 1990a), and baboon (see below) models of endotoxemia and sepsis. In contrast to rodents and similar to humans, sheep are very sensitive to small doses of endotoxin. Indeed, infusion of either endotoxin or bacteria into sheep leads to a hyperdynamic circulation with a fall in peripheral vascular resistance, and an increase in cardiac output and organ blood flow associated with reduced oxygen extraction. In this model, prolonged periods of endotoxemia or gram-negative bacteremia are also associated with increased total renal blood flow and the development of precapillary arteriovenous shunting, resulting in regional maldistribution of renal blood flow and reduced glomerular filtration pressure and rate. Interestingly, administration of L-NMMA 24 hr after the onset of endotoxemia increases urine output and reverses the impairment in creatinine clearance caused by the infusion of bacteria, without causing a significant fall in renal blood flow below baseline. In addition to these beneficial effects on renal blood flow and function, NOS inhibition results in increased oxygen extraction, a restoration of organ blood flow (in brain, heart, jejunum, ileum) to normal levels, and increased peripheral vascular resistance without a significant increase in lactate, indicating a normalization of hemodynamic parameters in the absence of excessive vasoconstriction (Booke *et al.*, 1996a,b). In conscious baboons, administration of live *Escherichia coli* bacteria results in a significant increase in the serum levels of biopterin, neopterin, and nitrate, suggesting induction of GTP cyclohydrolase I and iNOS (Strohmeier *et al.*, 1995). In this model, infusion of L-NMMA (5 mg/kg per hr) attenuates the rise in the serum nitrate and creatinine levels, hypotension, decreased peripheral vascular resistance, and the substantial 7-day mortality caused by severe sepsis in this species (Daryl Rees and Heinz Redl, personal communication). These findings clearly document that the circulatory failure caused by septic shock in baboons is largely mediated by an enhanced formation of NO by iNOS, and that inhibition of iNOS with L-NMMA improves outcome in this model. In summary, L-NMMA (currently developed by Glaxo Wellcome as 546C88 for use in septic shock; see below) is a nontoxic (e.g., LD_{50} in the rat: >1–2 g/kg) inhibitor of NOS activity that exerts beneficial hemodynamic effects in animals and humans with septic shock.

4.2. N^G-Nitro-L-arginine Methyl Ester

Following early findings in 1990 that L-NMMA exerted beneficial hemodynamic effects in animal models of endotoxemia, many studies aimed at elucidating the role of NO in septic shock utilized the NOS inhibitor N^G-nitro-L-arginine methyl ester (L-NAME) rather than L-NMMA, as L-NAME is inexpensive and readily available. However, in contrast to L-NMMA, L-NAME is a relatively selective inhibitor of eNOS rather than iNOS activity (Southan *et al.*, 1995), hence higher doses of this agent may cause excessive vasoconstriction (particularly in the pulmonary, renal, and myocardial vascular beds) and enhance the incidence of both microvascular thrombosis and neutrophil adhesion to the endothelium. This probably accounts for the reduction in oxygen delivery (Walker *et al.*, 1995; Waurick *et al.*, 1997) and exacerbation of organ injury induced by L-NAME in many (though not all) animal models of endotoxic or septic shock (Thiemermann, 1998). These results are not necessarily attributable to the use of very large amounts of L-NAME, but rather reflect the greater selectivity of L-NAME for eNOS rather than iNOS activity. In rats with endotoxemia, infusion of very low doses of L-NAME (e.g., 0.03 to 0.3 mg/kg per hr) results in a dose-related increase in blood pressure via inhibition of eNOS activity, without reducing the rise in plasma levels of nitrite/nitrate (an indicator of iNOS activity) or organ injury caused by endotoxin (Wu *et al.*, 1996). The notion that L-NAME is a very potent and fairly selective inhibitor of eNOS activity is highlighted by the observation that infusions of very low doses (30–50 (g/kg per min) of L-NAME (1) cause a reduction in renal cortical blood flow without causing an increase in blood pressure in the rat (Walder *et al.*, 1991), and (2) significantly enhance the increased pulmonary vascular resistance caused by endotoxin in the pig (Robertson *et al.*, 1994). In summary, L-NAME is a relatively selective inhibitor of eNOS activity that, with very few exceptions (Meyer *et al.*, 1994), has been shown to exert detrimental effects in animals with septic shock. In our opinion, this compound is an inappropriate pharmacological tool for the modulation of NO biosynthesis to improve organ dysfunction or survival in septic shock.

4.3. Aminoguanidine and Derivatives

Aminoguanidine was the first relatively selective inhibitor of iNOS activity to be discovered (Corbett *et al.*, 1992). Nevertheless, although aminoguanidine is a more potent inhibitor of iNOS than eNOS activity *in vitro* and *in vivo*, it is not a very potent inhibitor of iNOS activity ($IC_{50} \approx 100-150\ \mu M$) (Thiemermann, 1998). The inhibition of NOS by aminoguanidine becomes greater with increasing incubation time, indicating that it is a mechanism-based inhibitor (Wolff and Lubeskie, 1995). Aminoguanidine attenuates the delayed hypotension observed in rats (Wu *et al.*, 1995) and rabbits (Seo *et al.*, 1996) with endotoxic shock, and improves the survival of mice challenged with endotoxin (Wu *et al.*, 1995). Aminoguanidine and its analogue 1-hydroxy-2-guanidine also attenuate the liver injury and hepatocellular dysfunction caused by endotoxin in the rat (Ruetten *et al.*, 1996; Wu *et al.*, 1996). In rats with endotoxic shock, aminoguanidine decreases the degree of bacterial translocation (presumably by preventing injury to the gut mucosal barrier), attenuates disruption of the blood–brain barrier (Boje, 1996), and reduces the increase in pulmonary capillary leakage (Arkovitz *et al.*, 1996). Interpreting the mechanism(s) by which aminoguanidine exerts these beneficial effects is difficult, as aminoguanidine is not exclusively an inhibitor of iNOS activity. Indeed, it has many other pharmacological properties including inhibition of (1) histamine metabolism, (2) polyamine catabolism, (3) the formation of advanced glycosylation end products, and (4) catalase activity (as well as other copper- or iron-containing enzymes). Interestingly, aminoguanidine also prevents the expression of iNOS protein by an unknown mechanism (Thiemermann, 1998). Thus, aminoguanidine has to be regarded as an agent that (1) is a relatively selective but not very potent inhibitor of iNOS activity, (2) reduces the formation of NO by two distinct mechanisms, namely, prevention of iNOS expression and inhibition of iNOS activity, and (3) exerts many other effects that appear to be unrelated to the inhibition of iNOS activity.

4.4. Aminoethyl-isothiourea and Other *S*-Substituted Isothioureas

S-substituted isothioureas (ITUs) are non-amino acid analogues of L-arginine and potent inhibitors of iNOS activity with variable isoform selectivity (Garvey *et al.*, 1994; Szabo *et al.*, 1994; Southan *et al.*, 1995). The most potent isothioureas are those with only short alkyl chains on the sulfur atom and no substituents on the nitrogen atoms. For instance, *S*-ethyl-ITU is a potent competitive inhibitor of all isoforms of human NOS, while aminoethyl-ITU and *S*-methyl-ITU are more selective inhibitors of iNOS than of eNOS activity (Southan *et al.*, 1995). In 1994, we demonstrated that *S*-methyl-ITU reverses the circulatory failure caused by endotoxin in the rat. This beneficial hemodynamic effect of *S*-methyl-ITU is associated with an attenuation of the liver injury and hepatocellular dysfunction caused by endotoxin in rats, as well as an increase in the survival rate of mice challenged with a high dose of endotoxin (Szabo *et al.*, 1994). Similarly, administration of aminoethyl-ITU (1 mg/kg per hr commencing 2 hr after injection of endotoxin) results in beneficial hemodynamic effects and attenuates the degree of liver injury/dysfunction caused by endotoxin in the rat (Thiemermann *et al.*, 1995). In pigs with endotoxemia, injection of aminoethyl-ITU (10 mg/kg i.v. administered 3 hr after endotoxin) restores hepatic arterial blood flow to normal levels and increases hepatic oxygen consumption, without affecting cardiac output (Saetre *et al.*, 1998). Having emphasized that some of the beneficial effects of aminoguanidine in shock may be related to its ability to inhibit iNOS activity, it should be noted that *S*-substituted ITUs are also likely to have effects that are unrelated to inhibition of NOS activity. For instance, aminoethyl-ITU is a scavenger of peroxynitrite and exerts beneficial effects in models of disease known to be mediated by oxygen-derived free radicals (Thiemermann, 1998). Interestingly, dimethyl-ITU, which does *not* inhibit iNOS activity, is a weak radical scavenger and inhibits the activation of the transcription factor NF-κB. In rats challenged with either endotoxin or live *Salmonella typhimurium*, dimethyl-ITU attenuates the formation of TNFα and improves survival (Sprong *et al.*, 1997). It is conceivable that other *S*-substituted ITUs will also prevent the activation of NF-κB. This may well explain the ability of aminoethyl-ITU to prevent the endotoxin-enhanced expression of iNOS protein in cultured macrophages and the rat *in vivo* (Ruetten and Thiemermann, 1996).

4.5. Highly Selective Inhibitors of iNOS Activity: 1400W and L-NIL

S-Substituted ITUs and guanidines contain the amidine function [$-CH(=NH)NH_2$], a feature they share with *O*-substituted isoureas and amidines themselves. In 1996, we reported that certain amidines (e.g., 2-iminopiperidine, butyramine, 2-aminopyridine, propioamidine, and acetamidine) inhibit NOS activity (Southan *et al.*, 1996). Recently, 1400W [*N*-(3-(aminomethyl)benzyl)-

acetamidine], an analogue of acetamidine, has been reported to be a slow, tightly binding inhibitor of human iNOS. The inhibition by 1400W of the activity of human iNOS is potent ($K_d \sim 7$ nM), dependent on the cofactor NADPH, and either irreversible or extremely slowly reversible. Most notably, 1400W is an approximately 5000-fold more potent inhibitor of human iNOS activity than eNOS activity. In a rat model of vascular injury caused by endotoxin, 1400W is 50-fold more potent as an inhibitor of iNOS than eNOS activity, and attenuates the vascular leak syndrome (Garvey *et al.*, 1997). We have recently shown that selective inhibition of iNOS activity with 1400W attenuates the circulatory failure, but not the liver injury caused by endotoxin in the rat (Wray *et al.*, 1998).

L-NIL is another highly selective and potent inhibitor of iNOS activity in the rat (Faraci *et al.*, 1996) and mouse (Moore *et al.*, 1994). Like 1400W, L-NIL (3 mg/kg i.v., 2 hr after administration of endotoxin) attenuates the delayed hypotension, but does not reduce the degree of renal dysfunction, liver dysfunction, or hepatocellular injury caused by endotoxin in the rat (Fig. 3). These findings support the view that selective inhibition of iNOS activity might be a useful approach for the restoration of blood pressure in patients with shock. Most notably, however, our data are also consistent with the notion that enhanced formation of NO by iNOS primarily contributes to circulatory failure but not to the liver injury and dysfunction caused by endotoxin, as suggested by earlier studies of iNOS knockout mice challenged with endotoxin (MacMicking *et al.*, 1995).

5. NOS Inhibition in Humans with Septic Shock

Although our understanding of the role of NO in animal models of circulatory shock has improved substantially in recent years, our knowledge regarding the biosynthesis and importance of NO in the pathophysiology of patients with septic shock is still very limited. There is evidence that endotoxin and cytokines in combination induce the expression of iNOS as well as the formation of NO in various primary human cells or cell lines, including hepatocytes, mesangial cells, retinal pigmented epithelial cells, and lung epithelial cells (Morris and Billiar, 1994; Preiser and Vincent, 1996). Elevated plasma and urine levels of nitrite/nitrate have been reported in adults and children with severe septic shock, as well as in patients with burn injuries who subsequently developed sepsis (Ochoa *et al.*, 1991; Preiser *et al.*, 1996; Wong *et al.*, 1996). Moreover, elevated plasma levels of nitrite/nitrate have been demonstrated in patients receiving IL-2 chemotherapy (Hibbs *et al.*, 1992; Preiser and Vincent, 1996). Interestingly, increased iNOS activity in leukocytes obtained from patients with sepsis appears to correlate with the number of failing organs, but not with blood pressure. Nevertheless, these studies together support the view that septic shock in humans is associated with an enhanced formation of NO. It should, however, be stressed that the increase in

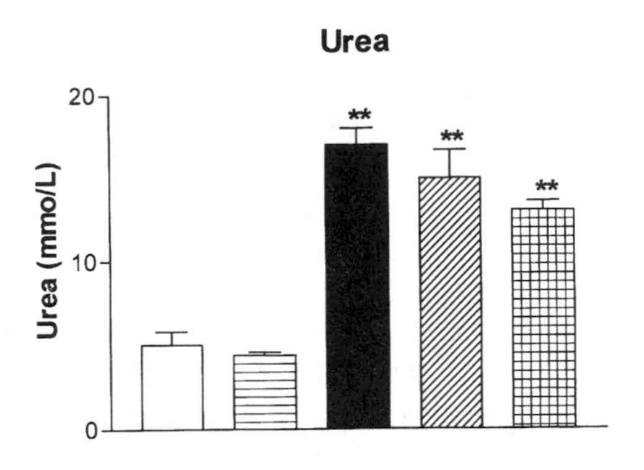

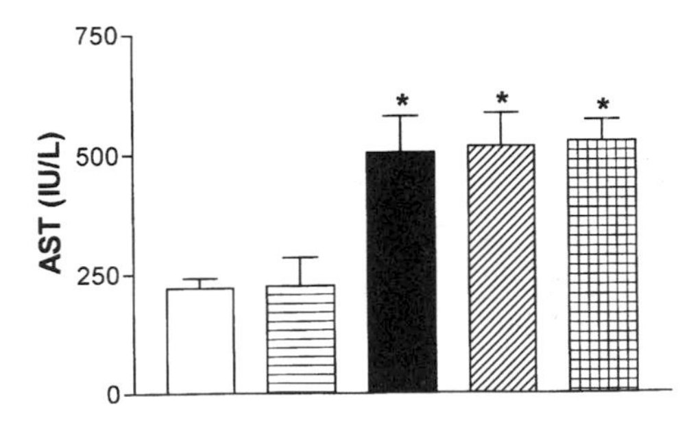

FIGURE 3. Effects of NOS inhibition on endotoxin-induced organ dysfunction. The selective iNOS inhibitor 1400W attenuates the rise in the serum levels of urea (an indicator of the development of renal dysfunction) and aspartate aminotransferase (an indicator of hepatocellular injury) in anesthetized rats challenged with endotoxin (LPS, 6 mg/kg i.v.). Animals received injections of saline rather than LPS and were treated with infusions of either saline (vehicle for 1400W, open columns, $n = 10$) or 1400W (10 mg/kg bolus plus 10 mg/kg per hr, (horizontal stripes, $n = 3$). Different groups of LPS-treated rats were treated with (starting 2 hr after LPS) (1) vehicle (saline control, black columns, $n = 10$), (2) 1400W 3 mg/kg bolus + 3 mg/kg per hr (diagonal stripes, $n = 8$), or (3) 1400W 10 mg/kg bolus + 10 mg/kg per hr (checked columns, $n = 5$). $*p < 0.01$, $**p < 0.001$ when compared by ANOVA to rats that had received vehicle rather than LPS. There were no differences in urea or AST between the LPS controls and 1400W-treated rats.

plasma levels of nitrite/nitrate elicited by endotoxin, cytokines, or bacteria in rodents is substantially (10-fold) higher than the observed increases in the plasma levels of these NO metabolites in other animal species (e.g., pig, sheep) including humans. Moreover, our understanding of (1) the biosynthesis of NO, (2) the regulation of iNOS expression, and (3) the role of NO in MODS in shock are largely based on rodent models of endotoxic shock. By comparison, we understand little about the role of NO in humans with septic and other forms of circulatory shock.

Early reports of beneficial hemodynamic effects of L-NMMA in humans with septic shock (Petros *et al.*, 1991, 1994; Lorente *et al.*, 1993; Schilling *et al.*, 1993) stimulated a phase I, multicenter, open-label, dose escalation (1, 2.5, 5, 10, or 20 mg/kg per hr for up to 8 hr) study using L-NMMA (546C88) in 32 patients with septic shock. In this study, L-NMMA sustained blood pressure and enabled a reduction in vasopressor (norepinephrine) support. The cardiac index fell to baseline values (possibly because of an increase in peripheral vascular resistance), and left ventricular function was well maintained. Moreover, L-NMMA increased oxygen extraction, while pulmonary shunting was not worsened. A recent placebo-controlled multicenter study involving 312 patients with septic shock has evaluated the effects of L-NMMA on the resolution of shock at 72 hr (primary endpoint). The severity of illness according to the SAPS (simplified acute physiology score) II score was similar between the groups receiving placebo and L-NMMA. Infusion of L-NMMA enhanced mean arterial blood pressure and systemic vascular resistance index, while decreasing cardiac output toward normal levels. L-NMMA had no effect on left ventricular systolic work index, indicating that the fall in cardiac output was not related to an impairment in cardiac contractility. In patients treated with L-NMMA, there was a transient increase in mean pulmonary artery pressure. Interestingly, L-NMMA did not affect the thrombocytopenia or the renal dysfunction caused by sepsis. Most notably, 41% of patients treated with L-NMMA, but only 21% of patients treated with placebo, recovered from shock within 72 hr. There was also a strong trend toward a reduction in mortality at day 14 in patients treated with L-NMMA.

6. Concluding Remarks

Since 1990, numerous studies have documented enhanced formation of NO in various animal models of endotoxic/septic shock. Similarly, patients with septic shock or those receiving IL-2 immunotherapy exhibit elevated plasma levels of nitrite/nitrate. Although the enhanced formation of NO in animals and humans with septic shock contributes to hypotension and hyporeactivity of the vasculature to vasoconstrictor agents (vasoplegia), it is still unclear whether NO from iNOS contributes to the organ dysfunction/failure syndrome associated with severe septic shock. Our finding that highly selective inhibitors of iNOS activity (such as 1400W or L-NIL) attenuate the delayed hypotension but do not affect the multiple organ dysfunction caused by endotoxin in the rat, supports the view that enhanced formation of NO within the vasculature contributes to sepsis-associated circulatory failure (vasodilatation, vasoplegia, and possibly vascular leak), but does not directly mediate the development of organ injury. This is corroborated by the finding that mice with a targeted interruption of the iNOS gene experience less endotoxin-induced hypotension, but remain susceptible to liver injury (MacMicking *et al.,* 1995).

Although it is evident that human cells and tissues can be induced to produce iNOS protein and activity when challenged with endotoxin and cytokines, the degree of iNOS activation in patients with septic shock appears to be substantially lower than in some other animal species (e.g., rodents). Nevertheless, inhibition of NOS activity with L-NMMA exerts beneficial hemodynamic effects (e.g., resolution of shock) without causing significant side effects in septic patients. Whether the beneficial hemodynamic effects of L-NMMA in patients with septic shock will be sufficient to attenuate 28-day mortality is currently being determined in a large phase III, multicenter trial involving more than 2000 patients.

References

Arkovitz, M. S., Wispe, J. R., Garcia, V. F., and Szabo, C., 1996, Selective inhibition of the inducible isoform of nitric oxide synthase prevents pulmonary transvascular flux during acute endotoxemia, *J. Pediatr. Surg.* **31:**1009–1015.

Baue, A. E., 1993, The multiple organ or systems failure syndrome, in: *Pathophysiology of Shock, Sepsis and Organ Failure* (G. Schlag and H. Redl, eds.), Springer, Berlin, pp. 1004–1018.

Billiar, T. R., Curran, R. D., Harbrecht, B. G., Stuehr, D. J., Demetris, A. J., and Simmons, R. L., 1990, Modulation of nitrogen oxide synthesis in vivo: N^G-monomethyl-L-arginine inhibits endotoxin-induced nitrite/nitrate biosynthesis while promoting hepatic damage, *J. Leukoc. Biol.* **48:**565–569.

Boje, K. M., 1996, Inhibition of nitric oxide synthase attenuates blood–brain barrier disruption during experimental meningitis, *Brain Res.* **720:**75–83.

Booke, M., Hinder, F., McGuire, R., Traber, L. D., and Traber, D. L., 1996a, Nitric oxide synthase inhibition versus norepinephrine for the treatment of hyperdynamic sepsis in sheep, *Crit. Care Med.* **24:**835–844.

Booke, M., Hinder, F., McGuire, R., Traber, L. D., and Traber, D. L., 1996b, Nitric oxide synthase inhibition versus norepinephrine in ovine sepsis: Effects on regional blood flow, *Shock* **5:**362–370.

Carnigie, P. R., Fellows, F. C. I., and Symington, G. R., 1977, Urinary excretion of methylarginine in human disease, *Metabolism,* **26:**531–537.

Cobb, J. P. and Danner, R. L., 1996, Nitric oxide and septic shock, *J. Am. Med. Assoc.* **275:**1192–1196.

Corbett, J. A., Tilton, R. G., Chang, K., Hasan, K. S., Ido, Y., Wang, J. L., Sweetland, M. A., Lancaster, J. R., Williamson, J. R., and McDaniel, M. L., 1992, Aminoguanidine, a novel inhibitor of nitric oxide formation, prevents diabetic vascular dysfunction, *Diabetes* **41:**552–558.

Dinerman, J. L., Lowenstein, C. J., and Snyder, S. H., 1993, Molecular mechanism of nitric oxide regulation: Potential relevance to cardiovascular disease, *Circ. Res.* **73:**217–222.

Faraci, W. S., Nagel, A. A., Verdries, K. A., Vincent, L. A., Xu, H., Nichols, L. E., Labasi, J. M., Salter, E. D., and Pettipher, E. R., 1996, 2-Amino-4-methylpyridine as a potent inhibitor of inducible NO synthase activity *in vitro* and *in vivo*, *Br. J. Pharmacol.* **119:**1101–1108.

Garvey, P. E., Oplinger, J. A., Tanoury, G. J., Sherman, P. A., Fowler, M., Marshall, S., Marmon, M. F., Paith, J. E., and Furfine, E. S., 1994, Potent and selective inhibition of human nitric oxide synthases. Inhibition by non-amino acid isothioureas, *J. Biol. Chem.* **269:**26669–26676.

Garvey, P. E., Oplinger, J. A., Furfine, E. S., Kiff, R. J., Laszlo, F., Whittle, B. J. R., and Knowles, R. G.,

1997, 1400W is a slow, tight binding, and highly selective inhibitor of inducible nitric oxide synthase *in vitro* and *in vivo*, *J. Biol. Chem.* **272**:4959–4963.

Gross, S. S., Stuehr, D. J., Aisaka, K., Jaffe, E. A., Levi, R., and Griffith, O. W., 1990, Macrophage and endothelial nitric oxide synthesis: Cell-type selective inhibition by N^G-aminoarginine, N^G-nitroarginine and N^G-methyl-arginine, *Biochem. Biophys. Res. Commun.* **170**:96–103.

Hibbs, J. B., Jr., Westenfelder, C., Taintor, R., Vavrin, Z., Kablitz, C., Baranowski, R. L., Ward, J. H., Menlove, R. L., McMurry, M. P., Kushner, J. P., and Samlowski, W. E., 1992, Evidence for cytokine-inducible nitric oxide synthesis from L-arginine in patients receiving interleukin-2 therapy, *J. Clin. Invest.* **89**:867–877.

Julou-Schaeffer, G., Gray, G. A., Fleming, I., Schott, C., Parratt, J. R., and Stoclet, J. C., 1990, Loss of vascular responsiveness induced by endotoxin involves the L-arginine pathway, *Am. J. Physiol.* **259**:H1038–H1043.

Kilbourn, R. G., Gross, S. S., Jubran, A., Adams, J., Griffith, O. W., Levi, R., and Lodato, R. F., 1990a, N^G-methyl-L-arginine inhibits tumour necrosis factor-induced hypotension: Implications for the involvement of nitric oxide, *Proc. Natl. Acad. Sci. USA* **87**:3629–3632.

Kilbourn, R. G., Jubran, A., Gross, S. S., Griffith, O. W., Levi, R., and Adams, J., 1990b, Reversal of endotoxin-mediated shock by N^G-monomethyl-L-arginine, an inhibitor of nitric oxide synthesis, *Biochem. Biophys. Res. Commun.* **172**:1132–1138.

Kilbourn, R. G., Szabo, C., and Traber, D. L., 1997, Beneficial versus detrimental effects of nitric oxide synthase inhibitors in circulatory shock: Lessons learned from experimental and clinical studies, *Shock* **7**:235–246.

Lorente, J. A., Landin, L., De Pablo, R., Renes, E., and Liste, D., 1993, L-Arginine pathway in the sepsis syndrome, *Crit. Care Med.* **21**:1287–1295.

Loscalzo, J., and Welch, G., 1995, Nitric oxide and its role in the cardiovascular system, *Prog. Cardiovasc. Dis.* **38**:87–104.

MacMicking, J. D., Nathan, C., and Hom, G., 1995, Altered responses to bacterial infection and endotoxic shock in mice lacking inducible nitric oxide synthase, *Cell* **82**:641–650.

Meyer, J., Lentz, C. W., Stothert, J. C., Traber, L. D., Herndon, D. N., and Traber, D. L., 1994, Effects of nitric oxide synthesis inhibition in hyperdynamic endotoxemia, *Crit. Care Med.* **22**:306–312.

Moncada, S., and Higgs, A., 1993, The L-arginine–nitric oxide pathway, *N. Engl. J. Med.* **329**:2202–2212.

Moore, W. M., Webber, R. K., Jerome, G. M., Tjoeng, F. S., Misko, T. P., and Currie, M. D., 1994, L-*N*6-(1-iminoethyl)lysine: A selective inhibitor of inducible nitric oxide synthase, *J. Med. Chem.* **37**:3886–3888

Morris, S. M., and Billiar, T. R., 1994, New insights into the regulation of inducible nitric oxide synthase, *Am. J. Physiol.* **266**:E829–E839.

Natanson, C., Hoffmann, W. D., Suffredini, E. F., Eichacker, P. Q., and Danner, R. L., 1994, Selected treatment strategies for septic shock based on proposed mechanism of pathogenesis, *Ann. Intern. Med.* **120**:771–783.

Nathan, C., 1992, Nitric oxide as a secretory product of mammalian cells, *FASEB J.* **6**:3051–3064.

Ochoa, J. B., Udekwu, A. O., Billiar, T. R., Curran, R. D., Cerra, F. B., Simmons, R. L., and Peitzman, A. B., 1991, Nitrogen oxide levels in patients after trauma and during sepsis, *Ann. Surg.* **214**:621–626.

Parrillo, J. E., 1990, Septic shock in humans. Advances in the understanding of pathogenesis, cardiovascular dysfunction and therapy, *Ann. Intern. Med.* **113**:227–242.

Park, K. S., Lee, H. W., and Hong, S. Y., 1988, Determination of methylated amino acids in human serum by high-performance liquid chromatography, *J. Chromatogr.* **440**:225–230.

Petros, A., Bennett, D., and Vallance, P., 1991, Effect of nitric oxide synthase inhibitors on hypotension in patients with septic shock, *Lancet* **338**:1557–1558.

Petros, A., Lamb, G., Leone, A., Moncada, S., Bennett, D., and Vallance, P., 1994, Effects of a nitric oxide synthase inhibitor in humans with septic shock, *Cardiovasc. Res.* **28:**34–39.

Preiser, J. C., and Vincent, J. L., 1996, Nitric oxide involvement in septic shock: Do human beings behave like rodents? in: *1996 Yearbook of Intensive Care and Emergency Medicine* (J. L. Vincent, ed.), Springer, Berlin, pp. 358–365.

Preiser, J. C., Reper, P., Vlasselaer, D., Vray, B., Zhang, H., Metz, G., Vanderkelen, A., and Vincent, J. L., 1996, Nitric oxide production is increased in patients after burn injury, *J. Trauma* **40:**368–371.

Rees, D. D., Cellek, S., Palmer, R. M. J., and Moncada, S., 1990, Dexamethasone prevents the induction of nitric oxide synthase and the associated effects on the vascular tone: An insight into endotoxic shock, *Biochem. Biophys. Res. Commun.* **173:**541–547.

Robertson, F. M., Offner, P. J., Ciceri, D. P., Becker, W. K., and Pruitt, B. A., Jr., 1994, Detrimental hemodynamic effects of nitric oxide synthase inhibition in septic shock, *Arch. Surg.* **129:**149–155.

Ruetten, H., Southan, G. J., Abate, A., and Thiemermann, C., 1996, Attenuation of the multiple organ dysfunction caused by endotoxin by 1-amino-2-hydroxy-guanidine, a potent inhibitor of inducible nitric oxide synthase, *Br. J. Pharmacol.* **118:**261–270.

Saetre, T., Gundersen, Y., Thiemermann, C., Lilleansen, P., and Aasen, A. O., 1998, Aminoethyl-isothiourea, a selective inhibitor of inducible nitric oxide synthase activity, improves liver circulation and oxygen metabolism in a porcine model of endotoxaemia, *Shock* **9:**109–115.

Schilling, J., Cakmakci, M., Battig, U., and Geroulanos, S., 1993, A new approach in the treatment of hypotension in human septic shock by N^{G}-monomethyl-L-arginine, an inhibitor of the nitric oxide synthetase, *Intensive Care Med.* **19:**227–231.

Seo, H. G., Fujiwara, N., Kaneto, H., Asashi, M., Fujii, J., and Taniguchi, N., 1996, Effect of the nitric oxide synthase inhibitor, *S*-ethyl-isothiourea, on cultured cells and cardiovascular functions of normal and lipopolysaccharide-treated rabbits, *J. Biochem.* **119:**553–558.

Southan, G. J., and Szabo, C., 1996, Selective pharmacological inhibition of distinct nitric oxide synthase isoforms, *Biochem. Pharmacol.* **51:**383–394.

Southan, G., Szabo, C., and Thiemermann, C., 1995, Isothioureas: Potent inhibitors of nitric oxide synthases with variable isoform selectivity, *Br. J. Pharmacol.* **114:**510–516.

Southan, G. J., Szabo, C., O'Conner, M. P., Salzman, A. C., and Thiemermann, C., 1996, Amidines are potent inhibitors of constitutive and inducible nitric oxide synthases: Preferential inhibition of the inducible isoform, *Eur. J. Pharmacol.* **291:**311–318.

Sprong, R. C., Aarsman, C. J. M., Oirschot, J. F. L. M., and Asbeck, B. S., 1997, Dimethylthiourea protects rats against gram-negative sepsis and decreases tumour necrosis factor and nuclear factor κB activity, *J. Lab. Clin. Med.* **129:**470–481.

Strohmeier, W., Werner, E. R., Redl, H., Wachter, H., and Schlag, G., 1995, Plasma nitrate and pteridine levels in experimental bacteremia in baboons, *Pteridines* **6:**8–11.

Szabo, C., and Thiemermann, C., 1995, Regulation of the expression of the inducible isoform of nitric oxide synthase, *Adv. Pharmacol.* **34:**113–154.

Szabo, C., Southan, G., and Thiemermann, C., 1994, Beneficial effects and improved survival in rodent models of septic shock with *S*-methyl-isothiourea sulfate, a novel, potent and selective inhibitor of inducible nitric oxide synthase, *Proc. Natl. Acad. Sci. USA* **91:**12472–12476.

Thiemermann, C., 1994, The role of L-arginine:nitric oxide pathway in circulatory shock, *Adv. Pharmacol.* **28:**45–79.

Thiemermann, C., 1998, The use of selective inhibitors of inducible nitric oxide synthase in septic shock, *Sepsis* **1:**123–129.

Thiemermann, C., and Vane, J. R., 1990, Inhibition of nitric oxide synthesis reduces the hypotension induced by bacterial lipopolysaccharide in the rat, *Eur. J. Pharmacol.* **182:**591–595.

Thiemermann, C., Ruetten, H., Wu, C. C., and Vane, J. R., 1995, The multiple organ dysfunction syndrome caused by endotoxin in the rat: Attenuation of liver dysfunction by inhibitors of nitric oxide synthase, *Br. J. Pharmacol.* **116:**2845–2851.

Walder, C. E., Thiemermann, C., and Vane, J. R., 1991, The involvement of endothelium-derived relaxing factor in the regulation of renal cortical blood flow in the rat, *Br. J. Pharmacol.* **102:**967–973.

Walker, T. A., Curtis, S. E., King-VanVlack, C. E., Chapler, C. K., Vallet, B., and Cain, S. M., 1995, Effects of nitric oxide synthase inhibition on regional hemodynamics and oxygen transport in endotoxic dogs, *Shock* **4:**415–420.

Waurick, R., Bone, H. G., Meyer, J., Booke, M., Meissner, A., Prien, T., and Van Aken, H., 1997, Haemodynamic effects of dopexamine and nitric oxide synthase inhibition in healthy and endotoxaemic sheep, *Eur. J. Pharmacol.* **333:**181–186.

Wei, X., Charles, I. G., and Smith, A., 1995, Altered immune responses in mice lacking inducible nitric oxide synthase, *Nature* **375:**408–411.

Wolff, D. J., and Lubeskie, A., 1995, Aminoguanidine is an isoform-selective, mechanism-based inactivator of nitric oxide synthase, *Arch. Biochem. Biophys.* **316:**290–301.

Wong, H. R., Carcillo, J. A., Burckart, G., and Kaplan, S. S., 1996, Nitric oxide production in critically ill patients, *Arch. Dis. Child.* **74:**482–489.

Wray, G. M., Millar, C. G., Hinds, C. J., and Thiemermann, C., 1998, Selective inhibition of the activity of nitric oxide synthase prevents the circulatory failure, but not the organ injury/dysfunction, caused by endotoxin. *Shock* **9:**329–335.

Wu, C. C., Chen, S. J., Szabo, C., Thiemermann, C., and Vane, J. R., 1995, Aminoguanidine attenuates the delayed circulatory failure and improves survival in rodent models of endotoxic shock, *Br. J. Pharmacol.* **114:**1666–1672.

Wu, C. C., Ruetten, H., and Thiemermann, C., 1996, Comparison of the effects of aminoguanidine and N^{G}-nitro-L-arginine methyl ester on the multiple organ dysfunction caused by endotoxaemia in the rat, *Eur. J. Pharmacol.* **300:**99–104.

CHAPTER 14

Nitric Oxide in Mycobacterium tuberculosis *Infection*

JOHN CHAN and JOANNE FLYNN

1. Introduction

Tuberculosis continues to be the single greatest infectious cause of death in the world (Murray *et al.*, 1990; Bloom and Murray, 1992). Approximately 1.75 billion persons are infected with *Mycobacterium tuberculosis* worldwide, with 8 million new cases and 3 million deaths per year. The AIDS epidemic has been identified as the most important factor contributing to the recent resurgence of tuberculosis in the United States (Snider and Roper, 1992). Fueled by the untimely emergence of multidrug-resistant strains (WHO, 1997), the AIDS epidemic has escalated the threat of the tubercle bacillus to an alarming level not experienced in modern times.

Although *M. tuberculosis* is one of the earliest bacterial pathogens of humans to be discovered, mechanisms of tuberculosis host resistance and pathogenesis remain incompletely defined. Experimental evidence suggests that acquired resistance against *M. tuberculosis* depends primarily on macrophage activation via cytokines generated by specific T lymphocytes (reviewed in Barnes *et al.*, 1994; Chan and Kaufmann, 1994), and that IFNγ is among the major T-cell products that control macrophage antimycobacterial activity (reviewed in Barnes *et al.*, 1994; Chan and Kaufmann, 1994). Because the best understood cytotoxic mechanism of activated macrophages is the production of oxygen radicals (Iyer *et al.*, 1961; reviewed in Nathan, 1983; Chanock *et al.*, 1994; Wientjes and Segal, 1995), research activity to characterize the antimycobacterial function of cytokine-

JOHN CHAN • Departments of Medicine and Microbiology and Immunology, Albert Einstein College of Medicine, Bronx, New York 10467. *JOANNE FLYNN* • Departments of Molecular Genetics and Biochemistry and Medicine, University of Pittsburgh School of Medicine, Pittsburgh, Pennsylvania 15261.

Nitric Oxide and Infection, edited by Fang. Kluwer Academic / Plenum Publishers, New York, 1999.

stimulated phagocytes has focused until recently on the toxic effects of respiratory burst-generated reactive oxygen intermediates (ROI). Still, the role of ROI as effector molecules mediating antimycobacterial activity remains controversial (reviewed in Chan and Kaufmann, 1994).

As discussed earlier in this volume (see especially Chapter 2), work by various laboratories led to the discovery of the NO-generating, L-arginine-dependent cytotoxic mechanism of murine macrophages (reviewed in Nathan and Hibbs, 1991; Moncada, 1992; Nathan, 1992; Fang, 1997; MacMicking *et al.*, 1997b; Nathan, 1997). Although not invariably susceptible to reactive nitrogen intermediates (RNI) (Nathan, 1997), microbes as phylogenetically diverse as bacteria, fungi, helminths, protozoa, and viruses have been shown to be sensitive to the toxic effects of RNI (reviewed in Nathan and Hibbs, 1991; MacMicking *et al.*, 1997b; see also Chapter 12). Among the numerous microbial targets of RNI is *M. tuberculosis*, whose susceptibility to toxic nitrogen oxides has been well established *in vitro* (Denis, 1991; Chan *et al.*, 1992; O'Brien *et al.*, 1994; Rhoades and Orme, 1997; Yu *et al.*, in press) (Table I) as well as *in vivo* using various murine models (Flynn *et al.*, 1993, 1995; Chan and Kaufmann, 1994; Chan *et al.*, 1995; MacMicking *et al.*, 1997a) (Table II). Despite the apparent significance of the NO-generating pathway in host defense against *M. tuberculosis*, little is known about the interactions between RNI and the tubercle bacillus.

2. Antimycobacterial Effects of RNI: Evidence *in Vitro*

As a successful intracellular pathogen, *M. tuberculosis* can effectively evade antimicrobial defenses of the host in order to reside and multiply within macrophages (Haas and Prez, 1995). However, because most healthy people are able to control *M. tuberculosis* infection throughout their lifetimes, with only a minority of PPD skin test converters developing disease (Haas and Prez, 1995), the immune response of the host must be capable of killing or at least inhibiting the growth of the tubercle bacillus. RNI generated by immunologically activated iNOS make a critical contribution to this immune response.

Early *in vitro* experiments demonstrated that IFNγ-activated murine macrophages can prevent growth of intracellular BCG (a relatively avirulent mycobacterial strain derived from *M. bovis*) and virulent *M. tuberculosis* (reviewed in Chan and Kaufmann, 1994). However, the mechanisms by which these macrophages execute their antimycobacterial function were not identified until recently. Soon after the discovery of the L-arginine-dependent cytotoxic pathway (reviewed in Nathan and Hibbs, 1991; Nathan, 1992), Flesch and Kaufmann (1991) linked the *in vitro* anti-BCG activity of IFNγ-activated murine bone marrow macrophages to RNI. Evidence for a significant role of these toxic nitrogen oxides in controlling intracellular growth of *M. tuberculosis* emerged soon there-

TABLE I
Reactive Nitrogen Intermediates Are Effective Antimycobacterial Agents *in Vitro*

System	*Mycobacterium* spp.	Susceptibility	References
Murine macrophage/NOS inhibitors	BCG	+	Flesch and Kaufmann (1991)
	M. tuberculosis H37Rv	+	Denis (1991a)
	M. tuberculosis Erdman	+	Chan *et al. (1992)*
	M. tuberculosis, multiple strains	±	Rhoades and Orme (1997)
	M. avium complex	±	Doi *et al.* (1993)
Human macrophage/NOS inhibitors	*M. avium*	+	Denis (1991b)
	BCG	+	Nozaki *et al.* (1997)
Acidified NO_2^-	BCG	+	Flesch and Kaufmann (1991)
	M. tuberculosis Erdman	+	Chan *et al.* (1992)
	M. tuberculosis, multiple strains	±	Rhoades and Orme (1997)
	M. tuberculosis, multiple strains; *M. bovis*	±	O'Brien *et al.* (1994)
Authentic NO	*M. Smegmatis*, BCG, and *M. tuberculosis* Erdman	+	Yu *et al.* (in press)
Authentic NO_2	*M. smegmatis*, BCG, and *M. tuberculosis* Erdman	+	Yu *et al.* (in press)
$ONOO^-$	*M. smegmatis* and BCG	+	Yu *et al.* (in press)
	M. tuberculosis Erdman & strain C	−	Yu *et al.* (in press)
	M. bovis Ravenel	−	Yu *et al.* (in press)
NONOates: SPER/NO, DEA/NO, MOM-DEA/NO	*M. smegmatis* and BCG	+	Tsang and Chan (unpublished)
NOS2-transfected 293 cells	*M. tuberculosis* Erdman	+	Yu and Chan (unpublished)

TABLE II

In Vivo Evidence Supporting a Role for RNI in Host Defense against Mycobacteria

Host	Model[a]	*Mycobacterium* spp.	References
Mouse	AG in C57BL/6	*M. tuberculosis* Erdman	Chan *et al.* (1995), Flynn *et al.* (1998)
	NMMA in C57BL/6	*M. tuberculosis* Erdman	Chan *et al.* (1995)
	NIL in C57BL/6	*M. tuberculosis* Erdman	MacMicking *et al.* (1997a), Scanga and Flynn (unpublished)
	gko	*M. tuberculosis* Erdman	Flynn *et al.* (1993), Cooper *et al.* (1993)
		BCG	Dalton *et al.* (1993)
	IFNγ receptor k.o.	BCG	Kamijo *et al.* (1993)
	p55 TNF receptor k.o.	*M. tuberculosis* Erdman	Flynn *et al.* (1995)
	NOS2 k.o.	*M. tuberculosis* Erdman	MacMicking *et al.* (1997a)
	IRF-1 k.o.	BCG	Kamijo *et al.* (1994)
Human	Immunocytochemistry studies on lung macrophages from tuberculous patients	Primary infecting strains: *M. tuberculosis*	Nicholson *et al.* (1996)
	IFNγ receptor mutation	Primary infecting strains: *M. fortuitum, M. chelonei, M. avium*	Newport *et al.* (1996)
		Primary infecting strain: BCG	Jouanguy *et al.* (1996)

[a] AG, aminoguanidine; NMMA, N^G-monomethyl-L-arginine; NIL, N^6-(1-iminoethyl)-L-lysine; NOS2, macrophage inducible isoform of nitric oxide synthase; *gko*, interferon-gamma "knockout"; k.o., "knockout"; IRF-1; interferon regulatory factor-1; IFNγ, interferon-gamma.

after (Denis, 1991a; Chan *et al.*, 1992). By using the NOS inhibitor N^G-monomethyl-L-arginine (L-NMMA) or manipulating concentrations of L-arginine in the culture medium, investigators performing *in vitro* macrophage studies were able to demonstrate a positive correlation between nitrite (NO_2^-) production and killing or inhibition of *M. tuberculosis*. The mycobactericidal activity of acidified NO_2^- in cell-free systems (Chan *et al.*, 1992; O'Brien *et al.*, 1994; Rhoades and Orme, 1997) and the ability of iNOS-transfected 293 cells to inhibit the *M. tuberculosis* Erdman strain (Yu and Chan, unpublished) have further documented the antibacterial activity of RNI against *M. tuberculosis*. Comparative studies using the ROI-deficient D9 mouse macrophage cell line and its parental J774.16 line strongly suggest that generation of RNI is the primary antimycobacterial mechanism of macrophages (Chan *et al.*, 1992). However, a role of ROI in host defense against *M. tuberculosis* has not been ruled out. For example, children suffering from X-linked chronic granulomatous disease (CGD) characterized by mutations in the NADPH oxidase subunit *gp91-phox* (reviewed in Curnutte, 1993), appear to be at higher risk for developing tuberculosis (Lau *et al.*, 1996). Furthermore, disruption of *gp91-phox* in *M. tuberculosis*-infected mice results in a higher pulmonary bacillary burden compared with wild-type controls (Adams *et al.*, 1997). Interestingly, the same study reported that IFNγ- and LPS-stimulated, RNI-generating macrophages from CGD mice exhibit antimycobacterial function, while macrophages obtained from animals with disruption of *iNOS* do not (Adams *et al.*, 1997). The contribution of ROI to the control of *M. tuberculosis* deserves further evaluation.

3. The iNOS-Dependent Cytotoxic Pathway in Acute Murine Tuberculosis

Much of the *in vivo* evidence supporting a role for RNI in host defense against *M. tuberculosis* has been derived from studies using mice with disruption of genes involved in immune function (Table III). The majority of these mouse strains are defective, directly or indirectly, in the IFNγ–TNFα–iNOS axis leading to RNI production. Mice disrupted in the genes for IFNγ (gko mice) (Dalton *et al.*, 1993), the IFNγ receptor (Kamijo *et al.*, 1993), or IFNγ regulatory factor-1 (IRF-1) (Kamijo *et al.*, 1994) are more susceptible to BCG infection, and macrophages from these mice produce less NO compared with controls. IFNγ knockout mice are extremely susceptible to intravenously administered virulent *M. tuberculosis*, with a rapidly fatal course of infection, compared with wild-type littermates (Flynn *et al.*, 1993) (Table III). In this study, analysis of *M. tuberculosis*-infected gko mice revealed undetectable iNOS mRNA and a marked decrease in serum NO_2^-/NO_3^- levels compared with wild-type animals, while TNFα expression was unaffected. Immunohistochemical studies of lung granulomas from infected gko

TABLE III
Effect of Immune Deficiency on Survival of Acute Murine Tuberculosis

Mouse/treatment	Deficiency	Dose (i.v.)[a]	m.s.t. (days)[b]	References
C57BL/6	None	5×10^5	> 140	Flynn and Chan (1998)
gko	INFγ	2×10^4	22 ± 3	Flynn (unpublished)
gko	IFNγ	5×10^5	14 ± 1	Flynn *et al.* (1993)
TNFp55R$^{-/-}$	55-kDa TNF receptor	5×10^5	20 ± 2	Flynn *et al.* (1995)
C57BL/6 + anti-TNF-MAb	TNFα	5×10^5	22 ± 2	Flynn *et al.* (1995)
C57BL/6 + 2.5% AG[c]	RNI	1×10^6	29 ± 1	Chan *et al.* (1995)
C57BL/6 × NMMA[d]	RNI	1×10^6	31 ± 2	Chan *et al.* (1995)
NOS2$^{-/-}$	NOS2 activity	1×10^5	38 ± 2	MacMicking *et al.* (1997a)
NOS2$^{-/-}$	NOS2 activity	1×10^6	28 ± 1	MacMicking *et al.* (1997a)
C57BL/6 + HC[e]	Steroid-related immunodeficiency	1×10^6	27 ± 1	MacMicking *et al.* (1997a)
NOS2$^{-/-}$	NOS2 activity/steroid-related immunodeficiency	1×10^6	23 ± 1	MacMicking *et al.* (1997a)
C57BL/6 PCM[f]	Protein-deficient diet	1×10^6	67 ± 7	Chan *et al.* (1996)
C57BL/6 PCM	Protein-deficient diet	1×10^6	43 ± 14	Chan *et al.* (1996)
CD40L$^{-/-}$	CD40 ligand	2×10^5	> 210	Campos-Neto *et al.* (1998)
IL-12$^{-/-}$	IL-12	1×10^5	40-45	Cooper *et al.* (1997)

[a] All infections were performed with *M. tuberculosis* strain Erdman.
[b] Mean survival time, where calculated.
[c] AG, aminoguanidine.
[d] NMMA, N^G-monomethyl-L-arginine.
[e] HC, hydrocortisone acetate.
[f] PCM, protein-calorie malnourished.

mice demonstrate greatly diminished iNOS expression compared with controls, indicating that IFNγ is necessary for RNI production *in vivo* (Flynn, unpublished). Aerosol infection of gko mice with *M. tuberculosis* is also rapidly fatal (Cooper *et al.*, 1993). The overwhelming infections in gko mice support other evidence that RNI play a protective role against the tubercle bacillus *in vivo*.

TNFα can act synergistically with IFNγ to promote RNI production (Ding, 1988) and subsequent killing of intracellular mycobacteria by macrophages (reviewed in Chan and Kaufmann, 1994). Mice deficient in the p55 TNF receptor are highly susceptible to virulent *M. tuberculosis* infection compared with wild-type mice (Table III) (Flynn *et al.*, 1995). Neutralization of TNFα with a monoclonal antibody also increases the susceptibility of mice to *M. tuberculosis* infection (Flynn *et al.*, 1995) (Table III). Interestingly, iNOS expression and RNI production by macrophages were greatly diminished early in the TNFR $p55^{-/-}$ model (<10 days), but returned to wild-type levels by day 14 postinfection. However, infection remained uncontrolled despite the belated production of RNI. An analogous situation may exist during murine *Leishmania major* infection, in which delayed RNI production is associated with a poor outcome (Evans *et al.*, 1996). We hypothesize that a heavy bacterial burden may result in RNI production in the absence of TNFα, because mycobacterial components such as lipoarabinomannan can trigger RNI production by IFNγ-treated macrophages *in vitro* (Roach *et al.*, 1993, 1995; Anthony *et al.*, 1994; Schuller-Lewis *et al.*, 1994; Chan and Bloom, unpublished). TNFα appears to play an important role promoting RNI production early in infection, and early production of RNI by macrophages appears to be crucial to the outcome of infection.

The observations in IFNγ or TNFα receptor-deficient mice strongly suggest that RNI production is required for murine macrophages to control the replication of *M. tuberculosis*. This led us to test the specific importance of iNOS during *M. tuberculosis* infection *in vivo*. The NOS inhibitors aminoguanidine (AG) or L-NMMA were provided to mice in drinking water during acute tuberculosis infection (Chan *et al.*, 1995). AG administration impairs the ability of mice to control *M. tuberculosis* infection, resulting in a dramatic increase in mortality and tissue bacterial burden. Administration of L-NMMA to *M. tuberculosis*-infected mice produced similar results. Recently, a different relatively selective iNOS inhibitor, L-NIL [N^6-(1-iminoethyl) lysine], has been shown to exacerbate murine tuberculosis (MacMicking *et al.*, 1997a; Scanga and Flynn, unpublished). The ability of three chemically distinct NOS inhibitors to exert similar effects on the course of *M. tuberculosis* infection in mice provides compelling evidence that iNOS is a critical component of host defense in murine tuberculosis.

Definitive proof of the importance of RNI in resistance to acute tuberculosis in the mouse was provided by MacMicking *et al.* (1997a), when they infected mice carrying a disruption in the *iNOS* gene (MacMicking *et al.*, 1995; Wei *et al.*, 1995). $iNOS^{-/-}$ mice succumb quickly to infection with virulent *M. tuberculosis*, with

100- to 1000-fold greater pulmonary bacterial burdens than those in control mice at day 25 postinfection (MacMicking *et al.*, 1997a). Relative to controls, neither IFNγ nor TNFα expression was diminished in these mice. Consideration of these and other studies allows one to conclude that (1) $iNOS^{-/-}$ mice, compared with those deficient in IFNγ, β_2-microglobulin, or TNFR1, appear most vulnerable to *M. tuberculosis* Erdman during the first 2 weeks of infection, with bacillary doubling time used as a criterion for susceptibility; and (2) although severely immunodeficient SCID mice display residual, glucocorticoid-sensitive resistance to *M. tuberculosis*, glucocorticoids (which depress RNI production, see Chapter 5) do not further enhance the susceptibility of $iNOS^{-/-}$ mice to the tubercle bacillus.

M. tuberculosis-infected mice lacking $CD4^+$ T cells are also impaired in IFNγ and iNOS expression early in infection, and show reduced survival compared with controls (Myers and Flynn, in preparation). Mice with protein-calorie malnutrition, a condition known to be highly associated with increased susceptibility to tuberculosis (reviewed in McMurray, 1994), are unable to control acute tuberculosis; this susceptibility coincides with decreased expression of iNOS, IFNγ, and TNFα during the early phase of infection (Chan *et al.*, 1996). IL-12-deficient mice are compromised in their ability to mount a Th1 (IFNγ-producing) $CD4^+$ response and demonstrate unrestrained growth of *M. tuberculosis* organisms in conjunction with reduced early iNOS expression following intravenous mycobacterial challenge (Cooper *et al.*, 1997). Thus, it appears that any condition resulting in decreased iNOS expression or RNI production early in infection has a detrimental effect on the ability of mice to control acute tuberculosis. Conversely, RNI production is associated with resistance to *M. tuberculosis*. For example, mice deficient in CD40 are unaffected in their ability to control *M. tuberculosis* infection, and macrophages explanted from the spleens of these mice have only slightly diminished NO production compared with control mice (Campos-Neto *et al.*, 1998).

4. Persistent/Latent Tuberculosis: RNI and Beyond

It has long been recognized that certain intracellular pathogens can persist chronically in the host, well beyond the initial acute phase of infection (Mackowiak, 1984; Krueger and Ramon, 1988; Domingue and Woody, 1997). This unique host–parasite relationship has received various designations, including *persistence*, *chronic persistence*, *dormancy*, *latency*, and *premunition*. While some have proposed strict definitions for these conditions (McCune *et al.*, 1966a), others have used the terms to describe chronic persistence of pathogens rather interchangeably (for examples, see Krueger and Ramon, 1988; Beaman *et al.*, 1995) in the absence of clear molecular and biochemical criteria to define these disease states.

Among the most clinically important microbes capable of establishing latent infection are HIV (Pomerantz *et al.*, 1992), the herpesviruses (Jordan *et al.*, 1984; Krueger and Ramon, 1988; Steiner, 1996), the sporozoan *Toxoplasma gondii* (Beaman *et al.*, 1995), and bacteria including spirochetes (Coyle and Dattwyler, 1990), *Salmonella* (Watson, 1967), and the tubercle bacillus (Wayne, 1994; Flynn *et al.*, 1998). T-cell-mediated immunity is of the utmost importance in the maintenance of this tenuous host–parasite relationship, underscored by the reactivation of infections caused by these pathogens in individuals with AIDS (Blaser and Cohn, 1986; Lane *et al.*, 1994), a disease characterized by a severe quantitative and qualitative deficiency in the cellular immune response. Given the important role of reactivation in tuberculosis (Stead, 1965, 1967; Stead *et al.*, 1968), a disease afflicting 1.75 billion persons worldwide (Murray *et al.*, 1990; Bloom and Murray, 1992), an improved understanding of mechanisms involved in the establishment and reactivation of latent infection is of paramount importance.

4.1. RNI in Latent Tuberculosis

Granger *et al.* (1993) were perhaps the first to propose the potential import of the macrophage L-arginine-dependent cytotoxic pathway in the maintenance of dormant infections. However, several years elapsed before the publication of the first explicit experimental evidence that RNI contribute to the control of latent infection; treatment of *Leishmania major* latently infected mice with NOS inhibitors was found to result in prompt recrudescence of disease (Stenger *et al.*, 1996). Reports on the significance of RNI in experimental latent tuberculous infection appeared soon therafter.

Administration of the NOS inhibitor L-NIL to *M. tuberculosis*-infected mice 40 days after the initiation of infection leads to disease exacerbation during the chronic phase of tuberculosis (MacMicking *et al.*, 1997a). Our laboratory has recently conducted additional studies with the NOS inhibitor AG in two well-characterized murine models (Flynn *et al.*, 1998). Inhibition of RNI production in a low-dose chronic persistent tuberculosis model described by Orme (1988) results in disease recrudescence. Reactivation is manifested by marked hepatosplenomegaly, a rigorous granulomatous response, and increased bacterial burden, especially in the lungs. Importantly, expression of three critical factors for RNI production—IFNγ, TNFα, and iNOS—can be demonstrated throughout the latent phase of infection. The role of RNI in preventing reactivation was confirmed in a second murine tuberculosis model involving treatment with antituberculous drugs (McCune and Tompsett, 1956; McCune *et al.*, 1956, 1966a,b; Grosset, 1978). As in the low dose model, the increase in bacillary burden observed in the drug-treatment model is also most apparent in the lungs. Intriguingly, the kinetics of reactivation-associated bacterial proliferation in the lungs observed in the two latency models are remarkably different (Flynn *et al.*, 1998). The potential significance of this observation will be discussed in the following section.

The expression of IFNγ and TNFα throughout the persistent phase of murine tuberculosis (Flynn *et al.*, 1998) suggests that these cytokines are important in preventing reactivation, at least in part by maintaining iNOS activity. TNFα has been shown previously to play a significant role in the control of chronic murine tuberculosis (Adams *et al.*, 1995). In those studies, adenovirus expressing the 55-kDa TNFα receptor was used to functionally neutralize TNFα *in vivo*. Administration of this recombinant adenovirus to mice infected with *M. tuberculosis* 6 months earlier resulted in striking increases in bacterial numbers in the lungs. Although neither RNI production nor iNOS expression was measured in these mice, it seems likely that functional neutralization of TNFα allowed reactivation of infection by compromising macrophage RNI production. Glucocorticoids inhibit iNOS induction (Radomski *et al.*, 1990; Kunz *et al.*, 1996) (see Chapter 5), and tuberculous reactivation associated with corticosteroids in murine experimental models (McCune *et al.* 1966a; North and Izzo, 1993; Brown *et al.*, 1995; MacMicking *et al.*, 1997a) or in humans (Rook *et al.*, 1987; references in Flynn *et al.*, 1998) might also result from attenuation of RNI production.

In sum, recent studies have provided compelling evidence that RNI contribute significantly to the control of chronic persistent tuberculous infection. However, attempts to extrapolate from these results to human tuberculosis must be viewed with caution, as mechanisms of latency and reactivation in murine models may differ in significant ways from mechanisms of tuberculous persistence in humans. In this regard, a recently developed experimental tuberculosis model using cynomolgus monkeys may eventually help to clarify the role of RNI in primate host defenses against latent *M. tuberculosis* (Walsh *et al.*, 1996).

4.2. iNOS-Independent Antimycobacterial Mechanisms

We have noted that AG treatment during the latent phase of the drug-treatment model of tuberculosis (Flynn *et al.*, 1998) causes an initial increase in pulmonary CFU from $\sim 10^2$/organ to $\sim 10^5$, but the bacillary burden subsequently remains stable for over 80 days despite continued treatment with the NOS inhibitor. This phenomenon was not observed in the low-dose model, in which the pulmonary bacterial load increases from $\sim 6 \times 10^5$ to 10^9 in less than 80 days after initiation of AG treatment. These observations suggest the existence of an iNOS-independent antimycobacterial mechanism(s) operative during the latent phase of tuberculous infection in the drug-treatment model (Flynn *et al.*, 1998). The ability of mice in this model to inhibit proliferation of *M. tuberculosis* in the lungs despite continuous treatment with the NOS inhibitor is probably not related to AG resistance, because this phenomenon is not apparent in the liver or spleen.

Moreover, RNI production by macrophages obtained from these mice during the plateau of pulmonary bacterial burden remains inhibitable by AG *in vitro* (Flynn, unpublished). NO-independent antimicrobial actions of IFNγ (Cooper *et al.*, 1993; Flynn *et al.*, 1993) and TNFα (Flynn *et al.*, 1995) may be responsible for this effect.

The existence of putative iNOS-independent antimycobacterial mechanisms is also suggested when patterns of disease progression in the various mouse models of acute tuberculosis, particularly those using "knockout" strains, are compared (Table III). Mice deficient in IFNγ are exquisitely susceptible to *M. tuberculosis* infection; even with low inocula (e.g., 2×10^4 bacilli i.v.), gko mice succumb to tuberculous infection more quickly than any other strains listed in Table III, including mice with disruption of *iNOS*. This suggests that IFNγ provides protection in part by a mechanism independent of iNOS induction. Tuberculous C57BL/6 mice treated with the NOS inhibitor L-NMMA, whose serum NO_2^-/NO_3^- levels are rendered virtually undetectable (Chan *et al.*, 1995), and whose mean survival time is almost identical to that of the *iNOS*$^{-/-}$ strain, display significantly greater resistance to *M. tuberculosis* compared with gko mice, using time to death as a criterion (Table III). Mice deficient in the 55-kDa TNFα receptor also succumb quickly to *M. tuberculosis* infection, with a mean survival time less than that reported for mice rendered RNI-deficient by gene disruption or treatment with NOS inhibitors (Table III). Thus, as in the case of latent tuberculosis, putative iNOS-independent mechanisms operative during acute infection may be related to IFNγ and TNFα, the two cytokines known to have vital roles in controlling *M. tuberculosis* infection. The principal caveat in making this comparison is that studies involving gko, p55 TNFα receptor-deficient, and NOS inhibitor-treated C57BL/6 mice were conducted in one laboratory (see Table III), while those using *iNOS*$^{-/-}$ mice were performed in another (MacMicking *et al.*, 1997a). Although there is little doubt that iNOS represents a major cytotoxic pathway against *M. tuberculosis*, it is also likely that RNI-independent mechanisms contribute to the control of tuberculosis. Elucidation of such mechanisms will further our understanding of host defense against the tubercle bacillus, and facilitate the design of novel therapeutic and preventive strategies.

5. Can *M. tuberculosis* Escape the Toxic Effects of RNI?*

Freely membrane-permeant (Denicola *et al.*, 1996) and reactive, NO or its various toxic derivatives (Stamler *et al.*, 1992; Stamler, 1994, 1996; Beckman and

* Since the writing of this chapter, the *ahpC* of *M. tuberculosis* Erdman has been reported to protect against RNI toxicity (Chen *et al.*, 1998). In addition, it has been shown that RNI induces the expression of the α-crystallin homolog by *M. tuberculosis* H37Rv (Garbe *et al.*, 1999).

Koppenol, 1996; Fang, 1997) (see also Chapters 3 and 12) can potentially target multiple vital components of microbes. However, microbial pathogens possess complex mechanisms of resistance to the toxic effects of RNI (Fang, 1997) (Chapter 12).

In theory, *M. tuberculosis* might escape the adverse effects of toxic nitrogen oxides by (1) decreasing uptake or increasing export of RNI, (2) reducing target reactivity with RNI or augmenting target synthesis, (3) manufacturing mycobacterial products to restrict the availability of essential components of the NO synthesis pathway, (4) producing NOS inhibitors, and (5) generating NO scavengers.

While evidence that the tubercle bacillus can resist RNI toxicity gradually emerges (O'Brien *et al.*, 1994; Ehrt *et al.*, 1997; Rhoades and Orme, 1997; Yu *et al.*, in press), specific evasion mechanisms have begun to be identified in other microbial species. In *Salmonella*, for example (Chapter 12), specific mechanisms of resistance to *S*-nitrosothiols have been identified (DeGroote *et al.*, 1995, 1996). *S*-Nitrosothiols such as GSNO (Gaston *et al.*, 1993) have apparent biological relevance, with glutathione being present in millimolar quantities in the host (Halliwell and Gutteridge, 1989), and substantial concentrations of *S*-nitrosothiols (Stamler, 1994, 1996) have been detected in various physiological settings. Using a genetic approach, defects in the transport of dipeptide derivatives of GSNO have been shown to impart RNI resistance (DeGroote *et al.*, 1995), suggesting that decreased RNI uptake could be an effective tactic to evade antimicrobial activity of toxic nitrogen oxides. Naturally occurring low-molecular-weight thiols such as glutathione (Fang, 1997) or homocysteine (DeGroote *et al.*, 1996) attenuate RNI toxicity in *Salmonella*, providing examples of resistance mediated by RNI scavengers. Other microbial products as chemically diverse as the phenazine-based pigment pyocyanin of *Pseudomonas aeruginosa* (Warren *et al.*, 1990) or the hemoglobin-derived hemozoin of *Plasmodium* spp. (Slater *et al.*, 1991; Prada *et al.*, 1996) may also function as RNI scavengers. It remains to be tested whether mycothiol [2(*N*-acetylcysteinyl)amido-2-deoxy-α-D-glucopyranosyl-(1 → 1)-myoinositol] (Newton *et al.*, 1996), a major thiol present in *Mycobacterium* spp. at concentrations ranging from 2.7 to 19 μmole per gram of residual dry weight, plays a role in defense against RNI toxicity. Finally, the intriguing observation that *Streptomyces* produces the NOS inhibitor *N*-iminoethyl-L-ornithine (Scannell *et al.*, 1972; Rees *et al.*, 1990) may indicate that microbes can evade RNI toxicity by directly restricting NOS activity.

While specific RNI resistance mechanisms employed by *M. tuberculosis* are yet to be defined, several lines of evidence strongly suggest that the tubercle bacillus possesses means to evade RNI toxicity (Table IV). The susceptibility of mycobacteria to RNI in the form of acidified nitrite varies significantly (O'Brien *et al.*, 1994). Nine strains of tubercle bacilli tested in this report included both clinical and laboratory *M. tuberculosis* isolates, as well as *M. bovis*. Strikingly,

TABLE IV

Evidence for Mycobacterial Resistance to Reactive Nitrogen Intermediates[a]

System	Gene	Characteristics	References
Acidified NO_2^-	Unknown	Wide spectrum of RNI susceptibility	O'Brien *et al.* (1994), Rhoades and Orme (1997)
Acidified NO_2^-	*noxR1*	1. Isolated from clinical strain C of *M. tuberculosis* 2. Confers resistance to the heterologous hosts *E. coli* and *M. smegmatis* 3. Present only in *M. tuberculosis* complex	Ehrt *et al.* (1997)
$ONOO^-$	Unknown	$ONOO^-$-resistant: *M. bovis*, *M. tuberculosis* strains Erdman and C $ONOO^-$-susceptible: *M. smegmatis*, BCG	Yu *et al.* (in press)

[a] See footnote on p. 291.

while the viability of *M. bovis* was not affected by 24-hr exposure to 250 μg/ml sodium nitrite (pH 5.0), the same treatment regimen was lethal to three of the eight *M. tuberculosis* strains examined. More importantly, the *in vitro* RNI resistance of these strains correlates positively with their virulence in guinea pigs, as determined by the rate of development of pathological lesions in organs of infected animals. Therefore, this study demonstrated that tubercle bacilli have variable susceptibility to the toxic effects of nitrogen oxides, and RNI resistance may be linked to virulence *in vivo*. The variability of *in vitro* susceptibility to acidified NO_2^- among different *M. tuberculosis* strains was later confirmed by another report (Rhoades and Orme, 1997). In this case, however, an *in vitro* system using IFNγ-stimulated primary murine bone marrow macrophages did not reveal a correlation between resistance to RNI and the ability of the *M. tuberculosis* to survive within NO-producing phagocytes. Because the two studies employed different systems to assess the biological significance of the RNI-resistance phenotype, reconciliation of these seemingly opposed observations will require further experimentation.

Molecular evidence supporting the existence of RNI-resistance mechanisms in *M. tuberculosis* began to emerge soon after the description of a positive correlation between mycobacterial resistance to RNI and virulence. Garbe *et al.* (1996) reported that *in vitro* treatment of *M. tuberculosis* H37Rv with *S*-nitroso-*N*-acetylpenicillamine (SNAP) induces the expression of eight polypeptides, as analyzed by two-dimensional protein gel electrophoresis. The function and the identity of these SNAP-inducible mycobacterial proteins remain largely uncharacterized,* with the exception of a 28-kDa species that has been designated Hsp (*h*eat *s*hock *p*rotein) 28. Significantly, the proteins elicited by RNI in this study were distinct from those triggered by $O_2^{-\bullet}$ produced by the redox-cycling agent menadione. In the latter case, each of the seven most conspicuously induced species belong to the Hsp family, including the 28-kDa species. This minimal overlap in SNAP- and menadione-induced genes in *M. tuberculosis* seems to suggest that the tubercle bacillus responds to reactive oxygen and nitrogen species via distinct regulatory elements. This finding is in apparent contrast to *E. coli*, in which both NO and $O_2^{-\bullet}$ are sensed via the *soxRS* regulon (Nunoshiba *et al.*, 1993), although additional regulatory systems induced by RNI may remain to be identified (Fang, 1997).

The most detailed molecular characterization of *M. tuberculosis* RNI-resistance mechanisms to date concerns *noxR1*, a novel antioxidant gene that confers resistance to the inhibitory effects of GSNO on the heterologous organisms *E. coli* and *Mycobacterium smegmatis* (Ehrt *et al.*, 1997). Expression of *noxR1* is also protective against the toxic effects of various reactive oxygen species in the same system (Ehrt *et al.*, 1997). Significantly, *noxR1* was cloned from a highly prevalent clinical *M. tuberculosis* isolate in New York City (Friedman *et al.*, 1997), and detected by Southern analysis only in members of the *M. tuberculosis* complex among 11 strains of mycobacteria examined. Molecular analysis predicts that *noxR1* encodes a 152-amino-acid protein. Evaluation of *noxR1*-expressing *E. coli* mutants deficient in the ROI- and RNI-responsive *oxyR* (reviewed in Storz *et al.*, 1990a,b) and *soxRS* loci (reviewed in Demple, 1991; Demple and Amabile-Cuevas, 1991) suggests that the antioxidant effects of NoxR1 are independent of *oxyR* and *soxRS*.

Recently, virulent mycobacteria including the *M. bovis* Ravanel strain, as well as the laboratory Erdman strain and the highly prevalent clinical C isolate of *M. tuberculosis* (Friedman *et al.*, 1997) have been found to be markedly resistant to peroxynitrite anion ($ONOO^-$) compared with relatively avirulent *M. smegmatis* and BCG (Yu *et al.*, in press). $ONOO^-$, a product of the reaction between $NO^\bullet$ and $O_2^{-\bullet}$ (rate constant: $6.7 \times 10^9\ M^{-1}\ sec^{-1}$; Huie and Padmaja, 1993), is a potent oxidant that has been implicated in a wide array of biochemical reactions and pathophysiological processes (reviewed in Beckman *et al.*, 1994a; Beckman and Koppenol, 1996) (see Chapter 8). Notably, rodent alveolar macrophages can be stimulated to produce significant amounts of $ONOO^-$ (Ischiropoulos *et al.*, 1992). As an antimicrobial effector molecule, $ONOO^-$ is toxic to *E. coli* (Zhu *et al.*, 1992), *Trypanosoma cruzi* (Denicola *et al.*, 1993; Rubbo *et al.*, 1994), and *Candida albicans* (Vazquez-Torres *et al.*, 1996) in *in vitro* studies. However, *Leishmania*

*See footnote on p. 291.

major (Assreuy *et al.*, 1994) and virulent mycobacterial strains are resistant to the toxic effects of $ONOO^-$ (Yu *et al.*, in press).

Together, these studies provide evidence suggesting that *M. tuberculosis* has developed mechanisms to evade or resist toxic nitrogen oxides. The significance of such evasion mechanisms in *M. tuberculosis* virulence remains to be rigorously tested. This will probably not be a straightforward task, given the intricacy of the host defense mechanisms against *M. tuberculosis* and the complex array of possible RNI-related reactions in biological systems.

6. RNI–*M. tuberculosis* Interactions

Given the significance of the L-arginine-dependent cytotoxic pathway in host defense against the tubercle bacillus, identifying distinct reactive nitrogen species that react with specific bacterial targets would seem likely to illuminate both pathogenetic and antimycobacterial mechanisms in tuberculosis. Such investigations could also yield novel therapeutic agents. However, biochemical and molecular details of RNI–*M. tuberculosis* interactions presently remain obscure. Our ignorance in this area largely results from intrinsic technical difficulties in working with *M. tuberculosis* and RNI. The complexity of RNI chemistry has been the subject of many recent reviews (reviewed in Stamler *et al.*, 1992; Stamler, 1994, 1996; Beckman and Koppenol, 1996) (see also Chapter 3). Less touted is the arduous nature of research with the tubercle bacillus, an airborne human pathogen that requires high-level biosafety containment. Moreover, its propensity to clump presents a considerable obstacle to biological experimentation, including genetic manipulation, and assessment of growth inhibition and killing. The long doubling time (~24 hr) of *M. tuberculosis* (Harshey and Ramakrishnan, 1976; Hiriyanna and Ramakrishnan, 1986) imposes a hiatus of 3 to 4 weeks between experiments that require growth of visible colonies. The complex cell wall of the tubercle bacillus (Brennan *et al.*, 1990; Besra and Chatterjee, 1994; Brennan and Draper, 1994) has considerably delayed the biochemical (reviewed in Wheeler and Ratledge, 1994) and genetic (Jacobs *et al.*, 1987) characterization of this pathogen.

6.1. RNI-Generating Systems

The discovery in the late 1980s of mammalian NO synthesis by cells of the neurological, immunological, and vascular systems (reviewed in Nathan, 1992) triggered intensive investigation of the reaction of RNI in biological systems. But research activity aimed at characterizing the interaction between RNI and various microorganisms had existed years before the discovery of immunologically generated NO (see Tarr, 1941, and references therein). These research efforts were stimulated by the findings that NO_2^- is an effective antimicrobial agent in meat curing, and yielded much information concerning the characteristics of this nitrogen oxide as an RNI-generating system (Tarr, 1941; Castellani and Niven, 1955; Shank *et al.*, 1961). The bacteriostatic activity of NO_2^- was reported to be significantly altered by the nature and the composition of the medium in which the experiments were carried out. For example, these studies demonstrated that oxygen, acidity, glucose, buffering agents, and low-molecular-weight thiols each have the ability to modulate the antimicrobial effects of NO_2^-. In addition, it was also appreciated that NO_2^- acidification results in a series of complex chemical reactions that yield multiple reactive nitrogen species including NO, $NO_2^\bullet$, HNO_2, and N_2O_3 (Castellani and Niven, 1955; Shank *et al.*, 1961), each endowed with unique chemical properties. These observations indicate that the seemingly well-defined acidified nitrite system so widely employed in the study of RNI–microbe interactions is, in fact, confounded by a plethora of variables.

Another approach commonly used to examine the toxic effects of nitrogen oxides on microbes utilizes a class of compounds called *NO donors*. The carrier molecules for NO are, in general, nucleophiles (Hanson *et al.*, 1995; Smith *et al.*, 1996) or thiols (see Mathews and Kerr, 1989, and references therein). As in the case of nitrite, the antimicrobial effects of nitrosothiols have long been known (Incze *et al.*, 1974). These molecules are, however, not without their own confounding features, resulting from the presence of specific carrier components, the complexity of their decomposition chemistry, and the ability of donated nitrogen monoxide to exist in a variety of redox forms: NO^+ (nitrosonium ion), $NO^\bullet$, and NO^- (nitroxyl anion) (reviewed in Stamler *et al.*, 1992; Stamler, 1996). The implication of a role for γ-glutamyl transpeptidase in the internalization of GSNO (DeGroote *et al.*, 1995; Hogg *et al.*, 1997), as well as the possible existence of stereoselective nitrosothiol receptors (Davisson *et al.*, 1996, 1997; Travis *et al.*, 1997) in biological systems, further complicates matters. Finally, as in the acidified NO_2^- RNI-generating system, the uniformity of the experimental milieu generated by NO donors can also be under the influence of multiple factors including the chemical nature of the buffer used, the presence of thiols, contaminating trace metals, and oxygen (Singh *et al.*, 1996; Stamler, 1996), to name just a few. The complexity of NO-donor chemistry is well illustrated by the fate of GSNO in the presence of GSH (Singh *et al.*, 1996). This physiologically relevant interaction yields oxidized glutathione and multiple nitrogenous compounds, including $NO^\bullet$, N_2O_3, NH_3, NO_2^-, and N_2O. The quantities of these products are dependent on the concentrations of oxygen and glutathione, as well as phosphate (Singh *et al.*, 1996).

6.2. Whodunit?

The complexity of commonly used RNI-generating systems contributes significantly to uncertainty regarding the effector side of RNI–*M. tuberculosis* interactions. Nevertheless, the use of acidified NO_2^- and NO-donor systems to study effects of nitrogen oxides on the tubercle bacillus has been informative. As mentioned earlier, acidified NO_2^- can be used to demonstrate the *in vitro* antimycobacterial activity of RNI (Chan *et al.*, 1992; O'Brien *et al.*, 1994; Rhoades and Orme, 1997), and *M. tuberculosis* strains vary widely in their susceptibility (O'Brien *et al.*, 1994; Rhoades and Orme, 1997). The antimycobacterial effects of RNI have also been examined *in vitro* using various nucleophile-based NO donors (Hanson *et al.*, 1995; Smith *et al.*, 1996). Spermine–NO (SPER/NO) and diethylamine–NO (DEA/NO) adducts effectively kill *M. smegmatis* in Kreb Ringer's phosphate buffer (Tsang and Chan, unpublished), as assessed by CFU quantitation. Metabolic labeling studies using [^{3}H]uracil also indicate that methoxymethyl-DEA/NO (MOM-DEA/NO), SPER/NO, and diethylenetriamine–NO adducts each exhibit inhibitory effects against *M. smegmatis* and *M. bovis* BCG (Tsang and Chan, unpublished). Along with the *S*-nitrosothiol GSNO, acidified NO_2^- has been used to identify the *M. tuberculosis noxR1* gene that confers RNI resistance in otherwise susceptible heterologous hosts (Ehrt *et al.*, 1997). The *S*-nitrosothiol SNAP has been used to characterize RNI-responsive gene expression in *M. tuberculosis* (Garbe *et al.*, 1996).

One particular NO-derived reactive species that exhibits antimycobacterial activity is $ONOO^-$ (Yu *et al.*, in press). In a series of experiments, the short half-life of this potent oxidant [$\sim$1.9 sec in phosphate buffer, pH 7.4, 37°C (Beckman *et al.*, 1990; Zhu *et al.*, 1992)] was exploited to demonstrate that this effect is attributable to $ONOO^-$ itself. $ONOO^-$ effectively kills *M. smegmatis* and BCG in phosphate buffer *in vitro*. In contrast, the viability of mycobacteria exposed to equivalent amounts of $ONOO^-$ after spontaneous degradation in phosphate buffer is not affected. Metal chelators and hydroxyl radical scavengers had no effect on the antimycobacterial activity of $ONOO^-$, further indicating that $ONOO^-$ was responsible for the observed cytocidal effects. Contrasting with the susceptible phenotype of avirulent species, virulent *M. tuberculosis* and *M. bovis* isolates are $ONOO^-$ resistant; this warrants further investigation. Recently, we have obtained data suggesting that NO-saturated phosphate-buffered saline, authentic gaseous $NO^\bullet$ or $NO_2^\bullet$ exerts mycobactericidal activity against the virulent *M. tuberculosis* Erdman strain, as well as *M. smegmatis*, and BCG (Yu *et al.*, in press) under highly stringent anaerobic conditions, although even such pure nitrogen oxide preparations may represent complex mixtures of RNI (Yu *et al.*, in press).

Notwithstanding the useful information provided by studies to date, they have not revealed the specific effector RNI species responsible for antituberculous activity *in vivo*. As our knowledge in NO-based drug design strategies expands (Hanson *et al.*, 1995), identification of the specific RNI that mediate inhibition or killing of the tubercle bacillus could assume therapeutic importance.

6.3. The Targets

Reactive nitrogen oxides are a remarkably versatile class of reactants in biological systems (reviewed in Stamler *et al.*, 1992; Stamler, 1994, 1996; Beckman and Koppenol, 1996; Fang, 1997). By virtue of their abilities to mutagenize DNA, oxidize lipids, and react with biologically important functional groups including sulfhydryls, tyrosine residues, heme and nonheme iron, and iron–sulfur proteins, RNI can target virtually any class of biological macromolecules (reviewed in Chapters 3, 8, and 12). Little is presently known about the *M. tuberculosis* components targeted by reactive nitrogen oxides *in vivo*. Identification of such mycobacterial targets is likely to be a difficult task because of (1) the large number of potential targets, (2) the lack of knowledge of the specific RNI responsible for antimycobacterial activity, and (3) the complexity of disease mechanisms in tuberculous infection. The last point is particularly relevant to extending our understanding of RNI–*M. tuberculosis* interactions beyond mere chemistry, to include a perspective that emphasizes host defense and pathogenesis. It is likely that the biochemistry of the tubercle bacillus at various stages of disease—acute, persistent/latent, and reactivation—features unique reactions. New targets may become available for engaging in reactions with RNI as the tubercle bacillus makes transitions from one phase of infection to the next. This scenario, together with the relatively uncharacterized biochemistry of *M. tuberculosis* during various phases of infection and the lack of *in vitro* models to closely simulate specific disease states, poses a significant challenge to identifying mycobacterial targets of RNI.

Some RNI microbial targets have been identified in other systems (see Chapter 12). For example, *S*-nitrosothiols have been shown to prevent *Bacillus cereus* sporulation by targeting germination-specific, surface-associated, sulfhydryl groups via covalent modification (Morris and Hansen, 1981). These sulfhydryl groups were later shown to reside predominately in three membrane proteins (13, 28, and 29 kDa) of germinating *B. cereus*, and to be necessary for the optimal activity of lactose and dicarboxylic acid permeases (Morris *et al.*, 1984). These observations encouragingly suggest that the stage-specific microbial targets can be identified. In another study designed to examine the role of B-lymphocyte-derived RNI in maintaining Epstein–Barr virus (EBV) latency (Mannick *et al.*, 1994), Zta, a redox-regulated EBV transactivator controlling the transition from latent to lytic infection, was found to be downregulated by RNI. The investigators proposed that RNI may maintain viral latency via direct *S*-nitrosylation of Zta at a critical thiol contained in the DNA-binding domain, or by fostering the formation of the intermolecular disulfide bridges (Mannick *et al.*, 1994). This intriguing hypothesis underscores the possibility that highly specific interactions between RNI and microbial components could mediate biological phenomena of importance in the context of pathogenesis and host defense. iNOS is expressed in granulomata during persistent/latent murine tuberculosis and contributes to the prevention of disease reactivation (Flynn *et al.*, 1998); thus, *S*-nitrosylation of specific mycobacterial proteins during this important phase of infection could play a role in the maintenance of latency.

Nevertheless, while specific microbial targets of RNI have been implicated in a few elegant examples, the plethora of reactive nitrogen species and putative targets makes the establishment of biologically relevant targets difficult in most instances. For example, the inhibitory effects of RNI on *Clostridium* spp. have been variously attributed to (1) destruction of enzymatic iron–sulfur clusters , especially ferredoxin (Reddy *et al.*, 1983), (2) inhibition of the ATP-generating iron–sulfur center-containing phosphoroclastic system that oxidizes pyruvate to acetate (Woods *et al.*, 1981), and (3) attenuation of sulfhydryl-containing glycolytic enzymes such as glyceraldehyde-3-phosphate dehydrogenase and aldolase (O'Leary and Solberg, 1976). Each of the proposed targets is a legitimate reactant with RNI, but it is difficult to say whether any or all are biologically relevant. A simple mechanistic explanation for the antimycobacterial activity of RNI may not exist.

7. RNI in Human Tuberculosis

Although well established as a key antimicrobial factor in a wide variety of rodent infectious disease models, the role of iNOS in host defense against microorganisms in humans remains controversial.* This not unreasonable skepticism has been largely fueled by the inability to reproducibly demonstrate *in vitro* high-output iNOS activity in explanted human macrophages. Ironically, the first lead that prompted the detailed investigation of immunological induction of NO synthesis in rodents came from the observation that an individual with a febrile illness of probable infectious etiology produced large amounts of nitrate in the urine (Wagner and Tannenbaum, 1982; see references in MacMicking *et al.*, 1997b). Therefore, the pursuit of this lead, which has led to the discovery of macrophage iNOS and the establishment of RNI as a class of effective antimicrobials, has yet to come full circle. Nevertheless, the *in vitro* expression of human macrophage iNOS, associated on occasion with high-output enzymatic activity, has been reported by various laboratories (reviewed in Denis, 1994; MacMicking *et al.*, 1997b; Nathan, 1997; see also Chapter 6). A defined *in vitro* system conducive to high-output NO production by human macrophages remains elusive. Until the establishment of such a system, some controversy revolving around the human macrophage–NO issue is likely remain.

In contrast to the lack of a reliable *in vitro* human macrophage system that supports high-output RNI production, *in vivo* evidence suggesting that the

* Additional evidence for a role of NO in defense against *M. tuberculosis* in humans has been reported after the completion of this Chapter (Wang *et al.*, 1998)

induction of iNOS function occurs in certain human diseases, particularly those associated with infection and inflammation, has been more forthcoming (reviewed in Macmicking *et al.*, 1997b). Among the first of such evidence was the fact that individuals with sepsis and those receiving cytokine therapy produce RNI in abundance (Ochoa *et al.*, 1991, 1992; Hibbs *et al.*, 1992). The inference of these observations was further reinforced by the cloning of iNOS from a variety of human cell types (reviewed in Nathan, 1992; Nathan and Xie, 1994). More importantly, immunohistochemical studies have demonstrated the presence of iNOS in inflamed human pulmonary tissues (Kobzik *et al.*, 1993), as well as in macrophages or monocytes obtained from individuals afflicted with a variety of inflammatory diseases (reviewed in MacMicking *et al.*, 1997b). By the same approach, the detection of the NO-derived $ONOO^-$ using nitrotyrosine-specific antibodies in lung tissues of humans with sepsis, pneumonia, or adult respiratory disease syndrome (reviewed in Beckman and Koppenol, 1996), as well as in and around macrophages in atherosclerotic lesions (Beckman *et al.*, 1994b) infers the *in situ* expression of iNOS function in these pathological states. Directly relevant to *M. tuberculosis*, a high degree of iNOS expression in pulmonary macrophages obtained by bronchoalveolar lavage from individuals infected with the tubercle bacillus was detected using a highly specific antibody (Nicholson *et al.*, 1996): While an average of 65% of the alveolar macrophages obtained from the 11 tuberculous patients studied expressed readily detectable enzyme, only 10% of those from healthy donors stained positive. In addition, inflammatory human pulmonary macrophages have the ability to restrict BCG growth by the infection-induced iNOS activity (Nozaki *et al.*, 1997). A recent report (Bonecini-Almeida *et al.*, 1998) describes a correlation between antituberculous activity and human macrophages cocultured with lymphocytes and IFNγ. Finally, because IFNγ and IL-12 participate in the induction of iNOS activity, the observation that individuals with mutations in IFNγ receptor (Jouanguy *et al.*, 1996; Newport *et al.*, 1996) or abnormal IL-12 P40 production (Drysdale *et al.*, 1997) exhibit increased susceptibility to mycobacterial infection further suggests a role for RNI in host defense against human mycobacterial infection.

Collectively, these studies have provided compelling evidence that human macrophage iNOS activity is inducible, particularly in the setting of infection and inflammation, and that RNI play a role in defense against the tubercle bacillus in humans. Although there is little doubt that the macrophage iNOS expressed in these clinical settings is functional, formal proof for enzymatic activity at a capacity of high NO output is still lacking. For now, the induction of high-output iNOS activity in human macrophages remains a mystery. Given the large number of biological factors that can downregulate iNOS activity in the mouse (reviewed in MacMicking *et al.*, 1997b), and the significant differences between the cytokines required for optimal induction of iNOS in human hepatocytes (Nussler *et al.*, 1992; Geller *et al.*, 1993) or in murine macrophages (Ding *et al.*, 1988), it is not unreasonable to assume that this synthase can be functionally induced in human macrophages *in vitro* once the correct combination of triggering molecules is known. Functional analysis of the highly complex promoter of the human enzyme (de Vera *et al.*, 1996), which differs substantially from that of mouse iNOS (Lowenstein *et al.*, 1993; Xie *et al.*, 1993), may help to define this combination.

8. Beyond the Effector : Target Equation

Solving the effector:target equation of the RNI–*M. tuberculosis* reaction is likely to illuminate important mechanisms of host defense and pathogenesis in tuberculosis. Of practical importance, the solution may bear on new therapeutic strategies: With advances in NO-based drug design (Hanson *et al.*, 1995), it is not inconceivable that regional delivery of customized, nebulizable NO prodrugs into tuberculous lungs may be able to target the intracellular compartments where bacilli reside. In addition, understanding the mechanisms by which *M. tuberculosis* evades RNI toxicity and identifying mycobacterial components with which nitrogen oxides react can reveal novel drug targets. The impact of NO and its derivatives on tuberculosis as a disease goes beyond the RNI–*M. tuberculosis* reaction per se. For example, the mechanism by which NO production is regulated during the evolution of tuberculous infection *in vivo* is virtually unknown. Does the hypoxia-responsive element (Melillo *et al.*, 1995) in the iNOS promoter play a role in the maintenance of continuous RNI generation in the hypoxic environment (Rich, 1944) within a granuloma? Although direct measurement of the oxygen tension of a tubercle has yet to be undertaken, Loebel *et al.* (1933) calculated that the center of a tubercle with a radius of 0.075 to 0.35 mm should completely lack oxygen. Equally obscure is the role of NO in affecting the nature of the tuberculous granuloma: necrotizing versus fibrotic (reviewed in Dannenberg and Rook, 1994; Rook and Bloom, 1994), either of which may be fostered by NO (Heck *et al.*, 1992; Rojas *et al.*, 1997; Schaffer *et al.*, 1997; Shearer *et al.*, 1997; reviewed in Nathan and Xie, 1994; Nathan, 1997). If these opposing outcomes of tuberculous lesions represent NO wielding its notorious double-edged sword (Nathan and Xie, 1994; Nathan, 1997), mechanistic explanations behind the selective development of immunoprotection and immunopathology in tuberculosis [the latter so strikingly exemplified by the Koch phenomenon (Anderson, 1891; Koch, 1891)], and the genetic controls of NO production may be forthcoming. With respect to the last point, the role of *Nramp1* in modulating iNOS function (reviewed in Nathan, 1995) is of particular interest, in view of the recent demonstration that polymorphisms in the 3′ untranslated region and intron 4 of *Nramp1* are associated with increased susceptibility to *M. tuberculosis* in humans (Bellamy *et al.*, 1998).

Clearly, much remains to be learned about NO in *M. tuberculosis* infection. It is hoped that expanding our knowledge of the relationship between NO, the human host, and *M. tuberculosis* will ultimately impact on the prevention, control, and treatment of tuberculosis. This impact is not trivial when one considers that one-third of the world's population is infected with the tubercle bacillus (Murray *et al.*, 1990; Bloom and Murray, 1992).

ACKNOWLEDGMENT. The authors thank Larry Keefer for helpful comments.

References

Adams, L. B., Mason, C. M., Kolls, J. K., Scollard, D., Krahenbuhl, J. L., and Nelson, S., 1995, Exacerbation of acute and chronic murine tuberculosis by administration of a tumor necrosis factor receptor-expressing adenovirus, *J. Infect. Dis.* **171**:400–405.

Adams, L. B., Dinauer, M. C., Morgenstern, D., and Krahenbuhl, J. L., 1997, The role of reactive oxygen and nitrogen intermediates in the host response to *Mycobacterium tuberculosis*, in: *ASM Conference on Tuberculosis: Past, Present, and Future*, Copper Mountain, Colorado, p.22.

Anderson, M. C., 1891, On Koch's treatment, *Lancet* **1**:651–652.

Anthony, L. S., Chatterjee, D., Brennan, P. J., Nano, F. E., 1994, Lipoarabinomannan from Mycobacterium tuberculosis modulates the generation of reactive nitrogen intermediates by gamma interferon-activated macrophages, *FEMS Immunol. Medical Microbiol.* **8**:299–305.

Assreuy, J., Cunha, F.Q., Epperlein, M., Noronha-Dutra, A., O'Donnell, C. A., Liew, F. Y., and Moncada, S., 1994, Production of nitric oxide and superoxide by activated macrophages and killing of *Leishmania major*, *Eur. J. Immunol.* **24**:672–676.

Barnes, P. F., Modlin, R. L., and Ellner, J. J., 1994, T-cell responses and cytokines, in: *Tuberculosis: Pathogenesis, Protection, and Control* (B. R. Bloom, ed.), American Society for Microbiology, Washington, D.C., pp. 417–436.

Beaman, M. H., McCabe, R. E., Wong, S.-Y., and Remington, J. S., 1995, *Toxoplasma gondii*, in: *Principles and Practice of Infectious Diseases* (G. L. Mandell, J. E. Bennett, and R. Dolin, eds.), Churchill Livingstone, New York, pp. 2455–2475.

Beckman, J. S., and Koppenol, W. H., 1996, Nitric oxide, superoxide, and peroxynitrite: The good, the bad, and the ugly, *Am. J. Physiol.* **271**:C1424-C1437.

Beckman, J. S., Beckman, T. W., Chen, J., Marshall, P. A., and Freeman, B. A., 1990, Apparent hydroxyl radical production by peroxynitrite: Implications for endothelial injury from nitric oxide and superoxide, *Proc. Natl. Acad. Sci. USA* **87**:1620–1624.

Beckman, J. S., Chen, J., Ischiropoulos, H., and Crow, J. P., 1994a, Oxidative chemistry of peroxynitrite, *Methods Enzymol.* **233**:229–240.

Beckman, J.S., Ye, Y. Z., Anderson, P. G., Chen, J., Accavitti, M. A., Tarpey, M. M., and White, C. R., 1994b, Extensive nitration of protein tyrosines in human atherosclerosis detected by immunohistochemistry, *Biol. Chem. Hoppe-Seyler* **375**:81–88.

Bellamy, R., Ruwende, C., Corrah, T., McAdam, K. P. W. J., Hilton, C. W., and Hill, A. V. S., 1998, Variations in the *Nramp1* gene and susceptibility to tuberculosis in west Africans, *N. Engl. J. Med.* **338**:640–644.

Besra, G. S., and Chatterjee, D., 1994, Lipids and carbohydrates of *Mycobacterium tuberculosis*, in: *Tuberculosis: Pathogenesis, Protection, and Control* (B. R. Bloom, ed.), American Society for Microbiology, Washington, D.C., pp. 285–306.

Blaser, M. J., Cohn, D. L., 1986, Opportunistic infections in patients with AIDS: Clues to the epidemiology of AIDS and the relative virulence of pathogens, *Rev. Infect. Dis.* **8**:21–30.

Bloom, B. R., and Murray, C. J. L., 1992, Tuberculosis: Commentary on a reemergent killer, *Science* **257**:1055–1063.

Bonecini-Almeida, M. G., Chitale, S., Boutsikakis, I., Geng, J. Y., Doo, H., He, S. H., and Ho, J. L., 1998, Induction of *in vitro* human macrophage anti-*Mycobacterium tuberculosis* activity—Requirement for IFN-γ and primed lymphocytes, *J. Immunol.* **160:**4490–4499.

Brennan, P. J., and Draper, P., 1994, Ultrastructure of *Mycobacterium tuberculosis*, in: *Tuberculosis: Pathogenesis, Protection, and Control* (B. R. Bloom, ed.), American Society for Microbiology, Washington, D.C., pp. 271–284.

Brennan, P. J., Hunter, S. W., McNeil, M., Chatterjee, D., and Daffe, M., 1990, Reappraisal of the chemistry of mycobacterial cell walls, with a view to understanding the roles of individual entities in disease processes, in: *Microbial Determinants of Virulence and Host Response* (E. M. Ayoub, G. H. Cassell, W. C. Branche, Jr., and T. J. Henry, eds.), American Society for Microbiology, Washington, D.C., pp. 55–75.

Brown, D. H., Miles, B. A., and Zwilling, B. S., 1995, Growth of *Mycobacterium tuberculosis* in BCG-resistant and -susceptible mice: Establishment of latency and reactivation, *Infect. Immun.* **63:**2243–2247.

Campos-Neto, A., Ovendale, P., Bement, T., Koppi, T. A., Ranslow, W. C., Rossi, M. A., Alderson, M. R., 1998, CD40 ligand is not essential for the development of cell-mediated immunity and resistance to *Mycobacterium tuberculosis*, *J. Immunol.* **160:**2037–41.

Castellani, A. G., and Niven, C. F., Jr., 1955, Factors affecting the bacteriostatic action of sodium nitrite, *Appl. Microbiol.* **3:**154–159.

Chan, J., and Kaufmann, S. H. E., 1994, Immune mechanisms of protection, in: *Tuberculosis: Pathogenesis, Protection, and Control* (B. R. Bloom, ed.), American Society for Microbiology, Washington, D.C., pp. 389–415.

Chan, J., Xing, Y., Magliozzo, R. S., and Bloom, B. R., 1992, Killing of virulent *Mycobacterium tuberculosis* by reactive nitrogen intermediates produced by activated murine macrophages, *J. Exp. Med.* **175:**1111–1122.

Chan, J., Tanaka, K., Carroll, D., Flynn, J. L., and Bloom, B. R., 1995, Effects of nitric oxide synthase inhibitors on murine infection with *M. tuberculosis*, *Infect. Immun.* **63:**736–740.

Chan, J., Tian, Y., Tanaka, K., Tsang, M. S., Yu, K., Salgame, P., Carroll, D., Kress, Y., Teitelbaum, R., and Bloom, B. R., 1996, Effects of protein calorie malnutrition on tuberculosis in mice, *Proc. Natl. Acad. Sci. USA* **93:**14857–14861.

Chanock, S. J., el Benna, J., Smith, R. M., and Babior, B. M., 1994, The respiratory burst oxidase, *J. Biol. Chem.* **269:**24519–24522.

Chen, L., Xie, Q.-W., Nathan, C., 1998, Alkyl hydroperoxide reductase subunit C protects bacterial and human cells against reactive nitrogen intermediates. *Mol. Cell* **1:**795–805.

Coyle, P. K., and Dattwyler, R., 1990, Spirochetal infection of the central nervous system, *Infect. Dis. Clin. North Am.* **4:**731–746.

Cooper, A. M., Dalton, D. K., Stewart, T. A., Griffin, J. P., Russell, D. G., and Orme, I. M., 1993, Disseminated tuberculosis in IFN-γ gene-disrupted mice, *J. Exp. Med.* **178:**2243–2248.

Cooper, A. M., Magram, J., Ferrante, J., Orme, I. M., 1997, Interleukin 12 (IL-12) is crucial to the development of protective immunity in mice intravenously infected with mycobacterium tuberculosis, *J. Exp. Med.* **186:**39–45.

Curnutte, J. T., 1993, Chronic granulomatous disease: The solving of a clinical riddle at the molecular level, *Clinical Immunol. Immunopathol.* **67:**S2–S15.

Dalton, D., Pitts-Meek, S., Keshav, S., Figari, I. S., Bradley, A., and Stewart, T. A., 1993, Multiple defects of immune cell function in mice with disrupted interferon-γ genes, *Science* **259:**1739–1742.

Dannenberg, A. M., Jr., and Rook, G. A. W., 1994, Pathogenesis of pulmonary tuberculosis: An interplay of tissue-damaging and macrophage-activating immune responses—dual mechanisms that control bacillary multiplication, in: *Tuberculosis: Pathogenesis, Protection, and Control* (B. R. Bloom, ed.), American Society for Microbiology, Washington, D.C., pp. 459–484.

Davisson, R.L., Travis, M. D., Bates, J. N., and Lewis, S. J., 1996, Hemodynamic effects of L-and D-*S*-nitrosocysteine in the rat. Stereoselective *S*-nitrosothiol recognition sites, *Circ. Res.* **79:**256–262.

Davisson, R. L., Travis, M. D., Bates, J. N., Johnson, A. K., and Lewis, S. J., 1997, Stereoselective actions of *S*-nitrosocysteine in central nervous system of conscious rats, *Am. J. Physiol.* **272:**H2361–H2368.

DeGroote, M. A., Granger, D., Xu, Y., Campbell, G., and Prince, R., 1995, Genetic and redox determinants of nitric oxide cytotoxicity in a *Salmonella typhimurium* model, *Proc. Natl. Acad. Sci. USA* **92:**6399–6403.

DeGroote, M. A., Testerman, T., Xu, Y., Stauffer, G., and Fang, F. C., 1996, Homocysteine antagonism of nitric oxide-related cytostasis in *Salmonella typhimurium*, *Science* **272:**414–417.

Demple, B., 1991, Regulation of bacterial oxidative stress genes, *Annu. Rev. Genet.* **25:**315–337.

Demple, B., and Amabile-Cuevas, C. F., 1991, Redox redux: The control of oxidative stress responses, *Cell* **67:**837–839.

Denicola, A., Rubbo, H., Rodriguez, D., and Radi, R., 1993, Peroxynitrite-mediated cytotoxicity to *Trypanosoma cruzi*, *Arch. Biochem. Biophys.* **304:**279–286.

Denicola, A., Souza, J. M., Radi, R., and Lissi, E., 1996, Nitric oxide diffusion in membranes determined by fluorescence quenching, *Arch. Biochem. Biophys.* **328:**208–212.

Denis, M., 1991a, Interferon-gamma-treated murine macrophages inhibit growth of tubercle bacilli via the generation of reactive nitrogen intermediates, *Cell. Immunol.* **132:**150–157.

Denis, M., 1991b, Tumor necrosis factor and granulocyte macrophage colony-stimulating factor stimulate human macrophages to restrict growth of virulent *Mycobacterium avium* and to kill avirulent *M. avium:* Killing effector mechanism depends on the generation of reactive nitrogen intermediates, *J. Leukoc. Biol.* **49:**380–387.

Denis, M., 1994, Human monocytes/macrophages: NO or no NO? *J. Leukoc. Biol.* **55:**682–684.

de Vera, M. E., Shapiro, R. A., Nussler, A. K., Mudgett, J. S., Simmons, R. L., Morris, S. M., Billiar, T. R., and Geller, D. A., 1996, Transcriptional regulation of human inducible nitric oxide synthase (NOS2) gene by cytokines: Initial analysis of the human NOS2 promoter, *Proc. Natl. Acad. Sci. USA* **93:**1054–1059.

Ding, A.H., Nathan, C. F., and Stuehr, D. J., 1988, Release of reactive nitrogen intermediates and reactive oxygen intermediates from mouse peritoneal macrophages. Comparison of activating cytokines and evidence for independent production, *J. Immunol.* **141:**2407–2412.

Doi, T., Ando, M., Akaike, T., Suga, M., Sato, K., and Maeda, H., 1993, Resistance to nitric oxide in *Mycobacterium avium* complex and its implication in pathogenesis, *Infect. Immun.* **61:**1980–1989.

Domingue, G. J., Sr., and Woody, H.B., 1997, Bacterial persistence and expression of disease, *Clin. Microbiol. Rev.* **10:**320–344.

Drysdale, P., Lammas, D. A., Harris, J., Quibell, K., Kumararantne, D. S., Schoel-Tollner, D., Girdlestone, J., Wadhwa, M., Bilger, P., Dockrell, H., Britten, K., and Segal, A., 1997, An abnormality in IL–12 P40 production in three patients with disseminated intracellular infections, in: *ASM Conference on Tuberculosis: Past, Present, and Future*, Copper Mountain, Colorado, p. 32.

Ehrt, S., Shiloh, M. U., Ruan, J., Choi, M., Gunzburg, S., Nathan, C., Xie, Q.-W., and Riley, L.W., 1997, A novel antioxidant gene from *Mycobacterium tuberculosis*, *J. Exp. Med.* **186:**1885–1896.

Evans, T. G., Reed, S. S., and Hibbs, J. B., Jr., 1996, Nitric oxide production in murine leishmaniasis: Correlation of progressive infection with increasing systemic synthesis of nitric oxide, *Am. J. Trop. Med. Hyg.* **54:**486–489.

Fang, F.C., 1997, Mechanisms of nitric oxide-related antimicrobial activity, *J. Clin. Invest.* **99:**2818–2825.

Flesch, I. E. A., and Kaufmann, S. H. E., 1991, Mechanisms involved in mycobacterial growth inhibition by gamma interferon-activated bone marrow macrophages: Role of reactive nitrogen intermediates, *Infect. Immun.* **59:**3213–3218.

Flynn, J. L., Chan, J., Triebold, K. J., Dalton, D. K., Stewart, T. A., and Bloom, B. R., 1993, An essential role for IFN-γ in resistance to *Mycobacterium tuberculosis* infection, *J. Exp. Med.* **178:**2249–2254.

Flynn, J. L., Goldstein, M. M., Chan, J., Triebold, K. J., Pfeffer, K., Lowenstein, C. J., Schreiber, R., Mak, T. W., and Bloom, B. R., 1995, Tumor necrosis factor is required in the protective immune response against *Mycobacterium tuberculosis* in mice, *Immunity* **2:**561–572.

Flynn, J., Scanga, C. A., Tanaka, K., and Chan, J., 1998, Reactivation of latent murine tuberculosis by inhibition of inducible nitric oxide synthase, *J. Immunol.* **160:**1796–1803.

Friedman, C. R., Quinn, G. C., Kreiswirth, B. N., Perlman, D. C., Salomon, N., Schluger, N., Lutfey, M., Berger, J., Poltoratskaia, N., and Riley, L. W., 1997, Widespread dissemination of a drug-susceptible strain of *Mycobacterium tuberculosis*, *J. Infect. Dis.* **176:**478–484.

Garbe, T. R., Hibler, N. S., and Deretic, V., 1996, Response of *Mycobacterium tuberculosis* to reactive oxygen and nitrogen intermediates, *Mol. Med.* **2:**134–142.

Garbe, T. R., Hibler, N. S., Deretic, V., 1999, Response to reactive nitrogen intermediates in *Mycobacterium tuberculosis*: Induction of the 16-kilodalton α-crystallin homolog by exposure to nitric oxide donors, *Infect. Immun.* **67:**460–465.

Gaston, B., Reilly, J., Drazen, J. M., Fackler, J., Ramdev, P., Arnelle, D., Mullins, M., Sugarbaker, D. J., Chee, C., Singel, D. J., Loscalzo, J., and Stamler, J. S., 1993, Endogenous nitrogen oxides and bronchodilator *S*-nitrosothiols in human airways, *Proc. Natl. Acad. Sci. USA* **90:**10957–10961.

Geller, D. A., Lowenstein, C. J., Shapiro, R. A., Nussler, A. K., Di Silvio, M., Wang, S. W., Nakayama, D. K., Simmons, R. L., Snyder, S. H., and Billiar, T. R., 1993, Molecular cloning and expression of inducible nitric oxide synthase from human hepatocytes, *Proc. Natl. Acad. Sci. USA* **90:**3491–3495.

Granger, D. L., Cameron, M. L., Lee-See, K., and Hibbs, J. B., Jr., 1993, Role of macrophage-derived nitrogen oxides in antimicrobial function, in: *Mononuclear Phagocytes in Cell Biology* (G. Lopez-Berenstein and J. Klostergaard, eds.), CRC Press, Boca Raton, pp. 7–30.

Grosset, J., 1978, The sterilizing value of rifampicin and pyrazinamide in experimental short course chemotherapy, *Tubercle* **59:**287–297.

Haas, D. W., and des Perez, R. M., 1995, *Mycobacterium tuberculosis*, in: *Principles and Practice of Infectious Diseases* (G. L. Mandell, J. E. Bennett, and R. Dolin, eds.), Churchill Livingstone, New York, pp. 2213–2243.

Halliwell, B., and Gutteridge, J. M., 1989, Protection against oxidants in biological systems: The superoxide theory of oxygen toxicity, in: *Free Radicals in Biology and Medicine*, Clarendon Press, Oxford, p. 98.

Hanson, S. R., Hutsell, T. C., Keefer, L. K., Mooradian, D. L., and Smith, D. J., 1995, Nitric oxide donors: A continuing opportunity in drug design, *Adv. Pharmacol.* **34:**383–398.

Harshey, R. M., and Ramakrishnan, T., 1976, Purification and properties of DNA-dependent RNA polymerase from *Mycobacterium tuberculosis* H37Rv, *Biochim. Biophys. Acta* **432:**49–59.

Heck, D. E., Laskin, D. L., Gardner, C. R., and Laskin, J. D., 1992, Epidermal growth factor suppresses nitric oxide and hydrogen peroxide production by keratinocytes. Potential role for nitric oxide in the regulation of wound healing, *J. Biol. Chem.* **267:**21277–21280.

Hibbs, J. B., Jr., Westenfelder, C., Taintor, R., Vavrin, Z., Kablitz, C., Baranowski, R. L., Ward, J. H., Menlove, R. L., McMurry, M. P., Kushner, J. P., and Samlowski, W. E., 1992, Evidence for cytokine-inducible nitric oxide synthesis from L-arginine in patients receiving interleukin-2 therapy, *J. Clin. Invest.* **89:**867–877.

Hiriyanna, K.T., and Ramakrishnan, T., 1986, DNA replication time in *Mycobacterium tuberculosis* H37Rv, *Arch. Microbiol.* **144:**105–109.

Hogg, N., Singh, R. J., Konorev, E., Joseph, J., and Kalyanaraman, B., 1997, *S*-Nitrosoglutathione as a substrate for gamma-glutamyl transpeptidase, *Biochem. J.* **323:**477–481.

Huie, R. E., and Padmaja, S., 1993, The reaction rate of nitric oxide with superoxide, *Free Radical Res. Commun.* **18:**195–199.

Incze, K., Parkes, J., Mihalyi, V., and Zukal, E., 1974, Antibacterial effect of cysteine-nitrosothiol and possible precursors thereof, *Appl. Microbiol.* **27:**202–205.

Ischiropoulos, H., Zhu, L., and Beckman, J. S., 1992, Peroxynitrite formation from macrophage-derived nitric oxide, *Arch. Biochem. Biophys.* **298:**446–451.

Iyer, G. Y. N., Islam, M. F., and Quastel, J. H., 1961, Biochemical aspects of phagocytosis, *Nature* **192:**535–541.

Jacobs, W. R., Jr., Tuckman, M., and Bloom, B. R., 1987, Introduction of foreign DNA into mycobacteria using a shuttle phasmid, *Nature* **327:**532–535.

Jordan, M. C., Jordan, G. W., Stevens, J. G., and Miller, G., 1984, Latent herpesviruses of humans, *Ann. Intern. Med.* **100:**866–880.

Jouanguy, E., Altare, F., Lamhamedi, S., Revy, P., Emile, J.-F., Newport, M., Levin, M., Blanche, S., Seboun, E., Fischer, A., and Casanova, J.-L., 1996, Interferon-γ-receptor deficiency in an infant with fatal bacille Calmette-Guerin infection, *N. Engl. J. Med.* **335:**1956–1961.

Kamijo, R., Le, J., Shapiro, D., Havell, E. A., Huang, S., Aguet, M., Bosland, M., and Vilcek, J., 1993, Mice that lack the interferon-γ receptor have profoundly altered responses to infection with bacillus Calmette-Guerin and subsequent challenge with lipopolysaccharide, *J. Exp. Med.* **178:**1435–1440.

Kamijo, R., Harada, H., Matsuyama, T., Bosland, M., Gerecitano, J., Shapiro, D., Le, J., Koh, S. I., Kimura, T., Green, S. J., Mak, T. W., Taniguchi, T., and Vilcek, J., 1994, Requirement for transcription factor IRF-1 in NO synthase induction in macrophages, *Science* **263:**612–615.

Kobzik, L., Bredt, D. S., Lowenstein, C. J., Drazen, J., Gaston, B., Sugarbaker, D., and Stamler, J. S., 1993, Nitric oxide synthase in human and rat lung: Immunocytochemical and histochemical localization, *Am. J. Respir. Cell. Mol. Biol.* **9:**371–377.

Koch, R., 1891, Fortsetzung über ein Heilmittel gegen Tuberculose, *Dtsch. Med. Wochenschr.* **17:**101–102.

Krueger, G. R., and Ramon, A., 1988, Overview of immunopathology of chronic active herpesvirus infection, *J. Virol. Methods* **21:**11–18.

Kunz, D., Walker, G., Eberhardt, W., and Pfeilschifter, J., 1996, Molecular mechanisms of dexamethasone inhibition of nitric oxide synthase expression in interleukin 1β-stimulated mesangial cells: Evidence for the involvement of transcriptional and posttranslational regulation, *Proc. Natl. Acad. Sci. USA* **93:**255–259.

Lane, H. C., Laughon, B. E., Falloon, J., Kovacs, J. A., Davey, R. T., Jr., Polis, M. A., and Masur, H., 1994, Recent advances in the management of AIDS-related opportunistic infections, *Ann. Intern. Med.* **120:**945–955.

Lau, Y. L., Yuen, K. Y., Ha, S. Y., Chan, C. F., Hui, Y. F., 1996, *Mycobacterium tuberculosis* is a major pathogen in patients with chronic granulomatous disease, 7th International Congress for Infectious Diseases, Abstract 107.016, p. 261.

Loebel, R. O., Shorr, E., and Richardson, H. B., 1933, The influence of adverse conditions upon the respiratory metabolism and growth of human tubercle bacilli, *J. Bacteriol.* **26:**167–173.

Lowenstein, C. J., Alley, E., Raval, P., Snyder, S. H., Russel, S. W., and Murphy, W., 1993, Nitric oxide synthase gene: Two upstream regions mediate its induction by interferon-gamma and lipopolysaccharide, *Proc. Natl. Acad. Sci. USA* **90:**9730–9734.

Mackowiak, P. A., 1984, Microbial latency, *Rev. Infect. Dis.* **6:**649–668.

MacMicking, J. D., Nathan, C., Hom, G., Chartrain, N., Fletcher, D. S., Trumbauer, M., Stevens, K., Xie, Q.-W., Sokol, K., Hutchinson, N., Chen, H., and Mudgett, J. S., 1995, Altered responses to bacterial infection and endotoxic shock in mice lacking inducible nitric oxide synthase, *Cell* **81:**641–650.

MacMicking, J. D., North, R. J., LaCourse, R., Mudgett, J. S., Shah, S. K., and Nathan, C. F., 1997a,

Identification of nitric oxide synthase as a protective locus against tuberculosis, *Proc. Natl. Acad. Sci. USA* **94:**5243–5248.

MacMicking, J., Xie, Q.-W., and Nathan, C., 1997b, Nitric oxide and macrophage function, *Annu. Rev. Immunol.* **15:**323–350.

Mannick, J. B., Koichiro, A., Izumi, K., Kieff, E., and Stamler, J. S., 1994, Nitric oxide produced by human B lymphocytes inhibits apoptosis and Epstein–Barr virus reactivation, *Cell* **79:**1137–1146.

Mathews, W. R., and Kerr, S. W., 1993, Biological activity of *S*-nitrosothiols: The role of nitric oxide, *J. Pharmacol. Exp. Ther.* **267:**1529–1537.

McCune, R. M., and Tompsett, R., 1956, Fate of *Mycobacterium tuberculosis* in mouse tissues as determined by the microbial enumeration technique. I. The persistence of drug-susceptible tubercle bacilli in the tissues despite prolonged antimicrobial therapy, *J. Exp. Med.* **104:**737–762.

McCune, R. M., Tompsett, R., and McDermott, W., 1956, The fate of *Mycobacterium tuberculosis* in mouse tissues as determined by the microbial enumeration technique. II. The conversion of tuberculous infection to the latent state by the administration of pyrazinamide and a companion drug, *J. Exp. Med.* **104:**763–802.

McCune, R. M., Feldmann, F. M., Lambert, H. P., and McDermott, W., 1966a, Microbial persistence I. The capacity of tubercle bacilli to survive sterilization in mouse tissues, *J. Exp. Med.* **123:**445–468.

McCune, R. M., Feldmann, F. M., and McDermott, W., 1966b, Microbial persistence II. Characteristics of the sterile state of tubercle bacilli, *J. Exp. Med.* **123:**469–486.

McMurray, D.N., 1994, Guinea pig model of tuberculosis, in: *Tuberculosis. Pathogenesis, Protection, and Control* (B. R. Bloom, ed.), American Society for Microbiology, Washington, D.C., pp. 135–147.

Melillo, G., Musso, T., Sica, A., Taylor, L. S., Cox, G. W., and Varesio, L., 1995, A hypoxia-responsive element mediates a novel pathway of activation of the inducible nitric oxide synthase promoter, *J. Exp. Med.* **182:**1683–1693.

Moncada, S., 1992, The L-arginine:nitric oxide pathway, *Acta Physiol. Scand.* **145:**201–227.

Morris, S. L., and Hansen, N., 1981, Inhibition of *Bacillus cereus* spore outgrowth by covalent modification of a sulfhydryl group by nitrosothiol and iodoacetate, *J. Bacteriol.* **148:**465–471.

Morris, S. L., Walsh, R. C., and Hansen, J. N., 1984, Identification and characterization of some bacterial membrane sulfhydryl groups which are targets of bacteriostatic and antibiotic action, *J. Biol. Chem.* **259:**13590–13594.

Murray, C. J. L., Styblo, K., and Rouillon, A., 1990, Tuberculosis in developing countries: Burden, intervention, and cost, *Bull. Int. Union Tuberc.* **65:**6–24.

Nathan, C. F., 1983, Mechanisms of macrophage antimicrobial activity, *Trans. R. Soc. Trop. Med. Hyg.* **77:**620–630.

Nathan, C., 1992, Nitric oxide as a secretory product of mammalian cells, *FASEB J.* **6:**3051–3064.

Nathan, C., 1995, Natural resistance and nitric oxide, *Cell* **82:**873–876.

Nathan, C., 1997, Inducible nitric oxide synthase: What difference does it make? *J. Clin. Invest.* **100:**2417–2423.

Nathan, C. F., and Hibbs, J. B., Jr., 1991, Role of nitric oxide synthesis in macrophage antimicrobial activity, *Curr. Opin. Immunol.* **3:**65–70.

Nathan, C., and Xie, Q.-W., 1994, Nitric oxide synthases: Roles, tolls, and controls, *Cell* **78:**915–918.

Newport, M. J., Huxley, C. M., Huston, S., Hawrylowicz, C. M., Oostra, B. A., Williamson, R., and Levin, M., 1996, A mutation in the interferon-gamma-receptor gene and susceptibility to mycobacterial infection, *N. Engl. J. Med.* **335:**1941–1949.

Newton, G. L., Arnold, K., Price, M. S., Sherrill, C., Delcardayre, S. B., Aharonowitz, Y., Cohen, G., Davies, J., Fahey, R. C., and Davis, C., 1996, Distribution of thiols in microorganisms: Mycothiol is a major thiol in most Actinomycetes, *J. Bacteriol.* **178:**1990–1995.

Nicholson, S., Bonecini-Almeida, M. da G., Lapa e Silva, J. R., Nathan, C., Xie, Q.-W., Mumford, R., Weidner, J. R., Calaycay, J., Geng, J., Boechat, N., Linhares, C., Rom, W., and Ho, J. L., 1996, Inducible nitric oxide synthase in pulmonary alveolar macrophages from patients with tuberculosis, *J. Exp. Med.* **183:**2293–2302.

North, R. J., and Izzo, A., 1993, Mycobacterial virulence. Virulent strains of *Mycobacteria tuberculosis* have faster *in vivo* doubling times and are better equipped to resist growth-inhibiting functions of macrophages in the presence and absence of specific immunity, *J. Exp. Med.* **177:**1723–1733.

Nozaki, Y., Hasegawa, Y., Ichiyama, S., Nakashima, I., and Shimokata, K., 1997, Mechanism of nitric oxide-dependent killing of *Mycobacterium bovis* BCG in human alveolar macrophages, *Infect. Immun.* **65:**3644–3647.

Nunoshiba, T., DeRojas-Walker, T., Wishnok, J. S., Tannenbaum, S. R., and Demple, B., 1993, Activation by nitric oxide of an oxidative-stress response that defends *Escherichia coli* against activated macrophages, *Proc. Natl. Acad. Sci. USA* **90:**9993–9997.

Nussler, A. K., di Silvio, M., Billiar, T. R., Hoffman, R. A., Geller, D. A., Selby, R., Madariaga, J., and Simmons, R. L., 1992, Stimulation of nitric oxide synthase pathway in human hepatocytes by cytokines and endotoxin, *J. Exp. Med.* **176:**261–266.

O'Brien, L., Carmichael, J., Lowrie, D. B., and Andrew, P. W., 1994, Strains of *Mycobacterium tuberculosis* differ in susceptibility to reactive nitrogen intermediates *in vitro*, *Infect. Immun.* **62:**5187–5190.

Ochoa, J. B., Udekwu, A. O., Billiar, T. R., Curran, R. D., Cerra, F. B., Simmons, R. L., Peitzman, A. B., 1991, Nitrogen oxide levels in patients after trauma and during sepsis, *Annals of Surgery* **214:**621–626.

Ochoa, J. B., Curti, B., Peitzman, A. B., Simmons, R. L., Billiar, T. R., Hoffman, R., Rault, R., Longo, D. L., Urba, W. J., Ochoa, A. C., 1992, *J. Natl. Cancer Institute* **84:**864–867.

O'Leary, V., and Solberg, M., 1976, Effect of sodium nitrite inhibition on intracellular thiol groups and on the activity of certain glycolytic enzymes in *Clostridium perfringens*, *Appl. Environ. Microbiol.* **31:**208–212.

Orme, I. M., 1988, A mouse model of the recrudescence of latent tuberculosis in the elderly, *Am. Rev. Respir. Dis.* **137:**716–718.

Pomerantz, R. J., Bagasra, O., and Baltimore, D., 1992, Cellular latency of human immunodeficiency virus type 1, *Curr. Opin. Immunol.* **4:**475–480.

Prada, J., Malinowski, J., Muller, S., Bienzle, U., and Kremsner, P. G., 1996, Effects of *Plasmodium vinckei* hemozoin on the production of oxygen radicals and nitrogen oxides in murine macrophages, *Am. J. Trop. Med. Hyg.* **54:**620–624.

Radomski, M. W., Palmer, R. M., and Moncada, S., 1990, Glucocorticoids inhibit the expression of an inducible, but not the constitutive, nitric oxide synthase in vascular endothelial cells, *Proc. Natl. Acad. Sci. USA* **87:**10043–10047.

Reddy, D., Lancaster, J. R., Jr., and Cornforth, D. P., 1983, Nitrite inhibition of *Clostridium botulinum* electron spin resonance detection of iron–nitric oxide complexes, *Science* **221:**769–770.

Rees, D. D., Palmer, R. M. J., Schulz, R., Hodson, H. F., and Moncada, S., 1990, Characterization of three inhibitors of endothelial nitric oxide synthase *in vitro* and *in vivo*, *Br. J. Pharmacol.* **101:**746–752.

Rhoades, E. R., and Orme, I. M., 1997, Susceptibility of a panel of virulent strains of *Mycobacterium tuberculosis* to reactive nitrogen intermediates, *Infect. Immun.* **65:**1189–1195.

Rich, A., 1944, *The Pathogenesis of Tuberculosis*, Charles C. Thomas, Springfield, Ill., pp. 297–298.

Roach, T. I., Barton, C. H., Chatterjee, D., Blackwell, J. M., 1993, Macrophage activation: Lipoarabinomannan from avirulent and virulent strains of Mycobacterium tuberculosis differentially induces the early genes c-fos, KC, JE, and tumor necrosis factor-alpha, *J. Immunol.* **150:**1886–1896.

Roach, T. I., Barton, C. H., Chatterjee, D., Liew, F. Y., Blackwell, J. M., 1995, Opposing effects of interferon-gamma on iNOS and interleukin-10 lipopolysaccharide- and mycobacterial lipoarabinomannan-stimulated macrophages, *Immunol.* **85:**106–113.

Rojas, M., Barrera, L. F., Puzo, G., and Garcia, L. F., 1997, Differential induction of apoptosis by virulent *Mycobacterium tuberculosis* in resistant and susceptible murine macrophages: Role of nitric oxide and mycobacterial products, *J. Immunol.* **159:**1352–1361.

Rook, G. A. W., and Bloom, B. R., 1994, Mechanisms of pathogenesis in tuberculosis, in: *Tuberculosis: Pathogenesis, Protection, and Control* (B. R. Bloom, ed.), American Society for Microbiology, Washington, D.C., pp. 485–501.

Rook, G. A. W., Steele, J., Ainsworth, M., and Leveton, C., 1987, A direct effect of glucocorticoid hormones on the ability of human and murine macrophages to control the growth of *M. tuberculosis*, *Eur. J. Respir. Dis.* **71:**286–291.

Rubbo, H., Denicola, A., and Radi, R., 1994, Peroxynitrite inactivates thiol-containing enzymes of *Trypanosoma cruzi* energetic metabolism and inhibits cell respiration, *Arch. Biochem. Biophys.* **308:**96–102.

Scannell, J. P., Ax, H. A., Pruess, D. L., Williams, T., Demny, T. C., and Stempel, A., 1972, Antimetabolites produced by microorganisms. L-N^5-(1-iminoethyl) ornithine, *J. Antibiot.* **25:**179–184.

Schaffer, M. R., Efron, P. A., Thornton, F. J., Klingel, K., Gross, S. S., and Barbul, A., 1997, Nitric oxide, an autocrine regulator of wound fibroblast synthetic function, *J. Immunol.* **158:**2375–2381.

Schuller-Levis, G. B., Levis, W. R., Ammazzalorso, M., Nosrati, A., Park, E., 1994, Mycobacterial lipoarabinomannan induces nitric oxide and tumor necrosis factor alpha production in a macrophage cell line: down regulation by taurine chloramine, *Infect. Immun.* **62:**4671–4674.

Shearer, J. D., Richards, J. R., Mills, C. D., and Caldwell, M. D., 1997, Differential regulation of macrophage arginine metabolism: A proposed role in wound healing, *Am. J. Physiol.* **272:**E181-E190.

Shank, J. L., Silliker, J. H., and Harper, R. H., 1962, The effect of nitric oxide on bacteria, *Appl. Microbiol.* **10:**185–189.

Singh, S. P., Wishnok, J. S., Keshive, M., Deen, W. M., and Tannenbaum, S. R., 1996, The chemistry of the *S*-nitrosoglutathione/glutathione system, *Proc. Natl. Acad. Sci. USA* **93:**14428–14433.

Slater, A. F. G., Swiggard, W. J., Orton, B. R., Flitter, W. D., Goldberg, D. E., Gerami, A., and Henderson, G. B., 1991, An iron–carboxylate bond links the heme units of malaria pigment, *Proc. Natl. Acad. Sci. USA* **88:**325–329.

Smith, D. J., Chakravarthy, D., Pulfer, S., Simmons, M. L., Hrabie, J. A., Citro, M. L., Saavedra, J. E., Davies, K. M., Hutsell, T. C., Mooradian, D. L., Hanson, S. R., and Keefer, L. K., 1996, Nitric oxide-releasing polymers containing the $[N(O)NO]^-$ group, *J. Med. Chem.* **39:**1148–1156.

Snider, D. E., Jr., and Roper, W. L., 1992, The new tuberculosis, *N. Engl. J. Med.* **326:**703–705.

Stamler, J. S., 1994, Redox signaling: Nitrosylation and related target interactions of nitric oxide, *Cell* **78:**931–936.

Stamler, J.S., 1996, *S*-Nitrosothiols and the bioregulatory actions of nitrogen oxides through reactions with thiol groups, *Curr. Top. Microbiol. Immunol.* **196:**19–36.

Stamler, J. S., Singel, D. J., and Loscalzo, J., 1992, Biochemistry of nitric oxide and its redox-activated forms, *Science* **256:**1898–1902.

Stead, W. W., 1965, The pathogenesis of pulmonary tuberculosis among older persons, *Am. Rev. Respir. Dis.* **91:**811.

Stead, W. W., 1967, Pathogenesis of a first episode of chronic pulmonary tuberculosis in man: Recrudescence of residuals of the primary infection or exogenous reinfection? *Am. Rev. Respir. Dis.* **95:**729–745.

Stead, W. W., Kerby, G. R., Schleuter, D. P., and Jordahl, C. W., 1968, The clinical spectrum of primary tuberculosis in adults. Confusion with reinfection in the pathogenesis of chronic tuberculosis, *Ann. Intern. Med.* **68:**731–745.

Steiner, I., 1996, Human herpes viruses latent infection in the nervous system, *Immunol. Rev.* **152:**157–173.

Stenger, S., Donhaur, N., Thuring, H., Rollinghoff, M., and Bogdan, C., 1996, Reactivation of latent leishmaniasis by inhibition of inducible nitric oxide synthase, *J. Exp. Med.* **183:**1501–1514.

Storz, G., Tartaglia, L. A., and Ames, B. N., 1990a, Transcriptional regulator of oxidative stress-inducible gene: Direct activation by oxidation, *Science* **248:**189–194.

Storz, G., Tartaglia, L. A., Farr, S. B., and Ames, B. N., 1990b, Bacterial defenses against oxidative stress, *Trends Genet.* **6:**363–368.

Tarr, H. L. A., 1941, Bacteriostatic action of nitrites, *Nature* **147:**417–418.

Travis, M. D., Davisson, R. L., Bates, J. N., and Lewis, S. J., 1997, Hemodynamic effects of L- and D-S-nitroso-beta, beta-dimethylcysteine in rats, *Am. J. Physiol.* **273:**H1493–H1501.

Vazquez-Torres, A., Jones-Carson, J., and Balish, E., 1996, Peroxynitrite contributes to the candidacidal activity of nitric oxide-producing macrophages, *Infect. Immun.* **64:**3127–3133.

Wagner, D. A., and Tannenbaum, S. R., 1982, Enhancement of nitrate biosynthesis by *Escherichia coli* lipopolysaccharide, in: *Nitrosamines and Human Cancer*, Banbury Report 12 (P. N. Magee, ed.), Cold Spring Harbor Press, Cold Spring Harbor, N.Y., pp. 437–441.

Walsh, G. P., Tan, E. V., dela Cruz, E. C., Abalos, R. M., Billahermosa, L. G., Young, L. J., Cellona, R. V., Nazareno, J. B., and Horwitz, M. A., 1996, The Philippine cynomolgus monkey (*Macaca fascicularis*) provides a new nonhuman primate model of tuberculosis that resembles human disease, *Nature Med.* **2:**430–436.

Wang, C. H., Liu, C. Y., Lin, H. C., Yu, C. T., Chung, K. F., and Kuo, H. P., 1998, Increased exhaled nitric oxide in active pulmonary tuberculosis due to inducible NO synthase upregulation in alveolar macrophages, *European Respiratory J.* **11**:809–815.

Warren, J. B., Loi, R., Rendell, N. B., and Taylor, G. W., 1990, Nitric oxide is inactivated by the bacterial pigment pyocyanin, *Biochem. J.* **266:**921–923.

Watson, K. C., 1967, Intravascular *Salmonella typhi* as a manifestation of the carrier state, *Lancet* **2:**332–334.

Wayne, L. G., 1994, Dormancy of *Mycobacterium tuberculosis* and latency of disease, *Eur. J. Clin. Microbiol. Infect. Dis.* **13:**908–914.

Wei, X.-Q., Charles, I. G., Smith, A., Ure, J., Feng, G.-J., Huang, F.-P., Xu, D., Muller, W., Moncada, S., and Liew, F. Y., 1995, Altered immune responses in mice lacking inducible nitric oxide synthase, *Nature* **375:**408–411.

Wheeler, P. R., and Ratledge, C., 1994, Metabolism of *Mycobacterium tuberculosis*, in: *Tuberculosis: Pathogenesis, Protection, and Control* (B. R. Bloom, ed.), American Society for Microbiology, Washington, D.C., pp 353–385.

WHO, 1997, Anti-tuberculosis drug resistance in the world. The WHO/IUALTD. A global project on anti-tuberculosis drug resistance surveillence. WHO/TB/97.229. (WWW.WHO.CH/GTB/Publications/DRITW/index.HTML).

Wientjes, F. B., and Segal, A. W., 1995, NADPH oxidase and the respiratory burst, *Semin. Cell Biol.* **6:**357–365.

Woods, L. F. J., Wood, J. M., and Gibbs, P. A., 1981, The involvement of nitric oxide in the inhibition of the phosphoroclastic system in *Clostridium sporogenes* by sodium nitrite, *J. Gen. Microbiol.* **125:**399–406.

Xie, Q.-W., Whisnant, R., and Nathan, C., 1993, Promoter of the mouse gene encoding calcium-independent nitric oxide synthase confers inducibility by interferon gamma and bacterial lipopolysaccharide, *J. Exp. Med.* **177:**1779–1784.

Yu, K., Mitchell, C., Xing, Y., Magliozzo, R. S., Bloom, B. R., and Chan, J., (1999), Relative toxicity of nitrogen oxides and related oxidants on mycobacteria: *M. tuberculosis* is resistant to peroxynitrite anion. *Tubercle and Lung Dis.* In press.

Zhu, L., Gunn, C., and Beckman, J. S., 1992, Bactericidal activity of peroxynitrite, *Arch. Biochem. Biophys.* **208:**452–457.

CHAPTER 15

Nitric Oxide in Malaria

NICHOLAS M. ANSTEY, J. BRICE WEINBERG, and DONALD L. GRANGER

1. Malaria

Malaria is caused by infection with one of four species of the intracellular protozoan parasite, *Plasmodium* (see Knell, 1991; Wyler, 1993, for reviews), although only *P. falciparum* causes severe malaria. It is a major public health problem, causing more global mortality than any other parasitic disease, an estimated 1.5 to 2.7 million deaths each year (World Health Organization, 1996). The bulk of morbidity and mortality from malaria occurs in those populations from tropical developing countries least able to afford treatment and control measures (World Health Organization, 1996). Over the last century, studies of *Plasmodium* infection not only have formed the basis for our understanding of the pathogenesis, treatment, and control of malaria, but have also contributed greatly to the understanding of host–pathogen interactions in general. In recent years these studies have also provided significant insights into the role of NO in host defense. *In vitro* experiments using mouse and human cells and *in vivo* experiments in mice have shown that NO and NO-related species have both antiparasitic and antidisease effects. Studies of African children with malaria have shown an association between NO production/leukocyte iNOS expression and host-protective responses. In this chapter we provide an overview of the clinical spectrum, immunology, and pathogenesis of malaria, and review the *in vitro* and *in vivo*

NICHOLAS M. ANSTEY • Tropical Medicine and International Health Unit, Menzies School of Health Research and Royal Darwin Hospital, Casuarina, Darwin NT0811, Northern Territory, Australia. *J. BRICE WEINBERG* • Division of Hematology and Oncology, Veterans Affairs and Duke University Medical Centers, Durham, North Carolina 27705. *DONALD L. GRANGER* • Division of Infectious Diseases, Department of Medicine, University of Utah Medical Center, Salt Lake City 84132, Utah.

Nitric Oxide and Infection, edited by Fang. Kluwer Academic / Plenum Publishers, New York, 1999.

evidence for the antiplasmodial and disease-modulating effects of NO in this disease.

1.1. Clinical Epidemiology of Malaria

The clinical epidemiology of malaria within countries is complex and varied. In regions with intense malaria transmission, such as sub-Saharan Africa, children are infected repeatedly in infancy and early childhood, with most morbidity and mortality occurring before 7 years of age (World Health Organization, 1990; Marsh, 1992). Uncomplicated clinical malaria results in a nonspecific syndrome of fever, anorexia, vomiting, and/or tachypnea. Severe malaria in these children is characterized by coma (cerebral malaria), respiratory distress (metabolic acidosis), severe anemia, poor peripheral perfusion, and/or hypoglycemia (Marsh *et al.*, 1995; English and Marsh, 1997). Renal impairment is common (Waller *et al.*, 1995; Anstey *et al.*, 1996; English *et al.*, 1996b) but frank renal failure is rare. With the development of clinical immunity, the frequency of disease falls significantly. In semi-immune children, a common finding is subclinical and subpatent parasitemia (Mendis and Carter, 1995), frequently accompanied by anemia (Newton *et al.*, 1997). Clinical malaria in semi-immune African adults is infrequent and generally of mild severity, with the exception of pregnant women, who are at risk for severe malaria (World Health Organization, 1990).

In regions with low or unstable malaria transmission, such as South and Southeast Asia, clinical immunity is less easily achieved, and both children and adults are at risk for severe disease (Fig. 1). Severe malaria in such nonimmune adults is more commonly a multisystem disease with manifestations including unarousable coma (cerebral malaria), metabolic acidosis, renal failure, pulmonary edema, severe anemia, hypoglycemia, shock, repeated convulsions, disseminated intravascular coagulation, hemoglobinuria, and jaundice (World Health Organization, 1990). Maternal malaria is a distinct clinical entity that occurs in both high- and low-transmission regions, in both previously immune and nonimmune women. Maternal malaria is responsible for intrauterine growth retardation and preterm birth, maternal anemia, and increased maternal and perinatal mortality (World Health Organization, 1990; Fried and Duffy, 1996).

1.2. Malaria Life Cycle

The life cycle of malaria parasites in humans (Knell, 1991) (Fig. 2) commences during mosquito feeding, with inoculation of sporozoites from the salivary glands of the female *Anopheles* into human blood. Sporozoites invade hepatocytes within 30 min. Massive multiplication within infected liver cells then occurs, followed approximately 1 week later by hepatocyte rupture, releasing many thousands of merozoites into the circulation. This hepatic phase of infection

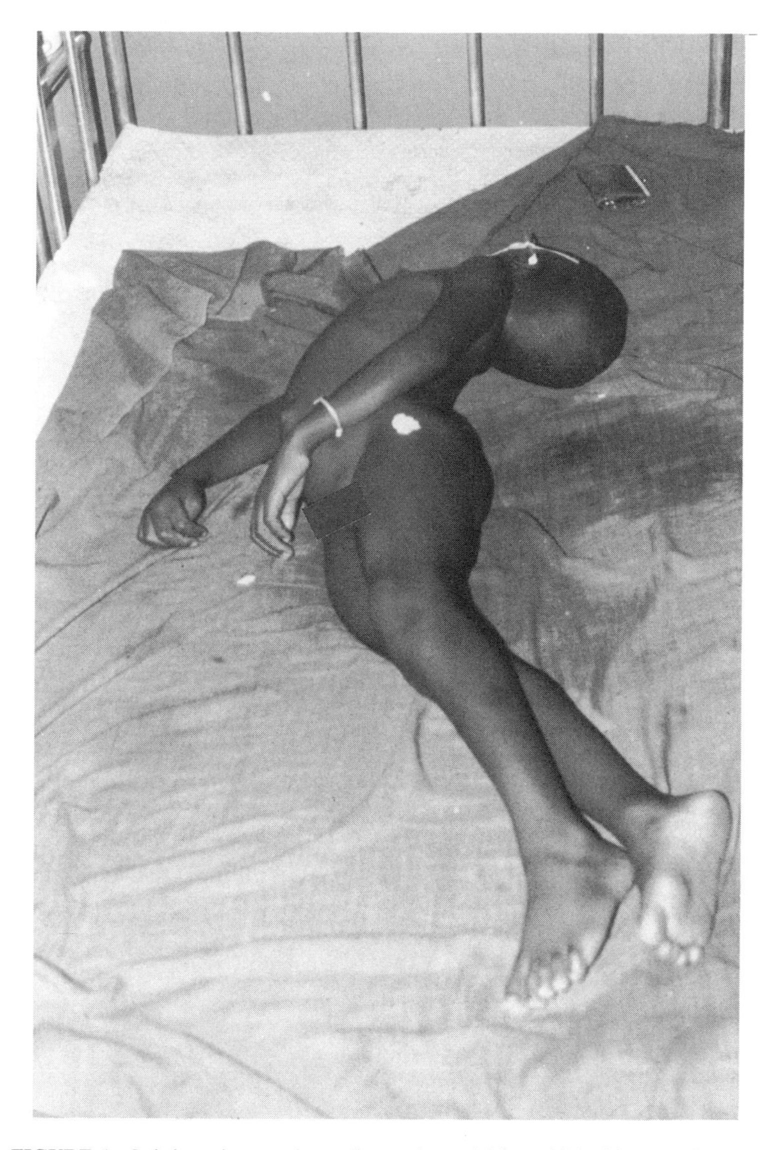

FIGURE 1. Opisthotonic posturing and coma in an African child with cerebral malaria.

(exoerythrocytic schizogony) is asymptomatic. Clinical disease results from the erythrocytic stage of the life cycle (erythrocytic schizogony), during which merozoites invade red cells to form intraerythrocytic trophozoites. Trophozoites ingest hemoglobin forming hemozoin or malaria pigment as they mature into the multinucleate schizont stage. Forty-eight hours after *P. falciparum* merozoite

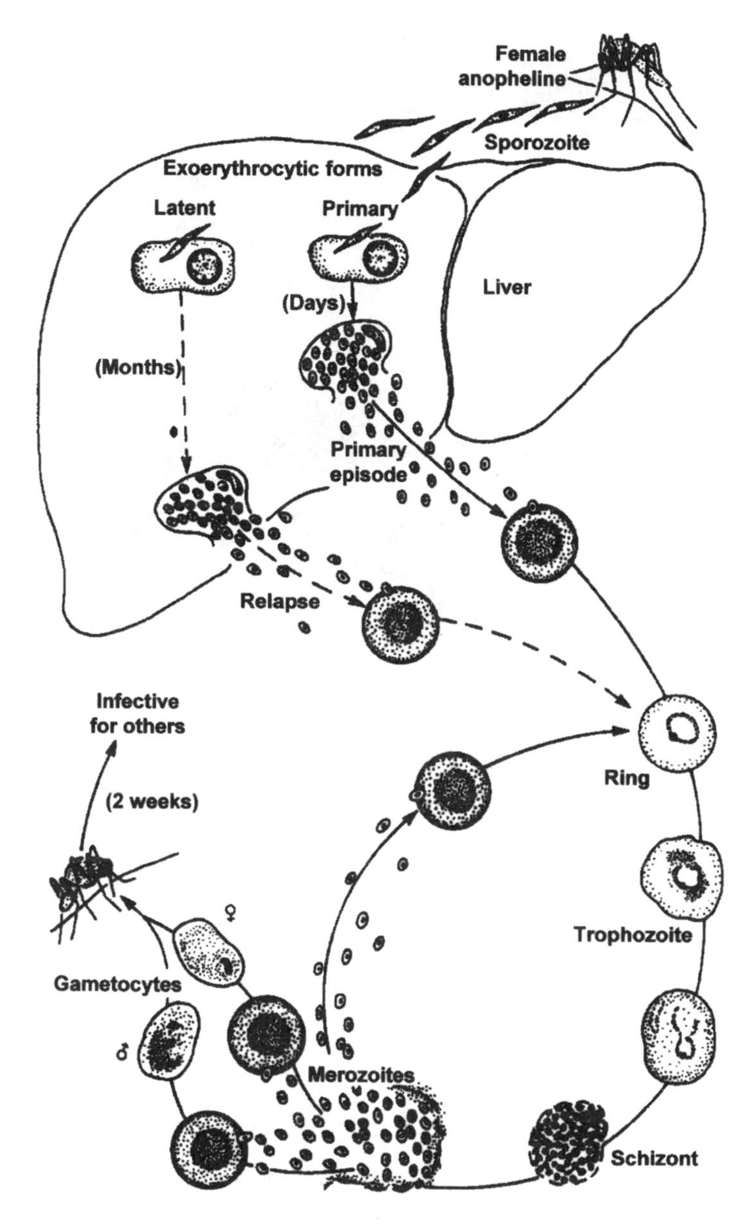

FIGURE 2. Life cycle of malaria parasites in humans. After mosquito inoculation of sporozoites, massive amplification of infection in the liver (exoerythrocytic schizogony) is followed by hepatocyte rupture and infection of red blood cells (erythrocytic stage). Red cell infection and rupture is responsible for clinical disease. After several cycles, sexual forms (gametocytes) are produced that can infect mosquitoes at the next blood meal. Reproduced from Wyler (1992) by copyright permission of W. B. Saunders Co.

invasion, the schizont-containing red cell ruptures releasing 8–16 merozoites, which then invade new red cells. Cyclical multiplication of parasites within red cells results in rising parasitemia, the level depending on both the host immune response and parasite factors. After several such blood cycles, a proportion of merozoites develop into male or female gametocytes. These sexual stages circulate, but only develop further if ingested by a mosquito during a blood meal. Fertilization occurs in the mosquito stomach. Oocysts then develop and rupture, releasing thousands of sporozoites that invade the mosquito salivary glands, from which they can infect humans and initiate a new cycle of infection.

1.3 Pathogenesis of Severe Malaria

Several pathogenic processes interact and overlap to cause severe malaria: sequestration of parasitized red cells within organ microvasculature, induction of disease-causing cytokines by parasite toxin(s), and the destruction and impaired production of red cells (see Pasvol *et al.*, 1995; English and Marsh, 1997; Marsh and Snow, 1997; Newton *et al.*, 1998; Newton and Krishna, 1998, for reviews) (Fig. 3).

Parasite sequestration in tissue microvascular beds is the predominant histopathologically evident process in severe malaria. Growth of asexual stages within the erythrocyte results in major structural and functional changes in the properties of the infected red cell. A unique feature of red cells parasitized with *P. falciparum* is their ability to adhere to human endothelial cells. This cytoadherence results from the binding of parasite proteins expressed on the surface of red cells such as PfEMP1 (Newbold *et al.*, 1997) and *clag* (Holt *et al.*, 1999) to a variety of endothelial ligands. These ligands include the constitutively expressed endothelial adhesion molecules CD36 and chondroitin sulfate A (CSA) (Barnwell *et al.*, 1989; Rogerson *et al.*, 1995), and also the cytokine-inducible ligands ICAM-1, E-selectin, and VCAM-1 (Berendt *et al.*, 1989; Ockenhouse *et al.*, 1992) (see Fig. 3). There is substantial evidence that endothelial cytoadherence and microvascular parasite sequestration contribute to the pathogenesis of severe falciparum malaria (particularly cerebral malaria) and maternal malaria (Fried and Duffy, 1996; Turner, 1997). The major autopsy finding in cerebral malaria is sequestration of parasitized red cells within postcapillary venules (Macpherson *et al.*, 1985; Pongponratn *et al.*, 1991) associated with widespread activation of endothelial receptors (Turner *et al.*, 1994; Turner, 1997). Such microvascular obstruction is thought to cause disease as a result of localized tissue hypoxia and metabolic derangements (Berendt *et al.*, 1994). Similar histopathological sequestration has been observed in a wide variety of other organs examined at autopsy including kidney, gut, muscle, lung, and placenta. Adhesion to CSA appears to select for a subpopulation of parasites that causes maternal malaria (Fried and Duffy, 1996).

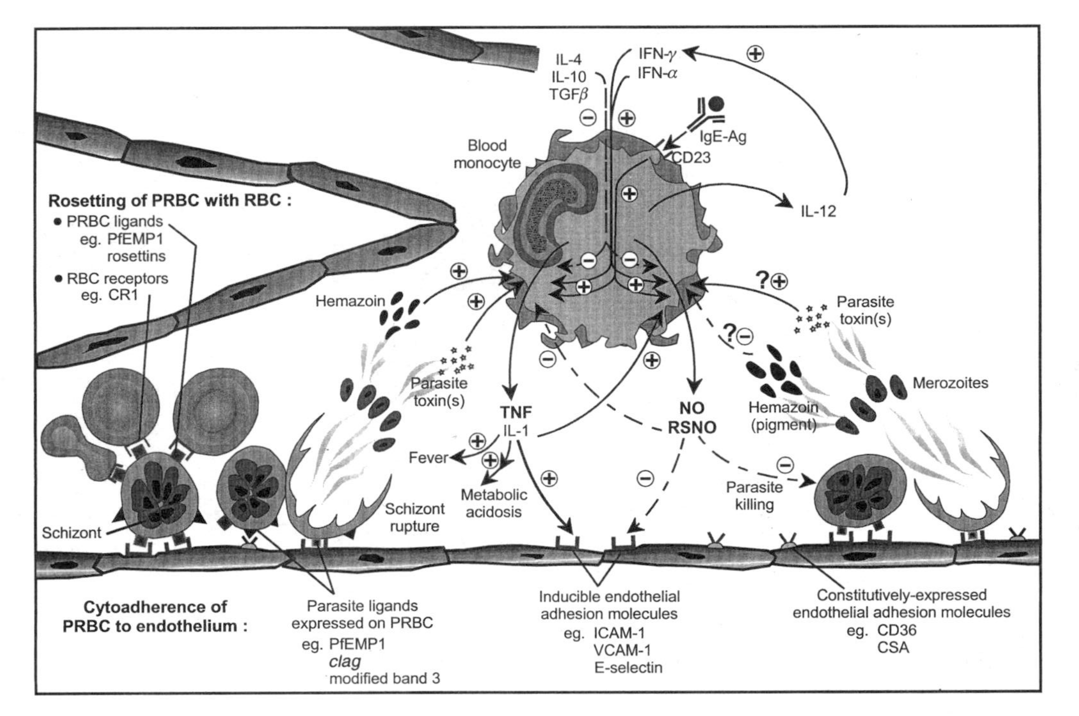

FIGURE 3. Pathogenesis of malaria. Central processes in malaria pathogenesis are (1) cytoadherence of parasitized red blood cells to (a) microvascular endothelium and (b) uninfected red cells (rosetting), resulting in microvascular obstruction and metabolic derangement, (2) excessive macrophage production of TNFα and other proinflammatory cytokines, and (3) red cell destruction and impaired production. TNFα contributes to fever, metabolic acidosis, and upregulation of the inducible endothelial adhesion molecules capable of binding parasitized red cells, thus contributing to parasite cytoadherence. Macrophage NO production appears to have a protective effect in malaria, downregulating expression of endothelial adhesion molecules, inhibiting TNFα production, and possibly killing or inhibiting parasites. Macrophage TNFα and NO production are both modulated by parasite products and host cytokines (see text).

Parasite products released at the time of schizont rupture include malaria pigment (hemozoin) (Arese and Schwarzer, 1997) and malaria toxin(s) (Bate *et al.*, 1992a–c; Schofield and Hackett, 1993; Bate and Kwiatkowski, 1994b; Schofield *et al.*, 1996; Kwiatkowski *et al.*, 1997; Schofield, 1997), important for their ability to induce production of fever-causing cytokines such as TNFα and IL-1 (Jakobsen *et al.*, 1995; Kwiatkowski *et al.*, 1997; Schofield, 1997). The nature of the parasite toxin(s) is not clear; glycosyl-phosphatidylinositol (GPI) (Schofield and Hackett, 1993; Schofield *et al.*, 1996; Schofield, 1997) and other lipid-modified polypeptides (Kwiatkowski *et al.*, 1997) have been implicated. Purification and identification of the *Plasmodium* toxin has been hampered by the recent finding of widespread contamination of *Plasmodium* cultures by mycoplasmas (Rowe *et al.*, 1998) products of which have similar cytokine-inducing ability (Turrini *et al.*, 1997).

Although part of the nonspecific protective antiparasitic immune response, there is also much evidence that *excessive* production of TNFα and other proinflammatory cytokines contributes to the pathogenesis of severe malaria. Circulating levels of TNFα and IL-1 correlate with malaria disease severity in both adults and children, and in studies of both nonimmune and semi-immune patients (Grau *et al.*, 1989; Kern *et al.*, 1989; Butcher *et al.*, 1990; Kwiatkowski *et al.*, 1990; Jakobsen *et al.*, 1995). Levels of TNFα are highest in fatal cerebral malaria (Kwiatkowski *et al.*, 1990). Further evidence for the importance of TNFα in the pathogenesis of severe malaria is provided by the sevenfold increased risk of death or severe neurological sequelae from cerebral malaria in Gambian children homozygous for the TNF2 allele (McGuire *et al.*, 1994). This variant of the TNFα gene promoter region is associated with higher constitutive and inducible levels of TNFα transcription compared with the TNF1 allele. TNFα is known to increase expression of a number of endothelial molecules involved in parasite cytoadherence (Berendt *et al.*, 1989) (see Fig. 3). In addition to parasite products, cross-linking of the CD23 macrophage low-affinity immunoglobulin E (IgE) receptor by IgE–anti-IgE antibody complexes can also induce TNFα (Perlmann *et al.*, 1997) (Fig. 3), providing another potential mechanism for increased TNFα production in severe disease, as plasma levels of both total and *Plasmodium*-specific IgE are highest in severe malaria (Perlmann *et al.*, 1997). High levels of Th2 cytokines such as IL-10 are also found in severe malaria (Peyron *et al.*, 1994; Anstey *et al.*, 1996). The timing and balance of Th1 and Th2 cytokine responses are thought to be possible determinants of cytokine-mediated pathology in malaria (Kwiatkowski, 1992).

Anemia is another major cause of morbidity and mortality from severe malaria. Two forms of anemia predominate: anemia associated with acute clinical episodes of malaria, and anemia associated with chronic low-grade parasitemia (Pasvol *et al.*, 1995), where the anemia is disproportionate to the low-level parasitemia. The pathogenesis of severe anemia is complex, multifactorial, and imperfectly understood (Pasvol *et al.*, 1995; Newton *et al.*, 1997). Although red cell destruction contributes to the anemia, the fall in hemoglobin with *Plasmodium* infection is often far greater than that attributable to loss of infected erythrocytes alone. Red cell destruction results from schizont rupture and phagocytosis of infected cells, as well as from autoimmune hemolysis and increased phagocytosis/splenic filtration of uninfected cells (Phillips and Pasvol, 1992). Decreased production of red cells is associated with bone marrow hypoplasia (Srichaikul and Siriasawakul, 1976) in acute infections, and dyserythropoiesis in chronic infections (Abdulla *et al.*, 1980), which may be cytokine mediated (Clark and Chaudhri, 1988). Recent evidence suggests that severe malarial anemia may be related to impaired production of the pro-erythropoietic cytokine IL-10 in response to the high TNF concentrations found in acute malaria (Kurtzhals *et al.*,1998).

In summary, severe malaria results from the complex interaction of several processes: microvascular sequestration of parasitized red cells, cytokine-mediated immunopathology, and the impaired production and increased destruction of red cells, with disease severity determined by the magnitude and tissue distribution of each process. Superimposed on these processes are hypoglycemia and an often profound metabolic acidosis, the latter being a major predictor of mortality (Taylor *et al.*, 1993; Krishna *et al.*, 1994; Marsh *et al.*, 1995). TNFα hyperproduction (Starnes *et al.*, 1988; Krishna *et al.*, 1994) and microvascular obstruction may play a role in the development of elevated lactate levels and metabolic acidosis, with important contributions from dehydration (English *et al.*, 1996b), hypovolemia (English *et al.*, 1996b), renal impairment (Waller *et al.*, 1995; Anstey *et al.*, 1996; English *et al.*, 1996b), and in many African children, salicylate poisoning (English *et al.*, 1996a).

1.4 Immune Response to Malaria

The immunology of malaria is complex, and the mechanisms of protection against each stage of infection remain incompletely understood. There are still no reliable *in vitro* correlates of a protective immune response (Miller *et al.*, 1997). Both specific and nonspecific immune responses are important; these have been reviewed in detail elsewhere (Ho and Sexton, 1995) and will be summarized only briefly. Specific immune responses against each stage are generally short-lived, and protection is limited by the enormous diversity and variation in antigens presented by *P. falciparum*, particularly the blood stages (Kemp *et al.*, 1996). The immune response to the brief sporozoite stage is antibody mediated. Protection against liver-stage parasites is mediated by T cells, predominantly $CD8^{+}$, but also $CD4^{+}$ cells. Two forms of immunity to blood stages develop following repeated exposure to malaria in endemic areas: "clinical immunity" (also known as anti-disease/antitoxic immunity or malaria "tolerance"), which ameliorates disease despite the persistence of circulating blood-stage parasites, and antiparasitic immunity in which parasites are cleared or reduced in number (Sinton *et al.*, 1931; Ho and Sexton, 1995; Mendis and Carter, 1995). Epidemiological studies suggest that clinical immunity develops earlier in endemic areas than does antiparasitic immunity, with high rates of asymptomatic parasitemia in childhood declining with increasing age. There is some evidence that clinical immunity is mediated by an antitoxin antibody (Bate and Kwiatkowski, 1994a; Kwiatkowski *et al.*, 1997), although chronic expression of leukocyte iNOS and NO production may also be involved (see sections 3.4 and 3.5). Both humoral and cell-mediated mechanisms are involved in antiparasitic immunity against blood-stage parasites. Humoral protection appears to occur through antibody-dependent phagocytosis of merozoites and intraerythrocytic parasites by monocytes (Bouharoun-Tayoun *et al.*, 1990) and neutrophils (Kumaratilake *et al.*, 1992). Antibody-independent cellular immunity to blood-stage parasites involves $CD4^+$ T cells (Fell *et al.*, 1994), cytolytic NK cells (Orago and Facer, 1991), and $\gamma\delta$ T cells (Elloso *et al.*, 1994).

Much recent interest has focused on the role of NO in nonspecific protective immune responses to both the liver and blood stages of malaria (James, 1995) and in protection against pathology in severe malaria. NO is a downstream mediator of cytokine activity, with iNOS expression and NO production positively regulated by Th1 cytokines and negatively regulated by Th2 cytokines (Nathan and Xie, 1994). NO downregulates TNFα production (Florquin *et al.*, 1994; Tiao *et al.*, 1994) and also has important effects on endothelial expression of the receptors used by parasitized red cells to adhere to vascular endothelium (Decaterina *et al.*, 1995; Khan *et al.*, 1996) (Fig. 3).

2. NO and Malaria: *In Vitro*, Mosquito and Animal Studies

2.1. Role of NO in the Immune Response to Exoerythrocytic Stages

NO has been shown to be an important mediator of the protective immune response to the exoerythrocytic stages of both rodent and human *Plasmodium* species. IFNγ inhibits the exoerythrocytic stage of *P. berghei* development via the induction of NO within murine hepatocytes *in vitro* (Mellouk *et al.*, 1991; Nussler *et al.*, 1991). IFNγ-stimulated human hepatocytes are also able to inhibit exoerythrocytic stages of *P. falciparum in vitro* through the induction of NO (Mellouk *et al.*, 1994). Similar mechanisms have been reported *in vivo* during rodent infections with *P. berghei* (Nussler *et al.*, 1993; Green *et al.*, 1994; Seguin *et al.*, 1994; Klotz *et al.*, 1995) and *P. yoelii* (Tsuji *et al.*, 1995). The protective immunity induced by irradiated *P. berghei* sporozoite vaccination of rodents is dependent on the production of IFNγ by $CD8^+$ T cells, which in turn inhibits and arrests parasite development through induction of iNOS in infected hepatocytes (Seguin *et al.*, 1994; Klotz *et al.*, 1995). Moreover, early production of NO in this model appears to enhance accumulation of these Plasmodium-specific $CD8^+$T cells (Scheller *et al.*, 1997). Maintenance of iNOS activity in infected hepatocytes is dependent on the intrahepatic persistence of the irradiation-attenuated parasite (Klotz *et al.*, 1995), suggesting a microbiostatic effect of NO. The protection induced by multigene DNA immunization with *P. yoelii* circumsporozoite protein is similarly dependent on secretion of IFNγ by $CD8^+$ T cells, resulting in hepatocyte induction of NO and inhibition of parasite development (Doolan *et al.*, 1996).

The same investigators have shown that IFNγ mediates the killing of exoerythrocytic stages of *P. yoelii* by recombinant IL-12 (Sedegah *et al.*, 1994) in part through production of NO. Administration of recombinant IL-12 has recently resulted in successful protection of rhesus monkeys against sporozoite challenge with *P. cynomolgi* (Hoffman *et al.*, 1997), in association with increased production of IFNγ production during the exoerythrocytic stage of infection. As in murine models, this successful immunoprophylaxis of primates may also be mediated by IFNγ induction of iNOS in hepatocytes. Immunoprophylaxis trials of IL-12 in humans may help clarify whether NO-dependent mechanisms of protection occur following sporozoite challenge with *P. falciparum.*

2.2. Role of NO in the Immune Response to Asexual Erythrocytic Stages

2.2.1. *In Vitro* Studies

In addition to its effects on exoerythrocytic stages of infection, NO also mediates inhibition of asexual blood stages of *Plasmodium* species *in vitro* (Siu, 1968; Rockett *et al.*, 1991; Gyan *et al.*, 1994; Taylor-Robinson, 1997a). NO donors *in vitro* have dose-dependent antiplasmodial activity, with cytostatic effects occurring at low concentrations and enhanced cytotoxicity at high concentrations (Taylor-Robinson, 1997a). NO-related cytotoxicity increases with decreasing oxygen tension (Taylor-Robinson and Looker, 1998). Later parasite stages, i.e., trophozoites and schizonts, are more susceptible than the earlier ring stages to NO donors (Taylor-Robinson, 1997a). NO-dependent activity of human monocytes also inhibits erythrocytic stages of *P. falciparum in vitro* (Gyan *et al.*, 1994).

The mechanism of NO-related antiplasmodial activity is not known. Potential microbial cellular targets of reactive nitrogen intermediates include DNA, membrane lipids, and multiple proteins, including thiols, heme proteins, metabolic enzymes, and tyrosine groups (Fang, 1997). NO congeners are known to vary in their antimicrobial activity (DeGroote and Fang, 1995; Fang, 1997). For *Plasmodium*, *S*-nitrosothiols such as *S*-nitrosocysteine and *S*-nitrosoglutathione inhibit *P. falciparum* at micromolar concentrations, but a saturated solution of $NO^\bullet$ itself has no inhibitory effect on *P. falciparum* at low millimolar concentrations (Rockett *et al.*, 1991). This suggests that a transfer of NO^+ species (Stamler, 1994) to parasite thiol groups may account for NO-related antiplasmodial activity.

Because of the potent NO-scavenging effect of hemoglobin (Lancaster, 1994) surrounding the intraerythrocytic parasite, investigators have questioned the likely importance of NO as an effector molecule against blood-stage parasites *in vivo* (Jones *et al.*, 1996). However, it has been recently recognized that *S*-nitrosothiols can transfer NO^+ groups to globin –SH groups without quenching by heme metal centers, resulting in the formation of *S*-nitrosohemoglobin (Jia *et al.*, 1996); this could conceivably provide an intraerythrocytic reservoir of nitrosating potential for the mediation of antiplasmodial activity.

2.2.2. Animal Studies

In contrast to the antiparasitic effects of NO found in the above *in vitro* studies, most rodent models of malaria suggest that *in vivo* NO production during blood-stage infection is associated more with host-protective effects than with direct antiplasmodial effects (Kremsner *et al.*, 1992; Taylor-Robinson *et al.*, 1993, 1996; Jacobs *et al.*, 1995; Stevenson *et al.*, 1995; Tsuji *et al.*, 1995; Jones *et al.*, 1996; Amante and Good, 1997; Favre *et al.*, 1997, 1999a) although this appears to depend on the rodent and parasite strains used. In *P. vinckei* malaria, pretreatment with the NOS inhibitor NMMA (*N*-monomethyl-L-arginine) results in increased and earlier IFNγ-induced mortality, with no change in the course of parasitemia (Kremsner *et al.*, 1992). Similar findings have been reported following infection with the normally nonlethal species *P. vinckei petteri*, in which inhibition of NO production has little effect on parasitemia but results in the death of all mice (Jones *et al.*, 1996). In *P. yoelii* malaria, primary infection in recipients of *P. yoelii*-specific T cells is associated with increased NO production. Inhibition of NO production in this setting does not alter the course of infection, but does result in increased mortality at low parasitemias (Amante and Good, 1997).

Although primary *P. chabaudi* parasitemia is associated with increased NO production (Taylor-Robinson *et al.*, 1993; Jacobs *et al.*, 1995; Favre *et al.*, 1997), there has been conflicting evidence regarding the role of NO in the killing of blood-stage parasites belonging to this species. An early study in inbred NIH mice suggested that control of primary *P. chabaudi chabaudi* parasitemia is dependent on NO (Taylor-Robinson *et al.*, 1993). This finding has not been confirmed in NOS2-deficient mice, where *P. chabaudi chabaudi* parasitemia and survival were not affected by the lack of NOS2 (Favre *et al.*, 1999b). However resistance to *P. chabaudi chabaudi* is lost in C57BL/6 mice treated with the NOS inhibitor aminoguanidine, suggesting that requirement for NO in blood stage killing may depend on the rodent strain used (Favre *et al.*, 1999b). The same group has also examined infection in IFNγ receptor-deficient mice which are incapable of generating NO in response to blood stage infection with *P. chabaudi chabaudi* (Favre *et al.*, 1997). Compared with wild-type mice, there is no difference in peak initial parasitemia following infection of IFNγ receptor-deficient mice with *P. chabaudi chabaudi*, although mortality is significantly increased (Favre *et al.*, 1997). Similarly, in both *P. chabaudi adami* and *P. yoelii yoelii* infections, peak parasitemias do not differ in IFNγ receptor-deficient mice (Tsuji *et al.*, 1995) when compared with wild-type mice, although the time to clearance is slightly prolonged. Finally, studies using the *P. chabaudi* AS strain also suggest that NO production in response to blood-stage infection is more important for host protection than for parasite killing. IL-12 treatment of susceptible mice increases NO production and reduces *P. chabaudi* AS parasitemia and mortality, but inhibition of NO production in IL-12-treated mice increases mortality without affecting parasitemia (Stevenson *et al.*, 1995). In mouse strains normally resistant to *P. chabaudi* AS infection, inhibition of NO production results in significant mortality without affecting parasitemia (Jacobs *et al.*, 1995). Because of the many differences between rodent and human malaria infections (Butcher, 1996), it is difficult to extrapolate from the host-protective and antiparasitic effects of NO in rodent models to human infection. The results of our human studies of *P. falciparum* infection in African children (Anstey *et al.*, 1996) also support the concept that NO production is host-protective (see below), but the importance of NO in blood-stage parasite killing in humans *in vivo* is not yet known.

2.3. Role of NO in the Immune Response to Gametocytes/Sexual Stages

Serum obtained during paroxysms of fever in human vivax malaria inhibits the ability of gametocytes to infect mosquitoes (Karunaweera *et al.*, 1992). There is evidence in both rodent and human malaria that NO is an important mediator of this gametocyte inactivation. Studies using direct mosquito feeding on rodents with *P. vinckei petteri* malaria have shown that gametocyte infectivity is significantly reduced following schizont rupture, but infectivity can be restored by pretreatment with an L-arginine analogue (Motard *et al.*, 1993). Cytokine-mediated *P. falciparum* gametocyte inactivation *in vitro* is dependent on the presence of human leukocytes and can be inhibited by L-NMMA, suggesting that gametocyte inactivation in humans is mediated at least in part by leukocyte-derived NO (Naotunne *et al.*, 1993).

2.4. Role of NO in the Mosquito Immune Response

The recent discovery of parasite-inducible NOS activity in *A. stephensi* (Luckhart *et al.*, 1998) and *A. gambiae* (Dimopoulos *et al.*, 1998), both major mosquito vectors of human malaria, suggests that mosquitoes share with vertebrates a conserved NO-mediated anti-*Plasmodium* defense. Expression of *A. stephensi* NO synthase (AsNOS), highly homologous to vertebrate neuronal NOS, increases soon after midgut invasion by *Plasmodium* (Luckhart *et al.*,

1998). The antiplasmodial effects of mosquito NOS are evident from a significant increase in parasite numbers in infected mosquitoes following inhibition of AsNOS activity with dietary L-NAME (Luckhart *et al.*, 1998).

3. NO in Human Malaria *In Vivo*

3.1. NO Methodology: Lessons from Malaria Field Studies

Studies attempting to measure NO production in human malaria have illustrated the range of methodological difficulties involved in clinical studies of NO biology, and the need for careful consideration of confounding variables. In the presence of oxygen, NO is rapidly converted to the stable metabolites nitrite and nitrate (Kosaka *et al.*, 1979; Westfelt *et al.*, 1995). Measurement of nitrate + nitrite (NO_x) in plasma and urine provides a valid and useful marker of NO production in rodents and humans in a variety of disease states (Hibbs *et al.*, 1992; Granger *et al.*, 1996, 1999; Anstey *et al.*, 1996), provided there is adequate control for the potentially confounding effects of dietary nitrate ingestion and nitrate retention resulting from renal impairment.

Although several studies have described plasma NO_x levels in human malaria (Cot *et al.*, 1994; Nussler *et al.*, 1994; Prada and Kremsner, 1995; Al Yaman *et al.*, 1996; Kremsner *et al.*, 1996; Agbenyega *et al.*, 1997), it has been difficult to extrapolate NO production from the NO_x levels reported in these studies because of absent or insufficient numbers of disease-free control subjects and inadequate control for the potential confounding effects of dietary nitrate ingestion (Mitchell *et al.*, 1916; Anonymous, 1981; Granger *et al.*, 1996), renal impairment (Mackenzie *et al.*, 1996; Anstey *et al.*, 1997c), decreased fractional excretion of NO_x (Anstey *et al.*, 1996), and altered volume of distribution of NO_x in malaria. Each of these confounders can act to increase plasma NO_x levels in malaria without reflecting increased NO production (Anstey *et al.*, 1997c). Some early studies (Al Yaman *et al.*, 1996; Kremsner *et al.*, 1996; Nussler *et al.*, 1994; Prada and Kremsner, 1995) used high uncorrected plasma nitrate levels to extrapolate increased NO production in severe and cerebral malaria, particularly those with a fatal outcome (Al Yaman *et al.*, 1996). However, more recent controlled studies have shown that while uncorrected plasma NO_x levels are higher in fatal compared with non fatal severe and cerebral malaria, this difference disappears when NO_x levels are corrected for renal impairment (Anstey *et al.*, 1996; Dondorp *et al.*, 1998; Taylor *et al.*, 1998), and at least in African children are markedly lower in cerebral malaria than in fasting malaria-exposed control children (Anstey *et al.*, 1996).

3.2. Controlled Studies in African Children

In semi-immune malaria-exposed Tanzanian children, we have recently demonstrated a close inverse correlation between NO production/ iNOS expression and malaria disease severity (Anstey *et al.*, 1996). Controlling for diet and renal function, we compared disease severity with markers of NO production (urinary and plasma NO_x), mononuclear cell (MNC) iNOS expression, and plasma TNFα and IL-10 levels. Urine NO_x excretion, plasma NO_x levels (corrected for renal impairment), and MNC iNOS expression were inversely related to disease severity, with levels highest in asymptomatic parasitemia and lowest in fatal cerebral malaria (Figs. 4 and 5). A very close association between NO production and leukocyte iNOS expression was noted among the disease categories. Results could not be explained by differences in dietary nitrate ingestion among the groups. In semi-immune malaria-exposed African children, NO production was thus correlated with *protective* rather than disease-producing responses. Because host immune responses and malaria disease phenotype are influenced by many variables including age (Baird, 1998), parasite polymorphism (Gupta and Hill, 1995), host genetics (Hill *et al.*, 1991), intensity of malaria transmission (Snow *et al.*, 1997), and prior exposure, it will be important to validate this finding in different age groups and in areas with differing malaria epidemiology.

Many investigators have shown high-level NO production by murine macrophages, but others have had difficulty showing that human monocytes or macrophages produce NO *in vitro* (Denis, 1994). A striking finding in the Tanzanian study was the increased NO production and expression of iNOS in circulating MNC from apparently healthy asymptomatic malaria-exposed children, with or without patent parasitemia. This finding and similar results during other inflammatory disease states such as pulmonary tuberculosis (Nicholson *et al.*, 1996), rheumatoid arthritis (St. Clair *et al.*, 1996), and IFNα treatment in hepatitis C infection (Sharara *et al.*, 1997) provide *in vivo* evidence that inducible high-output NO production does in fact occur in human macrophages (see Chapter 6).

3.3. Potential Mechanisms of Modulation of NO Production in Human Malaria

Because NO production and iNOS expression were higher in the group of Tanzanian children with subclinical patent parasitemia than in those with subpatent infections, it is likely that increased NO production/ iNOS expression in asymptomatic malaria-exposed children is related, at least in part, to infection with *Plasmodium*. *Plasmodium* itself is capable of inducing host leukocyte NO production directly, independent of TNFα production. In non-human primates, immunization with recombinant *P. vivax* and *P. falciparum* antigens results in increased expression of NOS2 in antigen-stimulated peripheral blood mononuclear

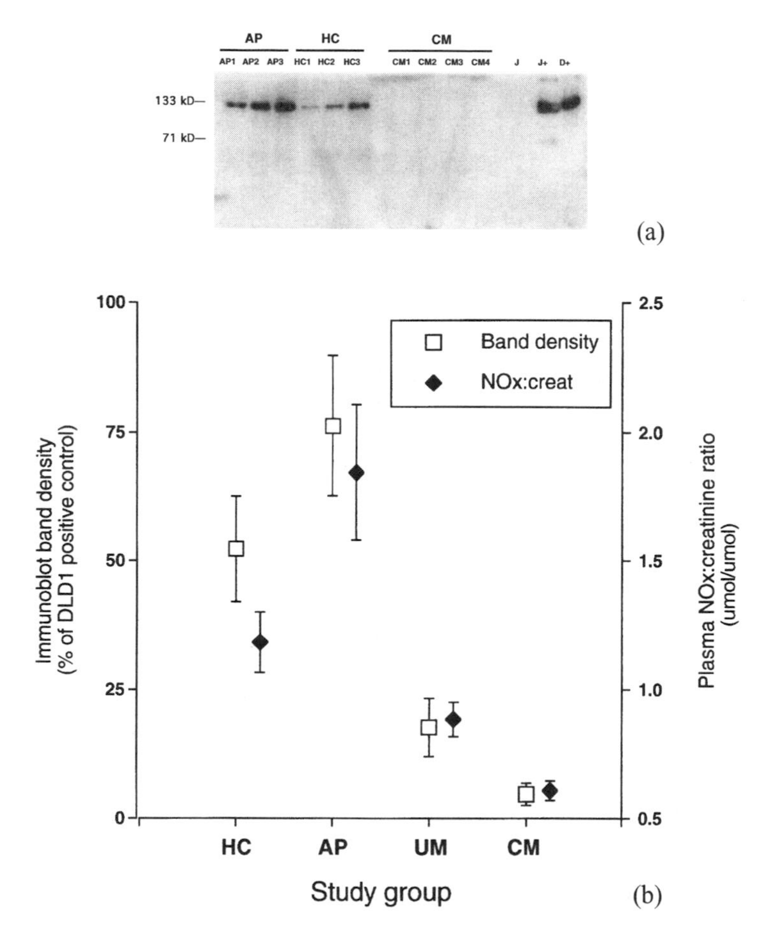

FIGURE 4. (a) Immunoblot analysis of mononuclear cells (MNC) from Tanzanian children. "Constitutive" expression of iNOS is seen in MNC from all asymptomatic malaria-exposed children with (AP) or without (HC) patent parasitemia. In contrast, MNC iNOS expression is suppressed in children with cerebral malaria (CM). (b) Inverse association between malaria severity and NO production in Tanzanian children. NO production was measured as plasma NO_x:creatinine ratio/MNC iNOS expression (by immunoblot). NO production/MNC iNOS expression is increased in asymptomatic malaria-exposed children with (AP) or without (HC) patent parasitemia, but suppressed in uncomplicated (UM) and cerebral (CM) malaria. Reproduced from *The Journal of Experimental Medicine*, 1996, Vol. 184, pp. 557–567 by copyright permission of The Rockefeller University Press.

cells. Blood-stage extracts of *P. vinckei* (Kremsner *et al.*, 1993) and *P. falciparum* (Naotunne *et al.*, 1993; Rockett *et al.*, 1996) induce NO production in mouse macrophages (Kremsner *et al.*, 1993; Rockett *et al.*, 1996) and human leukocytes (Naotunne *et al.*, 1993). However, as with earlier TNFα studies, the recent finding of widespread mycoplasmal contamination of *P. falciparum* cultures makes it

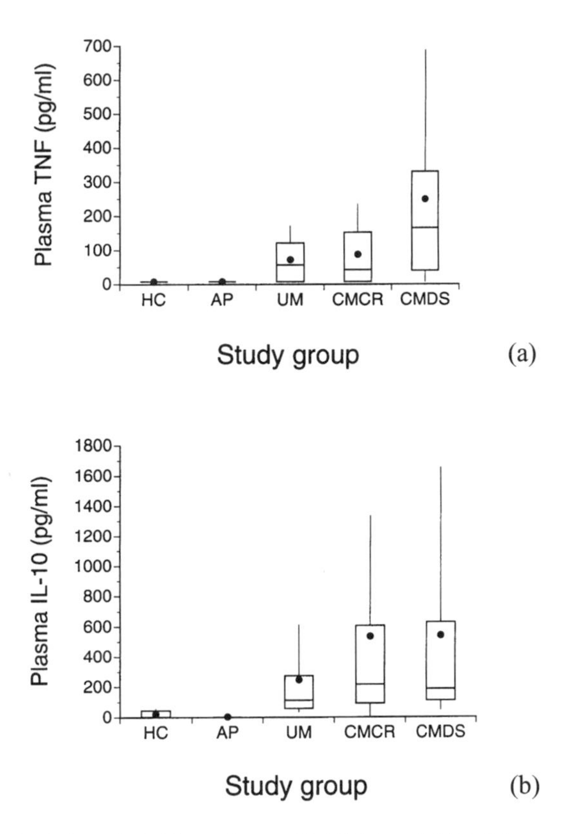

FIGURE 5. Cytokine levels in malaria. Plasma levels of (a) TNFα and (b) IL-10 are associated with malaria disease severity, being highest in patients with cerebral malaria with (CMDS) or without (CMCR) death or neurological sequelae. Reproduced from *The Journal of Experimental Medicine*, 1996, Vol. 184, pp. 557–567 by copyright permission of The Rockefeller University Press.

difficult to exclude an artifactual contribution from contaminants in these extracts (Turrini *et al.*, 1997). The extent to which such contaminants may have accounted for the apparent finding of NOS activity in erythrocytic stages of *P. falciparum* is also not clear (Ghigo *et al.*, 1995). An alternative mechanism for induction of iNOS in these malaria-exposed children is through cross-linking of macrophage CD23 by IgE–anti-IgE antibody complexes (Alonso *et al.*, 1995; Dugas *et al.*, 1995). Elevation in total and *Plasmodium*-specific IgE is nearly universal (Perlmann *et al.*, 1994) in malaria-endemic areas. Subclinical and subpatent infection with other microbial pathogens, including those causing increased IgE production (such as helminths), may also have contributed to the increased "basal" expression of iNOS in these Tanzanian children. Longitudinal *in vivo* studies are needed for more definitive analysis.

The mechanism of suppressed iNOS expression/NO production in uncomplicated and cerebral malaria in African children remains unclear. Determining the molecular basis of the deficient NO response in cerebral malaria may prove useful in identifying therapeutic and prophylactic strategies for severe disease. Deficient systemic NO production in cerebral malaria occurs despite high circulating levels of IFNγ, IL-1, and TNFα (Kwiatkowski *et al.*, 1990; Anstey *et al.*, 1996) (Fig. 5), each of which may increase iNOS expression (Chapter 5). Alternatively, host iNOS unresponsiveness may be mediated by excessive production of IL-10, TGFβ, and IL-4, known inhibitors of iNOS expression and NO production (Cunha *et al.*, 1992; Gazzinelli *et al.*, 1992; Bogdan and Nathan, 1993; Nathan and Xie, 1994; see also Chapter 5) (Fig. 3). Plasma levels of IL-10 are markedly increased in severe malaria (Peyron *et al.*, 1994; Ho *et al.*, 1995; Anstey *et al.*, 1996) (Fig. 5), but data on TGFβ and IL-4 are limited in African children. Inadequate host IL-12 production may also be important (Biron and Gazzinelli, 1995), though once again there are no available data.

Yet a further possibility is an iNOS gene or regulatory gene polymorphism associated with nonresponsiveness to Th1 cytokines in severe malaria. Two recent reports from Africa have shown an association of malaria disease severity with polymorphisms of the promoter region of NOS2. In Gabonese children, a single nucleotide polymorphism of the NOS2 promoter region was associated with protection from severe disease (Kun *et al.*, 1998). In Gambia children a pentanucleotide microsatellite polymorphism in a similar region was found to be associated with susceptibility to severe malaria (Burgner *et al.*, 1998). However, in our cohort of Tanzanian children with and without cerebral malaria (Anstey *et al.*, 1996), neither of these NOS2 promoter polymorphisms were associated with disease protection or susceptibility, and neither were associated with altered NO production (Levesque *et al.*, 1999). Such regional differences are not surprising in the light of the mounting evidence for complex heterogeneity in genetic susceptibility to severe and cerebral malaria in African populations (Bellamy *et al.*, 1998).

In addition to host factors, parasite products may be involved in suppression of macrophage NO production in clinical malaria (Fig. 3), as occurs in other protozoal infections (Green *et al.*, 1994). Infection by *Leishmania major* promastigotes or the addition of *L. major* glycoinositolphospholipid results in impaired macrophage NO production (Proudfoot *et al.*, 1995). β-Hematin, the heme moiety of malaria pigment (Arese and Schwarzer, 1997), has been shown to inhibit mouse macrophage NO production (Taramelli *et al.*, 1995), and ingestion of *P. vinckei* hemozoin by murine macrophages *in vitro* impairs NO production (Prada *et al.*, 1996). Similarly, hemozoin-laden liver macrophages from mice with *P. vinckei* blood-stage infections do not express iNOS, in contrast to adjacent hepatocytes (Prada *et al.*, 1996). Human macrophage functions, such as the oxidative burst and phagocytosis, are impaired after ingestion of *P. falciparum*-infected red cells or malaria pigment (hemozoin) (Schwarzer *et al.*, 1992; Leitner and Krzych, 1997).

However, it is not known whether hemozoin ingestion by human macrophages results in similar impairment of NO production.

3.4. NO and Host Protection *in Vivo*

The association between NO production and host protection in both animal and human malaria may in part be mediated by the ability of NO to downregulate production of TNFα (Florquin *et al.*, 1994; Tiao *et al.*, 1994) and other proinflammatory cytokines implicated in the pathogenesis of fever (Dinarello *et al.*, 1986; Kwiatkowski *et al.*, 1993) and clinical disease (Grau *et al.*, 1989; Kwiatkowski *et al.*, 1990). Although asymptomatic parasitemia, also known as clinical immunity or malaria "tolerance" (Mendis and Carter, 1995; Greenwood, 1996), appears to be mediated in part by an antitoxin antibody (Bate and Kwiatkowski, 1994a; Kwiatkowski *et al.*, 1997), chronic expression of leukocyte iNOS and NO production may also be involved. Clinical immunity or "tolerance" following repeated exposure to malaria resembles the diminution in febrile response seen following repeated exposure to endotoxin (Heyman and Beeson, 1949; Rubenstein *et al.*, 1965), and recent evidence suggests that endotoxin "tolerance" is mediated in part by NO (Fahmi and Chaby, 1993; Rojas *et al.*, 1993). Parasite-induced NO production seen in asymptomatic parasitemic children (Anstey *et al.*, 1996) may downregulate TNFα production and fever in response to parasitemia, and could explain (at least in part) the clinical immunity seen in these children (Anstey *et al.*, 1996, 1999b; Clark *et al.*, 1996).

Another potential mechanism of protection *in vivo* is the direct antiparasitic effect of NO shown to occur *in vitro*. This does not appear to be the mechanism of protection in most rodent models of malaria. Data from humans are limited. Although a number of human clinical studies have shown an inverse relationship between parasitemia and plasma NO_x levels (Agbenyega *et al.*, 1997; Al Yaman *et al.*, 1996; Kremsner *et al.*, 1996), this has not been a consistent finding and the correlations have been at best modest. Moreover, the NO_x levels reported in the studies showing this inverse correlation were not controlled for potential confounding effects of diet and renal impairment. NO may be involved in human blood stage antiparasitic immunity, but further studies are required.

3.5. Age Related Changes in NO Production and Clinical Immunity

Age appears to influence not only the acquisition of clinical immunity to malaria but also the susceptibility to and clinical manifestations of severe malaria (Marsh and Snow, 1997b; Baird 1998; Baird *et al.*, 1998). These observations may be related to age-related difference in NO production (Kissin *et al.*, 1997; Tsukahara *et al.*, 1997; Anstey *et al.*, 1999b). Our recent studies in Tanzanian children show that NO production is highest in infancy, falling after the first year of life, then rising again after 5 years of age (Anstey *et al.*, 1999b). This pattern of age-related NO production is the reverse of the pattern of age-related severe malarial morbidity in coastal Tanzanian children. Elevated production of NO in both infants and older children, may be related to age *per se* and malaria infection respectively, and may be one of the mediators of the anti-disease immunity found most commonly in these two age groups (Anstey *et al.*, 1999b).

3.6. Is NO Involved in Malaria Pathogenesis?

3.6.1. Cerebral Malaria

Suppression of systemic NO synthesis in cerebral malaria (Anstey *et al.*, 1996) may contribute to the excessive TNFα production implicated in the pathogenesis of this disease, through loss of the negative feedback of NO on TNFα production (Florquin *et al.*, 1994; Tiao *et al.*, 1994). Despite the *in vitro* and *in vivo* evidence for a protective role for NO in malaria, there has been speculation that organ-specific NO production may contribute to the pathogenesis of severe malaria (Clark *et al.*, 1991; Clark and Rockett, 1996). Some investigators have hypothesized that high concentrations of TNFα in cerebral malaria induce excessive local synthesis of NO by cerebrovascular endothelial cells, and that this NO alters neurotransmission, causing profound but reversible coma (Clark *et al.*, 1991, 1996). However, at present there are no *in vivo* data to support this hypothesis. Although mice provide an admittedly imperfect model of human cerebral malaria, inhibition of NOS activity in *P. berghei*-infected mice prevents neither neurological signs or death in infected wild-type (Arsenio *et al.*, 1993; Kremsner *et al.*, 1993) and NOS2-deficient (Favre *et al.*, 1999a) mice. As discussed earlier, systemic NO production is suppressed in human cerebral malaria in Tanzanian children (Anstey *et al.*, 1996). Recent data from Zambian children with cerebral malaria support these findings and cast more doubt on a role for excessive NO in malaria-associated coma. Desferrioxamine treatment, previously associated with more rapid resolution of coma (Gordeuk *et al.*, 1992), is associated with a *rise* in NO production compared with persistently low levels in controls (Weiss *et al.*, 1997).

Parasite products can induce NOS2 expression and NO production in human endothelial cells *in vitro* (Tachado *et al.*, 1996). Although the effects of organ-specific NO production are likely to be complex (Bogdan, 1998), it can be argued that any local NO production by cerebrovascular endothelium may well be protective in cerebral malaria. The cytokine-induced endothelial activation described in cerebral malaria (Turner *et al.*, 1994; Turner, 1997) is inhibited by NO. In human vascular endothelial cells, cytokine-induced expression of adhesion molecules such as VCAM-1, E-selectin, and ICAM-1 is decreased by exogenous and endogenous NO (Decaterina *et al.*, 1995; Khan *et al.*, 1996) via inhibition of NF-κB (Peng *et al.*, 1995). Any endothelial NO produced in cerebral malaria appears to be inadequate to prevent the increased expression of endothelial receptors for parasite cytoadherence, all of which are upregulated in human cerebral malaria (Turner *et al.*, 1994; Turner, 1997). Moreover, by increasing systemic TNFα production, the suppression of *systemic* NO in cerebral malaria may cause further TNFα-induced upregulation of cerebrovascular endothelial receptors and even greater parasite sequestration.

Measurements of NO_x in cerebrospinal fluid in human cerebral malaria have been inconsistent and difficult to interpret (Agbenyega *et al.*, 1997; Dondorp *et al.*, 1998; Weiss *et al.*, 1998). Levels are much lower than in plasma and it is not known how much is derived from NO production in adjacent brain parenchyma and how much from filtration of plasma nitrates through the choroid plexus (Granger *et al.*, 1999). Intravenous nitrate infusions in rabbits increase CSF nitrate concentrations (Nattie and Reeder, 1983). Uncorrected plasma and CSF NO_x levels are correlated in Thai adults with cerebral malaria (Dondorp *et al.*, 1998), but this has not been a universal finding. Detailed studies of postmortem cerebral malaria brain samples will be required to determine the extent of any *local* expression and activity of NOS2 in cerebrovascular endothelium, and expression/activity of all three NOS isoforms in brain parenchyma. However, because of increasing complexities in the interpretation of such studies, the results will not necessarily allow firm conclusions to be drawn about the functional role of local NO/NOS in neurological pathology versus protection (Bogdan, 1998).

3.6.2. Immunosuppression

Investigators have also postulated a role for NO (Rockett *et al.*, 1994; Ahvazi *et al.*, 1995; Taylor-Robinson, 1997b) in the suppression of nonspecific and specific immune responses found during and following acute falciparum malaria (Greenwood *et al.*, 1972; Williamson and Greenwood, 1978; MacDermott *et al.*, 1980; Riley *et al.*, 1988). There is some evidence to support this hypothesis in animal models, with suppression of lymphocyte proliferation in *P. vinckei* (Rockett *et al.*, 1994) and *P. chabaudi* (Rockett *et al.*, 1994; Ahvazi *et al.*, 1995) rodent malaria associated with NO production and reversed by the NOS inhibitor L-NMMA. This suppression of Th1 proliferation may be mediated by NO-mediated inhibition of IL-2 production (Taylor-Robinson, 1997b). However, there are no confirmatory data in human malaria. Because NO production is suppressed in uncomplicated falciparum malaria (Anstey *et al.*, 1996), NO is unlikely to be an important mediator of immunosuppression during acute malaria, but may prove to have such a role in convalescence.

3.6.3. Anemia

NO is known to decrease erythropoiesis, and is likely to be an important mediator of the anemia of chronic disease in humans (Maciejewski *et al.*, 1995; Domachowske, 1997). Bone marrow NO production is increased in *Trypanosoma brucei*-infected mice, and is thought to play a significant role in the anemia of *T. brucei* infection (Mabbott and Sternberg, 1996). Anemia is extremely common in malaria-endemic areas (Newton *et al.*, 1997), and its severity is frequently out of proportion to the low-level parasitemia found in such children. Because asymptomatic malaria-exposed children in malaria-endemic areas have increased NO production/leukocyte iNOS expression (Anstey *et al.*, 1996) which is likely to be sustained, we have hypothesized that chronic overproduction of NO in these asymptomatic children contributes to the anemia associated with subclinical/subpatent malaria. In our group of fasting, asymptomatic malaria-exposed Tanzanian children, NO production was inversely associated with hemoglobin concentration after controlling for age (Anstey *et al.*, 1999a). After controlling for age *and parasitemia* NO was no longer an independent predictor of anaemia, however one of the mechanisms of parasite-related anemia in such children may be through the adverse hematological effects of parasite-induced NO production (Anstey *et al.*, 1999a).

4. Conclusions

NO appears to be an important mediator of the protective immune response during all stages of *Plasmodium* infections. NO-related activity against liver stages and sexual blood stages is antiparasitic. NO production in asexual blood-stage infection *in vivo*, at least in rodent malaria, appears to be associated more with disease-ameliorating host-protective effects than with the direct antiplasmodial effects found *in vitro.* The association between NO production and disease protection shown in African children may be mediated by the ability of NO to downregulate TNFα, or to inhibit endothelial expression of receptors used by parasitized red cells to adhere to vascular endothelium. Further studies are required to explore mechanisms of NO-mediated antiplasmodial activity and host protection. Because host immune responses and malarial disease expression in humans are influenced by many variables, it will be important to validate the association between NO production and disease protection in other age groups, and in areas with differing malarial epidemiology. Such clinical studies must carefully control for potential confounding variables, if NO_x levels are used as markers of NO production. Further human studies are also required to clarify the role of NO (if any) in the pathogenesis of malaria, particularly organ-specific disease, immunosuppression, and anemia. Much has been learned since the life cycle of human malaria was discovered a century ago, but malaria mortality has not declined. The challenge will be to ensure that advances in our understanding of NO biology in malaria will inform new prophylactic, therapeutic, and vaccine strategies.

References

Abdulla, S., Weatherall, D., Wickramasinghe, S., and Hughes, M., 1980, The anaemia of *P. falciparum* malaria, *Br. J. Haematol.* **46:**171–183.

Agbenyega, T., Angus, B., Bedu-Addo, G., Baffoe-Bonnie, B., Griffin, G., Vallance, P., and Kriahna, S., 1997, Plasma nitrogen oxides and blood lactate concentrations in Ghanaian children with malaria, *Trans. R. Soc. Trop. Med. Hyg.* **91:**298–302.

Ahvazi, B. C., Jacobs, P., and Stevenson, M. M., 1995, Role of macrophage-derived nitric oxide in suppression of lymphocyte proliferation during blood-stage malaria, *J. Leukoc. Biol.* **58:**23–31.

Alonso, A., Carvalho, J., Alonso-Torre, S. R., Nunez, L., Bosca, L., and Sanchez Crespo, M., 1995, Nitric oxide synthesis in rat peritoneal macrophages is induced by IgE/DNP complexes and cyclic AMP analogues. Evidence in favor of a common signaling mechanism, *J. Immunol.* **154:**6475–6483.

Al Yaman, F. M., Mokela, D., Genton, B., Rockett, K. A., Alpers, M. P., and Clark, I. A., 1996, Association between serum levels of reactive nitrogen intermediates and coma in children with cerebral malaria in Papua New Guinea, *Trans. R. Soc. Trop. Med. Hyg.* **90:**270–273.

Amante, F. H., and Good, M. F., 1997, Prolonged Th1-like response generated by a *Plasmodium yoelii*-specific T cell clone allows complete clearance of infection in reconstituted mice, *Parasite Immunol.* **19:**111–126.

Anonymous, 1981, Nitrate, nitrite and nitrogen oxides: Environmental distribution and exposure of humans, in: *The Health Effects of Nitrate, Nitrite, and N-Nitroso Compounds,* Volume 1 (F. M. Peter, ed.), National Academy Press, Washington, D.C., pp. 3–52.

Anstey, N. M., Weinberg, J. B., Hassanali, M. Y., Mwaikambo, E. D., Manyenga, D., Misukonis, M. A., Arnelle, D. R., Hollis, D., McDonald, M. I., and Granger, D. L., 1996, Nitric oxide in Tanzanian children with malaria: Inverse relationship between malaria severity and nitric oxide production/nitric oxide synthase type 2 expression, *J. Exp. Med.* **184:**557–67.

Anstey, N., Granger, D., Hassanali, M., Duffy, P., Mwaikambo, E., and Weinberg, J., 1999a, Nitric oxide malaria and anaemia: Inverse relationship between NO production and haemoglobin concentration in asymptomatic malaria-exposed children, *Am. J. Trop. Med. Hyg.* **60:** in press.

Anstey, N., Weinberg, J., Wang, Z., Mwaikambo, E., Duffy, P., and Granger, D., 1999b, Effects of age and parasitemia on nitric oxide (NO) production/leukocyte nitric oxide synthase type 2 expression in asymptomatic, malaria-exposed children, *Am. J. Trop. Med. Hyg.* **60:** in press.

Anstey, N. M., Granger, D. L., and Weinberg, J. B., 1997, Nitrate levels in malaria [letter], *Trans. R. Soc. Trop. Med. Hyg.* **91:**238–240.

Arese, P., and Schwarzer, E., 1997, Malarial pigment (haemozoin)—a very active inert substance, *Ann. Trop. Med. Parasitol.* **91:**501–516.

Arsenio, V., Oshima, H., and Falanga, P., 1993, *Plasmodium berghei*: Is nitric oxide involved in the pathogenesis of mouse cerebral malaria? *Exp. Parasitol.* **77:**111–117.

Baird, J. K., 1998, Age-dependent characteristics of protection v. susceptibility to *Plasmodium falciparum, Ann. Trop. Med. Parasitol.* **92:**367–390.

Baird, J. K., Masbar, S., Basri, H., Tirtokusomo, S., Subianto, B., and Hoffman, S. L., 1998, Age-dependent susceptibility to severe disease with primary exposure to *Plasmodium falciparum, J. Infect. Dis.* **178:**592–595.

Barnwell, J., Asch, A., Nachman, R., Yamaya, M., Aikawa, M., and Ingravallo, P., 1989, A human 88-

kD membrane glycoprotein (CD36) functions in vitro as a receptor for a cytoadherence ligand on *Plasmodium falciparum*-infected erythrocytes, *J. Clin. Invest.* **84:**765–772.

Bate, C. A., and Kwiatkowski, D., 1994a, Inhibitory immunoglobulin M antibodies to tumor necrosis factor-inducing toxins in patients with malaria, *Infect. Immun.* **62:**3086–3091.

Bate, C. A., and Kwiatkowski, D., 1994b, A monoclonal antibody that recognizes phosphatidylinositol inhibits induction of tumor necrosis factor alpha by different strains of *Plasmodium falciparum*, *Infect. Immun.* **62:**5261–5266.

Bate, C. A., Taverne, J., Bootsma, H. J., Mason, R. C., Skalko, N., Gregoriadis, G., and Playfair, J. H., 1992a, Antibodies against phosphatidylinositol and inositol monophosphate specifically inhibit tumour necrosis factor induction by malaria exoantigens, *Immunology* **76:**35–41.

Bate, C. A., Taverne, J., and Playfair, J. H., 1992b, Detoxified exoantigens and phosphatidylinositol derivatives inhibit tumor necrosis factor induction by malarial exoantigens, *Infect. Immun.* **60:**1894–1901.

Bate, C. A., Taverne, J., Roman, E., Moreno, C., and Playfair, J. H., 1992c, Tumour necrosis factor induction by malaria exoantigens depends upon phospholipid, *Immunology* **75:**129–135.

Bellamy, R., Kwiatkowski, D., and Hill, A. V., 1998, Absence of an association between intercellular adhesion molecule 1, complement receptor 1 and interleukin 1 receptor antagonist gene polymorphisms and severe malaria in a West African population, *Trans. R. Soc. Trop. Med. HYg.* **92:**312–316.

Berendt, A. R., Simmons, D. L., Tansey, J., Newbold, C. I., and Marsh, K., 1989, Intercellular adhesion molecule is an endothelial cell adhesion receptor for *Plasmodium falciparum*, *Nature* **341:**57–59.

Berendt, A. R., Turner, G. D. H., and Newbold, C. I., 1994, Cerebral malaria: The sequestration hypothesis, *Parasitol. Today* **10:**412–414.

Biron, C., and Gazzinelli, R., 1995, Effects of IL–12 on immune responses to microbial infections: A key mediator in regulating disease outcome, *Curr. Opin. Immunol.* **7:**485–496.

Bogdan, C., 1998, The multiplex function of nitric oxide (NO) in (auto)immunity, *J. Exp. Med.* **187:**1361–1365.

Bogdan, C., and Nathan, C., 1993, Modulation of macrophage function by transforming growth factor β, interleukin-4, and interleukin-10, *Ann. N. Y. Acad. Sci.* **685:**713–739.

Bouharoun-Tayoun, H., Attanath, P., Sabchareon, A., Chongsuphajaisiddhi, T., and Druilhe, P., 1990, Antibodies that protect against *Plasmodium falciparum* blood stages do not on their own inhibit parasite growth and invasion *in vitro*, but act in cooperation with monocytes, *J. Exp. Med.* **172:**1633–1641.

Butcher, G., 1996, Models for malaria, *Parasitol. Today* **12:**378–382.

Butcher, G. A., Garland, T., Ajdukiewicz, A. B., and Clark, I. A., 1990, Serum tumor necrosis factor associated with malaria in patients in the Solomon Islands, *Trans. R. Soc. Trop. Med. Hyg.* **84:**658–661.

Clark, I. A., and Chaudhri, G., 1988, Tumour necrosis factor may contribute to the anaemia of malaria by causing dyserythropoiesis and erythrophagocytosis, *Br. J. Haematol.* **70:**99–103.

Clark, I. A., and Rockett, K. A., 1996, Nitric oxide and parasitic disease, *Adv. Parasitol.* **37:**1–56.

Clark, I. A., Rockett, K. A., and Cowden, W. B., 1991, Proposed link between cytokines, nitric oxide, and human cerebral malaria, *Parasitol. Today* **7:**205–207.

Clark, I. A., al-Yaman, F. M., Cowden, W. B., and Rockett, K. A., 1996, Does malarial tolerance, through nitric oxide, explain the low incidence of autoimmune disease in tropical Africa? *Lancet* **348:**1492–1494.

Cot, S., Ringwald, P., Mulder, B., Miailhes, P., Yap-Yap, J., Nussler, A. K., and Eling, W. M., 1994, Nitric oxide in cerebral malaria, *J. Infect. Dis.* **169:**1417–1418.

Cunha, F. Q., Moncada, S., and Liew, F. Y., 1992, Interleukin-10 (IL-10) inhibits the induction of nitric oxide synthase by interferon-γ in murine macrophages, *Biochem. Biophys. Res. Commun.* **186:**1155–1159.

Decaterina, R., Libby, P., Peng, H. B., Thannickal, V. J., Rajavashisth, T. B., Gimbrone, M. A., Shin, W. S., and Liao, J. K., 1995, Nitric oxide decreases cytokine-induced endothelial activation—Nitric oxide selectively reduces endothelial expression of adhesion molecules and proinflammatory cytokines, *J. Clin. Invest.* **96:**60–68.

DeGroote, M. A., and Fang, F. C., 1995, NO inhibitions. Antimicrobial properties of nitric oxide, *Clin. Infect. Dis.* **21:**S162-S165.

Denis, M., 1994, Human monocytes/macrophages: NO or no NO? *J. Leukoc. Biol.* **55:**682–684.

Dimopoulos, G., Seeley, D., Wolf, A., and Kafatos, F. C., 1998, Malaria infection of the mosquito *Anopheles gambiae* activates immune-responsive genes during critical transition stages of the parasite life cycle, *EMBO. J.* **17:**6115–6123.

Dinarello, C. A., Cannon, J. G., Wolff, S. M., Bernheim, H. A., Beutler, B., Cerami, A., Figari, I. S., Palladino, M. A. J., and O'Connor, J. V., 1986, Tumor necrosis factor (cachectin) is an endogenous pyrogen and induces production of interleukin 1, *J. Exp. Med.* **163:**1433–1450.

Domachowske, J. B., 1997, The role of nitric oxide in the regulation of cellular iron metabolism, *Biochem. Mol. Med.* **60:**1–7.

Dondorp, A. M., Planche, T., de Bel, E. E., Angus, B. J., Chotivanich, K. T., Silamut, K., Romijn, J. A., Ruangveerayuth, R., Hoek, F. J., Kager, P. A., Vreeken, J., and White, N. J., 1998, Nitric oxides in plasma, urine, and cerebrospinal fluid in patients with severe falciparum malaria, *Am. J. Trop. Med. Hyg.* **59:**497–502.

Doolan, D. L., Sedegah, M., Hedstrom, R. C., Hobart, P., Charoenvit, Y., and Hoffman, S. L., 1996, Circumventing genetic restriction of protection against malaria with multigene DNA immunization: CD8+ cell-, interferon gamma-, and nitric oxide-dependent immunity, *J. Exp. Med.* **183:**1739–1746.

Dugas, B., Mossalayi, D., Damais, C., and Kolb, J., 1995, Nitric oxide production by human monocytes: Evidence for a role for CD23, *Immunol. Today* **16:**574–580.

Duque, S., Montenegro-James, S., Arevalo-Herrera, M., Praba, A. D., Villinger, F., Herrera, S., and James, M. A., 1998, Expression of cytokine genes in Aotus monkeys immunized with synthetic and recombinant *Plasmodium vivax* and *P. falciparum* antigens, *Ann. Trop. Med. Parasitol.* **92:**553–559.

Elloso, M., van der Hyde, H., vande Waa, J., Manning, D., and Weidanz, W., 1994, Inhibition of *Plasmodium falciparum* in vitro by human γδ T cells, *J. Immunol.* **153:**1187–1194.

English, M., and Marsh, K., 1997, Childhood malaria—pathogenesis and treatment, *Curr. Opin. Infect. Dis.* **10:**221–225.

English, M., Marsh, V., Amukoye, E., Lowe, B., Murphy, S., and Marsh, K., 1996a, Chronic salicylate toxicity and severe malaria, *Lancet* **347:**1736–1737.

English, M., Waruiru, C., Lightowler, C., Murphy, S., Kirigha, G., and Marsh, K., 1996b, Hyponatraemia and dehydration in severe malaria, *Arch. Dis. Child.* **743:**210–205.

Fahmi, H., and Chaby, R., 1993, Desensitization of macrophages to endotoxin effects is not correlated with a down-regulation of lipopolysaccharide-binding sites, *Cell. Immunol.* **150:**219–229.

Fang, F. C., 1997, Mechanisms of nitric oxide-related antimicrobial activity, *J. Clin. Invest.* **99:**2818–2825.

Favre, N., Ryffel, B., Bordmann, G., and Rudin, W., 1997, The course of *Plasmodium chabaudi chabaudi* infections in interferon-gamma receptor deficient mice, *Parasite Immunol.* **19:**375–383.

Favre, N., Ryffel, B., and Rudin, W., 1999a, The development of murine cerebral malaria does not require nitric oxide production, *Parasitology* **118:**135–138.

Favre, N., Ryffel, B., and Rudin, W., 1999b, Parasite killing in murine malaria does not require nitric oxide production, *Parasitology* **118:**139–143.

Fell, A. H., Currier, J., and Good, M. F., 1994, Inhibition of *Plasmodium falciparum* growth in vitro by CD4+ and CD8+ T cells from non-exposed donors, *Parasite Immunol.* **16:**579–586.

Florquin, S., Amraoui, Z., Dubois, C., Decuyper, J., and Goldman, M., 1994, The protective role of endogenously synthesized nitric oxide in staphylococcal enterotoxin B-induced shock in mice, *J. Exp. Med.* **180:**1153–1158.

Fried, M., and Duffy, P., 1996, Adherence of *Plasmodium falciparum* to chondroitin sulphate A in the human placenta, *Science* **272:**1502–1504.

Gazzinelli, R. T., Oswald, I. P., James, S. L., and Sher, A., 1992, IL-10 inhibits parasite killing and nitrogen oxide production by IFNγ-activated macrophages, *J. Immunol.* **148:**1792–1796.

Ghigo, D., Todde, R., Ginsburg, H., Costamagna, C., Gautret, P., Bussolino, F., Ulliers, D., Giribaldi, G., Deharo, E., Gabrielli, G., Pescarmona, G., and Basia, A., 1995, Erythrocyte stages of *Plasmodium falciparum* exhibit a high nitric oxide synthase (NOS) activity and release an NOS-inducing soluble factor, *J. Exp. Med.* **182:**677–688.

Gordeuk, V., Thuma, P., Brittenham, G., McLaren, C., Parry, D., Backenstose, A., Biemba, G., Msiska, R., Holmes, L., McKinley, E., Vargas, L., Gilkeson, R., and Poltera, A. A., 1992, Effect of iron chelation therapy on recovery from deep coma in children with cerebral malaria, *N. Engl. J. Med.* **327:**1473–1477.

Granger, D., Anstey, N., Miller, W., and Weinberg, J., 1999, Measuring nitric oxide production in human clinical studies, *Meth. Enzymol.* **301:**49–61.

Granger, D. L., Miller, W. C., and Hibbs, J. B., Jr., 1996, Methods of analyzing nitric oxide production in the immune response, in: *Methods in Nitric Oxide (NO) Research* (M. Feelisch and J. S. Stamler, eds.), Wiley, New York, pp. 603–618.

Grau, G. E., Taylor, T. E., Molyneux, M. E., Wirima, J. J., Vassalli, P., Hommel, M., and Lambert, P. H., 1989, Tumor necrosis factor and disease severity in children with falciparum malaria, *N. Engl. J. Med.* **320:**1586–1591.

Green, S. J., Scheller, L. F., Marletta, M. A., Seguin, M. C., Klotz, F. W., Slayter, M., Nelson, B. J., and Nacy, C. A., 1994, Nitric oxide: Cytokine-regulation of nitric oxide in host resistance to intracellular pathogens, *Immunol. Lett.* **43:**87–94.

Greenwood, B., 1987, Asymptomatic malaria infections—do they matter? *Parasitology Today* **3:**206–214.

Greenwood, B., 1996, Fever and malaria, *Lancet* **348:**279–280.

Greenwood, B., Bradley-Moore, A., Palit, A., and Bryceson, A., 1972, Immunosuppression in children with malaria, *Lancet* **1:**169–172.

Gupta, S., and Hill, A. V. S., 1995, Dynamic interactions in malaria– Host heterogeneity meets parasite polymorphism, *Proc. R. Soc. London Ser. B* **261:**271–277.

Gyan, B., Troye-Blomberg, M., Perlmann, P., and Bjorkman, A., 1994, Human monocytes cultured with and without interferon-gamma inhibit *Plasmodium falciparum* parasite growth *in vitro* via secretion of reactive nitrogen intermediates, *Parasite Immunol.* **16:**371–375.

Heyman, A., and Beeson, P., 1949, Influence of various disease states upon the febrile response to intravenous injection of typhoid bacterial pyrogen, *J. Lab. Clin. Med.* **34:**1400–1403.

Hibbs, J. B., Jr., Westenfelder, C., Taintor, R., Vavrin, Z., Kablitz, C., Baranowski, R. L., Ward, J. H., Menlove, R. L., McMurry, M. P., Kushner, J. P., and Samlowski, W. E., 1992, Evidence for cytokine-inducible nitric oxide synthesis from L-arginine in patients receiving interleukin-2 therapy, *J. Clin. Invest.* **89:**867–877.

Hill, A. V. S., Allsopp, C. E. M., Kwiatkowski, D., Anstey, N. M., Twumasi, P., Rowe, P. A., Bennett, S., Brewster, D., McMichael, A. J., and Greenwood, B. M., 1991, Common West African HLA antigens are associated with protection from severe malaria, *Nature* **352:**595–600.

Ho, M., and Sexton, M. M., 1995, Clinical immunology of malaria, *Bailliere's Clin. Infect. Dis.* **2:**227–247.

Ho, M., Sexton, M. M., Tongtawe, P., Looareesuwan, S., Suntharasamai, P., and Webster, H. K., 1995, Interleukin-10 inhibits tumor necrosis factor production but not antigen-specific lymphoproliferation in acute *Plasmodium falciparum* malaria, *J. Infect. Dis.* **172:**838–844.

Hoffman, S., Crutcher, J., Puri, S., Ansari, A., Villinger, F., Franke, E., Singh, P., Finkelman, F., Gately, M., Dutta, G., and Sedegah, M., 1997, Sterile protection of monkeys against malaria after administration of interleukin 12, *Nature Med.* **3:**80–83.

Holt, D., Gardiner, D., Thomas, E., Mayo, M., Bourke, P., Sutherland, C., Carter, R., Myers, G., Kemp, D., and Trenholme, K., 1999, The *clag* gene family of *Plasmodium falciparum:* Are there roles other than cytoadherence? *Int. J. Parasitol.* in press.

Jacobs, P., Radzioch, D., and Stevenson, M. M., 1995, Nitric oxide expression in the spleen, but not in the liver, correlates with resistance to blood-stage malaria in mice, *J. Immunol.* **155:**5306–5313.

Jakobsen, P. H., Bate, C. A., Taverne, J., and Playfair, J. H., 1995, Malaria: Toxins, cytokines and disease, *Parasite Immunol.* **17:**223–231.

James, S. L., 1995, Role of nitric oxide in parasitic infections, *Microbiol. Rev.* **59:**533–547.

Jia, L., Bonaventura, C., Bonaventura, J., and Stamler, J. S., 1996, *S*-Nitrosohaemoglobin—A dynamic activity of blood involved in vascular control, *Nature* **380:**221–226.

Jones, I. W., Thomsen, L. L., Knowles, R., Gutteridge, W. E., Butcher, G. A., and Sinden, R. E., 1996, Nitric oxide synthase activity in malaria-infected mice, *Parasite Immunol.* **18:**535–538.

Karunaweera, N. D., Carter, R., Grau, G. E., Kwiatkowski, D., Del Giudice, G., and Mendis, K. N., 1992, Tumour necrosis factor-dependent parasite-killing effects during paroxysms in non-immune *Plasmodium vivax* malaria patients, *Clin. Exp. Immunol.* **88:**499–505.

Kemp, D., Sutherland, C., Holt, D., and Trenholme, K., 1996, The parasite's new clothes: Genetic aspects of antigenic diversity and variation in blood stages of *Plasmodium falciparum*, *Recent Adv. Microbiol.* **4:**161–200.

Kern, P., Hemmer, C. J., Van Damme, J., Gruss, H. J., and Dietrich, M., 1989, Elevated tumor necrosis factor alpha and interleukin-6 serum levels as markers for complicated *Plasmodium falciparum* malaria, *Am. J. Med.* **87:**139–143.

Khan, B., Harrison, D., Olbrych, M., Alexander, R., and Medford, R., 1996, Nitric oxide regulates vascular cell adhesion molecule 1 gene expression and redox-sensitive transcriptional events in human vascular endothelial cells, *Proc. Natl. Acad. Sci. USA* **93:**9114–9119.

Kissin, E., Tomasi, M., McCartneyfrancis, N., Gibbs, C. L., and Smith, P. D., 1997, Age-related decline in murine macrophage production of nitric oxide, *J. Infect. Dis.* **175:**1004–1007.

Klotz, F. W., Scheller, L. F., Seguin, M. C., Kumar, N., Marletta, M. A., Green, S. J., and Azad, A. F., 1995, Co-localization of inducible-nitric oxide synthase and *Plasmodium berghei* in hepatocytes from rats immunized with irradiated sporozoites, *J. Immunol.* **154:**3391–3395.

Knell, A. (ed.), 1991, *Malaria*, Oxford University Press, London.

Kosaka, H. K., Imaizumi, K., Imai, K., and Tyuma, I., 1979, Stoichiometry of the reaction of oxyhemoglobin with nitrite, *Biochim. Biophys. Acta* **581:**184–188.

Kremsner, P. G., Neifer, S., Chaves, M. F., Rudolph, R., and Bienzle, U., 1992, Interferon-gamma induced lethality in the late phase of *Plasmodium vinckei* malaria despite effective parasite clearance by chloroquine, *Eur. J. Immunol.* **22:**2873–2878.

Kremsner, P. G., Nussler, A., Neifer, S., Chaves, M. F., Bienzle, U., Senaldi, G., and Grau, G. E., 1993, Malaria antigen and cytokine-induced production of reactive nitrogen intermediates by murine macrophages: No relevance to the development of experimental cerebral malaria, *Immunology* **78:**286–290.

Kremsner, P. G., Winkler, S., Wildling, E., Prada, J., Bienzle, U., Graninger, W., and Nussler, A. K., 1996, High plasma levels of nitrogen oxides are associated with severe disease and correlate with rapid parasitological and clinical cure in *Plasmodium falciparum* malaria, *Trans. R. Soc. Trop. Med. Hyg.* **90:**44–47.

Krishna, S., Waller, D., ter Kuile, F., Kwiatkowski, D., Crawley, J., Craddock, C., Nosten, F., Chapman, D., Brewster, D., and Holloway, P., 1994, Lactic acidosis and hypoglycemia in children with severe malaria: Pathophysiological and prognostic significance, *Trans. R. Soc. Trop. Med. Hyg.* **88:**67–73.

Kumaratilake, L. M., Ferrante, A., Jaeger, T., and Rzepczyk, C. M., 1992, Effects of cytokines, complement, and antibody on the neutrophil respiratory burst and phagocytic response to *Plasmodium falciparum* merozoites, *Infect. Immun.* **60:**3731–3738.

Kun, J. F., Mordmuller, B., Lell, B., Lehman, L. G., Luckner, D., and Kremsner, P. G., 1998, Polymorphism in promoter region of inducible nitric oxide synthase gene and protection against malaria [letter], *Lancet* **351:**265–266.

Kurtzhals, J. A., Adabayeri, V., Goka, B. Q., Akanmori, B. D., Oliver-Commey, J. O., Nkrumah, F. K., Behr, C., and Hviid, L., 1998, Low plasma concentrations of interleukin 10 in severe malarial anaemia compared with cerebral and uncomplicated malaria, *Lancet* **351:**1768–1772.

Kwiatkowski, D., 1992, Malaria: Becoming more specific about non-specific immunity, *Curr. Opin. Immunol.* **4:**425–431.

Kwiatkowski, D., Hill, A. V., Sambou, I., Twumasi, P., Castracane, J., Manogue, K. R., Cerami, A., Brewster, D. R., and Greenwood, B. M., 1990, TNF concentration in fatal cerebral, non-fatal cerebral, and uncomplicated *Plasmodium falciparum* malaria, *Lancet* **336:**1201–1204.

Kwiatkowski, D., Molyneux, M. E., Stephens, S., Curtis, N., Klein, N., Pointaire, P., Smit, M., Allan, R., Brewster, D. R., Grau, G. E., and Greenwood, B. M., 1993, Anti-TNF therapy inhibits fever in cerebral malaria, *Q. J. Med.* **86:**91–98.

Kwiatkowski, D., Bate, C. A. W., Scragg, I. G., Beattie, P., Udalova, I., and Knight, J. C., 1997, The malarial fever response—Pathogenesis, polymorphism and prospects for intervention, *Ann. Trop. Med. Parasitol.* **91:**533–542.

Lancaster, J., 1994, Simulation of the diffusion and reaction of endogenously produced nitric oxide, *Proc. Natl. Acad. Sci. USA* **91:**8137–8141.

Leitner, W. W., and Krzych, U., 1997, *Plasmodium falciparum* malaria blood stage parasites preferentially inhibit macrophages with high phagocytic activity, *Parasite Immunol.* **19:**103–110.

Levesque, M., Hobbs, M., Anstey, N., Vaughn, T., Cancellor, J., Pole, A., Perkins, D., Misukonis, M., Chanock, S., Granger, D., and Weinberg, J., 1999, Lack of correlation of the NOS2 promoter G-954C and pentanucleotide microsatellite polymorphisms with disease severity and nitric oxide production in Tanzanian children with malaria, submitted.

Luckhart, S., Vodovotz, Y., Cui, L. W., and Rosenberg, R., 1998, The mosquito *Anopheles stephensi* limits malaria parasite development with inducible synthesis of nitric oxide, *Proc. Natl. Acad. Sci. USA* **95:**5700–5705.

Mabbott, N., and Sternberg, J., 1996, Bone marrow nitric oxide production and development of anemia in *Trypanosoma brucei*-infected mice, *Infect. Immun.* **63:**1563–1566.

MacDermott, R., Wells, R., Zolyomi, S., Pavanand, K., Phisphumvidhi, P., Permpanich, B., and Gilbreath, M., 1980, Examination of peripheral blood mononuclear cells and sera from Thai adults naturally infected with malaria in assays of blastogenic responsiveness to mitogenic lectins and allogenic cell surface antigens, *Infect. Immun.* **30:**781–785.

Maciejewski, J. P., Selleri, C., Sato, T., Cho, H. J., Keefer, L. K., Nathan, C. F., and Young, N. S., 1995, Nitric oxide suppression of human hematopoiesis in vitro—contribution to inhibitory action of interferon-gamma and tumor necrosis factor-alpha, *J. Clin. Invest.* **96:**1085–1092.

Mackenzie, I. M. J., Ekangaki, A., Young, J. D., and Garrard, C. S., 1996, Effect of renal function on serum nitrogen oxide concentrations, *Clin. Chem.* **42:**440–444.

Macpherson, G. G., Warrell, M. J., White, N. J., Looareesuwan, S., and Warrell, D. A., 1985, Human cerebral malaria. A quantitative ultrastructural analysis of parasitized erythrocyte sequestration, *Am. J. Pathol.* **119:**385–401.

Marsh, K., 1992, Malaria—a neglected disease? *Parasitology* **104:**S53-S69.

Marsh, K., and Snow, R., 1997, Host–parasite interaction and morbidity in malaria-endemic areas, *Philos. Trans. R. Soc. London Ser. B* **352:**1385–1394.

Marsh, K., Forster, D., Waruiru, C., Mwangi, I., Winstanley, M., Marsh, V., Newton, C., Winstanley, P.,

Warn, P., Peshu, N., Pasvol, G., and Snow, R., 1995, Indicators of life-threatening malaria in African children, *N. Engl. J. Med.* **332:**1399–1404.

McGuire, W., Hill, A. V., Allsopp, C. E., Greenwood, B. M., and Kwiatkowski, D., 1994, Variation in the TNF-alpha promoter region associated with susceptibility to cerebral malaria, *Nature* **371:**508–510.

Mellouk, S., Green, S. J., Nacy, C. A., and Hoffman, S. L., 1991, IFN-gamma inhibits development of *Plasmodium berghei* exoerythrocytic stages in hepatocytes by an L-arginine-dependent effector mechanism, *J. Immunol.* **146:**3971–3976.

Mendis, K., and Carter, R., 1995, Clinical disease and pathogenesis in malaria, *Parasitol. Today* **11:**1–11.

Miller, L., Good, M., and Kaslow, D., 1997, The need for assays predictive of protection in development of malaria bloodstage vaccines, *Parasitol. Today* **13:**46–47.

Mitchell, H. H., Shonle, H. A., and Grindley, H. S., 1916, The origin of nitrates in the urine, *J. Biol. Chem.* **24:**461–490.

Motard, A., Landau, I., Nussler, A., Grau, G., Baccam, D., Mazier, D., and Targett, G. A., 1993, The role of reactive nitrogen intermediates in modulation of gametocyte infectivity of rodent malaria parasites, *Parasite Immunol.* **15:**21–26.

Naotunne, T. S., Karunaweera, N. D., Mendis, K. N., and Carter, R., 1993, Cytokine-mediated inactivation of malarial gametocytes is dependent on the presence of white blood cells and involves reactive nitrogen intermediates, *Immunology* **78:**555–562.

Nathan, C., and Xie, Q.-W., 1994, Regulation of biosynthesis of nitric oxide, *J. Biol. Chem.* **269:**13725–13728.

Newbold, C. I., Craig, A. G., Kyes, S., Berendt, A. R., Snow, R. W., Peshu, N., and Marsh, K., 1997, Pfemp1, polymorphism and pathogenesis, *Ann. Trop. Med. Parasitol.* **91:**551–557.

Newton, C., Warn, P., Winstanley, P., Peshu, N., Snow, R., Pasvol, G., and Marsh, K., 1997, Severe anaemia in children living in a malaria endemic area of Kenya, *Trop. Med. Int. Health* **2:**165–178.

Newton, C. R., and Krishma, S., 1998, Severe falciparum malaria in children: Current understanding of pathophysiology and supportive treatment, *Pharmacol. Ther.* **79:**1–53.

Newton, C. R., Taylor, T. E., and Whitten, R. O., 1998, Pathophysiology of fatal falciparum malaria in African children, *Am. J. Trop. Med. Hyg.* **58:**673–683.

Nicholson, S., Bonecini-Almeida, M. da G., Lapa e Silva, J. R., Nathan, C., Xie, Q.-W., Mumford, R., Weidner, J. R., Calaycay, J., Geng, J., Boechat, N., Linhares, C., Rom, W., and Ho, J. L., 1996, Inducible nitric oxide synthase in pulmonary alveolar macrophages from patients with tuberculosis, *J. Exp. Med.* **183:**2293–2302.

Nussler, A., Drapier, J.-C., Renia, L., Pied, S., Miltgen, F., Gentilini, M., and Mazier, D., 1991, L-Arginine-dependent destruction of intrahepatic malaria parasites in response to tumor necrosis factor and/or interleukin 6 stimulation, *Eur. J. Immunol.* **21:**227–230.

Nussler, A. K., Renia, L., Pasquetto, V., Miltgen, F., Matile, H., and Mazier, D., 1993, *In vivo* induction of the nitric oxide pathway in hepatocytes after injection with irradiated malaria sporozoites, malaria blood parasites or adjuvants, *Eur. J. Immunol.* **23:**882–887.

Nussler, A. K., Eling, W., and Kremsner, P. G., 1994, Patients with *Plasmodium falciparum* malaria and *Plasmodium vivax* malaria show increased nitrite and nitrate plasma levels, *J. Infect. Dis.* **169:**1418–1419.

Ockenhouse, C. F., Tegoshi, T., Maeno, Y., Benjamin, C., Ho, M., Kan, K. E., Thway, Y., Win, K., Aikawa, M., and Lobb, R. R., 1992, Human vascular endothelial cell adhesion receptors for *Plasmodium falciparum*-infected erythrocytes: Roles for endothelial leukocyte adhesion molecule 1 and vascular cell adhesion molecule 1, *J. Exp. Med.* **176:**1183–1189.

Orago, A., and Facer, C., 1991, Cytotoxicity of human natural killer (NK) cell subsets for *Plasmodium falciparum* erythrocytic schizonts: Stimulation by cytokines and inhibition by neomycin, *Clin. Exp. Immunol.* **86:**23–29.

Pasvol, G., Clough, B., Carlsson, J., and Snounou, G., 1995, The pathogenesis of severe falciparum malaria, *Bailliere's Clin. Infect. Dis.* **2:**249–270.

Peng, H.-B., Libby, P., and Liao, J. K., 1995, Induction and stabilization of IκBα by nitric oxide mediates inhibition of NF-κB, *J. Biol. Chem.* **270:**14214–14219.

Perlmann, H., Helmby, H., Hagstedt, M., Carlson, J., Larsson, P. H., Troye-Blomberg, M., and Perlmann, P., 1994, IgE elevation and IgE anti-malarial antibodies in *Plasmodium falciparum* malaria: Association of high IgE levels with cerebral malaria, *Clin. Exp. Immunol.* **97:**284–292.

Perlmann, P., Perlmann, H., Flyg, B. W., Hagstedt, M., Elghazali, G., Worku, S., Fernandez, V., Rutta, A. S. M., and Troye-Blomberg, M., 1997, Immunoglobulin E, a pathogenic factor in *Plasmodium falciparum* malaria, *Infect. Immun.* **65:**116–121.

Peyron, F., Burdin, N., Ringwald, P., Vuillez, J. P., Rousset, F., and Banchereau, J., 1994, High levels of circulating IL-10 in human malaria, *Clin. Exp. Immunol.* **95:**300–303.

Phillips, R., and Pasvol, G., 1992, Anaemia of *Plasmodium falciparum* malaria, *Bailliere's Clin. Haematol.* **5:**315–330.

Pongponratn, E., Riganti, M., Punpoowong, B., and Aikawa, M., 1991, Microvascular sequestration of parasitized erythrocytes in human falciparum malaria: A pathological study, *Am. J. Trop. Med. Hyg.* **44:**168–175.

Prada, J., and Kremsner, P. G., 1995, Enhanced production of reactive nitrogen intermediates in human and murine malaria, *Parasitol. Today* **11:**409–410.

Prada, J., Malinowski, J., Muller, S., Bienzle, U., and Kremsner, P. G., 1996, Effects of *Plasmodium vinckei* hemozoin on the production of oxygen radicals and nitrogen oxides in murine macrophages, *Am. J. Trop. Med. Hyg.* **54:**620–624.

Proudfoot, L., Odonnell, C. A., and Liew, F. Y., 1995, Glycoinositolphospholipids of *Leishmania major* inhibit nitric oxide synthesis and reduce leishmanicidal activity in murine macrophages, *Eur. J. Immunol.* **25:**745–750.

Riley, E., Andersson, G., Otoo, L., Jepsen, S., and Greenwood, B., 1988, Cellular immune responses to *Plasmodium falciparum* antigens in Gambian children during and after an attack of falciparum malaria, *Clin. Exp. Immunol.* **73:**17–22.

Rockett, K. A., Awburn, M. M., Cowden, W. B., and Clark, I. A., 1991, Killing of *Plasmodium falciparum* in vitro by nitric oxide derivatives, *Infect. Immun.* **59:**3280–3283.

Rockett, K. A., Awburn, M. M., Rockett, E. J., Cowden, W. B., and Clark, I. A., 1994, Possible role of nitric oxide in malarial immunosuppression, *Parasite Immunol.* **16:**243–249.

Rockett, K. A., Kwiatkowski, D., Bate, C. A., Awburn, M. M., Rockett, E. J., and Clark, I. A., 1996, *In vitro* induction of nitric oxide by an extract of *Plasmodium falciparum*, *J. Infect.* **32:**187–196.

Rogerson, S. J., Chaiyaroj, S. C., Ng, K., Reeder, J. C., and Brown, G. V., 1995, Chondroitin sulfate A is a cell surface receptor for *Plasmodium falciparum*-infected erythrocytes, *J. Exp. Med.* **182:**15–20.

Rojas, A., Padron, J., Caveda, L., Palacios, M., and Moncada, S., 1993, Role of nitric oxide pathway in the protection against lethal endotoxemia afforded by low doses of lipopolysaccharide, *Biochem. Biophys. Res. Commun.* **191:**441–446.

Rowe, J. A., Scragg, I. G., Kwiatkowski, D., Ferguson, D. J., Carucci, D. J., and Newbold, C. I., 1998, Implications of mycoplasma contamination in *Plasmodium falciparum* cultures and methods for its detection and eradication, *Mol. Biochem. Parasitol.* **92:**177–180.

Rubenstein, M., Mulholland, J., Jeffery, G., and Wolff, S., 1965, Malaria induced endotoxin tolerance, *Proc. Soc. Exp. Biol. Med.* **118:**283–287.

Scheller, L. F., Green, S. J., and Azad, A. F., 1997, Inhibition of nitric oxide interrupts the accumulation of CD8+ T cells surrounding *Plasmodium berghei*-infected hepatocytes, *Infect. Immun.* **65:**3882–3888.

Schofield, L., 1997, Malaria toxins revisited, *Parasitol. Today* **13:**275–276.

Schofield, L., and Hackett, F., 1993, Signal transduction in host cells by a glycosylphosphatidylinositol toxin of malaria parasites, *J. Exp. Med.* **177:**145–153.

Schofield, L., Novakovic, S., Gerold, P., Schwarz, R. T., McConville, M. J., and Tachado, S. D., 1996, Glycosylphosphatidylinositol toxin of *Plasmodium* up-regulates intercellular adhesion molecule-1, vascular cell adhesion molecule-1, and E-selectin expression in vascular endothelial cells and increases leukocyte and parasite cytoadherence via tyrosine kinase-dependent signal transduction, *J. Immunol.* **156:**1886–1896.

Schwarzer, E., Turrini, F., Ulliers, D., Giribaldi, G., Ginsburg, H., and Arese, P., 1992, Impairment of macrophage functions after ingestion of *Plasmodium falciparum*-infected erythrocytes or isolated malaria pigment, *J. Exp. Med.* **176:**1033–1041.

Sedegah, M., Finkelman, F., and Hoffman, S. L., 1994, Interleukin 12 induction of interferon gamma-dependent protection against malaria, *Proc. Natl. Acad. Sci. USA* **91:**10700–10702.

Seguin, M. C., Klotz, F. W., Schneider, I., Weir, J. P., Goodbary, M., Slayter, M., Raney, J. J., Aniagolu, J. U., and Green, S. J., 1994, Induction of nitric oxide synthase protects against malaria in mice exposed to irradiated *Plasmodium berghei* infected mosquitoes: Involvement of interferon gamma and CD8 + T cells, *J. Exp. Med.* **180:**353–358.

Sharara, A. I., Misukonis, M. A., Chan, S., Shami, P. J., and Weinberg, J. B., 1997, Increased nitric oxide production as a potential mechanism of the alpha interferon anti-viral effect in chronic hepatitis C virus infection, *J. Exp. Med.* **186:**1495–1502.

Sinton, J., Harbhagwan, M., and Singh, J., 1931, The numerical prevalence of parasites in relation to fever in chronic benign tertian malaria, *Indian J. Med. Res.* **18:**871–874.

Siu, P. M., 1968, Antimalarial activity of 1-methyl-3-nitro-1-nitrosoguanidine, *Proc. Soc. Exp. Biol. Med.* **129:**753–756.

Snow, R. W., Omumbo, J. A., Lowe, B., Molyneux, C. S., Obiero, J. O., Palmer, A., Weber, M. W., Pinder, M., Nahlen, B., Obonyo, C., Newbold, C., Gupta, S., and Marsh, K., 1997, Relation between severe malaria morbidity in children and level of *Plasmodium falciparum* transmission in Africa, *Lancet* **349:**1650–1654.

Srichaikul, T., and Siriasawakul, T., 1976, Bone marrow changes in human malaria, *Ann. Trop. Med. Parasitol.* **8:**40–50.

Stamler, J. S., 1994, Redox signaling: Nitrosylation and related target interactions of nitric oxide, *Cell* **78:**931–936.

Starnes, H., Warren, R., Jeevanadam, M., Gabrilove, J., and Larchian, W., 1988, Tumor necrosis factor and the acute metabolic response to tissue injury in man, *J. Clin. Invest.* **82:**1321–1325.

St. Clair, E. W., Wilkinson, W. E., Lang, T., Sanders, L., Misukonis, M. A., Gilkeson, G. S., Pisetsky, D. S., Granger, D. L., and Weinberg, J. B., 1996, Increased expression of blood mononuclear cell nitric oxide synthase type 2 in rheumatoid arthritis patients, *J. Exp. Med.* **184:**1173–1178.

Stevenson, M. M., Tam, M. F., Wolf, S. F., and Sher, A., 1995, IL-12-induced protection against blood-stage *Plasmodium chabaudi* AS requires IFN-gamma and TNF-alpha and occurs via a nitric oxide-dependent mechanism, *J. Immunol.* **155:**2545–2556.

Tachado, S. D., Gerold, P., McConville, M. J., Baldwin, T., Quilici, D., Schwarz, R. T., and Schofield, L., 1996, Glycosylphosphatidylinositol toxin of *Plasmodium* induces nitric oxide synthase expression in macrophages and vascular endothelial cells by a protein tyrosine kinase-dependent and protein kinase C-dependent signaling pathway, *J. Immunol.* **156:**1897–1907.

Taramelli, D., Basilico, N., Pagani, E., Grande, R., Monti, D., Ghione, M., and Olliaro, P., 1995, The heme moiety of malaria pigment (beta-hematin) mediates the inhibition of nitric oxide and tumor necrosis factor-alpha production by lipopolysaccharide-stimulated macrophages, *Exp. Parasitol.* **81:**501–511.

Taylor, A. M., Day, N. P., Smith, D. X., Loc, P. P., Mai, T. T., Chau, T. T., Phu, N. H., Hien, T. T., and White, N. J., 1998, Reactive nitrogen intermediates and outcome in severe adult malaria, *Trans. R. Soc. Trop. Med. Hyg.* **92:**170–175.

Taylor, T. E., Borgstein, A., and Molyneux, M. E., 1993, Acid–base status in paediatric *Plasmodium falciparum* malaria, *Q. J. Med.* **86:**99–109.

Taylor-Robinson, A. W., 1997a, Antimalarial activity of nitric oxide—Cytostasis and cytotoxicity towards *Plasmodium falciparum*, *Biochem. Soc. Trans.* **25:**S262.

Taylor-Robinson, A. W., 1997b, Inhibition of IL-2 production by nitric oxide: A novel self-regulatory mechanism for Th1 cell proliferation, *Immunol. Cell Biol.* **75:**167–175.

Taylor-Robinson, A. W., and Looker, M., 1998, Sensitivity of malaria parasites to nitric oxide at low oxygen tensions [letter], *Lancet* **351:**1630.

Taylor-Robinson, A. W., Phillips, R. S., Severn, A., Moncada, S., and Liew, F. Y., 1993, The role of TH1 and TH2 cells in a rodent malaria infection, *Science* **260:**1931–1934.

Taylor-Robinson, A. W., Severn, A., and Phillips, R. S., 1996, Kinetics of nitric oxide production during infection and reinfection of mice with *Plasmodium chabaudi*, *Parasite Immunol.* **18:**425–430.

Tiao, G., Rafferty, J., Ogle, C., Fischer, J. E., and Hasselgren, P.-O., 1994, Detrimental effect of nitric oxide synthase inhibition during endotoxemia may be caused by high levels of tumor necrosis factor and interleukin-6, *Surgery* **116:**332–338.

Tsuji, M., Miyahira, Y., Nussenzweig, R. S., Aguet, M., Reichel, M., and Zavala, F., 1995, Development of antimalaria immunity in mice lacking IFN-gamma receptor, *J. Immunol.* **154:**5338–5344.

Tsukahara, H., Hiraoka, M., Hori, C., Miyanomae, T., Kikuchi, K., and Sudo, M., 1997, Age-related changes of urinary nitrite/nitrate excretion in normal children, *Nephron* **76:**307–309.

Turner, G., 1997, Cerebral malaria, *Brain Pathol.* **7:**569–582.

Turner, G. D., Morrison, H., Jones, M., Davis, T. M., Looareesuwan, S., Buley, I. D., Gatter, K. C., Newbold, C. I., Pukritayakamee, S., Nagachinta, B., White, N. J., and Berendt, A. R., 1994, An immunohistochemical study of the pathology of fatal malaria. Evidence for widespread endothelial activation and a potential role for intercellular adhesion molecule-1 in cerebral sequestration, *Am. J. Pathol.* **145:**1057–1069.

Turrini, F., Giribaldi, G., Valente, E., and Arese, P., 1997, *Mycoplasma* contamination of *Plasmodium* cultures—a case of parasite parasitism, *Parasitol. Today* **13:**367–368.

Waller, D., Krishna, S., Crawley, J., Miller, K., Nosten, F., Chapman, D., ter Kuile, F. O., Craddock, C., Berry, C., Holloway, P. A. H., Brewster, D., Greenwood, B. M., and White, N. J., 1995, Clinical features and outcome of severe malaria in Gambian children, *Clin. Infect. Dis.* **21:**577–587.

Weiss, G., Thuma, P. E., Biemba, G., Mabeza, G., Werner, E. R., and Gordeuk, V. R., 1998, Cerebrospinal fluid levels of biopterin, nitric oxide metabolites, and immune activation markers and the clinical course of human cerebral malaria, *J. Infect. Dis.* **177:**1064–1068.

Weiss, G., Thuma, P. E., Mabeza, G., Werner, E. R., Herold, M., and Gordeuk, V. R., 1997, Modulatory potential of iron chelation therapy on nitric oxide formation in cerebral malaria, *J. Infect. Dis.* **175:**226–230.

Westfelt, U. N., Benthin, G., Lundin, S., Stenqvist, O., and Wennmalm, A., 1995, Conversion of inhaled nitric oxide to nitrate in man, *Br. J. Pharmacol.* **114:**1621–1624.

Williamson, W., and Greenwood, B., 1978, Impairment of the immune response to vaccination after acute malaria, *Lancet* **1:**1328–1329.

World Health Organization, 1990, Severe and complicated malaria, *Trans. R. Soc. Trop. Med. Hyg.* **84** (Suppl 2):1–65.

World Health Organization, 1996, World malaria situation in 1993, *Weekly Epidemiol. Rec.* **71:**17–24.

Wyler, D., 1992, *Plasmodium* and *Babesia*, in: *Infectious Diseases* (S. L. Gorbach, J. G. Bartlett, and N. R. Blacklow, eds.), Saunders, Philadelphia, pp. 1967–1978.

Wyler, D., 1993, Malaria: Overview and update, *Clin. Infect. Dis.* **16:**449–458.

CHAPTER 16

Nitric Oxide in Schistosomiasis

ISABELLE P. OSWALD

1. Introduction

Schistosomiasis is a chronic and debilitating parasitic disease affecting 200 million people worldwide and responsible for at least 500,000 deaths each year. Schistosomes are complex metazoan parasites that require both a molluskan intermediate host and a mammalian definitive host for the completion of their life cycle. In the mammalian host, infection is initiated by the cercarial stage of the parasite, which then undergoes a remarkable transformation into a larval form called a *schistosomulum*. Schistosomula rapidly leave the skin and migrate via the blood and lymphatics to the lungs, through which they transit over a period of days. Eventually their journey through the bloodstream leads them to the liver, where male and female parasites mature and mate. Adult worm pairs live for years within the mesenteric veins, where they produce hundreds of eggs daily. The immune response to eggs that become lodged in host tissues is responsible for most of the pathological manifestations of schistosomiasis (Boros, 1989).

Although contrasting results can be obtained according to the experimental model used (rat or mouse, for instance), it is generally agreed that T cells participate in resistance to schistosomal infection (James and Sher, 1990; Capron, 1992). Exposure of mice to radiation-attenuated parasites induces a high level of resistance against challenge with virulent organisms. Treatment with antibodies against $CD4^+$ T cells or IFNγ abrogates protective immunity, while treatment with antibodies against the Th2 cytokines IL-4 and IL-5 fails to diminish resistance (Sher *et al.*, 1990; Smythies *et al.*, 1992). Investigation of the fate of challenge parasites indicates that they are eliminated as they migrate through the lungs of the

ISABELLE P. OSWALD • INRA, Laboratory of Pharmacology-Toxicology, 31931 Toulouse Cedex 9, France.

Nitric Oxide and Infection, edited by Fang. Kluwer Academic / Plenum Publishers, New York, 1999.

immunized host. Clearance of parasites from the lungs of vaccinated mice appears to be a protracted process lasting several weeks (Kassim *et al.*, 1992). Studies of histopathology indicate that larvae in transit through the lung become trapped in a mononuclear cell-rich inflammatory reaction histologically resembling a delayed hypersensitivity response, a Th1-associated phenomenon (Wilson and Coulson, 1989). A better understanding of the immune mechanisms operating in the experimental vaccine model will facilitate the design of an effective human vaccine.

As described in Chapter 2, pioneering work by Hibbs *et al.* (1988) indicated that macrophage cytotoxicity for tumor cells requires L-arginine, and subsequent work demonstrated that the active mediator of this cytotoxic pathway is nitric oxide (NO) (Lepoivre *et al.*, 1991; Karupiah *et al.*, 1993; Oswald and James, 1996). NO formation has now been demonstrated to participate in host defense against a diverse array of infectious agents including bacteria, fungi, and viruses, as well as parasites (Woods *et al.*, 1994) (see also Chapter 12). This chapter will focus on the role of NO during schistosomiasis, reviewing the mechanism of action of NO on helminthic parasites, *in vivo* antiparasitic effects of NO, types of NO-producing cells, and relevant immunological regulation of NO production.

2. Schistosome Targets of NO

NO modifies its targets in a bewildering variety of ways (reviewed in Nathan, 1992; Schmidt and Walter, 1994; see also Chapter 3). Because of its low molecular weight and lipophilic nature, NO rapidly diffuses across eukaryotic membranes and prokaryotic cell walls. NO can interact with the iron in the heme moiety of guanylyl cyclase, activating the enzyme to produce cyclic GMP. Alternatively, NO can bind to iron in iron–sulfur cluster-containing proteins, induce the ADP-ribosylation of a variety of enzymes (Dimmeler *et al.*, 1993), damage DNA by oxidation or deamination, and interact with superoxide to produce peroxynitrite (Beckman *et al.*, 1990), an oxidant species able to nitrate tyrosine residues of proteins. It has been postulated that the toxic effect of NO on helminths is related to the production of peroxynitrite (Brophy and Pritchard, 1992).

The principal targets of NO in the schistosome appear to be enzymes containing a catalytically active Fe–S group. Indeed, inactivation of aconitase or of the electron transport chain involved in respiration by various chemicals is toxic to newly transformed schistosomula (Ahmed *et al.*, 1997). The aconitase enzymes of the Krebs cycle, NADPH-ubiquinone oxidoreductase and succinate : ubiquinone reductase of the electron transport chain have each been identified as potential enzymatic targets of NO inhibition in eukaryotic cells (Woods *et al.*, 1994). Moreover, ultrastructural studies of newly transformed larvae cultured with activated macrophages show that, in contrast to antibody-dependent killing, macrophage-mediated cytotoxicity is not directed against the parasite surface.

Rather, progressive disintegration of internal structures beginning with perturbation of the mitochondria within subtegumental muscle cells and culminating in widespread vacuolation is observed (McLaren and James, 1985; Pearce and James, 1986). Such observations are consistent with the principal mechanisms of NO cytotoxicity proposed for mammalian cell targets, which involves inactivation of key enzymes (Woods *et al.*, 1994). The addition of excess iron and reducing agents, known to stabilize the activity of iron-containing enzymes in tumor cell targets, inhibits schistosomulum killing without decreasing nitrite production (James and Glaven, 1989), also supporting the interpretation that Fe-containing moieties serve as NO targets within the parasite.

Activated macrophages and NO-generating compounds have been shown to kill 2-week-old schistosomula *in vitro*, whereas younger lung-stage larvae are resistant (Pearce and James, 1986; Ahmed *et al.*, 1997). The effect of NO on metalloenzymes involved in DNA synthesis, such as ribonucleotide reductase, may be particularly destructive to the older growing and developing larvae, although this remains to be demonstrated. It must be emphasized that the biochemistry of schistosomes is incompletely understood, and these parasites may contain additional as yet unknown targets of NO.

3. Induced NO as an Antischistosome Effector

3.1. *In Vitro* Evidence

The antimicrobial capacity of NO was recognized in the food industry long before it was identified as a major effector molecule in host defense against microbial pathogens. An increasing number of *in vitro* studies demonstrate that NO possesses antimicrobial activity against a variety of pathogens including parasites, fungi, bacteria, and viruses (reviewed by Woods *et al.*, 1994; James, 1995; see also Chapter 12). However, one of the earliest demonstrations of microbicidal activity of NO was a study of the mechanism of macrophage cytotoxicity toward the larvae of the helminthic parasite *Schistosoma mansoni* (James and Glaven, 1989).

Today, evidence linking antischistosomal activity to production of NO and/or subsequent oxidation products includes (1) a correlation between NO production by cytokine-activated cell cultures and killing of parasites, (2) a requirement of L-arginine for antimicrobial activity and inhibition of this activity by iNOS inhibitors, and (3) the demonstration of a direct cytotoxic antiparasitic effect of NO-generating compounds (James and Glaven, 1989; Ahmed *et al.*, 1997).

3.2. *In Vivo* Evidence

While a strong correlation between antimicrobial activity and L-arginine-dependent production of NO by cytokine-activated cells has been readily demonstrated *in vitro*, the relationship between generation of NO *in vivo* and protection against bacterial or parasitic infection has only more recently been addressed (Granger *et al.*, 1991; Evans *et al.*, 1993).

As mentioned above, cell-mediated immunity participates in resistance to schistosomal infection. Examination of the levels of cytokine mRNA in the lungs of vaccinated mice during subsequent challenge with *S. mansoni* shows abundant production of Th1-type cytokines, including IFNγ, TNFα, and IL-2, which are known to activate NO-producing effector cells (Wynn *et al.*, 1994). Moreover, the demonstration of elevated iNOS mRNA levels in the lungs during the period of larval clearance and iNOS enzyme in inflammatory foci around parasites (Wynn *et al.*, 1994) gives further credence to a role of NO-mediated antiparasitic activity *in vivo.* Additional support is provided by the ability of the NOS inhibitor aminoguanidine to substantially reduce the resistance of vaccinated animals to a parasite challenge (Wynn *et al.*, 1994). Increased expression of iNOS mRNA has been found in the skin of vaccinated or multiply infected animals compared with naive mice during a challenge infection (Ramaswamy *et al.*, 1997). Local production of NO may participate in the early killing of the parasite, although this remains to be demonstrated.

During the past few years, biological functions of NO have been confirmed by the use of mice with a targeted inactivation of the iNOS gene. Macrophages from these animals fail to produce NO after stimulation with IFNγ and LPS (MacMicking *et al.*, 1995; Wei *et al.*, 1995). Likewise, these mice have markedly reduced defenses against parasites such as *Leishmania major* and *Toxoplasma gondii* (Wei *et al.*, 1995; Sharton-Kersten *et al.*, 1997). Infection of these iNOS knockout mice with *S. mansoni* have recently been performed (Coulson *et al.*, 1998; James *et al.*, 1998): these studies indicated that vaccine-induced protection is reduced but not eliminated in these mice. However, because iNOS knockout mice also displayed an immune response skewed toward type 1 reactivity with increased IFN-γ, TNF-α and IgG2a levels (James *et al.*, 1998) the exact contribution of NO in resistance to *S. mansoni* still remains to be determined.

4. NO-Producing Cells

Since the initial reports of NO production by murine macrophages (Stuehr and Marletta, 1985; Granger *et al.*, 1988), it has become apparent that the capacity to produce high quantities of NO is not limited to phagocytic cells. The potential for iNOS expression extends far beyond the classical immune system and includes tissues found at virtually every site within the body. The presence of the iNOS pathway in many different cell types suggests that NO may represent an intrinsic antimicrobial defense against pathogens sequestered in a variety of cells. Cells not traditionally considered to be part of the immune network might actually play a larger role in protective immunity against microbial pathogens than has been previously recognized.

4.1. Macrophages

The helminthic parasite, *S. mansoni*, remains extracellular throughout its interaction with the host. Several years ago it was observed that cytokine-activated macrophages are capable of killing larval schistosomula by a mechanism that closely resembles that of macrophage cytotoxicity for other extracellular targets, i.e., tumor cells (James and Glaven, 1989). Activation of inflammatory peritoneal macrophages by lymphokines induces the production of NO, which serves as the effector molecule of parasite killing (James and Glaven, 1989). Macrophage larvicidal activity does not require products of the respiratory burst (Scott *et al.*, 1985) but is arginine dependent and significantly inhibited in the presence of either N^{G}-monomethyl arginine, a competitive inhibitor of NOS, or arginase, an enzyme that degrades L-arginine into L-ornithine and urea (James and Glaven, 1989). More recently, an NO-generating compound has been shown to duplicate the effect of lymphokine-activated macrophages *in vitro* (Ahmed *et al.*, 1997).

4.2. Endothelial Cells

The observation that endothelial cells are responsive to cytokine signals for NO production suggested that these cells might participate in protection against microbial pathogens that possess an intravascular tropism (Kilbourn and Belloni, 1990). Because schistosomes remain intravascular throughout most of their life cycle in the mammalian host, schistosomiasis represented an ideal model to test this hypothesis. Studies employing murine endothelial cell lines showed that these cells are able to kill larval schistosomes via an arginine-dependent mechanism. Both newly transformed and older lung-stage parasites are susceptible to activated endothelial cells. Larval killing and NO production *in vitro* requires at least two cytokine signals with combinations of IFNγ, TNFα and IL-1 being most effective (Oswald *et al.*, 1994b).

It is tempting to speculate that endothelial cells might be especially important in the effector mechanism of protective immunity manifested against lung-stage schistosomula, since the parasites are in intimate contact with blood vessel walls during their migration through the lungs. It has been shown by measurement of mRNA levels that the necessary cytokine signals for endothelial cell activation are generated in the lungs of vaccinated mice (Wynn *et al.*, 1994). In addition, hypertrophy and hyperplasia of endothelial cells of the juxtabronchial arteries of mice vaccinated with irradiated cercariae suggest activation *in vivo* (Oswald *et al.*, 1994b). The absence of such endothelial cell changes in a strain of mice that fails to become protected as a result of vaccination provides strong circumstantial evidence for participation of iNOS activation in resistance (Oswald *et al.*, 1994b).

4.3. Other Cell Types

Cells including hepatocytes, epithelial cells, vascular smooth muscle cells, cardiac myocytes, osteoblasts, astrocytes, and fibroblasts can produce massive quantities of NO on cytokine stimulation (Oswald and James, 1996). It is easy to envision that such cells might play an important role in the host response to many pathogens. For example, IFNγ-treated hepatocytes are able to kill liver-stage forms of *Plasmodium berghei* by an NO-dependent mechanism (Mellouk *et al.*, 1991; Nussler *et al.*, 1991), and rats immunized with irradiated sporozoites demonstrate iNOS activity specifically in hepatocytes (Klotz *et al.*, 1995). In the case of schistosomiasis, epithelial cells may contribute to the NO synthesis observed in the skin of vaccinated animals (Ramaswamy *et al.*, 1997).

5. Cytokine Regulation of NO Production

As cytotoxic/cytostatic effector molecules, NO and its congeners participate in cell-mediated immunity (see above). However, generation of high levels of NO can also have detrimental effects on the host. Indeed, overproduction of NO can lead to the suppression of host immunity (Mills, 1991; Schleifer and Mansfield, 1993), metabolic failure, and eventual cardiovascular collapse. In addition, high NO production in noninfectious settings has been implicated in autoimmune diseases and inflammatory conditions such as arthritis and asthma (Barnes and Liew, 1995). NO production by iNOS is tightly regulated by cytokines (see also Chapter 5), and a major function of these pathways may be to prevent host toxicity. Regulation of iNOS activity has been most extensively studied in murine macrophages. Cytokines such as IFNγ and TNFα upregulate iNOS expression, whereas several other cytokines such as TGFβ, IL-4, IL-10, and IL-13 can block NO production, leading to an inhibition of cell-mediated antimicrobial activity.

5.1. IFNγ and TNFα

It has been recognized for several years that to become fully cytotoxic, macrophages must receive two consecutive signals, known as priming and triggering signals (Adams and Hamilton, 1987). IFNγ remains the prototypical priming signal for virtually all macrophage–monocyte functions including the production of NO, whereas TNFα is the major physiological triggering signal for macrophage activation (Ding *et al.*, 1988; Drapier *et al.*, 1988). TNFα production is induced in macrophages by LPS and other bacterial components such as muramyl dipeptide from the cell wall of mycobacteria. Endogenous production of TNFα can also be induced by infection of macrophages with intracellular pathogens such as *L. major*, *T. gondii*, or *M. bovis* (Green *et al.*, 1990; Langermans *et al.*, 1992; Flesch *et al.*, 1994). In thioglycollate-elicited inflammatory macrophages, which are among the most biochemically active macrophages, IFNγ treatment alone induces the endogenous production of TNFα, which acts as a second signal for full cytotoxic activation and NO production (Oswald *et al.*, 1992a). Other cytokines, notably IL-2 (Cox *et al.*, 1990), may also provide an accessory function for macrophage activation.

5.2. IL-4

Several cytokines, including IL-4, are known to suppress Th1 lymphocyte proliferation or production of IFNγ, thus limiting effector cell activation and indirectly inhibiting NO production. Production of NO can also be regulated more directly at the level of the effector cell by certain cytokines, leading to an inhibition of cell-mediated antimicrobial activity. *In vitro*, IL-4 blocks NO production and inhibits parasite killing by murine macrophages (Oswald *et al.*, 1992b; Bogdan *et al.*, 1993). *In vivo*, the production of IL-4 and IL-10 has been implicated in the failure of P strain mice to respond to vaccination against schistosomiasis (Oswald *et al.*, 1998).

The extent to which IL-4 and other counterregulatory cytokines modulate NO production varies under different experimental conditions. While IL-4, IL-10, and TGFβ block macrophage activation for NO production and parasite killing with equal potency, only IL-4 inhibits endothelial cell-mediated larvicidal activity (Oswald *et al.*, 1994b). Moreover, IL-4 may actually enhance macrophage cytotoxic activity under certain circumstances (Crawford *et al.*, 1987; Flesch and Kaufmann, 1990; Stenger *et al.*, 1991). The factors contributing to this alternative role of IL-4 remain poorly understood.

5.3. IL-10

IL-10 has been found by several different laboratories to inhibit NO secretion (Cuhna *et al.*, 1992; Gazzinelli *et al.*, 1992a), and this correlates with the inhibition of murine macrophage killing of extracellular (*S. mansoni*) as well as intracellular (*T. gondii* or *T. cruzi*) parasites (Cuhna *et al.*, 1992; Gazzinelli *et al.*, 1992a; Oswald *et al.*, 1992a). We have also established that blockade in NO production results from suppression of endogenous TNFα synthesis, and exogenous TNFα can restore NO expression in IL-10-treated inflammatory macrophages (Oswald *et al.*, 1992a).

The degree to which IL-10 inhibits NO production appears to depend on the macrophage population under investigation. For example, IL-10 blocks NO production and parasite killing in thioglycollate-elicited murine macrophages but has little effect on casein-elicited macrophages or resident cells (Oswald and James, 1996). Moreover, IL-10 increases iNOS mRNA and NO production in bone marrow-derived macrophages (Corradin *et al.*, 1993).

5.4. IL-13

IL-13 is a recently described cytokine that shares some activities with IL-4 (Zurawski and de Vries, 1994). IL-13 decreases the production of NO by activated macrophages, leading to a decrease in parasitacidal activity against *L. major* (Doherty *et al.*, 1993; Doyle *et al.*, 1994). Like IL-4, IL-13 appears to have dual effects. While IL-13 decreases NO production by LPS-stimulated GM-CSF-derived bone marrow macrophages, it enhances NO production by LPS-stimulated M-CSF-derived bone marrow macrophages (Doherty *et al.*, 1993; Doyle *et al.*, 1994).

5.5. TGFβ

In vitro, TGFβ blocks the NO-dependent cytocidal activity of activated macrophages for several parasites including *L. major* (Nelson *et al.*, 1991), *T. cruzi* (Gazinelli *et al.*, 1992b), and *S. mansoni* (Oswald *et al.*, 1992b). A recent study indicates that TGFβ suppresses iNOS expression by at least three distinct mechanisms: decreased stability and translation of iNOS mRNA, and increased degradation of iNOS protein (Bogdan and Nathan, 1993). *In vivo*, TGFβ aggravates *L. amazoniensis* infection (Barral-Netto *et al.*, 1992), and most likely reduces vaccination efficacy against *S. mansoni* (Williams *et al.*, 1995). In contrast to mice vaccinated by an intradermal route, animals vaccinated by either intramuscular or intravenous routes fail to develop protective immunity against *S. mansoni*. In both cases, vaccine failure is associated with elevated synthesis of TGFβ, which inhibits macrophage activation and NO production (Oswald *et al.*, 1992b; Williams *et al.*, 1995).

Ultimately, NO may regulate its own production by inactivating iNOS (Assreuy *et al.*, 1993; Griscavage *et al.*, 1993); such autoinhibition could serve to limit excessive production and resultant tissue damage. Another apparent control mechanism involves direct inhibition of Ia expression by NO in IFNγ-activated macrophages, which could limit antigen-presenting capability (Sicher *et al.*, 1994). Additional studies suggest that high-level NO production by antigen-presenting cells blocks T-cell proliferation (Roland *et al.*, 1994; Taylor-Robinson *et al.*, 1994). Finally experiments with iNOS knockout mice have revealed that NO down regulates type 1 cytokine responses (James *et al.*, 1998) which are ultimately required for NO production (see above).

6. Immune Evasion Mechanisms against NO

6.1. Manipulation of the Cytokine Response

Parasites have developed many methods for counteracting host defense mechanisms. Although the ability of some parasites to avoid the toxic effect of oxygen radicals has been postulated to be related to the production of scavenging enzymes (Hughes, 1988), analogous specific parasitic NO resistance mechanisms have not been described. Parasites may nevertheless escape NO-mediated damage by less direct means. For example, some life stages of a parasite may not depend on enzymes vulnerable to inactivation by NO (see below). Alternatively, parasites may manipulate the host's own immune apparatus to evade destruction by activated effector cells.

In fact, chronic *S. mansoni* infection is associated with massive stimulation of a Th2-type immune response, including production of IL-4, IL-10, IL-13, and/or TGFβ (Czaja *et al.*, 1989; Grzych *et al.*, 1991). These cytokines impede the development of an effective Th1 response and prevent effector cell activation and NO production (Oswald and James, 1996). Suppressive cytokines act synergistically to reduce macrophage function. Concentrations of TGFβ, IL-4, and IL-10 that alone are suboptimal for suppression of NO production or macrophage larvicidal activity exhibit potent effects when used in combination (Oswald *et al.*, 1992b). Therefore, these cytokines may exert potent inhibitory effects on host macrophage effector activity even when present in low quantities. Induction of multiple counterregulatory cytokines during schistosomal infection (Czaja *et al.*, 1989; Grzych *et al.*, 1991) may represent an important parasitic strategy for escaping protective host immune responses.

6.2. Schistosome Metabolic Transition

It is well known that parasitic helminths have a capacity to adapt to different environments by using different methods of energy respiration (Thompson *et al.*, 1984; Tielens, 1994; Komuniecki and Komuniecki, 1995). After infection of the mammalian host, schistosomes rapidly convert from an aerobic free-living stage to a form primarily dependent on fermentative metabolism (reviewed in Tielens, 1994). Recently we discovered that this early conversion is not permanent; rather, migrating schistosomal larvae pass through at least one additional stage in which they are more dependent on aerobic respiration for energy metabolism (Ahmed *et al.*, 1997). Newly transformed skin-stage schistosomula are highly susceptible to

NO-mediated killing by activated macrophages and activated endothelial cells (James and Glaven, 1989; Oswald *et al.*, 1994b), whereas lung-stage larvae appear to be totally resistant to this effector mechanism (Sher *et al.*, 1982). The second phase of vulnerability of schistosomal larvae to NO-mediated killing can be observed $2\frac{1}{2}$ weeks after infection when parasites trapped by the host inflammatory response are recovered from the livers of unimmunized mice or from the lungs of mice vaccinated with attenuated cercariae (Pearce and James, 1986). Thus, transitions in energy metabolism are accompanied by changes in susceptibility to activated cell-mediated killing, and the susceptibility to NO toxicity correlates with susceptibility to chemical inhibitors of aconitase and mitochondrial respiration (Ahmed *et al.*, 1997).

Transition to an alternative mechanism of energy metabolism to avoid dependence on enzymes inactivatable by NO is not restricted to helminthic parasites. Indeed, experiments have shown that conversion of *T. gondii* to a relatively quiescent bradyzoite stage is induced on treatment with exogenous NO or with an inhibitor of mitochondrial respiration (Bohne *et al.*, 1994). However, factors controlling the reversion of later stage migrating schistosomes to aerobic metabolism during a period of increased immune attack (James, 1995) remain uncertain.

7. Other Effects of NO in Schistosomiasis

7.1. Cachexia

C57BL/6 mice with targeted disruption of the IL-4 gene mount a Th1-type response during *S. mansoni* infection. At the onset of parasite egg deposition, these animals lose weight and subsequently succumb to infection (Brunet *et al.*, 1997). In contrast, wild-type C57BL/6 animals infected with *S. mansoni* mount a Th2-type response and survive, developing chronic disease without cachexia (Pearce *et al.*, 1996). The acute illness in C57BL/6 IL-$4^{-/-}$ mice is characterized by prominent parasite egg-associated inflammation in the ileum. Treatment of infected IL-$4^{-/-}$ mice with anti-TNFα mAb significantly lessens weight loss and prolongs life. Experimental observations support a possible role of NO in the pathophysiological process. Indeed, cells from infected IL-$4^{-/-}$ animals stimulated with Ag or Ag plus LPS produce significantly more NO than those from infected wild-type mice. Moreover, anti-TNFα treatment markedly reduces NO production by spleen cells from IL-$4^{-/-}$ animals (Brunet *et al.*, 1997). These findings suggest that IL-4 and possibly other type 2 cytokines such as IL-10 and IL-13 normally act to inhibit deleterious TNFα-dependent NO production during schistosomiasis.

7.2. Cancer

Schistosomal as well as other parasitic infections have been identified as risk factors for the subsequent development of cancer. These include a proposed association of *S. haematobium* infection with bladder cancer, as well as an association of *S. japonicum*, *S. mansoni*, *Opisthorchis viverrini*, and *Clonorchis sinensis* with liver cancer (Ishii *et al.*, 1994). NO and oxygen radicals produced in infected and inflamed tissues could contribute to the process of carcinogenesis by different mechanisms including increased DNA damage and the formation of carcinogenic nitrosamines from ingested precursors (Ohshima and Bartsch, 1994). In fact, individuals infected with *O. viverrini* or *Schistosoma* species have been found to excrete higher levels of nitrosoproline in their urine than uninfected subjects, and these differences are abolished by elimination of parasites with praziquantel treatment (Mostafa *et al.*, 1994; Satarug *et al.*, 1996). This suggests that infected individuals have an elevated endogenous nitrosylation potential. In related studies, NOS activity was immunohistochemically demonstrated in inflammatory cells surrounding parasite-containing bile ducts in the livers of *O. viverrini*-infected hamsters (Ohshima *et al.*, 1994). An analysis of the mutational spectra observed in the p53 gene from bladder tumors of patients with *S. haematobium* infection has revealed a high rate of G:C-to-A:T transitions, consistent with the deamination of 5-methylcytosine by NO (Warren *et al.,* 1995). Collectively, these observations support the hypothesis that *N*-nitroso compounds and other NO-derived mutagens have a role in human cancer etiology, particularly when exposure begins early in life and persists over a long period as occurs during parasitic infections.

8. Conclusion and Perspectives: NO Production by Human Monocytes/Macrophages

The production of NO and its antiparasitic effects in rodents is well established (see above). In contrast, the circumstances required for its release from human monocytes/macrophages and its potential role in human pathology remain controversial (review in Denis, 1994; Albina, 1995; see also Chap. 6).

Elevation of nitrate in the plasma and urine of individuals under cytokine treatment strongly suggest the existence of a cytokine-inducible NO pathway in humans (Hibbs *et al.*, 1992). Several human cell types including hepatocytes (Nussler *et al.*, 1992; Geller *et al.*, 1993), chondrocytes (Charles *et al.*, 1993), osteoblasts (Ralston *et al.*, 1994) epithelial cells (Adcock *et al.*, 1994) and eosinophils (del Pozo *et al.*, 1997) have been shown to express iNOS. The accumulation of iNOS mRNA and/or protein has also been documented in IFN-γ and LPS stimulated human monocytes and in those cultured with HIV (Reiling *et al.*, 1994; Weinberg *et al.*, 1995; Bukrinsky *et al.*, 1995). Despite this, the ability of human monocytes/macrophages to generate NO is very low (Schneemann *et al.*, 1993; Weinberg *et al.*, 1995) and a survey of the literature shows a mixture of positive results and numerous negative reports (review in Denis, 1994; Albina, 1995).

Several hypotheses have been put forward to explain why cytokine-inducible NOS activity has been so difficult to demonstrate in human macrophages. Difference in maturation and/or differentiation between the peritoneal rodent macrophages and human peripheral blood monocytes may explain, at least in part, the difficulties in inducing iNOS in the latter. When low NO production has been reported by human cells, a prolonged exposure to cytokines was generally required (Paul-Eugene *et al.*, 1994). Moreover, several papers suggest a need for other types of signals in the induction of NO production by human cells such as CD69 crosslinking (De Maria *et al.*, 1994) or HIV infection (Bukrinsky *et al.*, 1995). Nevertheless, human alveolar or peritoneal macrophages failed to produce NO when stimulated with cytokine and endotoxin or microorganisms (Cameron *et al.*, 1990; Schneemann *et al.*, 1993). The possibility must also be considered that human macrophages utilize microbicidal mechanisms other than NO production. Indeed, human monocytes are able to develop an antibacterial and antiparasitic activity through an NO-independent pathway (Cameron *et al.*, 1990; Woodman *et al.*, 1991; James *et al.*, 1990; Oswald *et al.*, unpublished results).

Thus, what emerges from the literature is that human monocytes/ macrophages appear to be particularly reluctant to produce NO, at least *in vitro*. In addition, neither rabbit monocytes/macrophages (Cameron *et al.*, 1990; Schneemann *et al.*, 1993) nor peripheral blood monocytes from new or old-world monkeys produce NO *in vitro* when stimulated with IFN-γ and/or LPS (Albina, 1995). In contrast, bovine monocyte/macrophages accumulate iNOS mRNA and secrete nitrite when stimulated with bacteria or with endotoxin, but not when treated with recombinant cytokines (Adler *et al.*, 1995). Further elucidation of some missing element in the *in vitro* production of NO by human macrophages and/or identification of some new effector pathway of microbicidal activity are still needed to elucidate the contribution of NO production in human immunity.

ACKNOWLEDGMENT. The author thanks Dr. C. M. Dozois for his help with the English text.

References

Adams, D. O., and Hamilton, T. A., 1987, Molecular transductional mechanism by which IFN-γ and other signals regulate macrophage development, *Immunol. Rev.* **97**:5–27.

Adcock, I. M., Brown, C. R., Kwon, O., and Barnes P. J., 1994, Oxidative stress induce NF kappa B

DNA binding and inducible NOS mRNA in human epithelial cells. *Biochem. Biophys. Res. Commun.* **199:**1518–1524.

Adler, H., Frech, B., Thöny, M., Pfister, H., Peterhans, E., and Jungi, T. W., 1995, Inducible nitric oxide synthase in cattle. Differential regulation of nitric oxide synthase in bovine and murine macrophages. *J. Immunol.* **154:**4710–4718.

Albina, J. E., 1995, On the expression of nitric oxide synthase by human macrophages: Why no NO? *J. Leuk. Biol.* **58:**643–649.

Assreuy, J., Cunha, F. Q., Liew, F. Y, and Moncada, S., 1993, Feedback inhibition of nitric oxide synthase activity by nitric oxide, *Br. J. Pharmacol.* **108:**833–837.

Barnes, P. J., and Liew, F. Y., 1995, Nitric oxide in asthmatic inflammation, *Immunol. Today* **16:**128–130.

Barral-Netto, M., Barral, A., Brownell, C. E., Skeiky, Y. A. W., Ellingsworth, L. R., Twardzik, D. R., and Reed, S. G., 1992, Transforming growth factor-β in leishmanial infection: A parasite escape mechanism, *Science* **257:**545–548.

Beckman, J. S., Beckman, T. W., Chen, J., Marshall, P. A., and Freeman, B. A., 1990, Apparent hydroxyl radical production by peroxynitrite: Implication for endothelial cell injury from nitric oxide and superoxide, *Proc. Natl. Acad. Sci. USA* **87:**1620–1624.

Bogdan, C., and Nathan, C., 1993, Modulation of macrophages function by transforming growth factor β, interleukin-4 and interleukin-10, *Ann. N.Y. Acad. Sci.* **685:**713–739.

Bogdan, C., Vodovotz, Y., Paik, J., Xie, Q. W., and Nathan, C., 1993, Mechanism of suppression of nitric oxide synthase expression by interleukin-4 in primary mouse macrophages, *J. Leukoc. Biol.* **55:**227–233.

Bohne, W., Heesemann, J., and Gross, U., 1994, Reduced replication of *Toxoplasma gondii* is necessary for induction of bradyzoite-specific antigens: A possible role for nitric oxide in triggering stage conversion, *Infect. Immun.* **62:**1761–1767.

Boros, D. L., 1989, Immunopathology of *Schistosoma mansoni* infection, *Clin. Med. Rev.* **2:**250–269.

Brophy, P. M., and Pritchard, D. I., 1992. Immunity to helminths: Ready to tip the biochemical balance, *Parasitol. Today* **8:**419–422.

Brunet, L. R., Finkelman, F. D., Cheever, A. W., Kopf, M. A., and Pearce, E. J., 1997, IL-4 protects against TNF-α mediated cachexia and death during acute schistosomiasis, *J. Immunol.* **159:**777–785.

Bukrinsky, M. I., Nottet, H. S. L. M., Schmidtmayerova, H., Dubrovsky, L., Flanagan, C. R., Mullins, M. E., Sipton, S. A., and Gendelman, H. E., 1995, Regulation of nitric oxide synthase activity in human immunodeficiency virus type 1 (HIV-1)-infected monocytes implications for HIV-associated neurological disease. *J. Exp. Med.* **181:**735–745.

Cameron, M. L., Granger, D. L., Weinberg, J. B., Kozumbo, W. J., and Koren, H. S., 1990, Human alveolar and peritoneal macrophages mediate fungistasis independently of L-arginine oxidation to nitrite or nitrate. *Am. Rev. Dis.* **142:**1313–1319.

Capron, A. R., 1992, Immunity to schistosomes, *Curr. Opin. Immunol.* **4:**419–424.

Charles, I. G., Palmer, R. M., Hickery, M. S., Bayliss, M. T., Chubb, A. P., Hall, V. S., Moss, D. W., and Moncada, S., 1993, Cloning, characterization, and expression of a cDNA encoding an inducible nitric oxide synthase from the human chondrocyte. *Proc. Natl. Acad. Sci. USA.* **90:**11419–11423.

Corradin, S. B., Fasel, N., Buchmüller-Rouiller, Y., Ransijn, A., Smith, J., and Mauël, J., 1993, Induction of macrophage nitric oxide production by interferon-γ and tumor necrosis factor-α is enhanced by interleukin-10, *Eur. J. Immunol.* **23:**2045–2058.

Coulson, P. S., Smythies, L. E., Betts, C., Mabbott, N. A., Sternberg, J. M., Wei, X.-G., Liew, F. Y., and Wilson, R. A., 1998. Nitric oxide produced in the lungs of mice immunized with the radiation attenuated schistosome vaccine is not the major agent causing challenge parasite elimination, *Immunology* **93:**55–63.

Cox, G. W., Mathieson, B. J., Giardina, S. L., and Varisio, L., 1990, Characterization of IL-2 receptor expression and function on murine macrophages, *J. Immunol.* **145:**1719–1726.

Crawford, R. M., Finbloom, D. S., Ohara, J., Paul, W. E., and Meltzer, M. S., 1987, B cell stimulatory factor-1 (interleukin-4) activates macrophages for increased tumoricidal activity and expression of Ia antigens, *J. Immunol.* **139:**135–141.

Cunha, F. Q., Moncada, S., and Liew, F. Y., 1992, Interleukin-10 (IL-10) inhibits the induction of nitric oxide synthase by interferon-gamma in murine macrophages, *Biochem. Biophys. Res. Commun.* **182:**1155–1159.

Czaja, M. J, Weiner, F. R., Flanders, K. C., Giambrone, M. A., Wind, R., Biempica, L., and Zern, M. A., 1989, *In vitro* and *in vivo* association of transforming growth factor-beta 1 with hepatic fibrosis, *J. Cell Biol.* **108:**2477–2482.

Del Pozo, V., de Arruda-Chaves E., De Andres, B., Cardaba B., Lopez-Farre A., Gallardo, S., Cortegana, I., Vidarte, L., Jurado, A., Sastre J., Palomino, P., and Lahoz, C., 1997, Eosinophils transcribe and translate messenger RNA for inducible Nitric Oxide Synthase. *J. Immunol.* **158:**859–864.

De Maria, R., Cifone, M. G., Trotta, R., Rippo, M. R., Festuccia, C., Santoni, A., and Testi, R., 1994, Triggering of human monocyte activation through CD69, a member of the natural killer gene complex family of signal transducting receptors. *J. Exp. Med.* **180:**1999–2004.

Denis, M., 1994, Human monocytes/macrophages: NO or no NO? *J. Leuk. Biol.* **55:**682–684.

Dimmeler, S., Ankarcrona, M., Nicotera, P., and Brune, B., 1993, Exogenous nitric oxide (NO) generation or IL-1β-induced intracellular NO production stimulates inhibitory auto-ADP-ribosylation of glyceraldehyde 3-phosphate dehydrogenase in RINm5F cells, *J. Immunol,* **150:**2964–2971.

Ding, A. H., Nathan, C., and Stuehr, D. J., 1988, Release of reactive nitrogen intermediates and reactive oxygen intermediate from mouse peritoneal macrophages, *J. Immunol.* **141:**2407–2414.

Doherty, T. M., Kastelein, R., Menon, S., Andrade, S., and Coffman, R. L., 1993, Modulation of murine macrophage function by IL-13, *J. Immunol.* **151:**7151–7160.

Doyle, A., Herbein, G., Montaner, L. J., Minty, A. J., Caput, D., Ferrara, P., and Gordon, S., 1994, Interleukin-13 alters the activation state of murine macrophages in vitro: Comparison with interleukin-4 and interferon-gamma, *Eur. J. Immunol.* **24:**1441–1445.

Drapier, J. C., Wietzerbin, J., and Hibbs, J. B., Jr., 1988, Interferon-γ and tumor necrosis factor induce the L-arginine-dependent cytotoxic effector mechanism in murine macrophages, *Eur. J. Immunol.* **18:**1587–1592.

Evans, T. G., Thai, L., Granger, D. L., and Hibbs, J. B., Jr., 1993, Effect of *in vivo* inhibition of nitric oxide production in murine leishmaniasis, *J. Immunol.* **151:**907–915.

Flesch, I. E., and Kaufmann, S. H., 1990, Activation of tuberculostatic macrophage function by gamma interferon, interleukin-4 and tumor necrosis factor, *Infect. Immun.* **58:**2675–2677.

Flesch, I. E. A., Hess, J. H., Oswald, I. P., and Kaufmann, S. H. E., 1994, Growth inhibition of *Mycobacterium bovis* by interferon-γ stimulated macrophages: Regulation by endogenous tumor necrosis factor and IL-10, *Int. Immunol.* **6:**693–700.

Fouad Ahmed, S., Oswald, I. P., Caspar, P., Heiny, S., Keefer, L., Sher, A., and James, S. L., 1997, Developmental differences determine susceptibility to nitric oxide-mediated killing in a murine model of vaccination against *Schistosoma mansoni*, *Infect. Immun.* **65:**219–226.

Gazzinelli, R. T., Oswald, I. P., James, S. L., and Sher, A., 1992a, IL-10 inhibits parasite killing and nitrogen oxide production by IFN-γ activated macrophages, *J. Immunol.* **148:**1792–1796.

Gazzinelli, R. T., Oswald, I. P., Heiny, S., James, S. L., and Sher, A., 1992b, Activated macrophage microbicidal functions against *Trypanosoma cruzi* are mediated by an L-arginine dependent mechanism and blocked by IFN-γ antagonist cytokines, *Eur. J. Immunol.* **22:**2501–2506.

Geller, D. A., Lowenstein, C. J., Shapiro, R. A., Nussler, A. K., Di Silvio, M., Wang, S. C., Nakayama, D. K., Simmons, R. L., Snyder, S. H., Billiar, T. R., 1993, Molecular cloning and expression of inducible nitric oxide synthase from human hepatocytes. *Proc. Natl. Acad. Sci. USA.* **90:**3491–3495.

Granger, D. L., Hibbs, J. B., Jr., Perfect, J. R., and Durack, D. T., 1988, Specific amino acid (L-arginine) requirement for the microbiostatic activity of murine macrophages, *J. Clin. Invest.* **81:**1129–1136.

Granger, D. L., Hibbs, J. B., Jr., and Broadnax, L. M., 1991, Urinary nitrate excretion in relation to macrophage activation: Influence of dietary Larginine and oral N^{G}-monomethyl-Larginine, *J. Immunol.* **146:**1294–1302.

Green, S. J., Crawford, R. M., Hockmeyer, J. T., Meltzer, M. S., and Nacy, C. A., 1990, *Leishmania major* amastigotes initiate the L-arginine-dependent killing mechanism in IFN-γ-stimulated macrophages by induction of tumor necrosis factor-α, *J. Immunol.* **145:**4290–4297.

Griscavage, J. M., Rogers, N. E., Sherman, M. P., and Ignaro, L. J., 1993, Inducible nitric oxide synthase from a rat alveolar macrophage cell line is inhibited by nitric oxide, *J. Immunol.* **151:**6329–6337.

Grzych, J. M., Pearce, E., Cheever, A., Caulada, Z. A., Caspar, P., Heiny, S., Lewis, F., and Sher, A., 1991, Egg deposition is the major stimulus for the production of Th2 cytokines in murine *Schistosomiasis mansoni*, *J. Immunol.* **146:**1322–1327.

Hibbs, J. B. Jr., Vavrin, Z., Taintor, R. R., and Rachelin, E. M., 1988, Nitric oxide: A cytotoxic activated macrophage effector molecule, *Biochem. Biophys. Res. Commun.* **157:**87–94.

Hibbs, J. B. Jr., Westenfelder, C., Taintor, R. R., Vavrin, Z., Kablitz, C., Baranowski, R.L., Ward, J. H., Menlove, R. L., Mc Murray, M. P., Kushner, J. P., and Samlowski, W., 1992, Evidence for cytokine-inducible nitric oxide synthesis from L-arginine in patient receiving interleukin-2 therapy. *J. Clin. Invest.* **89:**867–877.

Hughes, H. P. A., 1988, Oxidative killing of intracellular parasite mediated by macrophages, *Parasitol. Today* **4:**340–347.

Ishii, A., Matsuoka, H., Aji, T., Otha, N., Arimoto, S., Wataya, Y., and Hayatsu, H., 1994, Parasite infection and cancer: With special emphasis on *Schistosoma japonicum* infection (Trematoda). A review, *Mutat. Res.* **305:**273–281.

James, S. L., 1995, The role of nitric oxide in parasitic infections, *Microbiol. Rev.* **59:**533–547.

James, S. L., and Glaven, J., 1989, Macrophage cytotoxicity against schistosomula of *Schistosoma mansoni* involves arginine-dependent production of reactive nitrogen intermediates, *J. Immunol.* **143:**4208–4212.

James, S. L., and Sher, A., 1990, Cell-mediated immune response to schistosomiasis, *Curr. Top. Microbiol. Immunol.* **155:**21–31.

James, S. L., Cook, K. W., and Lazdins, J. K., 1990, Activation of human monocyte-derived macrophages to kill schistosomula of *Schistosoma mansoni in vitro*, *J. Immunol.* **145:**2686–2690.

James, S. L., Cheever, A. W., Caspar, P., and Wynn, T. A., 1998, Inducible nitric oxide synthase-deficient mice develop enhanced type 1 cytokine-associated cellular and humoral immune response after vaccination with attenuated *Schistosoma mansoni cercariae* but display partially reduced resistance, *Infect. Immun.* **66:**3510–3518.

Karupiah, G., Kie, Q. W., Buller, M. L., Nathan, C., Duarte, C., and MacMicking, J. D., 1993, Inhibition of viral replication by interferon-γ-induced nitric oxide synthase, *Science* **261:**1445–1448.

Kassim, O., Dean, D. A., Mangold, B. L., and Von Lichtenberg, F., 1992, Combined autoradiographic and histopathologic analysis of the fate of challenge *Schistosoma mansoni* schistosomula in mice immunized with irradiated cercariae, *Am. J. Trop. Med. Hyg.* **47:**231–237.

Kilbourn, R. G., and Belloni, P., 1990, Endothelial cell production of nitrogen oxides in response to interferon-γ in combination with tumor necrosis factor, interleukin-1 or endotoxin, *J. Natl. Cancer Inst.* **82:**772–776.

Klotz, F. W., Scheller, L. F., Segin, M. C., Kumar, N., Marletta, M. A., Green, S. J., and Azad, A. F., 1995, Co-localization of inducible nitric oxide synthase and *Plasmodium berghei* in hepatocytes from rats with irradiated sporozoites, *J. Immunol.* **154:**3391–3395.

Komuniecki, R., and Komuniecki, P.R., 1995, Aerobic–anaerobic transitions in energy metabolism during the development of the parasite nematode *Ascaris suum*, in: *Molecular Approaches to Parasitology* (J. C. Boothroyd and R. Komuniecki, eds.), Wiley–Liss, New York, pp. 109–121.

Langermans, J. A. M., Van der Hulst, M. E. B., Nibbering, P. H., Hiemstra, P. S., Fransen, L., and Van Furth, R., 1992, IFN-γ induced L-arginine-dependent toxoplasmastatic activity in murine peritoneal macrophages is mediated by endogenous tumor necrosis factor, *J. Immunol.* **148:**568–574.

Lepoivre, M., Raddassi, K., Oswald, I., Tenu, J. P., and Lemaire, G., 1991, Antiproliferative effects of NO synthase products, *Res. Immunol.* **142:**580–583.

MacMicking, J. D., Nathan, C., Hom, G., Chartrain, N., Flechter, D. S., Trumbauer, M., Stevens, K., Xie, Q., Sokol, K., Hutchinson, N., Chen, H., and Mudgett, J. S., 1995, Altered response to bacterial infection and endotoxic shock in mice lacking inducible nitric oxide synthase, *Cell* **81:**641–651.

McLaren, D. J., and James, S. L., 1985, Ultrastructural studies of the killing of schistosomula of *Schistosoma mansoni* by activated macrophages *in vitro*, *Parasite Immunol.* **7:**315–331.

Mellouk, S., Green, S. J., Nacy, C. A., and Hoffman, S. L., 1991, IFN-γ inhibits development of *Plasmodium berghei* exoerythrocytic stage in hepatocytes by an Larginine-dependent effector mechanism, *J. Immunol.* **146:**3971–3976.

Mills, C. D., 1991, Molecular basis of suppressor macrophages: Arginine metabolism via the NO synthase pathway, *J. Immunol.* **146:**2719–2723.

Mostafa, M. H., Helmi, H., Badawi, A. F., Tricker, A. R., Spiegelhalder, B., and Preussmann, R., 1994, Nitrate, nitrite and volatile *N*-nitroso compounds in the urine of *Schistosoma haematobium* and *Schistosoma mansoni* infected patients, *Carcinogenesis* **15:**619–625.

Nathan, C., 1992, Nitric oxide as a secretory product of mammalian cells, *FASEB J.* **6:**3051–3064.

Nelson, B. J., Ralph, P., Green, S. J., and Nacy, C. A., 1991, Differential susceptibility of activated macrophage cytotoxic effector reactions to the suppressive effects of transforming growth factor-beta 1, *J. Immunol.* **146:**1849–1857.

Nussler, A. K., Drapier, J. C., Renia, L., Pied, S., Miltgen, F., Gentilini, M., and Mazier, D., 1991, L-arginine dependent destruction of intrahepatic malaria parasites in response to tumor necrosis factor and/or interleukin stimulation, *Eur. J. Immunol.* **22:**227–230.

Ohshima, H., and Bartsch, H., 1994, Chronic infections and inflammatory processes as cancer risk factors: Possible role of nitric oxide in carcinogenesis, *Mutat. Res.* **305:**253–264.

Ohshima, H., Bandaletova, T. Y., Brouet, I., Bartsch, H., Kirby, G., Ogunbiyi, F., Vatanasapt, V., and Pitigool, V., 1994, Increased nitrosamine and nitrate biosynthesis mediated by nitric oxide induced in hamster infected with liver fluke (*Opisthorchis viverrini*), *Carcinogenesis* **15:**271–275.

Oswald, I. P., and James, S. L., 1996, Nitrogen oxide in host defense against parasites, *Methods* **10:**8–14.

Oswald, I. P., Wynn, T. A., Sher, A., and James, S. L., 1992a, IL-10 inhibits macrophage microbicidal activity by blocking the endogenous production of tumor necrosis factor-α required as a costimulatory factor for interferon γ-induced activation, *Proc. Natl. Acad. Sci. USA* **89:**8676–8680.

Oswald, I. P., Gazzinelli, R. T., Sher, A., and James, S. L., 1992b, IL-10 synergizes with IL-4 and transforming growth factor-β to inhibit macrophage cytotoxic activity, *J. Immunol.* **148:**3578–3582.

Oswald, I. P., Wynn, T. A., Sher, A., and James, S. L., 1994a, NO as an effector molecule of parasite killing: Modulation of its synthesis by cytokines, *Comp. Biochem. Physiol.* **108**C:11–18.

Oswald, I. P., Eltoum, I., Wynn, T. A., Schartz, B., Paulin, D., Sher, A, and James, S. L., 1994b, Endothelial cells are activated by cytokine to kill an intravascular parasite, *Schistosoma mansoni*, through the production of nitric oxide, *Proc. Natl. Acad. Sci. USA* **91:**999–1003.

Oswald, I. P., Caspar, P., Sharton-Kersten, T., Williams, M. E., Heiny, S., Sher, A. and James, S. L., 1998, Failure of P strain mice to respond to vaccination against schistosomiasis correlates with impaired production of IL-12 and up-regulation of Th2 cytokines that inhibit macrophage activation, *Eur. J. Immunol.* **28:**1762–1772.

Paul-Eugene, N., Kolb, J. P., Damais, C., and Dugas, B., 1994, Heterogeneous nitrite production by IL-4 stimulated human monocytes and peripheral blood mononuclear cells. *Immunol. Lett.* **42:**31–34.

Pearce, E. J., and James, S. L., 1986, Post lung stage schistosomula of *Schistosoma mansoni* exhibit transient susceptibility to macrophage mediated cytotoxicity *in vitro* that may relate to the late phase killing *in vivo*, *Parasite Immunol.* **8:**513–527.

Pearce, E. J., Cheever, A., Leonard, S., Covalesky, M., Fernandez-Botran, R., Kohler, G., and Kopf, M., 1996, *Schistosoma mansoni* in IL-4 deficient mice, *Int. Immunol.* **8:**453.

Ralston, S. H., Todd, D., Helfrich, M., Benjamin, N., Grabowski, P. S., 1994, Human osteoblast-like cells produce nitric oxide and express inducible nitric oxide synthase, *Endocrinology.* **135:**330–336

Ramaswamy, K., He, Y. X., and Salafsky, B., 1997, ICAM-1 and iNOS expression increased in the skin of mice after vaccination with γ-irradiated cercaria of *Schistosoma mansoni*, *Exp. Parasitol.* **86:**118–132.

Reiling, N., Ulmer, A. J., Duchrow, M., Ernst, M., Flad, H. D., and Hauschildt, S., 1994, Nitric oxide synthase: mRNA expression of different isoforms in human monocyte/macrophage. *Eur. J. Immunol.* **24:**1941–1944.

Roland, C. R., Walp, L., Stack, R. M., and Flye, M. W., 1994, Outcome of Kupffer cell antigen presentation to a cloned murine Th1 lymphocyte depends on the inducibility of nitric oxide synthase by IFN-γ, *J. Immunol.* **153:**5453–5464.

Satarug, S., Haswell-Elkins, M. R., Tsuda, M., Mairiang, P., Sithithaworn, P., Mairiang, E., Eumi, H., Sukprasert, S., Yongvanit, P., and Elkins, D. B., 1996, Thiocyanate-independent nitrosation in human carcinogenic parasite infection, *Carcinogenesis* **17:**1075–1081.

Schleifer, K. W., and Mansfield, J. M., 1993, Suppressor macrophages in African trypanosomiasis inhibit T cell proliferative responses by nitric oxide and prostaglandins, *J. Immunol.* **151:**5492–5503.

Schmidt, H. H. H. W., and Walter, U., 1994, NO at work, *Cell* **78:**919–925.

Schneemann, M., Schoedon, G., Hofer, S., Blau, N., Guerrero, L., and Schaffner, A., 1993, Nitric oxide is not a constituent of the antimicrobial armature of hum•n mononuclear phagocytes. *J. Infect. Diseases.* **167:**1358–1363.

Scott, P., James, S. L., and Sher, A., 1985, The respiratory burst is not required for killing of intracellular and extracellular parasites by lymphokine-activated macrophage cell line, *Eur. J. Immunol.* **15:**553–558.

Sharton-Kersten, T. M., Yap, G., Magram, J., and Sher, A., 1997, Inducible nitric oxide is essential for host control of persistent but not acute infection with the intracellular pathogen *Toxoplasma gondii*, *J. Exp. Med.* **185:**1261–1273.

Sher, A., James, S. L., Simpson, A. J. G., Ladzin, J. K., and Meltzer, M. S., 1982, Macrophage as effector cells of protective immunity in murine schistosomiasis. III Loss of susceptibility to macrophage mediated killing during maturation of *S. mansoni* schistosomula from the skin to the lung stage, *J. Immunol.* **128:**1876–1879.

Sher, A., Coffman, R. L., Heiny, S., and Cheever, A. W., 1990, Ablation of eosinophil and IgE responses with anti IL-5 or anti IL-4 antibodies fails to affect immunity against *Schistosoma mansoni* in the mouse, *J. Immunol.* **145:**3911–3916.

Sicher, S. C., Vazquez, M. A., and Lu, C. Y., 1994, Inhibition of macrophage Ia expression by nitric oxide, *J. Immunol.* **153:**1293–1300.

Smythies, L. S., Coulson, P. S., and Wilson, R. A., 1992, Monoclonal antibody to IFN-γ modifies pulmonary inflammatory responses and abrogates immunity to *Schistosoma mansoni* in mice vaccinated with irradiated cercariae, *J. Immunol.* **149:**3654–3658

Stenger, S., Solbach, W., Röllinghoff, M., and Bogdan, C., 1991, Cytokine interactions in experimental cutaneous leishmaniasis. II. Endogenous tumor necrosis factor-α production by macrophages is induced by the synergistic action of interferon (IFN-γ) and interleukin (IL)-4 and accounts for the antiparasitic effect mediated by IFN-γ and IL-4, *Eur. J. Immunol.* **21:**1669–1675.

Stuer, D. J., and Marletta, M. A., 1985, Mammalian nitrate biosynthesis: mouse macrophages produce nitrite and nitrate in response to *Escherichia coli* lipopolysaccharide. *Proc. Natl. Acad. Sci. USA* **22:**7738–7742.

Taylor-Robinson, A. W., Liew, F. Y., Severn, A., Xu, D., McSorley, S. J., Garside, P., Padron, J., and Phillips, R. S., 1994, Regulation of the immune response by nitric oxide differentially produced by T helper type 1 and T helper type 2 cells, *Eur. J. Immunol.* **24:**980–984.

Thompson, D. P., Morrison, D. D., Pax, A., and Bennett, J. L., 1984, Changes in glucose metabolism and cyanide sensitivity in *Schistosoma mansoni* during development, *Mol. Biochem. Parasitol.* **13:**39–51.

Tielens, A. G. M., 1994, Energy generation in parasitic helminths, *Parasitol. Today* **10:**346–352.

Warren, W., Biggs, P. J., el-Baz, M., Ghoneim, M. A., Stratton, M. R., and Venitt, S., 1995, Mutations in the p53 gene in schistosomal bladder cancer: A study of 92 tumours from Egyptian patients and a comparison between mutational spectra from schistosomal and non-schistosomal urothelial tumours, *Carcinogenesis* **16:**1181–1189.

Wei, X., Charles, I. G., Smith, A., Ure, J., Feng, G., Huang, F., Xu, D., Muller, W., Moncada, S., and Liew, F. Y., 1995, Altered immune responses in mice lacking inducible nitric oxide synthase, *Nature* **375:**408–411.

Weinberg, J. B., Misukonis, M. A., Shami, P. J., Mason, S. N., Sauls, D. L., Dittman, W. A., Wood, E. R., Smith, G. K., McDonald, B., Bachus, K. E., Haney, A. F., Granger, D. L., 1995, Human mononuclear phagocyte inducible nitric oxide synthase (iNOS): analysis of iNOS mRNA, iNOS protein, biopterin and nitric oxide production by blood monocytes and peritoneal macrophages. *Blood* **86:**1184–1195.

Williams, M. A, Caspar, P., Oswald, I., Sharma, H. K., Pankewycz, O., Sher, A., and James, S. L., 1995, Vaccination routes that fail to elicit protective immunity against *Schistosoma mansoni* induce the production of TGF-β which down-regulates macrophage antiparasitic activity, *J. Immunol.* **154:**4693–4700.

Wilson, R. A., and Coulson, P. S., 1989, Lung phase immunity to schistosomes: A new perspective to an old problem, *Parasitol. Today* **5:**274–278.

Woodman, J. P., Dimier, I. H., and Bout, D. T., 1991, Human endothelial cells are activated by interferon gamma to inhibit *Toxoplasma gondii* replication, *J. Immunol.* **137:**2019.

Woods, M. S., Mayer, J., Evans, T. G., and Hibbs, J. B., Jr., 1994, Antiparasitic effects of nitric oxide in an *in vitro* model of *Chlamydia trachomatis* infection and an *in vivo* murine model of *Leishmania major* infection, *Immunol. Ser.* **60:**179–195.

Wynn, T. A., Oswald, I. P., Eltoum, I., Caspar, P., Lowenstein, C. J., Lewis, F. A., James, S. L., and Sher, A., 1994, Elevated expression of Th1 cytokines and NO synthase in the lungs of vaccinated mice after challenge infection with *Schistosoma mansoni*, *J. Immunol.* **153:**5200–5209.

Zurawski, G., and de Vries, J. E., 1994, Interleukin 13, an interleukin 4-like cytokine that acts on monocytes and B cells, but not on T cells, *Immunol. Today* **15:**19–26.

CHAPTER 17

Nitric Oxide in Leishmaniasis

From Antimicrobial Activity to Immunoregulation

CHRISTIAN BOGDAN, MARTIN RÖLLINGHOFF, and ANDREAS DIEFENBACH

1. Introduction

Long before the L-arginine/nitric oxide (NO) pathway in eukaryotic cells was discovered, nitrates and nitrites have been used to prevent bacterial contamination of food and to preserve the red color of meat (Kuschel, 1902; Tanner and Evans, 1934; Tarr, 1941). Today, we not only understand much of the chemistry behind the protective effect of exogenously added nitrogen oxides, but also have learned that endogenous NO or NO derivatives synthesized by the type 2 (or inducible) isoform of NO synthase (NOS2, iNOS) serve a very similar function in mammalian hosts, and form an important part of our defense system against microbial pathogens. In fact, over the past 10 years the analysis of the antimicrobial activity of NO has proved to be one of the cornerstones of NO research, and has yielded an extensive list of infectious agents that appear to be susceptible to the static and/or cidal effects of NO (Bogdan, 1997; MacMicking *et al.*, 1997a) (see also Chapter 12).

The first microbial organism shown to be controlled by NO *in vivo* was the protozoan parasite *Leishmania major* (Liew *et al.*, 1990b). *Leishmania* parasites exist in an extracellular flagellated form (promastigote), and an intracellular nonflagellated (amastigote) form that typically resides in phagocytic cells such

---

CHRISTIAN BOGDAN, MARTIN RÖLLINGHOFF, and ANDREAS DIEFENBACH • Institute of Clinical Microbiology, Immunology, and Hygiene, Friedrich-Alexander-University of Erlangen-Nuremberg, D-91054, Erlangen, Germany.

Nitric Oxide and Infection, edited by Fang. Kluwer Academic / Plenum Publishers, New York, 1999.

TABLE I
Experimental (Mouse) and Human Leishmaniasis: Examples of Clinical Syndromes

Leishmania species	Course of infection in mice	Course of infection in humans
L. major	Self-healing cutaneous lesions (e.g., C57BL/6, C3H, CBA mice)	Cutaneous ulcer (oriental sore; mostly self-healing)
	Visceral, fatal disease (BALB/c)	Kala-azar-like illness (rare)
L. amazonensis	Mostly chronic, nonhealing primary and metastatic skin lesions	Localized cutaneous lesions (New World; self-healing)
		diffuse cutaneous leishmaniasis (non-healing; rare)
L. donovani	Mostly self-limiting visceral disease (e.g., BALB/c, C57)	Visceral disease (kala-azar; fatal if untreated)

as macrophages, granulocytes, and dendritic cells. *Leishmania* transmitted between mammalian hosts by sandflies can cause a wide spectrum of diseases, which is influenced by the parasite species and the immune status of the host (Table I) (Pearson and de Queiroz Sousa, 1996). In both humans and mice, nonhealing cutaneous or visceral disease is associated with a predominance of type 2 (Th2) T-helper cells producing IL-4 and IL-5, and a lack of the macrophage-activating cytokine IFNγ. In contrast, control and ultimate cure of *Leishmania* infections requires the expansion of $CD4^+$ type 1 T-helper lymphocytes (Th1), which is dependent on stimulation by IL-12. Th1 cells interact with macrophages via the CD40/CD40L receptor/ligand pair, produce IFNγ, and thereby activate infected host cells to kill intracellular amastigotes (Reiner and Locksley, 1995; Bogdan *et al.*, 1996). In this chapter we will summarize the evidence that NO is a critical effector and regulatory molecule during this process.

2. Antileishmanial Activity of NO *in Vitro*

2.1. NO and Extracellular *Leishmania*

In host cell-free systems, promastigotes and amastigotes of various *Leishmania* species have been treated with $NO^{\bullet}$ gas, NO-nucleophile adducts donating $NO^{\bullet}$ (so-called NONOates or diazenium diolates), *S*-nitroso compounds (which transfer NO^+ or release $NO^{\bullet}$), acidified sodium nitrite ($NaNO_2$), 3-morpholino-sydnonimine (SIN-1, which decomposes in aqueous solution to $O_2^{-\bullet}$ and $NO^{\bullet}$ with the subsequent formation of peroxynitrite), sodium nitroprusside (SNP, which releases both NO and the toxic anion cyanide), and authentic peroxynitrite ($ONOO^-$, which on homolytic or heterolytic fission can yield $HO^{\bullet}$

and $NO_2^{\bullet}$ or HO^- and NO_2^+, respectively). Each of these compounds has been reported to exert cidal effects on *Leishmania* parasites, although peroxynitrite failed to do so in one study (Table II, Fig. 1). In a direct comparison of promastigote and amastigote forms of *L. amazonensis*, *L. chagasi*, and *L. major* parasites, the amastigote form turned out to be relatively more resistant to $NO^{\bullet}$ gas or *S*-nitroso-*N*-acetylpenicillamine (SNAP) than the promastigote stage (Lemesre *et al.*, 1997). The antileishmanial mechanism of action has not been studied in detail for different NO-generating substances, but some evidence suggests that nitrogen oxides can trigger iron loss from enzymes possessing iron–sulfur prosthetic groups, such as cytosolic *cis*-aconitase (Lemesre *et al.*, 1997), and *S*-nitrosothiols can inhibit glyceraldehyde-3-phosphate dehydrogenase activity of *Leishmania* parasites (Bourguignon *et al.*, 1997). In addition to its leishmanicidal activity, NO exerts leishmaniastatic effects on both promastigotes and amastigotes of *L. major* (C. Bogdan, unpublished observation), *L. chagasi*, *L. mexicana*, and *L. amazonensis*, and blocks the amastigote-to-promastigote transformation (Lemesre *et al.*, 1997).

TABLE II
Leishmanicidal and/or Growth-Inhibitory Effects of NO on Extracellular *Leishmania* Parasites

NO agent	*Leishmania* species tested	References	Static/cidal effect of NO
$NO^{\bullet}$ gas	*L. major* PM[a]	Liew *et al.* (1990b)	+
	L. chagasi PM and AM	Lemesre *et al.* (1997)	+
	L. mexicana PM and AM		+
	L. amazonensis PM and AM		+
NONOate (DETA-NO)[b]	L. major PM and AM	Diefenbach *et al.* (in preparation)	+
$NaNO_2^- + H^+$ (pH 5.0)	*L. enrietti* PM	Mauël, *et al.* (1991)	+
S-Nitroso-*N*-acetylpenicillamine (SNAP)	*L. major* PM	Assreuy *et al.* (1994), Vouldoukis *et al.* (1997), Diefenbach *et al.* (1999a)	+
	L. infantum PM	Vouldoukis *et al.* (1997)	+
	L. braziliensis PM	Vouldoukis *et al.* (1997)	+
S-Nitrosoglutathione (GSNO)	*L. major*	Diefenbach *et al.* (1999a)	+
3-Morpholino-sydnonimine (SIN-1)	*L. major* PM	Assreuy *et al.* (1994)	+
Peroxynitrite	*L. major* PM	Assreuy *et al.* (1994)	−
	L. amazonensis PM	Augusto *et al.* (1996)	+
Sodium nitroprusside (SNP)	*L. major* PM	Liew *et al.* (1990b), Vouldoukis *et al.* (1997)	+

[a] PM, promastigotes; AM, amastigotes.
[b] DETA-NO, diethylenetriamine–nitric oxide adduct.

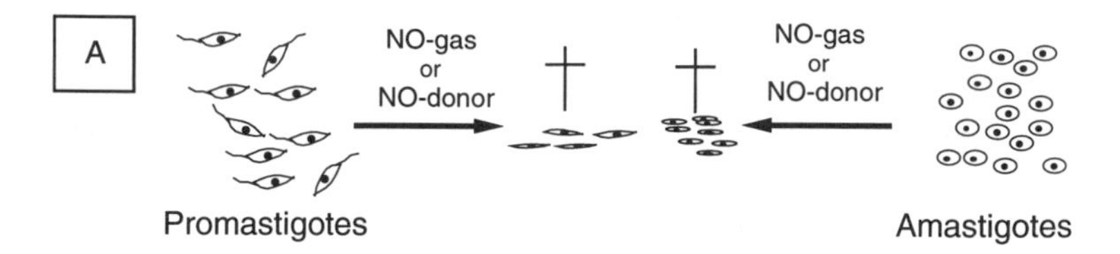

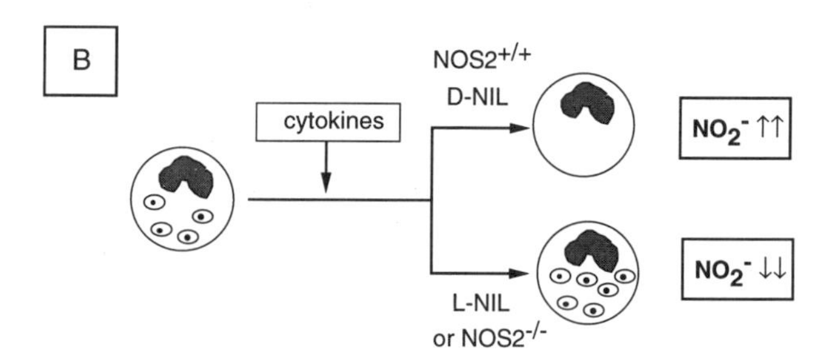

FIGURE 1. Experimental evidence for the antileishmanial function of NO *in vitro*. (A) Extracellular *Leishmania* pro- and amastigotes are killed by NO$^{\bullet}$ gas or NO donors. (B) The elimination of intracellular *Leishmania* by cytokine-activated macrophages is dependent on iNOS activity. Genetic deletion or pharmacological inhibition of iNOS abolishes the leishmanicidal activity of macrophages.

2.2. NO and Intracellular Leishmania

2.2.1. Control of Intracellular *Leishmania* by Endogenous iNOS

Macrophages, as classical host cells for *Leishmania*, are able to kill or restrict the growth of intracellular amastigotes if they are appropriately activated by T-helper lymphocytes or by soluble mediators. Activation of macrophages via cell–cell contact with Th1 cells is likely to involve membrane TNFα, the CD40/CD40L pair, and LFA-1/ICAM-1 interactions (Sypek and Wyler, 1991; Tian *et al.*, 1995; Soong *et al.*, 1996; Stout *et al.*, 1996). The key soluble cytokine in this respect is IFNγ, which alone or in combination with other mediators (e.g., TNFα, IL-4, IL-7, LPS) has been shown to strongly increase the leishmanicidal activity of murine and human macrophages (Murray, 1990, 1994; Bogdan *et al.*, 1993; Titus *et al.*, 1993). There is now compelling evidence that NO produced by iNOS is required for macrophages to kill *Leishmania* (Fig. 2): (1) *Leishmania* killing by activated macrophages is paralleled by the production of NO (measured as NO_2^- accumulating in culture supernatants); (2) suppression of macrophage NO release by L-arginine analogues such as L-N^G-monomethylarginine (L-NMMA, which inhibits all NOS isoforms) or L-N^6-iminoethyl-lysine (L-NIL, which has relative selectivity for iNOS) renders cytokine-activated macrophages unable to kill intracellular

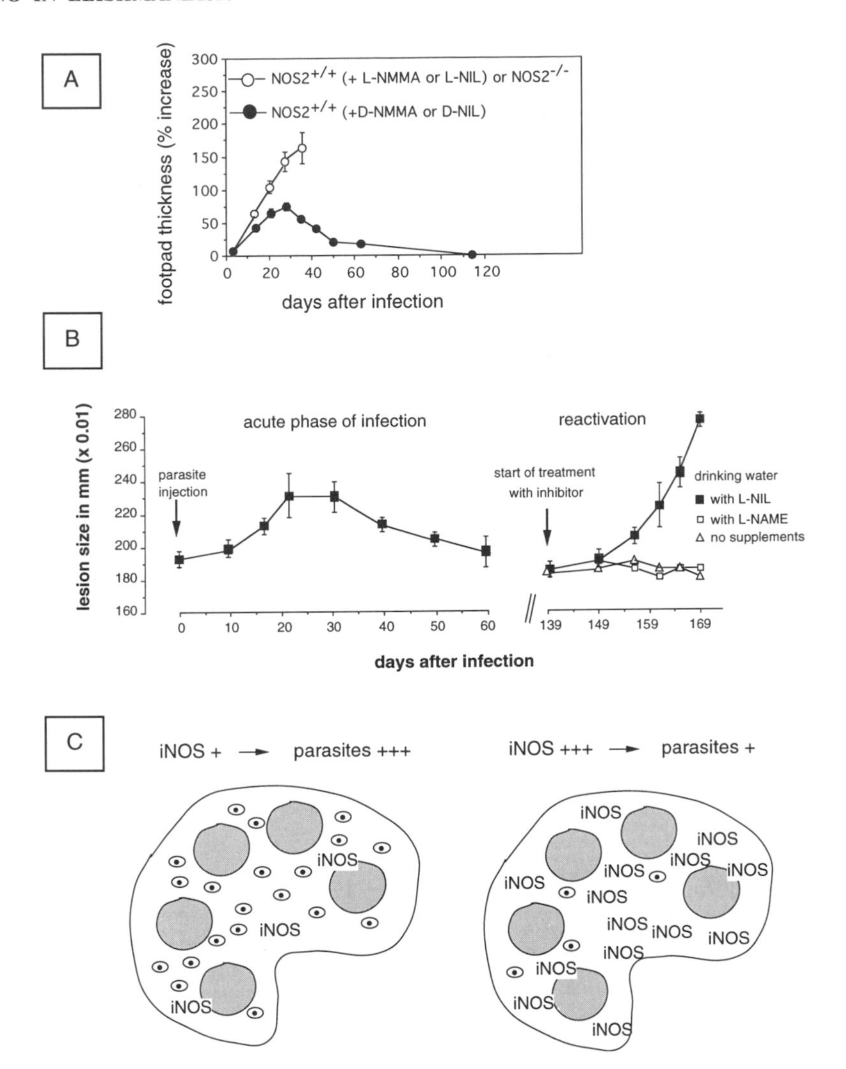

FIGURE 2. Experimental evidence for the antileishmanial function of NO *in vivo*. Genetic deletion or functional inactivation of iNOS exacerbated the acute phase of cutaneous leishmaniasis (A) or reactivated latent leishmaniasis (B) in mice. *In situ*, the expression of iNOS protein in the lymph node tissue correlated inversely with the number of parasites (C).

Leishmania; (3) macrophages genetically deficient in iNOS are devoid of leishmanicidal activity; and (4) cytokines that inhibit the production of NO by macrophages (e.g., TGFβ, IL-4, IL-10) rescue the survival of intracellular *Leishmania* amastigotes according to some, but not all, reports.

These correlations have been demonstrated for various *Leishmania* species (*L. major*, *L. enriettii*, *L. donovani*, *L. infantum*), for different murine macrophage populations (from the peritoneum or bone marrow) (Green *et al.*, 1990a; Liew *et al.*, 1990a,b, 1991; Bogdan *et al.*, 1991; Mauël *et al.*, 1991; Nelson *et al.*, 1991; Roach *et al.*, 1991; Vieth *et al.*, 1994; Soong *et al.*, 1996; Diefenbach *et al.*, 1998), and for human blood monocyte-derived macrophages (Vouldoukis *et al.*, 1995, 1997). In infected mouse peritoneal macrophages, IFNγ is a potent inducer of iNOS and antileishmanial activity, which can be synergistically enhanced by exogenous or endogenous TNFα (Bogdan *et al.*, 1990; Green *et al.*, 1990a; Liew *et al.*, 1990a). In the case of human macrophages, iNOS has been reported to be induced by IFNγ or after ligation of CD23 (by IgE–anti-IgE immune complexes or anti-CD23 antibody) (Vouldoukis *et al.*, 1995, 1997), a finding awaiting confirmation amid many frustrating attempts of other investigators to induce iNOS in human macrophages *in vitro* (Albina, 1995) (see Chapter 6 for an extensive review of this topic). Interestingly, the presence of intracellular *Leishmania* facilitates the induction of iNOS by IFNγ in both murine and human macrophages (Green *et al.*, 1990b; Vouldoukis *et al.*, 1997), which might provide a mechanism for the restriction of iNOS expression to infected cells *in vivo*.

2.2.2. Control of Intracellular *Leishmania* by Exogenous NO

NO-releasing compounds have been used to target *Leishmania* amastigotes residing within macrophages. Vouldoukis *et al.* (1997) reported that SNAP or SNP added to infected human macrophages induces effective killing of intracellular parasites. The interpretation of this type of experiment, however, can be complicated because exogenous NO donors can exert toxic effects on host cells. In fact, we have been unable to induce killing of intracellular *L. major* amastigotes in mouse macrophages by concentrations of SNAP, GSNO, or DETA-NO that do not simultaneously damage the host cells (Diefenbach *et al.*, 1999a, submitted). This problem can be circumvented by using alternative NO-generating systems. Mauël *et al.* (1991) used sodium nitrite, which was very efficient in inducing parasite killing when added to macrophages infected with *L. enriettii*, without affecting the viability or function of the host cells. A possible explanation for the lack of toxicity is that conversion of nitrite to NO occurs only within the acidic phagolysosomal compartment. In a different approach, we have demonstrated NO-dependent killing of *L. major* amastigotes within *iNOS*$^{-/-}$ macrophages by cocultured cytokine-activated *iNOS*$^{+/+}$ macrophages located in the vicinity of the infected host cells (Diefenbach *et al.*, 1999a, submitted).

2.2.3. The Contribution of Reactive Oxygen Intermediates versus NO to the Killing of Leishmania

Prior to the discovery of L-arginine-dependent generation of NO as an antimicrobial effector mechanism, reactive oxygen intermediates generated by the NADPH oxidase pathway were believed to be critical for the destruction of *Leishmania* by mouse and human macrophages. Published data correlated cytokine-induced killing of intracellular *Leishmania* with the generation of H_2O_2 and/or $O_2^{-\bullet}$, or with priming for the production of these intermediates in response to phorbol esters or zymosan (Murray, 1981b, 1982; Haidaris and Bonventre, 1982; Pearson *et al.*, 1983; Passwell *et al.*, 1986; Lehn *et al.*, 1989). Catalase was among the most effective scavengers in inhibiting the leishmanicidal activity of macrophages, which led to the conclusion that H_2O_2 rather than superoxide or hydroxyl radical is necessary for killing. Furthermore, in a cell-free xanthine oxidase microbicidal system, catalase but not scavengers of $O_2^{-\bullet}$, $HO^{\bullet}$, or $^1O_2$ protected *L. donovani* or *L. major* promastigotes (Murray, 1982). One way to reconcile these findings with more recent observations implicating iNOS is to postulate a convergence of the NADPH oxidase and the iNOS pathways, which might lead to the formation of a more toxic product than either pathway alone (e.g., generation of peroxynitrite from $O_2^{-\bullet}$ and $NO^{\bullet}$), as has been described in *E. coli* or in superoxide dismutase-deficient *Salmonella* (Pacelli *et al.*, 1995; DeGroote *et al.*, 1997). This hypothesis is somewhat at odds with the inability of superoxide dismutase to diminish the antimicrobial activity of macrophages against *Leishmania* (Murray, 1981a, 1982), although synergistic interactions between $NO^{\bullet}$ and H_2O_2 would still be conceivable. The temporal segregation of $NO^{\bullet}$ and $O_2^{-\bullet}$ production within a single cell (Assreuy *et al.*, 1994), at least *in vitro*, also argues somewhat against such an interaction. A number of additional observations obtained with mouse macrophages suggest that reactive oxygen intermediates do not have a principal role in the control of intracellular *Leishmania*: (1) the cell line IC-21, which is incapable of producing reactive oxygen intermediates, can still be activated by cytokines for the killing of intracellular *L. donovani* amastigotes (Scott *et al.*, 1985); (2) macrophage treatment with L-NMMA inhibits iNOS and prevents parasite killing despite an increase in production of H_2O_2 in response to phorbol ester (Ding *et al.*, 1988; Liew *et al.*, 1990a), and priming for high-output production of $O_2^{-\bullet}$ in the absence of $NO^{\bullet}$ release leaves intracellular *Leishmania* unimpaired (Mauël and Buchmüller-Rouiller, 1987; Mauël *et al.*, 1991); and (3) catalase has been found to inhibit the NO synthesis and leishmanicidal activity of cytokine-activated mouse peritoneal exudate macrophages (Li *et al.*, 1992). The latter effect can be reversed by the addition of tetrahydrobiopterin, a NOS cofactor, and offers an alternative explanation for the aforementioned parasite-protective function of catalase when added to infected host macrophages. As these experiments have not been performed with human monocytes/macrophages, the possibility that $O_2^{-\bullet}$ and/or H_2O_2 have an antileishmanial function in human phagocytes has not yet been excluded.

3. Antileishmanial Activity of NO *in Vivo*

3.1. Acute Phase of Infection

In genetically resistant inbred mice, infection with *L. major* causes self-healing cutaneous lesions. Intralesional or oral treatment of these mice with NOS inhibitors that are nonselective (e.g., L-NMMA) or selective (e.g., L-NIL) for iNOS leads to the development of ulcerating, nonhealing lesions with dramatically increased parasite burdens at the original site of infection, as well as within draining lymph nodes (Liew *et al.*, 1990b; Evans *et al.*, 1993; Stenger *et al.*, 1995). Similar observations have been made with two different strains of iNOS-deficient mice (129/Sv × MF1, Wei *et al.*, 1995; 129/SvEv × C57BL/6, Diefenbach *et al.*, 1998) (Fig. 2A). These findings clearly demonstrate that high-level production of NO is required for the control of *L. major in vivo*. A recent analysis with *iNOS*$^{-/-}$ mice on a pure 129/Sv background, however, suggests that infection with *L. major* can ultimately be overcome even in the absence of iNOS activity. The mechanism of lesion resolution in this model remains to be elucidated, but may involve Fas/FasL-mediated macrophage apoptosis (Huang *et al.*, 1998b). Comparative analysis of skin and lymph node tissue from *L. major*-infected genetically resistant (e.g., C57BL/6, C3H/HeN) or susceptible (BALB/c) mice has revealed a more rapid and marked upregulation of iNOS protein in the tissues of resistant mice during the acute phase of infection ($<$day 41) (Stenger *et al.*, 1994). This difference is also reflected by the urinary nitrate excretion, which is much higher in resistant mice up to day 25 of infection (Evans *et al.*, 1993). Furthermore, iNOS-positive areas in the lymph node contain far fewer parasites than iNOS-negative areas (Fig. 2C). The expression of iNOS protein is confined to macrophages and not observed in keratinocytes, Langerhans cells, endothelial cells, B cells, T cells, or granulocytes (Stenger *et al.*, 1994; Thüring *et al.*, 1995; Blank *et al.*, 1996).

More recently, we performed a detailed analysis of the very early phase of infection with *L. major* in genetically resistant mice. In contrast to the predictions of some investigators, iNOS expression was readily detectable on day 1 of infection in the skin lesion. Although iNOS-positive cells were only focally distributed within the dermis, and the majority of *L. major*-infected cells were negative for iNOS, we observed a striking antileishmanial function of iNOS. Mice lacking the iNOS gene or those treated with the iNOS inhibitor L-NIL experienced dissemination of parasites to all visceral organs examined (spleen, liver, bone marrow, lung) within

24 hr of infection, whereas the parasites remained confined to the original cutaneous site of infection and the initial draining lymph node in control mice. The early induction of iNOS is dependent on IFNα/β and not on IFNγ (Diefenbach *et al.*, 1998). Thus, the expression and function of iNOS is not a late event, but part of the immediate innate response to *L. major* (see also Fig. 3).

In the *L. donovani* mouse model, inhibition of iNOS has also been found to increase parasite burdens in the liver. Interestingly, iNOS, the antileishmanial effector mechanism, and the formation of liver granulomata can be induced by exogenous IL-12 in the absence of endogenous IFNγ, which argues for the existence of an IFNγ-independent pathway for *Leishmania* control (Taylor and Murray, 1997). As yet, there is no evidence for or against iNOS involvement in the control of *Leishmania* in infected humans.

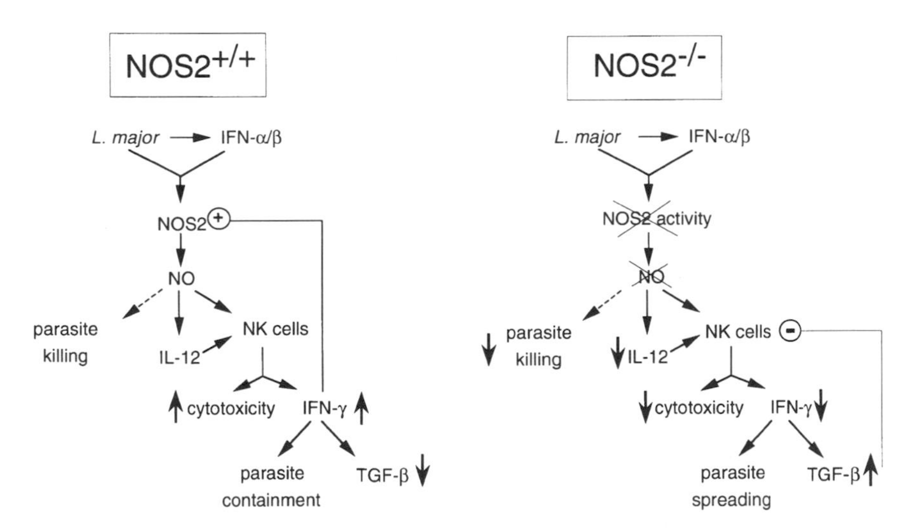

FIGURE 3. Summary of the possible cascade of events during the innate response to *L. major* at day 1 of infection (Diefenbach *et al.*, 1998). After cutaneous infection with *L. major*, IFNα/β is induced, which synergizes with the parasites for the focal expression of iNOS in the dermis of $iNOS^{+/+}$ mice at day 1 of infection. The NO produced by iNOS might exert a direct leishmanicidal effect (dashed arrows), but, more importantly, it is critical for the baseline expression of IL-12, mediates the production of IFNγ (via maintaining the responsiveness to IL-12), and is required for the functional maturation of natural killer (NK) cells (cytotoxic activity). IFNγ, in turn, leads to parasite containment, suppresses the production of TGFβ, and later further enhances the synthesis of iNOS (positive feedback loop). In $iNOS^{-/-}$ mice, IFNα/β is induced, but because of the genetic deletion iNOS protein is not expressed. Consequently, there is a decreased baseline level of IL-12, no responsiveness to IL-12 and no early upregulation of IFNγ, an almost complete absence of NK-cell cytotoxic activity, and hyperexpression of TGFβ, which might further deactivate the NK cells (negative feedback loop).

3.2. Chronic Phase of Infection

During our analysis of the acute phase of *L. major* infection, we happened to observe that mice cured of their cutaneous disease continued to express high amounts of iNOS mRNA and protein in the draining lymph node and at the former site of infection in the skin. At the same time, a small number of parasites persisted in these tissues as reported previously (Aebischer *et al.*, 1993). This raised the question whether the parasites are under the continuous control of iNOS-derived NO, and if so, why they are not eliminated in the first place. Treatment of long-term-infected, clinically cured mice with the iNOS inhibitor L-NIL had a striking effect and provided an answer to the first part of the question: Cutaneous disease was promptly reactivated leading to high parasite burdens in the skin and lymph node tissue, and eventually to ulcerated lesions (Fig. 2B). Thus, the continuous expression of iNOS was not just a consequence of an ongoing inflammatory response, but unmistakably exhibited a (direct or indirect) antimicrobial activity. Immunohistological analyses provided some clues as to why the parasites managed to persist despite the presence of iNOS. In the lymph node, approximately 70% of the organisms were located in iNOS-negative areas, densely surrounded by reticular fibroblasts and matrix deposits (Stenger *et al.*, 1996; Bogdan *et al.*, 1999a, submitted). Based on *in vitro* results, these cells appear to function as "safe targets" (Mirkovich *et al.*, 1986), but surrounding iNOS-positive macrophages prevent net replication of the *Leishmania* and thereby maintain a clinically stable balance (Diefenbach *et al.*, 1999a, submitted). These studies provided the first evidence for a NO-dependent control of an intracellular microorganism in a latently infected host organism and have recently been corroborated in a *Mycobacterium tuberculosis* model (MacMicking *et al.*, 1997b; Flynn *et al.*, 1998).

4. NO Limitations

The results summarized so far may lead the reader to believe that NO/iNOS is a perfect antileishmanial effector mechanism. This, however, is not the case. A number of observations underscore its limitations and provide a reminder of the "double-edged sword" character of the NO molecule (Schmidt and Walter, 1994).

1. When amastigotes of *L. major* are exposed to high concentrations of SNAP, GSNO, or DETA-NO (500–1000 μM) *in vitro*, 1–5% will survive and resume replication as soon as the NO source is removed (Diefenbach *et al.*, 1999a, submitted). This leishmaniostatic rather than leishmanicidal effect might contribute to the long-term persistence of *Leishmania in vivo*.

2. *Leishmania* have evolved mechanisms to suppress the induction of iNOS in macrophages. Macrophages that have already been infected with *L. major* promastigotes for 4–18 hr, exhibit strongly reduced expression of iNOS after stimulation with IFNγ. Various known surface molecules of *Leishmania* (glycoinositolphospholipids, phosphoglycan backbone of lipophosphoglycan) appear to account for this effect (Proudfoot *et al.*, 1995, 1996). In addition, the saliva of the sandfly vector *Phlebotomus papatasi* has been found to reduce NO production by macrophages (Hall and Titus, 1995; Waitumbi and Warburg, 1998).
3. NO controls *L. major* in the skin and in the lymph node, but probably not in the spleen. First, genetic deletion or functional inactivation of iNOS causes only a minor increase of the parasite numbers in the spleen (Stenger *et al.*, 1996; Diefenbach *et al.*, 1998). Second, genetically resistant mice show only scant expression of iNOS mRNA and protein in the spleen (Stenger *et al.*, 1994). Third, nonhealing BALB/c mice upregulate the expression of iNOS in the spleen and the urinary excretion of nitrate at late time points of *L. major* infection, but nevertheless harbor enormous numbers of parasites in the spleen and other visceral organs (Evans *et al.*, 1996c; Bogdan *et al.*, unpublished observation). Similar observations have been made in *L. amazonensis*-infected BALB/c mice, and peroxynitrite has been suggested to be at least partially responsible for damage to parasitized tissue (Giorgio *et al.*, 1996). These results argue for a pathological function of NO under certain circumstances (see also Chapter 8).

Currently, it appears that NO/iNOS is indispensable for the killing of *Leishmania*, but additional factors are required for the ultimate control of parasite replication and/or clinical cure. One of these factors might be TNFα. *L. major*-infected mice lacking the type I (or p55) receptor TNFα (TNFRp55$^{-/-}$) are able to reduce the parasite burden in the tissue to the same degree as wild-type mice, but fail to resolve their cutaneous lesions (Vieira *et al.*, 1996; Nashleanas *et al.*, 1998). The critical role of TNFα is also emphasized by a related study with *Listeria monocytogenes*, in which the TNFRp55$^{-/-}$ mice succumbed to the bacterial infection despite an unimpaired production of NO and reactive oxygen intermediates (Endres *et al.*, 1997).

5. Regulation of the Antileishmanial Immune Response by NO

NO research in infectious diseases has principally focused on the antimicrobial activity of NO (Bogdan, 1997), although *in vitro* studies have pointed to important immunoregulatory functions of NO for quite some time (see Chapter 10).

Of these, the inhibitory effect of NO on T-lymphocyte proliferation (Albina *et al.*, 1991) and the stimulation as well as suppression of cytokine production by macrophages, T cells, B cells, and endothelial cells have been best documented (Magrinat *et al.*, 1992; Lander *et al.*, 1993; Marcinkiewicz *et al.*, 1995; Peng *et al.*, 1995; Remick and Villarete, 1996).

In *L. major*-infected mice lacking the iNOS gene, Wei *et al.* (1995) reported an expansion of $CD4^+$ T lymphocytes and a high production of IFNγ by antigen-stimulated spleen cells harvested at day 70 of infection. They argued that this is a consequence of deficient feedback suppression of Th1 cells by NO, because NO was found to (a) inhibit the proliferation and cytokine production of malaria-specific Th1, but not Th2, clones (Wei *et al.*, 1995) and (b) to reduce the production of IL-12, a potent stimulus of Th1 cells, by macrophages (Huang *et al.*, 1998). Alternatively, however, the expansion of Th1 cells could simply reflect the attempt of the host to control the strongly increased parasite burden in $iNOS^{-/-}$ mice. More recently, we have obtained strong evidence that NO regulates the innate response to *Leishmania* at a time of infection when differences in the parasite load between wild-type and $iNOS^{-/-}$ mice are small and, thus, unlikely to account for regulatory differences (Diefenbach *et al.*, 1998). On day 1 of infection, genetic deletion or functional inactivation of iNOS lowered the baseline expression of IL-12 and increased the expression of TGF-β. Most importantly, there was no early IFN-γ production and NK cell activity in the absence of iNOS. *In vitro* experiments demonstrated that iNOS-derived NO was critically required for the response of NK cells to IL-12 and the subsequent release of IFN-γ (Diefenbach *et al.*, 1998; Diefenbach *et al.*, 1999b). Neutralization of IFNα/β, which we identified as the principal inducer of iNOS on day 1, caused the same alterations as observed in $iNOS^{-/-}$ mice. *In vitro*, IFNγ suppresses the production of TGFβ by macrophages (Schindler *et al.*, 1998), but not in the absence of iNOS, offering a possible explanation for the hyperexpression of TGFβ in $iNOS^{-/-}$ mice (Diefenbach *et al.*, 1998). Finally, in $iNOS^{-/-}$ or anti-IFNα/β-treated wild-type mice, *Leishmania* parasites spread to all visceral organs within 24 hr of infection as detailed in Section 3.1. Similar observations have been made in mice lacking the IFNγ gene, although the expression of iNOS was unimpaired in that model (Diefenbach *et al.*, 1998). Thus, parasite containment is not purely a function of NO, but does require the expression of IFNγ. Our current view of the cascade of NO-dependent events on day 1 of infection with *L. major* is summarized in Fig. 3.

6. Therapeutic Approaches: A Perspective

To date, therapy of leishmanial infections with chemotherapeutic agents is far from satisfactory. Several of the established drugs used for treatment of patients with visceral leishmaniasis have severe side effects. In patients with diffuse cutaneous leishmaniasis, therapy often completely fails. Furthermore, even after successful therapy, i.e., when clinical cure of the disease has been achieved, small numbers of *Leishmania* are known to persist in mammalian hosts, presumably lifelong. Numerous case reports document that these residual parasites form a potential threat for patients, as they can be reactivated and cause fulminant leishmaniasis during periods of immunosuppression. Therefore, it is highly desirable to develop strategies for the ultimate elimination of persisting *Leishmania*. Considering the impressive susceptibility of *Leishmania* to NO, one possible approach would be the design of tissue-specific NO donors that become activated in infected host cells, thereby minimizing toxic side effects. The synthesis of such compounds is being intensively pursued by a number of researchers, and certainly will be further stimulated by a recent study of Lopez-Jaramillo *et al.* (1998), who have obtained promising results in the treatment of patients with cutaneous leishmaniasis with the topical application of the "conventional" NO donor SNAP.

ACKNOWLEDGMENT. The preparation of this review and the conduct of part of the studies reviewed were supported by the Deutsche Forschungsgemeinschaft (Sonderforschungsbereich 263, project A5).

References

Aebischer, T., Moody, S. F., and Handman, E., 1993, Persistence of virulent *Leishmania major* in murine cutaneous leishmaniasis: A possible hazard for the host, *Infect. Immun.* **61:**220–226.

Albina, J. E., 1995, On the expression of nitric oxide synthase by human macrophages. Why no NO? *J. Leukoc. Biol.* **58:**643–649.

Albina, J. E., Abate, J. A., and Henry, W. L., Jr., 1991, Nitric oxide production is required for murine resident peritoneal macrophages to suppress mitogen-stimulated T cell proliferation. Role of IFN-γ in the induction of the nitric oxide-synthesizing pathway, *J. Immunol.* **147:**144–148.

Assreuy, J., Cunha, F. Q., Epperlein, M., Noronha-Dutra, A., O'Donnell, C. A., Liew, F. Y., and Moncada, S., 1994, Production of nitric oxide and superoxide by activated macrophages and killing of *Leishmania major*, *Eur. J. Immunol.* **24:**672–676.

Augusto, O., Linares, E., and Giorgio, S., 1996, Possible roles of nitric oxide and peroxynitrite in murine leishmaniasis, *Braz. J. Med. Biol. Res.* **29:**853–862.

Blank, C., Bogdan, C., Bauer, C., Erb, K., and Moll, H., 1996, Murine epidermal Langerhans cells do not express inducible nitric oxide synthase, *Eur. J. Immunol.* **26:**792–796.

Bogdan, C., 1997, Of microbes, macrophages and NO, *Behring Inst. Res. Commun.* **99:**58–72.

Bogdan, C., Moll, H., Solbach, W., and Röllinghoff, M., 1990, Tumor necrosis factor-α in combination with interferon-γ, but not with interleukin 4 activates murine macrophages for elimination of *Leishmania major* amastigotes, *Eur. J. Immunol.* **20:**1131–1135.

Bogdan, C., Stenger, S., Röllinghoff, M., and Solbach, W., 1991, Cytokine interactions in experimental cutaneous leishmaniasis. Interleukin 4 synergizes with interferon-γ to activate murine macrophages for killing of *Leishmania major* amastigotes, *Eur. J. Immunol.* **21:**327–333.

Bogdan, C., Gessner, A., and Röllinghoff, M., 1993, Cytokines in leishmaniasis: A complex network of stimulatory and inhibitory interactions, *Immunobiology* **189:**356–396.

Bogdan, C., Gessner, A., Solbach, W., and Röllinghoff, M., 1996, Invasion, control, and persistence of *Leishmania* parasites, *Curr. Opin. Immunol.* **8:**517–525.

Bogdan, C., Donhauser, N., Lorenz, E., Stenger, S., Röllinghoff, M., and Diefenbach, A., 1999b, Fibroblasts as safe targets for *Leishmania in vivo*, submitted for publication.

Bourguignon, S. C., Alves, C. R., and Giovanni-de-Simone, S., 1997, Detrimental effect of nitric oxide on *Trypanosoma cruzi* and *Leishmania major* like cells, *Acta Trop.* **66:**109–118.

DeGroote, M. A., Ochsner, U. A., Shiloh, M. U., Nathan, C., McCord, J. M., Dinauer, M. C., Libby, S. J., Vazquez-Torres, A., and Fang, F. C., 1997, Periplasmic superoxide dismutase protects *Salmonella* from products of phagocyte NADPH-oxidase and nitric oxide synthase, *Proc. Natl. Acad. Sci. USA* **94:**13997–14001.

Diefenbach, A., Schindler, H., Donhauser, N., Lorenz, E., Laskay, T., MacMicking, J., Röllinghoff, M., Gresser, I., and Bogdan, C., 1998, Type I interferon (IFN-α/β) and type 2 nitric oxide synthase regulate the innate immune response to a protozoan parasite, *Immunity* **8:**77–87.

Diefenbach, A., Döring, R., Röllinghoff, M., and Bogdan, C., 1999a, An *in vitro* model for the nitric oxide-dependent control of *Leishmania major* in the chronically infected host, submitted for publication.

Diefenbach, A., Schindler, H., Röllinghoff, M., Yokoyama, W. M., and Bogdan, C., 1999b, Requirement for type 2 NO synthase for IL-12 responsiveness in innate immunity, *Science* (in press).

Ding, A. H., Nathan, C. F., and Stuehr, D. J., 1988, Release of reactive nitrogen intermediates and reactive oxygen intermediates from mouse peritoneal macrophages. Comparison of activating cytokines and evidence for independent production, *J. Immunol.* **141:**2407–2412.

Endres, R., Luz, A., Schulze, H., Neubauer, H., Fütterer, A., Holland, S. M., Wagner, H., and Pfeffer, K., 1997, Listeriosis in p47phox$^{-/-}$ and TRp55$^{-/-}$ mice: Protection despite absence of ROI and susceptibility despite presence of RNI, *Immunity* **7:**419–432.

Evans, T. G., Thai, L., Granger, D. L., and Hibbs, J. B., Jr., 1993, Effect of *in vivo* inhibition of nitric oxide production in murine leishmaniasis, *J. Immunol.* **151:**907–915.

Evans, T. G., Reed, S. S., and Hibbs, J. B., 1996, Nitric oxide production in murine leishmaniasis: Correlation of progressive infection with increasing systemic synthesis of nitric oxide, *Am. J. Trop. Med. Hyg.* **54:**486–489.

Flynn, J. L., Scanga, C. A., Tanaka, K. E., and Chan, J., 1998, Effects of aminoguanidine on latent murine tuberculosis, *J. Immunol.* **160:**1796–1803.

Giorgio, S., Linares, E., Capurro, M. d. L., de Bianchi, A. G., and Augusto, O., 1996, Formation of nitrosyl hemoglobin and nitrotyrosine during murine leishmaniasis, *Photochem. Photobiol.* **63:**750–754.

Green, S. J., Crawford, R. M., Hockmeyer, J. T., Meltzer, M. S., and Nacy, C. A., 1990a, *Leishmania major* amastigotes initiate the L-arginine-dependent killing mechanism in IFN-γ stimulated macrophages by induction of tumor necrosis factor-α, *J. Immunol.* **145:**4290–4297.

Green, S. J., Meltzer, M. S., Hibbs, J. B., Jr., and Nacy, C. A., 1990b, Activated macrophages destroy intracellular *Leishmania major* amastigotes by an L-arginine-dependent killing mechanism, *J. Immunol.* **144:**278–283.

Haidaris, C. G., and Bonventre, P. F., 1982, A role for oxygen-dependent mechanisms in killing of *Leishmania donovani* tissue forms by activated macrophages, *J. Immunol.* **129:**850–855.

Hall, L. R., and Titus, R. G., 1995, Sand fly vector saliva selectively modulates macrophage functions that inhibit killing of *Leishmania major* and nitric oxide production. *J. Immunol.* **155:**3501–3506.

Huang, F.-P., Niedbala, W., Wei, X.-O., Xu, D., Feng, G.-J., Robinson, J. H., Lam, C., and Liew, F. Y., 1998a, Nitric oxide regulates Th1 cell development through inhibition of IL-12 synthesis by macrophages. *Eur. J. Immunol.* **28:**4062–4070.

Huang, F.-P., Xu, D., Esfandiari, E.-O., Sands, W., Wei, X.-Q., and Liew, F. Y., 1998b, Mice defective in Fas are highly susceptible to *Leishmania major* infection despite elevated IL-12 synthesis, strong Th1 responses, and enhanced nitric oxide production, *J. Immunol.* **160:**4143–4147.

Kuschel, F., 1902, Über die Wirkung des Einlegens von Fleisch in verschiedene Salze, *Arch. Hyg.* **43:**134–150.

Lander, H. M., Sehajpal, P. K., and Novogrodsky, A., 1993, Nitric oxide signaling: A possible role for G proteins, *J. Immunol.* **151:**7182–7187.

Lehn, M., Weiser, W. Y., Engelhorn, S., Gillis, S., and Remold, H. G., 1989, IL-4 inhibits H_2O_2 production and antileishmanial capacity of human cultured monocytes mediated by IFN-γ, *J. Immunol.* **143:**3020–3024.

Lemesre, J.-L., Sereno, D., Daulouède, S., Veyret, B., Brajon, N., and Vincendeau, P., 1997, *Leishmania* spp.: Nitric oxide-mediated inhibition of promastigote and axenically grown amastigote forms, *Exp. Parasitol.* **86:**58–68.

Li, Y., Severn, A., Rogers, M. V., Palmer, R. M. J., Moncada, S., and Liew, F. Y., 1992, Catalase inhibits nitric oxide synthesis and the killing of intracellular *Leishmania major* in murine macrophages, *Eur. J. Immunol.* **22:**441–446.

Liew, F. Y., Li, Y., and Millott, S., 1990a, Tumor necrosis factor-α synergizes with IFN-γ in mediating killing of *Leishmania major* through the induction of nitric oxide, *J. Immunol.* **145:**4306–4310.

Liew, F. Y., Millott, S., Parkinson, C., Palmer, R., M. J., and Moncada, S., 1990b, Macrophage killing of *Leishmania* parasite *in vivo* is mediated by nitric oxide from L-arginine, *J. Immunol.* **144:**4794–4797.

Liew, F. Y., Li, Y., Severn, A., Millott, S., Schmidt, J., Salter, M., and Moncada, S., 1991, A possible novel pathway of regulation by murine T helper type-2 (Th2) cells of a Th1 cell activity via the modulation of the induction of nitric oxide synthase in macrophages, *Eur. J. Immunol.* **21:**2489–2494.

Lopez-Jaramillo, P., Ruano, C., Rivera, J., Teran, E., Salazar-Irigoyen, R., Esplugues, J. V., and Moncada, S., 1998, Treatment of cutaneous leishmaniosis with nitric-oxide donor, *Lancet* **351:**1176–1177.

MacMicking, J., Xie, Q.-W., and Nathan, C., 1997a, Nitric oxide and macrophage function, *Annu. Rev. Immunol.* **15:**323–350.

MacMicking, J. D., North, R. J., LaCourse, R., Mudgett, J. S., Shah, S. K., and Nathan, C. F., 1997b, Identification of nitric oxide synthase as a protective locus against tuberculosis, *Proc. Natl. Acad. Sci. USA* **94:**5243–5248.

Magrinat, G., Mason, S. N., Shami, P. J., and Weinberg, J. B., 1992, Nitric oxide modulation of human leukemia cell differentiation and gene expression, *Blood* **80:**1880–1884.

Marcinkiewicz, J., Grabowska, A., and Chain, B., 1995, Nitric oxide up-regulates the release of inflammatory mediators by mouse macrophages, *Eur. J. Immunol.* **25:**947–951.

Mauël, J., and Buchmüller-Rouiller, Y., 1987, Effect of lipopolysaccharide on intracellular killing of *Leishmania enriettii* and correlation with macrophage oxidative metabolism, *Eur. J. Immunol.* **17:**203–208.

Mauël, J., Ransijn, A., and Buchmüller-Rouiller, Y., 1991, Killing of *Leishmania* parasites in activated murine macrophages is based on an L-arginine-dependent process that produces nitrogen derivatives, *J. Leukoc. Biol.* **49:**73–82.

Mirkovich, A. M., Galelli, A., Allison, A. C., and Modabber, F. Z., 1986, Increased myelopoiesis during *Leishmania major* infection in mice: Generation of "safe targets", a possible way to evade the effector immune mechanism, *Clin. Exp. Immunol.* **64:**1–7.

Murray, H. W., 1981a, Interaction of *Leishmania* with a macrophage cell-line. Correlation between intracellular killing and the generation of oxygen intermediates, *J. Exp. Med.* **153:**1690–1695.

Murray, H. W., 1981b, Susceptibility of *Leishmania* to oxygen intermediates and killing by normal macrophages, *J. Exp. Med.* **153:**1302–1315.

Murray, H. W., 1982, Cell-mediated immune response in experimental visceral leishmaniosis. II. Oxygen-dependent killing of intracellular *Leishmania donovani* amastigotes, *J. Immunol.* **129:**351–357.

Murray, H. W., 1990, Gamma interferon, cytokine-induced macrophage activation, and antimicrobial host defense *in vitro*, in animal models, and in humans, *Diagn. Microbiol. Infect. Dis.* **13:**411–421.

Murray, H. W., 1994, Blood monocytes: Differing effector role in experimental visceral versus cutaneous leishmaniasis, *Parasitol. Today* **10:**220–223.

Nashleanas, M., Kanaly, S., and Scott, P., 1998, Control of *Leishmania major* infection in mice lacking TNF receptors, *J. Immunol.* **160:**5506–5513.

Nelson, B. J., Ralph, P., Green, S. J., and Nacy, C. A., 1991, Differential susceptibility of activated macrophage cytotoxic reactions to the suppressive effects of transforming growth factor-β1, *J. Immunol.* **146:**1849–1857.

Pacelli, R., Wink, D. A., Cook, J. A., Krishna, M. C. W. D., Friedman, N., Tsokos, M., Samuni, A., and Mitchell, J. B., 1995, Nitric oxide potentiates hydrogen peroxide-induced killing of *Escherichia coli*, *J. Exp. Med.* **182:**1469–1479.

Passwell, J. H., Shor, R., and Shoham, J., 1986, The enhancing effect of interferon-β and -γ on the killing of *Leishmania tropica major* in human mononuclear phagocytes *in vitro*, *J. Immunol.* **136:**3062–3066.

Pearson, R. D., and de Queiroz Sousa, A., 1996, Clinical spectrum of leishmaniasis, *Clin. Infect. Dis.* **22:**1–13.

Pearson, R. D., Harcus, J. L., Roberts, D., and Donowitz, G. R., 1983, Differential survival of *Leishmania donovani* amastigotes in human monocytes, *J. Immunol.* **131:**1994–1999.

Peng, H.-P., Rajavashisth, T. B., Libby, P., and Liao, J. K., 1995, Nitric oxide inhibits macrophage-colony stimulating factor gene transcription in vascular endothelial cells, *J. Biol. Chem.* **270:**17050–17055.

Proudfoot, L., O'Donnell, C. A., and Liew, F. Y., 1995, Glycoinositolphospholipids of *Leishmania major* inhibit nitric oxide synthesis and reduce leishmanicidal activity in murine macrophages, *Eur. J. Immunol.* **25:**745–750.

Proudfoot, L., Nikolaev, A. V., Feng, G.-J., Wei, X.-Q., Ferguson, M. A. J., Brimacombe, J. S., and Liew, F. Y., 1996, Regulation of the expression of nitric oxide synthase and leishmanicidal activity by glycoconjugates of *Leishmania* lipophosphoglycan in murine macrophages, *Proc. Natl. Acad. Sci. USA* **93:**10984–10989.

Reiner, S. L., and Locksley, R. M., 1995, The regulation of immunity to *Leishmania major*, *Annu. Rev. Immunol.* **13:**151–177.

Remick, D. G., and Villarete, L., 1996, Regulation of cytokine gene expression by reactive oxygen and reactive nitrogen intermediates, *J. Leukoc. Biol.* **59:**471–475.

Roach, T. I. A., Kiderlen, A. F., and Blackwell, J. M., 1991, Role of inorganic nitrogen oxides and tumor necrosis factor alpha in killing *Leishmania donovani* amastigotes in gamma interferon/lipopolysaccharide-activated macrophages from Lsh^s and Lsh^r congenic mouse strains, *Infect. Immun.* **59:**3935–3944.

Schindler, H., Diefenbach, A., Röllinghoff, M., and Bogdan, C., 1998, IFN-γ inhibits the production of latent transforming growth factor-β1 by mouse inflammatory macrophages, *Eur. J. Immunol.* **28:**1181–1188.

Schmidt, H. H. H. W., and Walter, U., 1994, NO at work, *Cell* **78:**919–925.

Scott, P., James, S., and Sher, A., 1985, The respiratory burst is not required for killing of intracellular and extracellular parasites by a lymphokine-activated macrophage cell-line, *Eur. J. Immunol.* **15:**553–558.

Soong, L., Xu, J.-C., Grewal, I. S., Kima, P., Sun, J., Longley, B. J., Ruddle, N. H., McMahon-Pratt, D., and Flavell, R. A., 1996, Disruption of CD40–CD40-ligand interactions results in an enhanced susceptibility to *Leishmania amazonensis* infection, *Immunity* **4:**263–273.

Stenger, S., Thüring, H., Röllinghoff, M., and Bogdan, C., 1994, Tissue expression of inducible nitric oxide synthase is closely associated with resistance to *Leishmania major*, *J. Exp. Med.* **180:**783–793.

Stenger, S., Thüring, H., Röllinghoff, M., Manning, P., and Bogdan, C., 1995, L-N^6-(1-iminoethyl)-lysine potently inhibits inducible nitric oxide synthase and is superior to N^G-monomethyl-arginine *in vitro* and *in vivo*, *Eur. J. Pharmacol.* **294:**703–712.

Stenger, S., Donhauser, N., Thüring, H., Röllinghoff, M., and Bogdan, C., 1996, Reactivation of latent leishmaniasis by inhibition of inducible nitric oxide synthase, *J. Exp. Med.* **183:**1501–1514.

Stout, R. D., Suttles, J., Xu, J., Grewal, I., and Flavell, R. A., 1996, Impaired T cell-mediated macrophage activation in CD40 ligand-deficient mice, *J. Immunol.* **156:**8–11.

Sypek, J. P., and Wyler, D. J., 1991, Antileishmanial defense in macrophages triggered by tumor necrosis factor expressed on $CD4^+$ T lymphocyte plasma membrane, *J. Exp. Med.* **174:**755–759.

Tanner, F. W., and Evans, F. L., 1934, Effect of meat curing solutions on anaerobic bacteria. III. Sodium nitrite, *Zbl. Bakteriol. II* **91:**1–14.

Tarr, H. L. A., 1941, Bacteriostatic action of nitrates, *Nature* **147:**417–418.

Taylor, A. P., and Murray, H. W., 1997, Intracellular antimicrobial activity in the absence of interferon-γ: Effect of interleukin-12 in experimental visceral leishmaniasis in interferon-γ gene-disrupted mice, *J. Exp. Med.* **185:**1231–1239.

Thüring, H., Stenger, S., Gmehling, D., Röllinghoff, M., and Bogdan, C., 1995, Lack of inducible nitric oxide synthase in T cell clones and T lymphocytes from naive and *Leishmania major*-infected mice, *Eur. J. Immunol.* **25:**3229–3234.

Tian, L., Noelle, R. J., and Lawrence, D. A., 1995, Activated T cells enhance nitric oxide production by murine splenic macrophages through gp39 and LFA–1, *Eur. J. Immunol.* **25:**306–309.

Titus, R. G., Theodos, C. M., Shankar, A., and Hall, L. R., 1993, Interactions between *Leishmania major* and macrophages, in: *Macrophage–Pathogen Interactions* (B. Zwilling and T. Eisenstein, eds.), Dekker, New York, pp. 437–459.

Vieira, L. Q., Goldschmidt, M., Nashleanas, M., Pfeffer, K., Mak, T., and Scott, P., 1996, Mice lacking the TNF receptor p55 fail to resolve lesions caused by infection with *Leishmania major*, but control parasite replication, *J. Immunol.* **157:**827–835.

Vieth, M., Will, A., Schröppel, K., Röllinghoff, M., and Gessner, A., 1994, Interleukin–10 inhibits antimicrobial activity against *Leishmania major* in murine macrophages, *Scand. J. Immunol.* **40:**403–409.

Vouldoukis, I., Riveros-Moreno, V., Dugas, B., Quaaz, F., Bécherel, P., Debré, P., Moncada, S., and Mossalayi, M. D., 1995, The killing of *Leishmania major* by human macrophages is mediated by nitric oxide induced after ligation of the FcεRII/CD23 surface antigen, *Proc. Natl. Acad. Sci. USA* **92:**7804–7808.

Vouldoukis, I., Bécherel, P.-A., Riveros-Moreno, V., Arock, M., da Silva, O., Debré, P., Mazier, D., and Mossalayi, M. D., 1997, Interleukin-10 and interleukin-4 inhibit intracellular killing of *Leishmania infantum* and *Leishmania major* by human macrophages by decreasing nitric oxide generation, *Eur. J. Immunol.* **27:**860–865.

Waitumbi, J., and Warburg, A., 1998, *Phlebotomus papatasi* saliva inhibits protein phosphatase activity and nitric oxide production by murine macrophages, *Infect. Immun.* **66:**1534–1537.

Wei, X.-Q., Charles, I. G., Smith, A., Ure, J., Feng, G.-J., Huang, F.-P., Xu, D., Müller, W., Moncada, S., and Liew, F. Y., 1995, Altered immune responses in mice lacking inducible nitric oxide synthase, *Nature* **375:**408–411.

CHAPTER 18

Nitric Oxide in Viral Myocarditis

CHARLES J. LOWENSTEIN, MARTA SAURA, and CARLOS ZARAGOZA

1. Introduction

The natural history of viral myocarditis is highly variable. A given viral agent can cause mild myocarditis in some patients and fulminant infection in others. Some patients with severe myocarditis develop a permanent dilated cardiomyopathy, while others recover without residual evidence of cardiac dysfunction. This broad spectrum of illness is attributable in part to variable host factors.

The immune mechanisms that are activated by viral infection of the heart are not completely understood. Still less is known about the variations in host immune function that can lead to strikingly different clinical syndromes. Observational studies of humans and experiments with mice have shown that the host response to viral myocarditis can be divided into specific and nonspecific defenses. Macrophages and IFNγ are important mediators of the nonspecific antiviral response. Recent work has shown that macrophages play a pivotal role in the defense against viral infection by producing the potent antiviral effector molecule nitric oxide (NO) when stimulated with cytokines such as IFNγ.

This chapter will first summarize the clinical course of myocarditis in humans, then describe what is known about the immune response to viral myocarditis based on observations in a murine model. After a brief description of coxsackieviruses, the most common etiological agents of human myocarditis, the discussion will focus on recent research that has established NO to be a key inhibitor of viral

CHARLES J. LOWENSTEIN, MARTA SAURA, and CARLOS ZARAGOZA • Division of Cardiology, Department of Medicine, Johns Hopkins University School of Medicine, Baltimore, Maryland 21205.

Nitric Oxide and Infection, edited by Fang. Kluwer Academic / Plenum Publishers, New York, 1999.

replication in myocarditis. These studies have defined a critical role for NO in the innate immune system.

2. Viral Myocarditis in Humans

Myocarditis is defined as an inflammatory infiltrate of the heart associated with myocyte necrosis (Aretz *et al.,* 1987). The differential diagnosis of myocarditis is broad, including toxins, hypersensitivity to various drugs, collagen-vascular diseases, giant cell myocarditis, hypereosinophilic syndrome, and infections. A wide variety of infectious agents have been reported to cause myocarditis, including viruses, bacteria, fungi, and parasites. Although myocarditis is associated with a variety of viruses (Table I), coxsackieviruses account for more than 50% of viral myocarditis cases (Ray, 1994).

TABLE I
Viruses Associated with Myocarditis

Family / Virus
Adenoviridae
Adenovirus
Flaviviridae
Yellow fever virus
Hepadnaviridae
Hepatitis B virus
Herpesviridae
Cytomegalovirus
Epstein–Barr virus
Orthomyxoviridae
Influenza
Paramyxoviridae
Measles virus
Mumps virus
Respiratory syncytial virus
Picornaviridae
Coxsackievirus
Echovirus
Encephalomyocarditis
Poliovirus
Poxviridae
Vaccinia virus
Variola virus
Retroviridae
Human immunodeficiency virus
Rhabdoviridae
Rabies
Togaviridae
Rubella

The clinical and laboratory features of viral myocarditis vary markedly, and the host factors responsible for this variation are not well understood (Woodruff, 1980). Coxsackievirus infection is more virulent in children than in adults. Symptoms can be limited to an acute flulike syndrome with fever, myalgias, and diarrhea, or be followed by myocarditis or pericarditis with palpitations and chest pain. Coxsackievirus infection can also cause meningitis, ataxia, paralysis, exanthems, and enanthems. Myocarditis can be associated with a variety of electrocardiographic abnormalities, including ST segment and T-wave abnormalities, AV (atrioventricular) conduction disturbances, and ventricular dysrhythmias. Cardiac myocyte necrosis is demonstrated by serum creatine kinase elevations. Echocardiograms demonstrate regional or diffuse wall motion abnormalities. An endomyocardial biopsy may be necessary to confirm the diagnosis in patients presenting with a clinical syndrome consistent with myocarditis: Only 10% of patients referred for biopsy to exclude the diagnosis of myocarditis actually had histological evidence of myocarditis in the Myocarditis Treatment Trial (Mason *et al.,* 1995). Biopsy findings include an inflammatory infiltrate with myocyte degeneration or necrosis, and the severity of myocarditis can be graded according to the "Dallas criteria" (Aretz *et al.,* 1987).

If patients survive the initial period of infection, they usually recover from viral myocarditis within months, with eventual resolution of symptoms and echocardiographic abnormalities. The reversibility of ventricular dysfunction may provide an important clue to the nature of the host response in viral myocarditis. Left ventricular dysfunction caused by viral myocarditis is usually transient; dysfunction in approximately 50% of patients with biopsy-proven myocarditis improves spontaneously (Quigley *et al.,* 1987; O'Connell and Mason, 1989).

However, between 10 and 30% of patients with viral myocarditis have persistent ventricular dysfunction that slowly progresses to a dilated cardiomyopathy (O'Connell and Mason, 1989). The mortality of patients with myocarditis and dilated cardiomyopathy is surprisingly high, with approximately 20% of patients dead within 1 year, rising to 56% after 4 years (Mason *et al.,* 1995). Host factors that affect survival are poorly understood, but a stronger immune response is associated with patient survival, as discussed in greater detail below (Mason *et al.,* 1995).

Several longstanding controversies about myocarditis remain unresolved. The true prevalence of viral myocarditis is unknown but is probably greater than reported, and dilated cardiomyopathy of unknown cause may often result from undiagnosed viral myocarditis. Approximately 30% of patients with dilated cardiomyopathy have enteroviral RNA in their myocardium (Kandolf *et al.,* 1987), although enteroviral RNA is also detectable in a lower percentage of normal heart tissue.

Another controversy in myocarditis concerns whether the immune response to viral infection is beneficial or detrimental to the host. Coxsackievirus infection of mice can produce an early myocarditis caused by viral injury of the heart, followed by a prolonged myocarditis characterized by autoimmune damage to the heart (Rose *et al.*, 1986). This autoimmune response is mediated in part by lymphocytes; T-cell-deficient mice do not develop myocarditis after infection unless T lymphocytes are harvested from infected normal mice and adoptively transferred (Hashimoto *et al.,* 1983). B lymphocytes produce autoreactive antibodies in certain strains of infected mice (Wolfgram *et al.,* 1985), and may also contribute to autoimmune cardiac injury. Finally, autoimmune myocarditis can be duplicated by injecting myosin into mice (Neu *et al.,* 1987; Rose *et al.,* 1987; Huber and Cunningham, 1996). Perhaps human myocarditis also has an autoimmune component. However, the Myocarditis Treatment Trial showed that routine immunosuppressive therapy of myocarditis does not improve left ventricular function or mortality (Mason *et al.*, 1995). Thus, the contribution of autoimmunity to human myocarditis is not well understood. In fact, little is known about the human immune response to viral myocarditis. Most knowledge on this subject has been extrapolated from animal models.

3. Viral Myocarditis in Mice

The severity of coxsackievirus myocarditis in mice depends both on the strain of virus, and on host characteristics such as age, gender, and genetic background (Wolfgram *et al.*, 1986; Rose *et al.*, 1988; Huber and Pfaeffle, 1994). Coxsackievirus infection causes an acute myocarditis, and in certain strains of mice is followed by a chronic myocarditis.

The murine host response to coxsackievirus is mediated by humoral and cellular factors, which can additionally be classified as specific or nonspecific. Specific cellular responses include T and B lymphocytes. Cytotoxic T lymphocytes participate in the specific cellular response to coxsackievirus group B type 3 (CVB3) infection, and lyse infected cardiac myocytes (Woodruff and Woodruff, 1974; Huber *et al.*, 1980). However, these cytotoxic T lymphocytes are $CD4^+$, rather than the $CD8^+$ cells that appear to be involved in the autoimmune phase of experimental myocarditis in mice. The role of T lymphocytes depends on the genetics of the host (Huber and Pfaeffle, 1994). The effector mechanism by which T lymphocytes kill virally infected cells has not been established, but presumably involves perforin or granzyme B. B lymphocytes produce not only neutralizing anticoxsackievirus antibodies within 2 days of infection (Wolfgram *et al.*, 1985, 1986), but also anti-cardiac myocyte antibodies which may play a role in the autoimmune phase of myocarditis (Wolfgram *et al.*, 1985; Alvarez *et al.*, 1987; Neumann *et al.*, 1994). Mice that lack B and T lymphocytes develop severe and extensive myocardial necrosis when infected with coxsackievirus (Chow *et al.*, 1992).

Nonspecific responses to infection are mediated by natural killer (NK) cells and macrophages. Coxsackievirus activates NK cells for 2–4 days after infection of mice (Godeny and Gauntt, 1986, 1987b). Mice that lack NK cells have extensive myocardial necrosis following infection, and lesions are characterized by dystrophic calcification instead of inflammatory infiltrates and fibrosis (Godeny and Gauntt, 1986). The antiviral effect of NK cells appears to be mediated at least in part by direct lysis of infected cells; NK cells obtained from CVB3-infected mice can lyse CVB3 target cells *in vitro* (Godeny and Gauntt, 1987b).

Macrophages are another component of the nonspecific immune response believed to be required for the host to limit CVB3 infection (Woodruff, 1979; Godeny and Gauntt, 1987a; Hiraoka *et al.*, 1995). However, macrophages may also exert effects detrimental to the infected host; mice unable to produce the chemokine macrophage inflammatory protein-1 alpha (MIP-1α) are resistant to coxsackievirus myocarditis (Cook *et al.*, 1995). Thus, the role of the macrophage is not clearly defined in the host response to coxsackievirus infection in mice.

IFNγ is a regulator of nonspecific antiviral defense, and can limit the replication of coxsackievirus *in vivo* and *in vitro* (Kandolf *et al.*, 1985; Godeny and Gauntt, 1987b; Heim *et al.*, 1992). IFNγ reduces the number of focal cardiac lesions produced by coxsackievirus in infected mice (Godeny and Gauntt, 1987b) and reduces the replication of coxsackievirus in cultured cardiac myocytes (Kandolf *et al.*, 1985). However, the mechanism by which interferon exerts its antiviral effect was not understood at the time these observations were first made.

In summary, studies during the 1970s and 1980s revealed that T and B lymphocytes produce a specific defense against murine coxsackievirus infection. In addition to this slowly developing specific immunity, there is a rapid nonspecific response to coxsackievirus involving macrophages and IFNγ. The connection between interferon and macrophages was revealed by the discovery of NO and NO synthase (NOS).

4. NO and the Transcriptional Regulation of iNOS

As discussed in Chapters 3 and 4, NO is a small, radical molecule synthesized by the three NOS isoforms (Marletta, 1989; Ignarro, 1990; Moncada *et al.*, 1991; Lowenstein and Snyder, 1992; Nathan and Xie, 1994). The inducible iNOS (NOS2) isoform produces large amounts of NO relative to the other isoforms, and therefore is ideally suited to respond to pathogens. While many cytokines can induce iNOS expression (Chapter 5), IFNγ can act synergistically with other stimuli to promote expression in murine cells via interferon regulatory factor-1 (IRF-1), which binds to a response element upstream of the iNOS gene (Lowenstein *et al.*, 1993; Xie *et al.*,

1993; Kamijo *et al.*, 1994; Martin *et al.*, 1994). When transcription factors NF-κB and IRF-1 each bind to their respective regulatory sites, the iNOS promoter is rendered more accessible to the transcriptional apparatus (Xie *et al.,* 1994). Previous work has strongly implicated IFNγ in host defense against viral infection, focusing the attention of several investigators on the potential antiviral properties of NO.

5. Antiviral Properties of NO

NO has been demonstrated to protect cells from a variety of viral infections (see also Chapter 12). Exogenous NO reduces replication of Sindbis virus, vesicular stomatitis virus, and murine Friend leukemia virus (Akarid *et al.*, 1995; Bi and Reiss, 1995; Tucker *et al.*, 1996). When activated by IFNγ, macrophage-derived NO reduces replication of ectromelia, vaccinia, and herpes simplex-1 viruses (Croen, 1993; Karupiah *et al.*, 1993).

Infection with a wide variety of viruses induces iNOS in cells and in animals, presumably in part by inducing synthesis of IFNγ. However, the role of NO production in host defense against viruses is unclear. In some experimental models, NOS inhibition is detrimental, consistent with a role of NO as an antiviral effector. For example, mice infected with the retrovirus murine Friend leukemia virus have a greater viral load when treated with nitro-arginine (Akarid *et al.*, 1995). However, reduction of NO synthesis has no effect on other viral infections in mice. For example, NOS inhibitors have no effect on lymphocytic choriomeningitis or vaccinia infection of mice (Butz *et al.*, 1994; Rolph *et al.*, 1996). Indeed, in some experimental viral infections, inhibition of NOS actually improves survival. For example, NOS inhibition improves survival rates of mice with influenza pneumonitis (Akaike *et al.*, 1996) (Chapter 19) and mice infected with the flavivirus tick-borne encephalitis virus (Kreil and Eibl, 1996). Thus, although a variety of viruses induce iNOS and NO production in animals, the effect of NO may be pathogen specific.

Viral infection can induce iNOS expression in humans, according to a few case reports, but clinical data are scarce. Elevated serum levels of the NO metabolite nitrate have been reported in patients infected with hantavirus (Groeneveld *et al.*, 1995), and iNOS mRNA has been detected in brain tissue from patients infected with HIV (Bukrinsky *et al.*, 1995) (see Chapter 21).

6. Life Cycle of Coxsackievirus

Coxsackievirus infection is the most common cause of human viral myocarditis, and may in fact represent the most common cause of myocarditis overall. The

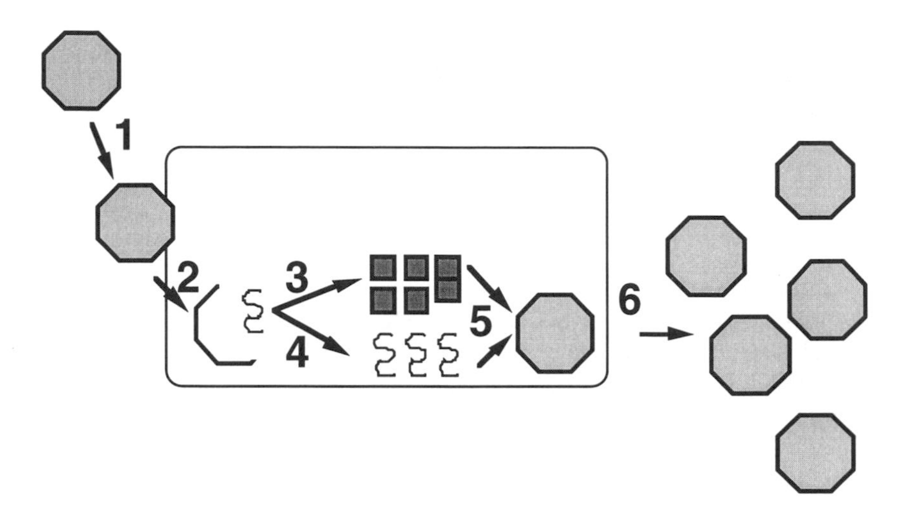

FIGURE 1. The coxsackievirus life cycle. Coxsackievirus binds to cell surface receptors (1) and penetrates the cell (2). Viral RNA is translated (3) and replicated (4), followed by virion assembly (5) and exit from the cell (6).

structure of coxsackievirus is similar to other members of the picornavirus family (Melnick, 1985; Rueckert, 1985), with a protein capsid coat surrounding a single positive strand RNA molecule. The life cycle of coxsackievirus is important for understanding the effects of NO on viral replication (Fig. 1). After binding to surface receptors such as the coxsackievirus adenovirus receptor, and possibly co-receptors such as decay-accelerating factor, nucleolin, and immunoglobulin, the virus penetrates the host cell and uncoats. The positive-strand RNA is translated into viral proteins. One of these coxsackievirus proteins, RNA-dependent RNA polymerase $3D^{pol}$, replicates the viral genome from positive-strand RNA into negative-strand RNA, and subsequently into multiple copies of positive-strand RNA. The coxsackievirus is assembled from positive-strand RNA and structural capsid proteins, and exits the cell. Each of the distinct life stages of coxsackieviruses could potentially be inhibited by NO.

7. Murine Model of Coxsackievirus Myocarditis

In general, mice infected with coxsackievirus develop a viremia that peaks 1 day after the onset of infection and resolves within 3 days. Virus is detected in the heart 2–15 days after infection, peaking on day 5 (Wolfgram *et al.,* 1986). Focal zones of myocyte necrosis and a mononuclear cellular infiltrate are seen in heart tissue 3–7 days after infection (Woodruff and Woodruff, 1974; Woodruff, 1979; Bendinelli *et al.*, 1982; Campbell *et al.*, 1982; Godeny and Gauntt, 1987a; Cook, 1996; Huber *et al.*, 1996).

7.1. Induction of iNOS Expression in Coxsackievirus Myocarditis

Infection with CVB3 induces iNOS expression in mice (Lowenstein *et al.*, 1996) (Fig. 2), and iNOS mRNA can be detected in cardiac tissue (Mikami *et al.*, 1996). iNOS activity is also increased in the spleens of infected mice. In the heart, iNOS mRNA is first detected 1 day after infection, peaks 5 days after infection, and declines to undetectable levels by 10–15 days after infection. Thus, iNOS is associated with the early immune response to CVB3 infection.

Macrophages appear to be the cells expressing iNOS in infected cardiac tissue (Lowenstein *et al.*, 1996; Mikami *et al.*, 1996). The time course of iNOS expression corresponds with the time course of macrophage infiltration, and immunohistochemical staining demonstrates iNOS within large mononuclear cells resembling macrophages. It is possible that other cells such as cardiac myocytes also express iNOS, but such expression has not been documented.

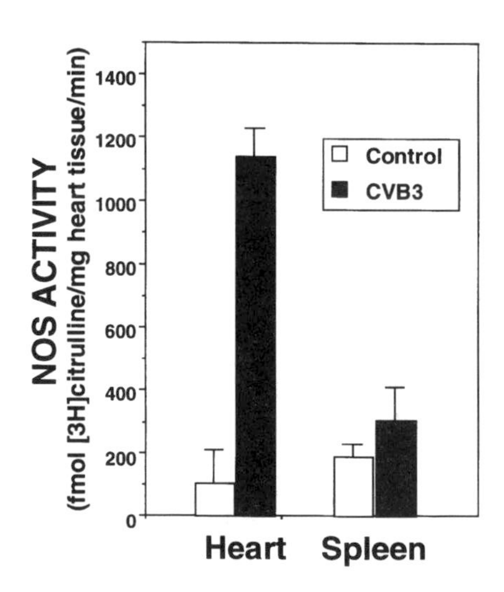

FIGURE 2. iNOS activity in infected hearts. Mice were infected with CVB3, and NOS enzyme activity was measured in the hearts and spleens harvested 7 days after infection. Reproduced from *The Journal of Clinical Investigation*, 1996, Vol. 97, pp. 1837–1843 by copyright permission of The American Society for Clinical Investigation.

7.2. NO-Dependent Inhibition of Viral Replication in Coxsackievirus Myocarditis

Initial observations suggested that NO plays a beneficial role during murine CVB3 infection. NOS inhibitors significantly increase the mortality of infected mice (Hiraoka *et al.,* 1996; Lowenstein *et al.*, 1996). However, the NOS inhibitors used in these initial experiments, L-NAME and L-NMMA, are not selective for iNOS. More definitive proof of the role of iNOS in viral infection has resulted from studies in animals that lack the iNOS gene.

The timing and extent of CVB3 replication are dramatically altered in a host lacking iNOS (Zaragoza *et al.,* 1998). Absence of iNOS permits CVB3 replication to increase, with peak cardiac CVB3 RNA levels in iNOS-deficient mice exceeding those in isogenic parental mice by at least 20-fold. Furthermore, CVB3 RNA is found earlier and persists longer in iNOS null murine hearts than in $iNOS^{+/+}$ controls. While CVB3 RNA is detected 5 days after infection and disappears 10 days after infection in normal mice, viral RNA is detected 1 day after infection and is still present 15 days after infection in iNOS null mice.

Viral titers essentially parallel RNA levels in iNOS null and control mice in the blood, heart, and all other organs studied (Zaragoza *et al.*, 1998). Depending on the initial inoculum, the quantity of CVB3 virus in the hearts of iNOS null mice is 50- to 500-fold higher than in the hearts of $iNOS^{+/+}$ control mice (Fig. 3). In noncardiac organs, CVB3 titers are elevated 10- to 100-fold in iNOS null mice compared with control mice (data not shown). Thus, the protective effect of NO during coxsackievirus infection does not seem to be confined to the heart.

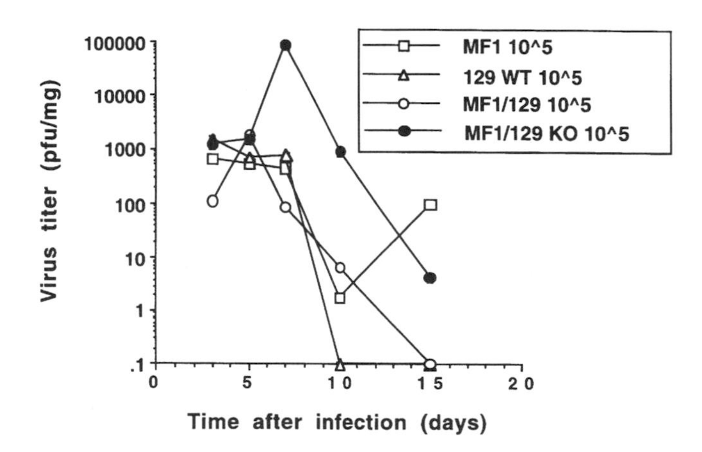

FIGURE 3. CVB3 titers in hearts of iNOS null mice. CVB3 10^5 pfu were injected into wild-type mice (MF1 or MF1/129 hybrids) and injected into iNOS null mice (MF1/129 KO). Hearts were harvested 3–15 days after infection, and viral particles were quantified by a plaque assay.

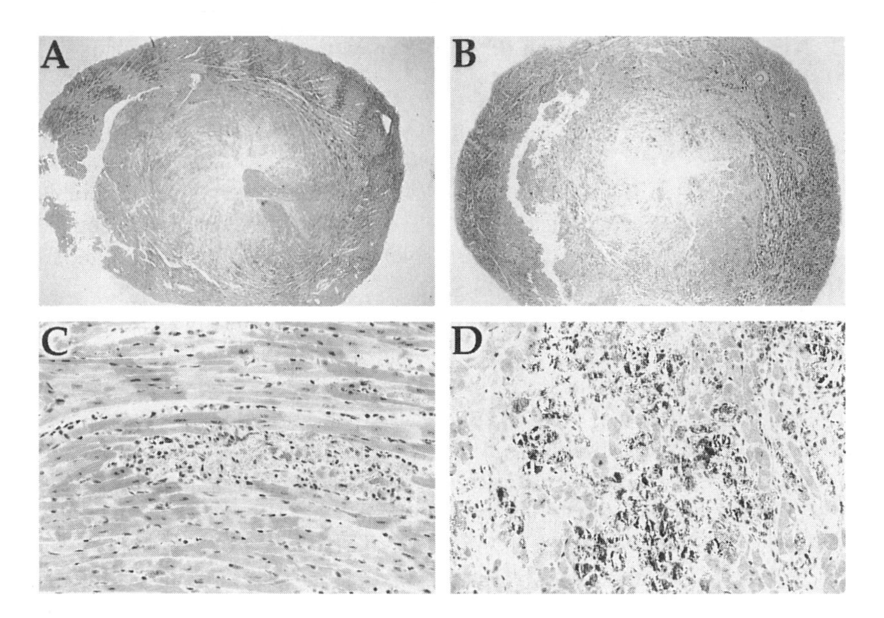

FIGURE 4. Pathology of experimental coxsackievirus myocarditis. Coxsackievirus infection causes focal inflammation in wild-type mice (A, C) but massive necrosis and calcification in the hearts of mice lacking iNOS (B, D).

7.3. NO-Dependent Reduction of Myocarditis in Coxsackievirus Infection

The result of excessive coxsackievirus growth is severe myocardial damage (Hiraoka *et al.*, 1996; Zaragoza *et al.*, 1998). The hearts of mice lacking iNOS not only have more lesions than do those of infected control mice, but the appearance of the inflammatory lesion is qualitatively different. Normal mice infected with coxsackievirus develop small foci in the heart with myocyte degeneration and mononuclear cell infiltration. Infected iNOS null mice develop larger lesions with more degeneration of myocytes and more extensive extracellular dystrophic calcification (Fig. 4), which may reflect rapid, widespread death of cardiac myocytes. This pattern of dystrophic calcification is strikingly similar to the pattern seen in coxsackievirus-infected mice lacking NK cells (Godeny and Gauntt, 1986) or $CD8^+$ lymphocytes (Huber and Lodge, 1984).

8. Mechanisms of NO-Dependent Inhibition of Coxsackievirus Replication

Although NO is capable of inhibiting the replication of a wide variety of RNA and DNA viruses, its molecular mechanism of action is unclear. NO has been reported to interfere with specific stages in the life cycle of viruses. For example, NO inhibits DNA synthesis, late protein translation, and virion assembly of vaccinia virus (Harris *et al.*, 1995; Karupiah and Harris, 1995; Melkova and Esteban, 1995). In some cases, NO may act on specific viral targets, such as the Epstein–Barr virus immediate early transactivator *Zta* (Mannick *et al.*, 1994). As NO can inhibit a variety of viruses, it is also likely that it inhibits general cellular processes necessary for viral replication, such as ribonucleotide reductase (Melkova and Esteban, 1994). The precise molecular targets of NO responsible for its antiviral properties have been incompletely defined.

Because the life cycle of coxsackievirus proceeds through well-defined stages, it may be possible to identify the specific stage(s) inhibited by NO. As discussed above, coxsackievirus attaches to and penetrates host cells, translates its RNA genome, replicates the genome, and assembles to exit the host cell. *In vitro* studies have begun to reveal the effects of NO on each stage of the coxsackievirus life cycle.

NO inhibits the replication of CVB3 in HeLa cells (Zaragoza *et al.*, 1997). The NO donor *S*-nitroso-*N*-acetylpenicillamine (SNAP) added 1 hr after infection progressively reduces the titers of CVB3 over time (Fig. 5). SNAP produces its maximal effects on viral replication 1 hr after infection, corresponding to the initiation of genome replication.

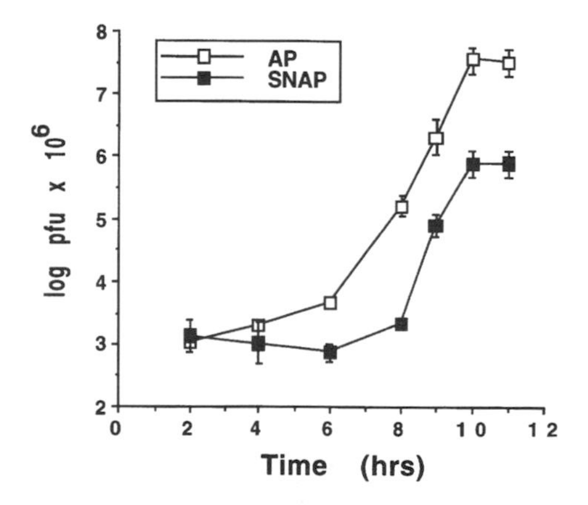

FIGURE 5. Effects of NO donor on coxsackievirus growth *in vitro*. The NO donor *S*-nitroso-*N*-acetylpenicillamine (SNAP) inhibits the replication of coxsackievirus *in vitro*, compared with the control compound acetylpenicillamine (AP). Reproduced from *The Journal of Clinical Investigation*, 1997, Vol. 100, pp. 1760–1767 by copyright permission of The American Society for Clinical Investigation.

8.1. Viral Attachment

NO does not appear to affect the first step in the coxsackievirus life cycle, viral attachment to the host cell (Zaragoza *et al.*, 1997). SNAP added to mixtures of HeLa cells and radioactively labeled virus does not affect viral binding to host cells.

8.2. Viral RNA Synthesis

However, NO does inhibit coxsackievirus RNA synthesis (Zaragoza *et al.*, 1997), required for viral replication. SNAP reduces incorporation of tritiated uridine during infection of HeLa cells (Fig. 6) and reduces viral RNA levels as measured by RT-PCR and Northern blot. Molecular targets involved in NO-mediated inhibition of RNA synthesis have not been determined, but could include the viral RNA-dependent RNA polymerase or viral protein VPg required to initiate RNA synthesis.

8.3. Viral Protein Synthesis

NO also inhibits coxsackievirus protein translation (Zaragoza *et al.*, 1997). Metabolic labeling and immunoblot techniques demonstrate an NO-dependent reduction in viral protein synthesis within 4 hr of infection. As translation of most viral proteins occurs after RNA synthesis, which is inhibitable by NO, it is perhaps not surprising that NO suppresses viral protein synthesis. However, it is possible that NO also exerts specific effects on the translation of viral proteins.

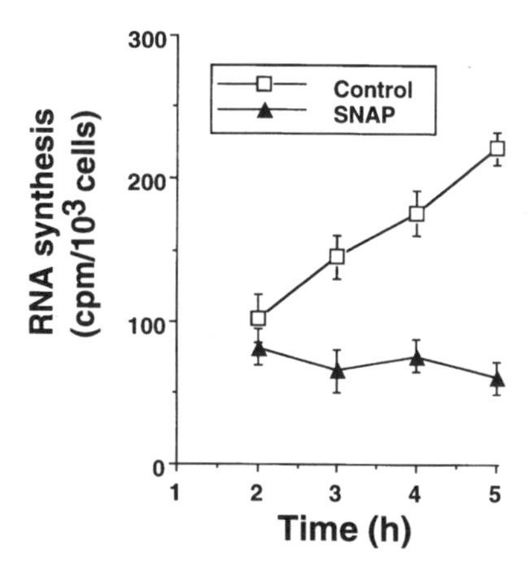

FIGURE 6. Inhibition of viral RNA synthesis by SNAP. The NO donor *S*-nitroso-*N*-acetylpenicillamine (SNAP) inhibits uridine incorporation by infected cells. Reproduced from *The Journal of Clinical Investigation*, 1997, Vol. 100, pp. 1760–1767 by copyright permission of The American Society for Clinical Investigation.

9. The Role of iNOS in Human Myocarditis

Although expression of iNOS in human viral myocarditis has not yet been demonstrated, the clinical aspects of human myocarditis suggest a possible role. One of the most intriguing aspects of human myocarditis is the transient nature of cardiac dysfunction. Some patients with fulminant myocarditis not only survive, but their left ventricular function is apparently normal within weeks of recovery. NO is a negative inotropic agent (Finkel *et al.*, 1992), and could be responsible for reversible depression of cardiac contractility in acute myocarditis. It is more difficult to conceive that myocyte destruction by viral or autoimmune mechanisms would be completely reversible. Another hallmark of coxsackievirus myocarditis is its variability. The murine model of coxsackievirus myocarditis reveals several factors correlated with clinical severity, including HLA type, gender, and age. Variability in the expression of iNOS provides another possible explanation for differences in clinical outcome. Different strains of mice differ dramatically in the amount and timing of cardiac iNOS expression during viral myocarditis, and the same may be true for humans (Zaragoza *et al.*, 1997).

Finally, a role of iNOS in antiviral host defense provides an explanation for the failure of immunosuppression to benefit myocarditis. Immunosuppressants are sometimes administered to patients with myocarditis, with the rationale that autoimmune mechanisms rather than direct viral injury are responsible for damage to cardiac myocytes. However, the Myocarditis Treatment Trial found that a prominent immune response to myocarditis (including greater numbers of macrophages and NK cells) is associated with *less* severe initial disease. Furthermore, immunosuppression did not improve left ventricular function or mortality. The immunosuppressive agents prednisone and cyclosporine have multiple anti-inflammatory effects, including suppression of iNOS expression (see Chapter 5). Immunosuppressive agents given to patients with active viral myocarditis may decrease expression of iNOS, leading to worsened infection.

Although iNOS has not yet been demonstrated in the hearts of patients with viral myocarditis, it has been detected in cardiac tissue of patients with other inflammatory disorders. For example, iNOS is found within human cardiac allografts, and the level of iNOS expression correlates with the extent of ventricular dysfunction (Lewis *et al.*, 1996). Also, iNOS activity has been found in endomyocardial biopsy specimens from patients with dilated cardiomyopathies.

10. NO Therapy for Myocarditis

Whether data from mouse studies can be extrapolated to human myocarditis remains to be seen. If NO is a major antiviral effector in humans, perhaps NO donors or even delivery of the iNOS gene can be used to treat viral myocarditis.

Inhaled NO has been successfully used to reduce pulmonary vascular resistance in neonates with primary pulmonary hypertension. However, with a short biological half-life, inhaled NO might not survive in the lungs long enough to diffuse into the blood and travel to the coronary arteries. Perhaps NO donors such as nitroglycerine or nitroprusside could be administered to patients with acute viral myocarditis, not only to reduce preload and afterload but also to inhibit viral growth. Several groups have incorporated different isoforms of NOS into viral vectors for gene therapy; however, none of these vectors is cardiotropic. Delivery of NO as pharmacological or gene therapy will probably be tested in the future.

11. Conclusions

NO is a critical component of the rapid, nonspecific response to viral infection of the heart. Coxsackievirus infection induces iNOS expression by macrophages, which in turn inhibits viral replication and prevents damage to cardiac myocytes. In the absence of iNOS and NO, coxsackievirus replicates to higher levels, grows more rapidly, and persists longer. However, even in the absence of NO, the host can eventually eliminate the virus by eliciting specific components of the immune system, B and T lymphocytes. Because this latter response is specific, it is also slower, so that infection causes extensive myocardial necrosis in a host lacking iNOS by the time viral clearance occurs.

Studies of NO in a murine myocarditis model not only point out the importance of NO in inhibiting coxsackievirus, but also suggest the general role of NO in the immune system. The host immune system includes both a rapid nonspecific component and a delayed but more specific response. NO is an important effector of the rapid nonspecific immune response (Nathan, 1995), with induction occurring within hours after viral infection. Although NO itself cannot eliminate virus, it controls viral replication until the specific immune system clears the infection. In the absence of NO, an infected host can succumb to overwhelming infection before the specific immune system can respond. Murine coxsackievirus myocarditis provides a vivid demonstration of these concepts.

References

Akaike, T., Noguchi, Y., Ijiri, S., Setoguchi, K., Suga, M., Zheng, Y. M., Dietzschold, B., and Maeda, H., 1996, Pathogenesis of influenza virus-induced pneumonia: Involvement of both nitric oxide and oxygen radicals, *Proc. Natl. Acad. Sci. USA* **93:**2448–2453.

Akarid, K., Sinet, M., Desforges, B., and Gougerot-Pocidalo, M. A., 1995, Inhibitory effect of nitric oxide on the replication of a murine retrovirus *in vitro* and *in vivo*, *J. Virol.* **69:**7001–7005.

Alvarez, F. L., Neu, N., Rose, N. R., Craig, S. W., and Beisel, K. W., 1987, Heart-specific autoantibodies induced by coxsackie virus B3: Identification of heart autoantigens, *Clin. Immunol. Immunopathol.* **43:**129–135.

Aretz, H. T., Billingham, M. E., Edwards, W. D., Factor, S. M., Fallon, J. T., Fenoglio, J. J., and Olsen, E. G. J., 1987, Myocarditis. A pathologic definition and classification, *Am. J. Cardiovasc. Pathol.* **1**:3–11.

Bendinelli, M., Matteucci, D., Toniolo, A., Patane, A. M., and Pistillo, M. P., 1982, Impairment of immunocompetent mouse spleen cell functions by infection with coxsackievirus B3, *J. Infect. Dis.* **146**:797–805.

Bi, Z., and Reiss, C. S., 1995, Inhibition of vesicular stomatitis virus infection by nitric oxide, *J. Virol.* **69**:2208–2213.

Bukrinsky, M. I., Nottet, H. S., Schmidtmayerova, H., Dubrovsky, L., Flanagan, C. R., Mullins, M. E., Lipton, S. A., and Gendelman, H. E., 1995, Regulation of nitric oxide synthase activity in human immunodeficiency virus type 1 (HIV-1)-infected monocytes: Implications for HIV-associated neurological disease, *J. Exp. Med.* **181**:735–745.

Butz, E. A., Hostager, B. S., and Southern, P. J., 1994, Macrophages in mice acutely infected with lymphocytic choriomeningitis virus are primed for nitric oxide synthesis, *Microb. Pathog.* **16**:283–295.

Campbell, A. E., Loria, R. M., Madge, G. E., and Kaplan, A. M., 1982, Dietary hepatic cholesterol elevation: Effects on coxsackievirus B infection and inflammation, *Infect. Immun.* **37**:307–317.

Chow, L. H., Beisel, K. W., and McManus, B. M., 1992, Enteroviral infection of mice with severe combined immunodeficiency. Evidence for direct viral pathogenesis of myocardial injury, *Lab. Invest.* **66**:24–31.

Cook, D. N., 1996, The role of MIP-1 alpha in inflammation and hematopoiesis, *J. Leukoc. Biol.* **59**:61–66.

Cook, D. N., Beck, M. A., Coffman, T. M., Kirby, S. L., Sheridan, J. F., Pragnell, I. B., and Smithies, O., 1995, Requirement of MIP-1 alpha for an inflammatory response to viral infection, *Science* **269**:1583–1585.

Croen, K. D., 1993, Evidence for antiviral effect of nitric oxide. Inhibition of herpes simplex virus type 1 replication, *J. Clin. Invest.* **91**:2446–2452.

Finkel, M. S., Oddis, C. V., Jacob, T. D., Watkins, S. C., Hattler, B. G., and Simmons, R. L., 1992, Negative inotropic effects of cytokines on the heart mediated by nitric oxide, *Science* **257**:387–389.

Godeny, E. K., and Gauntt, C. J., 1986, Involvement of natural killer cells in coxsackievirus B3 induced murine myocarditis, *J. Immunol.* **137**:1695–1702.

Godeny, E. K., and Gauntt, C. J., 1987a, *In situ* immune autoradiographic identification of cells in heart tissues of mice with coxsackievirus B3-induced myocarditis, *Am. J. Pathol.* **129**:267–276.

Godeny, E. K., and Gauntt, C. J., 1987b, Murine natural killer cells limit coxsackievirus B3 replication, *J. Immunol.* **139**:913–918.

Groeneveld, P. H., Colson, P., Kwappenberg, K. M., and Clement, J., 1995, Increased production of nitric oxide in patients infected with the European variant of hantavirus, *Scand. J. Infect. Dis.* **27**:453–456.

Harris, N., Buller, R. M., and Karupiah, G., 1995, Gamma interferon-induced, nitric oxide-mediated inhibition of vaccinia virus replication, *J. Virol.* **69**:910–915.

Hashimoto, I., Tatsumi, M., and Nakagawa, M., 1983, The role of T lymphocytes in the pathogenesis of coxsackie virus B3 heart disease, *Br. J. Exp. Pathol.* **64**:497–504.

Heim, A., Canu, A., Kirschner, P., Simon, T., Mall, G., Hofschneider, P. H., and Kandolf, R., 1992, Synergistic interaction of interferon-beta and interferon-gamma in coxsackievirus B3-infected carrier cultures of human myocardial fibroblasts, *J. Infect. Dis.* **166**:958–965.

Hiraoka, Y., Kishimoto, C., Takada, H., Suzaki, N., and Shiraki, K., 1995, Colony-stimulating factors and coxsackievirus B3 myocarditis in mice: Macrophage colony-stimulating factor suppresses acute myocarditis with increasing interferon-alpha, *Am. Heart J.* **130**:1259–1264.

Hiraoka, Y., Kishimoto, C., Takada, H., Nakamura, M., Kurokawa, M., Ochiai, H., and Shiraki, K., 1996, Nitric oxide and murine coxsackievirus B3 myocarditis: Aggravation of myocarditis by inhibition of nitric oxide synthase, *J. Am. Coll. Cardiol.* **28:**1610–1615.

Huber, S. A., and Cunningham, M. W., 1996, Streptococcal M protein peptide with similarity to myosin induces $CD4^+$ T cell-dependent myocarditis in MRL/++ mice and induces partial tolerance against coxsackieviral myocarditis, *J. Immunol.* **156:**3528–3534.

Huber, S. A., and Lodge, P. A., 1984, Coxsackievirus B-3 myocarditis in BALB/c mice. Evidence for autoimmunity to myocyte antigens, *Am. J. Pathol.* **116:**21–29.

Huber, S. A., and Pfaeffle, B., 1994, Differential Th1 and Th2 cell responses in male and female BALB/c mice infected with coxsackievirus group B type 3, *J. Virol.* **68:**5126–5132.

Huber, S. A., Job, L. P., and Woodruff, J. F., 1980, Lysis of infected myofibers by coxsackievirus B-3-immune T lymphocytes, *Am. J. Pathol.* **98:**681–694.

Huber, S. A., Mortensen, A., and Moulton, G., 1996, Modulation of cytokine expression by $CD4^+$ T cells during coxsackievirus B3 infections of BALB/c mice initiated by cells expressing the gamma delta+T-cell receptor, *J. Virol.* **70:**3039–3044.

Ignarro, L. J., 1990, Biosynthesis and metabolism of endothelium-derived nitric oxide, *Annu. Rev. Pharmacol. Toxicol.* **30:**535–560.

Kamijo, R., Harada, H., Matsuyama, T., Bosland, M., Gerecitano, J., Shapiro, D., Le, J., Koh, S. I., Kimura, T., and Green, S. J., 1994, Requirement for transcription factor IRF-1 in NO synthase induction in macrophages, *Science* **263:**1612–1615.

Kandolf, R., Canu, A., and Hofschneider, P. H., 1985, Coxsackie B3 virus can replicate in cultured human foetal heart cells and is inhibited by interferon, *J. Mol. Cell Cardiol.* **17:**167–181.

Kandolf, R., Kirschner, P., Ameis, D., Canu, A., and Hofschneider, P. H., 1987, *In situ* detection of enteroviral genomes in myocardial cells by nucleic acid hybridization: An approach to the diagnosis of viral heart disease, *Proc. Natl. Acad. Sci. USA* **84:**6272–6276.

Karupiah, G., and Harris, N., 1995, Inhibition of viral replication by nitric oxide and its reversal by ferrous sulfate and tricarboxylic acid cycle metabolites, *J. Exp. Med.* **181:**2171–2179.

Karupiah, G., Xie, Q. W., Buller, R. M., Nathan, C., Duarte, C., and MacMicking, J. D., 1993, Inhibition of viral replication by interferon-gamma-induced nitric oxide synthase, *Science* **261:**1445–1448.

Kreil, T. R., and Eibl, M. M., 1996, Nitric oxide and viral infection: NO antiviral activity against a flavivirus *in vitro*, and evidence for contribution to pathogenesis in experimental infection *in vivo*, *Virology* **219:**304–306.

Lewis, N. P., Tsao, P. S., Rickenbacher, P. R., Xue, C., Johns, R. A., Haywood, H., von der Leyen, G. A., Trindade, P. T., Cooke, J. P., Hunt, S. A., Billingham, M. E., Valantine, H. A., and Fowler, M. B., 1996, Induction of nitric oxide synthase in the human cardiac allograft is associated with contractile dysfunction of the left ventricle, *Circulation* **93:**720–729.

Lowenstein, C. J., and Snyder, S. H., 1992, Nitric oxide, a novel biologic messenger, *Cell* **70:**705–707.

Lowenstein, C. J., Alley, E. W., Raval, P., Snowman, A. M., Snyder, S. H., Russell, S. W., and Murphy, W. J., 1993, Macrophage nitric oxide synthase gene: Two upstream regions mediate induction by interferon gamma and lipopolysaccharide, *Proc. Natl. Acad. Sci. USA* **90:**9730–9734.

Lowenstein, C. J., Hill, S. L., Lafond-Walker, A., Wu, J., Allen, G., Landavere, M., Rose, N. R., and Herskowitz, A., 1996, Nitric oxide inhibits viral replication in murine myocarditis, *J. Clin. Invest.* **97:**1837–1843.

Mannick, J. B., Asano, K., Izumi, K., Kieff, E., and Stamler, J. S., 1994, Nitric oxide produced by human B lymphocytes inhibits apoptosis and Epstein–Barr virus reactivation, *Cell* **79:**1137–1146.

Marletta, M. A., 1989, Nitric oxide: Biosynthesis and biological significance, *Trends Biochem. Sci.* **14:**488–492.

Martin, E., Nathan, C., and Xie, Q. W., 1994, Role of interferon regulatory factor 1 in induction of nitric oxide synthase, *J. Exp. Med.* **180:**977–984.

Mason, J. W., O'Connell, J. B., Herskowitz, A., Rose, N. R., McManus, B. M., Billingham, M. E., and Moon, T. E., 1995, A clinical trial of immunosuppressive therapy for myocarditis. The Myocarditis Treatment Trial Investigators, *N. Engl. J. Med.* **333:**269–275.

Melkova, Z., and Esteban, M., 1994, Interferon-gamma severely inhibits DNA synthesis of vaccinia virus in a macrophage cell line, *Virology* **198:**731–735.

Melkova, Z., and Esteban, M., 1995, Inhibition of vaccinia virus DNA replication by inducible expression of nitric oxide synthase, *J. Immunol.* **155:**5711–5718.

Melnick, J. L., 1985, Enteroviruses: Polioviruses, coxsackieviruses, echoviruses, and newer enteroviruses, in: *Virology* (B. N. Fields, ed.), Raven Press, New York, pp. 739–794.

Mikami, S., Kawashima, S., Kanazawa, K., Hirata, K., Katayama, Y., Hotta, H., Hayashi, Y., Ito, H., and Yokoyama, M., 1996, Expression of nitric oxide synthase in a murine model of viral myocarditis induced by coxsackievirus B3, *Biochem. Biophys. Res. Commun.* **220:**983–989.

Moncada, S., Palmer, R. M., and Higgs, E. A., 1991, Nitric oxide: Physiology, pathophysiology, and pharmacology, *Pharmacol. Rev.* **43:**109–142.

Nathan, C., 1995, Natural resistance and nitric oxide, *Cell* **82:**873–876.

Nathan, C., and Xie, Q. W., 1994, Nitric oxide synthases: Roles, tolls, and controls, *Cell* **78:**915–918.

Neu, N., Craig, S. W., Rose, N. R., Alvarez, F., and Beisel, K. W., 1987, Coxsackievirus induced myocarditis in mice: Cardiac myosin autoantibodies do not cross-react with the virus, *Clin. Exp. Immunol.* **69:**566–574.

Neumann, D. A., Rose, N. R., Ansari, A. A., and Herskowitz, A., 1994, Induction of multiple heart autoantibodies in mice with coxsackievirus B3- and cardiac myosin-induced autoimmune myocarditis, *J. Immunol.* **152:**343–350.

O'Connell, J. B., and Mason, J. W., 1989, Diagnosing and treating active myocarditis, *West. J. Med.* **150:**431–437.

Quigley, P. J., Richardson, P. J., and Meaney, B. T., 1987, Long-term follow-up of acute myocarditis. Correlation of left ventricular function and outcome, *Eur. Heart J. Suppl.*: J39.

Ray, C. G., 1994, Enteroviruses and reoviruses, in: *Harrison's Principles of Internal Medicine* (K. J. Isselbacher, E. Braunwald, J. D. Wilson, J. B. Martin, A. S. Fauci, and D. L. Kasper, eds.), McGraw–Hill, New York, pp. 821–825.

Rolph, M. S., Ramshaw, I. A., Rockett, K. A., Ruby, J., and Cowden, W. B., 1996, Nitric oxide production is increased during murine vaccinia virus infection, but may not be essential for virus clearance, *Virology* **217:**470–477.

Rose, N. R., Wolfgram, L. J., Herskowitz, A., and Beisel, K. W., 1986, Postinfectious autoimmunity: Two distinct phases of coxsackievirus B3-induced myocarditis, *Ann. N.Y. Acad. Sci.* **475:**146–156.

Rose, N. R., Beisel, K. W., Herskowitz, A., Neu, N., Wolfgram, L. J., Alvarez, F. L., Traystman, M. D., and Craig, S. W., 1987, Cardiac myosin and autoimmune myocarditis, in: *Autoimmunity and Autoimmune Disease* (D. Evered and J. Whelan, eds.), Wiley, New York, pp. 3–24.

Rose, N. R., Neumann, D. A., Herskowitz, A., Traystman, M. D., and Beisel, K. W., 1988, Genetics of susceptibility to viral myocarditis in mice, *Pathol. Immunopathol. Res.* **7:**266–278.

Rueckert, R. R., 1985, Picornaviruses and their replication, in: *Virology* (B. N. Fields, ed.), Raven Press, New York, pp. 705–738.

Tucker, P. C., Griffin, D. E., Choi, S., Bui, N., and Wesselingh, S., 1996, Inhibition of nitric oxide synthesis increases mortality in Sindbis virus encephalitis, *J. Virol.* **70:**3972–3977.

Wolfgram, L. J., Beisel, K. W., and Rose, N. R., 1985, Heart-specific autoantibodies following murine coxsackievirus B3 myocarditis, *J. Exp. Med.* **161:**1112–1121.

Wolfgram, L. J., Beisel, K. W., Herskowitz, A., and Rose, N. R., 1986, Variations in the susceptibility to coxsackievirus B3-induced myocarditis among different strains of mice, *J. Immunol.* **136:**1846–1852.

Woodruff, J. F., 1979, Lack of correlation between neutralizing antibody production and suppression of coxsackievirus B3 replication in target organs: Evidence for the involvement of mononuclear inflammatory cells in defense, *J. Immunol.* **123:**31–35.

Woodruff, J. F., 1980, Viral myocarditis: A review, *Am. J. Pathol.* **101:**425–484.
Woodruff, J. F., and Woodruff, J. J., 1974, Involvement of T lymphocytes in the pathogeneisis of coxsackie virus B3 heart disease, *J. Immunol.* **113:**1726–1730.
Xie, Q. W., Whisnant, R., and Nathan, C., 1993, Promoter of the mouse gene encoding calcium-independent nitric oxide synthase confers inducibility by interferon gamma and bacterial lipopolysaccharide, *J. Exp. Med.* **177:**1779–1784.
Xie, Q. W., Kashiwabara, Y., and Nathan, C., 1994, Role of transcription factor NF-kappa B/Rel in induction of nitric oxide synthase, *J. Biol. Chem.* **269:**4705–4708.
Zaragoza, C., Ocampo, C. J., Saura, M., McMillan, A., and Lowenstein, C. J., 1997, Nitric oxide inhibition of coxsackievirus replication *in vitro*, *J. Clin. Invest.* **100:**1760–1767.
Zaragoza, C., Ocampo, C., Saura, M., Leppo, M., Wei, X. Q., Quick, R., Moncada, S., Liew, F. Y., and Lowenstein, C. J., 1998, The role of inducible nitric oxide synthase in the host response to coxsackie myocarditis, *Proc. Natl. Acad. Sci. USA* **95:**2469–2474.

CHAPTER 19

Nitric Oxide in Influenza

TAKAAKI AKAIKE and HIROSHI MAEDA

1. Introduction

Influenza is a highly contagious viral infection of the respiratory tract characterized by bronchitis, systemic illness, and sometimes, pneumonitis (Douglas, 1975; Murphy and Webster, 1990). Mice infected with a human influenza virus strain adapted to grow in the respiratory tract undergo severe and lethal tracheobronchitis and pneumonitis (Akaike *et al.*, 1989).

As in many infections, the pathogenesis of influenza is determined by a delicate balance of interactions between the host and pathogen. Free radical molecular species derived from the host have been a focus of considerable interest in recent studies of viral pathogenesis (Oda *et al.*, 1989; Akaike *et al.*, 1990, 1996, 1998; Maeda and Akaike, 1991; Hennet *et al.*, 1992; Ikeda *et al.*, 1993; Schwartz, 1993; Akaike and Maeda, 1994; Sato *et al.*, 1998). A series of studies have implicated superoxide anion radical ($O_2^{-}\cdot$) as a major pathological mediator in the experimentally induced influenza pneumonitis (Oda *et al.*, 1989; Akaike *et al.*, 1990; Maeda and Akaike, 1991; Akaike and Maeda, 1994). More recently, we have found that both nitric oxide radical ($NO^{\cdot}$) and $O_2^{-}\cdot$ are involved in the pathogenesis of influenza virus-induced pneumonitis in mice (Akaike *et al.*, 1996). In this chapter, we describe the biological relevance of overproduction of nitric oxide and superoxide in influenza pathogenesis from the perspective of the host–pathogen interaction, and discuss the implication of these observations for other viral infections.

TAKAAKI AKAIKE and HIROSHI MAEDA • Department of Microbiology, Kumamoto University School of Medicine, Kumamoto 860, Japan.

Nitric Oxide and Infection, edited by Fang. Kluwer Academic / Plenum Publishers, New York, 1999.

2. Overproduction of NO and Superoxide in Influenza Pneumonitis

When mice are infected with a lethal dose of influenza virus A (H2N2), a time-dependent induction of nitric oxide synthase (NOS) activity and inducible NOS (iNOS, NOS2) mRNA expression as assessed by RT-PCR is observed in infected lung tissue (Fig. 1A) (Akaike *et al.*, 1996). The iNOS induction becomes maximal on day 8 after infection, just before the infected animals become highly distressed and die of respiratory failure. The time course of iNOS induction in the lung parallels that of pulmonary consolidation, rather than the profile of virus replication in the lung (Fig. 1B).

To directly demonstrate $NO^\bullet$ overproduction in the mouse lung following infection with influenza virus, electron spin resonance (ESR) analysis of lung tissue has been performed using a dithiocarbamate and iron complex as a spin trap for $NO^\bullet$. $NO^\bullet$ generation is detectable through the formation of an NO–dithiocarbamate–iron adduct possessing a triplet hyperfine structure of *g* perpendicular 2.04 (Mordvintcev *et al.*, 1991; Yoshimura *et al.*, 1996), and the time course of nitric oxide production parallels that of iNOS induction (Fig. 2). These ESR signals are completely nullified by treatment with the NOS inhibitor N^G-monomethyl-L-arginine (L-NMMA), indicating that nitric oxide production in the virus-infected lung results from iNOS induction. Immunohistochemical studies using a specific anti-iNOS antibody reveal that iNOS is expressed in bronchial epithelial cells as well as in monocytes/macrophages infiltrating the interstitial tissue and alveolar spaces of virus-infected lung (Akaike *et al.*, unpublished observation).

Two major sources of $O_2^-\cdot$ generation are also markedly elevated in the influenza virus-infected lung (Oda *et al.*, 1989; Akaike *et al.*, 1990). First, the $O_2^-\cdot$-generating capacity of polymorphonuclear and mononuclear phagocytes recovered in bronchoalveolar lavage fluid (BALF) increases significantly after influenza virus infection. Second, the level of xanthine oxidase (XO) in BALF of virus-infected lung is elevated markedly compared with levels in BALF from noninfected mice. The conversion from xanthine dehydrogenase (XD) to XO is required for the efficient production of reactive oxygen from xanthine oxidoreductase (Amaya *et al.*, 1990). Therefore, it is of interest to note that XD-to-XO conversion was observed in the respiratory tract of virus-infected animals, while substrate (hypoxanthine and xanthine) availability was facilitated (Akaike *et al.*, 1990). The upregulation of XD (XO) during murine influenza virus infection has been further substantiated by Northern blotting for XD mRNA expression, as well as by Western blotting using a specific anti-XO antiserum (Akaike *et al.*, unpublished observation). $O_2^-\cdot$ generation by XO can be demonstrated by analysis of BALF from influenza virus-infected mice, and the time course parallels that of iNOS induction and NO production.

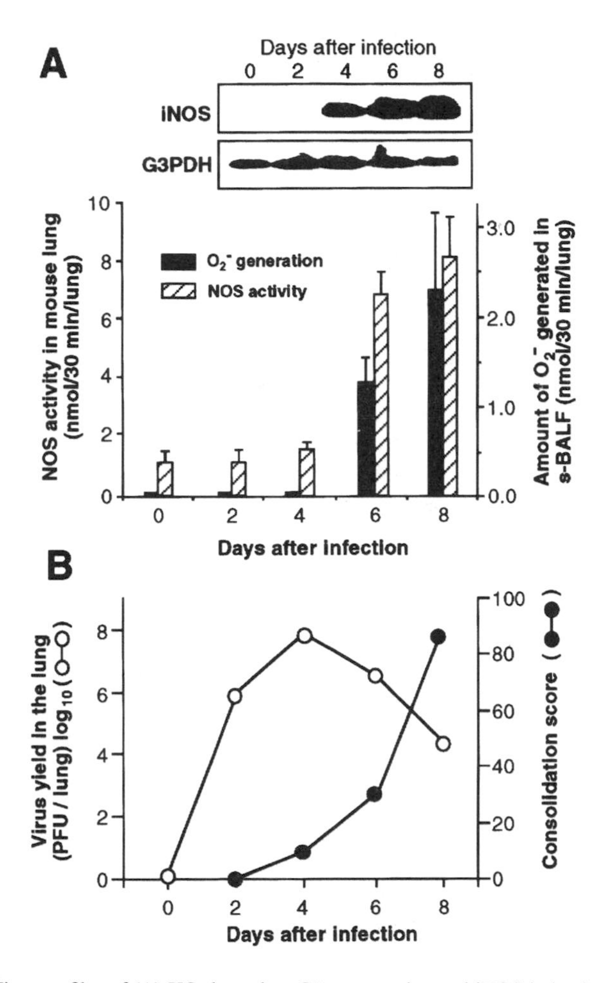

FIGURE 1. Time profiles of (A) XO-dependent $O_2^-\cdot$ generation and iNOS induction, and (B) virus yield and consolidation score, in mouse lung after influenza virus infection. Mice were infected with 2.0 LD_{50} of influenza virus [A/Kumamoto/Y5/67(H2N2)]. $O_2^-\cdot$ generation in the lung was assessed by measuring the amount of $O_2^-\cdot$ produced in brochoalveolar lavage fluid supernatant (s-BALF) obtained from infected animals. NOS activity and iNOS mRNA (upper panel in A) were determined radiochemically by using [^{14}C]L-arginine and RT-PCR/Southern blotting, respectively. Virus yield in the lung was quantified by the plaque-forming assay and was expressed as plaque-forming units (PFU). The consolidation score was measured by macroscopic observation of the pathological changes of the lung caused by the virus-induced pneumonia. Data in A are shown as means ± S.E.M. ($n = 4$), and those in B are mean values of three different experiments. G3PDH, glyceraldehyde-3-phosphate dehydrogenase. (A) Reproduced from Akaike *et al.* 1996, Pathogenesis of influenza virus-induced pneumonia: Involvement of both nitric oxide and oxygen radicals, *Proc. Natl. Acad. Sci. USA* **93:**2448–2453. Copyright 1996, National Academy of Sciences, U.S.A. (B) is from Akaike *et al.*, 1990, *The Journal of Clinical Investigation,* **85:**739–745, by Copyright permission of The American Society for Clinical Investigation.

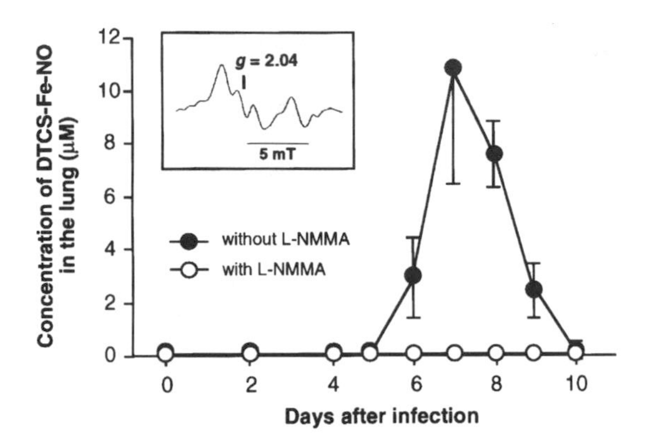

FIGURE 2. Time profile of NO production in the lung after influenza virus infection. Murine influenza infection was produced in the same manner as in Fig. 1. The amount of NO generated in the lung with or without L-NMMA treatment was quantified by ESR spectroscopy (110 K) using (*N*-dithiocarboxy)sarcosine $(DTCS)_2$–Fe^{2+} complex as a spin trap (Akaike *et al.*, 1996). A typical ESR spectrum of the NO–$(DTCS)_2$–Fe^{2+} adduct obtained with the virus-infected lung is shown in the inset. L-NMMA (2 mg/mouse) was given intraperitoneally to mice 2 hr before ESR measurements. Data are means ± S.E.M. ($n = 4$). Reproduced from Akaike *et al.* (1998) by copyright permission of Blackwell Science.

It is noteworthy that $O_2^-\cdot$ or $NO^\cdot$ per se are not particularly toxic for mammalian cells and many microbes. Earlier work suggested that $O_2^-\cdot$ might function as a reducing agent for ferric iron, forming ferrous iron to act as a catalyst for the formation of toxic hydroxyl radical ($HO^\cdot$) from hydrogen peroxides (Halliwell and Gutteridge, 1984). Because $HO^\cdot$ is a highly potent oxidizing radical species capable of mediating cell and tissue damage (Halliwell and Gutteridge, 1984; Sato *et al.*, 1992), we initially sought to identify $HO^\cdot$ generation in influenza virus-infected mouse lung by the ESR technique. However, evidence of $HO^\cdot$ generation could not be obtained from BALF of virus-infected animals.

Alternatively, the toxic effect of $O_2^-\cdot$ in combination with $NO^\cdot$ might be accounted for by the formation of peroxynitrite ($ONOO^-$), a reactive molecular species formed by rapid reaction of $O_2^-\cdot$ and $NO^\cdot$ (Beckman *et al.*, 1990; Huie and Padmaja, 1993; Pryor and Squadrito, 1995; Beckman and Koppenol, 1996; Rubbo *et al.*, 1996) that may contribute to diverse pathophysiological phenomena caused by simultaneous overproduction of $O_2^-\cdot$ and $NO^\cdot$.

3. Formation of Peroxynitrite in Influenza Pneumonitis

NO appears to have diverse molecular targets in biological systems (Moncada and Higgs, 1993; Rubbo *et al.*, 1996), including iron complex- or heme-containing proteins (Kosaka *et al.*, 1994; Henry *et al.*, 1997). Relatively stable NO–iron adducts can be formed *in vivo* when excess NO is produced (Doi *et al.*, 1996; Setoguchi *et al.*, 1996; Yoshimura *et al.*, 1996). The typical NO–hemoglobin signal is readily detectable and quantified by ESR spectroscopy in various tissues and blood.

The reaction of $O_2^{-}\cdot$ and $NO^{\cdot}$ is very rapid and diffusion-limited (rate constant $6.7 \times 10^9\ M^{-1}\ sec^{-1}$) resulting in the formation of $ONOO^-$ (Beckman *et al.*, 1990; Huie and Padmaja, 1993). Although the rate constant for the reaction of $O_2^{-}\cdot$ with superoxide dismutase (SOD) is slower ($1.9 \times 10^9\ M^{-1}\ sec^{-1}$) than that for the reaction with $NO^{\cdot}$, an excess of SOD might nevertheless limit the reaction of $O_2^{-}\cdot$ and $NO^{\cdot}$ by scavenging $O_2^{-}\cdot$.

To examine whether the reaction of $O_2^{-}\cdot$ and $NO^{\cdot}$ occurs in mouse lung during experimental influenza infection, we analyzed the formation of NO–hemoglobin in the virus-infected lung with or without SOD treatment (Akaike *et al.*, 1996). In this experiment, poly(vinylalcohol) (PVA)-conjugated Cu,Zn-SOD was used for more stable and effective drug delivery to the inflammatory site; the PVA-conjugated Cu,Zn-SOD has a prolonged plasma half-life and improved biocompatibility compared with native Cu,Zn-SOD (Kojima *et al.*, 1996). Removal of $O_2^{-}\cdot$ by SOD was predicted to yield a higher level of NO production.

In fact, the amount of NO–hemoglobin formed in mouse lung during influenza virus infection does increase significantly following treatment with polymer-conjugated SOD (Fig. 3). As expected, L-NMMA administration to virus-infected mice strongly suppresses NO–hemoglobin formation. The increase in NO–hemoglobin generation by the administration of SOD supports the notion that the reaction of $O_2^{-}\cdot$ with $NO^{\cdot}$ (and inferentially, the formation of $ONOO^-$) takes place during murine influenza pneumonitis.

A constant flux of $ONOO^-$ is very likely to cause pathophysiologically relevant effects on local tissues. It has been reported that tyrosine nitration mediated by $ONOO^-$ can be demonstrated using a specific antinitrotyrosine antibody (Beckman *et al.*, 1994). Accordingly, we performed immunohistochemical analysis of influenza virus-infected lung. Strong immunostaining for nitrotyrosine was most evident in macrophages and neutrophils infiltrating alveoli and interstitial spaces, as well as within inflammatory intraalveolar exudate (Akaike *et al.*, 1996). These observations provide strong support that $ONOO^-$ is produced and participates in biologically relevant reactions during experimental influenza pneumonitis.

4. Regulation of iNOS Expression in Viral Infections

Induction of iNOS has now been demonstrated during infection with a wide range of viruses with different tissue tropisms, including neuro-, pneumo-, and cardiotropic viruses such as Borna disease virus, herpes simplex virus type 1

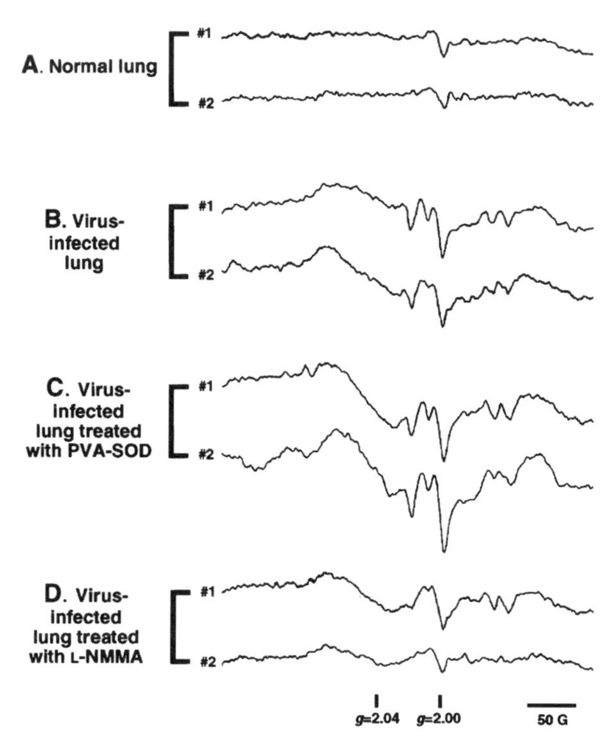

FIGURE 3. ESR spectra of NO–hemoglobin generated in the virus-infected lung. ESR study was performed with mouse lung obtained 7 days after influenza virus infection in the a same manner as in Fig. 2, without the use of a spin trapping agent for NO (Akaike *et al.*, 1996). PVA-SOD (3 mg, i.v.) and L-NMMA (2 mg, i.p.) were administered to mice 3 and 2 hr before ESR measurements, respectively. Two spectra observed with two different animals are shown for each experimental protocol.

(HSV-1), rabies virus, influenza virus, Sendai virus and coxsackievirus (Koprowski *et al.*, 1993; Zheng *et al.*, 1993; Campbell *et al.*, 1994; Akaike *et al.*, 1995, 1996; Bi *et al.*, 1995; Kreil and Eibl, 1996; Mikami *et al.*, 1996; Adler *et al.*, 1997; Akaike *et al.*, unpublished observation). iNOS expression has also been demonstrated within brain tissue of patients with HIV-1 encephalitis (Bukrinsky *et al.*, 1995) (see also Chapter 21). In experimental viral infections, iNOS expression seems to be related to the induction of proinflammatory cytokines, particularly IFNγ (see also Chapters 5 and 6).

We therefore examined the induction of IFNγ in the mouse lung during influenza virus infection using an enzyme immunoassay of BALF supernatant (Akaike *et al.*, 1996, 1998). The time courses of IFNγ and TNFα induction in the lung precede those of iNOS induction and NO overproduction (Fig. 4A,B), consistent with a causal relationship. Furthermore, the addition of BALF from

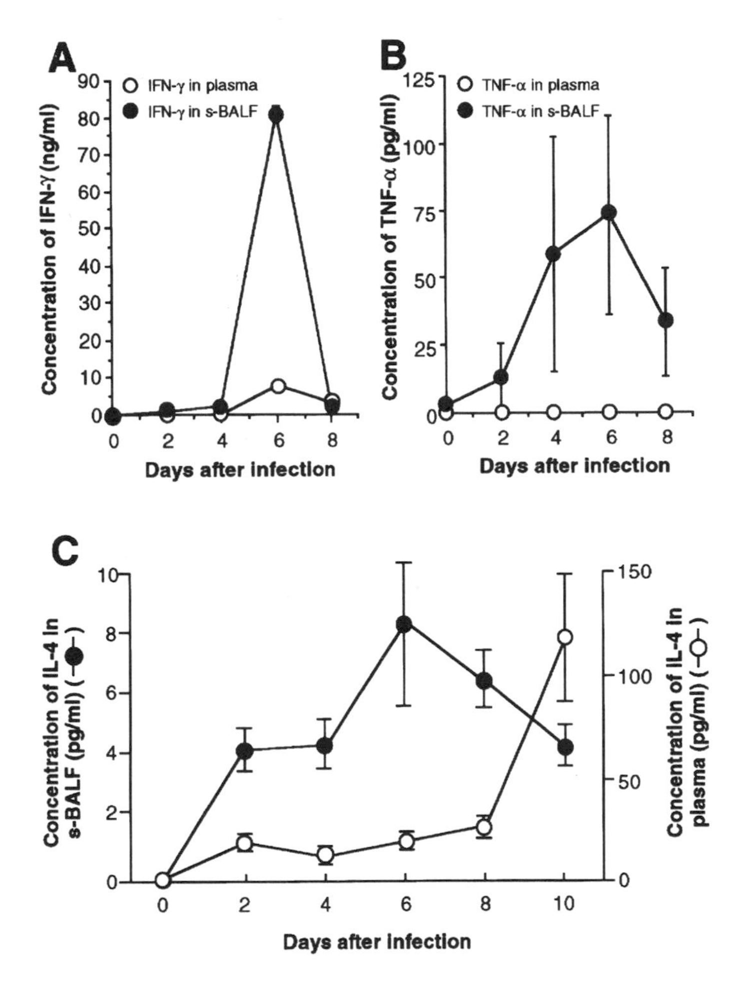

FIGURE 4. Induction of various cytokines during influenza virus infection in mice. (A, B) Time profiles of IFNγ and TNFα induction in bronchoalveolar lavage fluid supernatant (s-BALF) and plasma after influenza virus infection. (C) Induction of IL-4 in s-BALF and plasma after viral infection. Influenza infection was produced in the same manner as in Fig. 1. Each cytokine was measured using enzyme immunoassay kits (Endogen). Some of the data are from Akaike *et al.* (1996, 1998).

influenza virus-infected mice induces iNOS in a murine macrophage RAW 264 cell line. The iNOS-inducing activity of BALF can be almost completely nullified by treatment of the BALF with anti-murine IFNγ antibody (Fig. 5). From these results, IFNγ appears to be a major cytokine responsible for triggering iNOS expression in the influenza virus-infected murine lung.

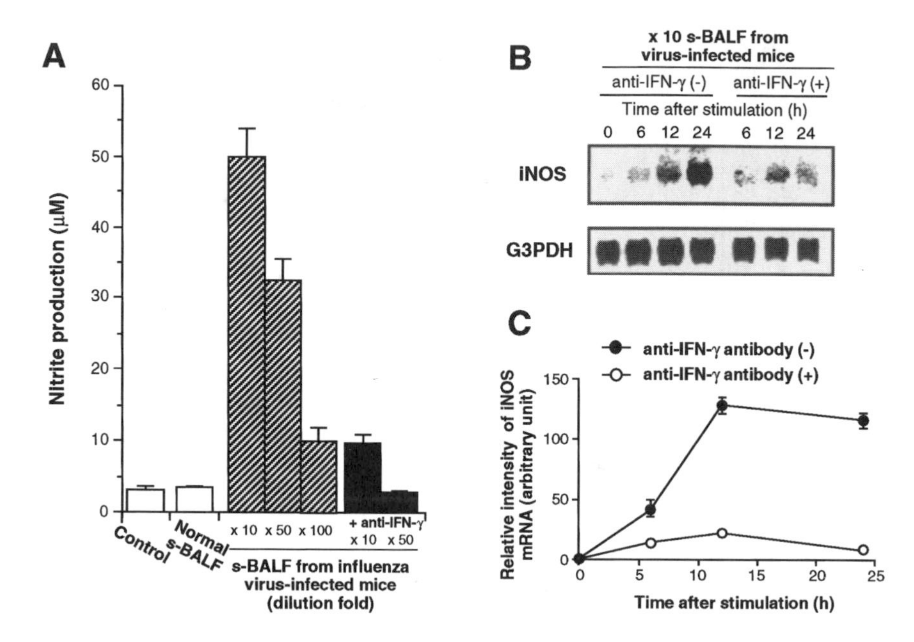

FIGURE 5. NOS induction by s-BALF in cultured RAW 264 cells. (A) NOS induction was assessed by measuring nitrite released in the culture during a 48-hr incubation period following stimulation with serially diluted bronchoalveolar lavage fluid supernatant (s-BALF) (Akaike *et al.*, 1996; Sato *et al.*, 1998). (B, C) iNOS mRNA expression was examined by Northern blotting (Setoguchi *et al.*, 1996; *Sato et al.*, 1998); the relative signal density of iNOS mRNA in B was quantified by comparison with G3PDH mRNA, and is shown in C. s-BALF treated with anti-murine IFNγ antibody was used in some assays as indicated. Data are means ± S.E.M. ($n = 4$). The data in A is from Akaike *et al.* (1996).

An interesting report by Kreil and Eibl (1995) observed that IFNα/β downregulates NO production in virus-infected murine macrophages in culture. Specifically, IFNα/β inhibited NO production by macrophages infected with tick-borne encephalitis (TBE) virus, in which iNOS expression was induced by IFNγ and TNFα. The downregulation of iNOS expression was most clearly observed in TBE virus-infected cells. This would suggest that NO production by virus-infected macrophages is antagonized by IFNα/β, an important effector molecule in the initial host response to viruses (Wright, 1997). However, divergent observations in other experimental systems (Zhang *et al.*, 1994; Zhou *et al.*, 1995; Sharara *et al.*, 1997) (see Chapter 6) suggest that these regulatory effects might be context-specific.

Other cytokines have also been associated with downregulation of iNOS expression, e.g., IL-4, IL-10, and TGFβ (Cunha *et al.*, 1992; Vodovotz *et al.*, 1993; Bogdan *et al.*, 1994); a suppressive effect of IL-4 and IL-10 on iNOS mRNA induction has been shown in murine macrophages. Furthermore, suppressive cytokines can reduce NO production indirectly via induction of arginase (Corraliza *et al.*, 1995; Gotoh *et al.*, 1996; Sonoki *et al.*, 1997), which diminishes the supply of substrate (L-arginine) for iNOS. In this regard, Xia and Zweier have reported the intriguing finding that effective $ONOO^-$ production is observed in L-arginine-depleted iNOS-expressing murine macrophages. However, appreciable $ONOO^-$ formation was not observed in L-arginine-supplemented cultures (Xia and Zweier, 1997). This suggests that an imbalance of various cytokines leading to insufficient L-arginine availability could result in preferential production of $ONOO^-$ rather than other NO congeners.

We have examined the time course of IL-4 and IFNγ production during influenza virus pneumonitis in mice (Fig. 4C) (Akaike *et al.*, 1998), and compared these data with the production of NO detected by ESR spectroscopy (Fig. 2). The induction of IL-4 becomes detectable in BALF as early as 2 days after viral infection, and increases steadily, attaining a maximum value 6 days after infection. In contrast, the level of IL-4 in plasma increases rapidly more than 8 days after infection. NO production in the lung is seen only 6 to 9 days after infection, corresponding with the appearance of pathological changes. Specifically, pulmonary consolidation appears after day 4 and persists up to 10 days after infection, when the animal becomes moribund. It is also important to note that induction of arginase ImRNA has been identified in virus-infected lung, paralleling IL-4 induction in the plasma (S. Fujii *et al.*, unpublished observation). This may indicate that IL-4 counteracts IFNγ actions on iNOS expression, attenuating the supply of L-arginine and limiting NO production. IL-4 and IL-10 are involved in the stimulation and differentiation of B cells as part of a Th2 response driven by the helper T-cell population (Wright, 1997). Therefore, suppressor cytokines down-regulating iNOS may shift host defense from an NO-dependent response to a humoral immune response directed against the intruding virus.

5. Pathophysiology of NO in Influenza Pneumonitis

NO has antimicrobial activity against bacteria, parasites and fungi (Granger *et al.*, 1988; Nathan and Hibbs, 1991; Doi *et al.*, 1993; James, 1995; Umezawa *et al.*, 1997) (see also Chapter 12). The antiviral action of NO is also known for some types of virus, typically DNA viruses such as a murine pox virus (ectromelia) and HSV-1 (Croen, 1993; Karupiah *et al.*, 1993). The antiviral effect, however, has not been observed with some RNA viruses (e.g., influenza virus, Sendai virus) that we have examined. In addition, a recent report shows a discrepancy between *in vitro* and *in vivo* effects of NO on a coronavirus (mouse hepatitis virus) (Lane *et al.*, 1997).

The antiviral activity of NO may be explained by the ability of NO to block DNA synthesis via inhibition of ribonucleotide reductase (Lepoivre *et al.*, 1991), and by effects on cellular energy metabolism by suppression of heme-containing mitochondrial electron transfer components (Cleeter *et al.*, 1994). Another interesting mechanism for NO-dependent antiviral action has been proposed from observations of Epstein–Barr virus (EBV) infection in cultured human B lymphocytes (Mannick *et al.*, 1994). A low level of NO production in EBV-transformed B lymphocytes results in inhibition of expression of an immediate-early EBV transactivator gene, possibly through regulation of the intracellular redox status.

In fact, inhibition of NO biosynthesis does not affect the titer of influenza virus in the lung during murine pneumonitis (Akaike *et al.*, 1996). The NOS inhibitor L-NMMA was administered daily to animals infected with influenza virus at lethal or sublethal doses. ESR analysis of virus-infected lung tissue with or without L-NMMA administration showed that NO production in the lung was strongly inhibited by the L-NMMA treatment protocol. However, the virus titers on days 4, 7, and 10 were not changed by L-NMMA treatment in either lethal or sublethal infections.

It is noteworthy that a significant improvement in survival rate was obtained with L-NMMA treatment of the influenza-virus infected animals (Akaike *et al.*, 1996). Similar results were obtained by Kreil and Eibl regarding the effect of NOS inhibition on TBE virus infection in mice (Kreil and Eibl, 1996). In their report, excessive NO generation in murine macrophages did not result in inhibition of TBE virus replication *in vitro*. Also, treatment of the TBE virus-infected mice with the NOS inhibitor aminoguanidine significantly prolonged survival.

We recently examined the effect of NOS inhibition with L-NMMA on HSV-1-induced encephalitis in rats. Although an antiproliferative action of NO against HSV was described for cells in culture (Croen, 1993; Karupiah *et al.*, 1993), our results *in vivo* indicate that L-NMMA suppression of excessive production of NO in the central nervous system (CNS) of HSV-1-infected animals led to improvement in neuronal damage, but suppression of NO generation did not affect viral replication in the CNS (Fujii *et al.*, 1999).

An important report by Adler *et al.* (1997) describes the effect of NOS inhibition during HSV-1-induced pneumonitis. L-NMMA treatment led to a significant improvement in histopathological changes in the lung, pulmonary compliance, and mortality despite increased viral proliferation. It is thus concluded that the tissue damage associated with HSV-1-induced pneumonia is more closely related to the NO-mediated inflammatory response of the host than to the direct effects of viral replication. This notion is also consistent with the role of NO in the pathogenesis of murine influenza pneumonitis.

6. Biological Effects of Peroxynitrite in Microbial Pathogenesis

6.1. Peroxynitrite as an Effector Molecule in Viral Pathogenesis

Based on the results described in this chapter, it is suggested that pathological effects resulting from overproduction of NO during viral infections, especially when accompanied by the production of $O_2^-\cdot$, may be more significant than the function of NO as a specific antiviral mediator, at least for some viral infections. This is supported by the known unique biochemical and biological properties of $ONOO^-$. $ONOO^-$ is much more reactive than either $NO^\cdot$ or $O_2^-\cdot$ (Beckman *et al.*, 1990; Pryor and Squadrito, 1995; Beckman and Koppenol, 1996; Rubbo *et al.*, 1996). $ONOO^-$ can have diverse actions in biological systems including nitration of protein tyrosine residues (Beckman *et al.*, 1994; Haddad *et al.*, 1994), lipid peroxidation (Radi *et al.*, 1991b; Haddad *et al.*, 1993), inactivation of aconitases (Castro *et al.*, 1994; Hausladen and Fridovich, 1994), inhibition of mitochondrial electron transport (Radi *et al.*, 1994), and oxidation of thiols (Radi *et al.*, 1991a). These reactions of $ONOO^-$ can have profound biological consequences including apoptotic and cytotoxic effects on various cells (Zhu *et al.*, 1992; Dawson *et al.*, 1993; Bonfoco *et al.*, 1995; Estevez *et al.*, 1995; Ischiropoulos *et al.*, 1995; Rubbo *et al.*, 1996; Troy *et al.*, 1996) (see also Chapter 8). The nitration of tyrosine residues in cells may compromise phosphorylation or adenylation modification of proteins, impairing intracellular signal transduction (Berlett *et al.*, 1996; Kong *et al.*, 1996). The biological relevance of $ONOO^-$ is further emphasized by the recent finding that $ONOO^-$ reactivity is modulated or potentiated by carbon dioxide or carbonate ion (Uppu *et al.*, 1996), which exists in physiological fluids at concentrations approximating 1.2 mM (Garrett and Grisham, 1995).

We have recently found that $ONOO^-$ activates human neutrophil procollagenase [matrix metalloproteinase 8 (MMP-8)], which has a critical role in tissue disintegration and remodeling under physiological as well as pathological conditions such as inflammation and infection (Okamoto *et al.*, 1997a,b). In addition to activation of MMP-8, $ONOO^-$ readily inactivates both tissue inhibitor for MMP (TIMP) and α_1-proteinase inhibitor, a major proteinase inhibitor in human plasma (Moreno and Pryor, 1992; Frears *et al.*, 1996; Whiteman *et al.*, 1996). This provides an additional mechanism by which $ONOO^-$ might accelerate tissue degradation and contribute to the pathogenesis of various inflammatory diseases. It is also reported that $ONOO^-$ activates cyclooxygenase, a key enzyme in the production of potent inflammatory prostaglandins (Landino *et al.*, 1996). Thus, $ONOO^-$ produced during virus-induced inflammation may promote tissue injury in numerous ways.

The involvement of $ONOO^-$ in influenza pathogenesis was indirectly shown by our earlier observations demonstrating improvement in the survival rate of the infected mice following injection of the pyran copolymer-conjugated SOD (Oda *et al.*, 1989; Akaike *et al.*, 1990), in which removal of $O_2^-\cdot$ would be predicted to suppress $ONOO^-$ production. More recently, the effect of recombinant human Mn-SOD was examined in mice infected with influenza virus (A or B) by Sidwell *et al.* (1996), who found a beneficial effect of SOD on both pulmonary function and mortality.

A protective effect of allopurinol, a potent inhibitor of XO, has similarly been observed in mice with influenza pneumonitis (Akaike *et al.*, 1990). In these studies, it is most likely that death of the infected animals resulted from elevated levels of $O_2^-\cdot$ produced by XO. In addition to the protective effect of either NO or $O_2^-\cdot$ inhibitors, we recently verified the therapeutic benefit of ebselen, a potent $ONOO^-$ scavenger (Matsumoto and Sies, 1996), during murine influenza pneumonitis (Akaike *et al.*, unpublished observation). $O_2^-\cdot$ generation by XO is also implicated in the pathogenesis of cytomegalovirus (CMV) infection in mice. Ikeda *et al.* (1993) have demonstrated elevated XO activity in the lung during CMV infection, and the number of pulmonary lesions was significantly reduced after treatment with either allopurinol or SOD.

6.2. Comparison of Toxic and Beneficial Effects of Peroxynitrite in Microbial Infections

The pathogenic action of nitric oxide and superoxide during the viral infections described in this chapter appears to be in contrast to the antimicrobial actions of reactive nitrogen and oxygen species observed during many bacterial, fungal, and parasitic infections (Chapter 12), although overproduction of NO has been implicated in pathogenesis of septic shock (Moncada and Higgs, 1993; Yoshida *et al.*, 1994) and neurological damage associated with bacterial meningitis (Kornellisse *et al.*, 1996) (see Chapter 20).

We recently examined the *in vivo* antimicrobial effects of $NO^\cdot$ and $O_2^-\cdot$ during *Salmonella typhimurium* infection in mice, during which XO and iNOS are strongly upregulated as in viral infections (Umezawa *et al.*, 1997). However, both mortality and bacterial burden were aggravated by treatment of infected animals with L-NMMA, allopurinol, or SOD (Umezawa *et al.*, 1995, 1997).

As depicted in Fig. 6, the different effects of $NO^\cdot$ and $O_2^-\cdot$ production in these bacterial and viral infections may relate to the contrasting nature of the host response to these pathogens. The host response to *S. typhimurium* results in physical containment of the pathogenic bacteria within a confined area, the abscesses or granulomata found in *Salmonella*-infected mice (Umezawa *et al.*, 1995, 1997). iNOS expression in the *Salmonella*-infected liver localizes mostly in microabscesses. As a result, reactive molecular species, such as $NO^\cdot$, $O_2^-\cdot$, and $ONOO^-$, directly affect invading pathogens in a limited area and primarily with intracellular compartments, minimizing tissue injury in the surrounding area. In contrast, viruses tend to involve tissues diffusely, although specific viruses may

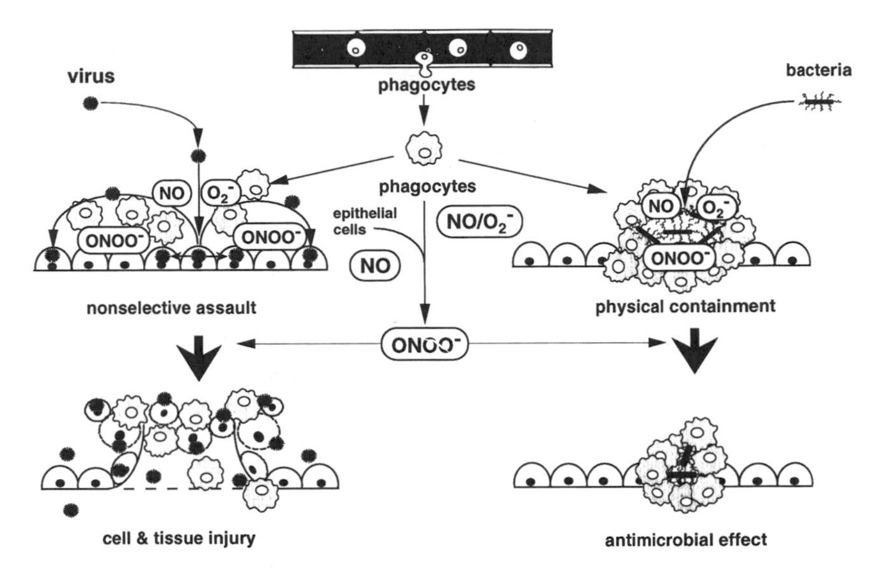

FIGURE 6. Biological effects of free radicals such as $O_2^-\cdot$ and NO$^\bullet$ and their product $ONOO^-$ in certain viral and bacterial infections. Modified from Akaike *et al.* (1998) by copyright permission of Blackwell Science.

exhibit selective tissue tropism (Akaike *et al.*, 1989; Wright, 1997). The ability of viruses to propagate and spread from cell to cell or through extracellular spaces prevents physical containment by host defenses, and allows free radical effector molecules such as NO$^\bullet$ and $O_2^-\cdot$ to exert cytotoxic effects on both normal and virus-infected tissues. This may help to account for the vastly different roles of free radical production in *Salmonella* and influenza virus infections.

7. Concluding Remarks

The free radicals $O_2^-\cdot$ and NO$^\bullet$ produced as effector molecules of host defense are not necessarily beneficial to the virus-infected host. The pathological consequence of free radical generation is determined by the intricate balance between the host and the microbial pathogen. In the case of influenza and certain other viruses, the detrimental effects of NO production and $ONOO^-$ formation appear to outweigh any benefits to the host.

Although this chapter did not discuss another biological aspect of $ONOO^-$, its mutagenetic potential (Ohshima and Bartsch, 1994; Liu and Hotchkiss, 1995; Yermilov *et al.*, 1996), it may be of future interest to explore a potential role of $ONOO^-$ and other nitrogen oxides as a missing link between viral infection and carcinogenesis, in view of the sustained and excessive generation of NO$^\bullet$ and $O_2^-\cdot$

during virus-induced inflammatory responses. An improved understanding of the pathophysiological function of NO and oxygen radicals during viral infection will provide profound insights into molecular mechanisms of viral pathogenesis, and help to identify novel therapeutic strategies.

ACKNOWLEDGMENTS. We thank Ms. Rie Yoshimoto for preparing the manuscript. Thanks are also due Drs. Moritaka Suga and Masayuki Ando for stimulating and critical discussion. This work is supported by a Grant-in-Aid from the Ministry of Education, Science, Sports and Culture of Japan (T.A. and H.M.), and a grant from the Ministry of Health and Welfare of Japan for surveys and research on specific diseases (T.A.).

References

Adler, H., Beland, J. L., Del-Pan, N. C., Kobzik, L., Brewer, J. P., Martin, T. R., and Rimm, I. J., 1997, Suppression of herpes simplex virus type 1 (HSV-1)-induced pneumonia in mice by inhibition of inducible nitric oxide synthase (iNOS, NOS2), *J. Exp. Med.* **185:**1533–1540.

Akaike, T., and Maeda, H., 1994, Molecular pathogenesis of influenza virus pneumonia: Impacts of proteases, kinins, and oxygen radicals derived from hosts, in: *Basic and Clinical Aspects of Pulmonary Fibrosis* (T. Takishima, ed.), CRC Press, Boca Raton, pp. 213–227.

Akaike, T., Molla, A., Ando, M., Araki, S., and Maeda, H., 1989, Molecular mechanism of complex infection by bacteria and virus analyzed by a model using serratial protease and influenza virus in mice, *J. Virol.* **65:**2252–2259.

Akaike, T., Ando, M., Oda, T., Doi, T., Ijiri, S., Araki, S., and Maeda, H., 1990, Dependence on O_2^- generation by xanthine oxidase of pathogenesis of influenza virus infection in mice, *J. Clin. Invest.* **85:**739–745.

Akaike, T., Weihe, E., Schaefer, M., Fu, Z. F., Zheng, Y. M., Vogel, W., Schmidt, H., Koprowski, H., and Dietzschold, B., 1995, Effect of neurotropic virus infection on neuronal and inducible nitric oxide synthase activity in rat brain, *J. Neurovirol.* **1:**118–125.

Akaike, T., Noguchi, Y., Ijiri, S., Setoguchi, K., Suga, M., Zheng, Y. M., Dietzschold, B., and Maeda, H., 1996, Pathogenesis of influenza virus-induced pneumonia: Involvement of both nitric oxide and oxygen radicals, *Proc. Natl. Acad. Sci. USA* **93:**2448–2453.

Akaike, T., Suga, M., and Maeda, H., 1998, Free radicals in viral pathogenesis: Molecular mechanisms involving superoxide and NO, *Proc. Soc. Exp. Biol. Med.* **217:**64–73.

Amaya, Y., Yamazaki, K., Sato, M., Noda, K., Nishino, T., and Nishino, T., 1990, Proteolytic conversion of xanthine dehydrogenase from the NAD-dependent type to the O_2-dependent type, *J. Biol. Chem.* **265:**14170–14175.

Beckman, J. S., and Koppenol, W. H., 1996, Nitric oxide, superoxide and peroxynitrite: The good, the bad, and the ugly, *Am. J. Physiol.* **271:**C1424–C1437.

Beckman, J. S., Beckman, T. W., Chen, J., Marshall, P. A., and Freeman, B. A., 1990, Apparent hydroxyl radical production by peroxynitrite: Implications for endothelial injury from nitric oxide and superoxide, *Proc. Natl. Acad. Sci. USA* **87:**1620–1624.

Beckman, J. S., Ye, Y. Z., Anderson, P. G., Chen, J., Accavitti, M. A., Tarpey, M. M., and White, C. R., 1994, Extensive nitration of protein tyrosines in human atherosclerosis detected by immunohistochemistry, *Biol. Chem. Hoppe Seyler* **375:**81–88.

Berlett, B. S., Friguet, B., Yim, M. B., Chock, P. B., and Stadtman, E. R., 1996, Peroxynitrite-mediated nitration of tyrosine residues in *Escherichia coli* glutamine synthetase mimics adenylylation: Relevance to signal transduction, *Proc. Natl. Acad. Sci. USA* **93:**1776–1780.

Bi, Z., Barna, M., Komatsu, T., and Reiss, C. S., 1995, Vesicular stomatitis virus infection of the central nervous system activates both innate and acquired immunity, *J. Virol.* **69:**6466–6472.

Bogdan, C., Vodovotz, Y., Paik, J., Xie, Q., and Nathan, C., 1994, Mechanism of suppression of nitric oxide synthase expression by interleukin-4 in primary mouse macrophages, *J. Leukoc. Biol.* **55:**227–233.

Bonfoco, E., Krainc, D., Ankarcrona, M., Nicotera, P., and Lipton, S. A., 1995, Apoptosis and necrosis: Two distinct events induced, respectively, by mild and intense insults with *N*-methyl-D-aspartate or nitric oxide/superoxide in cortical cell cultures, *Proc. Natl. Acad. Sci. USA* **92:**7162–7166.

Bukrinsky, M. I., Nottet, H. S. L. M., Schmidtmayerova, H., Dubrovsky, L., Flanagan, C. R., Mullins, M. E., Lipton, S. A., and Gendelman, H. E., 1995, Regulation of nitric oxide synthase activity in human immunodeficiency virus type 1 (HSV-1)-infected monocytes: Implications for HIV-associated neurological disease, *J. Exp. Med.* **181:**735–745.

Campbell, I. L., Samimi, A., and Chiang, C.-S., 1994, Expression of the inducible nitric oxide synthase. Correlation with neuropathology and clinical features in mice with lymphocytic choriomeningitis, *J. Immunol.* **153:**3622–3629.

Castro, L., Rodriguez, M., and Radi, R., 1994, Aconitase is readily inactivated by peroxynitrite, but not by its precursor, nitric oxide, *J. Biol. Chem.* **269:**29409–29415.

Cleeter, M. W. J., Cooper, J. M., Darley-Usmar, V. M., Moncada, S., and Schapiva, A. H. V., 1994, Reversible inhibition of cytochrome *c* oxidase, the terminal enzyme of the mitochondrial respiratory chain, by nitric oxide. Implications for neurodegenerative diseases, *FEBS Lett.* **345:**50–54.

Corraliza, I. M., Soler, G., Eichmann, K., and Modolell, M., 1995, Arginase induction by suppressors of nitric oxide synthesis (IL-4, IL-10 and PGE_2) in murine bone marrow-derived macrophages, *Biochem. Biophys. Res. Commun.* **206:**667–673.

Croen, K. D., 1993, Evidence for an antiviral effect of nitric oxide. Inhibition of herpes simplex virus type 1 replication, *J. Clin. Invest.* **91:**2446–2452.

Cunha, F. Q., Moncada, S., and Liew, F. Y., 1992, Interleukin-10 (IL-10) inhibits the induction of nitric oxide synthase by interferon-γ in murine macrophages, *Biochem. Biophys. Res. Commun.* **182:**1155–1159.

Dawson, V. L., Dawson, T. M., Uhl, G. R., and Snyder, S. H., 1993, Human immunodeficiency virus type 1 coat protein neurotoxicity mediated by nitric oxide in primary cortical cultures, *Proc. Natl. Acad. Sci. USA* **90:**3256–3259.

Doi, T., Ando, M., Akaike, T., Suga, M., Sato, K., and Maeda, H., 1993, Resistance to nitric oxide in *Mycobacterium avium* complex and its implication in pathogenesis, *Infect. Immun.* **61:**1980–1989.

Doi, K., Akaike, T., Horie, H., Noguchi, Y., Fujii, S., Beppu, T., Ogawa, M., and Maeda, H., 1996, Excessive production of nitric oxide in rat solid tumor and its implications in rapid tumor growth, *Cancer (Suppl.)* **77:**1598–1604.

Douglas, R. G., Jr., 1975, Influenza in man, in: *The Influenza Viruses and Influenza* (E. D. Kilbourne, ed.), Academic Press, New York, pp. 395–447.

Estevez, A. G., Radi, R., Barbeito, L., Shin, J. T., Thompson, J. A., and Beckman, J. S., 1995, Peroxynitrite-induced cytotoxicity in PC12 cells: Evidence for an apoptotic mechanism differentially modulated by neurotrophic factors, *J. Neurochem.* **65:**1543–1550.

Frears, E. R., Zhang, Z., Blake, D. R., O'Connell, J. P., and Winyard, P. G., 1996, Inactivation of tissue inhibitor of metalloproteinase-1 by peroxynitrite, *FEBS Lett.* **381:**21–24.

Fujii, S., Akaike, T., and Maeda, H., 1999, Role of nitric oxide in pathogenesis of herpes simplex virus encephalitis in rats, *Virology* (in press).

Garrett, R. H., and Grisham, C. M., 1995, Water, pH, and ionic equilibria, in: *Biochemistry*, Saunders College Publishing, Fort Worth, pp. 32–54.

Gotoh, T., Sonoki, T., Nagasaki, A., Terada, K., Takiguchi, M., and Mori, M., 1996, Molecular cloning of cDNA for nonhepatic mitochondrial arginase (arginase II) and comparison of its induction with nitric oxide synthase in a murine macrophage-like cell line, *FEBS Lett.* **395:**119–122.

Granger, D. L., Hibbs, J. B., Jr., Perfect, J. R., and Durack, D. T., 1988, Specific amino acid (L-arginine) requirement for microbiostatic activity of murine macrophages, *J. Clin. Invest.* **81:**1129–1136.

Haddad, I. Y., Ischiropoulos, B., Holm, B. A., Beckman, J. S., Baker, J. R., and Matalon, S., 1993, Mechanism of peroxynitrite-induced injury to pulmonary surfactant, *Am. J. Physiol.* **265:**L555–L564.

Haddad, I. Y., Pataki, G., Hu, P., Galliani, C., Beckman, J. S., and Matalon, S., 1994, Quantitation of nitrotyrosine levels in lung sections of patients and animals with acute lung injury, *J. Clin. Invest.* **94:**2407–2413.

Halliwell, B., and Gutteridge, J. M. C., 1984, Oxygen toxicity, oxygen radicals, transition metals and diseases, *Biochem. J.* **219:**1–14.

Hausladen, A., and Fridovich, I., 1994, Superoxide and peroxynitrite inactivate aconitases, but nitric oxide does not, *J. Biol. Chem.* **269:**29405–29408.

Hennet, T., Peterhans, E., and Stocker, R., 1992, Alterations in antioxidant defences in lung and liver of mice infected with influenza A virus, *J. Gen. Virol.* **73:**39–46.

Henry, Y. A., Guissani, A., and Ducastel, B., 1997, *Nitric Oxide Research from Chemistry to Biology: EPR Spectroscopy of Nitrosylated Compounds*, Molecular Biology Intelligence Unit, R. G. Landes Company, Austin.

Huie, R. E., and Padmaja, S., 1993, The reaction rate of nitric oxide with superoxide, *Free Radical Res. Commun.* **18:**195–199.

Ikeda, T., Shimokata, K., Daikoku, T., Fukatsu, T., Tsutsui, Y., and Nishiyama, Y., 1993, Pathogenesis of cytomegalovirus-associated pneumonitis in ICR mice: Possible involvement of superoxide radicals, *Arch. Virol.* **127:**11–24.

Ischiropoulos, H., Al-Mehdi, A., and Fisher, A. B., 1995, Reactive species in ischemic rat lung injury: Contribution of peroxynitrite, *Am. J. Physiol.* **269:**L158–L164.

James, S. L., 1995, Role of nitric oxide in parasitic infections, *Microbiol. Rev.* **59:**533–547.

Karupiah, G., Xie, Q., Buller, R. M. L., Nathan, C., Duarte, C., and MacMicking, J. D., 1993, Inhibition of viral replication by interferon-γ-induced nitric oxide synthase, *Science* **261:**1445–1448.

Kojima, Y., Akaike, T., Sato, K., Maeda, H., and Hirano, T., 1996, Polymer conjugation to Cu,Zn-SOD and suppression of hydroxyl radical generation on exposure to H_2O_2: Improved stability of SOD *in vitro* and *in vivo*, *J. Bioact. Compat. Polymers* **11:**169–190.

Kong, S. K., Yim, M. B., Stadtman, E. R., and Chock, P. B., 1996, Peroxynitrite disables the tyrosine phosphorylation regulatory mechanism: Lymphocyte-specific tyrosine kinase fails to phosphorylate nitrated cdc2(6-20)NH_2 peptide, *Proc. Natl. Acad. Sci. USA* **93:**3377–3382.

Koprowski, H., Zheng, Y. M., Heber-Katz, E., Fraser, N., Rorke, L., Fu, Z. F., Hanlon, C., and Dietzschold, B., 1993, *In vivo* expression of inducible nitric oxide synthase in experimentally induced neurologic diseases, *Proc. Natl. Acad. Sci. USA* **90:**3024–3027.

Kornellisse, R. F., Hoekman, K., Visser, J. J., Hop, W. C. J., Huijmans, J. G. M., van der Straaten, P. J. C., van der Heijden, A. J., Sukhai, R. N., Neijens, H. J., and de Groot, R., 1996, The role of nitric oxide in bacterial meningitis in children, *J. Infect. Dis.* **174:**120–126.

Kosaka, H., Sawai, Y., Sakaguchi, H., Kumura, E., Harada, N., Watanabe, M., and Shiga, T., 1994, ESR spectral transition by arteriovenous cycle in nitric oxide hemoglobin of cytokine-treated rats, *Am. J. Physiol.* **266:**C1400–C1405.

Kreil, T. R., and Eibl, M. M., 1995, Viral infection of macrophages profoundly alters requirements for induction of nitric oxide synthesis, *Virology* **212:**174–178.

Kreil, T. R., and Eibl, M. M., 1996, Nitric oxide and viral infection: NO antiviral activity against a flavivirus *in vitro*, and evidence for contribution to pathogenesis in experimental infection *in vivo*, *Virology* **219:**304–306.

Landino, L. M., Crews, B. C., Timmons, M. D., Morrow, J. D., and Marnett, L. J., 1996, Peroxynitrite, the coupling product of nitric oxide and superoxide, activates prostaglandin biosynthesis, *Proc. Natl. Acad. Sci. USA* **93:**15069–15074.

Lane, T. E., Paoletti, A. D., and Buchmeier, M. J., 1997, Dissociation between the *in vitro* and *in vivo* effects of nitric oxide on a neurotropic coronavirus, *J. Virol.* **71:**2202–2210.

Lepoivre, M., Fieschi, F., Coves, J., Thelander, L., and Fontecave, M., 1991, Inactivation of ribonucleotide reductase by nitric oxide, *Biochem. Biophys. Res. Commun.* **179:**442–448.

Liu, R. H., and Hotchkiss, J. H., 1995, Potential genotoxicity of chronically elevated nitric oxide: A review, *Mutat. Res.* **339:**73–89.

Maeda, H., and Akaike, T., 1991, Oxygen free radicals as pathogenic molecules in viral diseases, *Proc. Soc. Exp. Biol. Med.* **198:**721–727.

Mannick, J. B., Asano, K., Izumi, K., Kieff, E., and Stamler, J. S., 1994, Nitric oxide produced by human B lymphocytes inhibits apoptosis and Epstein-Barr virus reactivation, *Cell* **79:**1137–1146.

Matsumoto, H., and Sies, H., 1996, The reaction of ebselen with peroxynitrite, *Chem. Res. Toxicol.* **9:**262–267.

Mikami, S., Kawashima, S., Kanazawa, K., Hirata, K., Katayama, Y., Hotta, H., Hayashi, Y., Ito, H., and Yokoyama, M., 1996, Expression of nitric oxide synthase in a murine model of viral myocarditis induced by coxsackie virus B3, *Biochem. Biophys. Res. Commun.* **220:**983–989.

Moncada, S., and Higgs, A., 1993, The L-arginine–nitric oxide pathway, *N. Engl. J. Med.* **329:**2002–2012.

Mordvintcev, P., Mülsh, A., Busse, R., and Vanin, A., 1991, On-line detection of nitric oxide formation in liquid aqueous phase by electron paramagnetic resonance spectroscopy, *Anal. Biochem.* **199:**142–146.

Moreno, J. J., and Pryor, W., 1992, Inactivation of α_1-proteinase inhibitor by peroxynitrite, *Chem. Res. Toxicol.* **5:**425–431.

Murphy, B. R., and Webster, R. G., 1990, Orthomyxoviruses, in: *Virology*, Volume 2, 2nd ed. (B. N. Fields, D. M. Knipe, R. M. Chanock, M. S. Hirsh, J. L. Melnick, T. P. Monath, and B. Roizman, eds.), Raven Press, New York, pp. 1091–1152.

Nathan, C. F., and Hibbs, J. B., 1991, Role of nitric oxide synthesis in macrophage antimicrobial activity, *Curr. Opin. Immunol.* **3:**65–70.

Oda, T., Akaike, T., Hamamoto, T., Suzuki, F., Hirano, T., and Maeda, H., 1989, Oxygen radicals in influenza-induced pathogenesis and treatment with pyran polymer-conjugated SOD, *Science* **244:**974–976.

Ohshima, H., and Bartsch, H., 1994, Chronic infections and inflammatory processes as cancer risk factors: Possible role of nitric oxide in carcinogenesis, *Mutat. Res.* **305:**253–264.

Okamoto, T., Akaike, T., Suga, M., Tanase, S., Horie, H., Miyajima, S., Ando, M., Ichinose, Y., and Maeda, H., 1997a, Activation of human matrix metalloproteinases by various bacterial proteinases, *J. Biol. Chem.* **272:**6059–6066.

Okamoto, T., Akaike, T., Nagano, T., Miyajima, S., Suga, M., Ando, M., Ichimori, K., and Maeda, H., 1997b, Activation of human neutrophil procollagenase by nitrogen dioxide and peroxynitrite: A novel mechanism of procollagenase activation involving nitric oxide, *Arch. Biochem. Biophys.* **342:**261–274.

Pryor, W. A., and Squadrito, G. L., 1995, The chemistry of peroxynitrite: A product from the reaction of nitric oxide with superoxide, *Am. J. Physiol.* **268:**L699–L722.

Radi, R., Beckman, J. S., Bush, K. M., and Freeman, B. A., 1991a, Peroxynitrite oxidation of sulfhydryls, *J. Biol. Chem.* **266:**4244–4250.

Radi, R., Beckman, J. S., Bush, K. M., and Freeman, B. A., 1991b, Peroxynitrite-induced membrane lipid peroxidation: The cytotoxic potential of superoxide and nitric oxide, *Arch. Biochem. Biophys.* **288:**481–487.

Radi, R., Rodriguez, M., Castro, L., and Telleri, R., 1994, Inhibition of mitochondrial electron transport by peroxynitrite, *Arch. Biochem. Biophys.* **308:**89–95.

Rubbo, H., Darley-Usmar, V., and Freeman, B. A., 1996, Nitric oxide regulation of tissue free radical injury, *Chem. Res. Toxicol.* **9:**809–820.

Sato, K., Akaike, T., Kohno, M., Ando, M., and Maeda, H., 1992, Hydroxyl radical production by H_2O_2 plus Cu,Zn-superoxide dismutase reflects the activity of free copper released from the oxidatively damaged enzyme, *J. Biol. Chem.* **267:**25371–25377.

Sato, K., Suga, M., Akaike, T., Fujii, S., Muranaka, H., Doi, T., and Maeda, H., 1998, Therapeutic effect of erythromycin on influenza virus-induced lung injury in mice, *Am. J. Respir. Crit. Care Med.* **157:**853–857.

Schwartz, K. B., 1993, Oxidative stress during viral infection: A review, *Free Radical Biol. Med.* **21:**641–649.

Setoguchi, K., Takeya, M., Akaike, T., Suga, M., Hattori, R., Maeda, H., Ando, M., and Takahashi, K., 1996, Expression of inducible nitric oxide synthase and its involvement in pulmonary granulomatous inflammation in rats, *Am. J. Pathol.* **149:**2005–2022.

Sharara, A. I., Perkins, D. J., Misukonis, M. A., Chan, S. U., Dominitz, J. A., and Weinberg, J. B., 1997, Interferon-alpha activation of human mononuclear cells *in vitro* and *in vivo* for nitric oxide synthase type 2 mRNA and protein expression. Possible relationship of induced NOS2 to the anti-hepatitis C effects of IFN-α *in vivo*, *J. Exp. Med.* **186:**1495–1502.

Sidwell, R. W., Huffman, J. H., Bailey, K. W., Wong, M. H., Nimrod, A., and Panet, A., 1996, Inhibitory effects of recombinant manganese superoxide dismutase on influenza virus infections in mice, *Antimicrob. Agents Chemother.* **40:**2626–2631.

Sonoki, T., Nagasaki, A., Gotoh, T., Takiguchi, M., Takeya, M., Matsuzaki, H., and Mori, M., 1997, Coinduction of nitric oxide synthase and arginase I in cultured rat peritoneal macrophages and rat tissues *in vivo* by lipopolysaccharide, *J. Biol. Chem.* **272:**3689–3693.

Troy, C. M., Derossi, D., Prochiantz, A., Greene, L. A., and Shelanski, M. L., 1996, Downregulation of Cu/Zn superoxide dismutase leads to cell death via the nitric oxide–peroxynitrite pathway, *J. Neurosci.* **16:**253–261.

Umezawa, K., Ohnishi, N., Tanaka, K., Kamiya, S., Koga, Y., Nakazawa, H., and Ozawa, A., 1995, Granulation in livers of mice infected with *Salmonella typhimurium* is caused by superoxide released from host phagocytes, *Infect. Immun.* **63:**4402–4408.

Umezawa, K., Akaike, T., Fujii, S., Suga, M., Setoguchi, K., Ozawa, A., and Maeda, H., 1997, Induction of nitric oxide synthesis and xanthine oxidase and their role in the antimicrobial mechanism against *Salmonella typhimurium* in mice, *Infect. Immun.* **65:**2932–2940.

Uppu, R. M., Squadrito, G. L., and Pryor, W., 1996, Acceleration of peroxynitrite oxidations by carbon dioxide, *Arch. Biochem. Biophys.* **327:**335–343.

Vodovotz, Y., Bogdan, C., Paik, J., Xie, Q., and Nathan, C., 1993, Mechanisms of suppression of macrophage nitric oxide release by transforming growth factor β, *J. Exp. Med.* **178:**605–613.

Whiteman, M., Tritschler, H., and Halliwell, B., 1996, Protection against peroxynitrite-dependent tyrosine nitration and α_1-antiproteinase inactivation by oxidized and reduced lipoic acid, *FEBS Lett.* **379:**74–76.

Wright, P. F., 1997, Respiratory diseases, in: *Viral Pathogenesis* (N. Nathanson, R. Ahmed, F. Gonzalez-Scarano, D. E. Griffin, K. V. Holmes, F. A. Murphy, and H. L. Robinson, eds.), Lippincott–Raven Publishers, Philadelphia, pp. 703–711.

Xia, Y., and Zweier, J. L., 1997, Superoxide and peroxynitrite generation from inducible nitric oxide synthase in macrophages, *Proc. Natl. Acad. Sci. USA* **94:**6954–6958.

Yermilov, V., Yoshie, Y., Rubio, J., and Ohshima, H., 1996, Effects of carbon dioxide/bicarbonate on induction of DNA single-strand breaks and formation of 8-nitroguanine, 8-oxoguanine and base-propenal mediated by peroxynitrite, *FEBS Lett.* **399:**67–70.

Yoshida, M., Akaike, T., Wada, Y., Sato, K., Ikeda, K., Ueda, S., and Maeda, H., 1994, Therapeutic effect of imidazolineoxyl *N*-oxide against endotoxin shock through its direct nitric oxide-scavenging activity, *Biochem. Biophys. Res. Commun.* **202:**923–930.

Yoshimura, T., Yokoyama, H., Fuji, S., Takayama, F., Oikawa, K., and Kamada, H., 1996, *In vivo* EPR detection and imaging of endogenous nitric oxide in lipopolysaccharide-treated mice, *Nature Biotechnol.* **14:**992–994.

Zhang, X., Alley, E. W., Russell, S. W., and Morrison, D. C., 1994, Necessity and sufficiency of beta interferon for nitric oxide production in mouse peritoneal macrophages, *Infect. Immun.* **62:**33–40.

Zheng, Y. M., Schöfer, M. K. H., Weihe, E., Sheng, H., Corisdeo, S., Fu, Z. F., Koprowski, H., and Dietzschold, B., 1993, Severity of neurological signs and degree of inflammatory lesions in the brains of the rats with Borna disease correlate with the induction of nitric oxide synthase, *J. Virol.* **67:**5786–5791.

Zhou, A., Chen, Z., Rummage, J. A., Jiang, H., Kolosov, M., Stewart, C. A., and Leu, R. W., 1995, Exogenous interferon-gamma induces endogenous synthesis of interferon-alpha and -beta by murine macrophages for induction of nitric oxide synthase, *J. Interferon Cytokine Res.* **15:**897–904.

Zhu, L., Gunn, C., and Beckman, J. S., 1992, Bactericidal activity of peroxynitrite, *Arch. Biochem. Biophys.* **298:**452–457.

CHAPTER 20

Nitric Oxide in Bacterial Meningitis

GREGORY TOWNSEND and W. MICHAEL SCHELD

1. Introduction

Brain injury in bacterial meningitis is a multifactorial process, with significant contributions from the host inflammatory response. *In vitro* and *in vivo* research into the pathophysiology of bacterial meningitis indicates that leukocytes and host cell-derived proinflammatory mediators such as cytokines, prostaglandins, platelet-activating factor, leukocyte–endothelial cell adhesion molecules, and free radicals play a role in the disease. Among the latter, some of the more likely candidates include nitric oxide (NO) and its derivatives. This chapter will review the evidence supporting a role for NO in bacterial meningitis.

2. Roles of NO in Central Nervous System Function and Pathophysiology

As discussed in Chapter 3, NO is a water-soluble gaseous biological messenger that is reactive with atoms and free radicals. Inducible NO synthase (iNOS, NOS2) can produce cytotoxic concentrations of NO in response to various immunological stimuli, some of which are directly relevant to bacterial meningitis and central nervous system (CNS) inflammation, including LPS and the cytokines IFNα, TNFα, and IL-1 (Drapier *et al.*, 1988; Hibbs *et al.*, 1988; Knowles *et al.*, 1990; Murphy *et al.*, 1990; Pfeilschifter and Schwarzenbach, 1990; Kanno *et al.*, 1993; Lee *et al.*, 1993; Xie *et al.*, 1993; Geng *et al.*, 1994). Constitutive NOS enzymes are found in neurons (NOS1) and in vascular endothelial cells (NOS3) (see Chapter 4).

GREGORY TOWNSEND and W. MICHAEL SCHELD • Department of Medicine, University of Virginia, Charlottesville, Virginia 22908.

Nitric Oxide and Infection, edited by Fang. Kluwer Academic / Plenum Publishers, New York, 1999.

Cerebral endothelium-derived NO is involved in the maintenance of basal cerebrovascular vasodilatory tone (Kovach *et al.*, 1992), with a great deal of regional heterogeneity (Faraci, 1992). The primary stimulus for activation of neuronal NOS is glutamate-mediated activation of *N*-methyl-D-aspartate (NMDA) receptors (Bredt and Snyder, 1989; Garthwaite *et al.*, 1989). The localization of neuron-derived NO is also heterogeneous in the brain. Neuronal NO acts as an intercellular messenger with neighboring neurons and astrocytes as its main targets. Neuronal NOS may play a role in a number of important physiological functions in the CNS. For example, NO produced by neurons regulates changes in regional cerebral blood flow in response to local changes in neuronal activity. Neuronal NO also appears to be involved in modulation of the release and/or reuptake of several neurotransmitters, such as dopamine, acetylcholine, norepinephrine, γ-aminobutyric acid, and glutamate (Szabo, 1996).

NO may additionally contribute to the pathophysiology of bacterial meningitis indirectly through its derivatives. NO and superoxide ($O_2^{-\bullet}$), another oxygen-derived free radical produced by phagocytes and endothelial cells, may combine to form peroxynitrite ($ONOO^-$), a powerful oxidant (Blough and Zarifiou, 1985; Ischiropoulos *et al.*, 1992; Kooy and Royall, 1994). Peroxynitrite or one of its decomposition products with hydroxyl radical ($HO^\bullet$)-like reactivity (Beckman *et al.*, 1990; Hogg *et al.*, 1992) may cause oxidation of amino acid moieties or lipid peroxidation (Radi *et al.*, 1991a,b), thus interfering with normal cellular function and membrane integrity. Peroxynitrite may also inhibit mitochondrial respiration (Radi *et al.*, 1994). These effects may in turn lead to destabilization of the blood–brain barrier (BBB), as well as to neuronal dysfunction and/or death. Peroxynitrite and/or NO have been shown to cause necrosis or apoptotic cell death of neurons (Bonfoco *et al.*, 1995; Estevez *et al.*, 1995; Palluy and Rigaud, 1996; Leist *et al.*, 1997).

It is apparent, then, that NO may affect cerebral blood flow, cellular integrity, and neuronal function. Based on knowledge of its biological functions, it would be reasonable to postulate that NO might play a role in the pathophysiology of diseases of the CNS, including bacterial meningitis. A hypothetical scheme indicating the potential role of NO in the pathophysiology of bacterial meningitis is presented in Fig. 1.

There is evidence from nonmeningitis models of CNS injury to suggest that NO can contribute to the pathophysiology of CNS diseases. High concentrations of NO, such as are produced in response to the massive release of the excitatory amino acid glutamate during stroke, are neurotoxic (Boje and Arora, 1992; Peterson *et al.*, 1994), and NO appears to mediate the toxicity of excitatory amino acids (Uemura *et al.*, 1990; Dawson *et al.*, 1991). NOS inhibitors have been demonstrated to reduce neurological damage after stroke in some animal models (Nowicki *et al.*, 1991; Dawson *et al.*, 1992), although this benefit is not universally observed. It is also speculated that neurotoxic effects of NO may contribute to cell death in Parkinson's

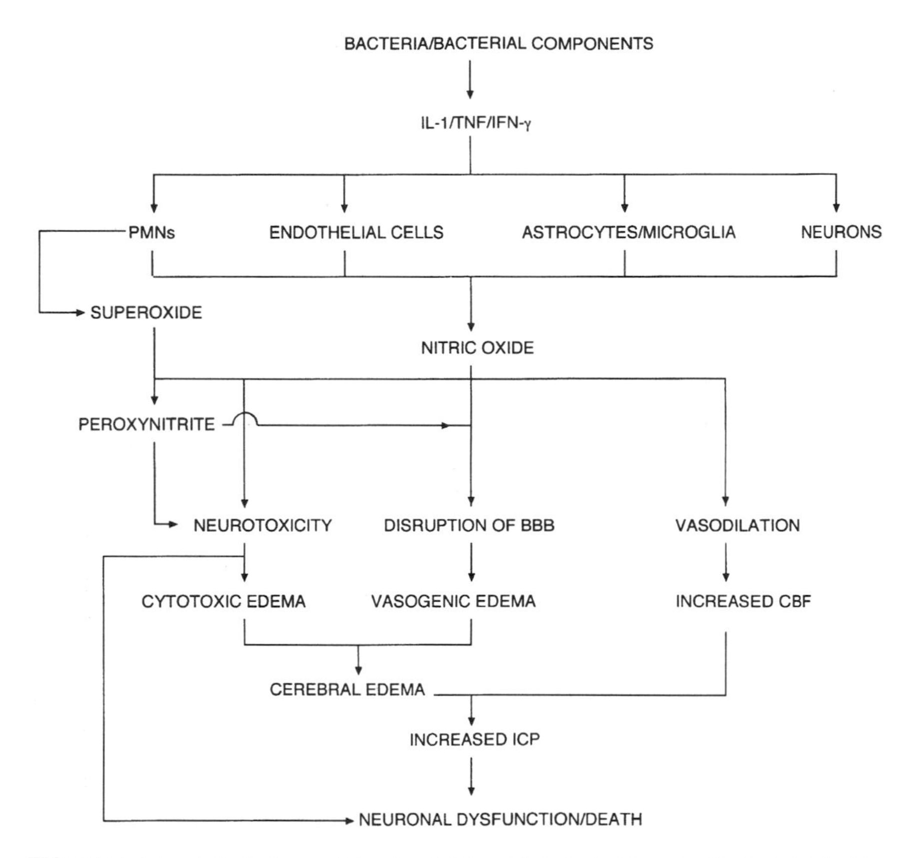

FIGURE 1. Hypothetical scheme of the role of nitric oxide in bacterial meningitis. IL-1, interleukin-1; TNF, tumor necrosis factor-alpha; IFN-γ, interferon-gamma; PMNs, polymorphonuclear leukocytes; BBB, blood–brain barrier; CBF, cerebral blood flow; ICP, intracranial pressure.

disease (Greenfield, 1992; Youdim *et al.*, 1993). NO-mediated inactivation of enzymes associated with mitochondrial respiration (Cleeter *et al.*, 1994) and inhibition of DNA replication by interaction with iron cofactors may lead to cellular dysfunction. This possibility is supported by the findings that breakdown of the BBB following cryogenic injury in the rat is prevented by an NMDA receptor antagonist, and stimulation of cerebral capillaries with NMDA results in increased capillary uptake of horseradish peroxidase, glucose, and calcium (Koenig *et al.*, 1992).

The cellular sources of NO relevant to bacterial meningitis and other diseases of the CNS are unknown, but several possibilities exist. As noted earlier, NOS is found in cerebral microvascular endothelial cells and in neurons and CNS macrophagelike cells (astrocytes and microglia). Neutrophils present in the subarachnoid space during bacterial meningitis may also generate NO. Thus, there are several potential sources of NO during bacterial meningitis.

3. *In Vitro* Studies

In vitro data suggest a potential role for NO in bacterial meningitis. In one study, primary astrocyte cultures from neonatal rat cortex were stimulated with heat-killed unencapsulated pneumococci. NO production was assessed by using the Griess reaction to measure concentrations of nitrite, one of the major oxidation products of NO, in the cell culture supernatant. Nitrite concentrations in astrocyte culture supernatants increased in a dose-dependent fashion in response to exposure to pneumococci. These increases were prevented by administration of the NOS inhibitors *N*-nitro-L-arginine (L-NA) and aminoguanidine. The inhibitory effect of L-NA was partially reversed by addition of L-arginine, a precursor of NO, but not by D-arginine. Rat cerebellar neurons and microglia were also demonstrated to produce NO when stimulated with pneumococci (Bernatowicz *et al.*, 1995). In another *in vitro* study, exposure of cultured rat astroglial cells to pneumococcal cell wall material resulted in a significant dose-dependent increase in nitrite concentrations in the supernatant after 24 hr. Again, this increase was inhibited by addition of the NOS inhibitor L-NA (Freyer *et al.*, 1996). Finally, L-NA has been used to prevent increases in nitrite concentrations after pneumococcal stimulation of rat cerebral endothelial cell cultures (Koedel *et al.*, 1995). Together, these results indicate that exposure of either resident CNS endothelial cells or macrophagelike cells to pneumococci stimulates NO production.

The ability of bacteria other than pneumococci to induce NO production has also been studied. In one experiment, murine macrophages were exposed to group B streptococci (GBS), the major etiological agents of neonatal meningitis, as well as to *Staphylococcus aureus* and *S. epidermidis*, leading causes of shunt-associated meningitis. All three bacterial species were found to cause a dose-dependent stimulation of NO production that could be abrogated by NOS inhibitors. Neither GBS nor *S. epidermidis* elicits NO production in cell lines deficient in complement receptor 3, which is necessary for nonopsonic phagocytosis of either species (Goodrum *et al.*, 1994).

Cultured rat neurons exposed to cell walls of *Streptococcus pneumoniae* or GBS do not appear to produce NO, but stimulation of astrocytes with extracts of either microorganism elicits cellular injury and NO production. In addition, cell walls are neurotoxic when neurons are cocultured with astrocytes or microglia; this effect can be prevented by the NOS inhibitor N^{G}-nitro-L-arginine methyl ester (L-NAME), but not by the excitatory amino acid antagonist MK801 (Kim and Tauber, 1996). These observations suggest that NO produced by astrocytes and microglia on exposure to gram-positive cell walls is toxic to neurons.

4. Animal Models

In vivo studies have supported a possible role of NO in the pathophysiology of bacterial meningitis. In a rat model of bacterial meningitis, increased NOS activity was induced by intracisternal administration of *Escherichia coli* LPS. NO synthesis occurred in the lateral and third ventricles and from the meninges, but not within peripheral leukocytes (Korytko and Boje, 1996). iNOS activity was also increased in the brains of rats after intracisternal inoculation with GBS (Leib *et al.*, 1996). These findings indicate that localized NO production occurs in the CNS during bacterial meningitis.

Animal models have demonstrated one of the key early steps in the pathophysiology of bacterial meningitis to be arteriolar dilatation accompanied by an increase in cerebral blood flow. In a rabbit model, flushing of cranial windows with artificial CSF containing LPS resulted in marked cerebral arteriolar vasodilatation after 4 hr. This effect was inhibited by the NOS inhibitor N^{G}-monomethyl-L-arginine, but the effects of the inhibitor were reversed by the presence of excess L-arginine, the precursor of enzymatically produced NO (Brian *et al.*, 1995). This study indicates that NO plays a role in mediating cerebral arteriolar dilatation induced by LPS during the early stages of bacterial meningitis.

In other studies, the NOS inhibitors L-NA or L-NAME have been administered intravenously in a rat model of pneumococcal meningitis (Haberl *et al.*, 1994; Koedel *et al.*, 1995). Treated rats demonstrated reduced CSF leukocyte concentrations, intracranial pressure, regional cerebral blood flow, and brain water content during the first 6 hr, when compared with rats that did not receive a NOS inhibitor. However, increased mortality was seen in at least some animals receiving the NOS inhibitor, possibly resulting from loss of beneficial NO-mediated effects on cerebral perfusion. L-NA was found to prevent dilatation of pial arterioles (Koedel *et al.*, 1995).

Another study examined the relative contributions of neuronal and inducible NOS to the cerebrovascular changes in the early phase of experimental pneumococcal meningitis in rats. In a rat model of meningitis induced by intracisternal inoculation of heat-killed pneumococci, the effects of the nNOS inhibitor 7-nitroindazole (NI) and the iNOS inhibitor *S*-methylisothiourea (SMT) on pial arteriolar diameter, intracranial pressure, brain water content, and CSF leukocyte concentrations were compared after 4 hr. Treatment with NI prevented pneumococcus-induced vasodilatation. Treatment with SMT 0.1 mg/kg administered intraperitoneally did not influence vasodilatation, but SMT 1.0 mg/kg attenuated the vasodilatation. The mean arterial pressure increased in rats given SMT at 1.0 mg/kg, but not at 0.1 mg/kg, indicating that the higher dose of SMT influenced NOS activity; this may have contributed to inhibition of pial vasodilatation seen at the higher dose. In contrast to other studies, the increase in brain water content, intracranial pressure, and CSF white blood cell counts in these pneumococcus-challenged rats was not attenuated by NI, nor by SMT at either dose (Paul *et al.*, 1997). Nevertheless, these findings suggest that nNOS is involved with pial arteriolar vasodilatation during the early phases of pneumococcal meningitis.

In a rabbit model examining the effects of TNFα and NO on cerebral circulation and metabolism, meningitis was induced by intracisternal inoculation of TNFα. Injection of recombinant human TNFα caused reductions in cerebral oxygen uptake and cerebral blood flow, as well as increases in intracranial pressure and in CSF lactate concentrations within the first 6 hr. Pretreatment with L-NAME resulted in reductions in oxygen uptake and increases in intracranial pressure and CSF lactate. Reduction in cerebral blood flow, which was related primarily to an increase in cerebrovascular resistance, was not affected by L-NAME (Tureen, 1995). These observations suggest that NO may mediate changes in cerebral circulation and metabolism during bacterial meningitis; increased intracranial pressure could result from an NO-induced fall in vascular tone.

In a study of the effects of iNOS inhibition at a later stage of bacterial meningitis, treatment with SMT had no effect on brain water content, intracranial pressure, or CSF leukocyte concentration at 24 hr in a rat model of pneumococcal meningitis (Koedel and Pfister, 1997). Although findings from other studies have demonstrated effects of NOS inhibition within 6 hr of inoculation, it is possible that NO produced by iNOS plays an important role during the early stages of bacterial meningitis, but not in later stages.

One of the more promising potential adjunctive treatments for bacterial meningitis is the cytokine IL-10. IL-10 deactivates cells of the macrophage/monocyte line, leading to decreased production of cytokines and reactive oxygen and nitrogen species (see also Chapter 5). In one study, intraperitoneal administration of IL-10 resulted in decreased CSF leukocyte concentrations, intracranial pressure, regional cerebral blood flow, and brain water content within 6 hr after intracisternal inoculation of pneumococci in a rat model of bacterial meningitis. *In vitro*, exposure of cultured rat cerebral endothelial cells to IL-10 attenuates the increase in nitrite concentration stimulated by heat-killed pneumococci (Koedel *et al.*, 1996). These results suggest that the salutary effects of IL-10 on parameters of CNS inflammation may at least in part be attributable to its effects on NO production.

Another important event in the pathophysiology of bacterial meningitis is an increase in the permeability of the blood–brain barrier (BBB). The BBB ordinarily serves to help maintain homeostasis in the CNS, particularly with regard to fluid balance and the passage of electrolytes and other solutes. Disruption of this barrier contributes to cerebral edema and neuronal dysfunction. In a rat model of meningitis induced by intracisternal inoculation of LPS, intravenous administration of the iNOS inhibitor aminoguanidine blocked NO production within the CNS. This was associated with a reduction in the pathological increase in BBB permeability (Boje, 1995, 1996).

Another study has examined the effects of NO on BBB permeability in a rat model of meningitis. In this study, meningitis was induced by intracisternal inoculation of live *Haemophilus influenzae* or bacterial endotoxin derived from *H. influenzae* or *E. coli*. Peak CSF nitrite concentrations were significantly higher in rats inoculated with live bacteria or with endotoxin, when compared with controls. Administration of the NOS inhibitor L-NAME resulted in significant reductions in mean CSF nitrite concentrations, CSF leukocyte concentrations, and BBB permeability in animals inoculated intracisternally with lipooligosaccharide. There was a significant correlation between increases in CSF nitrite concentrations over time and increases in BBB permeability (Buster *et al.*, 1995). These studies suggest strongly that NO or its metabolites contribute significantly to the increased permeability of the BBB observed in bacterial meningitis.

One of the more common and problematic sequelae of bacterial meningitis in children is sensorineural hearing loss; however, the mechanism(s) underlying this hearing loss remains undetermined. Pneumolysin, a toxin elaborated by *Streptococcus pneumoniae*, has been demonstrated to be cytotoxic to the guinea pig cochlea. Pretreatment of the cochlea with the NOS inhibitor N^G-methyl-L-arginine or the NMDA receptor antagonist MK801 protects against pneumolysin-induced damage (Amaee *et al.*, 1995). In another study, severe damage to cochlear cells, hair cells, and cells in the organ of Corti was elicited by perfusion of the scala tympani with molecules that enhance NO production (NMDA, sodium nitroprusside, and *S*-nitroso-*N*-acetylpenicillamine). This damage was associated with changes in electrophysiological parameters indicative of profound hearing loss. Pre-perfusion with either the NOS inhibitor L-methyl arginine or superoxide dismutase provided protection of the cochlea from such damage (Amaee *et al.*, 1997). These results indicate that NO, and perhaps peroxynitrite formed by the reaction of $NO^\bullet$ and $O_2^{-\bullet}$, could play an important role in the development of hearing loss associated with bacterial meningitis.

Peroxynitrite may also contribute to the pathophysiological sequelae of bacterial meningitis in other ways, for example, by mediating lipid peroxidation. In a rat model of meningitis, concentrations of by-products of lipid peroxidation (malonaldehyde and 4-hydroxyalkenal) in whole brain homogenates and CSF increased within 3 to 6 hr after intracisternal inoculation with *H. influenzae* or *S. pneumoniae*, and remained elevated through 18 hr when compared with controls. However, concentrations in brain homogenates were not increased in infected animals relative to controls (Rutgers *et al.*, 1997). This difference may reflect localization of lipid peroxidation to the meninges, subarachnoid space, and immediately adjacent brain parenchyma, but nonetheless indicates that lipid peroxidation is in fact increased during bacterial meningitis.

Although the majority of the published studies using NOS inhibitors in animal models of bacterial meningitis have demonstrated a benefit in one or more measures of CNS pathology, it should be noted that this trend has not been universal. In a rat model of GBS meningitis, administration of the iNOS inhibitor aminoguanidine resulted in *increased* neuronal injury (Leib *et al.*, 1996), higher bacterial titers, and worsening of cortical ischemia (Leib *et al.*, 1998). The contrast between the results of this study and the results of the other studies might reflect differences specific to the microorganism, NOS inhibitor, or animal model.

5. Clinical Observations

Along with the data from animal models, there is some evidence from clinical studies supporting a role for NO in the pathophysiology of bacterial meningitis. In one report, concentrations of nitrite and nitrate were compared between controls and groups of patients with a variety of neurological diseases, such as Huntington's and Alzheimer's disease, amyotrophic lateral sclerosis, HIV infection, and meningitis. In this study, CSF concentrations of quinolinic acid (a neurotoxin), neopterin (a marker of macrophage activation), and nitrite/nitrate were significantly increased in a small group of patients with bacterial and viral meningitis, when compared with controls (Milstien *et al.*, 1994). Another study compared CSF nitrite and nitrate concentrations in 35 patients with bacterial meningitis and 30 controls. CSF nitrate concentrations were increased in patients with bacterial meningitis in the absence of an increase in serum concentrations, suggesting local CNS production of NO, although increased CSF nitrate concentrations may also have resulted from increased permeability of the BBB. CSF concentrations of nitrite correlated positively with those of TNFα and negatively with CSF glucose concentrations. CSF concentrations of L-arginine were also lower in patients than in controls, again consistent with local production of NO (Kornelisse *et al.*, 1996).

In a study of CSF and serum from 94 patients with meningococcal meningitis compared with 44 controls with noninflammatory neurological diseases, increased mean CSF concentrations of nitrite and nitrate were demonstrated in CSF of patients with meningitis (Visser *et al.*, 1994). Increased CSF nitrite concentrations have also been demonstrated in children with meningitis caused by *H. influenzae* (Tsukahara *et al.*, 1996), and have been found to correlate with CSF concentrations of TNFα (van Furth *et al.*, 1996).

6. Conclusions

In summary, the physiological and pathophysiological effects of NO and its derivatives suggest that NO may play a significant role in the pathophysiology of bacterial meningitis. *In vitro* models have demonstrated that resident CNS cells are stimulated to increase NO production when exposed to bacteria and bacterial components, and that NO produced in response to bacterial components may be neurotoxic. In animal models of meningitis, local CNS NO production is increased, along with lipid peroxidation that may be mediated by peroxynitrite. Inhibition of NO synthesis is associated in most studies with inhibition of changes in CSF leukocyte and lactate concentrations, intracranial pressure, cerebral blood flow, intracranial pressure, pial arteriolar dilatation, cerebral oxygen uptake, BBB permeability, and inner ear structure and function. Some of these changes, such as pial arteriolar dilatation, appear to mediated by changes in nNOS activity, which may reflect the role played by neuronal NO in regulating local cerebral blood flow in response to changes in neuronal activity. Finally, clinical data demonstrate that local NO production is increased in patients with bacterial meningitis resulting from a variety of causes. These studies strongly suggest that CNS NO production is increased in bacterial meningitis, and that this increase contributes to pathophysiological sequelae of this disease. It is possible that further study will allow development of selective NOS inhibitors as adjunctive treatments in the management of bacterial meningitis, although the deleterious effects of NOS inhibition on cerebral perfusion mitigate against the use of currently available agents in this setting.

References

Amaee, F. R., Comis, S. D., and Osborne, M. P., 1995, N^G-Methyl-L-arginine protects the guinea pig cochlea from the cytotoxic effects of pneumolysin, *Acta Oto-Laryngol.* **115**:386–391.

Amaee, F. R., Comis, S. D., Osborne, M. P., Drew, S., and Tarlow, M. J., 1997, Possible involvement of nitric oxide in the sensorineural hearing loss of bacterial meningitis, *Acta Oto-Laryngol.* **117**:329–336.

Beckman, J. S., Beckman, T. W., Chen, J., Marshall, P. A., and Freeman, B. A., 1990, Apparent hydroxyl radical production by peroxynitrite: Implications for endothelial injury from nitric oxide and superoxide, *Proc. Natl. Acad. Sci. USA* **87**:1620–1624.

Bernatowicz, A., Kodel, U., Frei, K., Fontana, A., and Pfister, H. W., 1995, Production of nitrite by primary rat astrocytes in response to pneumococci, *J. Neuroimmunol.* **60**:53–61.

Blough, N., and Zafiriou, O., 1985, Reaction of superoxide with nitric oxide to form peroxynitrite in alkaline aqueous solution, *Inorg. Chem.* **24**:3502–3504.

Boje, K. M., 1995, Inhibition of nitric oxide synthase partially attenuates alterations in the blood–cerebrospinal fluid barrier during experimental meningitis in the rat, *Eur. J. Pharmacol.* **272**:297–300.

Boje, K. M., 1996, Inhibition of nitric oxide synthase attenuates blood–brain barrier disruption during experimental meningitis, *Brain Res.* **720**:75–83.

Boje, K., and Arora, P., 1992, Microglial-produced nitric oxide and reactive oxides mediate neuronal cell death, *Brain Res.* **587**:250–256.

Bonfoco, E., Krainc, D., Ankarcrona, M., Nicotera, P., and Lipton, S. A., 1995, Apoptosis and necrosis: Two distinct events induced, respectively, by mild and intense insults with *N*-methyl-D-aspartate or nitric oxide/superoxide in cortical cell cultures, *Proc. Natl. Acad. Sci. USA* **92**:7162–7166.

Bredt, D. S., and Snyder, S. H., 1989, Nitric oxide mediates glutamate-linked enhancement of cGMP levels in the cerebellum, *Proc. Natl. Acad. Sci. USA* **86**:9030–9033.

Brian, Y. E., Jr., Heistad, D. D., and Faraci, F. M., 1995, Dilatation of cerebral arterioles in response to lipopolysaccharide *in vivo*, *Stroke* **26:**277–280.

Buster, B. L., Weintrob, A. C., Townsend, O. C., and Scheld, W. M., 1995, Potential role of nitric oxide in the pathophysiology of experimental bacterial meningitis in rats, *Infect. Immun.* **63:**3835–3839.

Cleeter, M., Cooper, J., Darley-Usmar, V., Moncada, S., and Schapira, A., 1994, Reversible inhibition of cytochrome c oxidase, the terminal enzyme of the mitochondrial respiratory chain, by nitric oxide: Implications for neurodegenerative diseases, *FEBS Lett.* **345:**50–54.

Dawson, T., Dawson, V., and Snyder, S., 1992, A novel neuronal messenger molecule in brain: The free radical, nitric oxide, *Ann. Neurol.* **32:**297–311.

Dawson, V., Dawson, T., London, E., Brent, D., and Snyder, S., 1991, Nitric oxide mediates glutamate neurotoxicity in primary cortical cultures, *Proc. Natl. Acad. Sci. USA* **88:**6368–6371.

Drapier, J. C., Wietzerbin, J., and Hibbs, J. B., Jr., 1988, Interferon-gamma and tumor necrosis factor induce the L-arginine cytotoxic effector mechanism in murine macrophages, *Eur. J. Immunol.* **18:**1587–1592.

Estevez, A. G., Radi, R., Barbeito, L., Shin, J. T., Thompson, J. A., and Beckman, J. S., 1995, Peroxynitrite-induced cytotoxicity in PC12 cells: Evidence for an apoptotic mechanism differentially modulated by neurotrophic factors, *J. Neurochem.* **65:**1543–1550.

Faraci, F., 1992, Regulation of the cerebral circulation by endothelium, *Pharmacol. Ther.* **56:**1–22.

Freyer, D., Weih, M., Weber, J. R., Burger, W., Scholz, P., Manz, R., Ziegenhorn, A., Angestwurm, K., and Dirnagl, U., 1996, Pneumococcal cell wall components induce nitric oxide synthase and TNF-alpha in astroglial-enriched cultures, *Glia* **16:**1–6.

Garthwaite, J., Garthwaite, G., Palmer, R., and Moncada, S., 1989, NMDA receptor activation induces nitric oxide synthesis from arginine in rat brain slices, *Eur. J. Pharmacol.* **172:**413–416.

Geng, Y., Petersson, A., Wennmalm, A., and Hanson, G., 1994, Cytokine-induced expression of nitric oxide synthase results in nitrosylation of heme and nonheme iron proteins in vascular smooth muscle cells, *Exp. Cell Res.* **214:**418–428.

Goodrum, K. S., McCormick, L. L., and Schneider, B., 1994, Group B streptococcus-induced nitric oxide production in murine macrophages is CR3 (CD11b/CD18) dependent, *Infect. Immun.* **62:**3102–3107.

Greenfield, S., 1992, Cell death in Parkinson's disease, *Essays Biochem.* **27:**103–118.

Haberl, R. L., Anneser, F., Kodel, U., and Pfister, H. W., 1994, Is nitric oxide involved as a mediator of cerebrovascular changes in the early phase of experimental pneumococcal meningitis? *Neurol. Res.* **16:**108–112.

Hibbs, J. B., Jr., Taintor, R. R., Vavrin, Z., and Rachlin, E. M., 1988, Nitric oxide: A cytotoxic activated macrophage effector moleule, *Biochem. Biophys. Res. Commun.* **157:**87–94.

Hogg, N., Darley-Usmar, V. M., Wilson, M. T., and Moncada, S., 1992, Production of hydroxyl radicals from the simultaneous generation of superoxide and nitric oxide, *Biochem. J.* **281:**419–424.

Ischiropoulos, H., Zhu, L., and Beckman, J. S., 1992, Peroxynitrite formation from macrophage-derived nitric oxide, *Arch. Biochem. Biophys.* **298:**445–451.

Kanno, K., Hirata, Y., Imai, T., and Marumo, F., 1993, Induction of nitric oxide synthase gene by interleukin in vascular smooth muscle cells, *Hypertension* **22:**34–39.

Kim, Y. S., and Tauber, M. G., 1996, Neurotoxicity of glia activated by gram-positive bacterial products depends on nitric oxide production, *Infect. Immun.* **64:**3148–3153.

Knowles, R. G., Palacios, M., Palmer, R. M. J., and Moncada, S., 1990, Kinetic characteristics of nitric oxide synthase from rat brain, *Biochem. J.* **269:**207–210.

Koedel, U., and Pfister, H. W., 1997, Protective effect of the antioxidant *N*-acetyl-L-cysteine in pneumococcal meningitis in the rat, *Neurosci. Lett.* **225:**33–36.

Koedel, U., Bernatowicz, A., Paul, R., Frei, K., Fontana, A., and Pfister, H. W., 1995, Experimental pneumococcal meningitis: Cerebrovascular alterations, brain edema, and meningeal inflammation are linked to the production of nitric oxide, *Ann. Neurol.* **37:**313–323.

Koedel, U., Bernatowicz, A., Frei, K., Fontana, A., and Pfister, H. W., 1996, Systemically (but not intrathecally) administered IL-10 attenuates pathophysiologic alterations in experimental pneumococcal meningitis, *J. Immunol.* **157:**5185–5191.

Koenig, H., Trout, J., Goldstone, A., and Lu, C., 1992, Capillary NMDA receptors regulate blood–brain barrier function and breakdown, *Brain Res.* **588:**297–303.

Kooy, N., and Royall, J., 1994, Agonist-induced peroxynitrite production from endothelial cells, *Arch. Biochem. Biophys.* **310:**352–359.

Kornelisse, R. F., Hoekman, K., Visser, J. J., Hop, W. C., Huijmans, J. G., van der Straaten, P. J., van der Heijden, A. J., Sukhai, R. N., Neijens, H. J., and de Groot, R., 1996, The role of nitric oxide in bacterial meningitis in children, *J. Infect. Dis.* **174:**120–126.

Korytko, P. J., and Boje, K. M., 1996, Pharmacological characterization of nitric oxide production in a rat model of meningitis, *Neuropharmacology* **35:**231–237.

Kovach, A., Szabo, C., Benyo, Z., Csaki, C., Greenberg, J., and Reivich, M., 1992, Effects of N^G-nitro-L-arginine and L-arginine on regional blood flow in the cat, *J. Physiol. (London)* **449:**183–196.

Lee, S., Dickson, D., Liu, W., and Brosnan, C., 1993, Induction of nitric oxide synthase activity in human astrocytes by interleukin-1β and interferon-γ, *J. Neuroimmunol.* **46:**19–24.

Leib, S. L., Kim, Y. S., Black, S. M., Ferriero, D. M., and Tauber, M. G., 1996, Detrimental effect of nitric oxide inhibition in experimental bacterial meningitis, *Ann. Neurol.* **39:**555–556.

Leib, S. L., Kim, Y. S., Black, S. M., Tureen, J. H., and Tauber, M. G., 1998, Inducible nitric oxide synthase and the effect of aminoguanidine in experimental neonatal meningitis, *J. Infect. Dis.* **177:**692–700.

Leist, M., Fava, E., Montecucco, C., and Nicotera, P., 1997, Peroxynitrite and nitric oxide donors induce neuronal apoptosis by eliciting autocrine excitotoxicity, *Eur. J. Neurosci.* **9:**1488–1498.

Milstien, S., Sakai, N., Brew, B. J., Krieger, C., Vickers, J. H., Saito, K., and Heyes, M. P., 1994, Cerebrospinal fluid nitrite/nitrate levels in neurologic diseases, *J. Neurochem.* **63:**178–180.

Murphy, S., Minor, R. L., Jr., Welk, G., and Harrison, D. G., 1990, Evidence for an astrocyte-derived vasorelaxing factor with properties similar to nitric oxide, *J. Neurochem.* **55:**349–351.

Nowicki, J., Duval, D., Poignet, H., and Scatton, B., 1991, Nitric oxide mediates neuronal death after focal cerebral ischemia in the mouse, *Eur. J. Pharmacol.* **204:**339–340.

Palluy, O., and Rigaud, M., 1996, Nitric oxide induces cultured cortical neuron apoptosis, *Neurosci. Lett.* **208:**1–4.

Paul, K., Koedel, U., and Pfister, H. W., 1997, 7-Nitroimidazole inhibits pial arteriolar vasodilation in a rat model of pneumococcal meningitis, *J. Cereb. Blood Flow Metab.* **17:**985–991.

Peterson, P., Hu, S., Anderson, R., and Chao, C., 1994, Nitric oxide production and neurotoxicity mediated by activated microglia from human versus mouse brain, *J. Infect. Dis.* **170:**457–460.

Pfeilschifter, J., and Schwarzenbach, H., 1990, Interleukin 1 and tumor necrosis factor stimulate cGMP formation in rat mesangial cells, *FEBS Lett..* **273:**185–187.

Radi, R., Beckman, J. S., Bush, K. M., and Freeman, B. A., 1991a, Peroxynitrite-induced membrane lipid peroxidation: The cytotoxic potential of superoxide and nitric oxide, *Arch. Biochem. Biophys.* **288:**481–487.

Radi, R., Beckman, J. S., Bush, K. M., and Freeman, B. A., 1991b, Peroxynitrite oxidation of sulfhydryls, *J. Biol. Chem.* **266:**4244–4250.

Radi, R., Rodriguez, M., Castro, L., and Telleri, R., 1994, Inhibition of mitochondrial electron transport by peroxynitrite, *Arch. Biochem. Biophys.* **308:**89–95.

Rutgers, M. I. L., Yo, M. S. S., Buster, B. L., and Scheld, W. M., 1997, Increase of the lipidperoxidation byproducts malonaldehyde and 4-hydoxyalkenal in cerebrospinal fluid during experimental bacterial meningitis in the rat, in: *Proceedings and Abstracts of the 37th Interscience Conference on Antimicrobial Agents and Chemotherapy*, American Society for Microbiology, Washington, D.C., Abstr. B-71.

Szabo, C., 1996, Physiological and pathophysiological roles of nitric oxide in the central nervous system, *Brain Res. Bull.* **41:**131–141.

Tsukahara, H., Hara, Y., Tsuchida, S., Shigematsu, Y., Konishi, Y., Kikuchi, K., and Sudo, M., 1996, Nitrite concentration in cerebrospinal fluid of infants: Evidence for enhanced nitric oxide production in *Haemophilus influenzae* meningitis, *Acta Paediatr. Jpn.* **38:**420–422.

Tureen, J., 1995, Effect of recombinant human tumor necrosis factor-alpha on cerebral oxygen uptake, cerebrospinal fluid lactate, and cerebral blood flow in the rabbit: Role of nitric oxide, *J. Clin. Invest.* **95:**1086–1091.

Uemura, Y., Kowall, N., and Beal, M., 1990, Selective sparing of NADPH-diaphorase-somatostatin-neuropeptide Y neurons in ischemic gerbil striatum, *Ann. Neurol.* **27:**620–625.

van Furth, A. M., Seijmonsbergen, E. M., Groeneveld, P. H., van Furth, R., and Langermans, J. A., 1996, Levels of nitric oxide correlate with high levels of tumor necrosis factor alpha in cerebrospinal fluid samples from children with bacterial meningitis, *Clin. Infect. Dis.* **22:**876–878.

Visser, J. J., Scholten, R. J., and Hoekman, K., 1994, Nitric oxide synthesis in meningococcal meningitis, *Ann. Intern. Med.* **120:**345–346.

Xie, Q., Whisnant, R., and Nathan, C., 1993, Promoter of the mouse gene encoding calcium-independent nitric oxide synthase confers inducibility by interferon-gamma and bacterial lipopolysaccharide, *J. Exp. Med.* **177:**1779–1784.

Youdim, M., Ben-Shachar, D., Eshel, G., Finberg, J., and Riederer, P., 1993, The neurotoxicity of iron and nitric oxide: Relevance to the etiology of Parkinson's disease, *Adv. Neurol.* **60:**259–266.

CHAPTER 21

Nitric Oxide in AIDS-Associated Neurological Disease

STUART A. LIPTON

1. Introduction

Approximately one-quarter of adults and one-half of children with AIDS eventually suffer from neurological manifestations, including dysfunction of cognition, movement, and sensation, that are a direct consequence of HIV-1 infection of brain (Bacellar *et al.*, 1994; Lipton, 1994). These neurological problems can occur in the absence of superinfection with opportunistic pathogens or secondary malignancies (Price *et al.*, 1988). Clinical manifestations include difficulty with mental concentration and slowness of hand movements and gait. This malady was initially termed the *AIDS dementia complex* by Price *et al.* (1988), but more recently has been placed under the heading *HIV-1-associated cognitive/motor complex*. Pathologically, HIV-1 infection in the central nervous system (CNS) includes *HIV encephalitis* and is characterized by widespread reactive astrocytosis, myelin pallor, and infiltration by monocytoid cells, such as blood-derived macrophages, resident microglia, and multinucleated giant cells (Budka, 1991). In addition, most investigators have reported that particular subsets of neurons display a striking degree of injury, including dendritic pruning and simplification of synaptic contacts, as well as frank cell loss in some cases, which may herald the onset of cognitive and motor deficits in affected individuals (Ketzler *et al.*, 1990; Everall *et al.*, 1991; Wiley *et al.*, 1991; Masliah *et al.*, 1992; Tenhula *et al.*, 1992). Neuronal injury may result in reversible dysfunction rather than inevitable demise. In contrast to many

---

STUART A. LIPTON • CNS Research Institute, Harvard Medical School, and Brigham and Women's Hospital, Boston, Massachusetts 02115.

Nitric Oxide and Infection, edited by Fang. Kluwer Academic / Plenum Publishers, New York, 1999.

encephalidities, progressive clinical sequelae occur without direct infection of neurons by HIV-1 or significant autoimmune reactions triggered by virus. The mononuclear phagocytes (brain macrophages, microglia, and multinucleated giant cells) in the CNS represent the predominantly infected cell type (Koenig *et al.*, 1986). Although infection can occur in astrocytes, it is highly restricted (Saito *et al.*, 1994; Tornatore *et al.*, 1994). Hence, the number of HIV-infected cells in the brain is relatively small, consisting predominantly of macrophages and microglia, and the question of how a relatively small number of cells can produce so much dysfunction of uninfected neurons remains to be convincingly answered. Recent advances in our understanding of the pathogenic mechanisms underlying AIDS dementia raise the possibility that noxious substances released by HIV-infected or activated macrophages/microglia and astrocytes, including $NO^{\bullet}$, may play a role in the generation of the clinical dementia syndrome. Both *in vitro* and *in vivo* experiments from several different laboratories, including our group in collaboration with that of Howard Gendelman (University of Nebraska Medical Center), have lent support to the idea that these neurotoxins are largely responsible for the pathological alterations of brain tissue seen following HIV infection. These toxins may include HIV-1 proteins (e.g., gp120 and gp41, which together constitute the entire envelope protein gp160, Tat, Nef, and possibly others) as well as substances released from activated macrophages and astrocytes (e.g., glutamatelike neurotoxic molecules, amines, free radicals, cytokines, and eicosanoids) (Brenneman *et al.*, 1988; Wahl *et al.*, 1989; Dreyer *et al.*, 1990; Giulian *et al.*, 1990, 1993, 1996; Heyes *et al.*, 1991; Genis *et al.*, 1992; Hayman *et al.*, 1993; Dreyer and Lipton, 1994; Gelbard *et al.*, 1994; Lipton and Rosenberg, 1994; Toggas *et al.*, 1994; Yeh *et al.*, 1994; Bukrinsky *et al.*, 1995). The mechanism underlying this indirect form of neuronal injury has been shown to be related to excessive influx of Ca^{2+} into neurons in response to the noxious factors released from immune-activated, HIV-infected, or gp120-stimulated brain macrophages/microglia, which produce excessive stimulation of *N*-methyl-D-aspartate (NMDA) receptors (Dreyer *et al.*, 1990; Giulian *et al.*, 1990, 1993; Lipton, 1992c; Lo *et al.*, 1992; Savio and Levi, 1993; Diop *et al.*, 1994; Lannuzel *et al.*, 1995; Lipton and Gendelman, 1995). Although gp120 and Tat have been reported to *directly* induce neurotoxicity, the indirect form of neurotoxicity via macrophage toxins is thought to predominate (direct activation of chemokine receptors on neurons as well as on macrophages by gp120, for example, could contribute to neurotoxicity) (Meucci *et al.*, 1998). Interestingly, rising levels of excitatory amino acids during focal stroke (Benveniste *et al.*, 1984; Globus *et al.*, 1988; Silverstein *et al.*, 1991; Bustos *et al.*, 1992; Lombardi and Moroni, 1992; Louzada *et al.*, 1992; Perrson and Hillered, 1992) are thought to trigger pathways to neuronal cell death involving NMDA receptors that resemble those just described (for reviews, see Choi, 1988; Meldrum and Garthwaite, 1990; Lipton and Rosenberg, 1994). Moreover, neurotoxic (as well as trophic) factors from microglia or macrophages have been implicated in the pathogenesis of focal ischemia and a variety of neurodegenerative diseases.

2. Brain Macrophage- and Astrocyte-Mediated Neuronal Injury: Toxic Substances Released after HIV Infection or gp120 Stimulation

A paradoxical discordance exists between the small numbers of productively HIV-infected brain macrophages and microglia, and the severe clinical cognitive and motor deficits experienced by some patients with AIDS. This suggests that a mechanism of cellular amplification and/or activation is required for the generation of viral or cellular toxins that lead to tissue injury and sustained viral infection (Fig. 1). Indeed, there is ample evidence for diffuse CNS immune-related activation in HIV-1 associated neurological impairment (Tyor *et al.*, 1992; Wesselingh *et al.*, 1993; Griffin *et al.*, 1994). The secretion of neurotoxins by HIV-1-infected macrophages is likely to be regulated by a complex series of intracellular interactions between several different brain cell types including mononuclear phagocytes, astrocytes, and neurons (Lipton, 1992c). HIV-infected brain mononuclear phagocytes, especially after immune activation, secrete substances that are likely to contribute to neurotoxicity (Giulian *et al.*, 1990; Pulliam *et al.*, 1991; Genis *et al.*, 1992; Nottet *et al.*, 1995). These include but are not limited to eicosanoids such as arachidonic acid and its metabolites, platelet-activating factor (PAF), proinflammatory cytokines such as TNF-α and IL-1β, amines, free radicals such as nitric oxide (NO•) and superoxide anion (O_2^{-}•), and the glutamatelike agonist cysteine (Genis *et al.*, 1992; Gelbard *et al.*, 1994; Yeh *et al.*, 1994; Bukrinsky *et al.*, 1995; Giulian *et al.*, 1996). In a similar fashion, macrophages activated by HIV-1 envelope protein gp120 release arachidonic acid and its metabolites, TNF-α, IL-1β, and cysteine, which can lead to NMDA receptor-mediated neurotoxicity (Wahl *et al.*, 1989; Yeh *et al.*, 1994). Some eicosanoids and free radicals can lead to increased release or decreased reuptake of glutamate, which can also contribute to this type of neuronal damage (Lipton and Rosenberg, 1994). Additionally, PAF induces neuronal death in *in vitro* systems by a mechanism probably involving increased neuronal Ca^{2+} and the release of glutamate (Gelbard *et al.*, 1994).

Chronic immune stimulation of the brain, with widespread CNS (microglial and astroglial) activation, can result from IFN-γ production (Tyor *et al.*, 1992). This immune activation continues the process of neuronal injury initiated by HIV infection and its protein product, gp120. IFN-γ induces production of macrophage PAF (Valone and Epstein, 1988) and quinolinate, a tryptophan metabolite found in high concentrations in the cerebrospinal fluid of HIV-infected patients with dementia; quinolinate can also act as a glutamatelike agonist to injure neurons

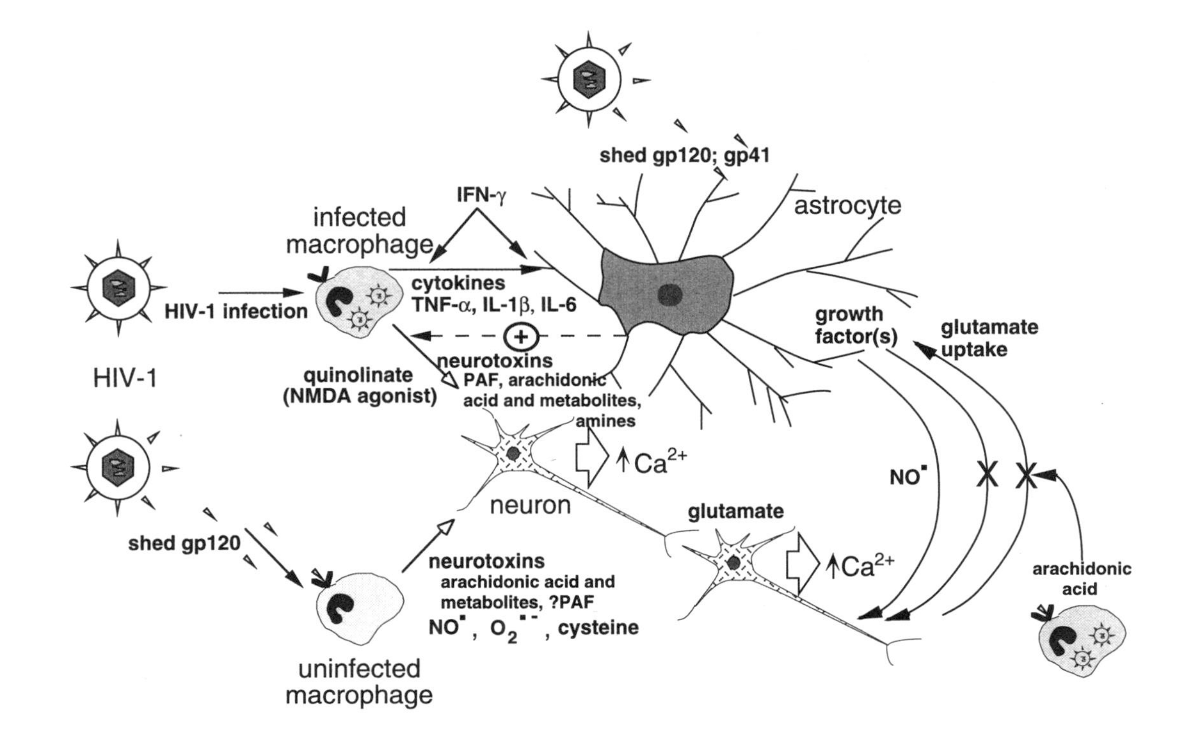

FIGURE 1. Model of immune activated and HIV-infected brain macrophages/microglia that release potentially neurotoxic substances. Substances from these macrophages and also possibly from reactive astrocytes contribute to neuronal injury and astrocytosis. Neuronal injury is primarily mediated by overactivation of NMDA receptors with a resultant increase in intracellular Ca^{2+} levels. This in turn leads to overactivation of a variety of potentially harmful enzyme systems, formation of free radicals ($NO^{\cdot} + O_2^{-\cdot}$ to form $ONOO^-$), and release of the neurotransmitter glutamate. Glutamate then overstimulates NMDA receptors on neighboring neurons, resulting in further injury. See text for additional details of the model.

(Heyes *et al.*, 1992). Cytokines participate in this cellular network in several additional ways. TNF-α may also increase voltage-dependent calcium currents in neurons (Soliven and Albert, 1992). TNF-α and IL-1β stimulate astrocytosis (Selmaj *et al.*, 1990). In conjunction with IL-1β, IFN-γ and TNF-α can induce immune nitric oxide synthase (NOS2, iNOS) expression with consequent NO production in cultured astrocytes (Simmons and Murphy, 1993), including human astrocytes (Lee *et al.*, 1993). Importantly, most of these factors (cytokines, quinolinate, PAF, and products of arachidonic acid metabolism) have been shown to be elevated in brain and/or cerebrospinal fluid of AIDS patients with clinical neurological deficits including dementia (Heyes *et al.*, 1991; Tyor *et al.*, 1992; Gelbard *et al.*, 1994; Griffin *et al.*, 1994).

The final common pathway for neuronal susceptibility appears to be similar to that observed in stroke and several neurodegenerative diseases. This mechanism involves overactivation of voltage-dependent Ca^{2+} channels and NMDA receptor-operated channels, which permit Ca^{2+} influx with resultant generation of free radicals (Lipton and Rosenberg, 1994). One pathway to NO generation in neurons is via constitutive neuronal NOS (NOS1, nNOS) activation by the rise in intraneuronal Ca^{2+} (Dawson *et al.*, 1992). Additionally, the increased levels of neuronal Ca^{2+} engendered by macrophage-synthesized toxins can lead to further release of glutamate. In turn, glutamate overexcites neighboring neurons leading to further increases in intracellular Ca^{2+}, neuronal injury, and more glutamate release. For many neurons, this cyclical pathway to toxicity can be blocked by antagonists of the NMDA receptor (Giulian *et al.*, 1990; Lipton *et al.*, 1990, 1991). For some neurons, this form of damage can also be ameliorated to some degree by calcium channel antagonists or non-NMDA receptor antagonists, perhaps depending on the repertoire of ion channel types in a specific population of neurons (Lipton, 1991). Thus, the elucidation of HIV-1-induced neurotoxins and their mechanism of action(s) offers hope for future pharmacological intervention (Lipton, 1992b; Lipton and Gendelman, 1995).

2.1. iNOS and nNOS in AIDS Dementia

Recently, iNOS has been shown to be activated in severe cases of AIDS dementia, most probably in astrocytes, by the HIV core protein gp41 (Adamson *et al.*, 1996). Increased serum and CSF concentrations of nitrate and nitrite have also been measured in patients with AIDS and CNS complications (Giovannoni *et al.*, 1998). These findings raise the interesting possibility that nitric oxide, likely via formation of peroxynitrite after reaction with superoxide anion (see below), can contribute to neuronal injury in AIDS brains and thus to the development of dementia. Additionally, overstimulation of NMDA receptors (via gp120-induced stimulation of macrophage toxins) can apparently lead to nNOS activation, and thus contribute to neurotoxicity via neuronal production of NO$^{\bullet}$ (Dawson *et al.*, 1993a).

3. Nature of the Neuronal Insult in AIDS Brains: Apoptosis versus Necrosis

Fulminant insults to the nervous system from excitotoxins or free radicals result in neuronal cell death from mitochondrial depolarization, energy failure, and necrosis, while less intense insults allow mitochondrial energy recovery, with toxin release inducing delayed apoptosis (Ankarcrona *et al.*, 1995; Bonfoco *et al.*, 1995). Apoptosis is an active process of cell destruction characterized by cell shrinkage, chromatin aggregation with extensive genomic fragmentation, and nuclear pyknosis (Kerr *et al.*, 1972; Wyllie *et al.*, 1980). *In vivo*, phagocytic cells normally sequester apoptotic cells, preventing inflammation and damage to the surrounding tissue (Duval *et al.*, 1985; Savill *et al.*, 1993). In contrast, necrosis is characterized by passive cell swelling, intense mitochondrial damage with rapid energy loss, and generalized disruption of internal homeostasis. This swiftly leads to membrane lysis, release of intracellular constituents to evoke a local inflammatory reaction, edema, and injury to the surrounding tissue (Schwartz *et al.*, 1993). Whether neuronal apoptosis occurs in AIDS dementia as it does in the penumbra of focal cerebral ischemic lesions, and how one might begin to develop treatment strategies for this problem, are subjects considered in the second half of this review.

3.1. Apoptotic Neuronal Cell Death from Mild Excitotoxic and Free Radical Insults

Increasing evidence suggests that in addition to the necrotic cell death occurring in the area of core ischemia during stroke, some neurons die in a delayed fashion by apoptosis; similarly, evidence for apoptosis has recently appeared for a variety of neurodegenerative disorders, including Alzheimer's disease and AIDS dementia (Dessi *et al.*, 1993; Linnik *et al.*, 1993; Loo *et al.*, 1993; MacManus *et al.*, 1993, 1994; Behl *et al.*, 1994a,b; Filipkowski *et al.*, 1994; Harvey *et al.*, 1994; Mitchell *et al.*, 1994; Pollard *et al.*, 1994a,b; Ratan *et al.*, 1994; Whittemore *et al.*, 1994; Dickson, 1995; Petito and Roberts, 1995).

With this increasing evidence of apoptosis in stroke and AIDS dementia, our laboratory has collaborated with colleagues in the laboratories of Pierluigi Nicotera (Konstanz, Germany) and Sten Orrenius (Stockholm, Sweden) to examine whether excitotoxic and free radical insults (consisting of $NO^{\bullet} + O_2^{-\bullet}$ leading to $ONOO^{-}$ formation) can result in either necrosis or apoptosis depending on the conditions of injury. We have found in cerebrocortical and cerebellar granule cell cultures that fulminant initial insults are associated with necrosis, while less intense injury results in apoptosis. Hence, the intensity of the initial insult determines whether neurons undergo early necrosis (because of loss of mitochondrial membrane potential with severe energy depletion, failure of ionic pumps, and lysis from osmotic swelling) or delayed apoptosis (associated with energy recovery in mitochondria, the generation of a mitochondrial signal to the nucleus to undergo a program of apoptosis, and condensation of the nucleus). Apoptotic neurons that are not phagocytosed in a timely fashion later undergo secondary necrosis, a phenomenon that has been observed previously in other tissues (Ankarcrona *et al.*, 1995; Bonfoco *et al.*, 1995; Dreyer *et al.*, 1995).

To demonstrate that neuronal apoptosis has taken place, no single criterion can be relied on exclusively, so multiple tests must be performed (although all features need not be present in any one cell type). We have monitored the following parameters: the presence of apoptotic nuclei by propidium iodide staining observed with confocal microscopy, the TUNEL (tdt mediated dUTP biotin nick-end-labeling) technique to demonstrate DNA damage, enzyme-linked immunosorbent assay (ELISA) with anti-histone/DNA monoclonal antibodies as evidence of DNA damage, agarose gel electrophoresis to detect chromatin fragmentation, amelioration of neuronal cell death by inhibition of transcription or translation, and ultrastructural criteria for nuclear condensation and apoptotic body formation. Lack of membrane integrity associated with necrosis can be documented by the leakage of the enzyme lactate dehydrogenase, the failure to exclude the dye trypan blue, or other similar techniques (Ankarcrona *et al.*, 1995; Bonfoco *et al.*, 1995, 1997; Dreyer *et al.*, 1995). The subsequent development of necrotic features, whether *in vitro* or *in vivo*, does not necessarily mitigate against initial death by apoptosis. It is well recognized that phosphatidylserine appears on the surface of apoptotic, but not necrotic, cells as a trigger for macrophages or other phagocytes to engulf the cell (Duval *et al.*, 1985); if phagocytic cells are not present, then necrotic features will eventually develop in the dead neurons, producing an ultrastructural phenotype with both apoptotic and necrotic characteristics (e.g., nuclear condensation but also mitochondrial swelling and plasma membrane leakage) (Ankarcrona *et al.*, 1995).

Thus, in a variety of neurological disorders including focal cerebral ischemia and AIDS dementia, mounting evidence suggests that apoptotic features in addition to necrosis can occur in neurons. However, of therapeutic significance, our data indicate that either necrotic or apoptotic events can be ameliorated by NMDA antagonists in the face of excitotoxic or free radical insults. This is of importance because current antiretroviral therapies, including the newer HAART (highly active antiretroviral therapy) regimens with protease inhibitors which do not penetrate well into the CNS, have not yet been shown to ameliorate AIDS dementia for greater than 12 weeks. Additionally, patients are living longer with AIDS due to these new systemic therapies and therefore the prevalence of the dementia, even though the incidence is decreasing, is thought to be on the rise (Sidtis *et al.*, 1993;

Lipton, 1997a,b). Hence, an adjunctive treatment might be beneficial (Lipton and Kieburtz, 1997). The next section highlights work in our laboratory on the development of NMDA antagonists using drugs known to be clinically tolerated by patients being treated for other disorders.

4. Potential Clinical Utility of NMDA Antagonists for AIDS Dementia and Stroke: Open-Channel Blockers and Redox Congeners of Nitric Oxide

As detailed elsewhere (Lipton, 1993b; Lipton and Rosenberg, 1994), many NMDA antagonists are not clinically tolerated, but some appear to be tolerated by humans at effective neuroprotectant concentrations. Several NMDA antagonists have been found to prevent neuronal injury associated with HIV-infected macrophages, gp120, PAF, cysteine, quinolinate, or amine neurotoxins (Giulian *et al.*, 1990, 1993, 1996; Lipton *et al.*, 1990, 1991; Lipton, 1992a–c, 1993a; Müller *et al.*, 1992; Dawson *et al.*, 1993a; Savio and Levi, 1993). Among these, two of the most promising (because of their extensive usage in patients with other diseases) are memantine and nitroglycerin (see Fig. 2 for mechanism and site of action). Memantine blocks the NMDA receptor-associated ion channel only when it is open. Unlike other NMDA open-channel blockers, such as dizocilpine (MK-801), memantine does not remain in the channel for an excessively long time interval, and this kinetic characteristic correlates with its safe use in humans for over a dozen years in Europe as a treatment for Parkinson's disease and spasticity (Chen *et al.*, 1992; Chen and Lipton, 1997). Increasing concentrations of glutamate or other NMDA agonists cause NMDA channels to remain open for a greater fraction of time on average. Under such conditions, an open-channel blocking drug such as memantine has a better chance to enter and block a channel. Because of this mechanism of action, the untoward effects of greater (pathological) glutamate concentrations are prevented to a greater extent than the effects of lower (physiological) concentrations (Lipton, 1993b; Lipton and Rosenberg, 1994; Chen and Lipton, 1997). Moreover, memantine can ameliorate neuronal injury associated with either focal cerebral ischemia or gp120 in model systems, both *in vitro* and *in vivo* (Seif el Nasr *et al.*, 1990; Erdö and Schäfer, 1991; Chen *et al.*, 1992; Keilhoff and Wolf, 1992; Lipton, 1992b; Lipton and Jensen, 1992; Müller *et al.*, 1992; Osborne and Quack, 1992; Pellegrini and Lipton, 1993; Stieg *et al.*, 1993; Toggas *et al.*, 1996).

As discussed earlier, $NO^{\bullet}$ can contribute to neuronal damage, and one of these pathways to neurotoxicity involves the reaction of $NO^{\bullet}$ with $O_2^{-\bullet}$ to form $ONOO^-$ (Beckman *et al.*, 1990; Dawson *et al.*, 1991, 1993a; Lipton *et al.*, 1993) (see also Chapter 8). Alternatively, $NO^{\bullet}$ can be converted to a chemical state that has precisely the opposite effect, i.e., one that protects neurons from injury caused by

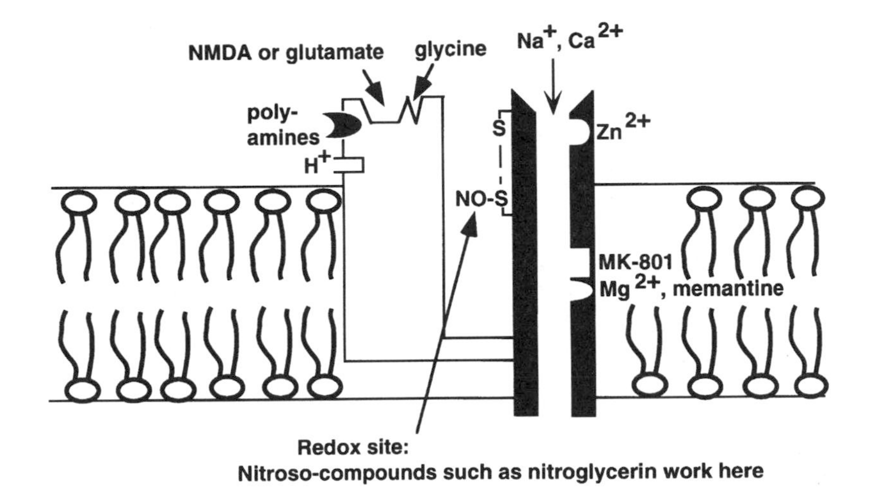

FIGURE 2. Sites of potential antagonist action on the NMDA receptor–channel complex. Competitive antagonists such as CGS-19755 can compete with NMDA or glutamate for binding to the agonist site. Several antagonists of the glycine coagonist site have been described, for example, chlorinated and sulfated derivatives of kynurenic acid. It is not yet definitively known if any will prove to be tolerated clinically at concentrations effective in combating neurotoxicity. H^+ effects are transmitted through a noncompetitive site; decreasing pH acts to downregulate channel activity. Other modulatory sites for polyamines and Zn^{2+} can also be used to affect receptor–channel function. Sites that inhibit channel activity by binding Mg^{2+} or drugs such as MK-801, phencyclidine, and memantine are within the electric field of the channel and are only exposed when the channel is previously opened by agonist (termed *uncompetitive antagonism*). A redox modulatory site(s), comprised of two or more thiol groups (based on our site-directed mutagenesis work on recombinant subunits), is affected by chemical reducing and oxidizing agents. Oxidation may possibly favor disulfide bond formation (S–S) over free thiol (–SH) groups and thus downregulate channel activity. At least one of the NMDA receptor's redox modulatory sites can also be downregulated by NO group transfer (*S*-nitrosylation with NO^+ equivalents to form RS–NO), which may facilitate disulfide bond formation (represented by the dashed line). This reaction, which leads to less NMDA-evoked Ca^{2+} influx and thus neuroprotection, involves NMDA receptor thiol groups because it is blocked by specific sulfhydryl alkylating agents such as *N*-ethylmaleimide.

NMDA receptor-mediated overstimulation. The change in chemical state is dependent on the removal or addition of an electron to $NO^{\bullet}$. This change in the chemical redox state can be influenced by the presence or absence of electron donors such as ascorbate and cysteine. With one less electron, $NO^{\bullet}$ becomes nitrosonium ion (NO^+), facilitating reaction with critical thiol group(s) [RSH or thiolate anion (RS^-)] comprising a redox modulatory site on the NMDA receptor–channel complex that decreases channel activity. This reaction can afford neuronal protection from overstimulation of NMDA receptors, which would otherwise result in an excessive Ca^{2+} influx (Lipton *et al.*, 1993; Lipton and Stamler, 1994). Free nitrosonium may not be present for such reactions under most physiological conditions, but both endogenous and exogenous NO donors may be capable of transferring nitrosonium equivalents. One such drug that can react with critical regulatory cysteine sulfhydryls on the NMDA receptor in a manner resembling nitrosonium is the common vasodilator nitroglycerin (Lei *et al.*, 1992; Lipton, 1993a; Lipton *et al.*, 1993; Lipton and Rosenberg, 1994). Chronic use of nitroglycerin induces tolerance to its effects on the cardiovascular system, but the drug appears to exert sustained effects in the brain to attenuate NMDA receptor-mediated neurotoxicity (Lipton, 1993a). The optimal dosing regimen has yet to be worked out for the neuroprotective effects of nitroglycerin in the brain; therefore, caution should be exercised before attempting to implement this form of therapy. In preliminary experiments, high concentrations of nitroglycerin have been found to be neuroprotective during various NMDA receptor-mediated insults, including focal ischemia (Sathi *et al.*, 1993; Lipton and Wang, 1996). Our *in vivo* data suggest that this effect of nitroglycerin may, at least in part, be a direct effect on neurons, consistent with an action at the NMDA receptor redox modulatory site(s). Additional targets of nitrosylation that ameliorate apoptosis have recently become apparent. These include intracellular signaling cascades downstream from the NMDA receptor, including caspases (cysteine proteases) and $p21^{ras}$ (Tenneti *et al.*, 1997; Yun *et al.*, 1998).

As the structural basis for redox modulation is further elucidated, it may become possible to design improved redox reactive reagents of clinical value, specifically targeting the NO group in appropriate redox state to the NMDA receptor. This strategy would avoid hypotensive and other side effects of systemically administered NO-donor drugs. One strategy to accomplish this goal might involve *S*-nitrosomemantine, exploiting NMDA channel blockade by memantine to target the NO group to the NMDA receptor. However, optimal design of a memantine-based NO donor will require more detailed knowledge of the ion pore and redox site(s) of the NMDA receptor–channel complex.

In recent years, the channel pore has been localized to the second membrane loop of the NMDA receptor subunit. Our laboratory, in collaboration with Stephen Heinemann's laboratory at The Salk Institute (La Jolla, California), and, independently, Peter Seeburg's laboratory (Heidelberg, Germany), have begun to characterize the redox modulatory sites of NMDA receptors at a molecular level using chimera formation and site-directed mutagenesis of recombinant subunits (NMDAR1, NMDAR2A–D) (Kohr *et al.*, 1994; Sullivan *et al.*, 1994). Two types of redox modulation can be distinguished. The first type gives rise to a persistent change in the functional activity of the receptor, and we have identified two cysteine residues on the NMDAR1 subunit (C744 and C798) that are responsible for this action (Sullivan *et al.*, 1994). The second type involves additional site(s), which also appear to be cysteine(s) based on the ability of ≤ 1 mM *N*-ethylmaleimide to block the effect in native neurons. The effect of NO interactions with the NMDA receptor at these sites is currently under intensive investigation (Omerovic *et al.*,

1995; Choi *et al.*, 1996; Lipton *et al.*, 1996; Sucher *et al.*, 1996; Stamler *et al.*, 1997). These studies should assist attempts to design NO donors that specifically target the NMDA receptor.

In the coming months, as these and other clinically tolerated NMDA antagonists are tested in clinical studies in an attempt to ameliorate AIDS dementia, we hope to be able to offer patients better adjunctive therapy to treat the cognitive and other neurological manifestations of AIDS. Interestingly, because focal cerebral ischemia (stroke) appears to be mediated to a large degree by overstimulation of NMDA receptors, these same drugs and several others are undergoing evaluation in clinical trials for stroke. Unlike many, if not all, of the other drugs currently under investigation, the agents discussed here are known to be clinically tolerated and have a high degree of safety at neuroprotective doses. These characteristics should expedite clinical studies for the use of these drugs in humans with AIDS dementia and stroke, as well as for other neurological disorders mediated at least in part by excessive NMDA receptor activity (Lipton and Rosenberg, 1994).

5. Conclusions

This article reviews an excitotoxic/free radical hypothesis accounting for the pathogenesis of focal cerebral ischemia (stroke) and AIDS dementia (HIV-1-associated cognitive/motor complex). The development of necrotic or apoptotic neuronal damage is dependent on the intensity of the excitotoxic/free radical insult. The free radical insults are generated, at least in part, by the $NO^{\bullet} + O_2^{-\bullet}$ reaction to form $ONOO^-$. Severe depletion of mitochondrial membrane potential and energy production is associated with necrosis, while recovery of mitochondrial energy levels allows an apoptotic program to proceed. Each of these forms of neuronal injury can be ameliorated in our models by NMDA antagonists. Two clinically tolerated NMDA antagonists, memantine and nitroglycerin, are mentioned here: (1) Memantine is an open-channel blocker of the NMDA-associated ion channel and a close congener of the antiviral and antiparkinsonian drug amantadine. Memantine blocks the effects of pathologically escalating levels of excitotoxins to a greater degree than lower physiological levels of these excitatory amino acids, thus sparing normal neuronal function to some extent. (2) Nitroglycerin acts at a redox modulatory site(s) of the NMDA receptor–channel complex to prevent excessive receptor activity and consequent neuronal damage caused by excessive Ca^{2+} influx and free radical formation. The neuroprotective action of nitroglycerin at the redox modulatory site(s) is mediated by a chemical species related to nitric oxide, but in a higher oxidation state, resulting in transfer of an NO^+ group to a critical cysteine sulfhydryl on the NMDA receptor (represented by the nitrosylation reaction: $NO^+ + RS^- \rightarrow RS{-}NO$). Because of the clinical safety of these drugs, they have the potential for expeditious trials in humans. Interestingly, one redox state of NO appears to contribute to neurotoxicity via $ONOO^-$ formation from $NO^{\bullet}$ and $O_2^{-\bullet}$ in a variety of neurodegenerative disorders including AIDS dementia, whereas another redox state of NO can decrease neuronal damage via nitrosylation of regulatory cysteine sulfhydryl groups on critical proteins.

ACKNOWLEDGMENTS. This work on the effects of NO on HIV infection in the brain was adapted from a longer work to appear in the *Annual Review of Pharmacology and Toxicology* by the same author. The research described herein was supported in part by National Institutes of Health grants P01 HD29587 and R01 EY09024, and by the American Foundation for AIDS Research and the Pediatric AIDS Foundation. S.A.L. is a consultant to and received sponsored research support from Neurobiological Technologies, Inc. (Richmond, Calif.) and Allergan, Inc. (Irvine, Calif.) for the clinical development of NMDA receptor antagonists.

References

Adamson, D. C., Wildemann, B., Sasaki, M., Glass, J. D., McArthur, J. C., Christov, V. I., Dawson, T. M., and Dawson, V. L., 1996, Immunologic NO synthase: Elevation in severe AIDS dementia and induction by HIV-1 gp41, *Science* **274:**1917–1921.

Ankarcrona, M., Dypbukt, J. M., Bonfoco, E., Zhivotovsky, B., Orrenius, S., Lipton, S. A., and Nicotera, P., 1995, Glutamate-induced neuronal death: A succession of necrosis or apoptosis depending on mitochondrial function, *Neuron* **15:**961–973.

Bacellar, H., Muñoz, A., Miller, E. N., Cohen, B. A., Besley, D., Selnes, O. A., Becker, J. T., and McArthur, J. C., 1994, Temporal trends in the incidence of HIV-1-related neurologic diseases: Multicenter AIDS cohort study, 1985–1992, *Neurology* **44:**1892–1900.

Beckman, J. S., Beckman, T. W., Chen, J., Marshall, P. A., and Freeman, B. A., 1990, Apparent hydroxyl radical production by peroxynitrite: Implications for endothelial injury from nitric oxide and superoxide, *Proc. Natl. Acad. Sci. USA* **87:**1620–1624.

Behl, C., Davis, J. B., Klier, F. G., and Schubert, D., 1994a, Amyloid beta peptide induces necrosis rather than apoptosis, *Brain Res.* **665:**253–264.

Behl, C., Davis, J. B., Lesley, R., and Schubert, D., 1994b, Hydrogen peroxide mediates amyloid β protein toxicity, *Cell* **77:**817–827.

Benveniste, H., Drejer, J., Schousboe, A., and Diemer, N. H., 1984, Elevation of the extracellular concentrations of glutamate and aspartate in rat hippocampus during transient cerebral ischemia monitored by intracerebral microdialysis, *J. Neurochem.* **43:**1369–1374.

Bonfoco, E., Krainc, D., Ankarcrona, M., Nicotera, P., and Lipton, S. A., 1995, Apoptosis and necrosis: Two distinct events induced respectively by mild and intense insults with NMDA or nitric oxide/superoxide in cortical cell cultures, *Proc. Natl. Acad. Sci. USA* **92:**7162–7166.

Bonfoco, E., Ankarcrona, M., Krainc, D., Nicotera, P., and Lipton, S. A., 1997, Techniques for distinguishing apoptosis from necrosis in cerebrocortical and cerebellar neurons, in: *Neuromethods: Apoptosis Techniques and Protocols* (J. Poirier, ed.), Humana Press, Totowa, N.J., pp. 237–253.

Brenneman, D. E., Westbrook, G. L., Fitzgerald, S. P., Ennist, D. L., Elkins, K. L., Ruff, M., and Pert, C. B., 1988, Neuronal cell killing by the envelope protein of HIV and its prevention by vasoactive intestinal peptide, *Nature* **335:**639–642.

Budka, H., 1991, Neuropathology of human immunodeficiency virus infection, *Brain Pathol.* **1:**163–175.

Bukrinsky, M. I., Nottet, H. S. L. M., Schmidtmayerova, H., Dubrovsky, L., Flanagan, C. R., Mullins, M. E., Lipton, S. A., and Gendelman, H. E., 1995, Regulation of nitric oxide synthase activity in human immunodeficiency virus type 1 (HIV-1)-infected monocytes: Implications for HIV-associated neurological disease, *J. Exp. Med.* **181:**735–745.

Bustos, G., Abarca, J., Forray, M. I., Gysling, K., Bradberry, C. W., and Roth, R. H., 1992, Regulation of excitatory amino acid release by *N*-methyl-D-aspartate receptors in rat striatum: *In vivo* microdialysis studies, *Brain Res.* **585:**105–115.

Chen, H.-S. V., and Lipton, S. A., 1997, Mechanism of memantine block of NMDA-activated channels in rat retinal ganglion cells: Uncompetitive antagonism, *J. Physiol. (London)* **499:**27–46.

Chen, H.-S. V., Pellegrini, J. W., Aggarwal, S. K., Lei, S. Z., Warach, S., Jensen, F. E., and Lipton, S. A., 1992, Open-channel block of NMDA responses by memantine: Therapeutic advantage against NMDA receptor-mediated neurotoxicity, *J. Neurosci.* **12:**4427–4436.

Choi, D. W., 1988, Glutamate neurotoxicity and diseases of the nervous system, *Neuron* **1:**623–634.

Choi, Y.-B., Chen, H.-S. V., Sucher, N. J., and Lipton, S. A., 1996, Redox modulatory site(s) of recombinant NMDA receptors: Subunit composition and conformation, *Soc. Neurosci. Abstr.* **22:**1761.

Dawson, T. M., Dawson, V. L., and Snyder, S. H., 1992, A novel neuronal messenger molecule in brain: The free radical, nitric oxide, *Ann. Neurol.* **32:**297–311.

Dawson, V. L., Dawson, T. M., London, E. D., Bredt, D. S., and Snyder, S. H., 1991, Nitric oxide mediates glutamate neurotoxicity in primary cortical cultures, *Proc. Natl. Acad. Sci. USA* **88:**6368–6371.

Dawson, V. L., Dawson, T. M., Bartley, D. A., Uhl, G. R., and Snyder, S. H., 1993a, Mechanisms of nitric oxide-mediated neurotoxicity in primary brain cultures, *J. Neurosci.* **13:**2651–2661.

Dawson, V. L., Dawson, T. M., Uhl, G. R., and Snyder, S. H., 1993b, Human immunodeficiency virus-1 coat protein neurotoxicity mediated by nitric oxide in primary cortical cultures, *Proc. Natl. Acad. Sci. USA* **90:**3256–3259.

Dessi, F., Charriaut-Marlangue, C., Khrestchatisky, M., and Ben-Ari, Y., 1993, Glutamate induced neuronal death is not a programmed cell death in cerebellar cultures, *J. Neurochem.* **60:**1953–1955.

Dickson, D. W., 1995, Apoptosis in the brain. Physiology and pathology, *Am. J. Pathol.* **146:**1040–1044.

Diop, A. G., Lesort, M., Esclaire, F., Sindou, P., Couratier, P., and Hugon, J., 1994, Tetrodotoxin blocks HIV coat protein (gp120) toxicity in primary neuronal cultures, *Neurosci. Lett.* **165:**187–190.

Dreyer, E. B., and Lipton, S. A., 1994, Toxic neuronal effects of the HIV coat protein gp120 may be mediated through macrophage arachidonic acid, *Soc. Neurosci. Abstr.* **20:**1049.

Dreyer, E. B., Kaiser, P. K., Offermann, J. T., and Lipton, S. A., 1990, HIV-1 coat protein neurotoxicity prevented by calcium channel antagonists, *Science* **248:**364–367.

Dreyer, E. B., Zhang, D., and Lipton, S. A., 1995, Transcriptional or translational inhibition blocks low dose NMDA-mediated cell death, *NeuroReport* **6:**942–944.

Duval, E., Wyllie, A. H., and Morris, R. G., 1985, Macrophage recognition of cells undergoing programmed cell death (apoptosis), *Immunology* **56:**351–358.

Erdö, S. L., and Schäfer, M., 1991, Memantine is highly potent in protecting cortical cultures against excitotoxic cell death evoked by glutamate and *N*-methyl-D-aspartate, *Eur. J. Pharmacol.* **198:**215–217.

Everall, I. P., Luthbert, P. J., and Lantos, P. L., 1991, Neuronal loss in the frontal cortex in HIV infection, *Lancet* **337:**1119–1121.

Filipkowski, R. K., Hetman, M., Kaminska, B., and Kaczmarek, L., 1994, DNA fragmentation in rat brain after intraperitoneal administration of kainate, *NeuroReport* **5:**1538–1540.

Gelbard, H. A., Nottet, H. S. L. M., Swindells, S., Jett, M., Dzenko, K. A., Genis, P., White, R., Wang, L., Choi, Y.-B., Zhang, D., Lipton, S. A., Tourtellotte, W. W., Epstein, L. G., and Gendelman, H. E., 1994, Platelet-activating factor: A candidate human immunodeficiency virus type 1-induced neurotoxin, *J. Virol.* **68:**4628–4635.

Genis, P., Jett, M., Bernton, E. W., Boyle, T., Gelbard, H. A., Dzenko, K., Keane, R. W., Resnick, L., Mizrachi, T., Volsky, D. J., Epstein, L. G., and Gendelman, H. E., 1992, Cytokines and arachidonic acid metabolites produced during human immunodeficiency virus (HIV)-infected macrophage–astroglia interactions: Implications for the neuropathogenesis of HIV disease, *J. Exp. Med.* **176:**1703–1718.

Giovannoni, G., Miller, R. F., Heales, S. J. R., Land, J. M., Harrison, M. J. G., and Thompson, E. J., 1998, Elevated cerebrospinal fluid and serum nitrate and nitrite levels in patients with central nervous system complications of HIV-1 infection—A correlation with blood-brain-barrier dysfunction, *J. Neurol. Sci.* **156:**53–58.

Giulian, D., Vaca, K., and Noonan, C. A., 1990, Secretion of neurotoxins by mononuclear phagocytes infected with HIV-1, *Science* **250:**1593–1596.

Giulian, D., Wendt, E., Vaca, K., and Noonan, C. A., 1993, The envelope glycoprotein of human immunodeficiency virus type 1 stimulates release of neurotoxins from monocytes, *Proc. Natl. Acad. Sci. USA* **90:**2769–2773.

Giulian, D., Yu, J., Li, X., Tome, D., Li, J., Wendt, E., Lin, S.-N., Schwarcz, R., and Noonan, C., 1996, Study of receptor-mediated neurotoxins released by HIV-1-infected mononuclear phagocytes found in human brain, *J. Neurosci.* **16:**3139–3153.

Globus, M. Y., Busto, R., Dietrich, W. D., Martinez, E., Valdes, I., and Ginsberg, M. D., 1988, Effect of ischemia on the in vivo release of striatal dopamine, glutamate, and gamma-aminobutyric acid studied by intracerebral microdialysis, *J. Neurochem.* **51:**1455–1464.

Griffin, D. E., Wesselingh, S. L., and McArthur, J. C., 1994, Elevated central nervous system prostaglandins in human immunodeficiency virus-associated dementia, *Ann. Neurol.* **35:**592–597.

Harvey, A. R., Cui, Q., and Robertson, D., 1994, The effect of cycloheximide and ganglioside GM1 on the viability of retinotectally projecting ganglion cells following ablation of the superior colliculus in rats, *Eur. J. Neurosci.* **5:**550–557.

Hayman, M., Arbuthnott, G., Harkiss, G., Brace, H., Filippi, P., Philippon, V., Thompson, D., Vigne, R., and Wright, A., 1993, Neurotoxicity of peptide analogies of the transactivating protein tat from Maedi-Visna virus and human immunodeficiency virus, *Neuroscience* **53:**1–6.

Heyes, M. P., Brew, B. J., Martin, A., Price, R. W., Salazqr, A. M., Sidtis, J. J., Yergey, J. A., Mouradian, M. M., Sadler, A., Keilp, J., Rubinow, D., and Markey, S. P., 1991, Quinolinic acid in cerebrospinal fluid and serum in HIV-1 infection: Relationship to clinical and neurological status, *Ann. Neurol.* **29:**202–209.

Heyes, M. P., Saito, K., and Markey, S. P., 1992, Human macrophages convert L-tryptophan to the neurotoxin quinolinic acid, *Biochem. J.* **283:**633–635.

Keilhoff, G., and Wolf, G., 1992, Memantine prevents quinolinic acid-induced hippocampal damage, *Eur. J. Pharmacol.* **219:**451–454.

Kerr, J. F. R., Wyllie, A. H., and Currie, A. R., 1972, Apoptosis: A basic biological phenomenon with wide ranging implications in tissue kinetics, *Br. J. Cancer* **26:**239–257.

Ketzler, S., Weis, S., Haug, H., and Budka, H., 1990, Loss of neurons in frontal cortex in AIDS brains, *Acta Neuropathol.* **80:**90–92.

Koenig, S., Gendelman, H. E., Orenstein, J. M., Dal Canto, M. C., Pozeshkpour, G. H., Yungbluth, M., Janotta, F., Kasmit, A., Martin, M. A., and Fauci, A. S., 1986, Detection of AIDS virus in macrophages in brain tissue from AIDS patients with encephalopathy, *Science* **233:**1089–1093.

Kohr, G., Eckardt, S., Lüddens, H., Monyer, H., and Seeburg, P. H., 1994, NMDA receptor channels: Subunit-specific potentiation by reducing agents, *Neuron* **12:**1031–1040.

Lannuzel, A., Lledo, P.-M., Lamghitnia, H. O., Vincent, J.-D., and Tardieu, M., 1995, HIV-1 envelope proteins gp120 and gp160 potentiate NMDA-induced $[Ca^{2+}]_i$ increase, alter $[Ca^{2+}]_i$ homeostasis and induce neurotoxicity in human embryonic neurons, *Eur. J. Neurosci.* **7:**2285–2293.

Lee, S. C., Dickson, D. W., Liu, W., and Brosnan, C. F., 1993, Induction of nitric oxide synthase activity in human astrocytes by IL-1β and IFN-γ, *J. Neuroimmunol.* **46:**19–24.

Lei, S. Z., Pan, Z.-H., Aggarwal, S. K., Chen, H.-S. V., Hartman, J., Sucher, N. J., and Lipton, S. A., 1992, Effect of nitric oxide production on the redox modulatory site of the NMDA receptor–channel complex, *Neuron* **8:**1087–1099.

Linnik, M. D., Zobrist, R. H., and Hatfield, M. D., 1993, Evidence supporting a role for programmed cell death in focal cerebral ischemia in rats, *Stroke* **24:**2002–2008.

Lipton, S. A., 1991, Calcium channel antagonists in the prevention of neurotoxicity, *Adv. Pharmacol.* **22:**271–291.

Lipton, S. A., 1992a, 7-Chlorokynurenate ameliorates neuronal injury mediated by HIV envelope protein gp120 in retinal cultures, *Eur. J. Neurosci.* **4:**1411–1415.

Lipton, S. A., 1992b, Memantine prevents HIV coat protein-induced neuronal injury *in vitro*, *Neurology* **42:**1403–1405.

Lipton, S. A., 1992c, Models of neuronal injury in AIDS: Another role for the NMDA receptor? *Trends Neurosci.* **15:**75–79.

Lipton, S. A., 1993a, Human immunodeficiency virus-infected macrophages, gp120, and *N*-methyl-D-aspartate receptor-mediated neurotoxicity, *Ann. Neurol.* **33:**227–228.

Lipton, S. A., 1993b, Prospects for clinically tolerated NMDA antagonists: Open-channel blockers and alternative redox states of nitric oxide, *Trends Neurosci.* **16:**527–532.

Lipton, S. A., 1994, Neurobiology: HIV displays its coat of arms, *Nature* **367:**113–114.

Lipton, S. A., 1997a, Neuropathogenesis of acquired immunodeficiency syndrome dementia, *Curr. Opin. Neurol.* **10:**247–253.

Lipton, S. A., 1997b, Treating AIDS dementia, *Science* **276:**1629–1630.

Lipton, S. A., and Gendelman, H. E., 1995, The dementia associated with the acquired immunodeficiency syndrome, *N. Engl. J. Med.* **332:**934–940.

Lipton, S. A., and Jensen, F. E., 1992, Memantine, a clinically-tolerated NMDA open-channel blocker, prevents HIV coat protein-induced neuronal injury *in vitro* and *in vivo*, *Soc. Neurosci. Abstr.* **18:**757.

Lipton, S. A., and Kieburtz, K., 1998, Development of adjunctive therapies for the neurologic manifestations of AIDS: Dementia and painful neuropathy, in: *The Neurology of AIDS* (H. E. Gendelman, S. A. Lipton, L. G. Epstein, and S. Swindells, eds.), Chapman & Hall, New York, pp. 377–381.

Lipton, S. A., and Rosenberg, P. A., 1994, Excitatory amino acids as a final common pathway for neurologic disorders, *N. Engl. J. Med.* **330:**613–622.

Lipton, S. A., and Stamler, J. S., 1994, Actions of redox-related congeners of nitric oxide at the NMDA receptor, *Neuropharmacology* **33:**1229–1233.

Lipton, S. A., and Wang, Y. F., 1996, NO-related species can protect from focal cerebral ischemia/reperfusion, in: *Pharmacology of Cerebral Ischemia* (J. Krieglstein, ed.), Medpharm Scientific Publishers, Stuttgart, pp. 183–191.

Lipton, S. A., Kaiser, P. K., Sucher, N. J., Dreyer, E. B., and Offermann, J. T., 1990, AIDS virus coat protein sensitizes neurons to NMDA receptor-mediated toxicity, *Soc. Neurosci. Abstr.* **16:** 289.

Lipton, S. A., Sucher, N. J., Kaiser, P. K., and Dreyer, E. B., 1991, Synergistic effects of HIV coat protein and NMDA receptor-mediated neurotoxicity, *Neuron* **7:**111–118.

Lipton, S. A., Choi, Y.-B., Pan, Z.-H., Lei, S. Z., Chen, H.-S. V., Sucher, N. J., Loscalzo, J., Singel, D. J., and Stamler, J. S., 1993, A redox-based mechanism for the neuroprotective and neurodestructive effects of nitric oxide and related nitroso-compounds, *Nature* **364:**626–632.

Lipton, S. A., Choi, Y.-B., Sucher, N. J., Pan, Z.-H., and Stamler, J. S., 1996, Redox state, NMDA receptors, and NO-related species, *Trends Pharmacol. Sci.* **17:**186–187.

Lo, T.-M., Fallert, C. J., Piser, T. M., and Thayer, S. A., 1992, HIV-1 envelope protein evokes intracellular calcium oscillations in rat hippocampal neurons, *Brain Res.* **594:**189–196.

Lombardi, G., and Moroni, F., 1992, GM_1 ganglioside reduces ischemia-induced excitatory amino acid output: A microdialysis study in the gerbil hippocampus, *Neurosci. Lett.* **134:**171–174.

Loo, D. T., Agata, C., Pike, C. J., Whittemore, E. R., Wulencewicz, A. J., and Cotman, C. W., 1993, Apoptosis is induced by beta-amyloid in cultured central nervous system neurons, *Proc. Natl. Acad. Sci. USA* **90:**7951–7955.

Louzada, J. P., Dias, J. J., Santos, W. F., Lachat, J. J., Bradford, H. F., and Coutinho, N. J., 1992, Glutamate release in experimental ischaemia of the retina: An approach using microdialysis, *J. Neurochem.* **59:**358–363.

MacManus, J. P., Buchan, A. M., Hill, I. E., Rasquinha, I., and Preston, E., 1993, Global ischemia can cause DNA fragmentation indicative of apoptosis in rat brain, *Neurosci. Lett.* **159:**89–92.

MacManus, J. P., Hill, I. E., Huang, Z.-G., Rasquinha, I., Xue, D., and Buchan, A. M., 1994, DNA damage consistent with apoptosis in transient focal ischemic neocortex, *NeuroReport* **5:**493–496.

Masliah, E., Achim, C. L., Ge, N., DeTeresa, R., Terry, R. D., and Wiley, C. A., 1992, Spectrum of human immunodeficiency virus-associated neocortical damage, *Ann. Neurol.* **32:**321–329.

Meldrum, B., and Garthwaite, J., 1990, Excitatory amino acid neurotoxicity and neurodegenerative disease, *Trends Pharmacol. Sci.* **11:**379–387.

Meucci, O., Fatatis, A., Simen, A. A., Bushell, T. J., Gray, P. W., and Miller, R. J., 1998, Chemokines regulate hippocampal neuronal signaling and gp120 neurotoxicity, *Proc. Natl. Acad. Sci. USA* **95:**14500–14505.

Mitchell, J., Lawson, S., Moser, B., Laidlaw, S. M., Cooper, A. J., Walkinshaw, G., and Waters, C. M., 1994, Glutamate-induced apoptosis results in a loss of striatal neurons in the parkinsonian rat, *Neuroscience* **63:**1–5.

Müller, W. E. G., Schröder, H. C., Ushijima, H., Dapper, J., and Bormann, J., 1992, gp120 of HIV-1 induces apoptosis in rat cortical cell cultures: Prevention by memantine, *Eur. J. Pharmacol. Mol. Pharm. Sect.* **226:**209–214.

Nottet, H. S. L. M., Jett, M., Flanagan, C. R., Zhai, Q.-H., Persidsky, Y., Rizzino, A., Genis, P., Baldwin, T., Schwartz, J., and Gendelman, H. E., 1995, A regulatory role for astrocytes in HIV-encephalitis: An overexpression of eicosanoids, platelet-activating factor and tumor necrosis factor-α by activated HIV-1-infected monocytes is attenuated by primary human astrocytes, *J. Immunol.* **154:**3567–3581.

Omerovic, A., Chen, S.-J., Leonard, J. P., and Kelso, S. R., 1995, Subunit-specific redox modulation of NMDA receptors expressed in Xenopus oocytes, *J. Recep. Sign. Transduc. Res.* **15:**811–827.

Osborne, N. N., and Quack, G., 1992, Memantine stimulates inositol phosphates production in neurones and nullifies *N*-methyl-D-aspartate-induced destruction of retinal neurones, *Neurochem. Int.* **21:**329–336.

Pellegrini, J. W., and Lipton, S. A., 1993, Delayed administration of memantine prevents *N*-methyl-D-aspartate receptor-mediated neurotoxicity, *Ann. Neurol.* **33:**403–407.

Perrson, L., and Hillered, L., 1992, Chemical monitoring of neurosurgical intensive care patients using intracerebral microdialysis, *J. Neurosurg.* **76:**72–80.

Petito, C. K., and Roberts, B., 1995, Evidence of apoptotic cell death in HIV encephalitis, *Am. J. Pathol.* **146:**1121–1130.

Pollard, H., Cantagrel, S., Charriaut-Marlangue, C., Moreau, J., and Ben-Ari, Y., 1994a, Apoptosis associated DNA fragmentation in epileptic brain damage, *NeuroReport* **5:**1053–1055.

Pollard, H., Charriaut-Marlangue, C., Cantagrel, S., Represa, A., Robain, O., Moreau, J., and Ben-Ari, Y., 1994b, Kainate-induced apoptotic cell death in hippocampal neurons, *Neuroscience* **63:**7–18.

Price, R. W., Brew, B., Sidtis, J., Rosenblum, M., Scheck, A. C., and Clearly, P., 1988, The brain and

AIDS: Central nervous system HIV-1 infection and AIDS dementia complex, *Science* **239:**586–592.

Pulliam, L., Herndler, B. G., Tang, N. M., and McGrath, M. S., 1991, Human immunodeficiency virus-infected macrophages produce soluble factors that cause histological and neurochemical alterations in cultured human brains, *J. Clin. Invest.* **87:**503–512.

Ratan, R. R., Murphy, T. H., and Baraban, J. M., 1994, Macromolecular synthesis inhibitors prevent oxidative stress-induced apoptosis in embryonic cortical neurons by shunting cysteine from protein synthesis to glutathione, *J. Neurosci.* **14:**4385–4392.

Saito, T., Sarer, L. R., Epstein, L. G., Michales, J., Mintz, M., Louder, M., Goldring, K., Cvetkovich, T. A., and Blumberg, B. M., 1994, Overexpression of *nef* as a marker for restricted HIV-1 infection of astrocytes in postmortem pediatric central nervous tissues, *Neurology* **44:**474–481.

Sathi, S., Edgecomb, P., Warach, S., Manchester, K., Donaghey, T., Stieg, P. E., Jensen, F. E., and Lipton, S. A., 1993, Chronic transdermal nitroglycerin (NTG) is neuroprotective in experimental rodent stroke models, *Soc. Neurosci. Abstr.* **19:**849.

Savill, J. S., Fadok, V., Henson, P., and Haslett, C., 1993, Phagocyte recognition of cells undergoing apoptosis, *Immunol. Today* **14:**131–136.

Savio, T., and Levi, G., 1993, Neurotoxicity of HIV coat protein gp120, NMDA receptors, and protein kinase C: A study with rat cerebellar granule cell cultures, *J. Neurosci. Res.* **34:**265–272.

Schwartz, L. M., Smith, S. W., Jones, M. E. E., and Osborne, B. A., 1993, Do all programmed cell deaths occur via apoptosis? *Proc. Natl. Acad. Sci. USA* **90:**980–984.

Seif el Nasr, M., Perucher, B., Rossberg, C., Mennel, H.-D., and Krieglstein, J., 1990, Neuroprotective effect of memantine demonstrated *in vivo* and *in vitro*, *Eur. J. Pharmacol.* **185:**19–24.

Selmaj, K. N., Farooq, M., Norton, T., Raine, C. S., and Brosman, C. F., 1990, Proliferation of astrocytes *in vitro* in response to cytokines, *J. Immunol.* **144:**129–135.

Sidtis, J. J., Gatsonis, C., Price, R. W., Singer, E. J., Collier, A. C., Richman, D. D., Hirsch, M. D., Schaerf, F. W., Fischl, M. A., Kieburtz, K., Simpson, D., Koch, M. A., Feinberg, J., and Dafni, U., 1993, Zidovudine treatment of the AIDS dementia complex: Results of a placebo-controlled trial, *Ann. Neurol.* **33:**343–349.

Silverstein, F. S., Naik, B., and Simpson, J., 1991, Hypoxia–ischemia stimulates hippocampal glutamate efflux in perinatal rat brain: An *in vivo* microdialysis study, *Pediatr. Res.* **30:**587–590.

Simmons, M. L., and Murphy, S., 1993, Cytokines regulateL-arginine-dependent cyclic GMP production in rat glial cells, *Eur. J. Neurosci.* **5:**825–831.

Soliven, B., and Albert, J., 1992, Tumor necrosis factor modulates Ca^{2+} currents in cultured sympathetic neurons, *J. Neurosci.* **12:**2665–2671.

Stamler, J. S., Toone, E. J., Lipton, S. A., and Sucher, N. J., 1997, (S)NO signals: Translocation, regulation, and a consensus motif, *Neuron* **18:**691–696.

Stieg, P. E., Sathi, S., Alvarado, S. P., Jackson, P. S., Pellegrini, J. W., Chen, H.-S. V., Lipton, S. A., and Jensen, F. E., 1993, Post-stroke neuroprotection by memantine minimally affects behavior and does not block LTP, *Soc. Neurosci. Abstr.* **19:**1503.

Sucher, N. J., Awobuluyi, M., Choi, Y.-B., and Lipton, S. A., 1996, NMDA receptors: From genes to channels, *Trends Pharmacol. Sci.* **17:**348–355.

Sullivan, J. M., Traynelis, S. F., Chen, H.-S. V., Escobar, W., Heinemann, S. F., and Lipton, S. A., 1994, Identification of two cysteine residues that are required for redox modulation of the NMDA subtype of glutamate receptor, *Neuron* **13:**929–936.

Tenhula, W. N., Xu, S. Z., Madigan, M. C., Heller, K., Freeman, W. R., and Sadun, A. A., 1992, Morphometric comparisons of optic nerve axon loss in acquired immunodeficiency syndrome, *Am. J. Ophthalmol.* **113:**14–20.

Tenneti, L., D'Emilia, D. M., and Lipton, S. A., 1997, Suppression of neuronal apoptosis by *S*-nitrosylation of caspases, *Neurosci. Lett.* **236:**139–142.

Toggas, S. M., Masliah, E., Rockenstein, E. M., Rall, G. F., Abraham, C. R., and Mucke, L., 1994,

Central nervous system damage produced by expression of the HIV-1 coat protein gp120 in transgenic mice, *Nature* **367:**188–193.

Toggas, S. M., Masliah, E., and Mucke, L., 1996, Prevention of HIV-1 gp120-induced neuronal damage in the central nervous system of transgenic mice by the NMDA receptor antagonist memantine, *Brain Res.* **706:**303–307.

Tornatore, C., Chandra, R., Berger, J. R., and Major, E. O., 1994, HIV-1 infection of subcortical astrocytes in the pediatric central nervous system, *Neurology* **44:**481–487.

Tyor, W. R., Glass, J. D., Griffin, J. W., Becker, S., McArthur, J. C., Bezman, L., and Griffin, D. E., 1992, Cytokine expression in the brain during the acquired immunodeficiency syndrome, *Ann. Neurol.* **31:**349–360.

Valone, F. H., and Epstein, L. B., 1988, Biphasic platelet-activating factor synthesis by human monocytes stimulated with IL-1β, tumor necrosis factor, or IFN-γ, *J. Immunol.* **141:**3945–3950.

Wahl, L. M., Corcoran, M. L., Pyle, S. W., Arthur, L. O., Harel-Bellan, A., and Farrar, W. L., 1989, Human immunodeficiency virus glycoprotein (gp120) induction of monocyte arachidonic acid metabolites and interleukin 1, *Proc. Natl. Acad. Sci. USA* **86:**621–625.

Wesselingh, S. L., Power, C., Glass, J. D., Tyor, W. R., McArthur, J. C., Farber, J. M., Griffin, J. W., and Griffin, D. E., 1993, Intracerebral cytokine messenger RNA expression in acquired immunodeficiency syndrome dementia, *Ann. Neurol.* **33:**576–582.

Whittemore, E. R., Loo, D. T., and Cotman, C. W., 1994, Exposure to hydrogen peroxide induces cell death via apoptosis in cultured rat cortical neurons, *NeuroReport* **5:**1485–1488.

Wiley, C. A., Masliah, E., Morey, M., Lemere, C., DeTeresa, R. M., Grafe, M. R., Hansen, L. A., and Terry, R. D., 1991, Neocortical damage during HIV infection, *Ann. Neurol.* **29:**651–657.

Wyllie, A. H., Kerr, J. F. R., and Currie, A. R., 1980, Cell death: The significance of apoptosis, *Int. Rev. Cytol.* **68:**251–306.

Yeh, M. W., Nottet, H. L. M., Gendelman, H. E., and Lipton, S. A., 1994, HIV-1 coat protein gp120 stimulates human macrophages to release L-cysteine, an NMDA agonist, *Soc. Neurosci. Abstr.* **20:**451.

Yun, H. Y., Gonzalez-Zulueta, M., Dawson, V. L., and Dawson, T. M., 1998, Nitric oxide mediates *N*-methyl-D-aspartate receptor-induced activation of p21ras, *Proc. Natl. Acad. Sci. USA* **95:**5773–5778.

CHAPTER 22

Nitric Oxide in Listeriosis

KENNETH S. BOOCKVAR, MITRA MAYBODI, REBECCA M. POSTON, ROGER L. KURLANDER and DONALD L. GRANGER

1. Introduction

Listeria monocytogenes is an uncommon but frequently fatal cause of meningitis and rhomboencephalitis in humans. The ease with which *Listeria* can be manipulated in the laboratory and the ability of mice to develop acquired antilisterial resistance after a brief period of illness and recovery have made listeriosis a useful model for the investigation of basic questions of mammalian immunity. From early clinical and histopathological observations, murine listeriosis was shown to involve cell-mediated defense against an intracellular pathogen. Neutrophils, macrophages, and lymphocytes are observed to congregate around the organism in the reticuloendothelial system. Classic experiments by Mackaness (1969) demonstrated that the anamnestic murine response against recurrent *Listeria* infection is mediated by cells, not serum. Later research into the cellular effector mechanism (Godfrey and Wilder, 1984; Rutherford and Schook, 1992; Tanaka *et al.*, 1995) showed that *Listeria* can be inhibited by activated macrophages in an oxygen radical-independent manner. With nitric oxide (NO) having been implicated as an alternative effector molecule produced by activated macrophages,

KENNETH S. BOOCKVAR • Department of Medicine, Cornell University Medical Center, New York, New York 10021. ***MITRA MAYBODI*** • Department of Ophthalmology, Washington University School of Medicine, St. Louis, Missouri 63110. ***REBECCA M. POSTON*** • Embrex Corporation, Research Triangle Park, North Carolina 27709. ***ROGER L. KURLANDER*** • Department of Clinical Pathology, National Institutes of Health, Bethesda, Maryland 20892. ***DONALD L. GRANGER*** • Department of Medicine, University of Utah Medical Center, Salt Lake City, Utah 84132.

Nitric Oxide and Infection, edited by Fang. Kluwer Academic / Plenum Publishers, New York, 1999.

the hypothesis that NO plays a role in defense against *Listeria* was subsequently tested. Experiments to date (Beckerman *et al.*, 1993; Boockvar *et al.*, 1994; MacMicking *et al.*, 1995) have documented the effect of manipulation of NO production on the severity of *Listeria* infection, and results *in vivo* indicate that NO synthesis is important for the clearance of *Listeria* from mice. However, the microbiological mechanism by which NO acts on *Listeria* is not known, and the potential roles of NO-mediated metabolic and regulatory effects on immune cell function have not been established.

2. Models of Listeriosis

2.1. *In Vivo*

The most thoroughly studied model of immune defense against *Listeria* involves systemic infection in mice, in which an intravenous inoculum triggers a multifaceted immunological response that includes the production of NO (Gregory *et al.*, 1993; Boockvar *et al.*, 1994). Visceral organism load provides a useful indicator of infection in studies of immune or NO modulation. The livers and spleens of infected mice are easily excised, homogenized, serially diluted, and plated onto medium selective for *Listeria*. After a sublethal inoculum, organisms can be isolated in increasing numbers from spleens and livers of nonimmune mice until the third day of infection, when they peak at 10^4-10^8 CFU per organ, depending on the strain and the size of the inoculum. The bacteria subsequently disappear over 7 days (Fig. 1A) (Huang *et al.*, 1993; Boockvar *et al.*, 1994). Visceral *Listeria* counts portray the dynamic balance between organism multiplication and immune inhibition over the course of infection. Because experimental listeriosis is an acute infection that has a turning point, studies commonly report organism loads from three time points (peak, prepeak, and postpeak) in order to show effects of immune or NO modification on all phases of infection. Other sources of organism isolation such as stool and blood cultures have been used to follow systemic listeriosis (Boockvar *et al.*, 1994), but the immune factors that affect these assays are not well-characterized and may differ from those that affect spleen and liver counts. However, stool cultures can be serially obtained from mice and thus provide longitudinal data for individual animals (in contrast to spleen and liver counts).

Histological examination of the liver provides another useful way to follow listeriosis in mice during studies of NO activity. After intravenous inoculation, approximately 99% of *Listeria* organisms can be found in the mouse liver (Gregory *et al.*, 1992). Mackaness (1962) documented the histology of primary infection over time: 24 hr after *Listeria* inoculation, lesions consisting primarily of neutrophils appear in the liver; at 48 hr, macrophages associate with bacteria at the

(A)

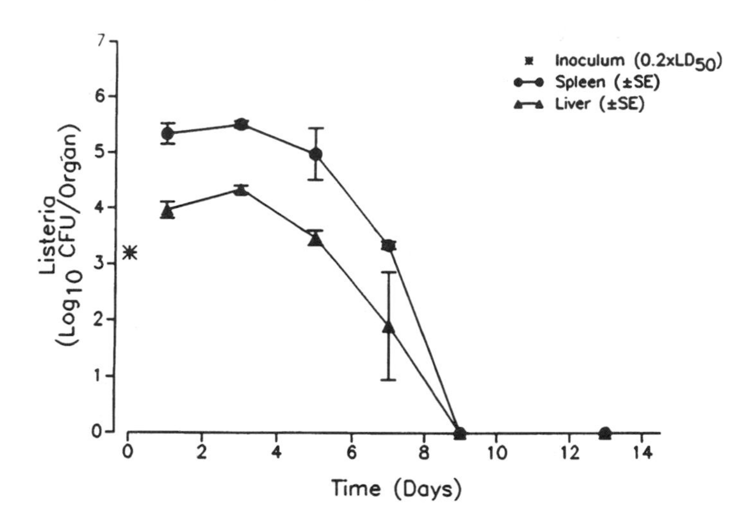

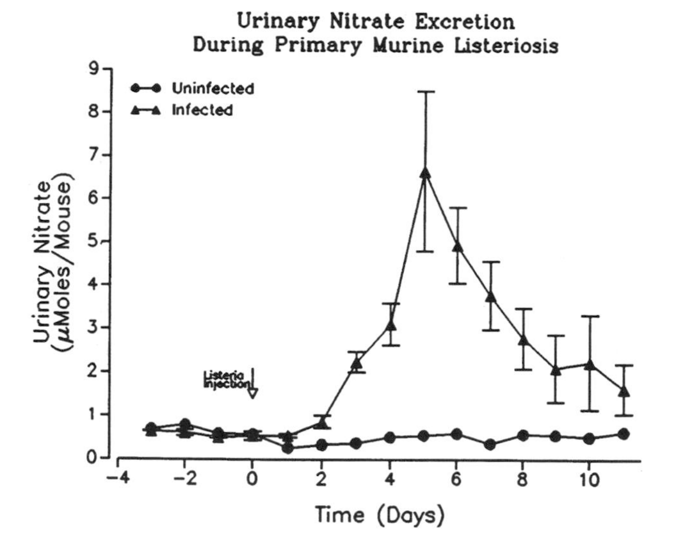

FIGURE 1. Visceral organism counts and urinary nitrate excretion during primary *Listeria monocytogenes* infection in C57BL/6 Mice. (A) Mice were injected intravenously with $\sim 0.1\,LD_{50}$ *L. monocytogenes* on day 0 and liver (▲) and spleen (●) CFU obtained on days 1, 3, 5, 7, and 9 after injection. The counts peak on day 3 after inoculation, after which the organism is rapidly eliminated. Each point represents the mean $\pm$S.E.M. for a group of four animals. (B) Mice were injected intravenously with $\sim 0.1\,LD_{50}$ *L. monocytogenes* (▲) or buffer (●) on day 0 and daily urinary nitrate excretion measured. Nitrate excretion rises to a peak on day 5 after inoculation, coinciding with the rapid elimination of *Listeria* from the mouse livers and spleens, after which it falls to baseline. Each point represents urinary nitrate per mouse per 24 hr averaged from a group of at least four mice. Adapted from Boockvar *et al.* (1994).

periphery of the lesions; at 3–4 days, macrophages dominate the lesions, and bacteria appear damaged; at 5–6 days, the liver lesions disperse. In contrast to primary infection, *Listeria* inoculated into mice convalescing from a previous *Listeria* infection (immune mice) do not multiply and are eliminated from the liver and spleen within 2 days. During such a secondary infection, macrophages congregate in the liver at 24 hr and disperse at 48 hr (Mackaness, 1962). Thus, depending on the stage of infection and whether the host is naive or immune to *Listeria*, different cell types and different anti-*Listeria* effector mechanisms appear to be involved in host defense. Variation in these histological patterns can be demonstrated after immune interventions, including NO modulation (Boockvar *et al.*, 1994), and are therefore useful to follow. However, histological studies require sacrifice of the animals, and thus cannot be used to obtain longitudinal data.

Morbidity and mortality are important measures of severity of *Listeria* infection. Measurement of the *Listeria* inoculum that would result in 50% mortality (LD_{50}) is the gold standard assay of mouse immunity against *Listeria*, although it is infrequently used because of resource limitations. Survival curves are often used as a substitute, but are not always accompanied by measures of statistical significance (Beckerman *et al.*, 1993; MacMicking *et al.*, 1995). In addition, mortality data do not provide information about the natural history of the infection. Clinical parameters such as mouse weight, food and water intake, and temperature have been reported (Boockvar *et al.*, 1994) and appear to correlate with visceral organism loads during murine listeriosis. In athymic mice, however, clinical parameters and mortality do not correlate well with organ *Listeria* counts. Athymic mice have higher LD_{50}s than controls and live for long periods of time with minimal signs of infection, yet these mice harbor higher organism counts in their livers and spleens (Emmerling *et al.*, 1977; Newborg and North, 1980; Sasaki *et al.*, 1990). Thus, organ *Listeria* counts, mortality, and clinical signs should be viewed as complementary parameters that together portray the immune effector capacity of mice during experimental listeriosis.

2.2. *In Vitro*

A variety of studies have been performed using both models of focal *Listeria* infection and *in vitro* cell culture infection. Peritoneal washing cultures can be used to follow mice after intraperitoneal (i.p.) *Listeria* inoculation (Miki and Mackaness, 1964), with subsequent removal of the mouse macrophages for *ex vivo* immunological and NO assays (Langermans *et al.*, 1992a). However, this technique is limited by the difficulty of standardizing *Listeria* counts from peritoneal cavity washings. In addition, there is a limited understanding of the immune response to i.p. listeriosis.

Because *Listeria* is a facultative intracellular organism, it is easily grown in cell culture, and many studies report measurements of NO in *Listeria*–cell culture systems, most commonly employing *ex vivo* murine peritoneal (Higginbotham *et al.*, 1992; Stokvis *et al.*, 1992; Higginbotham and Pruett, 1994) or bone marrow-derived (Rutherford and Schook, 1992) macrophages, a murine macrophagelike cell line (Inoue *et al.*, 1995), *ex vivo* murine liver cells (Curran *et al.*, 1989; Gregory *et al.*, 1993), or a murine hepatocytelike cell line (Szalay *et al.*, 1995). In these models, *Listeria* can be quantitated by plaque formation (Mackaness, 1962) or by bacterial culture of lysed cells. One study notes that intracellular organism viability may be adversely affected by washing cell cultures with antibiotics used to eradicate nonphagocytosed organisms, thereby artifactually lowering *in vitro Listeria* counts (Drevets *et al.*, 1994).

The efficacy of macrophage function under different experimental conditions can also be observed using electron microscopy, as activated macrophages inhibit *Listeria* function by retaining organisms in the phagocytic endosome and preventing them from entering the cytoplasm. The subcellular compartmentalization of ingested organisms reflects the activation state of the macrophages (Portnoy *et al.*, 1989).

Cell culture experiments have been useful in providing corroborating evidence regarding the potential roles of NO in *Listeria* infection. They also can be used to generate hypotheses regarding various immune effector mechanisms. However, because of the complexity of the mammalian immune system, *in vitro* observations cannot necessarily be generalized to *in vivo* systems.

3. NO Activity in Listeriosis

3.1. *In Vivo* Measurement

NO is the product of the oxidation of a guanidino nitrogen on L-arginine by NO synthase (see Chapter 4). Because of its reactivity, NO is short-lived and cannot be assayed directly in most biological systems, including models of *Listeria* infection. Methods for measuring NO depend on the fact that it quantitatively reacts with oxygen to produce nitrite, which can be assayed spectrophotometrically after reaction with Griess reagents. *In vivo*, NO reacts with iron atoms in proteins (Hibbs *et al.*, 1988), making it difficult to determine the rate of nitrite formation. Nitrite can be further oxidized to nitrate in the presence of oxyhemoglobin. Because nitrate does not react with the Griess reagents, it must be reduced to nitrite prior to the spectrophotometric assay. In models of *Listeria* infection, this has been accomplished using bacterial nitrate reductase (Granger *et al.*, 1996) or a cadmium column (Gregory *et al.*, 1993). Mice excrete 55 and 20% of a known ingested amount of nitrate in the urine and feces, respectively (Granger *et al.*, 1991). However, because the fraction and rate of NO conversion to nitrite and nitrate *in vivo* are unknown, researchers have had to assume that the metabolism of NO remains constant during different inflammatory states.

When mice are fed a nitrite- and nitrate-free diet, their rate of plasma and urinary nitrate production is found to be consistently low (Granger *et al.*, 1991). This allows the detection of increases in nitrate production related to various stimuli. As in other models of inflammation, nitrite and nitrate appear in the plasma during murine listeriosis, and are excreted in the urine. A sublethal *Listeria* inoculation has been shown to induce a rise in plasma nitrate by day 2 of infection (Samsom *et al.*, 1996), and in urinary nitrate by day 3 (Fig. 1B) (Boockvar *et al.*, 1994). Urinary nitrate peaks at 10- to 20-fold over baseline on day 5, then declines to baseline by recovery. Urinary nitrate represents a summation of NO activity over 24 hr, as the urine is collected once daily (Boockvar *et al.*, 1994). This technique fails to take into account other sources of nitrate loss such as stool excretion and metabolism by bacteria in the gut. Experiments measuring urinary nitrate also must assume that the fraction of nitrate excreted in the urine does not change during different inflammatory conditions, or over the course of *Listeria* infection (Granger *et al.*, 1996). Plasma nitrate rises earlier in infection than urinary nitrate and may give a more accurate picture of the time course of NO response to *Listeria* infection, but the utility of this variable is limited by the fact that measurement requires sacrifice of the animal and therefore cannot be followed longitudinally (Samsom *et al.*, 1996). Together, the techniques of urine and plasma nitrate analysis permit assessment of NO activity *in vivo*, while the clinical course of listeriosis is observed.

Of the several NO synthase isoenzymes, the inducible, calcium-independent NO synthase (iNOS, NOS2) is believed to be the principal source of high-level NO production during inflammation, based on cell culture models (Adler *et al.*, 1995). Because iNOS is induced by *Listeria* infection but is not constitutively present, assays for iNOS provide an alternative method to assess change in NO activity during the course of listeriosis. The cloned iNOS gene (Xie *et al.*, 1992) can be used as a probe for the presence of iNOS mRNA amplified by the polymerase chain reaction (PCR) *in situ* during *Listeria* infection. iNOS mRNA appears in the spleens of *Listeria*-infected mice by day 1 of infection and peaks by day 4, after which it declines to nearly undetectable levels by day 7 (Fig. 2) (Boockvar *et al.*, 1994; Flesch *et al.*, 1994). This pattern correlates closely with urinary nitrate excretion (Boockvar *et al.*, 1994).

The coincidence of NADPH diaphorase activity with macrophage nitrite production *in vitro*, and the ability of iNOS antisense oligonucleotides to suppress both activities, suggest that iNOS and NADPH colocalize in cells as is true for other forms of NOS (Flesch *et al.*, 1994). This allows NADPH diaphorase staining *in situ* to be used as a marker for cells with active iNOS. NADPH diaphorase activity is detectable within macrophages in the livers of mice only on day 4 after *Listeria* inoculation, corresponding to the greatest levels of iNOS mRNA (Flesch *et al.*,

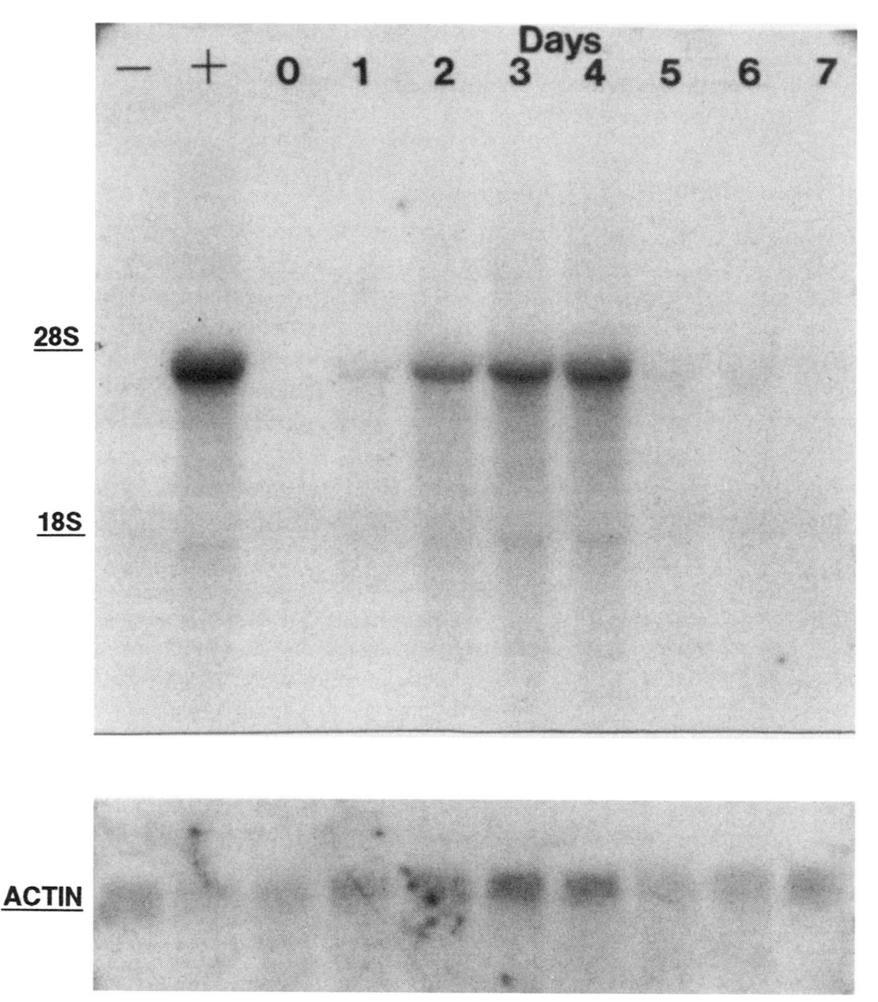

FIGURE 2. Northern blot for inducible nitric oxide synthase (iNOS) mRNA from spleens of C57BL/6 mice during primary *Listeria monocytogenes* infection. The probe is a radiolabeled 3.4-kb fragment. Negative control: RNA extract from unstimulated J774.1 macrophagelike cell culture. Positive control: RNA extract from J774.1 macrophagelike cell culture stimulated to produce iNOS by treatment with IFNγ and LPS. Numbered lanes: total splenic RNA extracted from mice injected intravenously with $\sim 0.1\ LD_{50}$ *L. monocytogenes* on day 0 and sacrificed on the indicated day after injection. iNOS mRNA appears in the spleens of mice by day 1 of infection, peaks in quantity by day 4, and by day 7 it declines to nearly undetectable levels. Each lane represents data from one animal. Adapted from Boockvar *et al.* (1994).

1994). Thus, both iNOS transcription and translation can be monitored as adjunctive means of detecting and following NO activity *in vivo*.

3.2. *In Vitro* Measurement

Assays for NO activity *in vitro* are based on the same biochemistry as those performed on animal samples. NO released in cell culture is rapidly oxidized to nitrite in the supernatant, which can be detected spectrophotometrically after treatment with Griess reagents. Because nitrate is produced *in vitro* in smaller quantities than nitrite, and in a relatively fixed proportion to nitrite (Ding *et al.*, 1988), cell culture supernatants can be analyzed without using a nitrate reducing agent. *Ex vivo* macrophage cultures derived from mouse bone marrow (Rutherford and Schook, 1992) or peritoneum (Langermans *et al.*, 1992b), *ex vivo* liver cell cultures (Curran *et al.*, 1989; Gregory *et al.*, 1993), murine macrophagelike cell lines (Cunha *et al.*, 1992), or hepatocytelike cell lines (Szalay *et al.*, 1995) can be stimulated to produce nitrite, usually with IFNγ as a primary stimulus, along with TNFα, LPS, or live or killed *L. monocytogenes*. PCR amplification has allowed detection of iNOS mRNA *in vitro* (Szalay *et al.*, 1995). iNOS protein has been assayed in macrophage lysates by Western blot with rabbit anti-iNOS IgG (MacMicking *et al.*, 1995). Such *in vitro* experiments have provided supportive data regarding the role of NO in listeriosis, and have stimulated hypotheses that can be tested *in vivo*.

3.3. Stimulation of NO Activity

Methods for modulating NO activity have been useful for investigating its significance in immune defense against *Listeria*. NO activity can be stimulated or inhibited at various levels of regulation, both *in vitro* and *in vivo*. Methods for stimulating NO activity *in vitro* have relied on the fact that IFNγ plus TNFα or LPS, among other stimuli, can activate cells infected with *Listeria* to produce NO. Interpretation of such experiments is limited by the fact that these compounds are not specific inducers of iNOS and have other immunostimulatory effects that may influence *Listeria* growth. A few studies have attempted to add NO to *in vitro* systems nonenzymatically, with dissolved NO gas (Gregory *et al.*, 1993) or NO donors such as nitroprusside (Xiong *et al.*, 1996). However, technical limitations and questions of *in vivo* relevance have limited the interpretation of these observations. *In vivo*, several studies have shown that treatment of mice with exogenous IFNγ or TNFα can influence the course of listeriosis. However, none have performed concurrent NO activity assays. No studies to date have been reported in which iNOS activity is targeted for genetic *up*regulation, but this may be possible with transgenic technology.

3.4. Inhibition of NO Activity

NO inhibition can be accomplished via several techniques. *In vitro*, the L-arginine substrate for iNOS can be omitted from the cell culture medium or hydrolyzed to urea and ornithine by arginase. Although either method will block iNOS activity, neither is effective *in vivo*, because arginine can be synthesized endogenously in the rodent liver and kidney (Granger *et al.*, 1991). iNOS activity has been inhibited in models of listeriosis by the antagonistic L-arginine analogues N^{G}-monomethyl-L-arginine (L-NMMA), N^{G}-nitro-L-arginine methyl ester (L-NAME), and aminoguanidine. Intravenous (Gregory *et al.*, 1993; Samsom *et al.*, 1996), intraperitoneal (Langermans *et al.*, 1992b; Samsom *et al.*, 1996), and oral (Boockvar *et al.*, 1994) routes of administration of these compounds have been used in *Listeria*-infected mice. Because the metabolism of iNOS inhibitors in rodents is largely unknown, experiments have been performed in which single and repeated administrations of the compounds have been given. Continuous subcutaneous administration of L-NMMA has failed because the experimental mice and their littermates chewed at the subcutaneous pumps until they fell out, while others became infected (K. Boockvar, unpublished data). L-NMMA and L-NAME are not specific for iNOS; they are believed to have effects on other NOS isoforms and on T-cell proliferation (Gregory *et al.*, 1994) *in vivo*, which may confound results of studies using these inhibitors. Other, more specific inhibitors of iNOS have been synthesized (Moore *et al.*, 1994; Nakane *et al.*, 1995), but have not yet been used in published studies of *Listeria* infection.

IL-10 (Bogdan *et al.*, 1991; Cunha *et al.*, 1992), anti-IFNγ antibodies (Beckerman *et al.*, 1993; Samsom *et al.*, 1996), or anti-TNFα antibodies (Langermans *et al.*, 1992b; Beckerman *et al.*, 1993; Samsom *et al.*, 1996) have been used to inhibit iNOS activity *in vitro*. Analogous *in vivo* models have included the infection of mice with targeted disruption of genes necessary for IFNγ (Kamijo *et al.*, 1994; Fehr *et al.*, 1997) or other immune activation pathways (Tanaka *et al.*, 1995). These systems suffer from lack of specificity, as many functions besides iNOS activity are influenced by the blockade of cytokine function. More specific inhibition of iNOS translation has been accomplished via the use of antisense mRNA *in vitro* (Flesch *et al.*, 1994). Targeted disruption of the iNOS gene itself (MacMicking *et al.*, 1995) has yielded particularly fruitful information on the role of NO in listeriosis *in vivo*, as will be discussed below.

4. NO-Dependent Response to Primary Listeriosis

As the course of murine listeriosis progresses, the immune response to the pathogen evolves. Direct and indirect evidence suggests that NO is most important early in listeriosis, when the multiplication rate of the bacterium is at its peak, the liver has its greatest burden of infection, and neutrophils and macrophages are most active. As already mentioned, inoculation of *Listeria* into mice induces a rise in plasma nitrate by day 2 of infection (Samsom *et al.*, 1996), and urinary nitrate rises by day 3. Urinary nitrate peaks at a level 10- to 20-fold greater than baseline on day 5, then declines to levels close to baseline by the time of recovery (Fig. 1B). iNOS mRNA appears in the spleens of *Listeria*-inoculated mice by day 1 of infection and quantitatively peaks by day 4, subsequently declining to nearly undetectable levels by day 7, a pattern closely correlating with urinary nitrate excretion in the same mice (Boockvar *et al.*, 1994).

Listeria-inoculated mice can be given sufficient doses of L-NMMA orally to suppress increases in nitrite excretion throughout infection, without observable clinical effects on uninfected mice (Fig. 3). *Listeria*-infected mice treated with

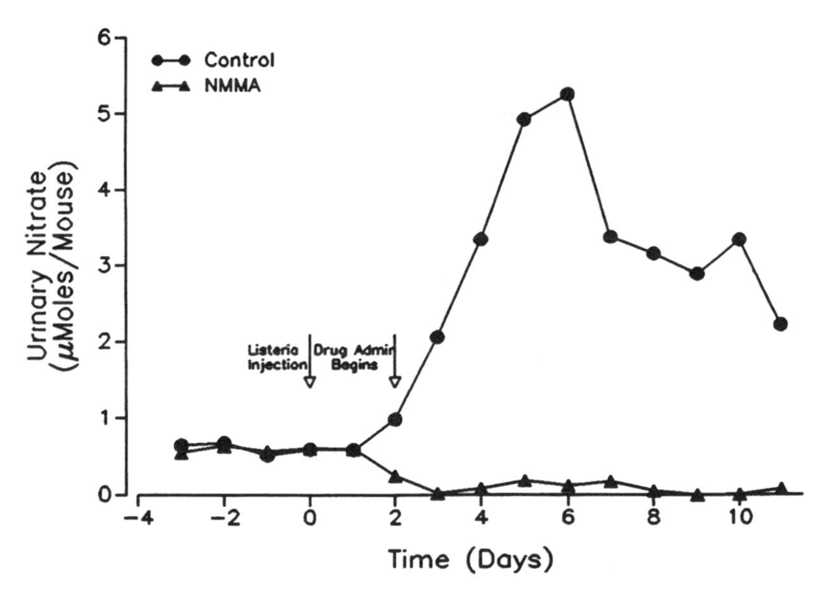

FIGURE 3. Effect of oral L-NMMA administration on urinary nitrate excretion during primary *Listeria monocytogenes* infection in C57BL/6 mice. Mice injected intravenously with ~0.1 LD_{50} *L. monocytogenes* on day 0 were treated with L-NMMA (▲) or L-arginine (●) by direct gastric installation beginning on day 1 after injection and continuing throughout the infection. Urinary nitrate excretion was completely suppressed in L-NMMA-treated infected mice, whereas L-arginine-treated infected control mice had a typical rise in urinary nitrate excretion which peaked on days 5 and 6. Each point represents urinary nitrate per mouse per 24 hr averaged from a group of at least four mice. Adapted from Boockvar *et al.* (1994).

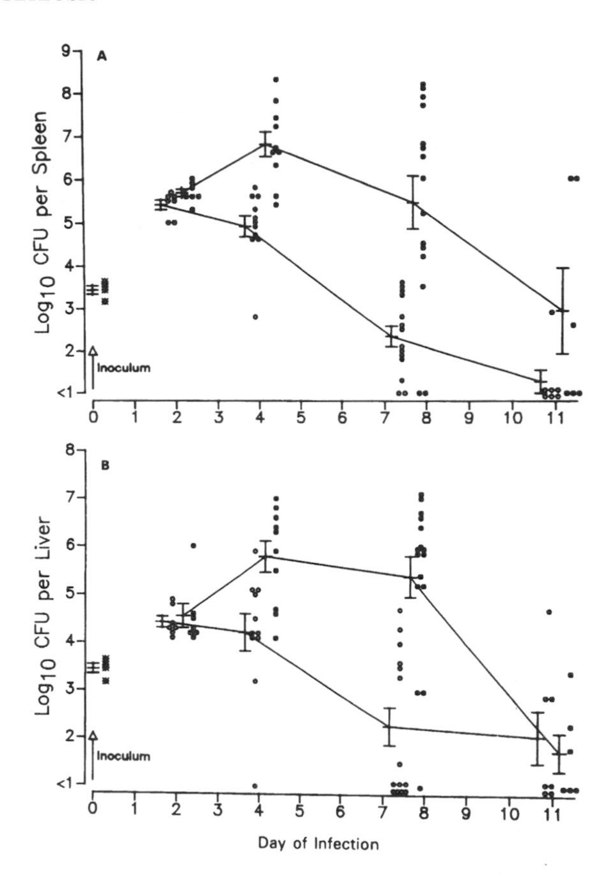

FIGURE 4. Effect of oral L-NMMA administration on visceral organism counts during primary *Listeria monocytogenes* infection in C57BL/6 mice. Mice injected intravenously with $\sim 0.1\,LD_{50}$ *L. monocytogenes* on day 0 were treated with L-NMMA (●) or L-arginine (○) by direct gastric installation beginning on day 1 after injection and continuing throughout the infection. Mice were sacrificed on days 2, 4, 7, 8, and 11 and per organ spleen (A) and liver (B) CFU obtained. L-NMMA-treated infected mice had statistically higher spleen and liver organism counts than L-arginine-treated infected control mice on days 4, 7, and 8, but not on day 11. There was one death in the L-NMMA-treated group and none in the L-arginine-treated group. Each point represents the value from a single animal. Each bar represents the standard error for the adjacent group. Adapted from Boockvar *et al.* (1994).

L-NMMA have 100-fold greater organism burdens in their spleens and livers compared with infected untreated controls by day 4 of infection, and 1000-fold increased organism burdens in their spleens and livers compared with controls by day 7 or 8 of infection (Fig. 4). On histological examination, the livers and spleens of infected L-NMMA-treated mice show more hepatocyte necrosis and monocytic infiltrate than controls, excluding an effect of L-NMMA on the migration of macrophages into infected foci. L-NMMA-treated infected mice also have at least 100-fold greater stool *Listeria* counts on days 5 through 9 after inoculation, and show worsened signs of illness including weight loss, anorexia, hypothermia, and death (Boockvar *et al.*, 1994). The influence of L-NMMA is dose dependent. Mice given a single i.v. dose of L-NMMA at the time of *Listeria* inoculation do not show significantly suppressed plasma nitrate levels on days 3 and 7 of infection compared with untreated infected controls. Correspondingly, mice treated with this regimen of L-NMMA have liver and spleen organism counts approximately equivalent to controls on day 3 after inoculation (Gregory *et al.*, 1993). These findings support the hypothesis that the effects of L-NMMA on listeriosis are related to its inhibition of NO production, and demonstrate the importance of NO in systemic immune defense against *Listeria*.

Additional *in vivo* experiments support an immune effector role of NO in listeriosis. Intraperitoneally inoculated mice treated with a regimen of i.p. aminoguanidine that inhibits NO activity by 50–100% have greater spleen organism counts and higher mortality than untreated mice. The effect of aminoguanidine is more pronounced if it is first administered on the day of *Listeria* inoculation rather than on day 2 after infection, causing 80 and 50% mortality by day 8, respectively (Beckerman *et al.*, 1993). As aminoguanidine is a more specific iNOS antagonist than L-NMMA (Nakane *et al.*, 1995), and is provided locally in this model, the potential confounding effects of the inhibitor on other organ and cellular function are minimized. The immunosuppressive effect of locally administered aminoguanidine suggests that locally produced NO is important for *Listeria* inhibition, supporting its role as an effector molecule *in vivo*. In addition, the fact that aminoguanidine is less immunosuppressive if withheld during the first 2 days indicates that NO begins its influence very early in infection.

A mouse strain with targeted disruption of the iNOS gene has been used to study listeriosis (MacMicking *et al.*, 1995), eliminating problems related to NOS inhibitor nonspecificity. iNOS-knockout mice have normal organ development, normal distributions of immune cells in their thymus and spleen, and a normal inflammatory response to i.p. stimuli. Although macrophages from these mice have preserved and perhaps even increased reactive oxygen species production, they do not produce NO in response to IFNγ and LPS, and have no iNOS mRNA or protein. When iNOS-knockout mice are inoculated with *Listeria*, they have 100-fold greater organism burdens in their livers and spleens by day 3, and succumb to *Listeria* inocula at least 10-fold lower than lethal doses in wild-type mice. Thus, specific systemic abrogation of iNOS results in a deficient immune response to *Listeria* as measured by morbidity, mortality, and microbiological assays. Perhaps the strongest evidence of a role for NO in listeriosis has been provided by the iNOS-deficient knockout mice, the contemporary standard for investigating the role of an enzyme *in vivo*.

Nevertheless, as might be expected for research involving a complex and redundant mammalian immune system, the experimental findings are sensitive to conditions under which NO and *Listeria* are manipulated and observed. For example, mice treated with a single, low i.v. dose of L-NMMA at the time of *Listeria* inoculation are observed to have 1000-fold *fewer* organisms in their spleens and livers on day 7 compared with untreated infected controls (Gregory *et al.*, 1993). This observation has been difficult to reconcile with the presumed short half-life of L-NMMA and the initial detection of iNOS mRNA in mouse spleens on the day *after Listeria* inoculation (Boockvar *et al.*, 1994; Flesch *et al.*, 1994). The authors of the study hypothesize that NO can inhibit T-cell function (see also Chapter 10) and thereby hinder *Listeria* elimination, an effect abrogated by L-NMMA. However, the purported immunosuppressive effect of NO in listeriosis has not been supported by other *in vivo* studies to date.

4.1. NO Production in Primary Listeriosis

4.1.1. Cellular Sources

Macrophages and hepatocytes are known to be capable of producing high fluxes of NO and are involved in the early elimination of *Listeria in vivo*. Following phagocytosis of *Listeria* by murine macrophagelike cells *in vitro*, the bacteria can be observed to escape from the endosome into the cytoplasm, traveling from cell to cell using host-derived actin filaments (Tilney and Portnoy, 1989). Inhibition of the early phagocyte response has been shown to exacerbate murine listeriosis (Rosen *et al.*, 1989; Conlan and North, 1992). Treatment with antibodies to the type 3 complement receptor prevents immigration of neutrophils and monocytes to sites of inflammation, and results in unrestricted replication of *Listeria* in the spleens, livers, and footpads of mice. Mice receiving such antibodies within 24 hr of a low *Listeria* inoculum succumb to lethal infection in 3–4 days. If given 3 days after *Listeria* inoculation, however, the antibodies cause a temporary increase in *Listeria* growth but no increase in mortality (Rosen *et al.*, 1989). Thus, phagocytes are required for *Listeria* inhibition very early in the course of infection, paralleling the requirement for iNOS activity.

Other studies have helped to confirm that macrophage-mediated inhibition of *Listeria* depends on NO production. When livers from mice infected with *Listeria* are stained using NADPH diaphorase as a marker for iNOS, only macrophages contain NADPH diaphorase activity, with temporal correlation to peak elevations of iNOS mRNA expression (Flesch *et al.*, 1994). This suggests that macrophages are the predominant cell type responsible for iNOS activity in *Listeria*-infected mice. *In vitro* data also support the role of macrophages (Rutherford and Schook, 1992; Szalay *et al.*, 1995). Murine bone marrow-derived macrophages can be stimulated to produce nitrite by combinations of IFNγ and LPS or IFNγ and live

Listeria, with concomitant inhibition of *Listeria* growth. This antilisterial effect can be abrogated by L-NMMA treatment or L-arginine depletion, but not by superoxide dismutase, suggesting that the antimicrobial activity depends on iNOS but not superoxide production. Likewise, the murine macrophagelike cell line J774A.1 can be stimulated by IFNγ to produce nitrite and inhibit *Listeria* growth (Frei *et al.*, 1993). Both effects can be abrogated by L-NMMA. Even though not all murine macrophagelike cell lines have yielded identical results (Leenen *et al.*, 1994; Inoue *et al.*, 1995), the majority of *in vivo* and *in vitro* evidence suggests that macrophages have a crucial role in defense against *Listeria,* linked to their ability to produce NO.

Similar data exist for hepatocytes in *Listeria* infection. After i.v. inoculation, approximately 99% of *Listeria* lodge in the liver, infecting both hepatocytes and Kupffer cells (Gregory *et al.*, 1992). During the first 24 hr, infected hepatocytes are lysed in the presence of neutrophils (Conlan and North, 1991), with some hepatocytes appearing to undergo apoptosis (Rogers *et al.*, 1996). Ninety percent of *Listeria* elimination during the first 24 hr is accomplished by this mechanism (Conlan and North, 1991). Hepatocytes have been shown to produce nitrite during listeriosis. In one system (Curran *et al.*, 1989), *ex vivo* hepatocytes from Sprague–Dawley rats produced nitrite when exposed to the supernatant of Kupffer cells stimulated by IFNγ and LPS, but not when exposed to IFNγ and LPS directly. In another system employing C57BL/6J mice (Gregory *et al.*, 1993), hepatocytes and Kupffer cells explanted on day 3 after *Listeria* inoculation and stimulated with a variety of cytokines produced significantly more nitrite than those obtained from uninfected mice. This nitrite production could be inhibited by L-NMMA, establishing that it results from NOS activity. Similarly, the murine hepatocytelike cell line TIB 75 can be stimulated by IFNγ to produce nitrite and inhibit bacterial growth when infected with *Listeria* (Szalay *et al.*, 1995). iNOS mRNA can be detected in the cell line under these conditions. When the TIB 75 cells are treated with L-NMMA, nitrite production is abolished and *Listeria* inhibition is partially reversed, suggesting that *Listeria* inhibition is achieved via NO production. Although hepatocytelike cell lines and *ex vivo* hepatocytes behave somewhat differently when infected with *Listeria* and treated with NOS inhibitors, the collective evidence suggests that hepatocytes, under the influence of stimulated phagocytic cells, produce NO during *Listeria* infection.

4.1.2. Cytokine Effects

Correlations between immune antilisterial activity and NO have also been observed in studies of cytokine modulation. *In vitro* IFNγ activates macrophages and prevents *Listeria* from escaping from the phagocytic vacuole into the cytoplasm (Portnoy *et al.*, 1989; Tilney and Portnoy, 1989). In separate experiments (Ding *et al.*, 1988), IFNγ was shown to be the only cytokine necessary and sufficient to stimulate nitrite production from explanted murine peritoneal macrophages. *In vivo*, IFNγ has been shown to be required for both NO synthesis and early *Listeria* elimination. IFNγ can be detected by ELISA in the spleens of mice by day 1 of *Listeria* infection, and peaks by day 2 (Nakane *et al.*, 1990; Poston and Kurlander, 1991) just before the maximal appearance of macrophages on histological sections. Early administration of exogenous IFNγ during a sublethal murine infection decreases splenic *Listeria* counts (Kiderlen *et al.*, 1984; Kurtz *et al.*, 1989; Langermans *et al.*, 1992b). Conversely, mice treated with antibody to IFNγ on day 1 of infection have 10^3-fold greater spleen *Listeria* counts by day 4 and 10^6-fold greater counts by day 6, respectively, as well as higher mortality (Buchmeier and Schreiber, 1985). Peritoneal macrophages removed from the anti-IFNγ-treated mice are not cytolytic for tumor cells, suggesting that the antibody abrogates macrophage activation *in vivo*. Thus, the influence of IFNγ on the course of infection parallels that of NO.

Stronger evidence that IFNγ is required to activate macrophages to produce NO is seen in mice with targeted disruption of the IFNγ receptor gene (Huang *et al.*, 1993). Peritoneal macrophages from IFNγ-receptor knockout mice do not produce nitrite in response to IFNγ, and knockout mice infected with *Listeria* have 10- to 100-fold higher visceral organism counts than wild-type mice on infection day 5. In addition, a *Listeria* inoculum sublethal for wild-type mice is lethal for knockout mice. Targeted disruption of genes of transcription factors dependent on the IFNγ signal has a similarly detrimental effect on host resistance to listeriosis (Fehr *et al.*, 1997), although the macrophages from some of these transgenic animals appear to retain the ability to produce NO when stimulated with LPS, with or without IFNγ. Thus, IFNγ signaling is required for early control of *Listeria* multiplication *in vivo*, and exacerbation of listeriosis correlates with a failure of explanted macrophages to produce NO in most models of IFNγ disruption. Differences between the background strains employed to generate transgenic mice as well as the *ex vivo* conditions used in each model, or differences in the effects of transcription factors on iNOS expression, may account for the preserved ability of macrophages from some mice with disruptions of the IFNγ signaling pathway to make NO.

Like IFNγ, TNFα has been well studied in murine listeriosis. *In vitro*, TNFα synergistically stimulates macrophages to produce nitrite when administered with IFNγ (Ding *et al.*, 1988), an observation that has not yet been evaluated *in vivo*. *In vivo*, TNFα can be detected in the spleens of mice by cytotoxic assay on days 1–3 of *Listeria* infection (Havell, 1989). TNFα mRNA, as detected by Northern blot of mouse spleen tissue, peaks on day 1 of *Listeria* infection and remains elevated until at least day 7 (Poston and Kurlander, 1992). In mice treated with anti-TNFα antibodies by day 1 after *Listeria* inoculation, organ bacterial counts are 100-fold greater than in controls by day 3, and all antibody-treated mice die (Havell, 1989; Langermans *et al.*, 1992b). If administered on infection days 3 or 5, anti-TNFα antibodies have only transient effects on organ *Listeria* counts and no effect on mortality. Exogenous administration of TNFα 1 to 24 hr before, but not 6 hr after, a large *Listeria* inoculation enhances organ *Listeria* clearance and mouse survival (Havell, 1989; Langermans *et al.*, 1992b). TNFα thus appears to exert its effects before day 3 and has little effect thereafter. Mice with targeted disruption of the type 1 TNFα receptor (TNFR1) whose cells do not bind TNFα have 10^3-10^5 greater organ *Listeria* counts by day 4 of infection (Rothe *et al.*, 1993) and universally succumb to *Listeria* challenge, even with inocula as low as 0.01 LD_{50} (Pfeffer *et al.*, 1993; Rothe *et al.*, 1993). Likewise, transgenic mice altered to constitutively express a TNFα inhibitor (the extracellular domain of TNFR1 fused to an immunoglobulin fragment) are more susceptible to *Listeria* infection (Garcia *et al.*, 1995). Histological examination of the liver of anti-TNFα-antibody-treated mice shows lesions containing heavily infected hepatocytes, but only a few mononuclear cells (Havell, 1989). These experiments demonstrate that TNFα like IFNγ, is required for the initial neutrophil and macrophage response to *Listeria* infection. In the absence of either cytokine, mice show an almost complete lack of myelomonocytic defense, with a phenotype similar to that produced by direct inhibition of iNOS.

In summary, NO production is required for normal murine immune defense against primary *Listeria* infection. Judging from corroborating *in vitro* data, the most likely source of this NO is macrophages under the influence of the cytokines IFNγ and TNFα. Hepatocytes may also be a significant source of NO in murine listeriosis. Depending on the model employed, the requirement for NO in primary listeriosis varies, which is perhaps not surprising as NO is an effector molecule of only one arm of a complex immune response. Studies to define the molecular actions of NO on *Listeria* in nonimmune mice, and to determine whether NO by itself is sufficient to mediate bacterial stasis or killing, have not yet been reported.

5. NO-Independent Response to Listeriosis

5.1. Primary Listeriosis

The murine immune response to listeriosis evolves from a nonspecific reaction dependent on neutrophil and macrophage effector molecules, including NO, to a reaction employing *Listeria*-specific T cells, generated after presentation of *Listeria* antigens. In antibody-mediated cell depletion experiments similar to those performed with neutrophils (Czuprynski *et al.*,1989; Sasaki *et al.*, 1990; Conlon *et al.*, 1993), T cells begin to have a demonstrable effect on murine listeriosis by day 3 of infection, and they are required for complete elimination of *Listeria*. In athymic mice, *Listeria* proliferation peaks at normal levels, but the organisms are never eliminated and establish a persistent infection resulting in the death of the mouse in 30 to 60 days (Newborg and North, 1980; Sasaki *et al.*, 1990). If the thymus is replaced surgically, the mice are able to eradicate the bacteria

(Newborg and North, 1980). In the only experiments examining the long-term effects of NO inhibition in listeriosis (Boockvar *et al.*, 1994), *Listeria*-infected mice treated with L-NMMA were found to have greater organism burdens on days 4, 7, and 8 after inoculation, and higher mortality than infected untreated controls; yet, the surviving animals eventually control bacterial proliferation, so that L-NMMA-treated animals have organism burdens similar to controls by day 11, despite complete inhibition of NO activity (Fig. 4). These data suggest that the influence of NO-producing cells wanes over the course of listeriosis, and that definitive clearance of the organism requires T cells. The later stage of *Listeria* elimination seems to be unaffected by iNOS inhibition.

As *Listeria* can grow both intracellularly and extracellularly, both $CD4^+$ and $CD8^+$ T cells appear to have a role in immune eradication. In mice treated with anti-CD4 antibodies, peak *Listeria* burdens are tenfold greater than in control mice, and anti-CD4-treated mice take 2 weeks longer to eliminate the organism (Czuprynski *et al.*,1989; Sasaki *et al.*, 1990). In mice treated with anti-CD8 antibodies, *Listeria* counts at day 4 are tenfold *lower* than in controls (Sasaki *et al.*, 1990), but counts at day 8 are tenfold greater than controls (Mielke *et al.*, 1988). Anti-CD8-treated mice take 3 weeks longer to eradicate the organism completely (Sasaki *et al.*, 1990). A detrimental effect of $CD8^+$ cells in the absence of $CD4^+$ cells is suggested by the higher mortality from listeriosis in athymic mice given spleen cells depleted of $CD4^+$ but not $CD8^+$ cells (Sasaki *et al.*, 1990). This may result from unregulated cytotoxic activity against host cells by an undefined mechanism. Thus, T cells seem to play a lesser role than neutrophils and macrophages in controlling early multiplication of *Listeria*, but are later required for mice to eradicate the organism. *Listeria* eradication is delayed in the absence of either $CD4^+$ or $CD8^+$ cells, and cannot occur at all if both subsets are absent.

5.2. Secondary Listeriosis

As in other infections that induce a potent T-cell response, specific immunity is established during listeriosis that enables convalescent mice to eliminate a secondary *Listeria* challenge much more efficiently. The components required for the secondary anamnestic response have been well studied, appear to depend on T cells, and, like the late stage of primary infection, seem to be independent of NO. *Listeria*-immune mice can withstand a secondary inoculum of ten LD_{50} organisms, clearing *Listeria* from the liver and spleen in 2 days, whereas control mice die on day 3 (Mackaness, 1962). In contrast to the histological appearance of primary infection, the liver lesions during secondary infection contain macrophages but no neutrophils, and the lesions begin to resolve by day 2.

This enhanced antilisterial activity can be isolated and transferred experimentally from immune mice to nonimmune mice, facilitating identification of the memory-containing factor. When nonimmune mice are injected with spleen cells from immune mice, recipient animals have greatly decreased mortality from systemic *Listeria* infection compared with uninjected controls (Mackaness, 1969; Mackaness and Hill, 1969; Lane and Unanue, 1972; North, 1973). Spleen cells are most protective if harvested from the donor at day 6 of *Listeria* infection (Mackaness, 1969), when spleen T and B cells are present in the greatest numbers (North, 1973). This transfer of secondary immunity can be abrogated by treating the spleen cells with antilymphocyte and anti-T-cell antibodies, but not anti-B-cell

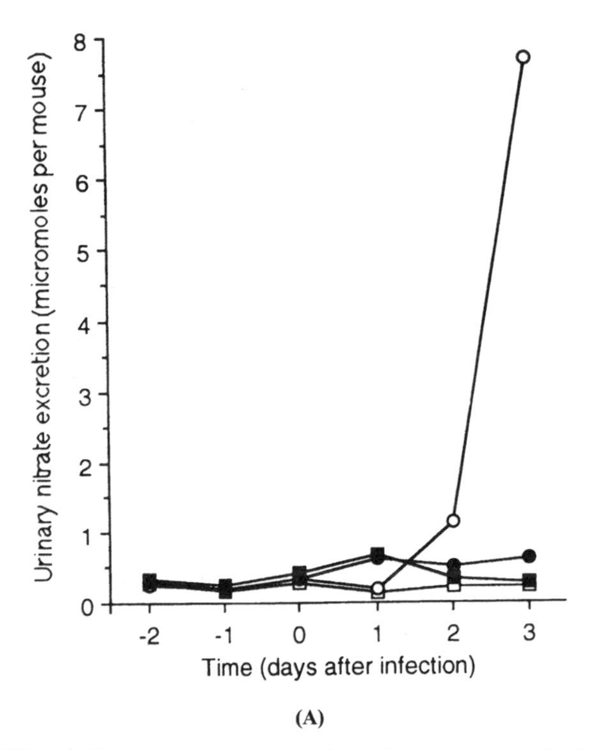

(A)

FIGURE 5. Effect of adoptive immunization on urinary nitrate excretion and splenic iNOS mRNA production during *Listeria monocytogenes* infection in C57BL/6 mice. (A) On day 0 adoptively immunized mice were injected intravenously with cells from the spleens of mice 7 days convalescent from a primary sublethal *L. monocytogenes* infection (closed circles and squares). Two groups of mice were not immunized (open circles and squares). One hour later, mice were injected intravenously with $\sim 2\,LD_{50}$ *L. monocytogenes* (circles) or buffer (squares). There was no detectable urinary nitrate excretion in adoptively immunized infected mice, whereas unimmunized infected mice had a steep rise in urinary nitrate excretion. Each point represents urinary nitrate per mouse per 24 hr averaged from a group of four mice. (B) Three adoptively immunized (AT lanes 1–3) and three unimmunized (1° lanes 1–3) mice were injected intravenously with $\sim 2\,LD_{50}$ *L. monocytogenes* and sacrificed on day 2 after injection. Total splenic RNA was probed for iNOS mRNA with a radiolabeled 3.4-kb fragment by Northern blot. iNOS mRNA appears in the spleens of unimmunized mice, but not adoptively immunized mice on day 2 of infection. Each lane represents data from one animal.

antibodies, suggesting that the protective capacity resides with T cells (Mackaness and Hill, 1969; Lane and Unanue, 1972; North, 1973). Furthermore, *Listeria*-immune mice depleted of $CD8^+$ but not $CD4^+$ cells using selective antibodies have a markedly diminished capacity to clear a secondary *Listeria* infection from liver and spleen. In $CD4^+$ cell-depleted *Listeria*-immune mice, clearance occurs normally despite the inability to form granulomas (Mielke *et al.*,1989). Serum transfer confers no antilisterial protection, showing that antibodies are not required for the secondary response (Miki and Mackaness, 1964). Thus, anamnestic immunity appears to be effected by $CD8^+$ cells.

Like the late response to primary infection, secondary eradication of *Listeria* from immune mice appears to occur independent of iNOS activity. Inoculation of

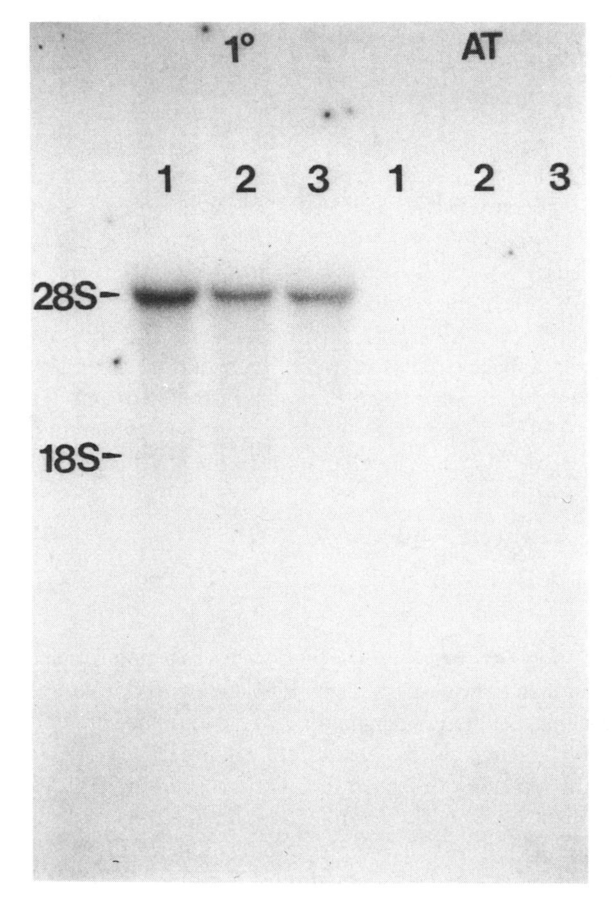

(B)

FIGURE 5. (*continued*)

Listeria into mice convalescing from a previous *Listeria* challenge induces 7-fold *lower* plasma nitrate levels than primary inoculation (Samsom *et al.*, 1996). Plasma nitrate peaks on day 2 of secondary listeriosis, in contrast to the later peak in primary infection. In addition, administering mice an L-arginine-deficient diet and L-NAME or aminoguanidine results in suppression of plasma nitrate, but no difference in visceral *Listeria* counts when compared with infected untreated mice. Nonimmune mice can be adoptively immunized by injection with spleen cells from immune mice, which confers the ability to efficiently clear large numbers of *Listeria* in an NO-independent fashion. Adoptively immunized mice injected with *Listeria* have 10^4- to 10^6-fold fewer organisms in their spleens and livers than infected unimmunized controls. *Listeria* infection causes no increase in urinary nitrate excretion and no detectable iNOS mRNA in splenic extracts of immunized animals, in contrast to controls (Fig. 5) (M. Maybodi, unpublished results). Oral administration of L-NMMA to adoptively immunized mice has no effect on hepatic or splenic *Listeria* counts, nor on mouse body weight, food intake, or mortality. Thus, multiple observations demonstrate that T cells do not depend on NO production for their antilisterial action.

More evidence that macrophage-derived NO is not required for immune defense against secondary listeriosis is provided by the observation that anti-IFNγ antibodies do not significantly exacerbate secondary infection in *Listeria*-immune mice, as measured by visceral *Listeria* counts on days 1–3 of infection (Samsom *et al.*, 1995, 1996). In contrast, anti-TNFα antibodies do exacerbate secondary infection (Samsom *et al.*, 1996), just as the enhanced toxoplasmastatic activity of peritoneal macrophages removed from mice during secondary *Toxoplasma gondii* infection can be abrogated by *in vivo* administration of anti-TNFα antibodies (Samsom *et al.*, 1995). These experiments suggest that cytotoxic $CD8^+$ T cells under the influence of TNFα and T-cell cytokines, but not IFNγ, mediate secondary immunity to listeriosis.

6. Conclusions

Murine listeriosis has been studied for decades as a model of acquired cellular immunity in mammals. The ease with which *Listeria* infection can be monitored in mice has generated a detailed understanding of the immune components needed at various stages of infection. Macrophages are acknowledged to be crucial for the initial nonspecific reaction to *Listeria* inoculation. Until the discovery of NO, the effector mechanism by which macrophages inhibit *Listeria* was unknown. The association of NO activity with macrophage activation and *Listeria* killing *in vitro*, and the readily demonstrable exacerbation of listeriosis in mice with impaired iNOS function, establish NO as an important effector molecule in this system. The increase in NO activity at the time of peak *Listeria* replication, prior to the development of an effective T-cell response, suggests that NO plays a distinct role in the early inflammatory response. The failure of NOS inhibition to affect the course of secondary listeriosis indicates that NO is a *non*specific effector molecule of macrophages, at least with regard to *Listeria*.

Further questions remain regarding the role of NO in listeriosis. How does NO affect *Listeria* organisms? Why is the interval of NO production during murine listeriosis limited in duration? Do *Listeria* become resistant to NO over time *in vivo*? Do *Listeria* induce NO production in humans? Research approaches to address some of these questions may include the engineering of NO-resistant mutants of *L. monocytogenes*, the creation of mice with controllable iNOS gene expression, and further studies of the role of NO production in human inflammation. Because *Listeria* and other NO-inhibited organisms cause significant morbidity and mortality in humans, an improved understanding of NO-dependent antilisterial activity will hopefully lead to advances in treatment.

References

Adler, H., Frech, B., Thony, M., Pfister, H., Peterhans, E., and Jungi, T. W., 1995, Inducible nitric oxide synthase in cattle: Differential cytokine regulation of nitric oxide synthase in bovine and murine macrophages, *J. Immunol.* **154:**4710–4718.

Beckerman, K. P., Rogers, H. W., Corbett, J. A., Schreiber, R. D., McDaniel, M. L., and Unanue, E. R., 1993, Release of nitric oxide during the T cell-independent pathway of macrophage activation: Its role in resistance to *Listeria monocytogenes, J. Immunol.* **150:**888–895.

Bogdan, C., Vodovotz, Y., and Nathan, C., 1991, Macrophage deactivation by interleukin 10, *J. Exp. Med.* **174:**1549–1555.

Boockvar, K. S., Granger, D. L., Poston, R. M., Maybodi, M., Washington, M. K., Hibbs, J. B., Jr., and Kurlander, R. L., 1994, Nitric oxide produced during murine listeriosis is protective, *Infect. Immun.* **62:**1089–1100.

Buchmeier, N. A., and Schreiber, R. D., 1985, Requirement of endogenous interferon-gamma production for resolution of *Listeria monocytogenes* infection, *Proc. Natl. Acad. Sci. USA* **82:**7404–7408.

Conlan, J. W., and North, R. J., 1991, Neutrophil-mediated dissolution of infected host cells as a defense strategy against a facultative intracellular bacterium, *J. Exp. Med.* **174:**741–744.

Conlan, J. W., and North, R. J., 1992, Monoclonal antibody NIMP-R10 directed against the CD11b chain of the type 3 complement receptor can substitute for monoclonal antibody 5C6 to exacerbate listeriosis by preventing the focusing of myelomonocytic cells at infectious foci in the liver, *J. Leukoc. Biol.* **52:**130–132.

Conlan, J. W., Dunn, P. L., and North, R. J., 1993, Leukocyte-mediated lysis of infected hepatocytes during listeriosis occurs in mice depleted of NK cells or CD4+ CD8+ Thy1.2+ T cells, *Infect. Immun.* **61:**2703–2707.

Cunha, F. Q., Moncada, S., and Liew, F. Y., 1992, Interleukin-10 (IL-10) inhibits the induction of nitric oxide synthase by interferon-gamma in murine macrophages, *Biochem. Biophys. Res. Commun.* **182:**1155–1159.

Curran, R. D., Billiar, T. R., Stuehr, D. J., Hofmann, K., and Simmons, R. L., 1989, Hepatocytes produce nitrogen oxides from L-arginine in response to inflammatory products of Kupffer cells, *J. Exp. Med.* **170:**1769–1774.

Czuprynski, C. J., Brown, J. F., Young, K. M., and Cooley, A. J., 1989, Administration of purified anti-L3T4 monoclonal antibody impairs the resistance of mice to *Listeria monocytogenes* infection, *Infect. Immun.* **57:**100–109.

Ding, A. H., Nathan, C. F., and Stuehr, D. J., 1988, Release of reactive nitrogen intermediates and reactive oxygen intermediates from mouse peritoneal macrophages: Comparison of activating cytokines and evidence for independent production, *J. Immunol.* **141:**2407–2412.

Drevets, D. A., Canono, B. P., Leenen, P. J., and Campbell, P. A., 1994, Gentamicin kills intracellular *Listeria monocytogenes, Infect. Immun.* **62:**2222–2228.

Emmerling, P., Finger, H., and Hof, H., 1977, Cell-mediated resistance to infection with *Listeria monocytogenes* in nude mice, *Infect. Immun.* **15:**382–385.

Fehr, T., Schoedon, G., Odermatt, B., Holtschke, T., Schneemann, M., Bachmann, M. F., Mak, T. W., Horak, I., and Zinkernagel, R. M., 1997, Crucial role of interferon consensus sequence binding protein, but neither of interferon regulatory factor 1 nor of nitric oxide synthesis for protection against murine listeriosis, *J. Exp. Med.* **185:**921–931.

Flesch, I. E., Hess, J. H., and Kaufmann, S. H., 1994, NADPH diaphorase staining suggests a transient and localized contribution of nitric oxide to host defence against an intracellular pathogen *in situ, Int. Immunol.* **6:**1751–1757.

Frei, K., Nadal, D., Pfister, H. W., and Fontana, A., 1993, *Listeria* meningitis: Identification of a cerebrospinal fluid inhibitor of macrophage listericidal function as interleukin 10, *J. Exp. Med.* **178:**1255–1261.

Garcia, I., Miyazaki, Y., Araki, K., Araki, M., Lucas, R., Grau, G. E., Milon, G., Belkaid, Y., Montixi, C., Lesslauer, W., and Vassalli, P., 1995, Transgenic mice expressing high levels of soluble TNF-R1 fusion protein are protected from lethal septic shock and cerebral malaria, and are highly sensitive to *Listeria monocytogenes* and *Leishmania major* infections, *Eur. J. Immunol.* **25:**2401–2407.

Godfrey, R. W., and Wilder, M. S., 1984, Relationships between oxidative metabolism, macrophage activation, and antilisterial activity, *J. Leukoc. Biol.* **36:**533–543.

Granger, D. L., Hibbs, J. B., Jr., and Broadnax, L. M., 1991, Urinary nitrate excretion in relation to murine macrophage activation: Influence of dietary L-arginine and oral N^{G}-monomethyl-L-arginine, *J. Immunol.* **146:**1294–1302.

Granger, D. L., Taintor, R. R., Boockvar, K. S., and Hibbs, J. B., Jr., 1996, Measurement of nitrate and nitrite in biological samples using nitrate reductase and Griess reaction, *Methods Enzymol.* **268:**142–151.

Gregory, S. H., Barczynski, L. K., and Wing, E. J., 1992, Effector function of hepatocytes and Kupffer cells in the resolution of systemic bacterial infections, *J. Leukoc. Biol.* **51:**421–424.

Gregory, S. H., Wing, E. J., Hoffman, R. A., and Simmons, R. L., 1993, Reactive nitrogen intermediates suppress the primary immunologic response to *Listeria, J. Immunol.* **150:**2901–2909.

Gregory, S. H., Sagnimeni, A. J., and Wing, E. J., 1994, Arginine analogues suppress antigen-specific and -nonspecific T lymphocyte proliferation, *Cell. Immunol.* **153:**527–532.

Havell, E. A., 1989, Evidence that tumor necrosis factor has an important role in antibacterial resistance, *J. Immunol.* **143:**2894–2899.

Hibbs, J. B., Jr., Taintor, R. R., Vavrin, Z., and Rachlin, E. M., 1988, Nitric oxide: A cytotoxic activated macrophage effector molecule, *Biochem. Biophys. Res. Commun.* **157:**87–94.

Higginbotham, J. N., and Pruett, S. B., 1994, Assessment of the correlation between nitrite concentration and listericidal activity in cultures of resident and elicited murine macrophages, *Clin. Exp. Immunol.* **97:**100–106.

Higginbotham, J. N., Lin, T. L., and Pruett, S. B., 1992, Effect of macrophage activation on killing of *Listeria monocytogenes:* Roles of reactive oxygen or nitrogen intermediates, rate of phagocytosis, and retention of bacteria in endosomes, *Clin. Exp. Immunol.* **88:**492–498.

Huang, S., Hendriks, W., Althage, A., Hemmi, S., Bluethmann, H., Kamijo, R., Vilcek, J., Zinkernagel, R. M., and Aguet, M., 1993, Immune response in mice that lack the interferon-gamma receptor, *Science* **259:**1742–1745.

Inoue, S., Itagaki, S., and Amano, F., 1995, Intracellular killing of *Listeria monocytogenes* in the J774.1 macrophage-like cell line and the lipopolysaccharide (LPS)-resistant mutant LPS1916 cell line defective in the generation of reactive oxygen intermediates after LPS treatment, *Infect. Immun.* **63:**1876–1886.

Kamijo, R., Harada, H., Matsuyama, T., Bosland, M., Gerecitano, J., Shapiro, D., Le, J., Koh, S. I., Kimura, T., Green, S. J., Mak, T. W., Taniguchi, T., and Vilcek, J., 1994, Requirement for transcription factor IRF-1 in NO synthase induction in macrophages, *Science* **263:**1612–1615.

Kiderlen, A. F., Kaufmann, S. H., and Lohmann-Matthes, M. L., 1984, Protection of mice against the intracellular bacterium *Listeria monocytogenes* by recombinant immune interferon, *Eur. J. Immunol.* **14:**964–967.

Kurtz, R. S., Young, K. M., and Czuprynski, C. J., 1989, Separate and combined effects of recombinant interleukin-1 alpha and gamma interferon on antibacterial resistance, *Infect. Immun.* **57:**553–558.

Lane, F. C., and Unanue, E. R., 1972, Requirement of thymus (T) lymphocytes for resistance to listeriosis, *J. Exp. Med.* **135:**1104–1112.

Langermans, J. A., van der Hulst, M. E., Nibbering, P. H., van der Meide, P. H., and van Furth, R., 1992a, Intravenous injection of interferon-gamma inhibits the proliferation of *Listeria monocytogenes* in the liver but not in the spleen and peritoneal cavity, *Immunology* **77:**354–361.

Langermans, J. A., van der Hulst, M. E., Nibbering, P. H., and van Furth, R., 1992b, Endogenous tumor necrosis factor alpha is required for enhanced antimicrobial activity against *Toxoplasma gondii* and *Listeria monocytogenes* in recombinant gamma interferon-treated mice, *Infect. Immun.* **60:**5107–5112.

Leenen, P. J., Canono, B. P., Drevets, D. A., Voerman, J. S., and Campbell, P. A., 1994, TNF-alpha and IFN-gamma stimulate a macrophage precursor cell line to kill *Listeria monocytogenes* in a nitric oxide-independent manner, *J. Immunol.* **153:**5141–5147.

Mackaness, G. B., 1962, Cellular resistance to infection, *J. Exp. Med.* **116:**381–406.

Mackaness, G. B., 1969, The influence of immunologically committed lymphoid cells on macrophage activity *in vivo, J. Exp. Med.* **129:**973–992.

Mackaness, G. B., and Hill, W. C., 1969, The effect of anti-lymphocyte globulin on cell-mediated resistance to infection, *J. Exp. Med.* **129:**993–1012.

MacMicking, J. D., Nathan, C., Hom, G., Chartrain, N., Fletcher, D. S., Trumbauer, M., Stevens, K., Xie, Q.-W., Sokol, K., Hutchinson, N., Chen, H., and Mudgett, J. S., 1995, Altered responses to bacterial infection and endotoxic shock in mice lacking inducible nitric oxide synthase, *Cell* **81:**641–650.

Mielke, M. E., Ehlers, S., and Hahn, H., 1988, T-cell subsets in delayed-type hypersensitivity, protection, and granuloma formation in primary and secondary *Listeria* infection in mice: Superior role of Lyt-2+ cells in acquired immunity, *Infect. Immun.* **56:**1920–1925.

Mielke, M. E., Niedobitek, G., Stein, H., and Hahn, H., 1989, Acquired resistance to *Listeria monocytogenes* is mediated by Lyt-2+ T cells independently of the influx of monocytes into granulomatous lesions, *J. Exp. Med.* **170:**589–594.

Miki, K., and Mackaness, G. B., 1964, The passive transfer of acquired resistance to *Listeria monocytogenes, J. Exp. Med.* **120:**93–103.

Moore, W. M., Webber, R. K., Jerome, G. M., Tjoeng, F. S., Misko, T. P., and Currie, M. G., 1994, L-N^6-(1-Iminoethyl)lysine: A selective inhibitor of inducible nitric oxide synthase, *J. Med. Chem.* **37:**3886–3888.

Nakane, A., Numata, A., Asano, M., Kohanawa, M., Chen, Y., and Minagawa, T., 1990, Evidence that endogenous gamma interferon is produced early in *Listeria monocytogenes* infection, *Infect. Immun.* **58:**2386–2388.

Nakane, M., Klinghofer, V., Kuk, J. E., Donnelly, J. L., Budzik, G. P., Pollock, J. S., Basha, F., and Carter, G. W., 1995, Novel potent and selective inhibitors of inducible nitric oxide synthase, *Mol. Pharmacol.* **47:**831–834.

Newborg, M. F., and North, R. J., 1980, On the mechanism of T cell-independent anti-*Listeria* resistance in nude mice, *J. Immunol.* **124:**571–576.

North, R. J., 1973, Cellular mediators of anti-*Listeria* immunity as an enlarged population of short lived, replicating T cells. Kinetics of their production, *J. Exp. Med.* **138:**342–355.

Pfeffer, K., Matsuyama, T., Kundig, T. M., Wakeham, A., Kishihara, K., Shahinian, A., Wiegmann, K., Ohashi, P. S., Kronke, M., and Mak, T. W., 1993, Mice deficient for the 55 kD tumor necrosis factor receptor are resistant to endotoxic shock, yet succumb to *L. monocytogenes* infection, *Cell* **73:**457–467.

Portnoy, D. A., Schreiber, R. D., Connelly, P., and Tilney, L. G., 1989, Gamma interferon limits access of *Listeria monocytogenes* to the macrophage cytoplasm, *J. Exp. Med.* **170:**2141–2146.

Poston, R. M., and Kurlander, R. J., 1991, Analysis of the time course of IFN-γ and protein production during primary murine listeriosis, *J. Immunol.* **146:** 4333–4337.

Poston, R. M., and Kurlander, R. J., 1992, Cytokine expression *in vivo* during murine listeriosis, *J. Immunol.* **149:**3040–3044.

Rogers, H. W., Callery, M. P., Deck, B., and Unanue, E. R., 1996, *Listeria monocytogenes* induces apoptosis of infected hepatocytes, *J. Immunol.* **156:**679–684.

Rosen, H., Gordon, S., and North, R. J., 1989, Exacerbation of murine listeriosis by a monoclonal antibody specific for the type 3 complement receptor of myelomonocytic cells. Absence of monocytes at infective foci allows *Listeria* to multiply in nonphagocytic cells, *J. Exp. Med.* **170:**27–37.

Rothe, J., Lesslauer, W., Lotscher, H., Lang, Y., Koebel, P., Kontgen, F., Althage, A., Zinkernagel, R., Steinmetz, M., and Bluethmann, H., 1993, Mice lacking the tumour necrosis factor receptor 1 are resistant to TNF-mediated toxicity but highly susceptible to infection by *Listeria monocytogenes, Nature* **364:**798–802.

Rutherford, M. S., and Schook, L. B., 1992, Differential immunocompetence of macrophages derived using macrophage or granulocyte-macrophage colony-stimulating factor, *J. Leukoc. Biol.* **51:**69–76.

Samsom, J. N., Langermans, J. A., Savelkoul, H. F., and van Furth, R., 1995, Tumour necrosis factor, but not interferon-gamma, is essential for acquired resistance to *Listeria monocytogenes* during a secondary infection in mice, *Immunology* **86:**256–262.

Samsom, J. N., Langermans, J. A., Groeneveld, P. H., and van Furth, R., 1996, Acquired resistance against a secondary infection with *Listeria monocytogenes* in mice is not dependent on reactive nitrogen intermediates, *Infect. Immun.* **64:**1197–1202.

Sasaki, T., Mieno, M., Udono, H., Yamaguchi, K., Usui, T., Hara, K., Shiku, H., and Nakayama, E., 1990, Roles of $CD4^+$ and $CD8^+$ cells, and the effect of administration of recombinant murine interferon gamma in listerial infection, *J. Exp. Med.* **171:**1141–1154.

Stokvis, H., Langermans, J. A., de Backer-Vledder, E., van der Hulst, M. E., and van Furth, R., 1992, Hydrocortisone treatment of BCG-infected mice impairs the activation and enhancement of antimicrobial activity of peritoneal macrophages, *Scand. J. Immunol.* **36:**299–305.

Szalay, G., Hess, J., and Kaufmann, S. H., 1995, Restricted replication of *Listeria monocytogenes* in a gamma interferon-activated murine hepatocyte line, *Infect. Immun.* **63:**3187–3195.

Tanaka, T., Akira, S., Yoshida, K., Umemoto, M., Yoneda, Y., Shirafuji, N., Fujiwara, H., Suematsu, S., Yoshida, N., and Kishimoto, T., 1995, Targeted disruption of the NF-IL6 gene discloses its essential role in bacteria killing and tumor cytotoxicity by macrophages, *Cell* **80:**353–361.

Tilney, L. G., and Portnoy, D. A., 1989, Actin filaments and the growth, movement, and spread of the intracellular bacterial parasite, *Listeria monocytogenes, J. Cell Biol.* **109:**1597–1608.

Xie, Q.-W., Cho, H. J., Calaycay, J., Mumford, R. A., Swiderek, K. M., Lee, T. D., Ding, A., Troso, T., and Nathan, C., 1992, Cloning and characterization of inducible nitric oxide synthase from mouse macrophages, *Science* **256:**225–228.

Xiong, H., Kawamura, I., Nishibori, T., and Mitsuyama, M., 1996, Suppression of IFN-gamma production from *Listeria monocytogenes*-specific T cells by endogenously produced nitric oxide, *Cell. Immunol.* **172:**118–125.

Part E

Future Directions

CHAPTER 23

Therapeutic Applications of Nitric Oxide in Infection

ANDRÉS VAZQUEZ-TORRES and FERRIC C. FANG

1. Introduction

Nitric oxide (NO) has been an object of intensive investigation as a possible therapeutic agent or target for the treatment of multiple disease conditions ever since it was discovered to be a product of eukaryotic cell metabolism. Indeed, numerous clinical trials have demonstrated that NO, NO donors, or NO scavengers can be used to treat a vast array of circulatory and respiratory ailments. To cite just a few examples, *S*-nitrosoglutathione, nitrate, L-arginine, sodium nitroprusside, and NO gas have been administered orally, topically, parentally, or inhalationally to treat disorders as varied as interstitial cystitis, heart failure, preeclampsia, penile erectile dysfunction, respiratory distress syndrome, and angina pectoris (Berrazueta *et al.*, 1994; Karamanoukian *et al.*, 1994; Langford *et al.*, 1994; Pedrinelli *et al.*, 1995; Wegner and Knispel, 1995). Moreover, several established pharmacological agents (e.g., aspirin, corticosteroids, tetracyclines, cyclosporin) have been only recently discovered to have significant effects on endogenous NO production (DiRosa *et al.*, 1990; Aeberhard *et al.*, 1995; Amin *et al.*, 1995, 1996; Conde *et al.*, 1995; Wu *et al.*, 1995; Walker *et al.*, 1997), suggesting that many longstanding treatment modalities may work, at least in part, via their effects on NO.

Preceding chapters in this volume have documented numerous examples in which NO overproduction can be detrimental during infection, resulting in vascular collapse or tissue injury (Chapters 8, 13, 19, 21). Yet, we have also seen that NO and its derivatives are potent mediators of cellular immunity and constitute an integral

ANDRÉS VAZQUEZ-TORRES and FERRIC C. FANG • Departments of Medicine, Pathology, and Microbiology, University of Colorado Health Sciences Center, Denver, Colorado 80262.

Nitric Oxide and Infection, edited by Fang. Kluwer Academic / Plenum Publishers, New York, 1999.

component of the host's antimicrobial arsenal against many helminths, protozoans, fungi, bacteria, and viruses (reviewed in Chapter 12) (James, 1995), although at the present time there are only a few examples in which the antimicrobial potential of NO has been therapeutically exploited for the treatment of infections. A rapidly growing understanding of the role of NO in infectious processes and the development of an expanding variety of pharmacological NO agonists and antagonists make prospects for NO-based therapy of infection increasingly feasible. In this chapter we present evidence indicating that manipulation of NO can indeed provide therapeutic benefit in infectious diseases.

2. NO Antagonism in the Treatment of Infection

The first application of NO as an antimicrobial agent was probably the addition of nitrites to food products. Because nitrites generate NO, *S*-nitrosothiols, and other reactive nitrogen intermediates, they inhibit microbial multiplication and impart an appealing color to meat via reaction with the heme group of myoglobin. However, despite this well-recognized antimicrobial activity, most NO-related therapeutic interventions in infectious diseases have actually focused on the elimination of pathological side effects arising from NO overproduction.

2.1. Septic Shock

Septic shock (see also Chapters 7 and 13), a syndrome characterized by fever, hypotension, heart failure, tachycardia, tachypnea, respiratory insufficiency, central and peripheral hypoxemia, oliguria, and disseminated intravascular coagulation, typically results from the massive stimulation of monocytes and endothelial cells by microbial cell wall constituents such as LPS, peptidoglycan, or lipoteichoic acid. Excessive quantities of proinflammatory cytokines including IL-1β, IL-6, IL-8, IFNγ, and TNFα are detectable systemically, along with nitrogen oxides. Numerous observations in experimental animal models of septic shock and in infected humans suggest that NO is responsible for many of the hemodynamic alterations that characterize this syndrome (Wright *et al.*, 1992; Gomez-Jimenez *et al.*, 1995) (see also Chapters 7 and 13). Ever since the initial association of NO with the pathophysiology of septic shock, tremendous attention has been focused on the potential therapeutic benefit of NO synthase (NOS) inhibition in this setting (reviewed in Palmer, 1993; Thiemermann, 1994; Evans and Cohen, 1995; Kilbourn *et al.*, 1997a,b).

In support of the hypothesis that NO contributes to the pathology of septic shock, the NOS inhibitor N^{G}-monomethyl-L-arginine (L-NMMA) increases the blood pressure and systemic vascular resistance in endotoxin-induced septic shock in dogs; similar observations have been made in a limited number of patients with sepsis (Kilbourn *et al.*, 1990; Petros *et al.*, 1991). However, NOS inhibition normalizes pulmonary arterial pressure in only a minority of septic patients suffering from acute respiratory distress syndrome (Krafft *et al.*, 1996), and the salutary rise in blood pressure is typically accompanied by an undesirable fall in cardiac index (Petros *et al.*, 1994; Mitaka *et al.*, 1995; Jourdain *et al.*, 1997). Although normal blood pressure can be restored in most cases, inhibition of NOS has not increased survival (Wright *et al.*, 1992; Krafft *et al.*, 1996; Park *et al.*, 1996).

One reason for these disappointing preliminary results may be the protective role of low-level NO production by endothelial NOS. Therefore, nonspecific inhibition of all NOS isoforms by inhibitors such L-NMMA or N^G-nitro-L-arginine (L-NNA) may produce both beneficial and detrimental effects. In support of this interpretation, relatively specific inducible NOS (iNOS) inhibitors such as L-canavanine or *S*-methylisothiourea both stabilize blood pressure and increase survival in endotoxemic rats or mice, effects that are not observed when nonselective inhibitors are used (Szabo *et al.*, 1994; Teale and Atkinson, 1994; Liaudet *et al.*, 1998) . Also, mice genetically deficient in iNOS have reduced mortality and hypotension induced by LPS (Nathan, 1995; Wei *et al.*, 1995). Thus, selective inhibition of the iNOS isoform may be a more appropriate therapeutic approach to septic shock. The recent structural resolution of the iNOS oxygenase domain and inhibitor complexes should expedite the development of novel selective inhibitors (Crane *et al.*, 1997). As discussed in Chapter 7, the dosage and timing of drug administration and the patient's fluid status appear to be additional important factors determining the benefit of NOS inhibition during sepsis. Another possible confounding factor is the role of NO as an antimicrobial mediator (Chapter 12); beneficial hemodynamic effects of NOS inhibition could be counterbalanced by enhanced microbial proliferation. However, the combination of effective antimicrobial therapy and iNOS inhibition may circumvent this problem. Indeed, Teale and Atkinson (1992) found that NOS inhibition is beneficial in an experimental model of bacterial peritonitis when effective antibiotics are coadministered.

An alternative to inhibiting NO synthesis is the removal of NO from the circulation. Maeda *et al.* (1995) have shown that imidazolineoxyl *N*-oxide, an effective NO scavenger, can prolong the survival of LPS-treated rats. Increased survival coincided with an improvement in the mean arterial pressure. Iron chelates have been used in a murine model to bind NO and decrease sepsis-associated mortality (Kazmierski *et al.*, 1996). Hemoglobin is another NO scavenger that has been used to treat the sepsis syndrome. Hemoglobin can undergo *S*-nitrosylation as well as heme–NO interactions (Gow and Stamler, 1998). In rat or ovine models of endotoxic shock, polymerized hemoglobin has been shown to restore mean arterial pressure and heart rate without interfering with renal function, in contrast to the NOS inhibitor L-NNA (De Angelo, 1997; Heneka *et al.*, 1997). Polymerized hemoglobin remains in the circulation as a consequence of its high molecular weight, which may permit NO in interstitial and intracellular compartments to continue mediating physiologic NO actions. Preliminary observations in healthy volunteers, patients with septic shock, or patients receiving adjunctive cytokine therapy for cancer encourage further evaluation of polymerized hemoglobin for the treatment of sepsis-related hypotension (De Angelo, 1997; Kilbourn, 1997; Reah *et al.*, 1997), although the enhanced mortality following coadministration of LPS and hemoglobin to mice emphasizes the need for caution (Su *et al.*, 1997).

Another alternative therapeutic approach in settings of NO overproduction is the downregulation of the NOS enzyme itself. For example, the protection against LPS-induced shock conferred by tetracycline or doxycycline appears to result from a decrease in IL-1β, TNFα and iNOS expression (Milano *et al.*, 1997). Abnormally high TNFα and IL-1β production are hallmarks of septic shock; therefore, considerable effort has been spent in developing anticytokine-based therapies (Dinarello, 1995) that can indirectly reduce NO production. However, treatment with certain cytokines may also be beneficial in septic shock. In a murine model of posttraumatic sepsis, increased survival conferred by GM-CSF correlated with decreased macrophage NO-producing capacity (Austin *et al.*, 1995).

Because of the diverse etiologies of septic shock, the complex interactions of the multiple inflammatory mediators produced, and the complex roles of NO in this syndrome, it is unlikely that NO inhibition alone will provide a panacea for sepsis. Nevertheless, selective iNOS inhibitors may well become an important component of a multifaceted therapeutic approach in the future.

2.2. Other Infections

Although not as extensively investigated as septic shock, NO also contributes to the immunopathology of many other infectious diseases (e.g., Chapters 8, 19–21) (Khan *et al.*, 1997). The pathology of whooping cough can be mimicked *in vitro* by tracheal cytotoxin, a muramyl peptide produced by *Bordetella pertussis*. Goldman and collaborators have reported that tracheal cytotoxin triggers epithelial NO production, leading to autodestruction of the epithelium (Heiss *et al.*, 1994; Flak and Goldman, 1996). The NOS inhibitors L-NMMA and aminoguanidine can attenuate the ciliostasis and epithelial cell death caused by tracheal cytotoxin (Heiss *et al.*, 1994), raising the possibility that NO inhibition might ameliorate the clinical manifestations of whooping cough *in vivo*.

The therapeutic potential of NOS inhibition has also been investigated in animal models of acute viral pneumonitis. Excessive production of NO elicited during influenza virus infection appears to play a crucial role in the associated respiratory tract pathology (Akaike *et al.*, 1996; see Chapter 18). Akaike and colleagues have provided evidence that nitrotyrosine, an oxidative signature of peroxynitrite or certain other NO congeners, accumulates in macrophages, neutrophils, and intraalveolar exudate from influenza-infected lungs. Treatment with L-NMMA improved the survival of the mice with influenza pneumonitis, without affecting viral replication. In a murine cytomegalovirus-associated immune-mediated pneumonitis model, NO antagonism was found to be beneficial despite the absence of tyrosine nitration, suggesting that NO overproduction can be detrimental for lung tissue even in the absence of peroxynitrite formation (Tanaka *et al.*, 1997). Inhibition of NO synthesis has been shown to decrease lethality in a murine model of herpes simplex virus (HSV) pneumonitis (Adler *et al.*, 1997), despite *in vitro* evidence that NO is a potent inhibitor of HSV replication (Croen, 1993; Karupiah *et al.*, 1993; Komatsu *et al.*, 1996). In contrast to the absence of an effect on viral replication seen in the murine influenza model, NOS inhibition *in vivo* coincided with a significant augmentation of the HSV viral burden (Adler *et al.*, 1997). Nevertheless, L-NMMA treatment resulted in increased survival, increased pulmonary compliance, and decreased lymphocyte infiltration. Upregulation of iNOS is also observed in HSV encephalitis (Meyding-Lamade *et al.*, 1998), a devastating condition in which antiviral therapy has limited efficacy and immunomodulatory intervention is highly attractive. Khan and co-workers have recently shown that NOS inhibition reduces early mortality and tissue injury associated with acute toxoplasmosis in mice, despite an associated enhancement of parasite replication (Khan *et al.*, 1997). Together, these results illustrate that the importance of NO's immunopathological effects can supersede its antimicrobial actions in certain infections.

Deleterious effects of NO overproduction in chronic infections should not be overlooked. Infectious agents including *Helicobacter pylori, Schistosoma haematobium,* hepatitis C virus, and *Opisthorchis viverrini* have been strongly correlated with both NO overproduction and carcinogenesis (Ohshima and Bartsch, 1994; Warren *et al.*, 1995; Satarug *et al.*, 1996; Tsuji *et al.*, 1996; Kane *et al.*, 1997). Eradication of *H. pylori* from gastric lesions using a combination of antimicrobial agents and antioxidants can reduce iNOS expression and nitrotyrosine formation in the gastric mucosa (Mannick *et al.*, 1996). Because the overproduction of NO has been proposed to be a genotoxic mechanism leading to the development of cancer, reduction of NO synthesis related to chronic infection could have far-reaching clinical implications (Bartsch *et al.*, 1992; De Koster *et al.*, 1994; Fox, 1994).

3. Nutritional Modulation of NO-Mediated Host Resistance

The disproportionally high incidence of tuberculosis in developing countries and in immunosuppressed individuals may be partly attributable to malnutrition. An iNOS-dominated immune response correlates with a favorable prognosis in mice and humans suffering from tuberculosis (Nicholson *et al.*, 1996; MacMicking *et al.*, 1997b). In an interesting report, Chan *et al.* (1996) recently demonstrated that malnourished mice infected with *M. tuberculosis* exhibit a reduced granulomatous reaction, low expression of iNOS in pulmonary tissue, and an increased mycobacterial burden. These signs were reversed after nutritional supplementation, suggesting that proper nutrition can boost NO-mediated antimicrobial immunity. Such noninvasive measures might dramatically reduce the incidence of tuberculosis and other "opportunistic" pathogens in impoverished populations throughout the world.

The risk of microbial translocation from the gastrointestinal tract to systemic sites may also be amenable to nutritional NO-related intervention. Dietary supplementation with L-arginine can improve the survival of mice suffering from sepsis-related experimentally induced peritonitis or extensive burns (Gianatti *et al.*, 1993; Gennari and Alexander, 1997; Horton *et al.*, 1998). Although the enhanced resistance observed in L-arginine-supplemented mice could be related to a nonspecific stimulation of a T-cell-mediated immunity (Barbul *et al.*, 1980; Kirk *et al.*, 1992), reversal of L-arginine's salutary effects by the NOS inhibitor L-NNA strongly suggests that NO is involved (Gianatti *et al.*, 1993).

4. Indirect NO Antimicrobial Therapy

Many cytokines as well as some transduction pathways regulating expression of iNOS have been identified (MacMicking *et al.*, 1997a) (Chapter 5). Although a comprehensive review of the modulation of cytokines for the treatment of disease is beyond the scope of this chapter, it must be considered that certain effects of these therapies are likely to be mediated by NO. Some investigators have shown that genetic or immunological depletion of cytokines can abrogate iNOS expression and increase susceptibility to infection (Kimura *et al.*, 1994). Similarly, cytokine therapy can enhance resistance to infection. To cite just a few examples, GM-CSF, IFNγ, or IL-12 therapy can stimulate a robust NO response and increase host resistance to *Candida albicans*, *Histoplasma capsulatum*, *Cryptococcus neoformans*, *Leishmania donovani*, or *L. major* in animal models (Hill *et al.*, 1995; Lovchik *et al.*, 1995; Kawakami *et al.*, 1997; Taylor and Murray, 1997; Zhou *et al.*, 1997). IFNα has been demonstrated to induce iNOS expression in human mononuclear cells during treatment of patients with hepatitis C infection (Sharara *et al.*, 1997) (Chapter 6), and the beneficial effects of IFNα in this infectious condition could be attributable to its effects on endogenous NO production. Xu *et al.* (1998) have recently used the novel approach of expressing migration inhibitor factor, IL-2, IFNγ, and TNFα from recombinant attenuated *Salmonella* strains; these constructs were able to enhance endogenous iNOS expression and reduce parasite burden in *Leishmania*-infected mice.

The action of amphotericin B, the antifungal agent of choice for many opportunistic systemic infections including candidiasis, cryptococcosis, blastomycosis, and histoplasmosis (Abu-Salah, 1996), has traditionally been associated with its capacity to interfere with the synthesis of ergosterol, a cholesterol-like constituent of the fungal membrane. More recently, it has become evident that amphotericin B also possesses immunomodulatory properties. Amphotericin B-dependent stimulation of macrophage TNFα and IL-1 production (Yamagushi *et al.*, 1993; Louie *et al.*, 1994; Tohyama *et al.*, 1996) along with enhancement of iNOS expression appears to be required for its anticryptococcal activity in an *in vitro* macrophage model (Tohyama *et al.*, 1996). Ironically, these immunomodulatory actions may partially account both for amphotericin B's antifungal actions and for its undesirable systemic side effects.

5. Direct NO Antimicrobial Therapy

Although *in vitro* and *in vivo* studies have demonstrated that reactive nitrogen intermediates possess broad-spectrum antimicrobial activity (Chapter 12), few researchers have yet investigated the direct therapeutic potential of these compounds. NO-related antimicrobial activity can potentiate the effects of other antimicrobial agents. For example, in an *in vitro* system, diazenium diolate NO-donors were shown to synergize with fluconazole, miconazole, or ketoconazole against strains of *C. albicans*, *C. krusei*, *C. parapsilosis*, and *C. tropicalis* (McElhaney-Feser *et al.*, 1997).

By inhibiting the enzyme ribonucleotide reductase, hydroxyurea blocks deoxynucleotide synthesis and interferes with HIV-1 replication (Lori *et al.*, 1994). This observation has prompted the evaluation of hydroxyurea in combination with other antiretroviral agents for the treatment of HIV-infected patients, with encouraging preliminary results (Rossero *et al.*, 1997). Hydroxyurea has been shown to eradicate Epstein–Barr virus episomes in *in vitro* experimental models (Chodosh *et al.*, 1998), suggesting that this agent might also be useful in other viral infections. Noting the structural similarity with N^{ω}-hydroxy-L-arginine, Kwon *et al.* (1991) have investigated whether hydroxyurea might be generating NO, a known potent inhibitor of ribonucleotide reductase. In fact, catalyzed either by hydrogen peroxide and a transition metal or by hemoproteins, NO can be formed from hydroxyurea (Kwon *et al.*, 1991; Pacelli *et al.*, 1996). These studies suggest that at least some of the antiviral activity of hydroxyurea might be mediated by NO, and should prompt the investigation of additional NO-based antiviral therapeutic strategies.

An encouraging multinational clinical study conducted by Drs. Patricio Lopez-Jaramillo, Salvador Moncada, and collaborators has tested the therapeutic potential of the NO donor *S*-nitroso-*N*-acetyl-penicillamine (SNAP) used topically in cutaneous infectious diseases. Application of a cream containing 200 μM SNAP to cutaneous lesions caused by the fungi *Trichophyton tonsurans*, *T. mentagraphytes*, *Epidermophyton floccosum*, or *C. albicans*, or by the protozoan *Leishmania mexicana*, resulted in both an improvement in clinical signs and the resolution of infection as demonstrated by sterilization of the affected site (Lopez-Jaramillo *et al.*, 1995, 1998). A double-blind and more extensive study assessing the therapeutic applications of topical SNAP is now under way in Ecuador. Acidified nitrite cream has also been used for the treatment of tinea pedis (Weller *et al.*, 1998). The accessibility of skin lesions makes cutaneous infection a particularly attractive setting in which to test the feasibility of NO-based antimicrobial therapy. Treatment of infection at other tissue sites may need to await the development of more sophisticated drug delivery strategies.

6. Conclusions

Our understanding of the complex roles of NO in infection has advanced remarkably during the past decade. However, the practical application of this knowledge to the prevention or treatment of infection has only scratched the surface. A better understanding of the NO-producing host cells and tissues that participate in the immune response to infection, the NO congeners that mediate immunopathology or host resistance to infectious diseases, the microbial species and critical molecular targets of specific NO metabolites, and the mechanisms that microbes use to avoid or resist NO congeners will ultimately contribute to the rational utilization of NO-based antimicrobial therapies. Emerging problems with resistant or refractory infections (Neu, 1992; Ash, 1996; Gold and Moellering, 1996; Nicolle *et al.*, 1996) make novel NO-based approaches attractive as therapeutic alternatives. Utilization of NO-based therapies will be further facilitated by the development of new NO donors, NO scavengers, and selective NOS inhibitors.

Delivery systems to target specific organs or tissues may also expedite the use of NO-modulating drugs for the therapy of localized infectious diseases, and might lessen many of the unwanted side effects associated with NO therapies. One step toward this objective has been achieved by Saavedra and collaborators, who successfully engineered a drug, 1-(pyrrolidin-1-yl)diazen-1-ium-1,2-diolate, to deliver NO specifically to the liver (Saavedra *et al.*, 1997).

Initial applications of NO-based therapies have focused on cardiovascular and respiratory conditions. It is exciting to contemplate the expansion of this therapeutic revolution to the realm of infectious disease in the near future.

References

Abu-Salah, K. M., 1996, Amphotericin B: An update, *Br. J. Biomed. Sci.* **53:**122–133.

Adler, H., Beland, J. L., Del-Pan, N. C., Kobzik, L., Brewer, J. P., Martin, T. R., and Rimm, I. J., 1997, Suppression of herpes simplex virus type 1 (HSV-1)-induced pneumonia in mice by inhibition of inducible nitric oxide synthase (iNOS, NOS2), *J. Exp. Med.* **185:**1533–1540.

Aeberhard, E. E., Henderson, S. A., Arabolos, N. S., Griscavage, J. M., Castro, F. E., Barrett, C. T., and Ignarro, L. J., 1995, Nonsteroidal anti-inflammatory drugs inhibit expression of the inducible nitric oxide synthase gene, *Biochem. Biophys. Res. Commun.* **208:**1053–1059.

Akaike, T., Noguchi, Y., Ijiri, S., Setoguchi, K., Suga, M., Zheng, Y. M., Dietzschold, B., and Maeda, H., 1996, Pathogenesis of influenza virus-induced pneumonia: Involvement of both nitric oxide and oxygen radicals, *Proc. Natl. Acad. Sci. USA* **93:**2448–2453.

Amin, A. R., Vyas, P., Attur, M., Leszczynska-Piziak, J., Patel, I. R., Weissmann, G., and Abramson, S. B., 1995, The mode of action of aspirin-like drugs: Effect on inducible nitric oxide synthase, *Proc. Natl. Acad. Sci. USA* **92:**7926–7930.

Amin, A. R., Attur, M. G., Thakker, G. D., Patel, P. D., Vyas, P. R., Patel, R. N., Patel, I. R., and Abramson, S. B., 1996, A novel mechanism of action of tetracyclines: Effects on nitric oxide synthases, *Proc. Natl. Acad. Sci. USA* **93:**14014–14019.

Ash, C., 1996, Antibiotic resistance: The new apocalypse? *Trends Microbiol.* **4:**371–372.

Austin, O. M. B., Redmond, H. P., Watson, R. W. G., Cunney, R. J., Grace, P. A., and Boudhier-Hayes, D., 1995, The beneficial effects of immunostimulation in posttraumatic sepsis, *J. Surg. Res.* **59:**446–449.

Barbul, A., Wasserkrug, H. L., Seifter, E., Rettura, G., Levenson, S. M., and Efron, G., 1980, Immunostimulatory effects of arginine in normal and injured rats, *J. Surg. Res.* **29:**228–235.

Bartsch, H., Ohshima, H., Pignatelli, B., and Calmels, S., 1992, Endogenously formed *N*-nitroso compounds and nitrosating agents in human cancer etiology, *Pharmacogenetics* **2:**272–277.

Berrazueta, J. R., Fleitas, M., Salas, E., Amado, J. A., Poveda, J., Ochoteco, A., Sanchez de Vega, M. J., and Ruiz de Celis, G., 1994, Local transdermal glyceryl trinitrate has an antiinflammatory action on thrombophlebitis induced by sclerosis of leg varicose veins, *Angiology* **45:**347–351.

Chan, J., Tian, Y., Tanaka, K. E., Tsang, M. S., Yu, K., Salgame, P., Carroll, D., Kress, Y., Teitelbaum, R., and Bloom, B. R., 1996, Effects of protein calorie malnutrition on tuberculosis in mice, *Proc. Natl. Acad. Sci. USA* **93:**14857–14861.

Chodosh, J., Holder, V. P., Gan, Y. J., Belgaumi, A., Sample, J., and Sixbey, J. W., 1998, Eradication of latent Epstein–Barr virus by hydroxyurea alters the growth-transformed cell phenotype, *J. Infect. Dis.* **177:**1194–1201.

Conde, M., Andrade, J., Bedoya, F. J., Santa Maria, C., and Sobrino, F., 1995, Inhibitory effect of cyclosporin A and FK506 on nitric oxide production by cultured macrophages. Evidence of a direct effect on nitric oxide synthase activity, *Immunology* **84:**476–481.

Crane, B. R., Arvai, A. S., Gachhui, R., Wu, C., Ghosh, D. K., Getzoff, E. D., Stuehr, D. J., and Tainer, J. A., 1997, The structure of nitric oxide synthase oxygenase domain and inhibitor complexes, *Science* **278:**425–431.

Croen, K. D., 1993, Evidence for antiviral effect of nitric oxide. Inhibition of herpes simplex virus type 1 replication, *J. Clin. Invest.* 91:2446–2452.

De Angelo, J., 1997, PHP in the treatment of nitric oxide dependent shock, in: *Nitric Oxide: Novel Therapeutics for Clinical Application,* International Business Communications, Philadelphia.

De Koster, E., Buset, M., Fernandes, E., and Deltenre, M., 1994, *Helicobacter pylori*: The link with gastric cancer, *Eur. J. Cancer Prevent.* **3:**247–257.

Dinarello, C., 1995, Cytokines as mediators in the pathogenesis of septic shock, *Curr. Top. Microbiol. Immunol.* **196:**133–165.

DiRosa, M., Radomski, M., Carnuccio, R., and Moncada, S., 1990, Glucocorticoids inhibit the induction of nitric oxide synthase in macrophages, *Biochem. Biophys. Res. Commun.* **172:**1246–1252.

Evans, T. J., and Cohen, T., 1995, Nitric oxide and other toxic oxygen species, *Curr. Top. Microbiol. Immunol.* **216:**189–207.

Flak, T. A., and Goldman, E., 1996, Autotoxicity of nitric oxide in airway disease, *Am. J. Respir. Crit. Care Med.* **154:**S202-S206.

Fox, J. G., 1994, Gastric disease in ferrets: Effects of *Helicobacter mustelae*, nitrosamines and reconstructive gastric surgery, *Eur. J. Gastroenterol. Hepatol.* **6**(Suppl 1)**:**S57–S65.

Gennari, R., and Alexander, J. W., 1997, Arginine, glutamine, and dehydroepiandrosterone reverse the immunosuppressive effect of prednisone during gut-derived sepsis, *Crit. Care Med.* **25:**1207–1214.

Gianatti, L., Alexander, J. W., Pyles, T., and Fukushima, R., 1993, Arginine-supplemented diets improve survival in gut-derived sepsis and peritonitis by modulating bacterial clearance, *Ann. Surg.* **217:**644–654.

Gold, H. S., and Moellering, R. C., Jr., 1996, Antimicrobial-drug resistance, *N. Engl. J. Med.* **335:**1445–1453.

Gomez-Jimenez, J., Salgado, A., Mourelle, M., Martin, M. C., Segura, R. M., Peracaula, R., and Moncada, S., 1995, L-Arginine: Nitric oxide pathway in endotoxemia and human septic shock, *Crit. Care Med.* **23:**253–258.

Gow, A. J., and Stamler, J. S., 1998, Reaction between nitric oxide and haemoglobin under physiological conditions, *Nature* **391:**169–173.

Heiss, L. N., Lancaster, J. R., Jr., Corbett, J. A., and Goldman, W. E., 1994, Epithelial autotoxicity of nitric oxide: Role in the respiratory cytopathology of pertussis, *Proc. Natl. Acad. Sci. USA* **91:**267–270.

Heneka, M. T., Loschmann, P. A., and Osswald, H., 1997, Polymerized hemoglobin restores cardiovascular and kidney function in endotoxin-induced shock in the rat, *J. Clin. Invest.* **99:**47–54.

Hill, A. D., Naama, H., Shou, J., Calvano, S. E., and Daly, J. M., 1995, Antimicrobial effects of granulocyte-macrophage colony-stimulating factor in protein-energy malnutrition, *Arch. Surg.* **130:**1273–1277.

Horton, J. W., White, J., Maass, D., and Sanders, B., 1998, Arginine in burn injury improves cardiac performance and prevents bacterial translocation, *J. Appl. Physiol.* **84**:695–702.

James, S. L., 1995, Role of nitric oxide in parasitic infections, *Microbiol. Rev.* **59:**533–547.

Jourdain, M., Tounoys, A., Leroy, X., Mangalaboyi, J., Fourrier, F., Goudemand, J., Gosselin, B., Vallet, B., and Chopin, C., 1997, Effects of N^{ω}-nitro-L-arginine methyl ester on the endotoxin-induced disseminated intravascular coagulation in porcine septic shock, *Crit. Care Med.* **25:**452–459.

Kane, J. M., Shears, L. L., Hierholzer, C., Ambs, S., Billiar, T. R., and Posner, M. C., 1997, Chronic hepatitis C virus infection in humans: Induction of hepatic nitric oxide synthase and proposed mechanisms for carcinogenesis, *J. Surg. Res.* **69:**321–324.

Karamanoukian, J. L., Glick, P. L., Zayek, M., Steinharn, R. M., Zwass, J. S., Fineman, J. R., and Morin, F. C. R., 1994, Inhaled nitric oxide in congenital hypoplasia of the lungs due to diaphragmatic hernia or oligohydramnios, *Pediatrics* **94:**715–718.

Karupiah, G., Xie, Q. W., Buller, R. M., Nathan, C., Duarte, C., and MacMicking, J. D., 1993, Inhibition of viral replication by interferon-gamma-induced nitric oxide synthase, *Science* **261:**1445–1448.

Kawakami, K., Tohyama, M., Qifeng, X., and Saito, A., 1997, Expression of cytokines and inducible nitric oxide synthase mRNA in the lungs of mice infected with *Cryptococcus neoformans:* Effects of interleukin-12, *Infect. Immun.* **65:**1307–1312.

Kazmierski, W. M., Wolberg, G., Wilson, J. G., Smith, S. R., Williams, D. S., Thorp, H. H., and Molina, L., 1996, Iron chelates bind nitric oxide and decrease mortality in an experimental model of septic shock, *Proc. Natl. Acad. Sci. USA* **93:**9138–9141.

Khan, I. A., Schwartzman, J. D., Matsuura, T., and Kasper, L. H., 1997, A dichotomous role for nitric oxide during acute *Toxoplasma gondii* infection in mice, *Proc. Natl. Acad. Sci. USA* **94:**13955–13960.

Kilbourn, R.G., 1997, Nitric oxide overproduction in septic shock—Methemoglobin concentrations and blockade with diaspirin cross-linked hemoglobin, *Crit. Care Med.* **25:**1446–1447.

Kilbourn, R. G., Jubran, A., Gross, S. S., Griffith, O. W., Levi, R., Adams, J., and Lodato, R. F., 1990, Reversal of endotoxin-mediated shock by N^G-methyl-L-arginine, an inhibitor of nitric oxide synthesis, *Biochem. Biophys. Res. Commun.* **172:**1132–1138.

Kilbourn, R. G., Szabo, C., and Traber, D. L., 1997a, Beneficial versus detrimental effects of nitric oxide synthase inhibitors in circulatory shock: Lessons learned from experimental and clinical studies, *Shock* **7:**235–246.

Kilbourn, R., Traber, D., and Szabo, C., 1997b, Nitric oxide and shock, *Dis. Mon.* **43:**279–348.

Kimura, T., Nakayama, K., Penninger, J., Kitagawa, M., Harada, H., Matsuyama, T., Tanaka, N., Kamijo, R., Vilcek, J., Mak, T. W., and Taniguchi, T., 1994, Involvement of the IRF-1 transcription factor in antiviral responses to interferons, *Science* **264:**1921–1924.

Kirk, S. J., Regan, N. C., Wasserkrug, H. L., Sodeyama, M., and Barbul, A., 1992, Arginine enhances T cell responses in athymic nude mice, *J. Parenter. Enterol. Nutr.* **16:**429–432.

Komatsu, T., Bi, Z., and Reiss, C. S., 1996, Interferon-gamma induced type I nitric oxide synthase activity inhibits viral replication in neurons, *J. Neuroimmunol.* **68:**101–108.

Krafft, P., Fridich, P., Fitzgerald, R. D., Koc, D., and Steltzer, H., 1996, Effectiveness of nitric oxide inhalation in septic ARDS, *Chest* **109:**486–493.

Kwon, N. S., Stuehr, D. J., and Nathan, C. F., 1991, Inhibition of tumor cell ribonucleotide reductase by macrophage-derived nitric oxide, *J. Exp. Med.* **174:**761–767.

Langford, E. J., Brown, A. S., Wainwright, R. J., de Belder, A. J., Thomas, M. R., Smith, R. E., Radomski, M. W., Martin, J. F., and Moncada, S., 1994, Inhibition of platelet activity by *S*-nitrosoglutathione during coronary angioplasty, *Lancet* **344:**1458–1460.

Liaudet, L., Rosselet, A., Schaller, M. D., Markert, M., Perret, C., and Feihl, F., 1998, Nonselective versus selective inhibition of inducible nitric oxide synthase in experimental endotoxic shock, *J. Infect. Dis.* **177:**127–132.

Lopez-Jaramillo, P., Ruano, C., Rivera, J., Teran, E., and Moncada, S., 1995, Treatment of cutaneous mycosis with the NO donor *S*-nitroso-*N*-acetylpenicillamine (SNAP), *Endothelium* **3(**Suppl.1**):**s13.

Lopez-Jaramillo, P., Ruano, C., Rivera, J., Teran, E., Salazar-Irigoyen, R., Esplugues, J. V., and Moncada, S., 1998, Treatment of cutaneous leishmaniasis with nitric-oxide donor, *Lancet* **351**:1176–1177.

Lori, F., Malykh, A., Cara, A., Sun, D., Weinstein, J. N., Lisziewicz, J., and Gallo, R. C., 1994, Hydroxyurea as an inhibitor of human immunodeficiency virus-type 1 replication, *Science* **266:**801–805.

Louie, A., Baltch, A. L., Franke, M. A., Smith, R. P., and Gordon, M. A., 1994, Comparative capacity of four antifungal agents to stimulate murine macrophages to produce tumour necrosis factor-alpha: An effect that is attenuated by pentoxifylline, liposomal vesicles and dexamethasone, *J. Antimicrob. Chemother.* **34:**975–987.

Lovchik, J. A., Lyons, C. R., and Lipscomb, M. F., 1995, A role for gamma interferon-induced nitric oxide in pulmonary clearance of *Cryptococcus neoformans*, *Am. J. Respir. Cell. Mol. Biol.* **13:**116–124.

MacMicking, J., Xie, Q. W., and Nathan, C., 1997a, Nitric oxide and macrophage function, *Annu. Rev. Immunol.* **15:**323–350.

MacMicking, J. D., North, R. J., LaCouse, R., Mudgett, J. S., Shah, S. K., and Nathan, C. F., 1997b, Identification of nitric oxide synthase as a protective locus against tuberculosis, *Proc. Natl. Acad. Sci. USA* **94:**5243–5248.

Maeda, H., Akaike, T., Yoshida, M., Sato, K., and Noguchi, Y., 1995, A new nitric oxide scavenger, imidazolineoxyl *N*-oxide derivative, and its effects in pathophysiology and microbiology, *Curr. Top. Microbiol. Immunol.* **196:**37–50.

Mannick, E. E., Bravo, L. E., Zarama, G., Realpe, J. L., Zhang, X. J., Ruiz, B., Fonham, E. T., Mera, R., Miller, M. J., and Correa, P., 1996, Inducible nitric oxide synthase, nitrotyrosine, and apoptosis in *Helicobacter pylori* gastritis: Effect of antibiotics and antioxidants, *Cancer Res.* **56:**3238–3243.

McElhaney-Feser, G. E., Raulli, R. E., and Cihlar, R. L., 1998, Synergy of nitric oxide releasing compounds and azoles against *Candida* species in vitro, *Antimicrob. Agents Chemother.* **42:**2342–2346.

Meyding-Lamade, U., Haas, J., Lamade, W., Stingele, K., Kehm, R., Fath, A., Heinrich, K., Hagenlocher, B. S., and Wildemann, B., 1998, Herpes simplex virus encephalitis—Long-term comparative study of viral load and the expression of immunologic nitric oxide synthase in mouse brain tissue, *Neurosci. Lett.* **244**:9–12.

Milano, S., Arcoleo, F., D'Agostino, P., and Cillari, E., 1997, Intraperitoneal injection of tetracyclines protects mice from lethal endotoxemia downregulating inducible nitric oxide synthase in various organs and cytokine and nitrate secretion in blood, *Antimicrob. Agents Chemother.* **41:**117–121.

Mitaka, C., Hirata, Y., Ichikawa, K., Uchida, T., Yokoyama, K., Nagura, T., Tsunoda, Y., and Amaha, K., 1995, Effects of nitric oxide synthase inhibitor on hemodynamic change and O_2 delivery in septic dogs, *Am. J. Physiol.* **268:**H2017-H2023.

Nathan, C., 1995, Natural resistance and nitric oxide, *Cell* **82:**873–876.

Neu, H. C., 1992, The crisis in antibiotic resistance, *Science* **257:**64–73.

Nicholson, S., Bonecini-Almeida, M. G., Lapa e Silva, J. R., Nathan, C., Xie, Q. W., Mumford, R., Weidner, J. R., Calaycay, J., Geng, J., Boechat, N., Linhares, C., Rom, W., and Ho, J. L., 1996, Inducible nitric oxide synthase in pulmonary alveolar macrophages from patients with tuberculosis, *J. Exp. Med.* **183:**2293–2302.

Nicolle, L. E., Strausbaugh, L. J., and Garibaldi, R. A., 1996, Infections and antibiotic resistance in nursing homes, *Clin. Microbiol. Rev.* **9:**1–17.

Ohshima, H., and Bartsch, H., 1994, Chronic infections and inflammatory processes as cancer risk factors: Possible role of nitric oxide in carcinogenesis, *Mutat. Res.* **305:**253–264.

Pacelli, R., Taira, J., Cook, J. A., Wink, D. A., and Krishna, M. C., 1996, Hydroxyurea reacts with heme proteins to generate nitric oxide, *Lancet* **347:**900.

Palmer, R. M. J., 1993, The discovery of nitric oxide in the vessel wall: A unifying concept in the pathogenesis of sepsis, *Arch. Surg.* **128:**396–401.

Park, J. H., Chang, S. H., Lee, K. M., and Shin, S. H., 1996, Protective effect of nitric oxide in an endotoxin-induced septic shock, *Am. J. Surg.* **171:**340–345.

Pedrinelli, R., Ebel, M., Catapano, G., Dell'Omo, G., Ducci, M., Del Chicca, M., and Clevico, A., 1995, Pressor, renal and endocrine effects of L-arginine in essential hypertensives, *Eur. J. Clin. Pharmacol.* **48:**195–201.

Petros, A., Bennet, D., and Vallance, P., 1991, Effect of nitric oxide synthase inhibitors on hypotension in patients with septic shock, *Lancet* **338:**1557–1558.

Petros, A., Lamb, G., Leone, A., Moncada, S., Bennett, D., and Vallance, P., 1994, Effects of a nitric oxide synthase inhibitor in humans with septic shock, *Cardiovasc. Res.* **28:**34–39.

Reah, G., Bodenham, A. R., Mallick, A., Daily, E. K., and Przybelski, R. J., 1997, Initial evaluation of diaspirin cross-linked hemoglobin (DCLHb) as a vasopressor in critically ill patients, *Crit. Care Med.* **25:**1480–1488.

Rossero, R., McKinsey, D., Green, S., Andron, L., and Pollard, R., 1997, Open label combination therapy with stavudine, didanosine, and hydroxyurea in nucleoside experienced HIV-1 infected patients, in: *Abstracts of the 5th Conference on Retroviruses and Opportunistic Infections*, Chicago, #519.

Saavedra, J. E., Billiar, T. R., Williams, D. L., Kim, Y. M., Watkins, S. C., and Keefer, L. K., 1997, Targeting nitric oxide (NO) in vivo. Design of a liver-selective NO donor prodrug that blocks tumor necrosis factor-alpha-induced apoptosis and toxicity in the liver, *J. Med. Chem.* **40:**1947–1954.

Satarug, S., Haswell-Elkins, M., Tsuda, M., Mairiang, P., Sithithaworn, P., Mairiang, E., Esumi, H., Sukpraser, S., Yongvanit, P., and Elkins, D. B., 1996, Thiocyanate-independent nitrosation in humans with carcinogenic parasite infection, *Carcinogenesis* **17:**1075–1081.

Sharara, A. I., Perkins, D. J., Misukonis, M. A., Chan, S. U., Dominitz, J. A., and Weinberg, J. B., 1997, Interferon (IFN)-α activation of human blood mononuclear cells *in vitro* and *in vivo* for nitric oxide synthase (NOS) type 2 mRNA and protein expression: Possible relationship of induced NOS2 to the anti-hepatitis C effects of IFN-α *in vivo*, *J. Exp. Med.* **186:**1495–1502.

Su, D., Roth, R. I., Yoshida, M., and Levin, J., 1997, Hemoglobin increases mortality from bacterial endotoxin, *Infect. Immun.* **65:**1258–1266.

Szabo, C., Southan, G. J., and Thiemermann, C., 1994, Beneficial effects and improved survival in rodent models of septic shock with *S*-methylisothiourea sulfate, a potent and selective inhibitor of inducible nitric oxide synthase, *Proc. Natl. Acad. Sci. USA* **91:**12472–12476.

Tanaka, K., Nakazawa, H., Okada, K., Umezawa, K., Fukuyama, N., and Koga, Y., 1997, Nitric oxide mediates murine cytomegalovirus-associated pneumonitis in lungs that are free of the virus, *J. Clin. Invest.* **100:**1822–1830.

Taylor, A. P., and Murray, H. W., 1997, Intracellular antimicrobial activity in the absence of interferon-gamma: Effect of interleukin-12 in experimental visceral leishmaniasis in interferon-gamma gene-disrupted mice, *J. Exp. Med.* **185:**1231–1239.

Teale, D. M., and Atkinson, A. M., 1992, Inhibition of nitric oxide synthase improves survival in a murine peritonitis model of sepsis that is not cured by antibiotics alone, *J. Antimicrob. Chemother.* **30:**839–842.

Teale, D. M., and Atkinson, A. M., 1994, L-Canavanine restores blood pressure in a rat model of endotoxic shock, *Eur. J. Pharmacol.* **271:**87–92.

Thiemermann, C., 1994, Role of the L-arginine–nitric oxide pathway in circulatory shock, *Adv. Pharmacol.* **28:**45–79.

Tohyama, M., Kawakami, K., and Saito, A., 1996, Anticryptococcal effect of amphotericin B is mediated through macrophage production of nitric oxide, *Antimicrob. Agents Chemother.* **40:**1919–1923.

Tsuji, S., Kawano, S., Tsujii, M., Takei, Y., Tanaka, M., Sawaoka, H., Nagano, K., Fusamoto, H., and Kamada, T., 1996, *Helicobacter pylori* extract stimulates inflammatory nitric oxide production, *Cancer Lett.* **108:**195–200.

Walker, G., Pfeilschifter, J., and Kunz, D., 1997, Mechanisms of suppression of inducible nitric-oxide synthase (iNOS) expression in interferon (IFN)-gamma-stimulated RAW 264.7 cells by dexamethasone. Evidence for glucocorticoid-induced degradation of iNOS protein by calpain as a key step in post-transcriptional regulation, *J. Biol. Chem.* **272:**16679–16687.

Warren, W., Biggs, P. J., el-Baz, M., Ghoneim, M. A., Stratton, M. R., and Venitt, S., 1995, Mutations in the p53 gene in schistosomal bladder cancer: A study of 92 tumours from Egyptian patients and a comparison between mutational spectra from schistosomal and nonschistosomal urothelial tumours, *Carcinogenesis* **16:**1181–1189.

Wegner, H. E., and Knispel, H. H., 1995, Effect of nitric oxide donor, linsidomine chlorhydrate, in treatment of human erectile dysfunction caused by venous leakage, *Urology* **42:**409–411.

Wei, X. Q., Charles, I. G., Smith, A., Ure, J., Feng, G. J., Huang, F. P., Xu, D., Muller, W., Moncada, S., and Liew, F. Y., 1995, Altered immune responses in mice lacking inducible nitric oxide synthase, *Nature* **375:**408–411.

Weller, R., Ormerod, A. D., Hobson, R. P., and Benjamin, N. J., 1998, A randomized trial of acidified nitrite cream in the treatment of tinea pedis, *J. Am. Acad. Dermatol.* **38:**559–563.

Wright, C. E., Rees, D. D., and Moncada, S., 1992, Protective and pathological roles of nitric oxide in endotoxin shock, *Cardiovasc. Res.* **26:**48–57.

Wu, C. C., Croxtall, J. D., Perretti, M., Bryant, C. E., Thiemermann, C., Flower, R. J., and Vane, J. R., 1995, Lipocortin 1 mediates the inhibition by dexamethasone of the induction by endotoxin of nitric oxide synthase in the rat, *Proc. Natl. Acad. Sci. USA* **92:**3473–3477.

Xu, D. M., McSorley, S. J., Tetley, L., Chatfield, S., Dougan, G., Chan, W. L., Satoskar, A., David, J. R., and Liew, F. Y., 1998, Protective effect on *Leishmania major* infection of migration inhibitory factor, TNF-alpha, and IFN-gamma administered orally via attenuated *Salmonella typhimurium*, *J. Immunol.* **160:**1285–1289.

Yamagushi, H. S., Abe, S., and Tokuda, Y., 1993, Immunomodulatory activity of antifungal drugs, *Ann. N.Y. Acad. Sci.* **685:**447–457.

Zhou, P., Sieve, M. C., Tewari, R. P., and Seder, R. A., 1997, Interleukin-12 modulates the protective immune response in SCID mice infected with *Histoplasma capsulatum*, *Infect. Immun.* **65:**936–942.

Index